Lecture Notes in Computer Science 5350

Commenced Publication in 1973
Founding and Former Series Editors:
Gerhard Goos, Juris Hartmanis, and Jan van Leeuwen

Editorial Board

David Hutchison
Lancaster University, UK
Takeo Kanade
Carnegie Mellon University, Pittsburgh, PA, U.
Josef Kittler
University of Surrey, Guildford, UK
Jon M. Kleinberg
Cornell University, Ithaca, NY, USA
Alfred Kobsa
University of California, Irvine, CA, USA
Friedemann Mattern
ETH Zurich, Switzerland
John C. Mitchell
Stanford University, CA, USA
Moni Naor
Weizmann Institute of Science, Rehovot, Israel
Oscar Nierstrasz
University of Bern, Switzerland
C. Pandu Rangan
Indian Institute of Technology, Madras, India
Bernhard Steffen
University of Dortmund, Germany
Madhu Sudan
Massachusetts Institute of Technology, MA, USA
Demetri Terzopoulos
University of California, Los Angeles, CA, USA
Doug Tygar
University of California, Berkeley, CA, USA
Gerhard Weikum
Max-Planck Institute of Computer Science, Saarbruecken, Germany

Josef Pieprzyk (Ed.)

Advances in Cryptology - ASIACRYPT 2008

14th International Conference on the Theory
and Application of Cryptology and Information Security
Melbourne, Australia, December 7-11, 2008
Proceedings

Volume Editor

Josef Pieprzyk
Macquarie University, Department of Computing
Sydney, NSW 2109, Australia
E-mail: josef@ics.mq.edu.au

Library of Congress Control Number: 2008939141

CR Subject Classification (1998): E.3, D.4.6, F.2.1-2, K.6.5, C.2, J.1, G.2

LNCS Sublibrary: SL 4 – Security and Cryptology

ISSN 0302-9743
ISBN-10 3-540-89254-0 Springer Berlin Heidelberg New York
ISBN-13 978-3-540-89254-0 Springer Berlin Heidelberg New York

This work is subject to copyright. All rights are reserved, whether the whole or part of the material is concerned, specifically the rights of translation, reprinting, re-use of illustrations, recitation, broadcasting, reproduction on microfilms or in any other way, and storage in data banks. Duplication of this publication or parts thereof is permitted only under the provisions of the German Copyright Law of September 9, 1965, in its current version, and permission for use must always be obtained from Springer. Violations are liable to prosecution under the German Copyright Law.

Springer is a part of Springer Science+Business Media

springer.com

© Springer-Verlag Berlin Heidelberg 2008
Printed in Germany

Typesetting: Camera-ready by author, data conversion by Scientific Publishing Services, Chennai, India
Printed on acid-free paper SPIN: 12562770 06/3180 5 4 3 2 1 0

Preface

The 14th International Conference on the Theory and Applications of Cryptology and Information Security—ASIACRYPT 2008—was held in Melbourne during December 7–11, 2008. The conference was sponsored by the International Association for Cryptologic Research (IACR) in cooperation with the Center for Advanced Computing – Cryptography and Algorithms (ACAC), Macquarie University, Deakin University, the Research Network for a Secure Australia (RNSA) and SECIA. ASIACRYPT 2008 was chaired by Lynn Batten and I had the honor of serving as the Program Chair.

There were 208 submissions from which 12 papers were withdrawn. Each paper got assigned to at least three referees. Papers submitted by the members of the Program Committee got assigned to five referees. In the first stage of the review process, the submitted papers were read and evaluated by the Program Committee members and then in the second stage, the papers were scrutinized during an extensive discussion. Finally, the Program Committee chose 33 papers to be included in the conference program. The authors of the accepted papers had three weeks for revision and preparation of final versions. The revised papers were not subject to editorial review and the authors bear full responsibility for their contents.

The Program Committee selected three best papers. They were: "Speeding up Pollard Rho Method on Prime Fields" by Jung Hee Cheon, Jin Hong, and Minkyu Kim, "A Modular Security Analysis of the TLS Handshake Protocol" by Paul Morrissey, Nigel P. Smart and Bogdan Warinschi and "Breaking the F-FCSR-H Stream Cipher in Real Time" by Martin Hell and Thomas Johansson. The authors of the three papers were invited to submit the full versions of their papers to the *Journal of Cryptology*. The authors of the first paper, Jung Hee Cheon, Jin Hong and Minkyu Kim, were recipients of the Best Paper Award.

The conference program included two invited lectures by Andrew Chi-Chih Yao and John Cannon. Andrew Chi-Chih Yao spoke about "Some Perspectives on Complexity-Based Cryptography" and an abstract has been included in the proceedings.

There are many people who contributed to the success of ASIACRYPT 2008. First I would like to thank the authors of all papers (both accepted and rejected) for submitting their papers to the conference. A special thanks go to the members of the Program Committee and the external referees who gave their time, expertise and enthusiasm in order to ensure that each paper received a thorough and fair review. I am grateful to Andy Clark, Helena Handschuh, Arjen Lenstra and Bart Preneel for their support and advice; I thank Vijayakrishnan Pasupathinathan for taking care of the iChair server and Michelle Kang and Judy Chow for maintenance of the conference website. Shai Halevi deserves our thanks for the registration site. Judy Chow, the conference secretary is warmly thanked for her enormous contribution responding to participant queries and on site at the conference registration. I would like to thank Matthieu Finiasz and Thomas Baignères from EPFL, LASEC, Switzerland for letting us use their iChair software that was used not only as the submission server but also facilitated the review and discussion process. Finally, I would like to thank Ed Dawson for organizing a traditional Rump Session.

December 2008 Josef Pieprzyk

ASIACRYPT 2008

December 7–11, 2008, Melbourne, Australia

Sponsored by
The International Association for Cryptologic Research

in co-operation with
School of Engineerring and Information Technology, Deakin University
Center for Advanced Computing – Cryptography and Algorithms, Macquarie University
Research Network for a Secure Australia
SECIA

General Chair

Lynn Batten	Deakin University, Australia

Program Chair

Josef Pieprzyk	Macquarie University, Australia

Program Committee

Masayuki Abe	NTT, Japan
Josh Benaloh	Microsoft, USA
Daniel Bernstein	University of Illinois at Chicago, USA
Colin Boyd	QUT, Australia
Claude Carlet	University of Paris 8, France
Nicolas Courtois	UCL, UK
Claus Diem	Leipzig University, Germany
Christophe Doche	Macquarie University, Australia
Stefan Dziembowski	La Sapienza University, Italy
Serge Fehr	CWI, The Netherlands
Jovan Golić	Telecom, Italy
Jonathan Katz	University of Maryland, USA
Kaoru Kurosawa	Ibaraki University, Japan
Xuejia Lai	Jiao Tong University, China
Tanja Lange	Technische Universiteit Eindhoven, The Netherlands
Byoungcheon Lee	Joongbu University, Korea
Arjen Lenstra	EPFL, Switzerland, and Alcatel-Lucent Bell Laboratories
Keith Martin	RHUL, UK
Tsutomu Matsumoto	Yokohama National University, Japan

Mitsuru Matsui	Mitsubishi, Japan
Siguna Mueller	University of Wyoming, USA
Kaisa Nyberg	Helsinki University of Technology and Nokia, Finland
Eiji Okamoto	Tsukuba University, Japan
Tatsuaki Okamoto	NTT, Japan
Pascal Paillier	Gemplus, France
Jacques Patarin	University of Versailles, France
David Pointcheval	ENS, France
Manoj Prabhakaran	University of Illinois, USA
Bart Preneel	Katholieke Universiteit Leuven, Belgium
C. Pandu Rangan	IIT, India
Vincent Rijmen	Katholieke Universiteit Leuven, Belgium and Graz University of Technology, Austria
Phillip Rogaway	UC Davis, USA
Bimal Roy	ISI, India
Rei Safavi-Naini	University of Calgary, Canada
Palash Sarkar	ISI, India
Ron Steinfeld	Macquarie University, Australia
Huaxiong Wang	NTU, Singapore
Yuliang Zheng	University of North Carolina at Charlotte, USA

External Reviewers

Michel Abdalla
Hadi Ahmadi
Kennichiro Akai
Jesus Almansa
Elena Andreeva
Kazumaro Aoki
François Arnault
Mina Askari
Mandal Avradip
Joonsang Baek
Endre Bangerter
Gregory Bard
Rana Barua
Lejla Batina
Aurélie Bauer
Amos Beimel
Côme Berbain
Thierry Berger
Florent Bernard
Joppe Bos
Billy Bob Brumley
Sébastien Canard
Yaniv Carmeli
David Cash
Maximilien Chagon
Hao Chen
Jung Hee Cheon
Céline Chevalier
Benoît Chevallier-Mâmes
Joo Yeon Cho
Raymond Choo
Ashish Choudhary
Cheng-Kang Chu
Carlos Cid
Jean-Sébastien Coron
Yang Cui
Paolo D'Arco
Alex Dent
Cunsheng Ding
Yevgeniy Dodis
Gwenael Doerr
Ling Dong
Ratna Dutta
Reza Rezaeian Farashahi
Jean-Charles Faugère
Sebastian Faust
Anna Lisa Ferrara
Matthias Fitzi
Pierre-Alain Fouque
Georg Fuchsbauer
Eiichiro Fujisaki
Jun Furukawa
Philippe Gaborit
Steven Galbraith
Craig Gentry
Marc Girault
Zheng Gong
Juan Gonzalez
Choudary Gorantla
Louis Goubin
Aline Gouget
Tim Gueneysu
Risto Hakala

Mike Hamburg
Safuat Hamdy
Danny Harnik
Swee-Huay Heng
Miia Hermelin
Alain Hiltgen
Jason Hinek
Dennis Hofheinz
Michal Hojsik
Qiong Huang
Laurent Imbert
Sebastiaan Indesteege
Yuval Ishai
Tetsu Iwata
Kimmo Järvinen
Mahabir P. Jhanwar
Guo Jian
Shaoquan Jiang
Ellen Jochemsz
Antoine Joux
Marc Joye
Ari Juels
Pascal Junod
Sumit K. Pandey
Özgül Küçük
Marcelo Kaihara
Alexandre Karlov
Emilia Käsper
Shahram Khazaei
Aggelos Kiayias
Eike Kiltz
Markulf Kohlweiss
Ted Krovetz
Małgorzata Kupiecka
Mario Lamberger
Donghoon Lee
Jin Li
Qiang Li
Wei Li
Helger Lipmaa
Xianhui Lu
Yi Lu
Yiyuan Luo
Vadim Lyubashevsky
Changshe Ma
Hemanta Maji
Krystian Matusiewicz
Alexander May
Florian Mendel
Renato Menicocci
Miodrag Mihaljevic
Seyed Mohammadhassanzadeh
David Molnar
Guglielmo Morgari
David Naccache
Tomislav Nad
Michael Naehrig
Yusuke Naito
Gregory Neven
Cedric Ng
Phong Nguyen
Ryo Nishimaki
Wakaha Ogata
Frédérique Oggier
Alina Oprea
Rafail Ostrovsky
Raphael Overbeck
Onur Ozen
Kenny Paterson
Arpita Patra
Goutam Paul
Souradyuti Paul
Christiane Peters
Slobodan Petrovic
Thomas Peyrin
Duong-Hieu Phan
Benny Pinkas
Norbert Pramstaller
Emmanuel Prouff
Bartosz Przydatek
Michael Quisquater
Ragavendran
Dominik Raub
Christian Rechberger
Tom Ristenpart
Mike Rosulek
Kazue Sako
Juraj Sarinay
Martin Schläffer
Berry Schoenmakers
Sharmila Devi Selvi
Pouyan Sepherdad
Yannick Seurin
Siamak Shahandashti
Hongsong Shi
Igor Shparlinski
Francesco Sieca
Michal Sramka
Martijn Stam
Marc Stevens
Hung-Min Sun
Willy Susilo
Daisuke Suzuki
Katsuyuki Takashima
Christophe Tartary
Jean-Pierre Tillich
Elmar Tischhauser
Tomas Toft
Toshio Tokita
Joana Treger
Toyohiro Tsurumaru
Ashraful Tuhin
Vesselin Velichkov
Frederik Vercauteren
Damien Vergnaud
Ivan Visconti
Zhongmei Wan
Peishun Wang
Bogdan Warinschi
Brent Waters
Douglas Wikström
Christopher Wolf
Zhongming Wu
Jürg Wullschleger
Guomin Yang
Rui Zhang
Zhifang Zhang
Yunlei Zhao
Jinmin Zhong
Sébastien Zimmer
Benne de Weger

Table of Contents

Multi-Party Computation

Invited Talk

Cryptographic Protocols I

Cryptographic Hash Functions I

Cryptographic Protocols II

Cryptographic Hash Functions II

Public-Key Cryptography I

Lattice-Based Cryptography

Private-Key Cryptography

Public-Key Cryptography II

Analysis of Stream Ciphers

MPC vs. SFE : Unconditional and Computational Security*

Martin Hirt, Ueli Maurer, and Vassilis Zikas

Department of Computer Science, ETH Zurich, 8092 Zurich, Switzerland
{hirt,maurer,vzikas}@inf.ethz.ch

Abstract. In secure computation among a set $\mathcal{P}$ of players one considers an adversary who can corrupt certain players. The three usually considered types of corruption are active, passive, and fail corruption. The adversary's corruption power is characterized by a so-called adversary structure which enumerates the adversary's corruption options, each option being a triple (A, E, F) of subsets of $\mathcal{P}$, where the adversary can actively corrupt the players in A, passively corrupt the players in E, and fail-corrupt the players in F.

This paper is concerned with characterizing for which adversary structures general secure function evaluation (SFE) and secure (reactive) multi-party computation (MPC) is possible, in various models. This has been achieved so far only for the very special model of perfect security, where, interestingly, the conditions for SFE and MPC are distinct. Such a separation was first observed by Ishai *et al.* in the context of computational security. We give the exact conditions for general SFE and MPC to be possible for information-theoretic security (with negligible error probability) and for computational security, assuming a broadcast channel, with and without setup. In all these settings we confirm the strict separation between SFE and MPC. As a simple consequence of our results we solve an open problem for computationally secure MPC in a threshold model with all three corruption types.

1 Introduction

Secure Function Evaluation and Secure Multi-Party Computation. Secure function evaluation (SFE) allows a set $\mathcal{P} = \{p_1, \ldots, p_n\}$ of n players to compute an arbitrary agreed function f of their inputs $x_1, \ldots, x_n$ in a secure way. (Reactive) secure multi-party computation (MPC) is a generalization of SFE where the function to be computed is "reactive": players can give inputs and get outputs several times during the computation. If one models SFE and MPC as ideal functionalities, then the main difference is that in MPC (but not in SFE) the functionality must be able to keep state.

The potential dishonesty of players is modeled by a central adversary corrupting players, where players can be actively corrupted (the adversary takes full control over them), passively corrupted (the adversary can read their internal state), or fail-corrupted

* This research was partially supported by the Swiss National Science Foundation (SNF), project no. 200020-113700/1 and by the Zurich Information Security Center (ZISC). The full version of this paper is available at http://www.crypto.ethz.ch/pubs/HiMaZi08.

J. Pieprzyk (Ed.): ASIACRYPT 2008, LNCS 5350, pp. 1–18, 2008.
© International Association for Cryptologic Research 2008

(the adversary can make them crash at any suitable time). A crashed player stops sending any messages, but the adversary cannot read the internal state of the player (unless he is actively or passively corrupted at the same time).

Summary of Known Results. SFE (and MPC) was introduced by Yao [Yao82]. The first general solutions were given by Goldreich, Micali, and Wigderson [GMW87]; these protocols are secure under some intractability assumptions. Later solutions [BGW88, CCD88] provide information-theoretic security. In particular, it is remarkable that if a (physical) broadcast channel is assumed, strictly more powerful adversaries can be tolerated [RB89, Bea91].

In the seminal papers solving the general SFE and MPC problems, the adversary is specified by a single corruption type (active or passive) and a threshold t on the tolerated number of corrupted players. Goldreich, Micali, and Wigderson [GMW87] proved that, based on cryptographic intractability assumptions, general secure MPC is possible if and only if $t < n/2$ players are actively corrupted, or, alternatively, if and only if $t < n$ players are passively corrupted. In the information-theoretic model, Ben-Or, Goldwasser, and Wigderson [BGW88] and independently Chaum, Crépeau, and Damgård [CCD88] proved that unconditional security is possible if and only if $t < n/3$ for active corruption and $t < n/2$ for passive corruption. Finally, in [GMW87, GL02, Gol04] it was shown that, based on cryptographic intractability assumptions, any number of active cheaters ($t < n$) can be tolerated for SFE, but only if we sacrifice fairness and guaranteed delivery of the output [Cle86]. Some of the above results were unified, and extended to include fail-corruption, in [FHM98]: perfectly secure MPC (and SFE) is achievable if and only if $3t_a+2t_p+t_f < n$, and unconditionally secure MPC (SFE) (without a trusted setup or a broadcast channel) is achievable if and only if $2t_a + 2t_p + t_f < n$ and $3t_a + t_f < n$, where t_a, t_p, and t_f denote the upper bounds on the number of actively, passively, and fail-corrupted players, respectively. These results consider an adversary who can perform all three corruption types *simultaneously*. For the computational-security case, Ishai *et al.* [IKLP06] gave a protocol for SFE which tolerates an adversary who can either corrupt $t_a < n/2$ players actively, or, *alternatively*, $t_p < n$ players passively. They also showed that such an adversary cannot be tolerated for MPC.

Generalizing threshold models, the adversary's corruption power can be characterized by a so-called adversary structure which enumerates the adversary's corruption options, each option being a triple (A, E, F) of subsets of $\mathcal{P}$, where the adversary can actively corrupt the players in A, passively corrupt the players in E, and fail-corrupt the players in F. Of course, the adversary's choice of the option is secret and a protocol must tolerate any choice by the adversary.

General adversary structures were first considered in [HM97, HM00] for active-only and passive-only corruption. General mixed-corruption (active and passive) adversary structures were considered in [FHM99]. The full generality, including fail-corruption, was first considered in [BFH$^+$08], where only the perfect-security case could be solved, both for SFE and MPC. An interesting aspect of those results is the separation between SFE and MPC: the condition for SFE is strictly weaker than the condition for MPC. This can also be seen as a justification for the most general mixed corruption models. Such a separation was previously observed for the perfect-security case [Alt99] and, as already mentioned, for the computational-security case [IKLP06].

Contributions of this Paper. We prove the exact conditions for general SFE and MPC to be possible, in the most general mixed adversary model, with synchronous communication, and where a broadcast channel is assumed. We consider the most natural and desirable security notion, where full security (including fairness and guaranteed output delivery) is required. We solve the two cases of general interest: unconditional (information-theoretic with negligible error probability) security and computational security, both with and without setup. We show a strict separation between SFE and MPC.

Our results imply that for the threshold model with all three corruption types simultaneously, and for computational security, SFE and MPC are possible if and only if $2t_a + t_p + t_f < n$. As in [FHM98] there is no separation in this model.

Outline of this paper. In Section 2 we describe the model. In Sections 3,4,5 and 6 we handle the unconditional-security case; in particular, in Sections 3 and 4 we describe techniques and sub-protocols that are used for the construction of MPC and SFE protocols described in Sections 5 and 6, respectively. Finally, in Section 7 we handle the computational-security case.

2 The Model

We consider a set $\mathcal{P} = \{p_1, \ldots, p_n\}$ of players. Some of these players can be corrupted by the adversary. We consider active corruption (the adversary takes full control), passive corruption (the adversary can read the internal state), and fail-corruption (the adversary can make the player crash). We use the following characterizations for players: a player that is not corrupted is called *uncorrupted*, a player that (so far) has followed the protocol instructions is called *correct*, and a player that has deviated from the protocol (e.g., has crashed or has sent wrong messages) is called *incorrect*. The adversary's corruption capability is characterized by an adversary structure $\mathcal{Z} = \{(A_1, E_1, F_1), \ldots, (A_m, E_m, F_m)\}$ (for some m) which is a monotone set of triples of player sets. At the beginning of the protocol, the adversary chooses a triple $Z^\star = (A^\star, E^\star, F^\star) \in \mathcal{Z}$ and actively corrupts the players in $A^\star$, passively corrupts the players in $E^\star$ (eavesdropping), and fail-corrupts the players in $F^\star$;[1] this triple is called the *actual adversary class* or simply the actual adversary. Note that $Z^\star$ is not known to the honest players and appears only in the security analysis. A protocol is called $\mathcal{Z}$*-secure* if it is secure against an adversary with corruption power characterized by $\mathcal{Z}$. For notational simplicity we assume that $A \subseteq E$ and $A \subseteq F$ for any $(A, E, F) \in \mathcal{Z}$, since an actively corrupted player can behave as being passively or fail-corrupted. Furthermore, as many constructions only need to consider the maximal classes of a structure, we define the maximal structure $\overline{\mathcal{Z}}$ as the smallest subset of $\mathcal{Z}$ such that $\forall (A, E, F) \in \mathcal{Z}\ \exists (\bar{A}, \bar{E}, \bar{F}) \in \overline{\mathcal{Z}} : A \subseteq \bar{A}, E \subseteq \bar{E}, F \subseteq \bar{F}$.

Communication takes place over a complete network of secure channels. Furthermore, we assume authenticated broadcast channels, which allow every $p_i \in \mathcal{P}$ to consistently send an authenticated message to all players in $\mathcal{P}$. All communication is synchronous, i.e., the delays in the network are upper-bounded by a known constant.

[1] We focus on static security, although our results could be generalized to adaptive corruption.

In the computational model (Section 7), the secrecy of the bilateral channels can be implemented by using encryption, where the public keys are distributed using the authenticated broadcast channels. We mention that in a model with simultaneous active and passive corruption, the authenticity cannot easily be implemented using setup, as the adversary can forge signatures of passively corrupted players. Also implementing the authenticated broadcast channels by point-to-point communication seems non-trivial, as it must be guaranteed that fail-corrupted players send either the right value or no value (but not a wrong value), and that passively corrupted players always send the right value.

To simplify the description, we adopt the following convention: Whenever a player does not receive an expected message (over a bilateral or a broadcast channel), or receives a message outside of the expected range, then the special symbol $\perp \notin \mathbb{F}$ is taken for this message. Note that after a player has crashed, he only sends $\perp$.

The function to be computed is described as an arithmetic circuit over some finite field $\mathbb{F}$, consisting of addition (or linear) gates and multiplication gates. Our protocols take as input the player's inputs and additionally the maximal adversary structure. The running time of the suggested protocols is polynomial in the size of their input,[2] and the error probability is negligible.

3 Information Checking

An actively corrupted player might send a value to another player and then deny that the value was sent by him. To deal with such behavior, we need a mechanism which binds a player to the messages he sends. In [RB89, CDD$^+$99, BHR07] the Information Checking (IC) method was developed for this purpose, and used to design unconditionally secure protocols tolerating up to $t < n/2$ active cheaters. In this section, we extend the IC method to the setting of general adversaries with active, passive, and fail-corruption.

The IC-authentication scheme involves three players, a sender p_s, a recipient p_r, and a verifier p_v, and consists of three protocols, called IC-Setup, IC-Distr, and IC-Reveal. Protocol IC-Distr allows p_s to send a value v to p_r in an authenticated way, so that p_r can, by invoking IC-Reveal, open v to p_v and prove that v was received from p_s. Both IC-Distr and IC-Reveal assume a secret key α known exclusively to p_s and p_v (but not to p_r). This key is generated and distributed in IC-Setup. Note that the same key can be used to authenticate multiple messages.

Informally, the three protocols can be described as follows: In IC-Setup, p_s generates a uniformly random key α and sends it to p_v over the bilaterally secure channel. In IC-Distr, α is used to generate an authentication tag y and a verification tag z for the sent value v. The values (v, y) and z are given to p_r and p_v, respectively. In IC-Reveal, p_r sends (v, y) to p_v, who verifies that (y, z) is a valid authentication/verification-tag pair for v with key α.

Ideally, an IC-authentication scheme should have the following properties: (1) Any value sent with IC-Distr is accepted in IC-Reveal, (2) in IC-Distr, p_v gets no information on v, and (3) only values sent with IC-Distr are accepted in IC-Reveal. However, these

[2] As the adversary structure might be exponentially large, our protocols' worst case running time can be exponential in the size of the player set. However, this is the best complexity one can hope to achieve for a protocol that tolerates *any* adversary structure [HM00].

properties cannot be (simultaneously) perfectly satisfied. In fact, Property 3 can only be achieved with negligible error probability, as the adversary might guess an authentication tag y' for a $v' \neq v$. Moreover, it can only be achieved when neither p_s nor p_v is passively corrupted, since otherwise the adversary knows α and z.

In our IC-authentication scheme the key α is chosen uniformly at random from $\mathbb{F}$ and the value v is also from $\mathbb{F}$. The authentication and verification tags, y and z, respectively, are such that for some degree-one polynomial $w(\cdot)$ over $\mathbb{F}$, $w(0) = v$, $w(1) = y$, and $w(\alpha) = z$. In other words, (y, z) is a valid IC-pair if $z = (y - v)\alpha + v$. Defining validity this way gives the IC-authentication scheme an additional *linearity* property. In particular, if (y, z) and (y', z') are valid IC-pairs for v and v', respectively, (for the same α) then $(y + y', z + z')$ is a valid IC-pair for $v + v'$. This implies that when some values have been sent with IC-Distr, then p_r and p_v can, without any interaction, compute valid authentication data for any linear combination of those values.

Due to space restrictions, the detailed description of the protocols IC-Setup, IC-Distr, and IC-Reveal, as well as the proof of the following lemma are deleted from this extended abstract.

Theorem 1. *Our IC-authentication scheme has the following properties. Correctness: When* IC-Distr *succeeds p_r learns a value v', where $v' = v$ unless p_s is actively corrupted.* IC-Distr *might abort only when p_s is incorrect. Completeness: If* IC-Distr *succeeds and p_r is correct then in* IC-Reveal *p_v accepts v'. Privacy:* IC-Distr *leaks no information on v to any player other than p_r. Unforgeability: When neither p_s nor p_v is passively corrupted, and the protocols* IC-Distr *and* IC-Reveal *have been invoked at most polynomially many times, then the probability that an adversary actively corrupting p_r makes p_v accept some v' which was not sent with* IC-Distr *is negligible.*

General IC-signatures. An IC-authentication scheme allows a sender $p_i \in \mathcal{P}$ to send a value v to a recipient $p_j \in \mathcal{P}$, so that p_j can later prove authenticity of v, but *only* towards a dedicated verifier $p_k \in \mathcal{P}$. In our protocols we want to use IC-authentication as a mechanism to bind the sender p_i to the messages he sends to p_j, so that p_j can prove to *every* $p_k \in \mathcal{P}$ that these messages originate from p_i. In [CDD+99], the IC-signatures where introduced for this purpose. These can be seen as semi "digital signatures" with information theoretic security. They do not achieve all properties of digital signatures, but enough to guarantee the security of our protocols.

The protocols used for generation and verification of IC-signatures are called ICS-Sign and ICS-Open, respectively. ICS-Sign allows a player $p_i \in \mathcal{P}$ to send a value v to $p_j \in \mathcal{P}$ signed with an IC-signature. The idea is the following: for each $p_k \in \mathcal{P}$, p_i invokes IC-Distr to send v to p_j with p_k being the verifier, where p_j checks that he receives the same v in all invocations. As syntactic sugar, we denote the resulting IC-signature by $\sigma_{i,j}(v)$. The idea in ICS-Open is the following: p_j announces v and invokes IC-Reveal once for each $p_k \in \mathcal{P}$ being the verifier. Depending on the outcomes of IC-Reveal the players decide to accept or reject v. As we want every $p_i \in \mathcal{P}$ to be able to send messages with ICS-Sign, we need a secret-key setup, where every $p_i, p_k \in \mathcal{P}$ hold a secret key $\alpha_{i,k}$. Such a setup can be easily established by appropriate invocations of IC-Setup.

The decision to accept or reject in ICS-Open has to be taken in a way which ensures that valid signatures are accepted (completeness), and forged signatures are rejected with overwhelming probability (unforgeability). To guarantee completeness, a signature must not be rejected when only actively corrupted players rejected in IC-Reveal. Hence, the players cannot reject the signature when there exists a class $(A_j, E_j, F_j) \in \mathcal{Z}$ such that all rejecting players are in A_j. Along the same lines, to guarantee unforgeability, the players cannot accept the signature when there exists a class $(A_i, E_i, F_i) \in \mathcal{Z}$ such that all accepting players are in E_i. To make sure that the above two cases cannot simultaneously occur, we require $\mathcal{Z}$ to satisfy the following property, denoted as $C_{\mathrm{IC}}(\mathcal{P}, \mathcal{Z})$:

$$C_{\mathrm{IC}}(\mathcal{P}, \mathcal{Z}) \iff \forall (A_i, E_i, F_i), (A_j, E_j, F_j) \in \mathcal{Z} : \; E_i \cup A_j \cup (F_i \cap F_j) \neq \mathcal{P}$$

We refer to the full version of this paper for a detailed description of the protocols ICS-Sign and ICS-Open and for a proof of the following lemma.

Lemma 1. *Assuming that $C_{\mathrm{IC}}(\mathcal{P}, \mathcal{Z})$ holds, our IC-signatures scheme has the following properties. Correctness: When* ICS-Sign *succeeds, then p_r learns a value v', where $v' = v$ unless p_s is actively corrupted.* ICS-Sign *might abort only when p_s is incorrect. Completeness: If* ICS-Sign *succeeds and p_r is correct then in* ICS-Open *all players accept v'. Privacy:* ICS-Sign *leaks no information on v to any player other than p_r. Unforgeability: When p_s is not passively corrupted, and the protocols* ICS-Sign *and* ICS-Open *have been invoked at most polynomially many times, then the probability that an adversary actively corrupting p_j can make the players accept some v' which was not sent with* ICS-Sign *is negligible.*

Linearity of IC-signatures. The linearity property of the IC-authentication scheme is propagated to the IC-signatures. In particular, when some values have be sent by p_i to p_j with ICS-Sign (using the same secret keys), then the players can locally, i.e., without any interaction, compute p_i's signature for any linear combination of those values, by applying the appropriate linear combination on the respective signatures. This process yields a signature which, when p_j is correct, will be accepted in ICS-Open.

4 Tools - Subprotocols

In this section we describe sub-protocols that are used as building blocks for MPC and SFE protocols. Some of the sub-protocols are non-robust, i.e., they might abort. When they abort then all (correct) players agree on a non-empty set $B \subseteq \mathcal{P}$ of incorrect players. The sub-protocols use IC-signatures to authenticate the sent values, therefore their security relies on the security of the IC-signatures. In particular, the security of the sub-protocols is guaranteed only when no signature is forged.[3] The secret-key setup, which is required for the IC-signatures, is established in a setup phase, before any of the sub-protocols is invoked. Due to space restrictions the security proofs and even the detailed descriptions of some of the sub-protocols are deleted from this extended abstract.

4.1 Share and Reconstruct

A secret-sharing scheme allows a player (called the dealer) to distribute a secret so that only qualified sets of players can reconstruct it. As secret-sharing scheme we employ a

[3] We use the term "forge" only for signatures corresponding to non-passively corrupted signers.

sum-sharing, i.e., the secret is split into summands that add up to the secret, where each summand might be given to several players. Additionally, for each summand all the players who hold it bilaterally exchange signatures on it. The sharing is characterized by a vector $\mathcal{S} = (S_1, \ldots, S_m)$ of subsets of $\mathcal{P}$, called the *sharing specification*. A value s is shared according to $\mathcal{S}$ if there exist summands $s_1, \ldots, s_m \in \mathbb{F}$ such that $\sum_{k=1}^{m} s_k = s$, and for each $k = 1, \ldots, m$ every $p_j \in S_k$ holds s_k along with IC-signatures on it from every $p_i \in S_k$. As syntactic sugar, we denote by $\sigma_{\mathcal{S}}(s)$ the set of all IC-signatures on the summands $s_1, \ldots, s_m$ held by the players. For each $p_j \in \mathcal{P}$ the vector $\langle s \rangle_j = (s_{j_1}, \ldots, s_{j_\ell})$ is considered to be p_j's *share* of s, where $s_{j_1}, \ldots, s_{j_\ell}$ are the summands held by p_j. The vector of all shares and the attached signatures, denoted as $\langle s \rangle = (\langle s \rangle_1, \ldots, \langle s \rangle_n, \sigma_{\mathcal{S}}(s))$, is a *sharing* of s. The vector of summands in $\langle s \rangle$ is denoted as $[s] = (s_1, \ldots, s_m)$. We say that $\langle s \rangle$ is a *consistent* sharing of s according to $\mathcal{S}$ if for each $k = 1, \ldots, m$ all (correct) players in S_k have the same view on the summands s_k and hold signatures on it from all other players in S_k, and $\sum_{k=1}^{m} s_k = s$.

For an adversary structure $\mathcal{Z}$, we say that a sharing specification $\mathcal{S}$ is *$\mathcal{Z}$-private* if for any sharing $\langle s \rangle$ according to $\mathcal{S}$ and for any adversary in $\mathcal{Z}$, there exists a summand s_k which this adversary does not know. Formally, $\mathcal{S}$ is $\mathcal{Z}$-private if $\forall (A, E, F) \in \mathcal{Z}\ \exists S \in \mathcal{S}: \ S \cap E = \emptyset$.[4] For an adversary structure $\mathcal{Z}$ with maximal classes $\overline{\mathcal{Z}} = \{(\cdot, E_1, \cdot), \ldots, (\cdot, E_m, \cdot)\}$, we denote the natural $\mathcal{Z}$-private sharing specification by $S_{\mathcal{Z}} = (\mathcal{P} \setminus E_1, \ldots, \mathcal{P} \setminus E_m)$.

Protocol Share (see below) allows a dealer p to share a value s among the players in $\mathcal{P}$ according to a sharing specification $\mathcal{S}$. The protocol is non-robust and might abort with a set $B \subseteq \mathcal{P}$ of incorrect players.

Protocol Share($\mathcal{P}, \mathcal{Z}, \mathcal{S}, p, s$)

1. Dealer p chooses summands $s_2, \ldots, s_{|\mathcal{S}|}$ randomly and sets $s_1 := s - \sum_{k=2}^{|\mathcal{S}|} s_k$.
2. For $k = 1, \ldots, |\mathcal{S}|$ the following steps are executed:
   (a) p sends s_k to each $p_j \in S_k$.
   (b) For each $p_i, p_j \in S_k$: ICS-Sign($\mathcal{P}, \mathcal{Z}, p_i, p_j, s_k$) is invoked to have p_i send s_k to p_j and attach an IC-signature on it. If ICS-Sign aborts, then Share aborts with $B := \{p_i\}$.
   (c) Each $p_j \in S_k$ broadcasts a complaint bit b, where $b = 1$ if p_j received a $\perp$ instead of s_k in Step 2a, or if he received some $s'_k \neq s_k$ from some p_i in Step 2b, and $b = 0$ otherwise.
   (d) If a complaint was reported p broadcasts s_k and the players in S_k create default signatures on it. If p broadcasts $\perp$ then Share aborts with set $B := \{p\}$.

Lemma 2. *If $\mathcal{S}$ is a $\mathcal{Z}$-private sharing specification, then protocol* Share($\mathcal{P}, \mathcal{Z}, \mathcal{S}, p, s$) *has the following properties. Correctness: It either outputs a consistent sharing of s' according to $\mathcal{S}$, where $s' = s$ unless the dealer p is actively corrupted, or it aborts with*

[4] Recall that for all $(A, E, F) \in \mathcal{Z}: \ A \subseteq E$.

a non-empty set $B \subseteq \mathcal{P}$ of incorrect players. Privacy: No information about s leaks to the adversary.

Reconstructing a shared value s is straightforward: The summands are announced one by one, and s is computed as the sum of the announced summands. To announce a summand s_k, each $p_i \in S_k$ broadcasts s_k and opens all the signatures on s_k which he holds (i.e., the signatures on s_k from all players in S_k). If all the signatures announced by p_i are accepted, then the value he announced is taken for s_k. If no $p_i \in S_k$ correctly announces all the signatures the announcing aborts with $B := S_k$. Protocols PubAnnounce and PubReconstruct invoked to publicly announce a summand and to publicly reconstruct a shared value are given in details in the full version of this paper. In the following two lemmas (also proved in the full version) we state their security.

Lemma 3. *Assume that $C_{\mathrm{IC}}(\mathcal{P}, \mathcal{Z})$ holds, the condition $\forall (A, E, F) \in \mathcal{Z} : S_k \not\subseteq E$ holds, and no signature is forged. Then protocol* PubAnnounce *either publicly announces the correct summand s_k, or it aborts with a non-empty set B of incorrect players. It might abort only if $S_k \subseteq F^\star$.*

Lemma 4. *Assume that $C_{\mathrm{IC}}(\mathcal{P}, \mathcal{Z})$ holds, the condition $\forall S \in \mathcal{S}$, $\forall (A, E, F) \in \mathcal{Z} : S \not\subseteq E$ holds, $\langle s \rangle$ is a consistent sharing according to $\mathcal{S}$, and no signature is forged. Then protocol* PubReconstruct *either publicly reconstructs s, or it aborts with a non-empty set $B \subseteq \mathcal{P}$ of incorrect players.*

Protocol PubReconstruct allows for public reconstruction of a shared value. However, in some of our protocols we need to reconstruct a shared value s privately, i.e., only towards some dedicated output player p. Such a private reconstruction protocol can be built using standard techniques (p shares a one-time pad used for perfectly blinding the output). We refer to the protocol for private reconstruction as Reconstruct, and point to the full version of this paper for a detailed description as well as for a proof of the following lemma.

Lemma 5. *Assume that $C_{\mathrm{IC}}(\mathcal{P}, \mathcal{Z})$ holds, $\mathcal{S}$ is a $\mathcal{Z}$-private sharing specification, the condition $\forall S \in \mathcal{S}$, $\forall (\cdot, E, \cdot) \in \mathcal{Z} : S \not\subseteq E$ holds, $\langle s \rangle$ is a consistent sharing according to $\mathcal{S}$, and no signature is forged. Then protocol* Reconstruct$(\mathcal{P}, \mathcal{Z}, \mathcal{S}, p, \langle s \rangle)$ *has the following properties. Correctness: Either it reconstructs s towards p, or it aborts with a non-empty set $B \subseteq \mathcal{P}$ of incorrect players. Privacy: No information about $\langle s \rangle$ leaks to the adversary.*

Addition. Due to the linearity of our secret sharing scheme, the players can locally compute a sharing of the sum of two shared values s and t as follows: each player adds his shares of s and t, and the corresponding signatures are also (locally) added. We refer to this sub-protocol as Add.

4.2 Multiplication

The goal of this section is to design a protocol for securely computing a sharing of the product of two shared values. Our approach combines techniques from [GRR98, Mau02, Mau06, BFH+08].

At a high level, the multiplication protocol for two shared values s and t works as follows: As s and t are already shared, we can use the summands $s_1, \ldots, s_m$ and $t_1, \ldots, t_m$ to compute the product as $st = \sum_{k,\ell=1}^{m} s_k t_\ell$. For each term $x_{k,\ell} = s_k t_\ell$, we have a player $p^{(k,\ell)} \in (S_k \cap S_\ell)$ share $x_{k,\ell}$ and prove that he shared the correct value. The sharing of st is computed as the sum of the sharings of the terms $x_{k,\ell}$.

For $p^{(k,\ell)} \in (S_k \cap S_\ell)$ to share $s_k t_\ell$ and prove that he did so properly the idea is the following: First, $p^{(k,\ell)}$ shares $s_k t_\ell$ by invoking Share. Denote by $x'_{k,\ell}$ the shared value.[5] Next, $p^{(k,\ell)}$ shares the summands s_k and t_ℓ by a protocol, called SumShare, which guarantees that he shares the correct summands. Finally, $p^{(k,\ell)}$ uses the sharings of s_k, t_ℓ, and $x'_{k,\ell}$ in a protocol, called MultProof, which allows him to prove that $x'_{k,\ell} = s_k t_\ell$. In the following we discuss the sub-protocols SumShare and MultProof, and then give a detailed description of the multiplication protocol.

Protocol SumShare (see full version) allows a player $p \in S_k$ to share a summand s_k of a sharing $\langle s \rangle$ according to $\mathcal{S}$, where $S_k \in \mathcal{S}$. The sharing specification of the output sharing can be some $\mathcal{S}' \neq \mathcal{S}$. In contrast to Share, protocol SumShare guarantees that p_i shares the correct value s_k. The idea is to have p share s_k, by Share, and then reconstruct the sharing (privately) towards each $p_j \in S_k$ who publicly approves or disapproves it. We refer to the full version of this paper for a proof of the following lemma.

Lemma 6. *Assume that $C_{\text{IC}}(\mathcal{P}, \mathcal{Z})$ holds, $\mathcal{S}'$ is a $\mathcal{Z}$-private sharing specification, the conditions $\forall(\cdot, E, \cdot) \in \mathcal{Z} : \;\; S_k \not\subseteq E$ and $\forall S' \in \mathcal{S}' \; \forall(\cdot, E, \cdot) \in \mathcal{Z} : \;\; S' \not\subseteq E$ hold, and no signature is forged. Then* SumShare$(\mathcal{P}, \mathcal{Z}, \mathcal{S}', S_k, p, s_k)$ *has the following properties. Correctness: Either it outputs a consistent sharing of s_k (p also outputs the vector $[s_k]$ of summands) according to $\mathcal{S}'$, or it aborts with a non-empty set $B \subseteq \mathcal{P}$ of incorrect players. Privacy: No information about s_k leaks to the adversary.*

Protocol MultProof (see full version) allows a player p, called the prover, who has shared three values a, b, and c (and knows the corresponding vectors $[a]$, $[b]$, and $[c]$ of summands) to prove that $c = ab$. The protocol can be seen as a distributed challenge-response protocol with prover p and verifier being all the players in $\mathcal{P}$. On a high level, it can be described as follows: First p shares some appropriately chosen values. Then the players jointly generate a uniformly random challenge r and expose it, and p answers the challenge. If p's answer is consistent with the sharings of a, b, and c and the sharings which he created in the first step, then the proof is accepted otherwise it is rejected. MultProof is non-robust and might abort with a set $B \subseteq \mathcal{P}$ of incorrect players. The proof of the following lemma is deleted from this extended abstract.

Lemma 7. *Assume that $C_{\text{IC}}(\mathcal{P}, \mathcal{Z})$ holds, $\mathcal{S}$ is a $\mathcal{Z}$-private sharing specification, the condition $\forall S \in \mathcal{S}, \; \forall(\cdot, E, \cdot) \in \mathcal{Z} : \;\; S \not\subseteq E$ holds, $\langle a \rangle, \langle b \rangle$, and $\langle c \rangle$ are consistent sharings according to $\mathcal{S}$, and no signature is forged. Then the protocol* MultProof *has the following properties. Correctness: If $c = ab$, then either the proof is accepted or* MultProof *aborts with a non-empty set $B \subseteq \mathcal{P}$ of incorrect players. Otherwise (i.e, if $c \neq ab$), with overwhelming probability, either the proof is rejected or* MultProof *aborts with a non-empty set $B \subseteq \mathcal{P}$ of incorrect players. Privacy: No information about $\langle a \rangle, \langle b \rangle$, and $\langle c \rangle$ leaks to the adversary.*

[5] Note that Share does not guarantee that $x'_{k,\ell} = s_k t_\ell$.

For completeness, we describe the multiplication protocol Mult (see below), which allows to compute a sharing of the product of two shared values. Mult is non-robust and might abort with a non-empty set $B \subseteq \mathcal{P}$ of incorrect players. When it succeeds, then with overwhelming probability it outputs a consistent sharing of the product.

Protocol Mult$(\mathcal{P}, \mathcal{Z}, \mathcal{S}, \langle s \rangle, \langle t \rangle)$

1. For every $(S_k, S_\ell) \in \mathcal{S} \times \mathcal{S}$, the following steps are executed, where $p^{(k,\ell)}$ denotes the player in $S_k \cap S_\ell$ with the smallest index:
   (a) $p^{(k,\ell)}$ computes $x_{k,\ell} := s_k t_\ell$ and shares it, by Share. Denote by $\langle x_{k,\ell} \rangle$ the resulting sharing.[a]
   (b) SumShare$(\mathcal{P}, \mathcal{Z}, \mathcal{S}, S_k, p^{(k,\ell)}, s_k)$ and SumShare$(\mathcal{P}, \mathcal{Z}, \mathcal{S}, S_\ell, p^{(k,\ell)}, t_\ell)$ are invoked. Denote by $\langle s_k \rangle$ and $\langle t_\ell \rangle$ the resulting sharings.
   (c) MultProof$(\mathcal{P}, \mathcal{Z}, \mathcal{S}, p^{(k,\ell)}, \langle s_k \rangle, \langle t_\ell \rangle, \langle x_{k,\ell} \rangle)$ is invoked. If the proof is rejected then Mult aborts with set $B = \{p^{(k,\ell)}\}$.
2. A sharing of the product st is computed as the sum of the sharings $\langle x_{k,\ell} \rangle$ by repeatedly invoking Add.
3. If any of the invoked sub-protocols aborts with B, then also Mult aborts with B.

[a] In addition to his share of $\langle x_{k,\ell} \rangle$, $p^{(k,\ell)}$ also outputs the vector of summands $[x_{k,\ell}]$.

Lemma 8. *Assume that $C_{\text{IC}}(\mathcal{P}, \mathcal{Z})$ holds, $\mathcal{S}$ is a $\mathcal{Z}$-private sharing specification, the conditions $\forall S \in \mathcal{S}, \forall(\cdot, E, \cdot) \in \mathcal{Z} : S \not\subseteq E$ and $\forall S_k, S_\ell \in \mathcal{S} : S_k \cap S_\ell \neq \emptyset$ hold, $\langle s \rangle$ and $\langle t \rangle$ are consistent sharings according to $\mathcal{S}$, and no signature is forged. Then protocol* Mult$(\mathcal{P}, \mathcal{Z}, \mathcal{S}, \langle s \rangle, \langle t \rangle)$ *has the following properties except with negligible probability. Correctness: It either outputs a consistent sharing of st according to $\mathcal{S}$ or it aborts with a non-empty set $B \subseteq \mathcal{P}$ of incorrect players. Privacy: No information about $\langle s \rangle$ and $\langle t \rangle$ leaks to the adversary.*

4.3 Resharing

In the context of MPC, we will need to reshare shared values according to a different sharing specification. To do that, each summand is shared by SumShare (see Section 4.2) according to the new sharing specification, and the players distributively add the sharings of the summands, resulting in a new sharing of the original value. A detailed description of the protocol Reshare as well as a proof of the following lemma can be found in the full version of this paper.

Lemma 9. *Assume that $C_{\text{IC}}(\mathcal{P}, \mathcal{Z})$ holds, $\mathcal{S}'$ is a $\mathcal{Z}$-private sharing specification, the conditions $\forall S \in \mathcal{S} \forall(\cdot, E, \cdot) \in \mathcal{Z} : S \not\subseteq E$, and $\forall S' \in \mathcal{S}' \forall(\cdot, E, \cdot) \in \mathcal{Z} : S' \not\subseteq E$ hold, and no signature is forged. Then* Reshare$(\mathcal{P}, \mathcal{Z}, \mathcal{S}, \mathcal{S}', \langle s \rangle)$ *has the following properties. Correctness: Either it outputs a consistent sharing of s according to $\mathcal{S}'$, or it aborts with a non-empty set $B \subseteq \mathcal{P}$ of incorrect players. Privacy: No information about $\langle s \rangle$ leaks to the adversary.*

5 (Reactive) Multi-party Computation

In this section we prove the necessary and sufficient condition on the adversary structure $\mathcal{Z}$ for the existence of unconditionally (i.e., i.t. with negligible error probability) $\mathcal{Z}$-secure multi-party computation protocols, namely, we prove the following theorem:

Theorem 2. *A set $\mathcal{P}$ of players can unconditionally $\mathcal{Z}$-securely compute any (reactive) computation, if and only if $C^{(2)}(\mathcal{P}, \mathcal{Z})$ and $C^{(1)}(\mathcal{P}, \mathcal{Z})$ hold, where*

$$C^{(2)}(\mathcal{P}, \mathcal{Z}) \iff \forall (A_i, E_i, F_i), (A_j, E_j, F_j) : \; E_i \cup E_j \cup (F_i \cap F_j) \neq \mathcal{P}$$
$$C^{(1)}(\mathcal{P}, \mathcal{Z}) \iff \forall (A_i, E_i, F_i), (A_j, E_j, F_j) : \; E_i \cup F_j \neq \mathcal{P}$$

The sufficiency of the above condition is proved by constructing an MPC protocol for any given circuit C consisting of input, addition, multiplication, and output gates.[6] The reactiveness of the computation is modeled by assigning to each gate a point in time when it should be evaluated.

The circuit is evaluated in a gate-by-gate fashion, where for input, addition, multiplication, and output gates, the corresponding sub-protocol Share, Add, Mult, and Reconstruct, respectively, is invoked.

The computation starts off with the initial player set $\mathcal{P}$ and adversary structure $\mathcal{Z}$, and with the sharing specification being $\mathcal{S} := \mathcal{S}_{\mathcal{Z}}$. Each time a sub-protocol aborts with set B of incorrect players, the players in B are deleted from the player set and from every set in the sharing specification, and the corresponding gate is repeated. Any future invocation of a sub-protocol is done in the updated player set $\mathcal{P}'$ and sharing specification $\mathcal{S}'$, and with the updated adversary structure $\mathcal{Z}'$, which contains only the classes in $\mathcal{Z}$ compatible with the players in $\mathcal{P} \setminus \mathcal{P}'$ being incorrect. Note that, as the players in $\mathcal{P} \setminus \mathcal{P}'$ are incorrect, any sharing according to $(\mathcal{P}, \mathcal{S})$ can be transformed, without any interaction, to a sharing according to $(\mathcal{P}', \mathcal{S}')$ by having the players delete all signatures of signers from $\mathcal{P} \setminus \mathcal{P}'$.

The delicate task is the multiplication of two shared values s and t. The idea is the following: First, we invoke Reshare to have both s and t shared according to the sharing specification $\mathcal{S}_{\mathcal{Z}'}$, i.e., the specification associated with the structure $\mathcal{Z}'$. Then we invoke Mult to compute a sharing of the product st according to $\mathcal{S}_{\mathcal{Z}'}$, and at the end we invoke Reshare once again to have the product shared back to the initial setting (i.e, according to $(\mathcal{P}', \mathcal{S}')$).

The security of the computation is guaranteed as long as no signature is forged. We argue that the forging probability is negligible. Observe that the total number of signatures in each sub-protocol invocation is polynomial in the input size; also, the total number of sub-protocol invocations is polynomial in the size of the circuit (since each time a sub-protocol aborts a new set B of incorrect players is identified, the total number of abortions is bounded by n). Hence, the total number of signatures in the computation is polynomial and, by the unforgeability property, the probability that a signature is forged is negligible.

We use the following operators on adversary structures, which were introduced in [BFH$^+$08]: For a set $B \subseteq \mathcal{P}$, we denote by $\mathcal{Z}|^{B \subseteq F}$ the sub-structure of $\mathcal{Z}$ that contains

[6] This does not exclude probabilistic circuits, as a random gate can be simulated by having each player input a random value and take the sum of those values as the input.

only adversaries who can fail-corrupt all the players in B, i.e., $\mathcal{Z}|^{B\subseteq F} = \{(A,E,F) \in \mathcal{Z} : B \subseteq F\}$. Furthermore, for a set $\mathcal{P}' \subseteq \mathcal{P}$, we denote by $\mathcal{Z}|_{\mathcal{P}'}$ the adversary structure with all classes in $\mathcal{Z}$ restricted to the player set $\mathcal{P}'$, i.e., $\mathcal{Z}|_{\mathcal{P}'} = \{(A\cap\mathcal{P}', E\cap\mathcal{P}', F\cap\mathcal{P}') : (A,E,F) \in \mathcal{Z}\}$. We also use the same operator on sharing specifications with similar semantics, i.e., for $\mathcal{S} = (S_1,\ldots,S_m)$ we denote $\mathcal{S}|_{\mathcal{P}'} = (S_1\cap\mathcal{P}',\ldots,S_m\cap\mathcal{P}')$. As syntactic sugar, we write $\mathcal{Z}|^{B\subseteq F}_{\mathcal{P}'}$ for $(\mathcal{Z}|^{B\subseteq F})|_{\mathcal{P}'}$.

It follows from the above definitions that when the players in $\mathcal{P}\setminus\mathcal{P}'$ have been detected to be incorrect, then the actual adversary $Z^\star$ is in $\mathcal{Z}|^{\mathcal{P}\setminus\mathcal{P}'\subseteq F}$. Furthermore, as the updated player set is $\mathcal{P}'$, the corresponding sharing specification and adversary structure are $\mathcal{S}' = \mathcal{S}|_{\mathcal{P}'}$ and $\mathcal{Z}' = \mathcal{Z}|^{\mathcal{P}\setminus\mathcal{P}'\subseteq F}_{\mathcal{P}'}$, respectively. One can easily verify that the conditions $C^{(2)}$ and $C^{(1)}$ hold in $(\mathcal{P}',\mathcal{Z}')$ when they hold in $(\mathcal{P},\mathcal{Z})$. This results in protocol MPC (see below).

Protocol MPC$(\mathcal{P},\mathcal{Z},C)$

0. Initialize $\mathcal{P}' := \mathcal{P}$, $\mathcal{Z}' := \mathcal{Z}$, and $\mathcal{S}' := \mathcal{S}_{\mathcal{Z}}$.
1. For every gate to be evaluated, do the following:
   - *Input gate for p:* If $p \in \mathcal{P}'$ invoke Share to have p share his input according to $(\mathcal{P}',\mathcal{S}')$. Otherwise, a default sharing of some pre-agreed default value is taken as the sharing of p's input.
   - *Addition gate:* Invoke Add to compute a sharing of the sum according to $\mathcal{S}'$.
   - *Multiplication gate:* Denote the sharings of the factors as $\langle s\rangle$ and $\langle t\rangle$, respectively, and the sharing specification corresponding to $\mathcal{Z}'$ as $\mathcal{S}_{\mathcal{Z}'}$. Invoke Reshare$(\mathcal{P}',\mathcal{Z}',\mathcal{S}',\mathcal{S}_{\mathcal{Z}'},\langle s\rangle)$ and Reshare$(\mathcal{P}',\mathcal{Z}',\mathcal{S}',\mathcal{S}_{\mathcal{Z}'},\langle t\rangle)$ to obtain the sharings $\langle s\rangle'$ and $\langle t\rangle'$ according to $(\mathcal{P}',\mathcal{S}_{\mathcal{Z}'})$, respectively. Invoke Mult$(\mathcal{P}',\mathcal{Z}',\mathcal{S}_{\mathcal{Z}'},\langle s\rangle',\langle t\rangle')$ to obtain a sharing $\langle st\rangle'$ of the product, according to $(\mathcal{P}',\mathcal{S}_{\mathcal{Z}'})$. Invoke Reshare$(\mathcal{P}',\mathcal{Z}',\mathcal{S}_{\mathcal{Z}'},\mathcal{S}',\langle st\rangle')$ to reshare this product according to $(\mathcal{P}',\mathcal{S}')$.
   - *Output gate for p:* If $p \in \mathcal{P}'$ invoke Reconstruct to have the output reconstructed towards p.
2. If any of the sub-protocols aborts with set B, then update $\mathcal{P}' := \mathcal{P}'\setminus B$, set $\mathcal{S}' := \mathcal{S}'|_{\mathcal{P}'}$ and $\mathcal{Z}' := \mathcal{Z}|^{\mathcal{P}\setminus\mathcal{P}'\subseteq F}_{\mathcal{P}'}$ and repeat the corresponding gate.

Lemma 10. *The protocol* MPC *is unconditionally $\mathcal{Z}$-secure if $C^{(2)}(\mathcal{P},\mathcal{Z})$ and $C^{(1)}(\mathcal{P},\mathcal{Z})$ hold.*

To complete this section, we give two lemmas that imply that unconditionally secure (reactive) MPC is not possible for some circuits when $C^{(2)}(\mathcal{P},\mathcal{Z})$ or $C^{(1)}(\mathcal{P},\mathcal{Z})$ is violated. The proofs of the lemmas are deleted from this extended abstract.

Lemma 11. *If $C^{(2)}(\mathcal{P},\mathcal{Z})$ is violated then there exist (even non-reactive) circuits which cannot be evaluated unconditionally $\mathcal{Z}$-securely.*

Lemma 12. *If $C^{(1)}(\mathcal{P},\mathcal{Z})$ is violated, then the players cannot hold a secret joint state with unconditional security.*

6 Secure Function Evaluation

In this section we prove the necessary and sufficient condition on the adversary structure $\mathcal{Z}$ for the existence of unconditionally $\mathcal{Z}$-secure function evaluation protocols. Note that the condition for SFE is weaker than the condition for MPC.

Theorem 3. *A set $\mathcal{P}$ of players can unconditionally $\mathcal{Z}$-securely compute any function if and only if $C^{(2)}(\mathcal{P},\mathcal{Z})$ and $C^{(1)}_{\mathsf{ORD}}(\mathcal{P},\mathcal{Z})$ hold, where*

$$C^{(2)}(\mathcal{P},\mathcal{Z}) \iff \forall (A_i,E_i,F_i),(A_j,E_j,F_j) \in \mathcal{Z}:\ E_i \cup E_j \cup (F_i \cap F_j) \neq \mathcal{P}$$

$$C^{(1)}_{\mathsf{ORD}}(\mathcal{P},\mathcal{Z}) \iff \begin{cases} \exists \text{ an ordering } ((A_1,E_1,F_1),\ldots,(A_m,E_m,F_m)) \text{ of } \overline{\mathcal{Z}} \text{ s.t.}^7 \\ \forall i,j \in \{1,\ldots,m\},\ i \leq j:\ E_j \cup F_i \neq \mathcal{P} \end{cases}$$

The sufficiency of the condition is proved by constructing an SFE protocol. Our approach is similar to the approach from [BFH$^+$08]: First all players share their inputs, then the circuit is evaluated gate-by-gate, and then the output is publicly reconstructed. However, our conditions do not guarantee robust reconstructibility. In fact, the adversary can break down the computation and cause all the sharings to be lost. As the circuit is non-reactive, we handle such an abortion by repeating the whole protocol, including the input gates. In each repetition, the adversary might choose new inputs for the actively corrupted players. By ensuring that the adversary gets no information on any secrets unless the full protocol succeeds (including the evaluation of output gates), we make sure that she chooses these inputs independently of the other players' inputs.

Termination is guaranteed, by the fact that whenever the protocol aborts, a new set B of incorrect players is identified, and the next iteration proceeds without them. Hence, the number of iterations is bounded by n. This implies also that the total number of signatures in the computation is polynomial, hence the forging probability is negligible.

Special care needs to be taken in the design of the output protocol. For simplicity, we describe the protocol for a single public output. Using standard techniques one can extend it to allow several outputs and, furthermore, private outputs.

The idea of the output protocol is the following: First observe that the privacy of our sharing scheme is protected by a particular summand which is not given to the adversary. In fact, such a summand s_k is guaranteed to exist for each $(A_k,E_k,F_k) \in \mathcal{Z}$ by the $\mathcal{Z}$-privacy of the sharing specification $\mathcal{S}_{\mathcal{Z}}$. As long as this summand is not published, an adversary of class (A_k,E_k,F_k) gets no information about the output (from the adversary's point of view, s_k is a perfect blinding of the output, and all other summands s_i are either known to the adversary or are distributed uniformly). Second, observe that whenever the publishing of some summand s_k fails (i.e., $\mathsf{PubAnnounce}$ aborts), the players get information about the actual adversary $(A^\star,E^\star,F^\star)$, namely that $S_k \subseteq F^\star$. The trick is to announce the summands in such an order, that if the announcing of a summand s_k aborts, then from the information that $S_k \subseteq F^\star$ the players can deduce that the summand associated with the actual adversary class has not been yet announced. In particular, if an adversary class $Z_i = (A_i,E_i,F_i)$ could potentially abort the announcing of the summand s_k (i.e., if $S_k \subseteq F_i$), then the summand s_k should be announced strictly before s_i, i.e., the summand associated with Z_i, is announced.

[7] Remember that $\overline{\mathcal{Z}}$ denotes the maximum classes in $\mathcal{Z}$. One can verify that such an ordering exists for $\overline{\mathcal{Z}}$ exactly if it exists for $\mathcal{Z}$.

Let $((A_1, E_1, F_1), \ldots, (A_m, E_m, F_m))$ denote an ordering of the maximal structure $\overline{\mathcal{Z}}$ satisfying: $\forall 1 \leq i \leq j \leq m : E_j \cup F_i \neq \mathcal{P}$, and let $\mathcal{S}$ denote the induced sharing specification $\mathcal{S} = (S_1, \ldots, S_m)$ with $S_k = \mathcal{P} \setminus E_k$. Then the protocol OutputGeneration (see below) either publicly reconstructs a sharing $\langle s \rangle$ according to $\mathcal{S}$ or it aborts with a non-empty set $B \subseteq \mathcal{P}$ of incorrect players. Privacy is guaranteed under the assumption that the summands of $\langle s \rangle$ not known to the adversary are uniformly distributed. As long as no signature is forged, this holds for all sharings in our protocols.

Protocol OutputGeneration($\mathcal{P}, \mathcal{Z}, \mathcal{S} = (S_1, \ldots, S_m), \langle s \rangle$)

1. For $k = 1, \ldots, m$, the following steps are executed *sequentially*:
   (a) PubAnnounce($\mathcal{P}, \mathcal{Z}, S_k, s_k, \sigma_{S_k}(s_k)$) is invoked to have the summand s_k published.
   (b) If PubAnnounce aborts with B, then OutputGeneration *immediately* aborts with B.
2. Every $p_j \in \mathcal{P}$ (locally) computes $s := \sum_{k=1}^{m} s_k$ and outputs s.

Lemma 13. *Assume that $C_{\text{IC}}(\mathcal{P}, \mathcal{Z})$ holds, $\mathcal{S}$ is a $\mathcal{Z}$-private sharing specification constructed as explained, the condition $\forall S_k \in \mathcal{S}, (\cdot, E, \cdot) \in \mathcal{Z} : S_k \not\subseteq E$ holds, $\langle s \rangle$ is a consistent sharing according to $\mathcal{S}$ with the property that those summands that are unknown to the adversary are randomly chosen, and no signature is forged. Then the protocol* OutputGeneration *either publicly reconstructs s, or it aborts with a non-empty set $B \subseteq \mathcal{P}$ of incorrect players. If* OutputGeneration *aborts, then the protocol does not leak any information on s to the adversary.*

For completeness, we also include a detailed description of the SFE protocol (see below) and state its security in the following lemma.

Protocol SFE($\mathcal{P}, \mathcal{Z}, C$)

0. Let $\mathcal{S} = (\mathcal{P} \setminus E_1, \ldots, \mathcal{P} \setminus E_m)$ for the assumed ordering $((A_1, E_1, F_1), \ldots, (A_m, E_m, F_m))$ of $\overline{\mathcal{Z}}$.
1. *Input stage:* For every input gate in C, Share is invoked to have the input player p_i share his input x_i according to $\mathcal{S}$.[a]
2. *Computation stage:* The gates in C are evaluated as follows:
   - *Addition gate:* Invoke Add to compute a sharing of the sum according to $\mathcal{S}$.
   - *Multiplication gate:* Invoke Mult to compute a sharing of the product according to $\mathcal{S}$.
3. *Output stage:* Invoke OutputGeneration($\mathcal{P}, \mathcal{Z}, \mathcal{S}, \langle s \rangle$) for the sharing $\langle s \rangle$ of the public output.
4. If any of the sub-protocols aborts with B, then set $\mathcal{P} := \mathcal{P} \setminus B$, and set $\mathcal{Z}$ to the adversary structure which is compatible with B being incorrect, i.e., $\mathcal{Z} := \mathcal{Z}|_{\mathcal{P}}^{B \subseteq F}$, and go to Step 1.

[a] If in a later iteration a player $p_i \notin \mathcal{P}$ should give input, then the players in $\mathcal{P}$ pick the default sharing of a default value.

Lemma 14. *The protocol* SFE *is unconditionally $\mathcal{Z}$-secure if $C^{(2)}(\mathcal{P},\mathcal{Z})$ and $C^{(1)}_{\mathsf{ORD}}(\mathcal{P},\mathcal{Z})$ hold.*

To complete the proof of Theorem 3 we need to show that unconditionally $\mathcal{Z}$-secure SFE is not possible for some circuits when $C^{(2)}(\mathcal{P},\mathcal{Z})$ or $C^{(1)}_{\mathsf{ORD}}(\mathcal{P},\mathcal{Z})$ is violated. The necessity of $C^{(2)}(\mathcal{P},\mathcal{Z})$ follows immediately from Lemma 11. The following lemma states the necessity of $C^{(1)}_{\mathsf{ORD}}(\mathcal{P},\mathcal{Z})$. The idea of the proof is that when $C^{(1)}_{\mathsf{ORD}}(\mathcal{P},\mathcal{Z})$ is violated then in any protocol evaluating the identity function, the adversary can break down the computation at a point where she has gained noticeable (i.e., not negligible) information about the output, although the correct players have only negligible information. For a more detailed proof the reader is referred to the full version of this paper.

Lemma 15. *If $C^{(1)}_{\mathsf{ORD}}(\mathcal{P},\mathcal{Z})$ is violated, then there are functions that cannot be unconditionally $\mathcal{Z}$-securely evaluated.*

7 Computational Security

In this section we show that conditions $C^{(1)}(\mathcal{P},\mathcal{Z})$ and $C^{(1)}_{\mathsf{ORD}}(\mathcal{P},\mathcal{Z})$ from Theorems 2 and 3 are sufficient and necessary for the existence of computationally $\mathcal{Z}$-secure MPC and SFE, respectively.

Theorem 4. *Assuming that enhanced trapdoor permutations exist, a set $\mathcal{P}$ of players can computationally $\mathcal{Z}$-securely compute any (reactive) computation (MPC) if and only if $C^{(1)}(\mathcal{P},\mathcal{Z})$ holds, and any non-reactive function (SFE) if and only if $C^{(1)}_{\mathsf{ORD}}(\mathcal{P},\mathcal{Z})$ holds.*

The proof of necessity is very similar to the proofs of Lemmas 12and 15 and, therefore, it is omitted. The sufficiency is proved by describing protocols that realize the corresponding primitive. Our approach is different than the one used in the previous sections. In particular, first, we design a protocol for SFE and then use it to design a protocol for MPC.

Note that the above bounds directly imply corresponding bounds for a threshold adversary who actively corrupts t_a players, passively corrupts t_p players, and fail-corrupts t_f players, simultaneously. Using the notation from [FHM98], we say that a protocol is (t_a,t_p,t_f)-*secure* if it tolerates such a threshold adversary.

Corollary 1. *Assuming that enhanced trapdoor permutations exist, a set $\mathcal{P}$ of players can computationally (t_a,t_p,t_f)-securely compute any computation (reactive or not) if and only if $2t_a+t_p+t_f<|\mathcal{P}|$.*

7.1 The SFE Protocol

Our approach to SFE uses ideas from [IKLP06]. The evaluation of the given circuit C proceeds in two stages, called the *computation stage* and the *output stage*. In the computation stage a uniformly random *sharing* of the output of C on inputs provided by the players is computed.[8] For this purpose we use the (non-robust) SFE protocol

[8] Without loss of generality (as in Section 6) we assume that the circuit C to be computed has one public output.

from [Gol04] for dishonest majority which achieves partial fairness and unanimous abort [GL02]. In the output stage the sharing of the output is publicly reconstructed, along the lines of the reconstruction protocol from Section 6. Both stages are non-robust and they might abort with a non-empty set $B \subseteq \mathcal{P}$ of incorrect players, but without violating privacy of the inputs. When this happens the whole evaluation is repeated among the players in $\mathcal{P} \setminus B$, where the inputs of the players in B are fixed to a default pre-agreed value, and the adversary structure $\mathcal{Z}$ is reduced to the structure $\mathcal{Z}|_{\mathcal{P}\setminus B}^{B\subseteq F}$, i.e., the structure which is compatible with the players in B being incorrect.

The secret-sharing scheme used here is similar to the one we use in the unconditional-security case. More precisely, the secret is split into uniformly random summands $s_1, \ldots, s_m \in \mathbb{F}$ that add up to the secret, where each player might hold several of those summands, according to some sharing specification $\mathcal{S} = (S_1, \ldots, S_m)$. The difference is that the players do not hold signatures on their summands, but they are committed to them (towards all players) by a perfectly hiding commitment scheme.[9] In particular, for each summand s_k, all players hold a commitment to s_k such that each $p_i \in S_k$ holds the corresponding decommitment information to open it.

The computation stage. In the computation stage, instead of C we evaluate the circuit C' which computes a uniformly random *sharing* $\langle y \rangle$ of the output y of C according to $\mathcal{S}_{\mathcal{Z}}$, i.e., the sharing specification associated with $\mathcal{Z}$. The circuit C' can be easily constructed from C [IKLP06]. To evaluate C' the players invoke the protocol for SFE from [Gol04] for the model where authenticated broadcast channels (but no bilateral point-to-point channels) are given, which tolerates any number of $t < n$ actively corrupted players. As proved in [Gol04], with this protocol we achieve the following properties: There is a $p \in \mathcal{P}$ (specified by the protocol), such that when p is uncorrupted the circuit C' is securely evaluated, otherwise the adversary can decide either to make *all* players abort the protocol or to allow C' to be securely evaluated. Note that the adversary can decide whether or not the protocol aborts even after having received the outputs of the passively corrupted players. Furthermore, by inspecting the protocol in [Gol04], one can verify that it actually satisfies some additional properties, which are relevant when all three corruption types are considered, namely (1) if p is *correct* then the protocol does not abort,[10] (2) a correct player always gives his (correct) input to the evaluation of C', and (3) a non-actively corrupted player does not give a wrong input (but might give no input if he crashes). By the above properties it is clear that the protocol can abort only if p is incorrect (i.e., $B = \{p\}$). Moreover, when it aborts privacy of the inputs is not violated as the outputs of passively corrupted players are their shares of $\langle y \rangle$ plus perfectly hiding commitments to all the summands of $\langle y \rangle$.

The output stage. The output stage is similar to the output stage of protocol SFE described in Section 6. The summands of $\langle y \rangle$ are announced sequentially in the order implied by $C_{\mathsf{ORD}}^{(1)}(\mathcal{P}, \mathcal{Z})$. This guarantees (as in protocol OutputGeneration) that when the announcing of a summand aborts, then the output stage can abort without violating

[9] Such commitment schemes are known to exist if (enhanced) trapdoor permutations exist [GMW86].

[10] Note that a correct player is not necessary uncorrupted.

privacy (the summand of $\langle y \rangle$ associated with the actual adversary has not been announced yet). To announce a summand, protocol CompPubAnnounce is invoked which is a trivially modified version of PubAnnounce to use openings of commitments instead of signatures. We refer to the abovely described SFE protocol as CompSFE.

Lemma 16. *Assuming that enhanced trapdoor permutations exist, the protocol* CompSFE *is computationally $\mathcal{Z}$-secure if $C^{(1)}_{\mathsf{ORD}}(\mathcal{P}, \mathcal{Z})$ holds.*

7.2 The MPC Protocol

A protocol for MPC can be built based on a (robust) general SFE protocol and a robustly reconstructible secret-sharing scheme, in a straightforward way: the SFE protocol is used to securely evaluate the circuit gate-by-gate, where each intermediary result is shared among the players. In fact, the secret-sharing scheme described is Section 7.1, for sharing specification $\mathcal{S}_{\mathcal{Z}}$, is robustly reconstructible if $C^{(1)}(\mathcal{P}, \mathcal{Z})$ holds. Indeed, condition $C^{(1)}(\mathcal{P}, \mathcal{Z})$ ensures that for any shared value each summand is known to at least one player who is not actively or fail-corrupted and will not change or delete it. Hence, the shared value is uniquely determined by the views of the players. Therefore, we can use protocol CompSFE to evaluate any (reactive) circuit as follows: For each input gate, invoke CompSFE to evaluate the circuit C_{input} which computes a sharing (according to $\mathcal{S}_{\mathcal{Z}}$) of the input value. For the addition and multiplication gate, invoke CompSFE to evaluate the circuits C_{add} and C_{mult} which on input the sharings of two values s and t output a sharing of the sum $s + t$ and of the product st, respectively. For output gates, invoke CompSFE to evaluate the circuit C_{output} which on input the sharing of some value s outputs s towards the corresponding player. We refer to the resulting MPC protocol as CompMPC.

Lemma 17. *Protocol* CompMPC *is computationally $\mathcal{Z}$-secure if $C^{(1)}(\mathcal{P}, \mathcal{Z})$ holds.*

8 Conclusions

We considered MPC and SFE in the presence of a general adversary who can actively, passively, and fail corrupt players, simultaneously. For both primitives we gave exact characterizations of the tolerable adversary structures for achieving unconditional (aka statistical) and computational security, when a broadcast channel is given. As in the case of threshold adversaries, the achieved bounds are strictly better than those required for perfect security, where no error probability is allowed. Our results confirm that in all three security models (perfect, unconditional, and computational) there are adversary structures that can be tolerated for SFE but not for MPC.

References

[Alt99] Altmann, B.: Constructions for efficient multi-party protocols secure against general adversaries. Diploma Thesis, ETH Zurich (1999)

[Bea91] Beaver, D.: Secure multiparty protocols and zero-knowledge proof systems tolerating a faulty minority. Journal of Cryptology 4(2), 370–381 (1991)

[BFH+08] Beerliová-Trubíniová, Z., Fitzi, M., Hirt, M., Maurer, U., Zikas, V.: MPC vs. SFE: Perfect security in a unified corruption model. In: Canetti, R. (ed.) TCC 2008. LNCS, vol. 4948, pp. 231–250. Springer, Heidelberg (2008)

[BGW88] Ben-Or, M., Goldwasser, S., Wigderson, A.: Completeness theorems for non-cryptographic fault-tolerant distributed computation. In: STOC 1988, pp. 1–10 (1988)

[BHR07] Beerliová-Trubíniová, Z., Hirt, M., Riser, M.: Efficient Byzantine agreement with faulty minority. In: Kurosawa, K. (ed.) ASIACRYPT 2007. LNCS, vol. 4833, pp. 393–409. Springer, Heidelberg (2007)

[CCD88] Chaum, D., Crépeau, C., Damgård, I.: Multiparty unconditionally secure protocols (extended abstract). In: STOC 1988, pp. 11–19 (1988)

[CDD+99] Cramer, R., Damgård, I., Dziembowski, S., Hirt, M., Rabin, T.: Efficient multiparty computations secure against an adaptive adversary. In: Stern, J. (ed.) EUROCRYPT 1999. LNCS, vol. 1592, pp. 311–326. Springer, Heidelberg (1999)

[Cle86] Cleve, R.: Limits on the security of coin flips when half the processors are faulty (extended abstract). In: STOC 1986, pp. 364–369 (1986)

[FHM98] Fitzi, M., Hirt, M., Maurer, U.: Trading correctness for privacy in unconditional multi-party computation. In: Krawczyk, H. (ed.) CRYPTO 1998. LNCS, vol. 1462, pp. 121–136. Springer, Heidelberg (1998)

[FHM99] Fitzi, M., Hirt, M., Maurer, U.: General adversaries in unconditional multi-party computation. In: Lam, K.-Y., Okamoto, E., Xing, C. (eds.) ASIACRYPT 1999. LNCS, vol. 1716, pp. 232–246. Springer, Heidelberg (1999)

[GL02] Goldwasser, S., Lindell, Y.: Secure computation without agreement. In: Malkhi, D. (ed.) DISC 2002. LNCS, vol. 2508, pp. 17–32. Springer, Heidelberg (2002)

[GMW86] Goldreich, O., Micali, S., Wigderson, A.: Proofs that yield nothing but their validity and a methodology of cryptographic protocol design (extended abstract). In: FOCS 1986, pp. 174–187 (1986)

[GMW87] Goldreich, O., Micali, S., Wigderson, A.: How to play any mental game — a completeness theorem for protocols with honest majority. In: STOC 1987, pp. 218–229 (1987)

[Gol04] Goldreich, O.: Foundations of Cryptography. Basic Applications, vol. 2. Cambridge University Press, New York (2004)

[GRR98] Gennaro, R., Rabin, M.O., Rabin, T.: Simplified VSS and fast-track multiparty computations with applications to threshold cryptography. In: PODC 1998, pp. 101–111 (1998)

[HM97] Hirt, M., Maurer, U.: Complete characterization of adversaries tolerable in secure multi-party computation. In: PODC 1997, pp. 25–34 (1997)

[HM00] Hirt, M., Maurer, U.: Player simulation and general adversary structures in perfect multiparty computation. Journal of Cryptology 13(1), 31–60 (2000)

[IKLP06] Ishai, Y., Kushilevitz, E., Lindell, Y., Petrank, E.: On combining privacy with guaranteed output delivery in secure multiparty computation. In: Dwork, C. (ed.) CRYPTO 2006. LNCS, vol. 4117, pp. 483–500. Springer, Heidelberg (2006)

[Mau02] Maurer, U.: Secure multi-party computation made simple. In: Cimato, S., Galdi, C., Persiano, G. (eds.) SCN 2002. LNCS, vol. 2576, pp. 14–28. Springer, Heidelberg (2003)

[Mau06] Maurer, U.: Secure multi-party computation made simple. Discrete Applied Mathematics 154(2), 370–381 (2006)

[RB89] Rabin, T., Ben-Or, M.: Verifiable secret sharing and multiparty protocols with honest majority. In: STOC 1989, pp. 73–85 (1989)

[Yao82] Yao, A.C.: Protocols for secure computations. In: FOCS 1982, pp. 160–164 (1982)

Strongly Multiplicative and 3-Multiplicative Linear Secret Sharing Schemes

Zhifang Zhang[1], Mulan Liu[1], Yeow Meng Chee[2], San Ling[2], and Huaxiong Wang[2,3]

[1] Key Laboratory of Mathematics Mechanization, Academy of Mathematics and Systems Science, Chinese Academy of Sciences, Beijing, China
{zfz,mlliu}@amss.ac.cn

[2] Division of Mathematical Sciences, School of Physical and Mathematical Sciences, Nanyang Technological University, Singapore
{ymchee,lingsan,hxwang}@ntu.edu.sg

[3] Centre for Advanced Computing - Algorithms and Cryptography
Department of Computing
Macquarie University, Australia

Abstract. Strongly multiplicative linear secret sharing schemes (LSSS) have been a powerful tool for constructing secure multi-party computation protocols. However, it remains open *whether or not there exist efficient constructions of strongly multiplicative LSSS from general LSSS*. In this paper, we propose the new concept of 3-*multiplicative LSSS*, and establish its relationship with strongly multiplicative LSSS. More precisely, we show that any 3-multiplicative LSSS is a strongly multiplicative LSSS, but the converse is not true; and that any strongly multiplicative LSSS can be efficiently converted into a 3-multiplicative LSSS. Furthermore, we apply 3-multiplicative LSSS to the computation of unbounded fan-in multiplication, which reduces its round complexity to four (from five of the previous protocol based on multiplicative LSSS). We also give two constructions of 3-multiplicative LSSS from Reed-Muller codes and algebraic geometric codes. We believe that the construction and verification of 3-multiplicative LSSS are easier than those of strongly multiplicative LSSS. This presents a step forward in settling the open problem of efficient constructions of strongly multiplicative LSSS from general LSSS.

Keywords: monotone span program, secure multi-party computation, strongly multiplicative linear secret sharing scheme.

1 Introduction

Secure multi-party computation (MPC) [16,9] is a cryptographic primitive that enables n players to jointly compute an agreed function of their private inputs in a secure way, guaranteeing the correctness of the outputs as well as the privacy of the players' inputs, even when some players are malicious. It has become a fundamental tool in cryptography and distributed computation. Linear secret sharing schemes (LSSS) play an important role in building MPC protocols.

J. Pieprzyk (Ed.): ASIACRYPT 2008, LNCS 5350, pp. 19–36, 2008.
© International Association for Cryptologic Research 2008

Cramer *et al.* [6] developed a generic method of constructing MPC protocols from LSSS. Assuming that the function to be computed is represented as an arithmetic circuit over a finite field, their protocol ensures that each player share his private input through an LSSS, and then evaluates the circuit gate by gate. The main idea of their protocol is to keep the intermediate results secretly shared among the players with the underlying LSSS. Due to the nature of linearity, secure additions (and linear operations) can be easily achieved. For instance, if player P_i holds the share x_{1i} for input x_1 and x_{2i} for input x_2, he can locally compute $x_{1i} + x_{2i}$ which is actually P_i's share for $x_1 + x_2$. Unfortunately, the above homomorphic property does not hold for multiplication. In order to securely compute multiplications, Cramer *et al.* [6] introduced the concept of *multiplicative* LSSS, where the product $x_1 x_2$ can be computed as a linear combination of the local products of shares, that is, $x_1 x_2 = \sum_{i=1}^{n} a_i x_{1i} x_{2i}$ for some constants $a_i, 1 \leq i \leq n$. Since $x_{1i} x_{2i}$ can be locally computed by P_i, the product can then be securely computed through a linear combination. Furthermore, in order to resist against an active adversary, they defined *strongly* multiplicative LSSS, where $x_1 x_2$ can be computed as a linear combination of the local products of shares by all players excluding any corrupted subset. Therefore, multiplicativity becomes an important property in constructing secure MPC protocols. For example, using strongly multiplicative LSSS, we can construct an error-free MPC protocol secure against an active adversary in the information-theoretic model [6]. Cramer *et al.* [7] also gave an efficient reconstruction algorithm for strongly multiplicative LSSS that recovers the secret even when the shares submitted by the corrupted players contain errors. This implicit "built-in" verifiability makes strongly multiplicative LSSS an attractive building block for MPC protocols.

Due to their important role as the building blocks in MPC protocols, efficient constructions of multiplicative LSSS and strongly multiplicative LSSS have been studied by several authors in recent years. Cramer *et al.* [6] developed a generic method of constructing a multiplicative LSSS from any given LSSS with a double expansion of the shares. Nikov *et al.* [14] studied how to securely compute multiplications in a dual LSSS, without blowing up the shares. For some specific access structures there exist very efficient multiplicative LSSS. Shamir's threshold secret sharing scheme is a well-known example of an ideal (strongly) multiplicative LSSS. Besides, self-dual codes give rise to ideal multiplicative LSSS [7], and Liu *et al.* [12] provided a further class of ideal multiplicative LSSS for some kind of graph access structure. We note that for strongly multiplicative LSSS, the known general construction is of exponential complexity. Käsper *et al.* [11] gave some efficient constructions for specific access structures (hierarchical threshold structures). It remains open whether there exists an efficient transformation from a general LSSS to a strongly multiplicative one.

On the other hand, although in a multiplicative LSSS, multiplication can be converted into a linear combination of inputs from the players, each player has to *reshare* the product of his shares, that is, for $1 \leq i \leq n$, P_i needs to reshare the product $x_{1i} x_{2i}$ to securely compute the linear combination $\sum_{i=1}^{n} a_i x_{1i} x_{2i}$. This resharing process involves costly interactions among the players. For example, if the players are to securely compute multiple multiplications, $\prod_{i=1}^{l} x_i$, the simple sequential multiplication requires interaction of round complexity proportional to l. Using the technique developed by Bar-Ilan and Beaver [1], Cramer *et al.* [4] recently showed that the round complexity can be significantly reduced to a constant of five for unbounded fan-in multiplications. However, the method does not seem efficient when l is small. For example, considering x_1x_2 and $x_1x_2x_3$, extra rounds of interactions seem unavoidable for computing $x_1x_2x_3$ even though we apply the method of Cramer *et al.* [4].

1.1 Our Contribution

In this paper, we propose the concept of 3-multiplicative LSSS. Roughly speaking, a 3-multiplicative LSSS is a generalization of multiplicative LSSS, where the product $x_1x_2x_3$ is a linear combination of the local products of shares. As one would expect, a 3-multiplicative LSSS achieves better round complexity for the computation of $\prod_{i=1}^{l} x_i$ compared to a multiplicative LSSS, if $l \geq 3$. Indeed, it is easy to see that computing the product $\prod_{i=1}^{9} x_i$ requires two rounds of interaction for a 3-multiplicative LSSS but four rounds for a multiplicative LSSS. We also extend the concept of a 3-multiplicative LSSS to the more general λ-multiplicative LSSS, for all integers $\lambda \geq 3$, and show that λ-multiplicative LSSS reduce the round complexity by a factor of $\frac{1}{\log \lambda}$ from multiplicative LSSS. In particular, 3-multiplicative LSSS reduce the constant round complexity of computing the unbounded fan-in multiplication from five to four, thus improving a result of Cramer *et al.* [4].

More importantly, we show that 3-multiplicative LSSS are closely related to strongly multiplicative LSSS. The latter is known to be a powerful tool for constructing secure MPC protocols against active adversaries. More precisely, we show the following:

(i) 3-multiplicative LSSS are also strongly multiplicative;
(ii) there exists an efficient algorithm that transforms a strongly multiplicative LSSS into a 3-multiplicative LSSS;
(iii) an example of a strongly multiplicative LSSS that is not 3-multiplicative.

Our results contribute to the study of MPC in the following three aspects:

- The 3-multiplicative LSSS outperform strongly multiplicative LSSS with respect to round complexity in the construction of secure MPC protocols.
- The 3-multiplicative LSSS are easier to construct than strongly multiplicative LSSS. First, the existence of an efficient transformation from a strongly multiplicative LSSS to a 3-multiplicative LSSS implies that efficiently constructing 3-multiplicative LSSS is not a harder problem. Second, verification of a strongly multiplicative LSSS requires checking the linear combinations for all possibilities of adversary sets, while the verification of a 3-multiplicative LSSS requires only one checking. We give two constructions of LSSS based on Reed-Muller codes and algebraic geometric codes that can be easily verified for 3-multiplicativity, but it does not seem easy to give direct proofs of their strong multiplicativity.

– This work provides two possible directions toward solving the open problem of determining the existence of efficient constructions for strongly multiplicative LSSS. On the negative side, if we can prove that in the information-theoretic model and with polynomial size message exchanged, computing $x_1x_2x_3$ inevitably needs more rounds of interactions than computing x_1x_2, then we can give a negative answer to this open problem. On the positive side, if we can find an efficient construction for 3-multiplicative LSSS, which also results in strongly multiplicative LSSS, then we will have an affirmative answer to this open problem.

1.2 Organization

Section 2 gives notations, definition of multiplicative LSSS, and general constructions for strongly multiplicative LSSS. Section 3 defines 3-multiplicative LSSS. Section 4 shows the relationship between 3-multiplicative LSSS and strongly multiplicative LSSS. Section 5 gives two constructions of 3-multiplicative LSSS from error-correcting codes, and Section 6 discusses the implications of 3-multiplicative LSSS in MPC. Section 7 concludes the paper.

2 Preliminaries

Throughout this paper, let $P = \{P_1, \ldots, P_n\}$ denote the set of n players and let $\mathcal{K}$ be a finite field. In a secret sharing scheme, the collection of all subsets of players that are authorized to recover the secret is called its *access structure*, and is denoted by AS. An access structure possesses the monotone ascending property: if $A' \in AS$, then for all $A \subseteq P$ with $A \supseteq A'$, we also have $A \in AS$. Similarly, the collection of subsets of players that are possibly corrupted is called the *adversary structure*, and is denoted as $\mathcal{A}$. An adversary structure possesses the monotone descending property: if $A' \in \mathcal{A}$, then for all $A \subseteq P$ with $A \subseteq A'$, we also have $A \in \mathcal{A}$. Owing to these monotone properties, it is often sufficient to consider the *minimum access structure* AS_{min} and the *maximum adversary structure* $\mathcal{A}_{max}$ defined as follows:

$$AS_{min} = \{A \in AS \mid \forall B \subseteq P, \text{ we have } B \subsetneq A \Rightarrow B \notin AS\},$$
$$\mathcal{A}_{max} = \{A \in \mathcal{A} \mid \forall B \subseteq P, \text{ we have } B \supsetneq A \Rightarrow B \notin \mathcal{A}\}.$$

In this paper, we consider the *complete* situation, that is, $\mathcal{A} = 2^P - AS$. Moreover, an adversary structure $\mathcal{A}$ is called Q^2 (respectively, Q^3) if any two (respectively, three) sets in $\mathcal{A}$ cannot cover the entire player set P. For simplicity, when an adversary structure $\mathcal{A}$ is Q^2 (respectively, Q^3) we also say the corresponding access structure $AS = 2^P - \mathcal{A}$ is Q^2 (respectively, Q^3).

2.1 Linear Secret Sharing Schemes and Monotone Span Programs

Suppose S is the secret-domain, R is the set of random inputs, and S_i is the share-domain of P_i, where $1 \leq i \leq n$. Let S and R denote random variables taking values in S and R, respectively. Then $\Pi : S \times R \rightarrow S_1 \times \cdots \times S_n$ is called a *secret sharing scheme* (SSS) with respect to the access structure AS, if the following two conditions are satisfied:

1. for all $A \in AS$, $H(\mathsf{S} \mid \Pi(\mathsf{S},\mathsf{R})|_A) = 0$;
2. for all $B \notin AS$, $H(\mathsf{S} \mid \Pi(\mathsf{S},\mathsf{R})|_B) = H(\mathsf{S})$,

where $H(\cdot)$ is the entropy function. Furthermore, the secret sharing scheme Π is called *linear* if we have $S = \mathcal{K}$, $R = \mathcal{K}^{l-1}$, and $S_i = \mathcal{K}^{d_i}$ for some positive integers l and d_i, $1 \leq i \leq n$, and the reconstruction of the secret can be performed by taking a linear combination of shares from the authorized players. The quantity $d = \sum_{i=1}^{n} d_i$ is called the *size* of the LSSS.

Karchmer and Wigderson [10] introduced monotone span programs (MSP) as a linear model for computing monotone Boolean functions. We denote an MSP by $\mathcal{M}(\mathcal{K}, M, \psi, \boldsymbol{v})$, where M is a $d \times l$ matrix over $\mathcal{K}$, $\psi : \{1, \ldots, d\} \rightarrow \{P_1, \ldots, P_n\}$ is a surjective labeling map, and $\boldsymbol{v} \in \mathcal{K}^l$ is a nonzero vector. We call d the *size* of the MSP and $\boldsymbol{v}$ the *target vector*. A monotone Boolean function $f : \{0,1\}^n \rightarrow \{0,1\}$ satisfies $f(\boldsymbol{\delta}') \geq f(\boldsymbol{\delta})$ for any $\boldsymbol{\delta}' \geq \boldsymbol{\delta}$, where $\boldsymbol{\delta} = (\delta_1, \ldots, \delta_n)$, $\boldsymbol{\delta}' = (\delta'_1, \ldots, \delta'_n) \in \{0,1\}^n$, and $\boldsymbol{\delta}' \geq \boldsymbol{\delta}$ means $\delta'_i \geq \delta_i$ for $1 \leq i \leq n$. We say that an MSP $\mathcal{M}(\mathcal{K}, M, \psi, \boldsymbol{v})$ *computes the monotone Boolean function* f if $\boldsymbol{v} \in span\{M_A\}$ if and only if $f(\boldsymbol{\delta_A}) = 1$, where A is a set of players, M_A denotes the matrix constricted to the rows labeled by players in A, $span\{M_A\}$ denotes the linear space spanned by the row vectors of M_A, and $\boldsymbol{\delta_A}$ is the characteristic vector of A.

Theorem 1 (Beimel [2]). *Suppose AS is an access structure over P and f_{AS} is the characteristic function of AS, that is, $f_{AS}(\boldsymbol{\delta}) = 1$ if and only if $\boldsymbol{\delta} = \boldsymbol{\delta}_A$ for some $A \in AS$. Then there exists an LSSS of size d that realizes AS if and only if there exists an MSP of size d that computes f_{AS}.*

Since an MSP computes the same Boolean function under linear transformations, we can always assume that the target vector is $\boldsymbol{e}_1 = (1, 0, \ldots, 0)$. From an MSP $\mathcal{M}(\mathcal{K}, M, \psi, \boldsymbol{e}_1)$ that computes f_{AS}, we can derive an LSSS realizing AS as follows: to share a secret $s \in \mathcal{K}$, the dealer randomly selects $\boldsymbol{\rho} \in \mathcal{K}^{l-1}$, computes $M(s, \boldsymbol{\rho})^\tau$ and sends $M_{P_i}(s, \boldsymbol{\rho})^\tau$ to P_i as his share, where $1 \leq i \leq n$ and τ denotes the transpose. The following property of MSP is useful in the proofs of our results.

Proposition 1 (Karchmer and Wigderson [10]). *Let $\mathcal{M}(\mathcal{K}, M, \psi, \boldsymbol{e}_1)$ be an MSP that computes a monotone Boolean function f. Then for all $A \subseteq P$, $\boldsymbol{e}_1 \notin span\{M_A\}$ if and only if there exists $\boldsymbol{\rho} \in \mathcal{K}^{l-1}$ such that $M_A(1, \boldsymbol{\rho})^\tau = \boldsymbol{0}^\tau$.*

2.2 Multiplicative Linear Secret Sharing Schemes

From Theorem 1, an LSSS can be identified with its corresponding MSP in the following way. Let $\mathcal{M}(\mathcal{K}, M, \psi, \boldsymbol{e}_1)$ be an LSSS realizing the access structure AS. Given two vectors $\boldsymbol{x} = (x_1, \ldots, x_d)$, $\boldsymbol{y} = (y_1, \ldots, y_d) \in \mathcal{K}^d$, we define $\boldsymbol{x} \diamond \boldsymbol{y}$ to be the vector containing all entries of the form $x_i \cdot y_j$ with $\psi(i) = \psi(j)$. More precisely, let

$$\boldsymbol{x} = (x_{11}, \ldots, x_{1d_1}, \ldots, x_{n1}, \ldots, x_{nd_n}),$$
$$\boldsymbol{y} = (y_{11}, \ldots, y_{1d_1}, \ldots, y_{n1}, \ldots, y_{nd_n}),$$

where $\sum_{i=1}^{n} d_i = d$, and $(x_{i1}, \ldots, x_{id_i})$, $(y_{i1}, \ldots, y_{id_i})$ are the entries distributed to P_i according to ψ. Then $\boldsymbol{x} \diamond \boldsymbol{y}$ is the vector composed of the $\sum_{i=1}^{n} d_i^2$ entries $x_{ij}y_{ik}$, where $1 \le j, k \le d_i, 1 \le i \le n$. For consistency, we write the entries of $\boldsymbol{x} \diamond \boldsymbol{y}$ in some fixed order. We also define $(\boldsymbol{x} \diamond \boldsymbol{y})^\tau = \boldsymbol{x}^\tau \diamond \boldsymbol{y}^\tau$.

Definition 1 (Multiplicativity). *Let $\mathcal{M}(\mathcal{K}, M, \psi, \boldsymbol{e}_1)$ be an LSSS realizing the access structure AS over P. Then $\mathcal{M}$ is called* multiplicative *if there exists a recombination vector $\boldsymbol{z} \in \mathcal{K}^{\sum_{i=1}^{n} d_i^2}$, such that for all $s, s' \in \mathcal{K}$ and $\boldsymbol{\rho}, \boldsymbol{\rho}' \in \mathcal{K}^{l-1}$, we have*

$$ss' = \boldsymbol{z}(M(s, \boldsymbol{\rho})^\tau \diamond M(s', \boldsymbol{\rho}')^\tau).$$

Moreover, $\mathcal{M}$ is strongly multiplicative *if for all $A \in \mathcal{A} = 2^P - AS$, $\mathcal{M}_{\overline{A}}$ is multiplicative, where $\mathcal{M}_{\overline{A}}$ denotes the MSP $\mathcal{M}$ constricted to the subset $\overline{A} = P - A$.*

Proposition 2 (Cramer *et al.* [6]). *Let AS be an access structure over P. Then there exists a multiplicative (respectively, strongly multiplicative) LSSS realizing AS if and only if AS is Q^2 (respectively, Q^3).*

2.3 General Constructions of Strongly Multiplicative LSSS

For all Q^2 access structure AS, Cramer *et al.* [6] gave an efficient construction to build a multiplicative LSSS from a general LSSS realizing the same AS. It remains *open* if we can *efficiently* construct a strongly multiplicative LSSS from an LSSS. However, there are general constructions with exponential complexity, as described below.

Since Shamir's threshold secret sharing scheme is strongly multiplicative for all Q^3 threshold access structure, a proper composition of Shamir's threshold secret sharing schemes results in a general construction for strongly multiplicative LSSS [6]. Here, we give another general construction based on multiplicative LSSS.

Let AS be any Q^3 access structure and $\mathcal{M}(\mathcal{K}, M, \psi, \boldsymbol{e}_1)$ be an LSSS realizing AS. For all $A \in \mathcal{A} = 2^P - AS$, it is easy to see that $\mathcal{M}_{\overline{A}}$ realizes the restricted access structure $AS_{\overline{A}} = \{B \subseteq \overline{A} \mid B \in AS\}$. The access structure $AS_{\overline{A}}$ is Q^2 over $\overline{A}$ because AS is Q^3 over $\overline{A} \cup A$. Thus, we can transform $\mathcal{M}_{\overline{A}}$ into a multiplicative LSSS following the general construction of Cramer *et al.* [6] to obtain a strongly multiplicative LSSS realizing AS. The example in Section 4.3 gives an illustration of this method.

We note that both constructions above give LSSS of exponential sizes, and hence are not *efficient* in general.

3 3-Multiplicative and λ-Multiplicative LSSS

In this section, we give an equivalent definition for (strongly) multiplicative LSSS. We then define 3-multiplicative LSSS and give a necessary and sufficient condition for its existence. The notion of 3-multiplicativity is also extended to λ-multiplicativity for all integer $\lambda > 1$. Finally, we present a generic (but inefficient) construction of λ-multiplicative LSSS.

Under the same notations used in Section 2.2, it is straightforward to see that we have an induced labeling map $\psi' : \{1, \ldots, \sum_{i=1}^{n} d_i^2\} \to \{P_1, \ldots, P_n\}$ on the entries of $\boldsymbol{x} \diamond \boldsymbol{y}$, distributing the entry $x_{ij}y_{ik}$ to P_i, since both x_{ij} and y_{ik} are labeled by P_i under ψ. For an MSP $\mathcal{M}(\mathcal{K}, M, \psi, \boldsymbol{e}_1)$, denote $M = (M_1, \ldots, M_l)$, where $M_i \in \mathcal{K}^d$ is the i-th column vector of M, $1 \le i \le l$. We construct a new matrix $M_\diamond$ as follows:

$$M_\diamond = (M_1 \diamond M_1, \ldots, M_1 \diamond M_l, M_2 \diamond M_1, \ldots, M_2 \diamond M_l, \ldots, M_l \diamond M_1, \ldots, M_l \diamond M_l).$$

For consistency, we also denote $M_\diamond$ as $M \diamond M$. Obviously, $M_\diamond$ is a matrix over $\mathcal{K}$ with $\sum_{i=1}^{n} d_i^2$ rows and l^2 columns. For any two vectors $\boldsymbol{u}, \boldsymbol{v} \in \mathcal{K}^l$, it is easy to verify that

$$(M\boldsymbol{u}^\tau) \diamond (M\boldsymbol{v}^\tau) = M_\diamond(\boldsymbol{u} \otimes \boldsymbol{v})^\tau,$$

where $\boldsymbol{u} \otimes \boldsymbol{v}$ denotes the tensor product with its entries written in a proper order. Define the induced labeling map ψ' on the rows of $M_\diamond$. We have the following proposition.

Proposition 3. *Let $\mathcal{M}(\mathcal{K}, M, \psi, \boldsymbol{e}_1)$ be an LSSS realizing the access structure AS, and let $M_\diamond$ be with the labeling map ψ'. Then $\mathcal{M}$ is multiplicative if and only if $\boldsymbol{e}_1 \in span\{M_\diamond\}$, where $\boldsymbol{e}_1 = (1, 0, \ldots, 0)$. Moreover, $\mathcal{M}$ is strongly multiplicative if and only if $\boldsymbol{e}_1 \in span\{(M_\diamond)_{\overline{A}}\}$ for all $A \in \mathcal{A} = 2^P - AS$.*

Proof. By Definition 1, $\mathcal{M}$ is multiplicative if and only if $ss' = \boldsymbol{z}(M(s, \boldsymbol{\rho})^\tau \diamond M(s', \boldsymbol{\rho}')^\tau)$ for all $s, s' \in \mathcal{K}$ and $\boldsymbol{\rho}, \boldsymbol{\rho}' \in \mathcal{K}^{l-1}$. Obviously,

$$M(s, \boldsymbol{\rho})^\tau \diamond M(s', \boldsymbol{\rho}')^\tau = M_\diamond((s, \boldsymbol{\rho}) \otimes (s', \boldsymbol{\rho}'))^\tau = M_\diamond(ss', \boldsymbol{\rho}'')^\tau, \tag{1}$$

where $(ss', \boldsymbol{\rho}'') = (s, \boldsymbol{\rho}) \otimes (s', \boldsymbol{\rho}')$. On the other hand, $ss' = \boldsymbol{e}_1(ss', \boldsymbol{\rho}'')^\tau$. Thus $\mathcal{M}$ is multiplicative if and only if

$$(\boldsymbol{e}_1 - \boldsymbol{z}M_\diamond)(ss', \boldsymbol{\rho}'')^\tau = 0. \tag{2}$$

Because of the arbitrariness of $s, s', \boldsymbol{\rho}$ and $\boldsymbol{\rho}'$, equality (2) holds if and only if $\boldsymbol{e}_1 - \boldsymbol{z}M_\diamond = \boldsymbol{0}$. Thus $\boldsymbol{e}_1 \in span\{M_\diamond\}$. The latter part of the proposition can be proved similarly. □

Now we are ready to give the definition of 3-multiplicative LSSS. We extend the diamond product "$\diamond$" and define $\boldsymbol{x} \diamond \boldsymbol{y} \diamond \boldsymbol{z}$ to be the vector containing all entries of the form $x_i y_j z_k$ with $\psi(i) = \psi(j) = \psi(k)$, where the entries of $\boldsymbol{x} \diamond \boldsymbol{y} \diamond \boldsymbol{z}$ are written in some fixed order.

Definition 2 (3-Multiplicativity). *Let $\mathcal{M}(\mathcal{K}, M, \psi, \boldsymbol{e}_1)$ be an LSSS realizing the access structure AS. Then $\mathcal{M}$ is called* 3-multiplicative *if there exists a recombination vector $\boldsymbol{z} \in \mathcal{K}^{\sum_{i=1}^{n} d_i^3}$ such that for all $s_1, s_2, s_3 \in \mathcal{K}$ and $\boldsymbol{\rho}_1, \boldsymbol{\rho}_2, \boldsymbol{\rho}_3 \in \mathcal{K}^{l-1}$, we have*

$$s_1 s_2 s_3 = \boldsymbol{z}(M(s_1, \boldsymbol{\rho}_1)^\tau \diamond M(s_2, \boldsymbol{\rho}_2)^\tau \diamond M(s_3, \boldsymbol{\rho}_3)^\tau).$$

We can derive an equivalent definition for 3-multiplicative LSSS, similar to Proposition 3: $\mathcal{M}$ is 3-multiplicative if and only if $\boldsymbol{e}_1 \in span\{(M \diamond M \diamond M)\}$. The following proposition gives a necessary and sufficient condition for the existence of 3-multiplicative LSSS.

Proposition 4. *For all access structures AS, there exists a 3-multiplicative LSSS realizing AS if and only if AS is Q^3.*

Proof. Suppose $\mathcal{M}(\mathcal{K}, M, \psi, \boldsymbol{e}_1)$ is a 3-multiplicative LSSS realizing AS, and suppose to the contrary, that AS is not Q^3, so there exist $A_1, A_2, A_3 \in \mathcal{A} = 2^P - AS$ such that $A_1 \cup A_2 \cup A_3 = P$. By Proposition 1, there exists $\boldsymbol{\rho}_i \in \mathcal{K}^{l-1}$ such that $M_{A_i}(1, \boldsymbol{\rho}_i)^\tau = \boldsymbol{0}^\tau$ for $1 \leq i \leq 3$. Since $A_1 \cup A_2 \cup A_3 = P$, we have $M(1, \boldsymbol{\rho}_1)^\tau \diamond M(1, \boldsymbol{\rho}_2)^\tau \diamond M(1, \boldsymbol{\rho}_3)^\tau = \boldsymbol{0}^\tau$, which contradicts Definition 2.

On the other hand, a general construction for building a 3-multiplicative LSSS from a strongly multiplicative LSSS is given in the next section, thus sufficiency is guaranteed by Proposition 2. □

A trivial example of 3-multiplicative LSSS is Shamir's threshold secret sharing scheme that realizes any Q^3 threshold access structure. Using an identical argument for the case of strongly multiplicative LSSS, we have a general construction for 3-multiplicative LSSS based on Shamir's threshold secret sharing schemes, with exponential complexity.

For any λ vectors $\boldsymbol{x}_i = (x_{i1}, \ldots, x_{id}) \in \mathcal{K}^d, 1 \leq i \leq \lambda$, we define $\diamond_{i=1}^{\lambda} \boldsymbol{x}_i$ to be the $\sum_{i=1}^{n} d_i^{\lambda}$-dimensional vector which contains entries of the form $\prod_{i=1}^{\lambda} x_{ij_i}$ with $\psi(j_1) = \cdots = \psi(j_\lambda)$.

Definition 3 (λ-Multiplicativity). *Let $\mathcal{M}(\mathcal{K}, M, \psi, \boldsymbol{e}_1)$ be an LSSS realizing the access structure AS, and let $\lambda > 1$ be an integer. Then $\mathcal{M}$ is* λ-multiplicative *if there exists a recombination vector $\boldsymbol{z}$ such that for all $s_1, \ldots, s_\lambda \in \mathcal{K}$ and $\boldsymbol{\rho}_1, \ldots, \boldsymbol{\rho}_\lambda \in \mathcal{K}^{l-1}$, we have*

$$\prod_{i=1}^{\lambda} s_i = \boldsymbol{z}(\diamond_{i=1}^{\lambda} M(s_i, \boldsymbol{\rho}_i)^\tau).$$

Moreover, $\mathcal{M}$ is strongly λ-multiplicative *if for all $A \notin AS$, the constricted LSSS $\mathcal{M}_{\overline{A}}$ is λ-multiplicative.*

Again, we can define a new matrix by taking the diamond product of λ copies of M. This gives an equivalence to (strongly) λ-multiplicative LSSS. Also, since Shamir's threshold secret sharing scheme is trivially λ-multiplicative and strongly λ-multiplicative, a proper composition of Shamir's threshold secret sharing schemes results in a general construction for both λ-multiplicative LSSS and strongly λ-multiplicative LSSS. Let Q^λ be a straightforward extension of Q^2 and Q^3, that is, an access structure AS is Q^λ if the player set P cannot be covered by λ sets in $\mathcal{A} = 2^P - AS$. The following corollary is easy to prove.

Corollary 1. *Let AS be an access structure over P. Then there exists a λ-multiplicative (respectively, strongly λ-multiplicative) LSSS realizing AS if and only if AS is Q^λ (respectively, $Q^{\lambda+1}$).*

Since a λ-multiplicative LSSS transforms the products of λ entries into a linear combination of the local products of shares, it can be used to simplify the secure computation of sequential multiplications. In particular, when compared to using only the multiplicative property (which corresponds to the case when $\lambda = 2$), a λ-multiplicative LSSS can lead to reduced round complexity by a factor of $\frac{1}{\log \lambda}$ in certain cases.

We also point out that Q^λ is not a necessary condition for secure computation. Instead, the necessary condition is Q^2 for the passive adversary model, or Q^3 for the active adversary model [6]. The condition Q^λ is just a necessary condition for the existence of λ-multiplicative LSSS which can be used to simplify computation. In practice, many threshold adversary structures satisfy the Q^λ condition for some appropriate integer λ, and the widely used Shamir's threshold secret sharing scheme is already λ-multiplicative. By using this λ-multiplicativity, we can get more efficient MPC protocols. However, since the special case $\lambda = 3$ shows a close relationship with strongly multiplicative LSSS, a fundamental tool in MPC, this paper focuses on 3-multiplicative LSSS.

4 Strong Multiplicativity and 3-Multiplicativity

In this section, we show that strong multiplicativity and 3-multiplicativity are closely related. On the one hand, given a strongly multiplicative LSSS, there is an *efficient* transformation that converts it to a 3-multiplicative LSSS. On the other hand, we show that any 3-multiplicative LSSS is a strongly multiplicative LSSS, but the converse is not true. It should be noted that strong multiplicativity, as defined, has a combinatorial nature. The definition of 3-multiplicativity is essentially algebraic, which is typically easier to verify.

4.1 From Strong Multiplicativity to 3-Multiplicativity

We show a general method to efficiently build a 3-multiplicative LSSS from a strongly multiplicative LSSS, for all Q^3 access structures. As an extension, the proposed method can also be used to efficiently build a $(\lambda + 1)$-multiplicative LSSS from a strongly λ-multiplicative LSSS.

Theorem 2. *Let AS be a Q^3 access structure and $\mathcal{M}(\mathcal{K}, M, \psi, \mathbf{e}_1)$ be a strongly multiplicative LSSS realizing AS. Suppose that $\mathcal{M}$ has size d and $|\psi^{-1}(P_i)| = d_i$, for $1 \leq i \leq n$. Then there exists a 3-multiplicative LSSS for AS of size $O(d^2)$.*

Proof. We give a constructive proof. Let $M_\diamond$ be the matrix defined in Section 3, and ψ' be the induced labeling map on the rows of $M_\diamond$. Then we have an LSSS $\mathcal{M}_\diamond(\mathcal{K}, M_\diamond, \psi', \boldsymbol{e}_1)$ that realizes an access structure $AS_\diamond$. Because $\mathcal{M}$ is strongly multiplicative, by Proposition 3 we have $\boldsymbol{e}_1 \in span\{(M_\diamond)_{\overline{A}}\}$ for all $A \notin AS$. Therefore $\overline{A} \in AS_\diamond$ and it follows that $AS^* \subseteq AS_\diamond$, where AS^* denotes the dual access structure of AS, defined by $AS^* = \{A \subseteq P \mid P - A \notin AS\}$.

The equality (1) in the proof of Proposition 3 shows that the diamond product of two share vectors equals sharing the product of the two secrets by the MSP $\mathcal{M}_\diamond(\mathcal{K}, M_\diamond, \psi', \boldsymbol{e}_1)$, that is,

$$(M(s_1, \boldsymbol{\rho}_1')^\tau) \diamond (M(s_2, \boldsymbol{\rho}_2')^\tau) = M_\diamond(s_1 s_2, \boldsymbol{\rho})^\tau, \quad \text{for some } \boldsymbol{\rho}_1', \boldsymbol{\rho}_2', \boldsymbol{\rho} \in \mathcal{K}^{l-1}.$$

Thus, using a method similar to Nikov *et al.* [14], we can get the product $(s_1 s_2) \cdot s_3$ by sharing s_3 through the dual MSP of $\mathcal{M}_\diamond$, denoted by $(\mathcal{M}_\diamond)^*$. Furthermore, since $(\mathcal{M}_\diamond)^*$ realizes the dual access structure $(AS_\diamond)^*$ and $(AS_\diamond)^* \subseteq (AS^*)^* = AS$, we can build a 3-multiplicative LSSS by the union of $\mathcal{M}$ and $(\mathcal{M}_\diamond)^*$, which realizes the access structure $AS \cup (AS_\diamond)^* = AS$. Now following the same method of Cramer *et al.* and Fehr [6,8], we prove the required result via the construction below.

Compute the column vector $\boldsymbol{v}_0$ as a solution to the equation $(M_\diamond)^\tau \boldsymbol{v} = \boldsymbol{e_1}^\tau$ for $\boldsymbol{v}$, and compute $\boldsymbol{v}_1, \ldots, \boldsymbol{v}_k$ as a basis of the solution space to $(M_\diamond)^\tau \boldsymbol{v} = \boldsymbol{0}^\tau$. Note that $(M_\diamond)^\tau \boldsymbol{v} = \boldsymbol{e_1}^\tau$ is solvable because $\boldsymbol{e}_1 \in span\{(M_\diamond)_{\overline{A}}\}$ for all $A \notin AS$, while $(M_\diamond)^\tau \boldsymbol{v} = \boldsymbol{0}^\tau$ may only have the trivial solution $\boldsymbol{v} = \boldsymbol{0}$ and $k = 0$. Let

$$M' = \begin{pmatrix} m_{11} & \cdots & m_{1l} & & & \\ \vdots & \ddots & \vdots & & & \\ m_{d1} & \cdots & m_{dl} & & & \\ \boldsymbol{v_0} & & & \boldsymbol{v_1} & \cdots & \boldsymbol{v_k} \end{pmatrix},$$

where $\begin{pmatrix} m_{11} & \cdots & m_{1l} \\ \vdots & \ddots & \vdots \\ m_{d1} & \cdots & m_{dl} \end{pmatrix} = M$ and the blanks in M' denote zeros. Define a labeling map ψ'' on the rows of M' which labels the first d rows of M' according to ψ and the other $\sum_{i=1}^n d_i^2$ rows according to ψ'.

As mentioned above, $\mathcal{M}'(\mathcal{K}, M', \psi'', \boldsymbol{e}_1)$ obviously realizes the access structure AS. We now verify its 3-multiplicativity.

Let $N = (\boldsymbol{v}_0, \boldsymbol{v}_1, \ldots, \boldsymbol{v}_k)$, a matrix over $\mathcal{K}$ with $\sum_{i=1}^n d_i^2$ rows and $k+1$ columns. For $s_i \in \mathcal{K}$ and $\boldsymbol{\rho}_i = (\boldsymbol{\rho}_i', \boldsymbol{\rho}_i'') \in \mathcal{K}^{l-1} \times \mathcal{K}^k$, $1 \leq i \leq 3$, denote $M'(s_i, \boldsymbol{\rho}_i)^\tau = (\boldsymbol{u}_i, \boldsymbol{w}_i)^\tau$, where $\boldsymbol{u}_i^\tau = M(s_i, \boldsymbol{\rho}_i')^\tau$ and $\boldsymbol{w}_i^\tau = N(s_i, \boldsymbol{\rho}_i'')^\tau$. We have

$$\boldsymbol{u}_1^\tau \diamond \boldsymbol{u}_2^\tau = (M(s_1, \boldsymbol{\rho}_1')^\tau) \diamond (M(s_2, \boldsymbol{\rho}_2')^\tau) = M_\diamond(s_1 s_2, \boldsymbol{\rho})^\tau,$$

where $(s_1 s_2, \boldsymbol{\rho}) = (s_1, \boldsymbol{\rho}_1') \otimes (s_2, \boldsymbol{\rho}_2')$. Then,

$$
\begin{aligned}
(\boldsymbol{u}_1 \diamond \boldsymbol{u}_2) \cdot \boldsymbol{w}_3^{\tau} &= (s_1 s_2, \boldsymbol{\rho})(M_{\diamond})^{\tau} \cdot N \begin{pmatrix} s_3 \\ \\ \boldsymbol{\rho}_3''^{\tau} \end{pmatrix} \\
&= (s_1 s_2, \boldsymbol{\rho}) \begin{pmatrix} 1\,0 \cdots 0 \\ 0\,0 \cdots 0 \\ \vdots\ \vdots \ddots \vdots \\ 0\,0 \cdots 0 \end{pmatrix} \begin{pmatrix} s_3 \\ \\ \boldsymbol{\rho}_3''^{\tau} \end{pmatrix} \\
&= s_1 s_2 s_3.
\end{aligned}
$$

It is easy to see that $(\boldsymbol{u}_1 \diamond \boldsymbol{u}_2) \cdot \boldsymbol{w}_3^{\tau}$ is a linear combination of the entries from $(\boldsymbol{u}_1 \diamond \boldsymbol{u}_2) \diamond \boldsymbol{w}_3$, and so is a linear combination of the entries from $M'(s_1, \boldsymbol{\rho}_1)^{\tau} \diamond M'(s_2, \boldsymbol{\rho}_2)^{\tau} \diamond M'(s_3, \boldsymbol{\rho}_3)^{\tau}$.

Hence $\mathcal{M}'$ is a 3-multiplicative LSSS for AS. Obviously, the size of $\mathcal{M}'$ is $O(d^2)$, since $d + \sum_{i=1}^{n} d_i^2 < d^2 + d$. □

If we replace the matrix $M_{\diamond}$ above by the diamond product of λ copies of M, using an identical argument, the construction from Theorem 2 gives rise to a $(\lambda+1)$-multiplicative LSSS from a strongly λ-multiplicative LSSS.

Corollary 2. *Let AS be a $Q^{\lambda+1}$ access structure and $\mathcal{M}(\mathcal{K}, M, \psi, \boldsymbol{e}_1)$ be a strongly λ-multiplicative LSSS realizing AS. Suppose the size of $\mathcal{M}$ is d and $|\psi^{-1}(P_i)| = d_i$, for $1 \le i \le n$. Then there exists a $(\lambda+1)$-multiplicative LSSS for AS of size $O(d^{\lambda})$.*

4.2 From 3-Multiplicativity to Strong Multiplicativity

Theorem 3. *Any 3-multiplicative LSSS is strongly multiplicative.*

Proof. Let $\mathcal{M}(\mathcal{K}, M, \psi, \boldsymbol{e}_1)$ be a 3-multiplicative LSSS realizing the access structure AS over P. For all $A \in \mathcal{A} = 2^P - AS$, by Proposition 1, we can choose a fixed vector $\boldsymbol{\rho}'' \in \mathcal{K}^{l-1}$ such that $M_A(1, \boldsymbol{\rho}'')^{\tau} = \boldsymbol{0}^{\tau}$. There exists a recombination vector $\boldsymbol{z} \in \mathcal{K}^{\sum_{i=1}^{n} d_i^3}$ such that for all $s, s' \in \mathcal{K}$ and $\boldsymbol{\rho}, \boldsymbol{\rho}' \in \mathcal{K}^{l-1}$, we have

$$ss' = \boldsymbol{z}(M(s, \boldsymbol{\rho})^{\tau} \diamond M(s', \boldsymbol{\rho}')^{\tau} \diamond M(1, \boldsymbol{\rho}'')^{\tau}).$$

Since $M_A(1, \boldsymbol{\rho}'')^{\tau} = \boldsymbol{0}^{\tau}$, and $M_{\overline{A}}(1, \boldsymbol{\rho}'')^{\tau}$ is a constant vector for fixed $\boldsymbol{\rho}''$, the vector $\boldsymbol{z}' \in \mathcal{K}^{\sum_{P_i \notin A} d_i^2}$ that satisfies

$$\boldsymbol{z}(M(s, \boldsymbol{\rho})^{\tau} \diamond M(s', \boldsymbol{\rho}')^{\tau} \diamond M(1, \boldsymbol{\rho}'')^{\tau}) = \boldsymbol{z}'(M_{\overline{A}}(s, \boldsymbol{\rho})^{\tau} \diamond M_{\overline{A}}(s', \boldsymbol{\rho}')^{\tau})$$

can be easily determined. Thus $ss' = \boldsymbol{z}'(M_{\overline{A}}(s, \boldsymbol{\rho})^{\tau} \diamond M_{\overline{A}}(s', \boldsymbol{\rho}')^{\tau})$. Hence, $\mathcal{M}$ is strongly multiplicative. □

Although 3-multiplicative LSSS is a subclass of strongly multiplicative LSSS, one of the advantages of 3-multiplicativity is that its verification admits a simpler process. For 3-multiplicativity, we need only to check that $\boldsymbol{e}_1 \in span\{(M \diamond M \diamond M)\}$, while strong multiplicativity requires the verification of $\boldsymbol{e}_1 \in span\{(M \diamond M)_{\overline{A}}\}$ for *all* $A \notin AS$.

Using a similar argument, the following results for $(\lambda+1)$-multiplicativity can be proved:

(i) A $(\lambda + 1)$-multiplicative LSSS is a strongly λ-multiplicative LSSS.
(ii) A λ-multiplicative LSSS is a λ'-multiplicative LSSS, where $1 < \lambda' < \lambda$.

4.3 An Example of a Strongly Multiplicative LSSS That Is Not 3-Multiplicative

We give an example of a strongly multiplicative LSSS that is not 3-multiplicative. It follows that 3-multiplicative LSSS are strictly contained in the class of strongly multiplicative LSSS. The construction process is as follows. Start with an LSSS that realizes a Q^3 access structure but is not strongly multiplicative. We then apply the general construction given in Section 2.3 to convert it into a strongly multiplicative LSSS. The resulting LSSS is however not 3-multiplicative.

Let $P = \{P_1, P_2, P_3, P_4, P_5, P_6\}$ be the set of players. Consider the access structure AS over P defined by

$$AS_{min} = \{(1,2), (3,4), (5,6), (1,5), (1,6), (2,6), (2,5), (3,6), (4,5)\},$$

where we use subscript to denote the corresponding player. For example, $(1,2)$ denotes the subset $\{P_1, P_2\}$. It is easy to verify that the corresponding adversary structure is

$$\mathcal{A}_{max} = \{(1,3), (1,4), (2,3), (2,4), (3,5), (4,6)\},$$

and that AS is a Q^3 access structure.

Let $\mathcal{K} = \mathbb{F}_2$. Define the matrix M over $\mathbb{F}_2$ with the labeling map ψ such that

$$M_{P_1} = \begin{pmatrix} 1\,0\,1\,0\,0 \\ 0\,0\,0\,1\,0 \\ 0\,0\,0\,0\,1 \end{pmatrix}, \; M_{P_2} = \begin{pmatrix} 0\,0\,1\,0\,0 \\ 0\,0\,0\,1\,0 \\ 0\,0\,0\,0\,1 \end{pmatrix}, \; M_{P_3} = \begin{pmatrix} 1\,1\,0\,0\,0 \\ 0\,0\,0\,0\,1 \end{pmatrix},$$

$$M_{P_4} = \begin{pmatrix} 0\,1\,0\,0\,0 \\ 0\,0\,0\,1\,0 \end{pmatrix}, \; M_{P_5} = \begin{pmatrix} 1\,1\,1\,0\,0 \\ 1\,0\,0\,1\,0 \end{pmatrix}, \; M_{P_6} = \begin{pmatrix} 0\,1\,1\,0\,0 \\ 1\,0\,0\,0\,1 \end{pmatrix}.$$

It can be verified that the LSSS $\mathcal{M}(\mathbb{F}_2, M, \psi, \boldsymbol{e}_1)$ realizes the access structure AS. Moreover, for all $A \in \mathcal{A} - \{(1,3), (1,4)\}$, the constricted LSSS $\mathcal{M}_{\overline{A}}$ is multiplicative. Thus in order to get a strongly multiplicative LSSS, we just need to expand $\mathcal{M}$ with multiplicativity when constricted to both $\{P_2, P_4, P_5, P_6\}$ and $\{P_2, P_3, P_5, P_6\}$.

Firstly, consider the LSSS $\mathcal{M}$ constricted to $P' = \{P_2, P_4, P_5, P_6\}$. Obviously, $\mathcal{M}_{P'}$ realizes the access structure $AS'_{min} = \{(5,6), (2,6), (2,5), (4,5)\}$, which is

Q^2 over P'. By the method of Cramer *et al.* [6], we can transform $\mathcal{M}_{P'}$ into the multiplicative LSSS $\mathcal{M}'_{P'}(\mathbb{F}_2, M', \psi', \boldsymbol{e}_1)$ defined as follows:

$$M'_{P_2} = \begin{pmatrix} 0\,0\,1\,0\,0 & \\ 0\,0\,0\,1\,0 & \\ 0\,0\,0\,0\,1 & \\ & 0\,1\,1\,1 \\ & 1\,1\,0\,0 \\ & 0\,0\,0\,1 \end{pmatrix}, \quad M'_{P_4} = \begin{pmatrix} 0\,1\,0\,0\,0 & \\ 0\,0\,0\,1\,0 & \\ & 0\,1\,1\,1 \\ & 1\,0\,0\,0 \end{pmatrix},$$

$$M'_{P_5} = \begin{pmatrix} 1\,1\,1\,0\,0 & \\ 1\,0\,0\,1\,0 & \\ 1 & 0\,1\,0\,1 \\ 0 & 0\,1\,0\,0 \end{pmatrix}, \quad M'_{P_6} = \begin{pmatrix} 0\,1\,1\,0\,0 & \\ 1\,0\,0\,0\,1 & \\ 1 & 0\,0\,1\,0 \\ 0 & 0\,0\,0\,1 \end{pmatrix},$$

where the blanks in the matrices denote zeros.

For consistency, we define

$$\begin{aligned} M'_{P_1} &= (M_{P_1}\ O_{3\times 4}), \\ M'_{P_3} &= (M_{P_3}\ O_{2\times 4}), \end{aligned}$$

where $O_{m\times n}$ denotes the $m \times n$ matrix of all zeros. It can be verified that for the subset $P'' = \{P_2, P_3, P_5, P_6\}$, the constricted LSSS $\mathcal{M}'_{P''}$ is indeed multiplicative. Therefore, $\mathcal{M}'(\mathbb{F}_2, M', \psi', \boldsymbol{e}_1)$ is a strongly multiplicative LSSS realizing the access structure AS. Furthermore, it can be verified that $\mathcal{M}'$ is not 3-multiplicative (the verification involves checking a 443×729 matrix using Matlab).

The scheme $\mathcal{M}(\mathbb{F}_2, M, \psi, \boldsymbol{v}_1)$ given above is the first example of an LSSS which realizes a Q^3 access structure but is not strongly multiplicative.

5 Constructions for 3-Multiplicative LSSS

It is tempting to find efficient constructions for 3-multiplicative LSSS. In general, it is a hard problem to construct LSSS with polynomial size for any specified access structure, and it seems to be an even harder problem to construct polynomial size 3-multiplicative LSSS with general Q^3 access structures. We mention two constructions for 3-multiplicative LSSS. These constructions are generally inefficient, which can result in schemes with exponential sizes. The two constructions are:

1. The Cramer-Damgård-Maurer construction based on Shamir's threshold secret sharing scheme [6].
2. The construction given in Subsection 4.1 based on strongly multiplicative LSSS.

There exist, however, some efficient LSSS with specific access structures that are multiplicative or 3-multiplicative. For instance, Shamir's t out of n threshold secret sharing schemes are multiplicative if $n \geq 2t + 1$, and 3-multiplicative if $n \geq 3t + 1$.

On the other hand, secret sharing schemes from error-correcting codes give good multiplicative properties. It is well known that a secret sharing scheme from a linear error-correcting code is an LSSS. We know that such an LSSS is multiplicative provided the underlying code is a self dual code [7]. The LSSS from a Reed-Solomon code is λ-multiplicative if the corresponding access structure is Q^λ. In this section, we show the multiplicativity of two other classes of secret sharing schemes from error-correcting codes:

(i) schemes from Reed-Muller codes are λ-multiplicative LSSS; and
(ii) schemes from algebraic geometric codes are λ-multiplicative ramp LSSS.

5.1 A Construction from Reed-Muller Codes

Let $\boldsymbol{v}_0, \boldsymbol{v}_1, \ldots, \boldsymbol{v}_{2^m-1}$ be all the points in the space $\mathbb{F}_2^m$. The binary Reed-Muller code $\mathcal{R}(r,m)$ is defined as follows:

$$\mathcal{R}(r,m) = \{(f(\boldsymbol{v}_0), f(\boldsymbol{v}_1), \ldots, f(\boldsymbol{v}_{2^m-1})) \mid f \in \mathbb{F}_2[x_1, \ldots, x_m],\ \deg f \leq r\}.$$

Take $f(\boldsymbol{v}_0)$ as the secret, and $f(\boldsymbol{v}_i)$ as the share distributed to player P_i, $1 \leq i \leq 2^m - 1$. Then $\mathcal{R}(r,m)$ gives rise to an LSSS for the set of players $\{P_1, \ldots, P_n\}$, with the secret-domain being $\mathbb{F}_2$, where $n = 2^m - 1$. For any three codewords

$$\boldsymbol{c}_i = (s_i, s_{i1}, \ldots, s_{in}) = (f_i(\boldsymbol{v}_0), f_i(\boldsymbol{v}_1), \ldots, f_i(\boldsymbol{v}_n)) \in \mathcal{R}(r,m), \quad 1 \leq i \leq 3,$$

it is easy to see that

$$\begin{aligned} \boldsymbol{c}_1 \diamond \boldsymbol{c}_2 \diamond \boldsymbol{c}_3 &= (s_1 s_2 s_3, s_{11} s_{21} s_{31}, \ldots, s_{1n} s_{2n} s_{3n}) \\ &= (g(\boldsymbol{v}_0), g(\boldsymbol{v}_1), \ldots, g(\boldsymbol{v}_n)) \in \mathcal{R}(3r, m), \end{aligned}$$

where $g = f_1 f_2 f_3 \in \mathbb{F}_2[x_1, \ldots, x_m]$ and $\deg g \leq 3r$. From basic results on Reed-Muller codes [15], we know that $\mathcal{R}(3r,m)$ has dual code $\mathcal{R}(m-3r-1,m)$ when $m > 3r$, and the dual code $\mathcal{R}(m-3r-1,m)$ trivially contains the codeword $(1,1,\ldots,1)$. It follows that $s_1 s_2 s_3 = \sum_{j=1}^n s_{1j} s_{2j} s_{3j}$, which shows that the LSSS from $\mathcal{R}(r,m)$ is 3-multiplicative when $m > 3r$. Certainly, this LSSS is strongly multiplicative. In general, we have the following result:

Theorem 4. *The LSSS constructed above from $\mathcal{R}(r,m)$ is λ-multiplicative, provided $m > \lambda r$.*

5.2 A Construction from Algebraic Geometric Codes

Chen and Cramer [3] constructed secret sharing schemes from algebraic geometric (AG) codes. These schemes are *quasi-threshold* (or *ramp*) schemes, which means that any t out of n players can recover the secret, and any fewer than t' players have no information about the secret, where $t' \leq t \leq n$. In this section, we show that ramp schemes from some algebraic geometric codes [3] are λ-multiplicative.

Let χ be an absolutely irreducible, projective, and nonsingular curve defined over $\mathbb{F}_q$ with genus g, and let $D = \{v_0, v_1, \ldots, v_n\}$ be the set of $\mathbb{F}_q$-rational points on χ. Let G be an $\mathbb{F}_q$-rational divisor with degree m satisfying $supp(G) \cap D = \emptyset$ and $2g - 2 < m < n + 1$. Let $\overline{\mathbb{F}}_q$ denote the algebraic closure of $\mathbb{F}_q$, let $\overline{\mathbb{F}}_q(\chi)$ denote the function field of the curve χ, and let $\Omega(\chi)$ denote all the differentials on χ. Define the linear spaces:

$$\mathcal{L}(G) = \{f \in \overline{\mathbb{F}}_q(\chi) \mid (f) + G \geq 0\},$$
$$\Omega(G) = \{\omega \in \Omega(\chi) \mid (\omega) \geq G\}.$$

Then the functional AG code $C_{\mathcal{L}}(D, G)$ and residual AG code $C_{\Omega}(D, G)$ are respectively defined as follows:

$$C_{\mathcal{L}}(D, G) = \{(f(v_0), f(v_1), \ldots, f(v_n)) \mid f \in \mathcal{L}(G)\} \subseteq \mathbb{F}_q^{n+1},$$
$$C_{\Omega}(D, G) = \{(Res_{v_0}(\eta), Res_{v_1}(\eta), \ldots, Res_{v_n}(\eta)) \mid \eta \in \Omega(G - D)\} \subseteq \mathbb{F}_q^{n+1},$$

where $Res_{v_i}(\eta)$ denotes the residue of η at v_i.

As above, $C_{\Omega}(D, G)$ induces an LSSS for the set of players $\{P_1, \ldots, P_n\}$, where for every codeword $(f(v_0), f(v_1), \ldots, f(v_n)) \in C_{\Omega}(D, G) = C_{\mathcal{L}}(D, D - G + (\eta))$, $f(v_0)$ is the secret and $f(v_i)$ is P_i's share, $1 \leq i \leq n$. For any λ codewords

$$\begin{aligned} \boldsymbol{c}_i &= (s_i, s_{i1}, \ldots, s_{in}) \\ &= (f_i(v_0), f_i(v_1), \ldots, f_i(v_n)) \in C_{\mathcal{L}}(D, D - G + (\eta)), \quad 1 \leq i \leq \lambda, \end{aligned}$$

it is easy to see that

$$\diamond_{i=1}^{\lambda} \boldsymbol{c}_i = \left(\prod_{i=1}^{\lambda} s_i, \prod_{i=1}^{\lambda} s_{i1}, \ldots, \prod_{i=1}^{\lambda} s_{in}\right) \in C_{\mathcal{L}}(D, \lambda(D - G + (\eta))).$$

If $2g - 2 < \deg(\lambda(D - G + (\eta))) < n$, then $C_{\mathcal{L}}(D, \lambda(D - G + (\eta)))$ has the dual code $C_{\Omega}(D, \lambda(D - G + (\eta))) = C_{\mathcal{L}}(D, \lambda G - (\lambda - 1)(D + (\eta)))$. When $\deg(\lambda G - (\lambda - 1)(D + (\eta))) \geq 2g$, $C_{\Omega}(D, \lambda(D - G + (\eta)))$ has a codeword with a nonzero first coordinate, implying $\prod_{i=1}^{\lambda} s_i = \sum_{j=1}^{n} a_j \prod_{i=1}^{\lambda} s_{ij}$ for some constants $a_j \in \mathbb{F}_q$. Thus, the LSSS induced by the AG code $C_{\Omega}(D, G)$ is λ-multiplicative. It is easy to see that if $\deg G = m \geq \frac{(\lambda-1)(n-1)}{\lambda} + 2g$ then we have $2g - 2 < \deg(\lambda(D - G + (\eta))) < n$ and $\deg(\lambda G - (\lambda - 1)(D + (\eta))) \geq 2g$. Therefore, we have the following theorem.

Theorem 5. *Let χ be an absolutely irreducible, projective, and nonsingular curve defined over $\mathbb{F}_q$ with genus g, let $D = \{v_0, v_1, \ldots, v_n\}$ be the set of $\mathbb{F}_q$-rational points on χ. Let G be an $\mathbb{F}_q$-rational divisor with degree m satisfying $supp(G) \cap D = \emptyset$ and $2g - 2 < m < n + 1$. Then the LSSS induced by the AG code $C_{\Omega}(D, G)$ is λ-multiplicative, provided $m \geq \frac{(\lambda-1)(n-1)}{\lambda} + 2g$.*

6 Implications of the Multiplicativity of LSSS

The property of 3-multiplicativity implies strong multiplicativity, and so is sufficient for building MPC protocols against active adversaries. The conditions for

3-multiplicativity are easy to verify, while verification for strong multiplicativity involves checking an exponential number of equations (each subset in the adversary structure corresponds to an equation).

With 3-multiplicative LSSS, or more generally λ-multiplicative LSSS, we can simplify local computation for each player and reduce the round complexity in MPC protocols. For example, using the technique of Bar-Ilan and Beaver [1], we can compute $\prod_{i=1}^{l} x_i$, $x_i \in \mathbb{F}_q$, in a constant number of rounds, independent of l. For simplicity, we consider passive adversaries in the information-theoretic model. Suppose for $1 \le i \le l$, the shares of x_i, denoted by $[x_i]$, have already been distributed among the players. To compute $\prod_{i=1}^{l} x_i$, $x_i \in \mathbb{F}_q$, we follow the process of Cramer *et al.* [4]:

(1) Generate $[b_0 \in_R \mathbb{F}_q^*], [b_1 \in_R \mathbb{F}_q^*], \ldots, [b_l \in_R \mathbb{F}_q^*]$ and $[b_0^{-1}], [b_1^{-1}], \ldots, [b_l^{-1}]$, where $b_i \in_R \mathbb{F}_q^*$ means that b_i is a random element in $\mathbb{F}_q^*$.
(2) For $1 \le i \le l$, each player computes $[b_{i-1} x_i b_i^{-1}]$ from $[b_{i-1}], [b_i^{-1}]$ and $[x_i]$.
(3) Recover $d_i = b_{i-1} x_i b_i^{-1}$ from $[b_{i-1} x_i b_i^{-1}]$ for $1 \le i \le l$, and compute $d = \prod_{i=1}^{l} d_i$.
(4) Compute $[d b_0^{-1} b_l]$ from $[b_0^{-1}], [b_l]$ and d.

It is easy to see that $d b_0^{-1} b_l = \prod_{i=1}^{l} x_i$. Using a multiplicative LSSS, the above process takes five rounds of interactions as two rounds are required in Step (2). However, if we use a 3-multiplicative LSSS instead, then only one round is needed for Step (2). Thus, 3-multiplicative LSSS reduce the round complexity of computing unbounded fan-in multiplication from five to four. This in turn simplifies the computation of many problems, such as polynomial evaluation and solving linear systems of equations.

In general, the relationship between λ-multiplicative LSSS and strongly λ-multiplicative LSSS can be described as follows:

$$\cdots \subseteq SMLSSS_{\lambda+1} \subsetneq MLSSS_{\lambda+1} \subseteq SMLSSS_{\lambda} \subsetneq MLSSS_{\lambda} \subseteq \cdots,$$

where $MLSSS_\lambda$ (respectively, $SMLSSS_\lambda$) denotes the class of λ-multiplicative (respectively, strongly λ-multiplicative) LSSS. It is easy to see that $SMLSSS_\lambda \subsetneq MLSSS_\lambda$ because they exist under the conditions $Q^{\lambda+1}$ and Q^λ, respectively. Since $SMLSSS_\lambda$ and $MLSSS_{\lambda+1}$ both exist under the same necessary and sufficient condition of $Q^{\lambda+1}$, it is not straightforward to see whether $MLSSS_{\lambda+1}$ is strictly contained in $SMLSSS_\lambda$. For $\lambda = 2$, we already know that $MLSSS_3 \subsetneq SMLSSS_2$ (Section 4.3). It would be interesting to find out if this is also true for $\lambda > 2$. We have also given an efficient transformation from $SMLSSS_\lambda$ to $MLSSS_{\lambda+1}$. It remains open whether an efficient transformation from $MLSSS_\lambda$ to $SMLSSS_\lambda$ exists when the access structure is $Q^{\lambda+1}$. When $\lambda = 2$, this is a well-known open problem [6].

7 Conclusions

In this paper, we propose the new concept of 3-multiplicative LSSS, which form a subclass of strongly multiplicative LSSS. The 3-multiplicative LSSS are easier to construct compared to strongly multiplicative LSSS. They can also simplify the computation and reduce the round complexity in secure multiparty computation protocols. We believe that 3-multiplicative LSSS are a more appropriate primitive as building blocks for secure multiparty computations, and deserve further investigation. We stress that finding efficient constructions of 3-multiplicative LSSS for general access structures remains an important open problem.

Acknowledgement

The work of M. Liu and Z. Zhang is supported in part by the open project of the State Key Laboratory of Information Security and the 973 project of China (No. 2004CB318000). Part of the work was done while Z. Zhang was visiting Nanyang Technological University supported by the Singapore Ministry of Education under Research Grant T206B2204.

The work of Y. M. Chee, S. Ling, and H. Wang is supported in part by the Singapore National Research Foundation under Research Grant NRF-CRP2-2007-03.

In addition, the work of Y. M. Chee is also supported in part by the Nanyang Technological University under Research Grant M58110040, and the work of H. Wang is also supported in part by the Australian Research Council under ARC Discovery Project DP0665035.

References

1. Bar-Ilan, J., Beaver, D.: Non-cryptographic fault-tolerant computing in constant number of rounds of interaction. In: PODC 1989, pp. 201–209 (1989)
2. Beimel, A.: Secure schemes for secret sharing and key distribution. PhD thesis, Technion - Israel Institute of Technology (1996)
3. Chen, H., Cramer, R.: Algebraic geometric secret sharing schemes and secure multi-party computations over small fields. In: Dwork, C. (ed.) CRYPTO 2006. LNCS, vol. 4117, pp. 521–536. Springer, Heidelberg (2006)
4. Cramer, R., Kiltz, E., Padró, C.: A note on secure computation of the Moore-Penrose pseudoinverse and its spplication to secure linear algebra. In: Menezes, A. (ed.) CRYPTO 2007. LNCS, vol. 4622, pp. 613–630. Springer, Heidelberg (2007)
5. Chen, H., Cramer, R., de Haan, R., Cascudo Pueyo, I.: Strongly multiplicative ramp schemes from high degree rational points on curves. In: Smart, N.P. (ed.) EUROCRYPT 2008. LNCS, vol. 4965, pp. 451–470. Springer, Heidelberg (2008)
6. Cramer, R., Damgård, I., Maurer, U.: General secure multi-party computation from any linear secret-sharing scheme. In: Preneel, B. (ed.) EUROCRYPT 2000. LNCS, vol. 1807, pp. 316–334. Springer, Heidelberg (2000)
7. Cramer, R., Daza, V., Gracia, I., Urroz, J., Leander, G., Martí-Farré, J., Padró, C.: On codes, matroids and secure multi-party computation from linear secret sharing schemes. In: Shoup, V. (ed.) CRYPTO 2005. LNCS, vol. 3621, pp. 327–343. Springer, Heidelberg (2005)
8. Fehr, S.: Efficient construction of the dual span program. Master Thesis, the Swiss Federal Institute of Technology (ETH) Zürich (1999), http://homepages.cwi.nl/~fehr/publications.html

9. Goldreich, O., Micali, S., Wigderson, A.: How to play ANY mental game. In: STOC 1987, pp. 218–219 (1987)
10. Karchmer, M., Wigderson, A.: On span programs. In: Proc. 8th Ann. Symp. Structure in Complexity Theory, pp. 102–111 (1993)
11. Käsper, E., Nikov, V., Nikova, S.: Strongly multiplicative hierarchical threshold secret sharing. In: 2nd International Conference on Information Theoretic Security - ICITS 2007. LNCS (to appear, 2007)
12. Liu, M., Xiao, L., Zhang, Z.: Multiplicative linear secret sharing schemes based on connectivity of graphs. IEEE Transactions on Information Theory 53(11), 3973–3978 (2007)
13. Massey, J.L.: Minimal codewords and secret sharing. In: Proc. 6th Joint Swedish-Russian Workshop on Information Theory, pp. 276–279 (1993)
14. Nikov, V., Nikova, S., Preneel, B.: On multiplicative linear secret sharing schemes. In: Johansson, T., Maitra, S. (eds.) INDOCRYPT 2003. LNCS, vol. 2904, pp. 135–147. Springer, Heidelberg (2003)
15. van Lint, J.H.: Introduction to coding theory, 3rd edn. Graduate Texts in Mathematics, vol. 86. Springer, Heidelberg (1999)
16. Yao, A.: Protocols for secure computation. In: FOCS 1982, pp. 160–164 (1982)

Graph Design for Secure Multiparty Computation over Non-Abelian Groups

Xiaoming Sun[1], Andrew Chi-Chih Yao[1], and Christophe Tartary[1,2]

[1] Institute for Theoretical Computer Science
Tsinghua University
Beijing, 100084
People's Republic of China
[2] Division of Mathematical Sciences
School of Physical and Mathematical Sciences
Nanyang Technological University
Singapore
{xiaomings,andrewcyao}@tsinghua.edu.cn,
ctartary@ntu.edu.sg

Abstract. Recently, Desmedt et al. studied the problem of achieving secure n-party computation over non-Abelian groups. They considered the passive adversary model and they assumed that the parties were only allowed to perform black-box operations over the finite group G. They showed three results for the n-product function $f_G(x_1, \ldots, x_n) := x_1 \cdot x_2 \cdot \ldots \cdot x_n$, where the input of party P_i is $x_i \in G$ for $i \in \{1, \ldots, n\}$. First, if $t \geq \lceil \frac{n}{2} \rceil$ then it is impossible to have a t-private protocol computing f_G. Second, they demonstrated that one could t-privately compute f_G for any $t \leq \lceil \frac{n}{2} \rceil - 1$ in exponential communication cost. Third, they constructed a randomized algorithm with $O(n\,t^2)$ communication complexity for any $t < \frac{n}{2.948}$.

In this paper, we extend these results in two directions. First, we use percolation theory to show that for any fixed $\epsilon > 0$, one can design a randomized algorithm for any $t \leq \frac{n}{2+\epsilon}$ using $O(n^3)$ communication complexity, thus nearly matching the known upper bound $\lceil \frac{n}{2} \rceil - 1$. This is the first time that percolation theory is used for multiparty computation. Second, we exhibit a deterministic construction having polynomial communication cost for any $t = O(n^{1-\epsilon})$ (again for any fixed $\epsilon > 0$). Our results extend to the more general function $\widetilde{f}_G(x_1, \ldots, x_m) := x_1 \cdot x_2 \cdot \ldots \cdot x_m$ where $m \geq n$ and each of the n parties holds one or more input values.

Keywords: Multiparty Computation, Passive Adversary, Non-Abelian Groups, Graph Coloring, Percolation Theory.

1 Introduction

In multiparty computation, a set of n parties $\{P_1, \ldots, P_n\}$ want to compute a function of some secret inputs held locally by these participants. Since its introduction by Yao [19], multiparty computation has been extensively studied. Most multiparty computation protocols rely on algebraic structures which are at least Abelian groups [14] as in

J. Pieprzyk (Ed.): ASIACRYPT 2008, LNCS 5350, pp. 37–53, 2008.
© International Association for Cryptologic Research 2008

[1, 3, 4, 8, 10, 11, 12] for instance. The usefulness of Abelian groups in cryptography is not restricted to multiparty computation as numerous cryptographic primitives are developed over such groups [6, 7, 17]. However, the construction of efficient quantum algorithms to solve the discrete logarithm problem as well as the factoring problem prevent the use of many of these primitives over those machines [18]. Since quantum algorithms seem to be less efficient over non-Abelian groups, there is increasingly a need for developing cryptographic constructions over such mathematical structures. The reader may be aware of the existence of public key cryptosystems for such groups [15, 16].

Recently, Desmedt et al. studied the problem of designing secure n-party protocol over non commutative finite groups for the *passive* (or *semi-honest*) adversary model [5]. Their goal is to guarantee unconditional security simply using a black-box representation of the finite non-Abelian group $(G,\cdot)$. This assumption means that the n parties can only perform three operations in $(G,\cdot)$: the group operation ($(x,y) \mapsto x \cdot y$), the group inversion ($x \mapsto x^{-1}$) and the uniformly distributed group sampling ($x \in_R G$).

Desmedt et al. focused on the existence and the design of t-private protocols for the n product function $f_G(x_1,\ldots,x_n) := x_1 \cdot \ldots \cdot x_n$ where the input of party P_i is $x_i \in G$ for $i \in \{1,\ldots,n\}$. In such a protocol, no colluding sets $\mathcal{C}$ of at most t participants learn anything about the data hold by any of the remaining members $\{P_1,\ldots,P_n\} \setminus \mathcal{C}$. Desmedt et al. obtained three important results. First, if $t \geq \lceil \frac{n}{2} \rceil$ (dishonest majority) then it is impossible to construct a t-private protocol to compute f_G. Second, if $t < \lceil \frac{n}{2} \rceil$ then one can always design a deterministic t-private protocol computing f_G with an exponential communication complexity of $O(n \binom{2t+1}{t}^2)$ group elements. Third, they built a probabilistic t-private protocol computing f_G with a polynomial communication complexity of $O(n\,t^2)$ group elements when $t < \frac{n}{2.948}$.

That work leads to two important questions. First, we would like to know if it is possible to construct a t-private protocol for values of $t \in \left[\frac{n}{2.948}, \lceil \frac{n}{2} \rceil - 1\right]$ with polynomial communication complexity. Second, Desmedt et al.'s construction shows that one can t-privately compute f_G with polynomial communication cost for any $t = O(\log n)$. A natural issue is to determine the existence and to construct a deterministic t-private protocol with polynomial communication complexity for other values t (ideally, up to the threshold $\lceil \frac{n}{2} \rceil - 1$).

In this article, we give a positive answer to these two questions. First, we demonstrate that the random coloring approach and the graph construction by Desmedt et al. can be used to guarantee t-privacy for any $t < \frac{n}{2+\epsilon}$ (for any fixed $\epsilon > 0$). The communication complexity of our construction is $O(n^3)$ group elements. This result is obtained using percolation theory. To the best of our knowledge, this is the first use of this theory in the context of multiparty computation. Second, we provide a deterministic construction for any $t = O(n^{1-\epsilon})$. This scheme has polynomial communication complexity as well.

This paper is organized as follows. In the next section, we will recall the different reductions performed in [5] to solve the t-privacy issue over non-Abelian groups. In Sect. 3, we present our randomized construction achieving t-privacy for any value $t \leq \frac{n}{2+\epsilon}$ which is closed to the theoretical bound $\lceil \frac{n}{2} \rceil - 1$. In Sect. 4, we show how to construct deterministic t-private protocols having polynomial communication cost for any $t = O(n^{1-\epsilon})$. In the last section, we conclude our paper with some remaining open problems for multiparty computation over non-Abelian black-box groups.

2 Achieving Secure Computation over Non-Abelian Groups

In this section, we present some of the results and constructions developed by Desmedt et al. which are necessary to understand our improvements from Sect. 3 and Sect. 4. First, we recall the definition of secure multiparty computation in the passive, computationally unbounded attack model, restricted to deterministic symmetric functionalities and perfect emulation as in [5].

We denote $[n]$ the set of integers $\{1, \ldots, n\}$, $\{0,1\}^*$ the set of all finite binary strings and $|A|$ the cardinality of the set A.

Definition 1. *We denote* $f : (\{0,1\}^*)^n \mapsto \{0,1\}^*$ *an n-input and single-output function. Let $\prod$ be a n-party protocol for computing f. We denote the n-party input sequence by* $\mathbf{x} = (x_1, \ldots, x_n)$, *the joint protocol view of parties in subset $I \subset [n]$ by* $\mathbf{VIEW}_I^{\prod}(\mathbf{x})$, *and the protocol output by* $\mathbf{OUT}^{\prod}(\mathbf{x})$. *For $0 < t < n$, we say that $\prod$ is a t-private protocol for computing f if there exists a probabilistic polynomial-time algorithm S, such that, for every $I \subset [n]$ with $|I| \leq t$ and every $\mathbf{x} \in (\{0,1\}^*)^n$, the random variables*

$$\langle S(I, \mathbf{x}_I, f(\mathbf{x})), f(\mathbf{x})\rangle \text{ and } \langle \mathbf{VIEW}_I^{\prod}(\mathbf{x}), \mathbf{OUT}^{\prod}(\mathbf{x})\rangle$$

are identically distributed, where $\mathbf{x}_I$ denotes the projection of the n-ary sequence $\mathbf{x}$ on the coordinates in I.

In the remaining of this paper, we assume that party P_i has a personal input $x_i \in G$ (for $i \in [n]$) and the function to be computed is the n-party product $f_G(x_1, \ldots, x_n) := x_1 \cdot \ldots \cdot x_n$.

Desmedt et al. first reduced the problem of constructing a t-private n-party protocol for f_G to the problem of constructing a *symmetric (strong) t-private protocol* $\prod'$ (see [5] for a detailed definition of symmetric privacy) to compute the shared 2-product function $f'_G(x, y) := x \cdot y$ where the inputs x and y are shared amongst the n parties. They demonstrated that iterating $(n-1)$ times the protocol $\prod'$ would give a t-private protocol to compute f_G.

The second reduction occurring in [5] consists of constructing a t-private n-party shared 2-product protocol $\prod'$ from a suitable coloring over particular directed graphs. We will detail the important steps of this reduction as they will serve the understanding of our own constructions.

Definition 2 ([5]). *We call graph $\mathcal{G}$ an* admissible Planar Directed Acyclic Graph (PDAG) *with share parameter ℓ and size parameter $m(\geq \ell)$ if it has the following properties:*

- *The nodes of $\mathcal{G}$ are drawn on a square $m \times m$ grid of points (each node of $\mathcal{G}$ is located at a grid point but some grid points may not be occupied by nodes). The rows of the grid are indexed from top to bottom and the columns from left to right by the integers $1, 2, \ldots, m$. A node of $\mathcal{G}$ at row i and column j is said to have index (i, j). $\mathcal{G}$ has $2\,\ell$ input nodes on the top row, and ℓ output nodes on the bottom row.*
- *The incoming edges of a node on row i only come from nodes on row $i-1$, and outgoing edges of a node on row i only go to nodes on row $i+1$.*

– *For each row i and column j, let $\eta_1^{(i,j)} < \cdots < \eta_{q(i,j)}^{(i,j)}$ denote the ordered column indices of the $q(i,j) > 0$ nodes on level $i+1$ which are connected to node (i,j) by an edge. Then, for each $j \in [m-1]$, we have:*

$$\eta_{q(i,j)}^{(i,j)} \leq \eta_1^{(i,j+1)}$$

which means that the rightmost node on level $i+1$ connected to node (i,j) is to the left of (or equal to) the leftmost node on level $i+1$ connected to node $(i,j+1)$.

An admissible PDAG has 2ℓ input nodes. The first ℓ ones (i.e. $(1,1),\ldots,(1,\ell)$) represent the x-input nodes while the remaining ones represent the y-input nodes. Let $C : [m] \times [m] \mapsto [n]$ be a n-coloring function that associates to each node (i,j) of $\mathcal{G}$ a color $C(i,j)$ chosen from a set of n possible colors. The following notion will be used to express the property we expect the graph coloring to have in order to build $\prod'$.

Definition 3 ([5]). *We say that $C : [m] \times [m] \mapsto [n]$ is a t-*reliable *n-*coloring *for the admissible* PDAG *$\mathcal{G}$ (with share parameter ℓ and size parameter m) if for each t-color subset $I \subset [n]$, there exist $j^* \in [\ell]$ and $j_y^* \in [\ell]$ such that:*

- *There exists a path* PATH_x *in $\mathcal{G}$ from the j^*th x-input node to the j^*th output node, such that none of the path node colors are in subset I (it is called an I-avoiding path), and*
- *There exists an I-avoiding path* PATH_y *in $\mathcal{G}$ from the j_y^*th y-input node to the j^*th output node.*

If $j_y^* = j^*$ for all I, we say that C is a symmetric *t-reliable n-coloring.*

Important Remark: Even if the graph $\mathcal{G}$ is directed, it is regarded as *non-directed* when building the I-avoiding paths in Definition 3.

Desmedt et al. built a protocol $\prod'(\mathcal{G}, C)$ taking as input a graph $\mathcal{G}$ and a n coloring C. We do not detail this protocol in our paper as its internal design does not have any influence in our work. The reader can find it in [5]. However, in order to ease the understanding of our work, we recall the relation between multiparty protocols over a non-Abelian group G and coloring of admissible PDAGs as it appear in [5].

The n participants $\{P_1,\ldots,P_n\}$ are identified by the n colors of the admissible PDAG $\mathcal{G}$. The input/output nodes of the graph $\mathcal{G}$ are labeled by the input/output elements of the group G. Each edge represents a group element sent from one participant to another one. Each internal node contains an intermediate value of the protocol. Those values are computed, at each node $\mathcal{N}$ of $\mathcal{G}$, as the group operation between the elements along all the incoming edges of $\mathcal{N}$ from the leftmost one to the rightmost one. This intermediate value is then redistributed along all the outgoing edges of $\mathcal{N}$ using the following $O_{\mathcal{N}}$-of-$O_{\mathcal{N}}$ secret sharing where $O_{\mathcal{N}}$ represents the number of outgoing edges of node $\mathcal{N}$.

Proposition 1 ([5]). *Let g be an element of the non-Abelian group G. Denote λ and μ two integers where $\mu \in [\lambda]$. We create a λ-of-λ sharing $(s_g(1),\ldots,s_g(\lambda))$ of g by picking the $\lambda - 1$ shares $\{s_g(\xi)\}_{\xi \in [\lambda] \setminus \{\mu\}}$ uniformly and independently at random from G, and computing $s_g(\mu)$ to be the unique element of G such that:*

$$g = s_g(1) \cdot s_g(2) \cdot \ldots \cdot s_g(\lambda)$$

Then, the distribution of the shares $(s_g(1), \ldots, s_g(\lambda))$ *is independent of* μ.

We recall the following important result:

Theorem 1 ([5]). *If* $\mathcal{G}$ *is an admissible* PDAG *and* C *is a symmetric* t*-reliable* n*-coloring for* $\mathcal{G}$ *then* $\prod'(\mathcal{G}, C)$ *achieves symmetric strong* t*-privacy.*

The last reduction is related to the admissible PDAG. Desmedt et al. only consider admissible PDAGs as defined below and represented in Fig. 1.

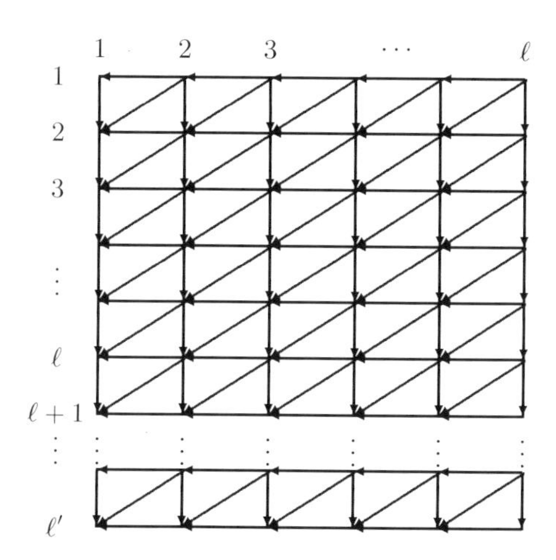

Fig. 1. The admissible PDAG $\mathcal{G}_{tri}(\ell', \ell)$

Definition 4 ([5]). *The admissible* PDAG $\mathcal{G}_{tri}(\ell', \ell)$ *is a* $\ell' \times \ell$ *directed grid such that:*

- *[horizontal edges] for* $i \in [\ell']$ *and for* $j \in [\ell - 1]$, *there is a directed edge from node* $(i, j+1)$ *to* (i, j),
- *[vertical edges] for* $i \in [\ell' - 1]$ *and for* $j \in [\ell]$, *there is a directed edge from node* (i, j) *to node* $(i+1, j)$,
- *[diagonal edges] for* $i \in [\ell' - 1]$ *and for* $j \in \{2, \ldots, \ell\}$, *there is a directed edge from node* (i, j) *to node* $(i+1, j-1)$.

According to Definition 2, an admissible PDAG has $2\,\ell$ input nodes and no horizontal edges. Desmedt et al. indicated that the y-input nodes could be arranged along a column on $\mathcal{G}_{tri}(\ell', \ell)$ instead of being along the same row as the x-input nodes. They also explained that $\mathcal{G}_{tri}(\ell', \ell)$ could also be drawn according the requirements of Definition 2. By rotating $\mathcal{G}_{tri}(\ell', \ell)$ by 45 degrees anticlockwise, the x-input nodes and y-input nodes of $\mathcal{G}_{tri}(\ell', \ell)$ are now on the same row and the horizontal edges of $\mathcal{G}_{tri}(\ell', \ell)$ have become diagonal edges which satisfies Definition 2.

A priori, $\mathcal{G}_{tri}(\ell', \ell)$ is a rectangular grid. In [5], Desmedt et al. considered square grids $\mathcal{G}_{tri}(\ell, \ell)$ for which they introduced the following notion.

Definition 5 ([5]). *We say that* $C : [\ell] \times [\ell] \mapsto [n]$ *is a weakly* t*-reliable* n*-coloring for* $\mathcal{G}_{tri}(\ell, \ell)$ *if for each* t*-color subset* $I \subset [n]$*:*

- *There exists an I-avoiding path $\mathcal{P}_x$ in $\mathcal{G}_{tri}(\ell,\ell)$ from a node on the top row to a node on the bottom row. Such a path is called an I-avoiding top-bottom path.*
- *There exists an I-avoiding path $\mathcal{P}_y$ in $\mathcal{G}_{tri}(\ell,\ell)$ from a node on the rightmost column to a node on the leftmost column. Such a path is called an I-avoiding right-left path.*

As said in [5], the admissible PDAG requirements (Definition 2) are still satisfied if we remove from $\mathcal{G}_{tri}$ some 'positive slope' diagonal edges and add some 'negative slope' diagonal edges (connecting a node (i,j) to node $(i+1,j+1)$, for some $i \in [\ell'-1]$ and $j \in [\ell-1]$). Such a generalized admissible PDAG is denoted $\mathcal{G}_{gtri}$.

Lemma 1 ([5]). *Let $C : [\ell] \times [\ell] \mapsto [n]$ be a weakly t-reliable n-coloring for square admissible* PDAG *$\mathcal{G}_{tri}(\ell,\ell)$. Then, we can construct a t-reliable n-coloring for a rectangular admissible* PDAG *$\mathcal{G}_{gtri}(2\ell-1,\ell)$.*

Thus, Desmedt et al. have demonstrated that it was sufficient to get a weakly t-reliable n coloring for some $\mathcal{G}_{tri}(\ell,\ell)$ in order to construct a t-private protocol for computing the n-product f_G. The cost communication cost of this protocol is $(n-1)$ times the number of edges of $\mathcal{G}_{gtri}(2\ell-1,\ell)$. Since that grid is obtained from $\mathcal{G}_{tri}(\ell,\ell)$ using a mirror, the communication cost of the whole protocol is $O(n\,\ell^2)$ group elements. The constructions that we propose in this paper are colorings of some grids $\mathcal{G}_{tri}(\ell,\ell)$.

3 A Randomized Construction Achieving Maximal Privacy

In this section, we present a randomized construction ensuring the t-privacy of the computation of f_G up to $\frac{n}{2+\epsilon}$. Our scheme has a linear share parameter $\ell = O(n)$.

We use the same random coloring C_{rand} for the grid $\mathcal{G}_{tri}(\ell,\ell)$ as in [5]. However, our analysis is based on percolation theory while Desmedt et al. used a counting-based argument. We first introduce the following definition which is illustrated in Fig. 2.

Algorithm 1. Coloring C_{rand}

Input: A grid $\mathcal{G}_{tri}(\ell,\ell)$.
1. For each $(i,j) \in [\ell] \times [\ell]$, choose the color $C(i,j)$ of node (i,j) independently and uniformly at random from $[n]$.

Output: A n-coloring of the grid.

Definition 6. *The* triangular lattice of depth ℓ *denoted $\mathcal{T}(\ell)$ is a directed graph drawn over a $\ell \times (3\,\ell-2)$ grid such that:*

- *[horizontal edges] for $i \in [\ell]$ and for $j \in [\ell-1]$, there is a directed edge from node $(i,i+2\,j)$ to $(i,i+2\,(j-1))$,*
- *[right downwards edges] for $i \in [\ell-1]$ and for $j \in \{0,\ldots,\ell-1\}$, there is a directed edge from node $(i,i+2\,j)$ to node $(i+1,i+2\,j+1)$,*
- *[left downwards edges] for $i \in [\ell-1]$ and for $j \in [\ell-1]$, there is a directed edge from node $(i,i+2\,j)$ to node $(i+1,i+2\,j-1)$.*

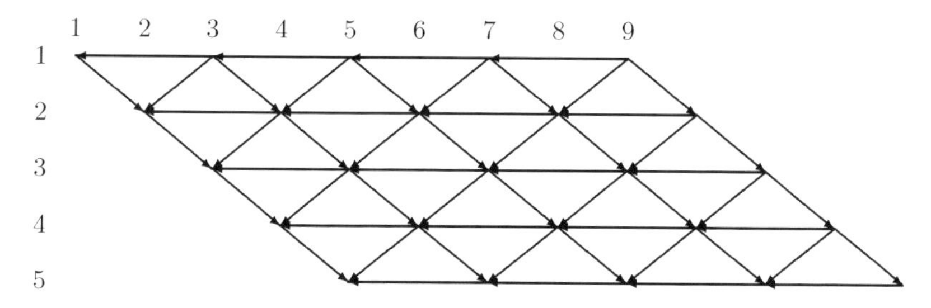

Fig. 2. The triangle $\mathcal{T}(5)$

Proposition 2. *For any positive integer ℓ, we have a graph isomorphism between $\mathcal{G}_{tri}(\ell, \ell)$ and $\mathcal{T}(\ell)$.*

Proof. Consider the mapping:

$$\begin{aligned} \mathcal{G}_{tri}(\ell, \ell) &\longrightarrow \mathcal{T}(\ell) \\ (i, j) &\longmapsto (i, i + 2(j-1)) \end{aligned}$$

It is easy to see that the nodes of the two graphs are in bijective correspondence while the direction of each edge is maintained. □

Theorem 2. *For any $\epsilon > 0$, there exists a constant c_ϵ such that if $t \leq \frac{n}{2+\epsilon}$ and $\ell \geq c_\epsilon n$, then there exists a weakly t-reliable n-coloring for $\mathcal{G}_{tri}(\ell, \ell)$.*

Proof. We prove that the coloring C_{rand} will work with high probability. Let $t_\epsilon = \left\lfloor \frac{n}{2+\epsilon} \right\rfloor$ where $\lfloor \cdot \rfloor$ denotes the floor function. Instead of considering the probability that C_{rand} is a weakly t_ϵ-reliable n-coloring for $\mathcal{G}_{tri}(\ell, \ell)$, we study the complementary event. A suitable value for ℓ will be given at the end of this demonstration.
The coloring C_{rand} is called *bad* if there exists a color set $I \subset [n]$ with $|I| = t_\epsilon$, such that either there are no I-avoiding top-bottom paths or there are no I-avoiding right-left paths. By the union bound, we obtain the following upper bound on $\Pr(C_{rand}$ is bad):

$$\begin{aligned} &2 \Pr(\exists I \subset [n], |I| = t_\epsilon, \text{there are no } I\text{-avoiding top-bottom paths in } \mathcal{G}_{tri}(\ell, \ell)) \\ &\leq 2 \sum_{I \subset [n], |I| = t_\epsilon} \Pr(\text{there are no } I\text{-avoiding top-bottom paths in } \mathcal{G}_{tri}(\ell, \ell)). \qquad (1) \end{aligned}$$

The factor 2 in (1) comes from the fact the top-bottom probability is equal to the right-left probability due to the symmetry of the grid $\mathcal{G}_{tri}(\ell, \ell)$ and the coloring C_{rand}.

Next, we demonstrate that for a fixed color set $I \subset [n]$ with $|I| = t_\epsilon$, the probability that there are no I-avoiding top-bottom paths in C_{rand} is exponentially small. Let us fix the color set I. We call a vertex *closed* if its color belongs to I. Otherwise, the vertex is called *open*. The random coloring C_{rand} of each vertex is equivalent to open it independently and randomly with probability $p := 1 - \frac{t_\epsilon}{n}$. An I-avoiding path is simply an *open path*. Therefore, we get:

$$\begin{aligned} &\Pr(\text{there are no } I\text{-avoiding top-bottom paths in } \mathcal{G}_{tri}(\ell, \ell)) \\ &= \Pr_p(\text{there are no open top-bottom paths in } \mathcal{G}_{tri}(\ell, \ell)) \\ &= 1 - \Pr_p(\text{there is an open top-bottom path in } \mathcal{G}_{tri}(\ell, \ell)) \qquad (2) \end{aligned}$$

We have the following result.

Lemma 2 ([2]). *The triangular lattice $\mathcal{T}(\ell)$ has the following property:*

$$\Pr_p(\text{there is an open top-bottom path in } \mathcal{T}(\ell)) + \Pr_p(\text{there is a closed right-left path in } \mathcal{T}(\ell)) = 1$$

When we combine Lemma 2, Proposition 2 and (2), we obtain the following:

$$\Pr(\text{there is no } I\text{-avoiding top-bottom path in } \mathcal{G}_{tri}(\ell,\ell)) = \Pr_p(\text{there is a closed right-left path in } \mathcal{T}(\ell)) = \Pr_{1-p}(\text{there is an open right-left path in } \mathcal{T}(\ell)) \quad (3)$$

In (3), $\Pr_{1-p}(\cdot)$ means that we open each vertex with probability $1-p$. We have the following result from percolation theory.

Lemma 3 ([13]). *Let T be the triangular lattice in the plane. Then, the critical probability of site percolation $p_c^s(T)$ is equal to $\frac{1}{2}$.*

When the open probability is less than the critical probability, the percolation has the following properties (see for example Chapter 4, Theorem 9 in [2]).

Lemma 4 ([9]). *If $p < p_c^s(T)$, then there is a constant $c = c(p)$,*

$$\Pr_p(0 \xrightarrow{n}) < e^{-cn}.$$

where $\{x \xrightarrow{n}\}$ is the event that there is an open path from x to a point in $S_n(x)$ with $S_n(x) := \{y : d(x,y) = n\}$ and $d(x,y)$ denotes the distance between x and y.

Remark: The value 0 from Lemma 4 represent the zero element of $\mathbb{Z} \times \mathbb{Z}$ when the graph is represented as a lattice over that set. In the case of the triangular lattice depicted as Fig. 2, the value 0 can be identified to the node $(1,1)$.

In our case, we have: $1 - p = \frac{t_\epsilon}{n} \leq \frac{1}{2+\epsilon} < p_c^s(T)$. Using Lemma 4, we get:

$$\Pr_{1-p}(\text{there is an open right-left path in } \mathcal{T}(\ell)) \leq \ell \, \Pr_{1-p}(0 \xrightarrow{\ell-1}) \leq \ell \, e^{-c\,(\ell-1)} \quad (4)$$

The first inequality is due to the fact that any right-left path has length at least $(\ell - 1)$ in $\mathcal{T}(\ell)$. Combining (1)-(4), we obtain:

$$\Pr(C_{rand} \text{ is bad}) \leq 2 \binom{n}{t_\epsilon} \ell \, e^{-c\,(\ell-1)}$$

Thus, if we choose $\ell := c_\epsilon \, n$ for some large enough constant c_ϵ, we have:

$$\Pr(C_{rand} \text{ is bad}) \leq \frac{1}{2^n}$$

which guarantees the fact that C_{rand} is a weakly t_ϵ-reliable n-coloring for $\mathcal{G}_{tri}(\ell,\ell)$ with overwhelming probability in n. □

Corollary 1. *There exists a black box t_ϵ-private protocol for f_G with communication complexity $O(n^3)$ group elements where $t_\epsilon = \lfloor \frac{n}{2+\epsilon} \rfloor$. Moreover, for any $\delta > 0$, we* *can construct a probabilistic algorithm, with run-time polynomial in n and $\log(\delta^{-1})$, which outputs a protocol $\prod$ for f_G such that the communication complexity of $\prod$ is $O(n^3 \log^2(\delta^{-1}))$ group elements and the probability that $\prod$ is not t_ϵ-private is at most δ.*

Proof. The existence of the protocol is a direct consequence of Theorem 2 as well as the different reductions exposed in Sect. 2. As our construction requires $\ell = O(n)$, we deduce that the communication cost of the protocol computing f_G is $O(n^3)$. The justification of the running time of the algorithm and the probability of failure δ is identical to what is done in [5]. □

We showed that it was possible to build a randomized algorithm to achieve $\left\lfloor \frac{n}{2+\epsilon} \right\rfloor$-private computation of f_G using $O(n^3)$ group elements. Even if the probability of failure of our previous construction is small, we would like to remove the randomized restriction so that we can get a (deterministic) protocol which is always guaranteed to succeed. In [5], Desmedt et al. only provided deterministic protocols to compute f_G in polynomial communication cost when $t = O(\log n)$. In the next section, we present a deterministic construction for any $t = O(n^{1-\epsilon})$ where ϵ is any positive constant. Our construction requires polynomial communication complexity as well.

4 A Deterministic Construction for Secure Computation

In this section, we show how to build a deterministic t-private protocol to compute f_G with polynomial complexity cost for any $t = O(n^{1-\epsilon})$. First, we will focus on particular pairs (t, n). Second, we generalize our result to any (t, n) with $t = O(n^{1-\epsilon})$.

We recursively construct our admissible PDAG $\mathcal{G}_{rec}$ and its coloring C_{rec}. Let $d \in \mathbb{N} \setminus \{0, 1\}$ be a constant. Denote $\mathcal{B}_d$ the binomial coefficient $\binom{2d-1}{d-1}$.

Theorem 3. *For any positive integer k, there is a weakly t_k-reliable n_k-coloring $C_{rec}(\ell_k)$ for the square admissible* PDAG *$\mathcal{G}_{rec}(\ell_k)$, where the parameters are: $t_k := d^k - 1$, $n_k := (2d-1)^k$ and $\ell_k = \mathcal{B}_d^k\, (\mathcal{B}_d + 1)^{k-1}$.*

Proof. We prove the theorem by induction on k.

$k = 1$: We have $t_1 = d - 1$, $n_1 = 2\,d - 1$ and $\ell_1 = \mathcal{B}_d$. We set $\mathcal{G}_{rec}(\ell_1) := \mathcal{G}_{tri}(\ell_1, \ell_1)$. We define $C_{rec}(\ell_1)$ as being the combinatorial coloring C_{comb} designed in [5] and recalled as Algorithm 2.

Algorithm 2. Coloring C_{comb}

Input: A $L \times L$ grid where $L = \binom{N}{T}$.

1. Let $I_1, \ldots, I_L$ denote the sequence of all T-color subsets of $[N]$ (in some ordering).
2. For each $(i, j) \in [L] \times [L]$, define the color $C(i, j)$ of node (i, j) in the grid to be any color in the set $S_{i,j} := [N] \setminus (I_i \cup I_j)$.

Output: A N-coloring of the grid.

Desmedt et al. noticed that, even if we removed the diagonal edges from $\mathcal{G}_{tri}(\ell_1, \ell_1)$, we still had the existence of I-avoiding top-bottom and right-left paths. Thus, we assume that $\mathcal{G}_{rec}(\ell_1)$ has no such edges so that $\mathcal{G}_{rec}(\ell_1)$ is a square grid the side length of which is ℓ_1 nodes. $\mathcal{G}_{rec}(\ell_1)$ is an admissible PDAG.

$k \geq 1$: Suppose we already have the construction and coloring for k, we recursively construct $\mathcal{G}_{rec}(\ell_{k+1})$ from $\mathcal{G}_{rec}(\ell_k)$.

We first build the block grid B by copying $(\mathcal{B}_d + 1) \times (\mathcal{B}_d + 1)$ times $\mathcal{G}_{rec}(\ell_1)$. The connections between two copies of $\mathcal{G}_{rec}(\ell_1)$ are as follows. Horizontally, we draw a directed edge from node $(i, 1)$ in the right-hand side copy to node (i, ℓ_1) in the left-hand side copy for $i \in [\ell_1]$ (i.e. we horizontally connect nodes at the same level). Vertically, we draw a directed edge from node (ℓ_1, j) in the top side copy to node $(1, j)$ in the bottom side copy for $j \in [\ell_1]$ (i.e. we vertically connect nodes at the same level).

The block B is a $(\mathcal{B}_d\,(\mathcal{B}_d + 1)) \times (\mathcal{B}_d\,(\mathcal{B}_d + 1))$ grid. It has the following property the proof of which can be found in Appendix A.

Proposition 3. *The block grid B admits a $(2\,d-1)$-coloring (just use the same C_{comb} for each copy of $\mathcal{G}_{rec}(\ell_1)$), such that for any $(d-1)$-color subset $I \subset [2\,d-1]$, there are $\mathcal{B}_d + 1$ horizontal (vertical) I-avoiding* **straight lines** *in B.*

Now, we construct $\mathcal{G}_{rec}(\ell_{k+1})$ and its coloring $C_{rec}(\ell_{k+1})$ as follows. We replace each node in $\mathcal{G}_{rec}(\ell_k)$ by a copy of B. If the node of $\mathcal{G}_{rec}(\ell_k)$ was colored by the color $c \in [n_k]$, then we color B with the set of colors $\{(2d-1)(c-1)+1, (2d-1)(c-1)+2, \ldots, (2\,d-1)\,c\}$, using C_{comb}. All the edges within each copy of B remain identical in $\mathcal{G}_{rec}(\ell_{k+1})$.

Now, we show how to connect two copies of B. We first focus on vertical connections. Consider an edge in $\mathcal{G}_{rec}(\ell_k)$ from a node in the i-th row to another node in the $(i+1)$-th row. Since these two nodes have been replaced by two copies of B, we denote the nodes on the top copy (i.e. those corresponding to the nodes of the i-th row in $\mathcal{G}_{rec}(\ell_k)$) as $v_{1,1}, \ldots, v_{1,\mathcal{B}_d}, v_{2,1}, \ldots, v_{\mathcal{B}_d+1,\mathcal{B}_d}$ and the nodes on the bottom copy as $w_{1,1}, \ldots, w_{1,\mathcal{B}_d}, w_{2,1}, \ldots, w_{\mathcal{B}_d+1,\mathcal{B}_d}$.

For each $(i, j) \in [\mathcal{B}_d] \times [\mathcal{B}_d]$, we add a directed edge $(v_{i,j}, w_{i,j+i-1})$ in $\mathcal{G}_{rec}(\ell_{k+1})$. If the index $(j+i-1)$ is greater than $\mathcal{B}_d$, $w_{i,j+i-1}$ is the node $w_{i+1,j+i-1-\mathcal{B}_d}$. Figure 3 gives the example for $d = 2$. The connection process works similarly for two consecutive columns where we replace each horizontal edge from $\mathcal{G}_{rec}(\ell_k)$ by $\mathcal{B}_d^2$ different edges in $\mathcal{G}_{rec}(\ell_{k+1})$.

It is clear that the number of nodes on each side of the square $\mathcal{G}_{rec}(\ell_{k+1})$ is:

$$\ell_{k+1} = \mathcal{B}_d\,(\mathcal{B}_d + 1) \cdot \ell_k = \mathcal{B}_d^{k+1}\,(\mathcal{B}_d + 1)^k$$

and the number of colors used in $C_{rec}(\ell_{k+1})$ is $n_{k+1} = (2\,d-1) \cdot n_k = (2\,d-1)^{k+1}$. The grid $\mathcal{G}_{rec}(\ell_{k+1})$ obtained by this recursive process is also an admissible PDAG due to the horizontal/vertical connection processes between two copies of B (as well as two copies of $\mathcal{G}_{rec}(\ell_1)$ inside B).

The last point to prove is that for any t_{k+1}-color subset $I \subset [n_{k+1}]$, there is an I-avoiding top-bottom (and right-left) path in $\mathcal{G}_{rec}(\ell_{k+1})$. We only prove the existence

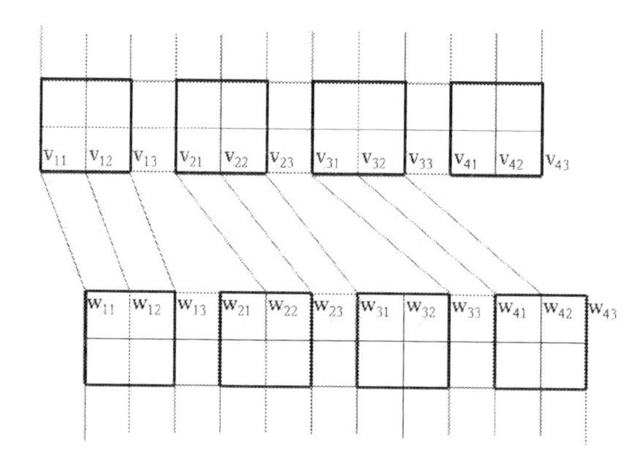

Fig. 3. How to vertically connect two copies of B when $d = 2$

of a top-bottom path in this paper as the demonstration of the existence for a right-left path is similar. For each $j \in [n_k]$, we define the set I_j as:

$$I_j := I \cap \{(2d-1)(j-1)+1, (2d-1)(j-1)+2, \ldots, (2d-1)j\}$$

Since

$$|I_1| + \cdots + |I_{n_k}| = |I| = t_{k+1} = d^{k+1} - 1 \tag{5}$$

and each $|I_j| \leq 2d-1$, there are at least $(n_k - t_k)$ subsets having at most $(d-1)$ elements. Indeed, in the opposite case, we would have:

$$|I_1| + \cdots + |I_{n_k}| \geq d\,(n_k - (n_k - t_k - 1)) = d \cdot d^k = d^{k+1},$$

which would contradict (5). Assume that $S \subseteq [n_k]$ is the set of these indices (i.e. for each $j \in S$, $|I_j| \leq d-1$). We have: $|[n_k] \setminus S| \leq t_k$. By the induction hypothesis, there is a $([n_k] \setminus S)$-avoiding top-bottom path in $\mathcal{G}_{rec}(\ell_k)$, i.e., the colors used on this path all belong to S. Let $v_1, \ldots, v_m$ be the vertices of the path and denote the color of node v_j as $c_j \in S$ $(j \in [m])$.

Now, we show there is an I-avoiding top-bottom path in $\mathcal{G}_{rec}(\ell_{k+1})$. In $\mathcal{G}_{rec}(\ell_{k+1})$, each node v_j has been replaced by a copy B_{v_j} with colors in $\{(2d-1)(c_j-1)+1, (2d-1)(c_j-1)+2, \ldots, (2\,d-1)\,c_j\}$. Since the color set I_{c_j} satisfies $|I_{c_j}| \leq d-1$, by Proposition 3 we deduce that there are $\mathcal{B}_d$ horizontal and $\mathcal{B}_d$ vertical I_{c_j}-avoiding paths in B_{v_j}.

One can show that this property involves the existence of an I-avoiding top-bottom path in $\mathcal{G}_{rec}(\ell_{k+1})$. This top-bottom path is the connection of an I_{c_1}-avoiding path (from B_{v_1}), an I_{c_2}-avoiding path (from B_{v_2}),. . ., an I_{c_m}-avoiding path (from B_{v_m}). The reader can find more details about this process in Appendix B. A similar demonstration leads to the existence of an I-avoiding right-left path in $\mathcal{G}_{rec}(\ell_{k+1})$ which achieves the demonstration of our theorem. □

The communication complexity of the protocol to t_k-privately compute the function $f_G(x_1, \ldots, x_{n_k})$ using the previous admissible PDAG is $O(n_k\,\ell_k^2)$ group elements where

$$\ell_k \leq \mathcal{B}_d^k(\mathcal{B}_d+1)^{k-1} \leq 2^{(2d-1)k} \times 2^{(2d-1)(k-1)} \leq 2^{2k(2d-1)} \leq n_k^{\frac{2(2d-1)}{\log_2(2d-1)}}$$

Note that the last inequality comes from $2^k = n_k^{\frac{1}{\log_2(2d-1)}}$.

Now, we generalize our result to any (t,n) where $t = O(n^{1-\epsilon})$ for any fixed positive ϵ. The class $O(n^{1-\epsilon})$ is the set of all functions f such that: $\exists \tau_f > 0\, \exists n_0 > 0 : \forall n \geq n_0\, f(n) \leq \tau_f\, n^{1-\epsilon}$. In our case, the function f is the privacy level t. Our main result is stated as follows.

Theorem 4. *For any fixed $\epsilon > 0$, for any fixed $\tau > 0$, there exists a constant $n_{\epsilon,\tau} \in \mathbb{N}$, such that for any $n \geq n_{\epsilon,\tau}$, if $t \leq \tau\, n^{1-\epsilon}$, then there exists a black-box t-private protocol to compute f_G with communication complexity polynomial in n. Moreover, there is a deterministic polynomial time algorithm to construct the protocol.*

Proof. We fix $\epsilon > 0$ and $\tau > 0$. We set $d = 2^{\lceil \frac{2}{\epsilon} \rceil - 1}$ and $k = \lfloor \log_{(2d-1)} n \rfloor$. We have $d \geq 2$. If $n \geq 2\,d-1$ then $k \geq 1$. In such a condition, we can apply Theorem 3 for the pair (k,d). There exists a t_k-private protocol to compute the value $f_G(x_1,\ldots,x_{n_k})$ using $O(n_k\,\ell_k^2)$ group elements where t_k, n_k, ℓ_k are defined as in Theorem 3. It is clear that the construction also t'-privately computes $f_G(x_1,\ldots,x_{n'})$ for any (t',n') such that $t' \leq t_k$ and $n' \geq n_k$. So, we only need to show $\tau\, n^{1-\epsilon} \leq t_k$, $n \geq n_k$ and $\ell_k = \mathrm{poly}(n)$. Due to our choice of d and k, we have:

$$n_k \leq (2d-1)^{\lfloor \log_{(2d-1)} n \rfloor} \leq (2d-1)^{\log_{(2d-1)} n} \leq n$$

And:

$$t_k \geq d^{\lfloor \log_{(2d-1)} n \rfloor} - 1 \geq d^{\log_{(2d-1)} n - 1} - 1 \geq \frac{n^{\frac{\log_2 d}{\log_2 (2d-1)}}}{d} - 1 \geq \frac{n^{\frac{\log_2 d}{\log_2 2d}}}{d} - 1$$

Since $d = 2^{\lceil \frac{2}{\epsilon} \rceil - 1}$, we get:

$$t_k \geq \frac{n^{\frac{\lceil \frac{2}{\epsilon} \rceil - 1}{\lceil \frac{2}{\epsilon} \rceil}}}{2^{\lceil \frac{2}{\epsilon} \rceil - 1}} - 1 \geq \frac{n^{1-\frac{\epsilon}{2}}}{2^{\lceil \frac{2}{\epsilon} \rceil - 1}} - 1 \geq \frac{n^{\frac{\epsilon}{2}}}{2^{\lceil \frac{2}{\epsilon} \rceil - 1}}\, n^{1-\epsilon} - 1$$

Since ϵ is a fixed positive constant, the mapping $n \mapsto \frac{n^{\frac{\epsilon}{2}}}{2^{\lceil \frac{2}{\epsilon} \rceil - 1}}$ has an infinite limit. Therefore: $\exists \widetilde{n}_{\epsilon,\tau} > 0 : \forall n \geq \widetilde{n}_{\epsilon,\tau} \quad \frac{n^{\frac{\epsilon}{2}}}{2^{\lceil \frac{2}{\epsilon} \rceil - 1}} \geq \tau + \frac{1}{n^{1-\epsilon}}$.

Remember that we early required $n \geq 2\,d-1$ in order to use Theorem 3. If we set $n_{\epsilon,\tau} := \max(2\,d-1, \widetilde{n}_{\epsilon,\tau})$ then:

$$\forall n \geq n_{\epsilon,\tau} \begin{cases} n_k \leq n \\ t_k \geq \tau\, n^{1-\epsilon} \geq t \end{cases}$$

It remains to argue about ℓ_k. Since $n_k \leq n$, we have: $\ell_k \leq n^{\frac{2\,(2\,d-1)}{\log_2 (2\,d-1)}}$. Since d is independent from n, ℓ_k is upper bounded by a polynomial in n. □

The previous theorem claims that for any fixed ϵ, if n is chosen large enough then we can t-privately compute f_G for any $t = O(n^{1-\epsilon})$. Such an asymptotic survey is also performed in [5]. However, in practical applications, the number of participants is not asymptotically large. The deterministic construction by Desmedt et al. has polynomial cost when $t = O(\log n)$. We now present a result valid for any group size n which guarantees privacy for larger t's than in [5] using polynomial communication as well.

Theorem 5. *For any positive integer n no smaller than 3, there exists a black-box protocol for f_G which is $(\lceil \frac{n^{\log_3 2}}{2} \rceil - 1)$-private. It requires the n participants to exchange $O(n^6)$ group elements. Moreover, there is a deterministic polynomial time algorithm to construct the protocol.*

Proof. We set $d = 2$ and $k := \lfloor \log_3(n) \rfloor$. The protocol obtained using Theorem 3 has parameter $t_k \geq \frac{n^{\log_3 2}}{2} - 1$ and $n_k \leq n$. We have: $\mathcal{B}_2 = 3$. Therefore: $\ell_k \leq \frac{n^{1+2\log_3 2}}{4}$. Thus, we obtain: $n_k \ell_k^2 = O(n^6)$. □

5 Conclusion and Open Problems

In this paper, we first demonstrated that we could construct a probabilistic t-private protocol computing the n-product function over any non-Abelian group for *any* t up to $\frac{n}{2+\epsilon}$ (for any fixed positive ϵ), thus nearly matching the known upper bound $\lceil \frac{n}{2} \rceil - 1$. As the communication complexity of our construction is $O(n^3)$ group elements, this result answers one of the questions asked by Desmedt et al. concerning the largest collision resistance achievable with an admissible PDAG of size polynomial in n. Note that Desmedt et al. indicated the discovery of a construction for $(n, t) = (24, 11)$ improving locally their own theoretical bound $\frac{n}{2.948}$ since $11 \approx \frac{24}{2.182}$. Our result demonstrates the existence of such a construction for any fixed positive ϵ (in [5], we have the particular case $\epsilon = 0.182$). Since the scheme developed in [5] (exclusively valid for $t < \frac{n}{2.948}$) only requires $O(n\,t^2)$ elements to be exchanged, a direction to further investigate is the existence of a (randomized) t-private protocol for any $t \leq \lceil \frac{n}{2} \rceil - 1$ having at most the cost of Desmedt et al.'s scheme.

Second, we showed that it was possible to construct a deterministic t-private n-party protocol to compute f_G having a polynomial communication cost for any $t = O(n^{1-\epsilon})$. For practical purpose, one may want to optimize the choice of parameters in our construction. For example, we have proved that one could t-privately compute f_G for any (t, n) satisfying $t \leq \left\lceil \frac{n^{\log_3 2}}{2} \right\rceil - 1$.

Desmedt et al. argued that the reduction from a protocol computing the n-product to a subroutine computing the shared 2-product extended to the more general function $\widetilde{f_G}(x_1, \ldots, x_m) := x_1 \cdot x_2 \cdot \ldots \cdot x_m$ where $m \geq n$ and each of the n parties holds one or more input values. This ensured the validity of their protocol to securely compute $\widetilde{f_G}$ as well. Since the constructions that we presented are particular admissible PDAGs, our results are also valid to compute $\widetilde{f_G}$.

Our work leads to the following two questions. First, is it possible to reduce the communication cost when $t = O(n^{1-\epsilon})$? Second, can we generalize this approach to design a deterministic polynomial communication cost algorithm for any t up to the threshold $\lceil \frac{n}{2} \rceil - 1$?

Apart from the previous points which constitute directions to improve the security for the passive adversary model, a problem which requires attention is the possibility of achieving secure computation of f_G against malicious parties. Indeed, even if multiparty computation can be used with small groups (as in the case of the Millionaires' problem [19]), the general purpose is to enable large communication groups to perform common computations and the larger the number of parties is, the more likely (at least) one of them will deviate from the given protocol.

Acknowledgments

The authors are grateful to the anonymous reviewers for their comments to improve the quality of this paper. The three authors' work was supported in part by the National Natural Science Foundation of China grant 60553001 and the National Basic Research Program of China grants 2007CB807900 and 2007CB807901. Xiaoming Sun's research was also funded by the National Natural Science Foundation of China under grant 60603005. Christophe Tartary's work was also financed by the Ministry of Education of Singapore under grant T206B2204.

References

[1] Ben-Or, M., Goldwasser, S., Wigderson, A.: Completeness theorems for non-cryptographic fault-tolerant distributed computation. In: 20th Annual ACM Symposium on Theory of Computing, Chicago, USA, pp. 1–10. ACM Press, New York (1988)
[2] Bollobàs, B., Riordan, O.: Percolation. Cambridge University Press, Cambridge (September 2006)
[3] Cramer, R., Damgård, I.B., Maurer, U.: General secure multi-party computation from any linear secret-sharing scheme. In: Preneel, B. (ed.) EUROCRYPT 2000. LNCS, vol. 1807, pp. 316–334. Springer, Heidelberg (2000)
[4] Damgård, I.B., Ishai, Y.: Scalable secure multiparty computation. In: Dwork, C. (ed.) CRYPTO 2006. LNCS, vol. 4117, pp. 501–520. Springer, Heidelberg (2006)
[5] Desmedt, Y., Pieprzyk, J., Steinfeld, R., Wang, H.: On secure multi-party computation in black-box groups. In: Menezes, A. (ed.) CRYPTO 2007. LNCS, vol. 4622, pp. 591–612. Springer, Heidelberg (2007)
[6] Diffie, W., Hellman, M.E.: New directions in cryptography. IEEE Transactions on Information Theory 22(6), 644–654 (1976)
[7] El Gamal, T.: A public-key cryptosystem and a signature scheme based on discrete logarithms. IEEE Transactions on Information Theory 31(4), 469–472 (1985)
[8] Goldreich, O., Vainish, R.: How to solve any protocol problem - an efficiency improvement. In: Pomerance, C. (ed.) CRYPTO 1987. LNCS, vol. 293, pp. 73–86. Springer, Heidelberg (1988)
[9] Hammersley, J.M.: Percolation processes: Lower bounds for the critical probability. The Annals of Mathematical Statistics 28(3), 790–795 (1957)
[10] Hirt, M., Maurer, U.: Robustness for free in unconditional multi-party computation. In: Kilian, J. (ed.) CRYPTO 2001. LNCS, vol. 2139, pp. 101–118. Springer, Heidelberg (2001)

[11] Hirt, M., Maurer, U., Przydatek, B.: Efficient secure multi-party computation. In: Okamoto, T. (ed.) ASIACRYPT 2000. LNCS, vol. 1976, pp. 143–161. Springer, Heidelberg (2000)
[12] Hirt, M., Nielsen, J.B.: Robust multiparty computation with linear communication complexity. In: Dwork, C. (ed.) CRYPTO 2006. LNCS, vol. 4117, pp. 463–482. Springer, Heidelberg (2006)
[13] Kesten, H.: Percolation Theory for Mathematicians. Birkhäuser, Basel (November 1982)
[14] Lang, S.: Algebra (Revised Third Edition). Springer, Heidelberg (November 2002)
[15] Magliveras, S.S., Stinson, D.R., van Trung, T.: New approaches to designing public key cryptosystems using one-way functions and trapdoors in finite groups. Journal of Cryptology 15(4), 285–297 (2002)
[16] Paeng, S.-H., Ha, K.-C., Kim, J.H., Chee, S., Park, C.: New public key cryptosystem using finite non Abelian groups. In: Kilian, J. (ed.) CRYPTO 2001. LNCS, vol. 2139, pp. 470–485. Springer, Heidelberg (2001)
[17] Rivest, R.L., Shamir, A., Adleman, L.M.: A method for obtaining digital signatures and public key cryptosystems. Communication of the ACM 21(2), 120–126 (1978)
[18] Shor, P.W.: Polynomial-time algorithms for prime factorization and discrete logarithms on a quantum computer. SIAM Journal on Computing 26(5), 1484–1509 (1997)
[19] Yao, A.C.-C.: Protocols for secure computations. In: 23rd Annual IEEE Symposium on Foundations of Computer Science, Chicago, USA, November 1982, pp. 80–91. IEEE Press, Los Alamitos (1982)

A Proof of Proposition 3

Let I be a $(d-1)$-color subset of $[2d-1]$. In [5], Desmedt et al. demonstrated that there were a I-avoiding top-bottom path and a I-avoiding right-left path in $\mathcal{G}_{tri}(\ell_1, \ell_1)$. They also showed that those two paths were straight lines. Thus, one can remove the diagonal edges of $\mathcal{G}_{tri}(\ell_1, \ell_1)$ while preserving those paths. This means that there exist a I-avoiding top-bottom path and a I-avoiding right-left path in $\mathcal{G}_{rec}(\ell_1)$ which are straight lines.

Since B is a-$(\mathcal{B}_d+1) \times (\mathcal{B}_d+1)$-copy of $\mathcal{G}_{rec}(\ell_1)$ and, due to the vertical/horizontal connections of these copies, we deduce that there are $(\mathcal{B}_d + 1)$ I-avoiding top-bottom paths and $(\mathcal{B}_d + 1)$ I-avoiding right-left paths in B. Moreover, each of these paths is a straight line.

B Connection of Color Avoiding Paths

It was shown in the proof of Theorem 3 that each block B_{c_i} had $\mathcal{B}_d$ horizontal and $\mathcal{B}_d$ vertical I_{c_i}-avoiding paths. In this appendix, we show how to construct a I-avoiding top-bottom path in $\mathcal{G}_{rec}(\ell_{k+1})$. Our path will start at the top of B_{v_1} and ends at the bottom of B_{v_m}.

Every grid from the family $(\mathcal{G}_{rec}(\ell_\lambda))_{\lambda \geq 1}$ is a square grid. Thus, the sequence of blocks $B_{v_1}, \ldots, B_{v_m}$ in $\mathcal{G}_{rec}(\ell_{k+1})$ is determined by the position of B_{v_1} as well as the m-tuple of letters from $\{\mathfrak{L}, \mathfrak{R}, \mathfrak{T}, \mathfrak{B}\}$ ($\mathfrak{L}$eft, $\mathfrak{R}$ight, $\mathfrak{T}$op, $\mathfrak{B}$ottom) indicating the output side of the block B_{v_i} for $i \in [m]$. Note that the last letter of the tuple is always $\mathfrak{B}$ since the I-avoiding top-bottom path ends at the bottom of B_{v_m}.

This tuple has the property the two consecutive letters cannot be opposite to each other (i.e, one cannot have $(\mathfrak{L}, \mathfrak{R})$, $(\mathfrak{R}, \mathfrak{L})$, $(\mathfrak{T}, \mathfrak{B})$ or $(\mathfrak{B}, \mathfrak{T})$). This means that you leave a block on a different side that you entered it. The reader can check the correctness of this claim by a simple recursive process on the parameter k. This property is trivially true for $k = 1$ since $\mathcal{G}_{rec}(\ell_1) = \mathcal{G}_{tri}(\ell_1)$. The recursion follows from the path construction that we will design below.

Proposition 4. *Let i be any element of $[m]$. Assume that $\mathcal{N}$ is any node on a side of B_{v_i} belonging to a I_{c_i}-avoiding straight line path. For each other side $\mathfrak{S}_i$ of B_{v_i}, we can construct a I_{c_i}-avoiding path from $\mathcal{N}$ to any of the $(\mathcal{B}_d + 1)$ nodes on $\mathfrak{S}_i$ belonging to a I_{c_i}-avoiding straight line path.*

Proof. We only provide a proof when $\mathcal{N}$ is on the top side of B_{v_i} (the three other cases are similar). The three possible output sides are $\mathfrak{B}$, $\mathfrak{L}$ and $\mathfrak{R}$. The block B_{v_i} is a-$(\mathcal{B}_d + 1) \times (\mathcal{B}_d + 1)$-copy of the original grid $\mathcal{G}_{rec}(\ell_1)$. Thus, B_{v_i} can be treated as a $(\mathcal{B}_d + 1) \times (\mathcal{B}_d + 1)$ array of grids $\mathcal{G}_{rec}(\ell_1)$. Based on this observation, we will use the terminology *grid-row* (respectively *grid-column*) to denote a set of $\mathcal{B}_d + 1$ horizontal (respectively vertical) grids $\mathcal{G}_{rec}(\ell_1)$ in B_{v_i}.

1. $\underline{\mathfrak{S}_i = \mathfrak{B}}$. The vertical I_{c_i}-avoiding path starting at node $\mathcal{N}$ intersects the **horizontal** I_{c_i}-avoiding path located within the bottom grid-row of B_{v_i} at node $\mathcal{I}$. That horizontal path intersects each of the $\mathcal{B}_d + 1$ **vertical** I_{c_i}-avoiding paths (one within each grid-column) at $\mathcal{I}_1, \ldots, \mathcal{I}_{\mathcal{B}_d+1}$. Note that $\mathcal{I} = \mathcal{I}_\mu$ for some $\mu \in [\mathcal{B}_d + 1]$. Once we are at one of the $\mathcal{I}_j$'s, we simply go vertically downwards to the node $\mathcal{N}'_j$ located at the bottom side of the block B_{v_i}.

Thus, we can construct a path from $\mathcal{N}$ to each of the $\mathcal{B}_d + 1$ output nodes on the bottom side of B_{v_j} belonging to the vertical I_{c_i}-avoiding paths. Those paths are $(\mathcal{N}, \mathcal{I}, \mathcal{I}_j, \mathcal{N}'_j)$ for $j \in [\mathcal{B}_d + 1]$.

2. $\underline{\mathfrak{S}_i = \mathfrak{R}}$. The vertical I_{c_i}-avoiding path starting at node $\mathcal{N}$ intersects the **horizontal** I_{c_i}-avoiding path located within the top grid-row of B_{v_i} at node $\mathcal{I}$. That horizontal path intersects the **vertical** I_{c_i}-avoiding path located within the rightmost grid-column of B_{v_i} at node $\widetilde{\mathcal{I}}$. This vertical path intersects each of the $\mathcal{B}_d + 1$ **horizontal** I_{c_i}-avoiding paths (one within each grid-row) at $\widetilde{\mathcal{I}}_1, \ldots, \widetilde{\mathcal{I}}_{\mathcal{B}_d+1}$. As before, we get: $\widetilde{\mathcal{I}} = \widetilde{\mathcal{I}}_\mu$ for some $\mu \in [\mathcal{B}_d + 1]$. Once we are at one of the $\widetilde{\mathcal{I}}_j$'s, we horizontally go rightwards to the node $\mathcal{N}'_j$ located on the right hand side of the block B_{v_i}.

Thus, we can construct a path from $\mathcal{N}$ to each of the $\mathcal{B}_d + 1$ output nodes on the right hand side of B_{v_j} belonging to the horizontal I_{c_i}-avoiding paths. Those paths are $(\mathcal{N}, \mathcal{I}, \widetilde{\mathcal{I}}, \widetilde{\mathcal{I}}_j, \mathcal{N}'_j)$ for $j \in [\mathcal{B}_d + 1]$.

3. $\underline{\mathfrak{S}_i = \mathfrak{L}}$. This is analogous to the previous case. □

We can finally construct a I-avoiding top-bottom path in $\mathcal{G}_{rec}(\ell_{k+1})$. We denote the m-tuple of output sides as $(\mathfrak{S}_1, \ldots, \mathfrak{S}_m)$. As previously said, we have: $\mathfrak{S}_m = \mathfrak{B}$.

We start at **any** node N_1 located on the top side of B_{v_1} and on a vertical I_{c_1}-avoiding path. Using Proposition 4, we can connect N_1 to any of the $\mathcal{B}_d + 1$ nodes on side $\mathfrak{S}_1$ of B_{v_1} using a I_{c_1}-avoiding path. An important remark is that each block of the whole grid $\mathcal{G}_{rec}(\ell_{k+1})$ is a set of $(\mathcal{B}_d + 1) \times (\mathcal{B}_d + 1)$ identical copies of $\mathcal{G}_{rec}(\ell_1)$ (including the coloring). As a consequence, these $\mathcal{B}_d + 1$ nodes have the same location in their respective copies of $\mathcal{G}_{rec}(\ell_1)$. Given the connection process between any pair of blocks within $\mathcal{G}_{rec}(\ell_{k+1})$, one of these $\mathcal{B}_d + 1$ nodes must be connected to a node N_2 from block B_{v_2} belonging to a I_{c_2}-avoiding straight line path. Similarly, N_2 is connected via a I_{c_2}-avoiding path in B_{v_2} to a node N_3 from B_{v_3} belonging to a I_{c_3}-avoiding straight line path. If we repeat this process for each of the remaining blocks, we obtain a set of $m-1$ nodes $N_1, \ldots, N_{m-1}$. The last node N_{m-1} can be connected to a node N_m on the bottom side of B_{v_m} using a I_{c_m}-avoiding path. Thus, N_1 (top side of $\mathcal{G}_{rec}(\ell_{k+1})$) is connected to N_m (bottom side of $\mathcal{G}_{rec}(\ell_{k+1})$) using a I-avoiding path which achieves the demonstration of our theorem.

Remark: As claimed above, this construction involves that the two consecutive side letters of the m-tuple cannot be opposite to each other.

Some Perspectives on Complexity-Based Cryptography

Andrew Chi-Chih Yao

Tsinghua University, Beijing, China
andrewcyao@tsinghua.edu.cn

Abstract. In the 1940's, Shannon applied his information theory to build a mathematical foundation for classical cryptography which studies how information can be securely encrypted and communicated. In the internet age, Turing's theory of computation has been summoned to augment Shannon's model and create new frameworks, under which numerous cryptographic applications have blossomed. Fundamental concepts, such as "information" and "knowledge transfer", often need to be re-examined and reformulated. The amalgamation process is still ongoing in view of the many unsolved security issues. In this talk we give a brief overview of the background, and discuss some of the recent developments in complexity-based cryptography. We also raise some open questions and explore directions for future work.

J. Pieprzyk (Ed.): ASIACRYPT 2008, LNCS 5350, p. 54, 2008.
© International Association for Cryptologic Research 2008

A Modular Security Analysis of the TLS Handshake Protocol

P. Morrissey, N.P. Smart, and B. Warinschi

Department Computer Science, University of Bristol,
Merchant Venturers Building, Woodland Road,
Bristol, BS8 1UB,
United Kingdom
paulm,nigel,bogdan@cs.bris.ac.uk

Abstract. We study the security of the widely deployed Secure Session Layer/Transport Layer Security (TLS) key agreement protocol. Our analysis identifies, justifies, and exploits the modularity present in the design of the protocol: the *application keys* offered to higher level applications are obtained from a *master key*, which in turn is derived, through interaction, from a *pre-master key*.

Our first contribution consists of formal models that clarify the security level enjoyed by each of these types of keys. The models that we provide fall under well established paradigms in defining execution, and security notions. We capture the realistic setting where only one of the two parties involved in the execution of the protocol (namely the server) has a certified public key, and where the same master key is used to generate multiple application keys.

The main contribution of the paper is a modular and generic proof of security for the application keys established through the TLS protocol. We show that the transformation used by TLS to derive master keys essentially transforms an *arbitrary* secure pre-master key agreement protocol into a secure master-key agreement protocol. Similarly, the transformation used to derive application keys works when applied to an arbitrary secure master-key agreement protocol. These results are in the random oracle model. The security of the overall protocol then follows from proofs of security for the basic pre-master key generation protocols employed by TLS.

1 Introduction

The SSL key agreement protocol, developed by Netscape, was made publicly available in 1994 [22] and after various improvements [20] has formed the bases for the TLS protocol [18, 19] which is nowadays ubiquitously present in secure communications over the internet. Surprisingly, despite its practical importance, this protocol had never been analyzed using the rigorous methods of modern cryptography. In this paper we offer one such analysis. Before describing our results and discussing their implications we recall the structure of the TLS protocol (Figure 1). The protocol proceeds in six phases. Through phases (1) and

J. Pieprzyk (Ed.): ASIACRYPT 2008, LNCS 5350, pp. 55–73, 2008.
© International Association for Cryptologic Research 2008

(2) parties confirm their willingness to engage in the protocol, exchange, and verify the validity of their identities and public keys (it is assumed that at least one party (the server) possess a long term public/private key pair $(\mathrm{PK}_B, \mathrm{SK}_B)$, as well as a certificate $\mathrm{sig}_{\mathrm{CA}}(\mathrm{PK}_B)$ issued by some certification authority CA). The next four phases, which are the focus of this paper, are as follows.

(3) A *pre-master secret* $s \in \mathcal{S}_{\mathrm{PMS}}$ is obtained using one of a number of protocols that include RSA based key transport and signed Diffie–Hellman key exchange (which we describe and analyze later in the paper).
(4) The pre-master secret key s is used to derive a *master secret* $m \in \mathcal{S}_{\mathrm{MS}}$, with $m = G(s, r_A, r_B)$. Here r_A, r_B are random nonces that the two parties exchange and G is a key derivation function. The obtained master secret key is confirmed by using it to compute two MACs of the transcript of the conversation which are then exchanged.
(5) In the next phase the master key m is used to obtain one or more application keys: for each application key, the parties exchange random nonces n_A and n_B and compute the shared application key via $k = k' \mid\mid k'' \leftarrow H(m, n_A, n_B)$. Here, H is a key derivation function. Notice, that each application key is actually two keys: one for securing communication from the client to the server, and one from the server to the client. This is important to prevent reflection attacks.
(6) Finally the application keys are used in an application (and we exhibit one possible use for encrypting some arbitrary messages). We emphasise that many applications can use the same master key by repeated application of Steps 5 and 6.

The proper use of keys in this last stage had been the object of previous studies [4, 25] and is not part of our analysis.

An interesting aspect of TLS is that the protocols used to obtain the pre-master secret in Step (3) are very simplistic and on their own insecure in the terms of modern cryptography. It is the combination of step (3) with those in (4) and (5) which leads (as we show in this paper) to secure key agreement protocol in the standard sense. Broadly speaking, our goal is to derive sufficient security conditions on the pre-master key agreement protocol which would ensure that the above combination indeed yields a secure key-agreement protocol in a standard cryptographic sense.

We caution that in our analysis we disregard steps (1) and (2), and therefore assume an existing PKI which authenticates all public keys in use in the system. In particular we do not take into account any so-called PKI attacks.

Models. Much of the previous work on key agreement protocols in the provable security community has focused on defining security models and then creating protocols which meet the security goals of the models. In some sense, we are taking the opposite approach: we focus on a particular existing protocol, namely TLS, and develop security models that capture the security levels that the various keys derived in one execution of the protocol enjoy. The path we take is also motivated by the lack of models that capture precisely the security of these keys.

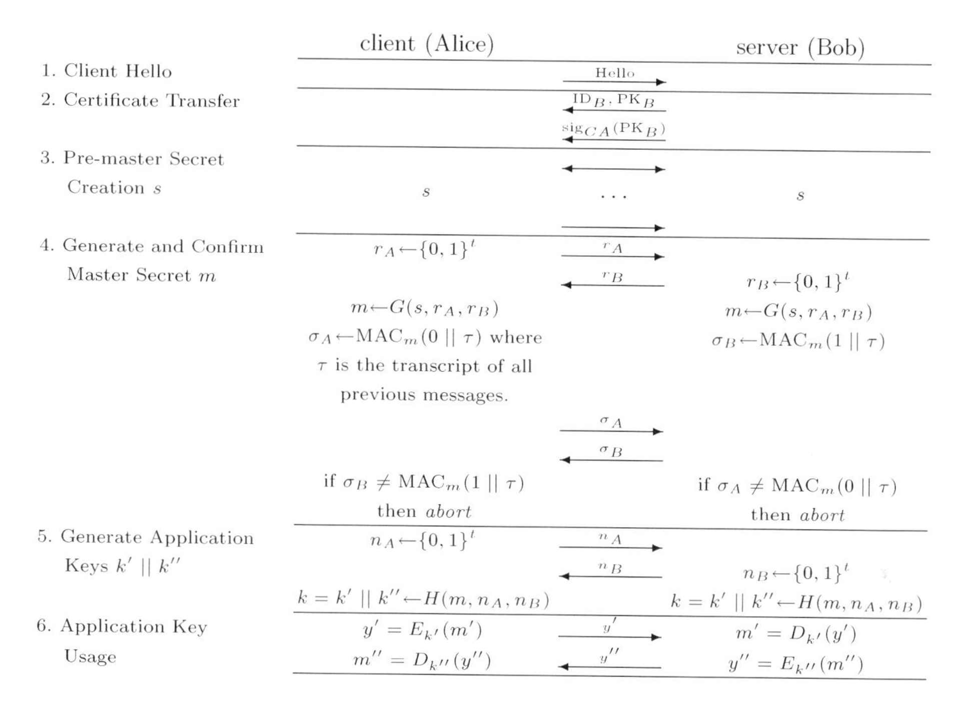

Fig. 1. A general TLS like protocol

A second important aspect of our approach is that unlike in prior work on key-agreement protocols, we do not regard the protocol as a monolithic structure. Instead, we identify the structure described above and give security models for each of the keys that are derived in the protocol. A benefit that follows from this modular approach is that we split the analysis of the overall protocol to the analysis of its components, thus making the task of proving security more manageable.

We first provide a model for pre-master key agreement protocols. The model is a weakened version of the Blake–Wilson, Johnson and Menezes (BJM) model [9]. In particular we only require that pre-master key agreement protocols are secure in a one-way sense (the adversary cannot recover the entire established key), and that the protocol is secure against man-in-the-middle attacks. In addition, unlike in prior work, we model the realistic setting where only one of the parties involved in the protocol is required to possess a certified public key.

Next, we give a security model for master-key agreement protocols which strengthens the one described above. We still only require secrecy for keys in the one-wayness sense, but now we ask for the protocol to also be secure against unknown-key-share attacks. In addition, we introduce key-confirmation as a requirement for master keys.

Finally, via a further extension, we obtain a model for the security of key agreement protocols. Our model for application key security is rather standard, and resembles the BJM model: we require for the established key to be indistinguishable from a randomly chosen one, and we give the adversary complete control over the network, and various corruption capabilities. Our model explicitly takes into consideration the possibility that the same master key is used to derive multiple application keys.

SECURITY ANALYSIS OF THE TLS HANDSHAKE PROTOCOL. Based on the models that we developed, we give a security proof for the TLS handshake protocol. In particular, we analyze a version where the MAC sent in step 4 is passed in the clear (and not encrypted under the application keys as in full TLS.) It is intuitively clear that the security of the full TLS protocol follows from our analysis. While a direct analysis of the latter may be desirable we choose to trade immediate applicability of our results to full TLS for the modularity afforded by our abstraction.

Our proof is modular and generic. Specifically, we show that the protocol $(\Pi; \mathsf{MKD}_{\mathsf{SSL}}(\mathsf{Mac}, G))$ obtained by appending to an arbitrary pre-master key agreement protocol Π the flows in phase (4) of TLS is a secure master-key agreement protocol in the sense that we define in this paper. The result holds provided that the message authentication code used in the transformation is secure and the hash function in the construction is modeled as a random oracle. Similarly, we show that starting from an arbitrary secure master-key agreement protocol Π, the protocol $(\Pi; \mathsf{AK}_{\mathsf{SSL}}(H))$ obtained by appending the flows in phase (5) of TLS is a secure application-key agreement protocol (provided that H is modeled as a random oracle).

An important benefit of the modular approach that we employ surfaces at this stage: to conclude the security of the overall protocol it is sufficient to show that the individual pre-master key agreement protocols of TLS are indeed secure (in the weak sense that we put forth in this paper). The analysis is thus more manageable, and avoids duplicating and rehashing proof ideas, which would be the case if one was to analyze TLS in its entirety for each distinct method for establishing pre-master keys.

IMPACT ON PRACTICE. An implication of practical consequence of our analysis concerns the use of encryption for implementing the pre-master key agreement protocol of TLS. Currently, the RSA key transport mode of TLS uses a randomized padding mechanism to avoid known problems with vanilla RSA. The original choice was the encryption scheme from PKCS-v1.0. The exact choice is historic, but in modern terms was made to attempt to create an $\mathsf{IND\text{-}CCA}$ encryption scheme. It turns out that the encryption scheme from PKCS-v1.0 is not in fact $\mathsf{IND\text{-}CCA}$ secure. This was exploited in the famous reaction attack by Bleichenbacher [11] on SSL, where invalid ciphertext messages were used to obtain pre-master secret keys. Our analysis implies that no randomized padding mechanism is actually needed, as deterministic encryption suffices to guarantee the security of the whole protocol.

Importantly, our models *do* capture security against reaction attacks as long as the full behaviour of the protocol is specified and analyzed. The key aspect is that the analysis should include the behaviour of the parties when the messages that they receive do not follow the protocol (e.g. are malformed). Our analysis of the premaster key agreement based on encryption schemes (e.g. that based on RSA) considers and thus justifies the validity of the patch proposed to cope with reaction attacks, i.e. by ensuring that the execution when malformed packages are received is indistinguishable from honest executions.

Our models can be used to explicitly capture one-way and mutual authentication via public-key certificate information. We do not model variants of the standard TLS protocol which can include password-based authentication or shared key-based techniques. We leave these extensions for future work.

It is important to observe that our model does not require that the application keys satisfy a notion of key-confirmation (as we require for the master-keys). Indeed, the TLS protocol does not ensure this property. However, one may obtain implicit key confirmation through the use of such keys in further applications. In some sense, this loss is a by-product of the way we have broken up the protocol. One of our goals was to show what security properties each of the stages provides, and therefore we modeled and analyzed the security of the application keys. However, if one considers Stages 1-4 as the key agreement protocol, and stages 5-6 as the application where the keys are used, then one does obtain an explicit notion of key confirmation. Hence, the loss of explicit key confirmation in Stage 5 should not be considered a design flaw in TLS.

On the use of the random oracle model. In our proofs we assume that the key derivation function is a random oracle, i.e. an idealized randomness extractor. As such, the typical disclaimer associated to proofs in the random oracle model certainly applies, and we caution against over optimism in their interpretation. A natural and important question is whether a standard model analysis is possible, ideally, assuming that the key derivation function is pseudorandom (as is the function based on HMAC used in the current specification of TLS). Unfortunately, indirect evidence indicates that such a result is extremely hard to obtain. As observed by Jonsson and Kaliski in their analysis of the use of RSA in TLS [23], the use of the key derivation function in TLS is akin to the use of such functions in deriving DEM keys under the KEM/DEM paradigm [16]. It is thus likely that a proof as above would immediately imply an efficient RSA-based encryption scheme secure in the standard model, thus solving a long-standing open question in cryptography.

Related Work. The work which is closest with ours is the analysis of the use of RSA in TLS by Jonsson and Kaliski [23]. They consider a very simplified security model for the master secret key, for the particular case when the protocol for premaster key is based on encryption. We share the modeling of the key derivation function as a random oracle, and the observation that deterministic encryption may suffice for a secure premaster key had also been made there. However, the present work uses a far more general and modular model for key-exchange, analyzes several pre-master key agreement protocols, including one based on DDH which is offered by TLS.

Other analyses of the TLS protocol used Dolev-Yao models, where ideal security of the underlying primitives is postulated, and thus no guarantees are offered for the more concrete world. Such analyses include the one carried out by Mitchel, Shmatikov, and Stern [28] using a model checker, and the one of Paulson who used the inductive method [30]. Wagner and Schneier analyze various security aspects of SSL 3.0 [32], but their treatment is informal. Finally, Bellare and Namprempre [4], and Krawczyk [25] study how to correctly use the application keys derived via TLS. Their treatment is focused exclusively on the use of keys, and is not concerned with the security of the entire key agreement protocol.

The first complexity theoretic model for key agreement was the Bellare-Rogaway (BR) model [6, 7]. The main driving forces of this model were the works of [8, 17]. Since the initial work of Bellare and Rogaway there have been a number of other models proposed for key-exchange in various applications and environments [1, 3, 5, 9, 10, 12, 13, 14, 27, 31]. These models can be loosely categorised into two main groups: those that use simulation based techniques [3, 14, 31], and those closer to the original BR model that use an indistinguishability based approach [9, 10, 13, 27]. As explained before, our analysis uses a model that falls in the latter category which, as argued elsewhere [13], has certain drawbacks but also several important benefits over the simulation based approach. Certainly, our general understanding of TLS would benefit from an analysis in a simulation based model, especially one that guarantees compositionality [14]. However, in such settings care must be taken on the use of the UC session identifiers which must be unique and predetermined. Furthermore, multiple sessions of TLS use the same long term secret keys which is a setting inherently difficult to handle in the UC framework. The joint state UC theorem [15] a technical tool sometimes useful in such situations does not apply to encryption (as used by encryption based pre-master key derivation). Furthermore, applying the JUC theorem to protocols that use signatures it requires signing messages/session identifier pairs, thus obtaining an analysis of a related but different protocol.

Some aspects of other indistinguishability-based models relevant to our work are the following. In [6] entity authentication and authenticated key distribution are considered in the two-party symmetric key case where users are modeled as message driven oracles. The adversary in this case acts as the communications channel between users. To define security, the notions of an "error-free history" of [8] and of "matching protocol runs" from [17] are made formal in [6] using the notion of a *matching conversation*. We use this notion in our definitions.

Various security attributes are then included in the definition of security by allowing the adversary to make corresponding queries such as Reveal queries. In [7] this was developed to model the three party symmetric key case for entity authentication and key distribution. The models most relevant to our work are the Blake–Wilson, Johnson and Menezes (BJM) based models [9, 10, 27]. The BJM model of [9] extended the BR model, to authenticated key agreement (AK) and authenticated key agreement with key confirmation (AKC) in the public key case. The work of [9] uses the notion of a No-Matching condition [6], to define a clearer separation between AK and AKC protocols and deals with Diffie–Hellman (DH) like protocols. Our execution models are inspired by the BJM model (while our security definitions are different.)

Following on from this [10] deals with the case of key transport using public key encryption (PKE) and key agreement using DH key agreement with digital signatures (DSS). In [27] a modular proof technique was used in a modified BJM model to prove security of key agreement protocols relative to a gap assumption. Indeed, the idea of transforming a one-way security definition into an indistinguishability definition occurs also in the generic transform proposed by Kudla and Paterson [26, 27] and our techniques are very similar to theirs.

Finally, an important security model that is related to ours is that of Canetti and Krawczyk (CK) [13]. In addition to the corruption capabilities that we consider, the CK model allows the adversary to obtain the entire internal state of a session and in particular the ephemeral secrets used in sessions. As pointed out by Choo et al. this type of query is the only essential difference between the adversarial capabilities in the model of Bellare and Rogaway and that of Canetti and Krawczyk (see Table 2 of [24]). Clearly, our analysis does not offer guarantees in the face of such extremely powerful types of adversaries and in fact it can be easily seen that under such attacks the TLS version that uses the DDH-based premaster secret key agreement is insecure. It may be possible that one can demonstrate security of TLS under such stronger attacks by assuming secure erasures as done for similar protocols [13, 14].

By adopting the style of the BR models over the style of the CK model we also avoid some of the idiosyncrasies of the latter related to the use of session identifiers (which need to be unique, and somehow agreed upon in advance by participating parties) [13, 24]. For a further discussion on the use of identifiers in the CK model versus the BR model see [24].

One other aspect of [13] which is somewhat related to our work is a modular framework for designing protocols. In the model of [13] one can first develop a secure protocol under the powerful assumption that all communication is authenticated. Then, a secure protocol in the more realistic setting with no authenticated communication is obtained by applying a generic transformation using an *authenticator*. Obviously, the modular structure of TLS that we observe and exploit is of a different nature. In particular it does not seem possible to regard TLS as the result of applying an authenticator to some other protocol.

2 A Generic Execution Model for Two-Party Protocols

The security models that we use in this paper are based on the earlier work of Bellare et al. [3, 5, 6, 7], as refined by BJM [9]. In this section we give a general description of the common features of these models, and recall some of the intuition behind them. Later, we specialise the general model for the different tasks that we consider in the paper.

Registered and unregistered users. We model a setting with two kinds of users: registered users (with identities in some set $\mathcal{U}$) and non-registered user (with identities in some set $\mathcal{U}'$). Each user $U \in \mathcal{U}$ has a long-term public key PK_U and a corresponding long term private key SK_U. The set $\mathcal{U}$ is intended to model the set of servers in the standard one-way authentication mode of TLS, the set of identities $\mathcal{U}'$ models users that do not have a long term public/private key pair.

Models for interactive protocols execution. We are concerned with two-party protocols: interactive programs in which an initiator and a responder communicate via some communication channel. Each of the two parties runs some reactive program: each program expects to receive a message from the communication channel, computes a response, and sends this back to the channel. We refer to one execution of the program for the initiator (respectively, responder) as an initiator session (respectively, a responder session). Each party may engage in multiple, concurrent, initiator and responder sessions.

As standard, we assume an adversary in absolute control of the communication network: the adversary intercepts all messages sent by parties, and may respond with whatever message it wants. This situation is captured by considering an adversary (an arbitrary probabilistic, polynomial-time algorithm) who has access to oracles that correspond to some (initiator or responder) sessions of the protocol which the oracle maintains internally. In particular, each oracle maintains an internal state which consists of the variables of the session to which it corresponds, and additional meta-variables used later to define security notions. In our descriptions we typically ignore the details of the local variables of the sessions, and we omit a precise specification of how these sessions are executed. Both notions are standard. The typical meta-variables of an oracle $\mathcal{O}$ include the following. Variable $\tau_{\mathcal{O}} \in \{0,1\}^* \cup \{\perp\}$ that maintains the transcript of all messages sent and received by the oracle, and occasionally, other data pertaining to the execution. Variable $\mathsf{role}_{\mathcal{O}} \in \{initiator, responder, \perp\}$ records the type of session to which the oracle corresponds. Variable $\mathsf{pid}_{\mathcal{O}} \in \mathcal{U}$ keeps track of the identity of the intended partner of the session maintained by $\mathcal{O}$. Variable $\delta_{\mathcal{O}}$ indicates whether the session had finished successfully, or unsuccessfully. We specify the values that this variable takes later in the paper. Finally, variable $\gamma_{\mathcal{O}} \in \{\perp, corrupted\}$ records whether or not the session had been corrupted by the adversary.

After an initialisation phase, in which long term keys for the parties are generated the adversary takes control of the execution which he drives forward using several types of queries. The adversary can create a new session of user U playing the role of the initiator/responder by issuing a query $\mathsf{NewSession}(U, role)$, with $role \in \{initiator, responder\}$. User U can be either registered or unregistered. We write Π_U^i for the i'th session of user U. To any oracle $\mathcal{O}$ the adversary can send a message msg using the query $\mathsf{Send}(\mathcal{O}, \mathrm{msg})$. In return the adversary receives an answer computed according to the session maintained by $\mathcal{O}$. The adversary may also corrupt oracles. Later in the paper when we specialise the general model, we also clarify the different versions of corruptions that can occur and how are they handled by the oracles. The execution halts whenever the adversary decides to do so.

To identify sessions that interact with each other we use the notion of matching conversations introduced by Bellare and Rogaway (which essentially states that the inputs to one session are outputs of the other sessions, and the other way around) [6].

3 Pre-master Key Agreement Protocols

In this section we specialise the general model described above for the case of pre-master key agreement protocols, and analyze the security of the pre-master key agreement protocols used in TLS.

As discussed in the introduction, the design of our models is guided by the security properties that the various subprotocols of TLS satisfy. In particular, we require extremely weak security properties for the pre-master secret key. Specifically, we demand that an adversary is not able to *fully* recover the key shared between two honest parties. In its attack the adversary is allowed to adaptively corrupt parties and obtain their long term secret key, and is allowed to check if a certain string s equals the pre-master secret key held by some honest session. The latter capability models an extremely limited form of reveal queries: our adversary is not allowed to obtain the pre-master secret key of any of the sessions, but can only guess (and then check) their values.

The formal model of security for pre-master key agreement protocols extends the general model in Section 2 and makes only mild assumptions regarding the syntax of such protocols. Specifically, we assume that the pre-master key belongs to some space $\mathcal{S}_{\mathrm{PMS}}$. This space is often the support set of some mathematical structure such as a group. We require that if t is the security parameter then $\#\mathcal{S}_{\mathrm{PMS}} \geq 2^t$. Furthermore, we assume that the initiator and responder programs use a variable $s \in \mathcal{S}_{\mathrm{PMS}} \cup \{\perp\}$ that stores the shared pre-master key. The corresponding variable stored by some oracle $\mathcal{O}$ is $s_{\mathcal{O}}$. For pre-master secret key agreement protocols the internal variable $\delta_{\mathcal{O}}$ stores one of the following values: $\perp$ (the session had not finished its execution), *accepted-pmk* (the session had finished its execution successfully (which in particular means that $s_{\mathcal{O}}$ holds some pre-master session key in $\mathcal{S}_{\mathrm{PMS}}$) or *rejected* (the session had finished its execution unsuccessfully). Unless $\delta_{\mathcal{O}} = \textit{accepted-pmk}$ we assume $s_{\mathcal{O}} = \perp$.

The corruption capabilities of the adversary discussed above are modeled using queries Corrupt and Check formally defined as follows. When the adversary issues a query Corrupt(U) the following actions take place. If $U \in \mathcal{U}$ then $\mathrm{SK}_{\mathcal{U}}$ is returned to the adversary, and we say that party U had been corrupted. In all sessions $\mathcal{O} = \Pi_U^i$ for some $i \in \mathbb{N}$ the value of $\gamma_{\mathcal{O}}$ is set to *corrupted* and no further interaction between these oracles and the adversary may take place. Additionally, no further queries NewSession($U, role$) are permitted.

When the adversary issues the query Check($\mathcal{O}, s$), for $\mathcal{O} = \Pi_U^i$, $i \in \mathbb{N}$, U some uncorrupted party, and $s \in \mathcal{S}_{\mathrm{PMS}}$, then the answer returned to the adversary is **true**, if $\delta_{\mathcal{O}} = \textit{accepted-pmk}$ and $s_{\mathcal{O}} = s$, and **false** otherwise. When a given oracle is initialized all values for the internal states are set to $\perp$. At the end of a protocol, the role, partner ID, and oracle state (but not the pre-master key) are recorded in the transcript.

The following definition captures the class of oracles which are valid targets for the attacker using the notion of "fresh oracles". These are uncorrupted oracles who have successfully finished their execution, and have a known intended partner who is also not corrupted.

Definition 1 (Fresh Pre-Master Secret Key Oracle). *A pre-master secret oracle $\mathcal{O}$ is said to be fresh if all of the following conditions are satisfied:*

(1) $\gamma_{\mathcal{O}} = \perp$, (2) $\delta_{\mathcal{O}} = \textit{accepted-pmk}$, and (3) $\exists\ V \in \mathcal{U}$ such that V is uncorrupted and $\mathrm{pid}_{\mathcal{O}} = V$.

SECURITY GAME FOR PRE-MASTER KEY AGREEMENT PROTOCOLS. We define the security of a pre-master key agreement protocol Π via the following game $\mathsf{Exec}^{\mathsf{OW\text{-}PMS}}_{\mathcal{A},\Pi}(t)$ between an adversary $\mathcal{A}$ and a challenger $\mathcal{C}$:

(1) The challenger, $\mathcal{C}$, generates public/secret key pairs for each user $U \in \mathcal{U}$ (by running the appropriate key-generation algorithm on the security parameter t), and returns the public keys to $\mathcal{A}$.

(2) Adversary $\mathcal{A}$, is allowed to make as many NewSession, Send, Check, and Corrupt queries as it likes.

(3) At some point $\mathcal{A}$ outputs a pair $(\mathcal{O}^*, s^*)$, where $\mathcal{O}^*$ is some pre-master secret oracle, and $s^* \in \mathcal{S}_{\mathrm{PMS}}$.

We say the adversary $\mathcal{A}$ wins if its output $(\mathcal{O}^*, s^*)$ is such that $\mathcal{O}^*$ is fresh, and $s^* = s_{\mathcal{O}^*}$. In this case the output of $\mathsf{Exec}^{\mathsf{OW\text{-}PMS}}_{\Pi,\mathcal{A}}(t)$ is set to 1. Otherwise the output of the experiment is set to 0. We write

$$\mathbf{Adv}^{\mathsf{OW\text{-}PMS}}_{\mathcal{A},\Pi}(t) = \Pr[\mathsf{Exec}^{\mathsf{OW\text{-}PMS}}_{\mathcal{A},\Pi}(t) = 1],$$

for the advantage of $\mathcal{A}$ in winning the $\mathsf{Exec}^{\mathsf{OW\text{-}PMS}}_{\mathcal{A},\Pi}(t)$ game. The probability is taken over all the random coins used in the game. We deem a pre-master secret key protocol secure if the adversary is not able to fully compute the key held by fresh oracles.

Definition 2 (Pre-Master Key Agreement Security). *A pre-master key agreement protocol is secure if it satisfies the following requirements:*

- **Correctness:** *If at the end of the execution of a benign adversary, who correctly relays messages, any two oracles which have had a matching conversation hold the same pre-master key, and the key should be distributed uniformly on the pre-master key space* $\mathcal{S}_{\mathrm{PMS}}$.
- **Key Secrecy:** *A pre-master key agreement protocol* Π *satisfies* OW-PMS *key secrecy if for any* p.p.t. *adversary* $\mathcal{A}$ *its advantage* $\mathbf{Adv}^{\mathsf{OW\text{-}PMS}}_{\mathcal{A},\Pi}(t)$ *is a negligible function.*

Before proceeding, we discuss the strength of our model for the security of pre-master secret keys, and several authentication issues.

REMARK 1. Our security requirements for pre-master secret key agreement are significantly weaker than the standard requirements for key exchange [6, 7]. In particular, we only require secrecy in the sense of one-wayness (not in the sense of indistinguishability from a random key). Furthermore, the corruption abilities of the adversary are severely limited: the adversary cannot obtain (or "reveal") pre-master secrets established by honest parties (even if these parties are not those under the attack).

REMARK 2. As a consequence of our security requirements our model may deem protocols that succumb to unknown-key-share attacks [17] secure. In such attacks, two sessions belonging to honest users U and V locally establish the same pre-master secret key, without intentional interaction with each other.

Remark 3. Security under our notion guarantees security against man-in-the-middle attacks: a situation where honest parties U and V believe they interact with each other but their pre-master key(s) is in fact shared with the adversary is a security break in our model.

Remark 4. Although the resulting security notion is very weak, it turns out that it suffices to obtain good master-key agreement protocols by appropriately designed protocols to derive such keys (e.g. the protocol in Step 4 of the TLS protocol – Figure 1.) More importantly, the weak notion also allows for many simple protocols to be proved secure. For example, in the next section we prove that deterministic encryption is sufficient to construct such protocols.

Remark 5. Our model is not concerned with secure establishment of pre-master secret keys between two unauthenticated parties (the oracle that is under attack always has $\mathrm{pid}_{\mathcal{O}} \neq \perp$). While treating this case is possible using the concept of matching conversations to pair sessions, the resulting definition would be heavier and not particularly illuminating. Instead, we concentrate on the situation more relevant to practice where at least one of the parties that take part in the protocol (the server) has a certified public key.

Remark 6. As usual, our security model can be easily adapted to the random oracle model by providing the adversary with access to the random oracle (whenever some hash function is modeled as a RO). The same holds true for the rest of the models that we develop in this paper.

We now discuss the security of the pre-master secret key agreement protocols used in TLS.

Protocols based on public-key encryption. A natural, intuitively appealing, construction for pre-master key agreement protocols is based on the following use of an arbitrary public-key encryption scheme Enc. A user selects a pre-master secret key s from an appropriate space, and sends to the server the encryption of s under the server's public-key. The server then obtains s as the decryption of the ciphertext that it receives. We write PMK(Enc) for the resulting protocol.

Theorem 1. *If* Enc *is a* OW-CPA *secure deterministic encryption or a* OW-CCA *secure randomized encryption scheme, then the pre-master secret key agreement protocol* $\Pi =$ PMK(Enc) *is a secure pre-master key transport protocol.*

The result of this theorem, like all theorems in this paper will be proved in the full version.

The weak security properties that we define for pre-master key agreement protocols enable us to show security of PMK(Enc) based on weak security requirements for Enc. Indeed, the one-wayness type secrecy for pre-master keys translates to the one-wayness of the encryption function of Enc. This result of our analysis implies, perhaps surprisingly, that one can avoid the use of full-fledged IND-CCA encryption schemes in favor of the much simpler *deterministic* OW-CPA schemes (*e.g.* textbook RSA). Of course, probabilistic encryption can also be used, but in this case we show security of the associated pre-master secret key protocol based on OW-CCA security. More generally our results holds under the assumption that the encryption scheme is secure against an attacker with access to a plaintext checking oracle. It is therefore not paradoxical that a deterministic scheme suffices but an IND-CPA scheme does not.

Finally, since IND-CCA implies OW-CCA, our security analysis *does* apply to the (correct) use of an IND-CCA secure public key encryption scheme within the TLS protocol. In particular, when Enc is RSA-OAEP, the pre-master secret key protocol PMK(Enc) is secure.

SIGNED DIFFIE-HELLMAN PRE-MASTER KEY AGREEMENT. The pre-master secret key in TLS can also be produced by exchanging a Diffie-Hellman key g^{xy}, for x and y randomly chosen by the two participants, who also sign the relevant message flow (either g^x or g^y) with their long term signing keys. It is known that this protocol, which we denote by $\mathsf{PMK}(\mathsf{Sig}, \mathbb{G})$, does not meet the requirements of an authenticated key agreement protocol, for example see [17] for a discussion of this protocol and various attacks on it. However, one can show.

Theorem 2. *Let $\mathbb{G}$ be cyclic group for which the gap-Diffie-Hellman assumption holds and let* Sig *be a secure digital signature scheme. Then $\Pi = \mathsf{PMK}(\mathsf{Sig}, \mathbb{G})$ is a secure pre-master key agreement protocol.*

4 Master Key Agreement Protocols

In this section we introduce a security model for master-key agreement protocols. We then show that master key agreement protocols obtained from secure pre-master key agreement protocols via the transformation used in TLS satisfy our notion of security.

Our security model for master key agreement protocols is similar to that for pre-master key agreement protocols. We again ask for the adversary not to be able to *fully* recover the master secret key of the session under attack. Moreover, we ask for a key confirmation guarantee: if a session of some user U accepts a certain master-key then there exists a unique session of its intended partner that had accepted the same key. In addition to the queries previously defined for the adversary, we also let the adversary obtain the master keys agreed in different sessions of the protocol, without corrupting the user to which this session belongs, i.e. we allow so-called Reveal queries.

In the formal model that we give below we make the following assumptions about the syntax of a master-key agreement protocol. We assume that the master key belongs to some space $\mathcal{S}_{\mathrm{MS}}$ for which we require that $\#\mathcal{S}_{\mathrm{MS}} \geq 2^t$, and assume that the programs that specify a master key agreement protocol use a variable m to store the agreed master key. For such protocols the variable $\delta_{\mathcal{O}}$ now takes values in $\{\perp, \mathit{accepted\text{-}mk}, \mathit{reject}\}$ with the obvious meaning. Furthermore, the variable $\gamma_{\mathcal{O}}$ can also take the value *revealed* to indicate that the stored master key has been given to the adversary (see below).

In addition to the queries allowed in the experiment for pre-master key security, the adversary is also allowed to issue queries of the form $\mathsf{Reveal}(\mathcal{O})$. This query is handled as follows: if $\delta_{\mathcal{O}}$ =*accepted-mk* then $m_{\mathcal{O}}$ is returned to $\mathcal{A}$ and $\gamma_{\mathcal{O}}$ is set to *revealed*, while if $\delta_{\mathcal{O}} \neq$*accepted-mk* then the query acts as a no-op. As before, when a given oracle is initialized all values for the internal states are set to $\perp$. At the end of a protocol the role, partner ID and oracle state (but not the master key) are recorded in the transcript. Unless $\delta_{\mathcal{O}}$ =*accepted-mk* we assume $m_U^i = \perp$.

The definition of freshness needs to be adapted to take into account the new adversarial capabilities. We call an oracle $\mathcal{O}$ fresh if it is uncorrupted, has successfully finished its execution, its intended partner V is uncorrupted, and none of the revealed oracles belonging to V has had a matching conversation with $\mathcal{O}$. The latter condition essentially says that the adversary can issue $\mathsf{Reveal}(\mathcal{Q})$ for any $\mathcal{Q}$ (including those that belong to the intended partner of $\mathcal{O}$), as long as $\mathcal{Q}$ is not the session with which $\mathcal{O}$ actually interacts.

Definition 3 (Fresh Master Secret Oracle). *A master secret oracle $\mathcal{O}$ is said to be fresh if all of the following conditions hold:*

(1) $\gamma_{\mathcal{O}} = \perp$, (2) $\delta_{\mathcal{O}}$ = accepted-mk, (3) $\exists$ $V \in \mathcal{U}$ such that V is uncorrupted and $\mathrm{pid}_{\mathcal{O}} = V$, *and*

(4) No revealed oracle Π_V^i has had a matching conversation with $\mathcal{O}$.

SECURITY GAME FOR MASTER-KEY AGREEMENT PROTOCOLS. The game, denoted by $\mathsf{Exec}_{\mathcal{A},\Pi}^{\mathsf{OW\text{-}MS}}(t)$, for defining the security of master-key agreement protocol Π in the presence of adversary $\mathcal{A}$ is similar to that for pre-master key, with the modification that $\mathcal{A}$ is also allowed to make any number of Reveal queries, in addition to the NewSession, Send, Corrupt, Reveal, and Check queries. Here, check queries are with respect to the master secret keys only. When the adversary stops, it outputs a pair $(\mathcal{O}^*, m^*)$, where $\mathcal{O}^*$ identifies one of its oracles, and m^* is some element of $\mathcal{S}_{\mathrm{MS}}$. We say that $\mathcal{A}$ wins if its output $(\mathcal{O}^*, m^*)$ is such that O^* is fresh and $m^* = m_{\mathcal{O}^*}$. In this case the output of $\mathsf{Exec}_{\mathcal{A},\Pi}^{\mathsf{OW\text{-}MS}}(t)$ is set to 1. Otherwise the output of the experiment is set to 0. We write

$$\mathbf{Adv}_{\mathcal{A},\Pi}^{\mathsf{OW\text{-}MS}}(t) = \Pr[\mathsf{Exec}_{\mathcal{A},\Pi}^{\mathsf{OW\text{-}MS}}(t) = 1]$$

for the advantage of $\mathcal{A}$ in winning the $\mathsf{Exec}_{\mathcal{A},\Pi}^{\mathsf{OW\text{-}MS}}(t)$ game. The probability is taken over all random coins used in the execution.

The following definition describes a situation where some party U had engaged in a session which terminated successfully with some party V, but no session of V has a matching conversation with U.

Definition 4 (No-Matching). *Let* $\mathsf{No\text{-}Matching}_{\mathcal{A},\Pi}(t)$ *be the event that at some point during the execution of* $\mathsf{Exec}_{\mathcal{A},\Pi}^{\mathsf{OW\text{-}MS}}(t)$ *for two uncorrupted parties $U \in \mathcal{U} \cup \mathcal{U}'$ and $V \in \mathcal{U}$ there exists an oracle $\mathcal{O} = \Pi_U^i$ with* $\mathrm{pid}_{\mathcal{O}} = V \in \mathcal{U}$, $\delta_{\mathcal{O}}$ = *accepted, and yet no oracle Π_V^i has had a matching conversation with $\mathcal{O}$.*

The following definition says that a protocol is a secure master-key agreement protocol if the key established in an honest session is secret (in the one-wayness sense) and no honest party can be coaxed into incorrectly accepting.

Definition 5 (Master Key Agreement Security). *A master key agreement protocol is secure if it satisfies the following requirements:*

- **Correctness:** *If at the end of the execution of a benign adversary, who correctly relays messages, any two oracles which have had a matching conversation hold the same master key, which is distributed uniformly over the master key space* $\mathcal{S}_{\mathrm{MS}}$.
- **Key Secrecy:** *A master key agreement protocol* Π *satisfies* OW-MS *key secrecy if for any* p.p.t. *adversary* $\mathcal{A}$, *its advantage* $\mathbf{Adv}_{\mathcal{A},\Pi}^{\mathsf{OW\text{-}MS}}(t)$ *is a negligible function.*
- **No Matching:** *For any* p.p.t. *adversary* $\mathcal{A}$, *the probability of the event* $\mathsf{No\text{-}Matching}_{\mathcal{A},\Pi}(t)$ *is a negligible function.*

REMARK 1. Our security requirements for master secret keys are still significantly weaker than the more standard requirements for key exchange [6, 7]. Although the adversarial powers are similar to those in existing models (*e.g.*[9]), we still require the adversary to recover the entire key. The weaker requirement is motivated by our use of TLS as guide in designing the security model. In this protocol, the master secret key is *not* indistinguishable from a random one since it is used to compute MACs that are sent over the network.

REMARK 2. The No Matching property we require is essentially the one based on matching conversations introduced by Bellare and Rogaway [6], adapted to our setting where only one of the parties involved in the execution is required to hold a certified key (and thus have a verifiable identity). One could potentially replace matching conversations with weaker versions of partnering, but only at the expense of making the definitions and results less clear. Bellare and Rogaway also show that if the No Matching property is satisfied, then agreement is injective. In our terms, with overwhelming probability it holds that if $\mathcal{O} = \Pi_U^i$ had accepted and has $\mathrm{pid}_{\mathcal{O}} = V \in \mathcal{U}$, then there exist precisely one session of V with which $\mathcal{O}$ has a matching conversation.

REMARK 3. Notice that, together, the first and third conditions in the above definitions imply a key confirmation guarantee: if one session has accepted a certain key, then there exists a unique session of the intended partner who has accepted the same key.

REMARK 4. The addition of Reveal queries implies security against "unknown-key-share" attacks: if parties U and V share a master-key without being aware that they interact with each other the adversary can obtain the key of U by performing a Reveal query on the appropriate session of V, thus breaking security in the sense defined above.

REMARK 5. Notice that an adversary against the master-secret key does not have any query that allows it to obtain information about the pre-master secret key. This is consistent with the SSL specification which states that the pre-master secret should be converted to the master secret immediately and that the pre-master secret should be securely erased from memory. In particular this means that the pre-master secret does not form part of the state of the master key agreement oracle, and so it does not get written on a transcript.

In this section we show that the master-key agreement protocol obtained from a secure pre-master key agreement protocol by using the transformation used in TLS is secure. Let Π be an arbitrary pre-master key agreement protocol, G a hash function, and $\mathsf{Mac} = (\mathcal{K}, \mathrm{MAC}, \mathrm{ver})$ a message authentication code. We write $(\Pi; \mathsf{MKD}_{\mathsf{SSL}}(\mathsf{Mac}, G))$ the master-key agreement protocol obtained by extending Π with the master-key derivation phase of TLS, i.e. by appending to the message flows of Π those in Step 4 of Figure 1. Starting from a secure pre-master key agreement protocol, the above transformation yields a secure master key agreement protocol.

Theorem 3. *Let Π be a secure pre-master agreement protocol,* Mac *a secure message authentication code, and G a random oracle. Then $(\Pi; \mathsf{MKD}_{\mathsf{SSL}}(\mathsf{Mac}, G))$ is a secure master-key agreement protocol.*

5 Application Key Agreement

In this section we extend the model developed so far to deal with application keys obtained from master-secret keys, and the analyze the security of the application keys obtained through the TLS protocol.

As discussed in the introduction we focus on protocols with a particular structure: first, a master-key is agreed by the parties via some master-key agreement protocol Π, and then this key is used as input to an application key derivation protocol, Σ. The same master-key can be used in multiple executions of the application key protocol which can take place in parallel and concurrently.

We capture this setting by modifying the model for master-key agreement protocols as follows. We consider two types of oracles: MK-oracles which correspond to sessions where the master secret key is derived (i.e. sessions of protocol Π), and AK-oracles, which correspond to sessions of the application key derivation protocol (i.e. sessions of Σ). The AK-oracles are spawned by MK-oracles that have established a master-secret key; spawning is done at the request of the adversary. The internal structure and behavior of MK-oracles are as defined in the previous section. To describe AK-oracles, we again impose some syntactic restrictions on the protocols (and thus on the oracles). We require that AK-oracle $\mathcal{Q}$ maintain variables $\tau_{\mathcal{Q}}, m_{\mathcal{Q}}, \mathrm{role}_{\mathcal{Q}}, \mathrm{pid}_{\mathcal{Q}}$ with the same roles as before. In addition, a new variable $k_{\mathcal{Q}} \in \mathcal{S}_{\mathrm{A}}$ holds the application key obtained in the session. (Here $\#\mathcal{S}_{\mathrm{A}} \geq 2^t$, where t is the security parameter). The state variable $\delta_{\mathcal{Q}}$ now assumes values in $\{\perp, \textit{accepted-ak}, \textit{rejected}\}$, with the obvious semantics. Finally, the corruption variable $\delta_{\mathcal{Q}}$ is either $\perp$ or *compromised* (we explain below when the latter value is set).

In addition to the powers previously granted to the adversary, now the adversary can also create new AK-oracles by issuing queries of the form $\mathsf{Spawn}(\mathcal{O})$, with $\mathcal{O}$ an MK-oracle that had successfully finished its execution. As a result, a new oracle $\mathcal{Q} = \Sigma^j_{\mathcal{O}}$ is created (where j indicates that $\mathcal{Q}$ is the j'th oracle spawned by $\mathcal{O}$.) Oracle $\mathcal{Q}$ inherits the variables $\tau_{\mathcal{Q}}$, $m_{\mathcal{Q}}$, $\mathrm{role}_{\mathcal{Q}}$, and $\mathrm{pid}_{\mathcal{Q}}$ from $\mathcal{O}$ in the obvious way. The adversary may also compromise AK-oracles: when a query $\mathsf{Compromise}(\mathcal{Q})$ is issued, if $\mathcal{Q}$ has accepted, then $k_{\mathcal{Q}}$ is returned to the adversary and $\delta_{\mathcal{Q}}$ is set to *compromised*. Notice that the Compromise queries are the analogue of Reveal queries for AK-oracles. We chose to have different names for clarity.

The security of keys is captured via a Test query. When $\mathsf{Test}(\mathcal{Q})$ is issued, a bit $b \in \{0,1\}$ is chosen at random. Then if $b = 0$ then $k_{\mathcal{Q}^*}$ is returned to the adversary, otherwise a randomly selected element from $\mathcal{S}_{\mathrm{A}}$ is returned to the adversary (who then has to guess b; see the game defined below).

An AK-oracle $\mathcal{Q}$ is a valid target for the adversary if the parent oracle of $\mathcal{Q}$ is fresh, $\mathcal{Q}$ has finished successfully its execution, its intended partner, say V, is not corrupt, and any session of V with which $\mathcal{Q}$ has a matching conversation is not compromised.

Definition 6 (Fresh Application Key Oracle). *Let $\mathcal{O}$ be a master key agreement oracle and $\mathcal{Q}$ denote one of its children. The oracle $\mathcal{Q}$ is said to be fresh if the following conditions hold:*

(1) $\mathcal{O}$ is a fresh master key agreement oracle, (2) $\gamma_{\mathcal{Q}} = \perp$, (3) $\delta_{O} =$ accepted-ak , (4) $\exists\, V \in \mathcal{U}$ such that $\mathrm{pid}_{\mathcal{Q}} = V$, and (5) No compromised session $\Sigma_{\mathcal{Q}'}$ that belongs to V has had a matching conversation with $\mathcal{Q}$.

Note that here, we are implicitly assuming that knowing a master key automatically gives the adversary all derived application keys. Whilst this will not be true of all protocols which one can think of, it is true for all application key derivation protocols that we consider here and in particular in Stage 5 of the protocol of Figure 1.

Security game for application-key agreement protocols. We define the security of an application-key protocol $\Pi;\Sigma$ via a game $\mathsf{Exec}^{\mathsf{IND\text{-}AK}}_{\mathcal{A},\Pi;\Sigma}(t)$ between an adversary $\mathcal{A}$ and a challenger $\mathcal{C}$.

(1) $\mathcal{C}$ generates public-secret key pairs for each user $U \in \mathcal{U}$, and returns the public keys to $\mathcal{A}$.
(2) $\mathcal{A}$ is allowed to make as many NewSession, Send, Spawn, Compromise, Reveal, Check, and Corrupt queries as it likes throughout the game.
(3) At any point during the game adversary $\mathcal{A}$ makes a single $\mathsf{Test}(\mathcal{Q}^*)$ query.
(4) The adversary outputs a bit b'.

We say that $\mathcal{A}$ wins if $\mathcal{Q}^*$ is fresh at the end of the game and its output bit b is such that $b = b'$ (where b is the bit internally selected during the Test query). In this case the result of $\mathsf{Exec}^{\mathsf{IND\text{-}AK}}_{\mathcal{A},\Pi;\Sigma}(t)$ is set to 1. Otherwise the output of the experiment is set to 0. We write

$$\mathbf{Adv}^{\mathsf{IND\text{-}AK}}_{\mathcal{A},(\Pi;\Sigma)}(t) = \left|\Pr[\mathsf{Exec}^{\mathsf{IND\text{-}AK}}_{\mathcal{A},\Pi;\Sigma}(t) = 1] - \frac{1}{2}\right|$$

for the advantage of $\mathcal{A}$ in winning the $\mathsf{Exec}^{\mathsf{IND\text{-}AK}}_{\mathcal{A},\Pi;\Sigma}(t)$ game. Using this security game we can now define the security of a application key agreement protocol.

Definition 7 (Application Key Agreement Security). *An application key agreement protocol is secure if it satisfies the following conditions:*

- **Correctness:** *In the presence of an adversary which faithfully relays messages, two oracles running the protocol accept holding the same application key and session ID, and the application key is distributed uniformly at random on the application key space.*
- **Key secrecy:** *An application key agreement protocol* $\Pi; \Sigma$ *satisfies* IND-AK *key secrecy if for any* p.p.t. *adversary* $\mathcal{A}$, *its advantage* $\mathbf{Adv}^{\mathsf{IND\text{-}AK}}_{\mathcal{A},\Pi;\Sigma}(t)$ *is negligible in* t.

REMARK 1. The model that we develop ensures strong security guarantees for the application keys, in the standard sense of indistinguishability against attackers with powerful corruption capabilities. In this sense our model is close to existing ones, but has the added feature that we explicitly consider the setting where more than one application-key can be derived from the same master key.

REMARK 2. Notice that at the application key layer we do not require key confirmation anymore. Indeed, a trivial attack on the standard notion of key confirmation can be mounted against application keys derived using the TLS protocol. However, implicit key confirmation for application keys may still be achieved, depending how the application key is actually used. (In the full version of the paper we discuss the composition of our application key agreement protocol with specific applications, especially confidentiality applications.)

The loss of this property is in some sense a result of how we chose to break down the protocol for analysis, since one of our goals was to identify what security properties each of the stages provides. However, if one considers Stages 1-4 as the key agreement protocol, and stages 5-6 as the application then one does obtain an explicit notion of key confirmation. Hence, the loss of explicit key confirmation in Stage 5 should not be considered a design flaw in TLS.

In this section we show that the application-key agreement protocol obtained from any secure master-key derivation protocol, and the application-key derivation protocol of TLS (Stage 5 of Figure 1) is secure.

For any master-key agreement protocol Π, and hash function H, we write $(\Pi; \mathsf{AK}_{\mathsf{SSL}}(H))$ for the application-key agreement protocol obtained by extending Π with the application-key derivation protocol of TLS. Informally, this means that we derive an application key agreement protocol from a master key agreement protocol using Stage 5 of Figure 1. We make no assumption as to whether the master key agreement protocol itself is derived from a pre-master key agreement protocol as in Figure 1. The following theorem says that starting with a master-key agreement protocol secure in the sense of Definition 5, the above transformation yields a secure application key protocol.

Theorem 4. *Let* Π *be a secure master-key agreement protocol and* H *a random oracle. Then* $(\Pi; \mathsf{AK}_{\mathsf{SSL}}(H))$ *is a secure application-key agreement protocol.*

The security of TLS follows from Theorems 1, 2, 3 and 4. For full details the reader should consult the full version of this paper.

Acknowledgements. The authors would like to thank Caroline Belrose for various discussions on key agreement protocols during the writing of this paper and Martin Abadi for interesting insights into various aspects of TLS. The work described in this paper has been supported in part by the EU FP6 project eCrypt and an EPSRC grant.

References

1. Abdalla, M., Chevassut, O., Pointcheval, D.: One–Time Verifier–based Encrypted Key Exchange. In: Vaudenay, S. (ed.) PKC 2005. LNCS, vol. 3386, pp. 47–64. Springer, Heidelberg (2005)
2. An, J.H., Dodis, Y., Rabin, T.: On the Security of Joint Signature and Encryption. In: Knudsen, L.R. (ed.) EUROCRYPT 2002. LNCS, vol. 2332, pp. 83–107. Springer, Heidelberg (2002)
3. Bellare, M., Canetti, R., Krawczyk, H.: A modular approach to the design and analysis of authentication and key exchange protocols. In: 30th Symposium on Theory of Computing – STOC 1998, pp. 419–428. ACM, New York (1998)
4. Bellare, M., Namprempre, C.: Authenticated encryption: Relations among notions and analysis of the generic composition paradigm. In: Okamoto, T. (ed.) ASIACRYPT 2000. LNCS, vol. 1976, pp. 531–545. Springer, Heidelberg (2000)
5. Bellare, M., Pointcheval, D., Rogaway, P.: Authenticated key exchange secure against dictionary attacks. In: Preneel, B. (ed.) EUROCRYPT 2000. LNCS, vol. 1807, pp. 139–155. Springer, Heidelberg (2000)
6. Bellare, M., Rogaway, P.: Entity authentication and key distribution. In: Stinson, D.R. (ed.) CRYPTO 1993. LNCS, vol. 773, pp. 232–249. Springer, Heidelberg (1994)
7. Bellare, M., Rogaway, P.: Provably secure session key distribution: The three party case. In: 27th Symposium on Theory of Computing – STOC 1995, pp. 57–66. ACM, New York (1995)
8. Bird, R., Gopal, I.S., Herzberg, A., Janson, P.A., Kutten, S., Molva, R., Yung, M.: Systematic Design of Two-Party Authentication Protocols. In: Feigenbaum, J. (ed.) CRYPTO 1991. LNCS, vol. 576, pp. 44–61. Springer, Heidelberg (1992)
9. Blake–Wilson, S., Johnson, D., Menezes, A.J.: Key agreement protocols and their security analysis. In: Darnell, M.J. (ed.) Cryptography and Coding 1997. LNCS, vol. 1355, pp. 30–45. Springer, Heidelberg (1997)
10. Blake–Wilson, S., Menezes, A.: Entity Authentication and Authenticated Key Transport Protocols Employing Asymmetric Techniques. In: Christianson, B., Lomas, M. (eds.) Security Protocols 1997. LNCS, vol. 1361, pp. 137–158. Springer, Heidelberg (1998)
11. Bleichenbacher, D.: Chosen ciphertext attacks against protocols based on the RSA encryption standard PKCS #1. In: Krawczyk, H. (ed.) CRYPTO 1998. LNCS, vol. 1462, pp. 1–12. Springer, Heidelberg (1998)
12. Bresson, E., Chevassut, O., Pointcheval, D.: Provably Authenticated Group Diffie–Hellman Key Exchange – The Dynamic Case. In: Boyd, C. (ed.) ASIACRYPT 2001. LNCS, vol. 2248, pp. 290–309. Springer, Heidelberg (2001)
13. Canetti, R., Krawczyk, H.: Analysis of Key-Exchange Protocols and Their Use for Building Secure Channels. In: Pfitzmann, B. (ed.) EUROCRYPT 2001. LNCS, vol. 2045, pp. 453–474. Springer, Heidelberg (2001)

14. Canetti, R., Krawczyk, H.: Universally Composable Notions of Key Exchange and Secure Channels. In: Knudsen, L.R. (ed.) EUROCRYPT 2002. LNCS, vol. 2332, pp. 337–351. Springer, Heidelberg (2002)
15. Canetti, R., Rabin, T.: Universal Composition with Joint State. In: Boneh, D. (ed.) CRYPTO 2003. LNCS, vol. 2729, pp. 265–281. Springer, Heidelberg (2003)
16. Cramer, R., Shoup, V.: Design and analysis of practical public-key encryption schemes secure against adaptive chosen ciphertext attack. SIAM Journal of Computing 33, 167–226 (2003)
17. Diffie, W., van Oorschot, P.C., Weiner, M.J.: Authentication and authenticated key exchange. Designs, Codes and Cryptography 2, 107–125 (1992)
18. Dierks, T., Allen, C.: The TLS Protocol Version 1.0. RFC 2246 (January 1999)
19. Dierks, T., Allen, C.: The TLS Protocol Version 1.2. RFC 4346 (April 2006)
20. Freier, A.O., Karlton, P., Kocher, P.C.: The SSL Protocol Version 3.0. Internet Draft (1996)
21. Fouque, P., Pointcheval, D., Zimmer, S.: HMAC is a Randomness Extractor and Applications to TLS. In: Symposium on Information, Computer and Communications Security, ASIACCS 2008 (2008)
22. Hickman, K.E.B.: The SSL Protocol Version 2.0. Internet Draft (1994)
23. Jonsson, J., Kaliski Jr., B.: On the Security of RSA Encryption in TLS. In: Yung, M. (ed.) CRYPTO 2002. LNCS, vol. 2442, pp. 127–142. Springer, Heidelberg (2002)
24. Choo, K.-K.R., Boyd, C., Hitchcock, Y.: Examining Indistinguishability-Based Proof Models for Key Establishment Protocols. In: Roy, B. (ed.) ASIACRYPT 2005. LNCS, vol. 3788, pp. 585–604. Springer, Heidelberg (2005)
25. Krawczyk, H.: The order of encryption and authentication for protecting communications (or: How secure is SSL?). In: Kilian, J. (ed.) CRYPTO 2001. LNCS, vol. 2139, pp. 310–331. Springer, Heidelberg (2001)
26. Kudla, C.: Special signature schemes and key agreement protocols. PhD Thesis, Royal Holloway University of London (2006)
27. Kudla, C., Paterson, K.: Modular security proofs for key agreement protocols. In: Roy, B. (ed.) ASIACRYPT 2005. LNCS, vol. 3788, pp. 549–565. Springer, Heidelberg (2005)
28. Mitchell, J.C., Shmatikov, V., Stern, U.: Finite-state analysis of SSL 3.0. In: SSYM 1998: Proceedings of the 7th conference on USENIX Security Symposium 1998 (1998)
29. Mazare, L., Warinschi, B.: On the security of encryption under adaptive corruptions (preprint, 2007)
30. Paulson, L.: Inductive analysis of the Internet protocol TLS. ACM Transations on Information and Systems Security 2(3), 332–351 (1999)
31. Shoup, V.: On formal models for secure key exchange (version 4) (preprint, 1999)
32. Wagner, D., Schneier, B.: Analysis of the SSL 3.0 protocol. In: 2nd USENIX Workshop on Electronic Commerce (1996)

Ambiguous Optimistic Fair Exchange

Qiong Huang[1], Guomin Yang[1], Duncan S. Wong[1], and Willy Susilo[2]

[1] City University of Hong Kong, Hong Kong, China
{csqhuang@student,csyanggm@cs,duncan@}cityu.edu.hk
[2] University of Wollongong, Australia
wsusilo@uow.edu.au

Abstract. Optimistic fair exchange (OFE) is a protocol for solving the problem of exchanging items or services in a fair manner between two parties, a signer and a verifier, with the help of an arbitrator which is called in only when a dispute happens between the two parties. In almost all the previous work on OFE, after obtaining a partial signature from the signer, the verifier can present it to others and show that the signer has indeed committed itself to something corresponding to the partial signature *even* prior to the completion of the transaction. In some scenarios, this capability given to the verifier may be harmful to the signer. In this paper, we propose the notion of *ambiguous optimistic fair exchange* (A-OFE), which is an OFE but also requires that the verifier cannot convince anybody about the authorship of a partial signature generated by the signer. We present a formal security model for A-OFE in the multi-user setting and chosen-key model. We also propose an efficient construction with security proven without relying on the random oracle assumption.

1 Introduction

Optimistic Fair Exchange (OFE) allows two parties to fairly exchange information in such a way that at the end of a protocol run, either both parties have obtained the complete information from one another or none of them has obtained anything from the counter party. In an OFE, there is a third party, called Arbitrator, which only gets involved when a dispute occurred between the two parties. OFE is a useful tool in practice, for example, it can be used for performing contract signing, fair negotiation and similar applications on the Internet. Since its introduction [1], there have been many OFE schemes proposed [2, 3, 4, 12, 13, 14, 18, 21, 23, 24]. For all recently proposed schemes, an OFE protocol for signature typically consists of three message flows. The initiator of OFE, Alice, first sends a message σ_P, called *partial signature*, to the responder, Bob. The partial signature σ_P acts as Alice's partial commitment to her full signature which is to be sent to Bob. But Bob needs to send his full signature to Alice first in the second message flow. After receiving Bob's full signature, Alice sends her full signature to Bob in the third message flow. If in the second message flow that Bob refuses to send his full signature back to Alice, Alice's partial signature σ_P should have no use to Bob, so that Alice has no concern about giving away σ_P. However, after Bob has sent his full signature to

J. Pieprzyk (Ed.): ASIACRYPT 2008, LNCS 5350, pp. 74–89, 2008.
© International Association for Cryptologic Research 2008

Alice while Alice refuses to send her full signature in the third message flow, then Bob can ask the Arbitrator to retrieve Alice's full signature from σ_P after sending both σ_P and Bob's full signature to the Arbitrator. To the best of our knowledge, among almost all the known OFE schemes, there is one common property about Alice's partial signature σ_P which has neither been captured in any of the security models for OFE nor been considered as a requirement for OFE. The property is that once σ_P is given out, at least one of the following statements is true.

1. Everyone can verify that σ_P must be generated by Alice because σ_P, similar to a standard digital signature, has the non-repudiation property with respect to Alice's public key;
2. Bob can show to anybody that Alice is the signer of σ_P.

For example, in the schemes proposed in [12, 18], the partial signature of Alice is a standard signature, which can only be generated by Alice. In many OFE schemes in the literature, Alice's signature is encrypted under the arbitrator's public key, and then a non-interactive proof is generated to show that the ciphertext indeed contains a signature of Alice. This is known as *verifiably encrypted signature*. However, this raises the question of whether a non-interactive proof that a signature is encrypted is really any different from a signature itself, since it alone is sufficient to prove to any third party that the signer has committed to the message [10].

This property may cause no concern in some applications, for example, in those where only the full signature is deemed to have some actual value to the receiving party. However, it may be undesirable in some other applications. Since σ_P is publicly verifiable and non-repudiative, in practice, σ_P may not be completely useless to Bob. Instead, σ_P has evidently shown Alice's commitment to the corresponding message. This may incur some unfair situation, to the advantage of Bob, if Bob does not send out his full signature. In contract signing applications, this could be undesirable because σ_P can already be considered as Alice's undeniable commitment to a contract in court while there is no evidence showing that Bob has committed to anything.

In another application, fair negotiation, the property above may also be undesirable. Suppose after obtaining σ_P from Alice on her offer, Bob may show it to Charlie, who is Alice's competitor, and ask Charlie for making a better offer. If Charlie's offer is better, then Bob may stop the OFE protocol run with Alice indicating that Bob is unwilling to conclude the negotiation with Alice, and instead carrying out a new OFE protocol run with Charlie. Bob can play the same game iteratively until that no one can give an even better offer. Then Bob can resolve the negotiation by sending his service (i.e. his full signature as the commitment to his service) to the highest bidder.

For making OFE be applicable to more applications and practical scenarios, in this paper, we propose to enhance the security requirements of OFE and construct a new OFE scheme which does not have the problems mentioned above. One may also think of this as an effort to make OFE more admissible as a viable fair exchange tool for real applications. We will build an OFE scheme which not only satisfies all the existing security requirements of OFE (with respect to the strongest security model available [18]), but in addition to that, will also have σ_P be not self-authenticating and unable for Bob to demonstrate to others that Alice has committed herself to something. We call this enhanced notion of OFE as *Ambiguous Optimistic Fair Exchange* (A-OFE). It inherits all the formalized properties of OFE [12, 18] and has a new property introduced: *signer ambiguity*. It requires that a partial signature σ_P generated by Alice or Bob should look alike and be indistinguishable even to Alice and Bob.

(Related Work): There have been many OFE schemes proposed in the past [2, 3, 4, 12, 13, 18, 21, 23, 24]. In the following, we review some recent ones by starting from 2003 when Park, Chong and Siegel [24] proposed an OFE based on sequential two-party multi-signature. It was later broken and repaired by Dodis and Reyzin [13]. The scheme is *setup-driven* [25, 26], which requires all users to register their keys with the arbitrator prior to any transaction. In [23], Micali proposed another scheme based on a CCA2 secure public key encryption with the property of *recoverable randomness* (i.e., both plaintext and randomness used for generating the ciphertext can be retrieved during decryption). Later, Bao et al. [4] showed that the scheme is not fair, where a dishonest party, Bob, can obtain the full commitment of another party, Alice, without letting Alice get his obligation. They also proposed a fix to defend against the attack.

In PKC 2007, Dodis, Lee and Yum [12] considered OFE in a multi-user setting. Prior to their work, almost all previous results considered the single-user setting only which consists of a single signer and a single verifier (along with an arbitrator). The more practical multi-user setting considers a system to have multiple signers and verifiers (along with the arbitrator), so that a dishonest party can collude with other parties in an attempt of cheating. Dodis et al. [12] showed that security of OFE in the single-user setting does not necessarily imply the security in the multi-user setting. They also proposed a formal definition of OFE in the multi-user setting, and proposed a generic construction, which is *setup-free* (i.e. no key registration is required between users and the arbitrator) and can be built in the random oracle model [5] if there exist one-way functions, or in the standard model if there exist trapdoor one-way permutations.

In CT-RSA 2008, Huang, Yang, Wong and Susilo [18] considered OFE in the multi-user setting and *chosen-key* model, in which the adversary is allowed to choose public keys arbitrarily without showing its knowledge of the corresponding private keys. Prior to their work, the security of all previous OFE schemes (including the one in [12]) are proven in a more restricted model, called *certified-key* (or *registered-key*) model, which requires the adversary to prove its knowledge of the corresponding private key before using a public key. In [18], Huang et al. gave a formal security model for OFE in the multi-user setting and chosen-key model, and proposed an efficient OFE scheme based on ring signature. In their scheme, a partial signature is a conventional signature and a full signature is a two-member ring signature in additional to the conventional signature. The security of their scheme was proven without random oracles.

Liskov and Micali [22] proposed an *online-untransferable signature* scheme, which in essence is an enhanced version of designated confirmer signature, with the extra property that a dishonest recipient, who is interacting with a signer, cannot convince a third party that the signature is generated by the signer. Their scheme is fairly complex and the signing process requires several rounds of interaction with the recipient. Besides, their scheme works in the certified-key model, and is not setup-free, i.e. there is a setup stage between each signer and the confirmer, and the confirmer needs to store a public/secret key pair for each signer, thus a large storage is required for the confirmer.

In [14], Garay, Jakobsson and MacKenzie introduced a similar notion for optimistic contract signing, named *abuse-freeness*. It requires that no party can ever prove to a third party that he is capable of choosing whether to validate or invalidate a contract. They also proposed a construction of abuse-free optimistic contract signing protocol. The security of their scheme is based on DDH assumption under the random oracle model. Besides they did not consider the multi-user setting for their contract signing protocol.

(Our Contributions): In this paper we make the following contributions.

1. We propose the notion of *Ambiguous Optimistic Fair Exchange* (Ambiguous OFE or A-OFE in short) which allows a signer Alice to generate a partial signature in such a way that a verifier Bob cannot convince anybody about the authorship of this partial signature, and thus cannot prove to anybody that Alice committed herself to anything prematurely. Realizing the notion needs to make the partial signature ambiguous with respect to Alice and Bob. We will see that this requires us to include both Alice and Bob's public keys into the signing and verification algorithms of A-OFE.
2. For formalizing A-OFE, we propose a strong security model in the multi-user setting and chosen-key model. Besides the existing security requirements for OFE, that is, resolution ambiguity[1], security against signers, security against verifiers and security against the arbitrator, A-OFE has an additional requirement: *signer ambiguity*. It requires that the verifier can generate partial signatures whose distribution is (computationally) indistinguishable from that of partial signatures generated by the signer. We also evaluate the relations among the security requirements and show that if a scheme has security against the arbitrator and (a weaker variant of) signer ambiguity, then it already has (a weaker variant of) security against verifiers.
3. We propose the first efficient A-OFE scheme and prove its security in the multi-user setting and chosen-key model without random oracle. It is based on Groth and Sahai's idea of constructing a fully anonymous group signature scheme [15, 16] and the security relies on the decision linear assumption and strong Diffie-Hellman assumption.

(Paper Organization): In the next section, we define A-OFE and propose a security model for it. We also show some relation among the formalized security

[1] Resolution ambiguity is just another name for the ambiguity considered in [12, 18].

requirements of A-OFE. In Sec. 3, we introduce some preliminaries which are used in our construction, which is described in Sec. 4. In Sec. 5, we prove the security of our scheme in the standard model, and compare our scheme with other two related work.

2 Ambiguous Optimistic Fair Exchange

In an A-OFE scheme, we require that after receiving a partial signature σ_P from Alice (the signer), Bob (the verifier) cannot convince others but himself that Alice has committed herself to σ_P. This property is closely related to the non-transferability of designated verifier signature [19] and the ambiguity of concurrent signature [11]. Similarly, we require that the verification algorithm in A-OFE should also take as the public keys of both signer and (designated) verifier as inputs, in contrast to that in the traditional definition of OFE [1, 2, 12, 18].

Definition 1 (Ambiguous Optimistic Fair Exchange). *An* ambiguous optimistic fair exchange *(*A-OFE *in short) scheme involves two users (a signer and a verifier) and an arbitrator, and consists of the following (probabilistic) polynomial-time algorithms:*

- PMGen*: On input 1^k where k is a security parameter, it outputs a system parameter* PM.
- $\mathsf{Setup}^{\mathsf{TTP}}$*: On input* PM*, the algorithm generates a public arbitration key APK and a secret arbitration key ASK.*
- $\mathsf{Setup}^{\mathsf{User}}$*: On input* PM *and (optionally) APK, the algorithm outputs a public/secret key pair (PK, SK). For user U_i, we use (PK_i, SK_i) to denote its key pair.*
- Sig *and* Ver*:* $\mathsf{Sig}(M, SK_i, PK_i, PK_j, APK)$ *outputs a (full) signature σ_F on M of user U_i with the designated verifier U_j, where message M is chosen by user U_i from the message space $\mathcal{M}$ defined under PK_i, while* $\mathsf{Ver}(M, \sigma_F, PK_i, PK_j, APK)$ *outputs* accept *or* reject*, indicating σ_F is U_i's valid full signature on M with designated verifier U_j or not.*
- PSig *and* PVer*: They are partial signing and verification algorithms respectively.* $\mathsf{PSig}(M, SK_i, PK_i, PK_j, APK)$ *outputs a partial signature σ_P, while* $\mathsf{PVer}(M, \sigma_P, \mathbf{PK}, APK)$ *outputs* accept *or* reject*, where $\mathbf{PK} = \{PK_i, PK_j\}$.*
- Res*: This is the resolution algorithm.* $\mathsf{Res}(M, \sigma_P, ASK, \mathbf{PK})$*, where $\mathbf{PK} = \{PK_i, PK_j\}$, outputs a full signature σ_F, or $\perp$ indicating the failure of resolving a partial signature.*

Note that we implicitly require that there is an efficient algorithm which given a a pair of (SK, PK), verifies if SK matches PK, i.e. (SK, PK) is an output of algorithm $\mathsf{Setup}^{\mathsf{User}}$. As in [12], PSig together with Res should be functionally equivalent to Sig.

For the correctness, we require that for any $k \in \mathbb{N}$, $\mathsf{PM} \leftarrow \mathsf{PMGen}(1^k)$, $(APK, ASK) \leftarrow \mathsf{Setup}^{\mathsf{TTP}}(\mathsf{PM})$, $(PK_i, SK_i) \leftarrow \mathsf{Setup}^{\mathsf{User}}(\mathsf{PM}, APK)$,

$(PK_j, SK_j) \leftarrow \mathsf{Setup}^{\mathsf{User}}(\mathsf{PM}, APK)$, and $M \in \mathcal{M}(PK_i)$, let $\mathbf{PK} = \{PK_i, PK_j\}$, we have the following

$$\mathsf{PVer}(M, \mathsf{PSig}(M, SK_i, PK_i, PK_j, APK), \mathbf{PK}, APK) = \mathsf{accept},$$
$$\mathsf{Ver}(M, \mathsf{Sig}(M, SK_i, PK_i, PK_j, APK), PK_i, PK_j, APK) = \mathsf{accept}, \text{ and}$$
$$\mathsf{Ver}(M, \mathsf{Res}(M, \mathsf{PSig}(M, SK_i, PK_i, PK_j, APK), ASK, \mathbf{PK}), PK_i, PK_j, APK) = \mathsf{accept}.$$

2.1 Security Properties

(Resolution Ambiguity): The *resolution ambiguity* property requires that any 'resolved signature' $\mathsf{Res}(M, \mathsf{PSig}(M, SK_i, PK_i, PK_j, APK), ASK, \{PK_i, PK_j\})$ is *computationally indistinguishable* from an 'actual signature' generated by the signer, $\mathsf{Sig}(M, SK_i, PK_i, PK_j, APK)$. It is identical to 'ambiguity' defined in [12, 18]. Here we just use another name, in order to avoid any confusion, as we will define another kind of ambiguity next.

(Signer Ambiguity): Informally, *signer ambiguity* means that given a partial signature σ_P from a signer A, a verifier B should not be able to convince others that σ_P was indeed generated by A. To capture this property, we use the idea of defining *ambiguity* in concurrent signature [11]. We require that B can generate partial signatures that look *indistinguishable* from those generated by A. This is also the reason why a verifier should also have a public/secret key pair, and the verifier's public key should be included in the inputs of PSig and Sig. Formally, we define an experiment in which D is a probabilistic polynomial-time distinguisher.

$$\begin{aligned}
\mathsf{PM} &\leftarrow \mathsf{PMGen}(1^k)\\
(APK, ASK) &\leftarrow \mathsf{Setup}^{\mathsf{TTP}}(\mathsf{PM})\\
(M, (PK_0, SK_0), (PK_1, SK_1), \delta) &\leftarrow D^{O_{\mathsf{Res}}}(APK)\\
b &\leftarrow \{0, 1\}\\
\sigma_P &\leftarrow \mathsf{PSig}(M, SK_b, PK_b, PK_{1-b}, APK)\\
b' &\leftarrow D^{O_{\mathsf{Res}}}(\delta, \sigma_P)\\
\text{success of } D &:= [b' = b \wedge (M, \sigma_P, \{PK_0, PK_1\}) \notin Query(D, O_{\mathsf{Res}})]
\end{aligned}$$

where δ is D's state information, oracle O_{Res} takes as input a valid[2] partial signature σ_P of user U_i on message M with respect to verifier U_j, i.e. $(M, \sigma_P, \{PK_i, PK_j\})$, and outputs a full signature σ_F on M under PK_i, PK_j, and $Query(D, O_{\mathsf{Res}})$ is the set of valid queries D issued to the resolution oracle O_{Res}. In this oracle query, D can arbitrarily choose a public key PK without knowing the corresponding private key. However, we do require that there exists a PPT algorithm to check the validity of the two key pairs output by D, i.e. if SK_b matches PK_b for $b = 0, 1$, or if (PK_b, SK_b) is a possible output of $\mathsf{Setup}^{\mathsf{User}}$. The advantage of D, $\mathtt{Adv}_D^{\mathrm{SA}}(k)$, is defined to be the gap between its success probability in the experiment above and 1/2, i.e. $\mathtt{Adv}_D^{\mathrm{SA}}(k) = |\Pr[b' = b] - 1/2|$.

[2] By 'valid', we mean that σ_P is a valid partial signature on M under public keys PK_i, PK_j, alternatively, the input $(M, \sigma_P, PK_i, PK_j)$ of O_{Res} satisfies the condition that $\mathsf{PVer}(M, \sigma_P, \{PK_i, PK_j\}, APK) = \mathsf{accept}$.

Definition 2 (Signer Ambiguity). *An OFE scheme is said to be* signer ambiguous *if for any probabilistic polynomial-time algorithm D,* $\mathtt{Adv}_D^{\mathrm{SA}}(k)$ *is* negligible *in k.*

Remark 1. We note that a similar notion was introduced in [14, 22]. It's required that the signer's partial signature can be simulated in an indistinguishable way. However, the '*indistinguishability*' in [14, 22] is defined in CPA fashion, giving the adversary no oracle that resolves a partial signature to a full one, while our definition of signer ambiguity is done in the CCA fashion, allowing the adversary to ask for resolving any partial signature except the challenge one to a full signature, which is comparable to the CCA security of public key encryption schemes.

(Security Against Signers): We require that no PPT adversary A should be able to produce a partial signature with non-negligible probability, which looks good to a verifier but cannot be resolved to a full signature by the honest arbitrator. This ensures the fairness for verifiers, that is, if the signer has committed to a message with respect to an (honest) verifier, the verifier should always be able to obtain the full commitment of the signer. Formally, we consider the following experiment:

$$\begin{aligned}
\mathsf{PM} &\leftarrow \mathsf{PMGen}(1^k)\\
(APK, ASK) &\leftarrow \mathsf{Setup}^{\mathsf{TTP}}(\mathsf{PM})\\
(PK_B, SK_B) &\leftarrow \mathsf{Setup}^{\mathsf{User}}(\mathsf{PM}, APK)\\
(M, \sigma_P, PK_A) &\leftarrow A^{O^B_{\mathsf{PSig}}, O_{\mathsf{Res}}}(APK, PK_B)\\
\sigma_F &\leftarrow \mathsf{Res}(M, \sigma_P, ASK, \{PK_A, PK_B\})\\
\text{success of } A &:= [\mathsf{PVer}(M, \sigma_P, \{PK_A, PK_B\}, APK) = \mathsf{accept}\\
&\quad \wedge \mathsf{Ver}(M, \sigma_F, PK_A, PK_B, APK) = \mathsf{reject}\\
&\quad \wedge (M, PK_A) \notin \mathit{Query}(A, O^B_{\mathsf{PSig}})]
\end{aligned}$$

where oracle O_{Res} is described in the previous experiment, O^B_{PSig} takes as input (M, PK_i) and outputs a partial signature on M under PK_i, PK_B generated using SK_B, and $\mathit{Query}(A, O^B_{\mathsf{PSig}})$ is the set of queries made by A to oracle O^B_{PSig}. In this experiment, the adversary can arbitrarily choose a public key PK_i, and it may not know the corresponding private key of PK_i. Note that the adversary is not allowed to corrupt PK_B, otherwise it can easily succeed in the experiment by simply using SK_B to produce a partial signature under public keys PK_A, PK_B and outputting it. The advantage of A in the experiment $\mathtt{Adv}_A^{\mathrm{SAS}}(k)$ is defined to be A's success probability.

Definition 3 (Security Against Signers). *An OFE scheme is said to be* secure against signers *if there is* no *PPT adversary A such that* $\mathtt{Adv}_A^{\mathrm{SAS}}(k)$ *is* non-negligible *in k.*

(Security Against Verifiers): This security notion requires that any PPT verifier B should not be able to transform a partial signature into a full signature with non-negligible probability if no help has been obtained from the signer or the arbitrator. This requirement has some similarity to the notion of *opacity* for verifiably encrypted signature [9]. Formally, we consider the following experiment:

$$\begin{aligned}
\mathsf{PM} &\leftarrow \mathsf{PMGen}(1^k) \\
(APK, ASK) &\leftarrow \mathsf{Setup}^{\mathsf{TTP}}(\mathsf{PM}) \\
(PK_A, SK_A) &\leftarrow \mathsf{Setup}^{\mathsf{User}}(\mathsf{PM}, APK) \\
(M, PK_B, \sigma_F) &\leftarrow B^{O_{\mathsf{PSig}}, O_{\mathsf{Res}}}(PK_A, APK) \\
\text{success of } B &:= [\mathsf{Ver}(M, \sigma_F, PK_A, PK_B, APK) = \mathsf{accept} \wedge \\
&\quad (M, \cdot, \{PK_A, PK_B\}) \notin Query(B, O_{\mathsf{Res}})]
\end{aligned}$$

where oracle O_{Res} is described in the experiment of signer ambiguity, $Query(B, O_{\mathsf{Res}})$ is the set of valid queries B issued to the resolution oracle O_{Res}, and oracle O_{PSig} takes as input a message M and a public key PK_j and returns a valid partial signature σ_F on M under PK_A, PK_j generated using SK_A. In the experiment, B can ask the arbitrator for resolving any partial signature with respect to any pair of public keys (adaptively chosen by B, probably without the knowledge of the corresponding private keys), with the limitation described in the experiment. The advantage of B in the experiment $\mathtt{Adv}_B^{\mathrm{SAV}}(k)$ is defined to be B's success probability in the experiment above.

Definition 4 (Security Against Verifiers). *An OFE scheme is said to be* secure against verifiers *if there is* no *PPT adversary B such that* $\mathtt{Adv}_B^{\mathrm{SAV}}(k)$ *is* non-negligible *in k.*

(Security Against the Arbitrator): Intuitively, an OFE is secure against the arbitrator if no PPT adversary C including the arbitrator, should be able to generate with non-negligible probability a full signature without explicitly asking the signer for generating one. This ensures the fairness for signers, that is, no one can frame the actual signer on a message with a forgery. Formally, we consider the following experiment:

$$\begin{aligned}
\mathsf{PM} &\leftarrow \mathsf{PMGen}(1^k) \\
(APK, ASK^*) &\leftarrow C(\mathsf{PM}) \\
(PK_A, SK_A) &\leftarrow \mathsf{Setup}^{\mathsf{User}}(\mathsf{PM}, APK) \\
(M, PK_B, \sigma_F) &\leftarrow C^{O_{\mathsf{PSig}}}(ASK^*, APK, PK_A) \\
\text{success of } C &:= [\mathsf{Ver}(M, \sigma_F, PK_A, PK_B, APK) = \mathsf{accept} \wedge \\
&\quad (M, PK_B) \notin Query(C, O_{\mathsf{PSig}})]
\end{aligned}$$

where the oracle O_{PSig} is described in the previous experiment, ASK^* is C's state information, which might not be the corresponding private key of APK, and $Query(C, O_{\mathsf{PSig}})$ is the set of queries C issued to the oracle O_{PSig}. The advantage of C in this experiment $\mathtt{Adv}_C^{\mathrm{SAA}}(k)$ is defined to be C's success probability.

Definition 5 (Security Against the Arbitrator). *An OFE scheme is said to be* secure against the arbitrator *if there is* no *PPT adversary C such that* $\mathtt{Adv}_C^{\mathrm{SAA}}(k)$ *is* non-negligible *in k.*

Remark 2. In A-OFE, both signer U_A and verifier U_B are equipped with public/secret key pairs (of the same structure), and U_A and U_B can generate indistinguishable partial signatures on the same message. If the security against the arbitrator holds for U_A (as described in the experiment above), it should also hold for U_B. That is, even when colluding with U_A (and other signers), the arbitrator should not be able to frame U_B for a full signature on a message, if it has not obtained a partial signature on the message generated by U_B.

Definition 6 (Secure Ambiguous Optimistic Fair Exchange). *An A-OFE scheme is said to be* secure *in the multi-user setting and chosen-key model if it is resolution ambiguous, signer ambiguous, secure against signers, secure against verifiers and secure against the arbitrator.*

2.2 Weaker Variants of the Model

In this section, we evaluate the relation between the signer ambiguity and security against verifiers. Intuitively, if an A-OFE scheme is not secure against verifiers, the scheme cannot be signer ambiguous because a malicious verifier can convert with non-negligible probability a signer's partial signature to a full one which allows the verifier to win the signer ambiguity game. For technical reasons, we first describe some weakened models before giving the proof for a theorem regarding the relation.

In our definition of signer ambiguity (Def. 2), the two public/secret key pairs are selected by the adversary D. In a weaker form, the key pairs can be selected by the challenger, and D is allowed to corrupt these two keys. This is comparable to the ambiguity definition for concurrent signature [11], or the strongest definition of anonymity of ring signature considered in [6], namely *anonymity against full key exposure.* We can also define an even weaker version of signer ambiguity, in which D is given two public keys, PK_A, PK_B, the oracle access of O_{PSig} which returns U_A's partial signatures, and is allowed to corrupt PK_B. We call this form of signer ambiguity as *weak signer ambiguity.*

In the definition of security against verifiers (Def. 4), the verifier's public key PK_B is adaptively selected by the adversary B. In a weaker model, PK_B can be generated by the challenger and the corresponding user secret key can be corrupted by B. The rest of the model remains unchanged. We call this as *weak security against verifiers.* Below we show that if an OFE scheme is weakly signer ambiguous and secure against the arbitrator, then it is also weakly secure against verifiers.

Theorem 1. *In A-OFE, weak signer ambiguity and security against the arbitrator (Def. 5) together imply weak security against verifiers.*

Proof. Suppose that an A-OFE scheme is not weakly secure against verifiers. Let B be the PPT adversary that has non-negligible advantage ϵ in the experiment of weak security against verifiers and B make at most q queries of the form $(\cdot, PK_B)$

to oracle O_{PSig}. Due to the security against the arbitrator, B must have queried O_{PSig} in the form $(\cdot, PK_B)$. Hence the value of q is at least one. Denote the experiment of weak security against verifiers by $\mathsf{Ex}^{(0)}$. Note that in $\mathsf{Ex}^{(0)}$ all queries to O_{PSig} are answered with partial signatures generated using SK_A. We now define a series of experiments, $\mathsf{Ex}^{(1)}, \cdots, \mathsf{Ex}^{(q)}$, so that $\mathsf{Ex}^{(i)}$ $(i \geq 1)$ is the same as $\mathsf{Ex}^{(i-1)}$ except that starting from the $(q+1-i)$-th query to O_{PSig} up to the q-th query of the form $(\cdot, PK_B)$, they are answered with partial signatures generated using SK_B. Let B's success probability in experiment $\mathsf{Ex}^{(i)}$ be ϵ_i. Note that $\epsilon_0 = \epsilon$, and in experiment $\mathsf{Ex}^{(q)}$ all queries of the form $(\cdot, PK_B)$ to O_{PSig} are answered with partial signatures generated using SK_B. Since B also knows SK_B (through corruption), it can use SK_B to generate partial signatures using SK_B on any message. Therefore, making queries of the form $(\cdot, PK_B)$ to O_{PSig} does not help B on winning the experiment if answers are generated using SK_B. It is equivalent to the case that B does not issue any query $(\cdot, PK_B)$ to O_{PSig}. Hence guaranteed by the security against the arbitrator, we have that B's advantage in $\mathsf{Ex}^{(q)}$ is negligible as B has to output a full signature without getting any corresponding partial signature.

Since the gap, $|\epsilon_0 - \epsilon_q|$, between B's advantage in $\mathsf{Ex}^{(0)}$ and that in $\mathsf{Ex}^{(q)}$ is non-negligible, there must exist an $1 \leq i \leq q$ such that $|\epsilon_{i-1} - \epsilon_i|$ is at least $|\epsilon_0 - \epsilon_q|/q$, which is non-negligible as well. Let i^* be such an i. We show how to make use of the difference of B's advantage in $\mathsf{Ex}^{(i^*-1)}$ and $\mathsf{Ex}^{(i^*)}$ to build a PPT algorithm D to break the weak signer ambiguity.

Given APK and PK_A, PK_B, D first asks its challenger for SK_B, and then invokes B on (APK, PK_A, PK_B). D randomly selects an i^* from $\{1, \cdots, q\}$, and simulates the oracles for B as follows. If B asks for SK_B, D simply gives it to B. The oracle O_{Res} is simulated by D using its own resolution oracle. If B makes a query (M, PK_j) to O_{PSig} where $PK_j \neq PK_B$, D forwards this query to its own partial signing oracle, and returns the obtained answer back to B. Now consider the ℓ-th query of the form (M, PK_B) made by B to O_{PSig}. If $\ell < q + 1 - i^*$, D forwards it to its own oracle, and returns the obtained answer. If $\ell = q + 1 - i^*$, D requests its challenger for the challenge partial signature σ_P^* on M and returns it to B. If $\ell > q + 1 - i^*$, D simply uses SK_B to produce a partial signature on M. At the end of the simluation, when B outputs (M^*, σ_F^*), if B succeeds in the experiment, D outputs 0; otherwise, D outputs 1.

It's easy to see that D guesses the correct i^* with probability at least $1/q$. Now suppose that D's guess of i^* is correct. If σ_P^* was generated by D's challenger using SK_A, i.e. $b = 0$, the view of B is identical to that in $\mathsf{Ex}^{(i^*-1)}$. On the other side, if σ_P^* was generated using SK_B, i.e. $b = 1$, the view of B is identical to that in $\mathsf{Ex}^{(i^*)}$. Let b' be the bit output by D. Since D outputs 0 only if B succeeds in the experiment, we have $\Pr[b' = 0|b = 0] = \epsilon_{i^*-1}$ and $\Pr[b' = 0|b = 1] = \epsilon_{i^*}$. Therefore, the advantage of D in attacking the weak signer ambiguity over random guess is

$$\begin{aligned}\left|\Pr[b' = b] - \frac{1}{2}\right| &= \left|\Pr[b' = 0 \wedge b = 0] + \Pr[b' = 1 \wedge b = 1] - \frac{1}{2}\right| \\ &= \left|\Pr[b' = 0 \wedge b = 0] + \left(\Pr[b = 1] - \Pr[b' = 0 \wedge b = 1]\right) - \frac{1}{2}\right| \\ &= \frac{1}{2}\left|\Pr[b' = 0|b = 0] - \Pr[b' = 0|b = 1]\right| \\ &\geq \frac{1}{2q}\left|\epsilon_{i^*-1} - \epsilon_{i^*}\right| \geq \frac{1}{2q^2}\left|\epsilon_0 - \epsilon_q\right|\end{aligned}$$

which is also non-negligible. This contradicts the weak signer ambiguity assumption. □

Corollary 1. *In A-OFE, signer ambiguity (Def. 2) and security against the arbitrator (Def. 5) together imply weak security against verifiers.*

Letting an adversary select the two challenge public keys gives the adversary more power in attacking signer ambiguity. Therefore, signer ambiguity defined in Sec. 2.1 is at least as strong as the weak signer ambiguity. Hence this corollary follows directly the theorem above.

3 Preliminaries

(Admissible Pairings): Let $\mathbb{G}_1$ and $\mathbb{G}_T$ be two cyclic groups of large prime order p. $\hat{e}$ is an *admissible pairing* if $\hat{e} : \mathbb{G}_1 \times \mathbb{G}_1 \to \mathbb{G}_T$ is a map with the following properties: (1) *Bilinear*: $\forall R, S \in \mathbb{G}_1$ and $\forall a, b \in \mathbb{Z}$, $\hat{e}(R^a, S^b) = \hat{e}(R, S)^{ab}$; (2) *Non-degenerate*: $\exists R, S \in \mathbb{G}_1$ such that $\hat{e}(R, S) \neq 1$; and (3) *Computable*: there exists an efficient algorithm for computing $\hat{e}(R, S)$ for any $R, S \in \mathbb{G}_1$.

(**Decision Linear Assumption (DLN)**[8]:) Let $\mathbb{G}_1$ be a cyclic group of large prime order p. The Decision Linear Assumption for $\mathbb{G}_1$ holds if for any PPT adversary $\mathcal{A}$, the following probability is negligibly close to 1/2.

$$\Pr[F, H, W \leftarrow \mathbb{G}_1; r, s \leftarrow \mathbb{Z}_p; Z_0 \leftarrow W^{r+s}; Z_1 \leftarrow \mathbb{G}_1; d \leftarrow \{0,1\} : \mathcal{A}(F, H, W, F^r, H^s, Z_d) = d]$$

(**q-Strong Diffie-Hellman Assumption (q-SDH)**[7]): The q-SDH problem in $\mathbb{G}_1$ is defined as follows: given a $(q+1)$-tuple $(g, g^x, g^{x^2}, \cdots, g^{x^q})$, output a pair $(g^{1/(x+c)}, c)$ where $c \in \mathbb{Z}_p^*$. The q-SDH assumption holds if for any PPT adversary $\mathcal{A}$, the following probability is negligible.

$$\Pr\left[x \leftarrow \mathbb{Z}_p^* : \mathcal{A}(g, g^x, \cdots, g^{x^q}) = (g^{\frac{1}{x+c}}, c)\right]$$

4 Ambiguous OFE without Random Oracles

In this section, we propose an A-OFE scheme, which is based on Groth and Sahai's idea of constructing a fully anonymous group signature scheme [15, 16]. Before describing the scheme, we first describe our construction in a high level.

4.1 High Level Description of Our Construction

As mentioned in the introduction part, many OFE schemes in the literature follows a generic framework: Alice encrypts her signature under the arbitrator's public key, and then provides a proof showing that the ciphertext indeed contains her signature on the message. To extend this framework to ambiguous optimistic fair exchange, we let Alice encrypt her signature under the arbitrator's public key and provide a proof showing that the ciphertext contains either her signature on the message or Bob's signature on it. Therefore, given Alice's partial signature, Bob cannot convince others that Alice was committed herself to something, as he can also generate this signature.

Our concrete construction below follows the aforementioned framework, which is based on the idea of Groth in constructing a fully anonymous group signature scheme [15]. In more details, Alice's signature consists of a weakly secure BB-signature [7] and a strong one-time signature. Since only the BB-signature is related to Alice's identity, we encrypt it under the arbitrator's public key using Kiltz' tag-based encryption scheme [20], with the one-time verification key as the tag. The non-interactive proof is based on a newly developed technique by Groth and Sahai [16], which is efficient and doesn't require any complex NP-reduction. The proof consists of two parts. The first part includes a commitment to Alice's BB-signature along with a non-interactive witness indistinguishable (NIWI) proof showing that either Alice's BB-signature or Bob's BB-signature on the one-time verification key is in the commitment. The second part is non-interactive zero-knowledge (NIZK) proof (of knowledge) showing that the commitment and the ciphertext contains the same thing. These two parts together imply that the ciphertext contains a BB-signature on the message generated by either Alice or Bob. Both the ciphertext and the proof are authenticated using the one-time signing key. Guaranteed by the strong unforgeability of the one-time signature, no efficient adversary can modify the ciphertext or the proof.

The NIWI proof system consists of four (PPT) algorithms, K_{NI}, P_{WI}, V_{WI} and X_{xk}, where K_{NI} is the key generation algorithm which outputs a common reference string $\mathtt{crs}$ and an extraction key xk; P_{WI} takes as input $\mathtt{crs}$, the statement to be proved x, and a corresponding witness w, and outputs a proof π; V_{WI} is the corresponding verification algorithm; and X_{xk} takes as input $\mathtt{crs}$ and a valid proof π, outputs a witness w'. The NIZK proof shares the same common reference string with the NIWI proof. P_{ZK} and V_{ZK} are the proving and verification algorithms of the NIZK proof system respectively. Due to the page limit, we refer readers to [16] for detained information about the non-interactive proofs and to [15] for an introduction to the building tools needed for our construction.

4.2 The Scheme

Now we propose our A-OFE scheme. It works as follows:

- PMGen takes 1^k and outputs $\mathsf{PM} = (1^k, p, \mathbb{G}_1, \mathbb{G}_T, \hat{\mathsf{e}}, g)$ so that $\mathbb{G}_1$ and $\mathbb{G}_T$ are cyclic groups of prime order p; g is a random generator of $\mathbb{G}_1$; $\hat{\mathsf{e}} : \mathbb{G}_1 \times \mathbb{G}_1 \rightarrow$

$\mathbb{G}_T$ is an admissible bilinear pairing; and group operations on $\mathbb{G}_1$ and $\mathbb{G}_T$ can be efficiently performed.
- $\mathsf{Setup}^{\mathsf{TTP}}$: The arbitrator runs the key generation algorithm of the non-interactive proof system to generate a common reference string $\mathtt{crs}$ and an extraction key xk, i.e. $(\mathtt{crs}, xk) \leftarrow K_{NI}(1^k)$, where $\mathtt{crs} = (F, H, U, V, W, U', V', W')$. It also randomly selects $K, L \leftarrow \mathbb{G}_1$, and sets $(APK, ASK) = ((\mathtt{crs}, K, L), xk)$, where F, H, K, L together form the public key of the tag-based encryption scheme [20], and xk is the extraction key of the NIWI proof system [15, 16], which is also the decryption key of the tag-based encryption scheme.
- $\mathsf{Setup}^{\mathsf{User}}$: Each user U_i randomly selects $x_i \leftarrow \mathbb{Z}_p$, and sets $(PK_i, SK_i) = (g^{x_i}, x_i)$.
- PSig: To partially sign a message m with verifier U_j, user U_i does the following:
  1. call the key generation algorithm of $\mathcal{S}$ to generate a one-time key pair $(otvk, otsk)$;
  2. use SK_i to compute a BB-signature $\overline{\sigma}$ on $\mathtt{H}(otvk)$, i.e. $\overline{\sigma} \leftarrow g^{\frac{1}{x_i + \mathtt{H}(otvk)}}$;
  3. compute an NIWI proof π_1 showing that $\overline{\sigma}$ is a valid signature under either PK_i or PK_j, i.e. $\pi_1 \leftarrow P_{WI}(\mathtt{crs}, (\hat{\mathsf{e}}(g,g), PK_i, PK_j, \mathtt{H}(otvk)), (\overline{\sigma}))$, which shows that the following holds:
$$\hat{\mathsf{e}}(\overline{\sigma}, PK_i \cdot g^{\mathtt{H}(otvk)}) = \hat{\mathsf{e}}(g,g) \vee \hat{\mathsf{e}}(\overline{\sigma}, PK_j \cdot g^{\mathtt{H}(otvk)}) = \hat{\mathsf{e}}(g,g)$$
  4. compute a tag-based encryption ([20]) y of $\overline{\sigma}$, i.e. $y = (y_1, y_2, y_3, y_4, y_5) \leftarrow \mathcal{E}.E_{pk}(\overline{\sigma}, \mathsf{tag})$, where $pk = (F, H, K, L)$ and $\mathsf{tag} = \mathtt{H}(otvk)$;
  5. compute an NIZK proof π_2 showing that y and the commitment C to $\overline{\sigma}$ in π_1 contain the same $\overline{\sigma}$, i.e. $\pi_2 \leftarrow P_{ZK}(\mathtt{crs}, (y, \pi_1), (r, s, t))$;
  6. use $otsk$ to sign the whole transcript and the message M, i.e. $\sigma_{ot} \leftarrow \mathcal{S}.S_{otsk}(M, \pi_1, y, \pi_2)$.

  The partial signature σ_P of U_i on message M then consists of $(otvk, \sigma_{ot}, \pi_1, y, \pi_2)$.
- PVer: After obtaining U_i's partial signature $\sigma_P = (otvk, \sigma_{ot}, \pi_1, y, \pi_2)$, the verifier U_j checks the following. If any one fails, U_j rejects; otherwise, it accepts.
  1. if σ_{ot} is a valid one-time signature on (M, π_1, y, π_2) under $otvk$;
  2. if π_1 is a valid NIWI proof, i.e. $V_{WI}(\mathtt{crs}, (\hat{\mathsf{e}}(g,g), PK_i, PK_j, \mathtt{H}(otvk)), \pi_1) \stackrel{?}{=} \mathsf{accept}$;
  3. if π_2 is a valid NIZK proof, i.e. $V_{ZK}(\mathtt{crs}, (y, \pi_1), \pi_2) \stackrel{?}{=} \mathsf{accept}$;
- Sig: To sign a message M with verifier U_j, user U_i generates a partial signature σ_P as in PSig, and set the full signature σ_F as $\sigma_F = (\sigma_P, \overline{\sigma})$.
- Ver: After receiving σ_F on M from U_i, user U_j checks if $\mathsf{PVer}(M, \sigma_P, \{PK_i, PK_j\}, APK) \stackrel{?}{=} \mathsf{accept}$, and if $\hat{\mathsf{e}}(\overline{\sigma}, PK_i \cdot g^{\mathtt{H}(otvk)}) \stackrel{?}{=} \hat{\mathsf{e}}(g,g)$. If any of the checks fails, U_j rejects; otherwise, it accepts.
- Res: After receiving U_i's partial signature σ_P on message M from user U_j, the arbitrator firstly checks the validity of σ_P. If invalid, it returns $\perp$ to U_j. Otherwise, it extracts $\overline{\sigma}$ from π_1 by calling $\overline{\sigma} \leftarrow X_{xk}(\mathtt{crs}, \pi_1)$. The arbitrator returns $\overline{\sigma}$ to U_j.

5 Security Analysis

Theorem 2. *The proposed A-OFE scheme is secure in the multi-user setting and chosen-key model (without random oracle) provided that DLN assumption and q-SDH assumption hold.*

Intuitively, the resolution ambiguity is guaranteed by the extractability and soundness of the NIWI proof of knowledge system. The signer ambiguity and security against verifiers are due to the CCA security of the encryption scheme. Security against signers and security against the arbitrator are guaranteed by the (weak) unforgeability of BB-signature scheme. Due to the page limit, we leave the detailed proof in the full version of this paper.

Remark 3. In our construction, the signer uses its secret key to generate a BB-signature on a fresh one-time verification key, while the message is signed using the corresponding one-time signing key. As shown by Huang et al. in [17], this combination leads to a strongly unforgeable signature scheme. It's not hard to see that our proposed A-OFE scheme actually achieves a stronger version of security against the verifier. That is, even if the adversary sees the signer U_A's full signature σ_F on a message M with verifier U_B, it cannot generate another σ'_F on M such that $\mathsf{Ver}(M, \sigma'_F, PK_A, PK_B, APK) = \mathsf{accept}$. The claim can be shown using the proof given in this paper without much modification.

(*Comparison*): We note that schemes proposed in [14, 22] have similar properties as our ambiguous OFE, i.e. (online, offline) *non-transferability*. Here we make a brief comparison with these two schemes. First of all, our A-OFE scheme is better than them in terms of the level of non-transferability. In [14, 22], the non-transferability is defined only in the CPA fashion. The adversary is not given an oracle for converting a partial signature to a full one. While in our definition of A-OFE, we define the ambiguity in the CCA fashion, allowing the adversary to ask for resolving a partial signature to a full one. Second, in terms of efficiency, our scheme outperforms the scheme proposed in [22], and is slightly slower than [14]. The generation of a partial signature of their scheme requires linear (in security parameter k) number of encryptions, and the size of a partial signature is also linear in k. While in our scheme both the computation cost and size of a partial signature are constant. The partial signature of ousr scheme includes about 41 group elements plus a one-time verification key and a one-time signature. Third, both our scheme and the scheme in [14] only require one move in generating a partial signature, while the scheme in [22] requires four moves. Fourth, in [22], there is a setup phase between each signer and the confirmer, in which the confirmer generates an encryption key pair for each signer. Therefore, the confirmer has to store a key pair for each signer, leading to a large storage. While our scheme and [14] don't need such a phase. Fifth, in terms of security, our scheme and [22] are provably secure without random oracles. But the scheme in [14] is only provably secure in the random oracle model.

Acknowledgements

We are grateful to the anonymous reviewers of Asiacrypt 2008 for their invaluable comments. The first three authors were supported by a grant from the Research Grants Council of the Hong Kong Special Administrative Region, China (RGC Ref. No. CityU 122107).

References

1. Asokan, N., Schunter, M., Waidner, M.: Optimistic protocols for fair exchange. In: CCS, pp. 7–17. ACM, New York (1997)
2. Asokan, N., Shoup, V., Waidner, M.: Optimistic fair exchange of digital signatures (extended abstract). In: Nyberg, K. (ed.) EUROCRYPT 1998. LNCS, vol. 1403, pp. 591–606. Springer, Heidelberg (1998)
3. Asokan, N., Shoup, V., Waidner, M.: Optimistic fair exchange of digital signatures. IEEE Journal on Selected Areas in Communication 18(4), 593–610 (2000)
4. Bao, F., Wang, G., Zhou, J., Zhu, H.: Analysis and improvement of Micali's fair contract signing protocol. In: Wang, H., Pieprzyk, J., Varadharajan, V. (eds.) ACISP 2004. LNCS, vol. 3108, pp. 176–187. Springer, Heidelberg (2004)
5. Bellare, M., Rogaway, P.: Random oracles are practical: A paradigm for designing efficient protocols. In: ACM CCS, pp. 62–73. ACM, New York (1993)
6. Bender, A., Katz, J., Morselli, R.: Ring signatures: Stronger definitions, and constructions without random oracles. In: Halevi, S., Rabin, T. (eds.) TCC 2006. LNCS, vol. 3876, pp. 60–79. Springer, Heidelberg (2006), `http://eprint.iacr.org/`
7. Boneh, D., Boyen, X.: Short signatures without random oracles. In: Cachin, C., Camenisch, J.L. (eds.) EUROCRYPT 2004. LNCS, vol. 3027, pp. 56–73. Springer, Heidelberg (2004)
8. Boneh, D., Boyen, X., Shacham, H.: Short group signatures. In: Franklin, M. (ed.) CRYPTO 2004. LNCS, vol. 3152, pp. 41–55. Springer, Heidelberg (2004)
9. Boneh, D., Gentry, C., Lynn, B., Shacham, H.: Aggregate and verifiably encrypted signatures from bilinear maps. In: Biham, E. (ed.) EUROCRYPT 2003. LNCS, vol. 2656, pp. 416–432. Springer, Heidelberg (2003)
10. Boyd, C., Foo, E.: Off-line fair payment protocols using convertible signatures. In: Ohta, K., Pei, D. (eds.) ASIACRYPT 1998. LNCS, vol. 1514, pp. 271–285. Springer, Heidelberg (1998)
11. Chen, L., Kudla, C., Paterson, K.G.: Concurrent signatures. In: Cachin, C., Camenisch, J.L. (eds.) EUROCRYPT 2004. LNCS, vol. 3027, pp. 287–305. Springer, Heidelberg (2004)
12. Dodis, Y., Lee, P.J., Yum, D.H.: Optimistic fair exchange in a multi-user setting. In: Okamoto, T., Wang, X. (eds.) PKC 2007. LNCS, vol. 4450, pp. 118–133. Springer, Heidelberg (2007)
13. Dodis, Y., Reyzin, L.: Breaking and repairing optimistic fair exchange from PODC 2003. In: DRM 2003, pp. 47–54. ACM, New York (2003)
14. Garay, J.A., Jakobsson, M., MacKenzie, P.: Abuse-free optimistic contract signing. In: Wiener, M. (ed.) CRYPTO 1999. LNCS, vol. 1666, pp. 449–466. Springer, Heidelberg (1999)
15. Groth, J.: Fully anonymous group signatures without random oracles. In: Kurosawa, K. (ed.) ASIACRYPT 2007. LNCS, vol. 4833, pp. 164–180. Springer, Heidelberg (2007)

16. Groth, J., Sahai, A.: Efficient non-interactive proof systems for bilinear groups. In: Smart, N.P. (ed.) EUROCRYPT 2008. LNCS, vol. 4965, pp. 415–432. Springer, Heidelberg (2008)
17. Huang, Q., Wong, D.S., Li, J., Zhao, Y.: Generic transformation from weakly to strongly unforgeable signatures. Journal of Computer Science and Technology 23(2), 240–252 (2008)
18. Huang, Q., Yang, G., Wong, D.S., Susilo, W.: Efficient optimistic fair exchange secure in the multi-user setting and chosen-key model without random oracles. In: Malkin, T.G. (ed.) CT-RSA 2008. LNCS, vol. 4964, pp. 106–120. Springer, Heidelberg (2008)
19. Jakobsson, M., Sako, K., Impagliazzo, R.: Designated verifier proofs and their applications. In: Maurer, U.M. (ed.) EUROCRYPT 1996. LNCS, vol. 1070, pp. 143–154. Springer, Heidelberg (1996)
20. Kiltz, E.: Chosen-ciphertext security from tag-based encryption. In: Halevi, S., Rabin, T. (eds.) TCC 2006. LNCS, vol. 3876, pp. 581–600. Springer, Heidelberg (2006)
21. Kremer, S.: Formal Analysis of Optimistic Fair Exchange Protocols. PhD thesis, Université Libre de Bruxelles (2003)
22. Liskov, M., Micali, S.: Online-untransferable signatures. In: Cramer, R. (ed.) PKC 2008. LNCS, vol. 4939, pp. 248–267. Springer, Heidelberg (2008)
23. Micali, S.: Simple and fast optimistic protocols for fair electronic exchange. In: PODC 2003, pp. 12–19. ACM, New York (2003)
24. Park, J.M., Chong, E.K., Siegel, H.J.: Constructing fair-exchange protocols for e-commerce via distributed computation of RSA signatures. In: PODC 2003, pp. 172–181. ACM, New York (2003)
25. Zhu, H., Bao, F.: Stand-alone and setup-free verifiably committed signatures. In: Pointcheval, D. (ed.) CT-RSA 2006. LNCS, vol. 3860, pp. 159–173. Springer, Heidelberg (2006)
26. Zhu, H., Susilo, W., Mu, Y.: Multi-party stand-alone and setup-free verifiably committed signatures. In: Okamoto, T., Wang, X. (eds.) PKC 2007. LNCS, vol. 4450, pp. 134–149. Springer, Heidelberg (2007)

Compact Proofs of Retrievability

Hovav Shacham[1] and Brent Waters[2,⋆]

[1] University of California, San Diego
hovav@cs.ucsd.edu
[2] University of Texas, Austin
bwaters@csl.sri.com

Abstract. In a proof-of-retrievability system, a data storage center convinces a verifier that he is actually storing all of a client's data. The central challenge is to build systems that are both efficient and *provably* secure – that is, it should be possible to extract the client's data from any prover that passes a verification check. In this paper, we give the first proof-of-retrievability schemes with full proofs of security against *arbitrary* adversaries in the strongest model, that of Juels and Kaliski. Our first scheme, built from BLS signatures and secure in the random oracle model, has the *shortest query and response* of any proof-of-retrievability with public verifiability. Our second scheme, which builds elegantly on pseudorandom functions (PRFs) and is secure in the standard model, has the *shortest response* of any proof-of-retrievability scheme with private verifiability (but a longer query). Both schemes rely on homomorphic properties to aggregate a proof into one small authenticator value.

1 Introduction

In this paper, we give the first proof-of-retrievability schemes with full proofs of security against *arbitrary* adversaries in the Juels-Kaliski model. Our first scheme has the shortest query and response of any proof-of-retrievability with public verifiability and is secure in the random oracle model. Our second scheme has the shortest response of any proof-of-retrievability scheme with private verifiability (but a longer query), and is secure in the standard model.

Proofs of storage. Recent visions of "cloud computing" and "software as a service" call for data, both personal and business, to be stored by third parties, but deployment has lagged. Users of outsourced storage are at the mercy of their storage providers for the continued availability of their data. Even Amazon's S3, the best-known storage service, has recently experienced significant downtime.[1]

In an attempt to aid the deployment of outsourced storage, cryptographers have designed systems that would allow users to verify that their data is still

⋆ Supported by NSF CNS-0749931, CNS-0524252, CNS-0716199; the US Army Research Office under the CyberTA Grant No. W911NF-06-1-0316; and the U.S. Department of Homeland Security under Grant Award Number 2006-CS-001-000001.

[1] See, e.g., `http://blogs.zdnet.com/projectfailures/?p=602`

J. Pieprzyk (Ed.): ASIACRYPT 2008, LNCS 5350, pp. 90–107, 2008.
© International Association for Cryptologic Research 2008

available and ready for retrieval if needed: Deswarte, Quisquater, and Saïdane [8], Filho and Barreto [9], and Schwarz and Miller [15]. In these systems, the client and server engage in a protocol; the client seeks to be convinced by the protocol interaction that his file is being stored. Such a capability can be important to storage providers as well. Users may be reluctant to entrust their data to an unknown startup; an auditing mechanism can reassure them that their data is indeed still available.

Evaluation: formal security models. Such proof-of-storage systems should be evaluated by both "systems" and "crypto" criteria. Systems criteria include: (1) the system should be as *efficient* as possible in terms of both computational complexity and communication complexity of the proof-of-storage protocol, and the storage overhead on the server should be as small as possible; (2) the system should allow *unbounded use* rather than imposing a priori bound on the number of audit-protocol interactions[2]; (3) verifiers should be *stateless*, and not need to maintain and update state between audits, since such state is difficult to maintain if the verifier's machine crashes or if the verifier's role is delegated to third parties or distributed among multiple machine.[3] Statelessness and unbounded use are required for proof-of-storage systems with *public verifiability*, in which anyone can undertake the role of verifier in the proof-of-storage protocol, not just the user who originally stored the file.[4]

The most important crypto criterion is this: Whether the protocol actually establishes that any server that passes a verification check for a file – even a malicious server that exhibits arbitrary, Byzantine behavior – is *actually storing the file*. The early cryptographic papers lacked a formal security model, let alone proofs. But provable security matters. Even reasonable-looking protocols could in fact be insecure; see Appendix C of the full paper [16] for an example.

The first papers to consider formal models for proofs of storage were by Naor and Rothblum, for "authenticators" [14], and by Juels and Kaliski, for "proofs of retrievability" [12]. Though the details of the two models are different, the insight behind both is the same: in a secure system if a server can pass an audit then a special extractor algorithm, interacting with the server, must be able (w.h.p.) to extract the file.[5]

---

[2] We believe that systems allowing a bounded number of interactions can be useful, but only as stepping stones towards fully secure systems. Some examples are bounded identity-based encryption [11] and bounded CCA-secure encryption [7]; in these systems, security is maintained only as long as the adversary makes at most t private key extraction or decryption queries.

[3] We note that the sentinel-based scheme of Juels and Kaliski [12], the scheme of Ateniese, Di Pietro, Mancini, and Tsudik [3], and the scheme of Shah, Swaminathan and Baker [17] lack both unbounded use and statelessness. We do not consider these schemes further in this paper.

[4] Ateniese et al. [1] were the first to consider public verifiability for proof-of-storage schemes.

[5] This is, of course, similar to the intuition behind proofs of knowledge.

A simple MAC-based construction. In addition, the Naor-Rothblum and Juels-Kaliski papers describe similar proof-of-retrievability protocols. The insight behind both is that checking that *most* of a file is stored is easier than checking that *all* is. If the file to be stored is first encoded redundantly, and each block of the encoded file is authenticated using a MAC, then it is sufficient for the client to retrieve retrieves a few blocks together with their MACs and check, using his secret key, that these blocks are correct. Naor and Rothblum prove their scheme secure in their model.[6] The simple protocol obtained here uses techniques similar to those proposed by Lillibridge et al. [13]. Signatures can be used instead of MACs to obtain public verifiability.

The downside to this simple solution is that the server's response consists of λ block-authenticator pairs, where λ is the security parameter. If each authenticator is λ bits long, as required in the Juels-Kaliski model, then the response is $\lambda^2 \cdot (s+1)$ bits, where the ratio of file block to authenticator length is $s:1$.[7]

Homomorphic authenticators. The proof-of-storage scheme described by Ateniese et al. [1] improves on the response length of the simple MAC-based scheme using *homomorphic authenticators.* In their scheme, the authenticators σ_i on each file block m_i are constructed in such a way that a verifier can be convinced that a linear combination of blocks $\sum_i \nu_i m_i$ (with arbitrary weights $\{\nu_i\}$) was correctly generated using an authenticator computed from $\{\sigma_i\}$.[8]

When using homomorphic authenticators, the server can combine the blocks and λ authenticators in its response into a single aggregate block and authenticator, reducing the response length by a factor of λ. As an additional benefit, the Ateniese et al. scheme is the first with public verifiability. The homomorphic authenticators of Ateniese et al. are based on RSA and are thus relatively long.

Unfortunately, Ateniese et al. do not give a rigorous proof of security for their scheme. In particular, they do not show that one can extract a file (or even a significant fraction of one) from a prover that is able to answer auditing queries convincingly. The need for rigor in extraction arguments applies equally to both the proof-of-retrievability model we consider and the weaker proof of data possession model considered by Ateniese et al.[9]

Our contributions. In this paper, we make two contributions.

1. We describe two new short, efficient homomorphic authenticators. The first, based on PRFs, gives a proof-of-retrievability scheme secure in the

[6] Juels and Kaliski do not give a proof of security against arbitrary adversaries, but this proof is trivial using the techniques we develop in this paper; for completeness, we give the proof in Appendix D of the full paper [16].

[7] Naor and Rothblum show that one-bit MACs suffice for proving security in their less stringent model, for an overall response length of $\lambda \cdot (s+1)$ bits. The Naor-Rothblum scheme is not secure in the Juels-Kaliski model.

[8] In the Ateniese et al. construction the aggregate authenticator is $\prod_i \sigma_i^{\nu_i} \bmod N$.

[9] For completeness, we give a correct and fully proven Ateniese-et-al.–inspired, RSA-based scheme, together with a full proof of security, in Appendix E of the full paper [16].

standard model. The second, based on BLS signatures [5], gives a proof-of-retrievability scheme with public verifiability secure in the random oracle model.
2. We prove both of the resulting schemes secure in a variant of the Juels-Kaliski model. Our schemes are the first with a security proof against arbitrary adversaries in this model.

The scheme with public retrievability has the shortest query and response of any proof-of-retrievability scheme: 20 bytes and 40 bytes, respectively, at the 80-bit security level. The scheme with private retrievability has the shortest response of any proof-of-retrievability scheme (20 bytes), matching the response length of the Naor-Rothblum scheme in a more stringent security model, albeit at the cost of a longer query. We believe that derandomizing the query in this scheme is the major remaining open problem for proofs of retrievability.

1.1 Our Schemes

In our schemes, as in the Juels-Kaliski scheme, the user breaks an erasure encoded file into n blocks $m_1, \ldots, m_n \in Z_p$ for some large prime p. The erasure code should allow decoding in the presence of *adversarial* erasure. Erasure codes derived from Reed-Solomon codes have this property, but decoding and encoding are slow for large files. In Appendix B of the full paper [16] we discuss how to make use of more efficient codes secure only against random erasures.

The user authenticates each block as follows. She chooses a random $\alpha \in \mathbb{Z}_p$ and PRF key k for function f. These values serve as her secret key. She calculates an authentication value for each block i as

$$\sigma_i = f_k(i) + \alpha m_i \in \mathbb{Z}_p \ .$$

The blocks $\{m_i\}$ and authenticators $\{\sigma_i\}$ are stored on the server. The proof of retrievability protocol is as follows. The verifier chooses a random challenge set I of l indices along with l random coefficients in Z_p.[10] Let Q be the set $\{(i, \nu_i)\}$ of challenge index–coefficient pairs. The verifier sends Q to the prover. The prover then calculates the response, a pair (σ, μ), as

$$\sigma \leftarrow \sum_{(i,\nu_i)\in Q} \nu_i \cdot \sigma_i \qquad \text{and} \qquad \mu \leftarrow \sum_{(i,\nu_i)\in Q} \nu_i \cdot m_i \ .$$

Now verifier can check that the response was correctly formed by checking that

$$\sigma \stackrel{?}{=} \alpha \cdot \mu + \sum_{(i,\nu_i)\in Q} \nu_i \cdot f_k(i) \ .$$

It is clear that our techniques admit short responses. But it is not clear that our new system admits a simulator that can extract files. Proving that it does is quite challenging, as we discuss below. In fact, unlike similar, seemingly correct schemes (see Appendix C of the full paper [16]), our scheme is provably secure in the standard model.

[10] Or, more generally, from a subset B of $\mathbb{Z}_p$ of appropriate size; see Section 1.1.

A scheme with public verifiability. Our second scheme is publicly verifiable. It follows the same framework as the first, but instead uses BLS signatures [5] for authentication values that can be publicly verified. The structure of these signatures allows for them to be aggregated into linear combinations as above. We prove the security of this scheme under the Computational Diffie-Hellman assumption over bilinear groups in the random oracle model.

Let $e\colon G \times G \to G_T$ be a computable bilinear map with group G's support being $\mathbb{Z}_p$. A user's private key is $x \in \mathbb{Z}_p$, and her public key is $v = g^x \in G$ along with another generator $u \in G$. The signature on block i is $\sigma_i = \left[H(i)u^{m_i}\right]^x$. On receiving query $Q = \{(i, \nu_i)\}$, the prover computes and sends back $\sigma \leftarrow \prod_{(i,\nu_i)\in Q} \sigma_i^{\nu_i}$ and $\mu \leftarrow \sum_{(i,\nu_i)\in Q} \nu_i \cdot m_i$. The verification equation is:

$$e(\sigma, g) \stackrel{?}{=} e\Big(\prod_{(i,\nu_i)\in Q} H(i)^{\nu_i} \cdot u^{\mu},\ v\Big) \ .$$

This scheme has public verifiability: the private key x is required for generating the authenticators $\{\sigma_i\}$ but the *public* key v is sufficient for the verifier in the proof-of-retrievability protocol.

Parameter selection. Let λ be the security parameter; typically, $\lambda = 80$. For the scheme with private verification, p should be a λ bit prime. For the scheme with public verification, p should be a 2λ-bit prime, and the curve should be chosen so that discrete logarithm is 2^λ-secure. For values of λ up to 128, Barreto-Naehrig curves [4] are the right choice; see the survey by Freeman, Scott, and Teske [10].

Let n be the number of blocks in the file. We assume that $n \gg \lambda$. Suppose we use a rate-ρ erasure code, i.e., one in which any ρ-fraction of the blocks suffices for decoding. (Encoding will cause the file length to grow approximately $(1/\rho)\times$.) Let l be the number of indices in the query Q, and $B \subseteq \mathbb{Z}_p$ be the set from which the challenge weights ν_i are drawn.

Our proofs – see Section 4.2 for the details – guarantee that extraction will succeed from any adversary that convincingly answers an ϵ-fraction of queries, provided that $\epsilon - \rho^l - 1/\#B$ is non-negligible in λ. It is this requirement that guides the choice of parameters.

A conservative choice is $\rho = 1/2$, $l = \lambda$, and $B = \{0, 1\}^\lambda$; this guarantees extraction against any adversary.[11] For applications that can tolerate a larger error rate these parameters can be reduced. For example, if a 1-in-1,000,000 error is acceptable, we can take B to be the set of 22-bit strings and l to be 22; alternatively, the coding expansion $1/\rho$ can be reduced.

A tradeoff between storage and communication. As we described our schemes above, each file block is accompanied by an authenticator of equal length. This gives a $2\times$ overhead beyond that imposed by the erasure code, and the server's

[11] The careful analysis in our proofs allows us to show that, for 80-bit security, the challenge coefficients ν_i can be 80 bits long, not 160 as proposed in [2, p. 17]. The smaller these coefficients, the more efficient the multiplications or exponentiations that involve them.

response in the proof-of-retrievability protocol is $2\times$ the length of an authenticator. In the full schemes of Section 3, we introduce a parameter s that gives a tradeoff between storage overhead and response length. Each block consists of s elements of $\mathbb{Z}_p$ that we call *sectors*. There is one authenticator per block, reducing the overhead to $(1 + 1/s)\times$. The server's response is one aggregated block and authenticator, and is $(1+s)\times$ as long as an authenticator. The choice $s = 1$ corresponds to our schemes as we described them above and to the scheme given by Ateniese et al. [1].[12]

Compressing the request. A request, as we have seen, consists of an l element subset of $[1, n]$ together with l elements of the coefficient set B, chosen uniformly and independently at random. In the conservative parametrization above, a request is thus $\lambda \cdot (\lceil \lg n \rceil + \lambda)$ bits long. One can reduce the randomness required to *generate* the request using standard techniques,[13] but this will not shorten the request itself. In the random oracle model, the verifier can send a short (2λ bit) seed for the random oracle from which the prover will generate the full query. Using this technique we can make the queries as well as responses compact in our publicly verifiable scheme, which already relies on random oracles.[14] Obtaining short queries in the standard model is the major remaining open problem in proofs of retrievability.

We note that, by techniques similar to those discussed above, a PRF can be used to generate the per-file secret values $\{\alpha_j\}$ for our privately verifiable scheme and a random oracle seed can be used to generate the per-file public generators $\{u_j\}$ in our publicly verifiable scheme. This allows file tags for both schemes to be short: $O(\lambda)$, asymptotically.

We also note that subsequent to our work Bowers, Juels, and Oprea [6] provided a framework, based on "inner and outer" error correcting codes, by which they describe parameterizations of our approach that trade off the cost of a single audit and the computational efficiency of extracting a file a series of audit requests. In our work we have chosen to put emphasis on reducing single audit costs. We envision an audit as a mechanism to ensure that a file is indeed available and that a file under most circumstances will be retrieved as a simple bytestream. In a further difference, the error-correcting codes employed by Bowers, Juels, and Oprea are optimized for the case where $\epsilon > 1/2$, i.e., for when the server answers correctly more than half the time. By contrast, our

---

[12] It would be possible to shorten the response further using knowledge-of-exponent assumptions, as Ateniese et al. do, but such assumptions are strong and nonstandard; more importantly, their use means that the extractor can never be implemented in the real world.

[13] For example, choose keys k' and k'' for PRFs with respective ranges $[1, n]$ and B. The query indices are the first l distinct values amongst $f'_{k'}(1), f'_{k'}(2), \ldots$; the query coefficients are $f''_{k''}(1), \ldots, f''_{k''}(l)$.

[14] Ateniese et al. propose to eliminate random oracles here by having the prover generate the full query using PRF keys sent by the verifier [2, p. 11], but it is not clear how to prove such a scheme secure, since the PRF security definition assumes that keys are kept secret.

techniques scale to any small (but nonnegligible) ϵ. We believe that this frees systems implementers from having to worry about whether a substantial error rate (for example, due to an intermitent connection between auditor and server) invalidates the assumptions of the underlying cryptography.

1.2 Our Proofs

We provide a modular proof framework for the security of our schemes. Our framework allows us to argue about the systems unforgeability, extractability, and retrievability with these three parts based respectively on cryptographic, combinatorial, and coding-theoretical techniques. Only the first part differs between the three schemes we propose. The combinatorial techniques we develop are nontrivial and we believe they will be of independent interest.

It is interesting to compare both our security model and our proof methodology to those in related work.

The proof of retrievability model has two major distinctions from that used by Naor and Rothblum [14] (in addition to the public-key setting). First, the NR model assumes a checker can request and receive specific memory locations from the prover. In the proof of retrievability model, the prover can consist of an arbitrary program as opposed to a simple memory layout and this program may answer these questions in an arbitrary manner. We believe that this realistically represents an adversary in the type of setting we are considering. In the NR setting the extractor needs to retrieve the file given the server's memory; in the POR setting the analogy is that the extractor receives the adversary's program.

Second, in the proof of retrievability model we allow the attacker to execute a polynomial number of proof attempts before committing to how it will store memory. In the NR model the adversary does not get to execute the protocol before committing its memory. This weaker model is precisely what allows for the use of 1-bit MACs with error correcting codes in one NR variant. One might argue that in many situations this is sufficient. If a storage server responds incorrectly to an audit request we might assume that it is declared to be cheating and there is no need to go further. However, this limited view overlooks several scenarios. In particular, we want to be able to handle setups where there are several verifiers that do not communicate or if there might be several storage servers handling the same encoded file that are audited independently. Only our stronger model can correctly reflect these situations. In general, we believe that the strongest security model allows for a system to be secure in the most contexts including those not previously considered.[15]

One of the distinctive and challenging parts of our work is to argue extraction from homomorphically accumulated blocks. While Ateniese et al. [1] proposed using homomorphic RSA signatures and proved what is equivalent to our unforgeability requirement, they did not provide an argument that one could extract individual blocks from a prover. The only place where extractability is

[15] We liken this argument to that for the strong definition currently accepted for chosen-ciphertext secure encryption.

addressed in their work is a short paragraph in Appendix A, where they provide some intuitive arguments. Here is one concrete example: Their constructions make multiple uses of pseudorandom functions (PRFs), yet the security properties of a PRF are never applied in a security reduction. This gives compelling evidence that a rigorous security proof was not provided. Again, we emphasize that extraction is needed in even the weaker proof of data possession model claimed by the authors.

Extractability issues arise in several natural constructions. Proving extraction from aggregated authenticator values can be challenging; in Appendix C of the full paper [16] we show an attack on a natural but incorrect system that is very similar to the "E-PDP" efficient alternative scheme given by Ateniese et al. (which they use in their performance measurements). For this scheme, Ateniese et al. claim only that the protocol establishes that a cheating prover has the sum $\sum_{i \in I} m_i$ of the blocks. We show that indeed this is all it can provide. Ateniese et al. calculate that a malicious server attacking the E-PDP scheme et al. that a malicious server attacking the E-PDP scheme would need to store 10^{140} blocks in order to cheat with probability 100%. By contrast, our attack, which allows the server to cheat with somewhat lower probability (almost 9% for standard parameters) requires no more storage than were the server faithfully storing the file.

Finally, we argue that the POR is the "right" model for considering practical data storage problems, since provides a successful audit guarantees that *all* the data can be extracted. Other work has advocated that a weaker Proof of Data Possession [1] model might be acceptable. In this model, one only wants to guarantee that a certain percentage (e.g., 90%) of data blocks are available. By offering this weaker guarantee one might hope to avoid the overhead of applying erasure codes. However, this weaker condition is unsatisfactory for most practical application demands. One might consider how happy a user would be were 10% of a file containing accounting data lost. Or if, for a compressed file, the compression tables were lost – and with them all useful data. Instead of hoping that there is enough redundancy left to reconstruct important data in an ad-hoc way, it is much more desirable to have a model that inherently provides this. We also note that Ateniese et al. [1] make an even weaker guarantee for their "E-PDP" system that they implement and use as the basis for their measurements. According to [1] their E-PDP system "only guarantees possession of the sum of the blocks." While this might be technically correct, it is even more difficult to discern what direct use could come from retrieving a sum of a subset of data blocks.

One might still hope to make use of systems proved secure under these models. For example, we might attempt to make a PDP system usable by adding on an erasure encoding step. In addition, if a system proved that one could be guaranteed sums of blocks for a particular audit, then it *might* be the case that by using multiple audit one could guarantee that individual file blocks could be extracted. However, one must prove that this is the case and account for the additional computational and communication overhead of multiple passes.

When systems use definitions that don't model full retrievability it becomes very difficult to make any useful security or performance comparisons.

2 Security Model

We recall the security definition of Juels and Kaliski [12]. Our version differs from the original definition in several details:

- we rule out any state ("α") in key generation and in verification, because (as explained in Section 1) we believe that verifiers in proof-of-retrievability schemes should be stateless;
- we allow the proof protocol to be arbitrary, rather than two-move, challenge-response; and
- our key generation emits a public key as well as a private key, to allow us to capture the notion of public verifiability.

Note that any stateless scheme secure in the original Juels-Kaliski model will be secure in our variant, and any scheme secure in our variant whose proof protocol can be cast as two-move, challenge-response protocol will be secure in the Juels-Kaliski definition. In particular, our scheme with private verifiability is secure in the original Juels-Kaliski model.[16]

A proof of retrievability scheme defines four algorithms, Kg, St, $\mathcal{V}$, and $\mathcal{P}$, which behave thus:

Kg(). This randomized algorithm generates a public-private keypair (pk, sk).

St(sk, M). This randomized file-storing algorithm takes a secret key sk and a file $M \in \{0,1\}^*$ to store. It processes M to produce and output M^*, which will be stored on the server, and a tag t. The tag contains information that names the file being stored; it could also contain additional secret information encrypted under the secret key sk.

$\mathcal{P}$, $\mathcal{V}$. The randomized proving and verifying algorithms define a protocol for proving file retrievability. During protocol execution, both algorithms take as input the public key pk and the file tag t output by St. The prover algorithm also takes as input the processed file description M^* that is output by St, and the verifier algorithm takes as input the secret key. At the end of the protocol run, $\mathcal{V}$ outputs 0 or 1, where 1 means that the file is being stored on the server. We can denote a run of two machines executing the algorithms as: $\{0,1\} \xleftarrow{\text{R}} \big(\mathcal{V}(pk, sk, t) \rightleftharpoons \mathcal{P}(pk, t, M^*)\big)$.

[16] In an additional minor difference, we do not specify the extraction algorithm as part of a scheme, because we do not expect that the extract algorithm will be deployed in outsourced storage applications. Nevertheless, the extract algorithm used in our proofs (cf. Section 4.2) is quite simple: undertake many random $\mathcal{V}$ interactions with the cheating prover; keep track of those queries for which $\mathcal{V}$ accepts the cheating prover's reply as valid; and continue until enough information has been gathered to recover file blocks by means of linear algebra. The adversary $\mathcal{A}$ could implement this algorithm by means of its proof-of-retrievability protocol access.

We would like a proof-of-retrievability protocol to be correct and sound. Correctness requires that, for all keypairs (pk, sk) output by Kg, for all files $M \in \{0,1\}^*$, and for all (M^*, t) output by $\mathsf{St}(sk, M)$, the verification algorithm accepts when interacting with the valid prover:

$$\big(\mathcal{V}(pk, sk, t) \rightleftharpoons \mathcal{P}(pk, t, M^*)\big) = 1 \ .$$

A proof-of-retrievability protocol is sound if any cheating prover that convinces the verification algorithm that it is storing a file M is actually storing that file, which we define in saying that it yields up the file M to an extractor algorithm that interacts with it using the proof-of-retrievability protocol. We formalize the notion of an extractor and then give a precise definition for soundness.

An extractor algorithm $\mathsf{Extr}(pk, sk, t, \mathcal{P}')$ takes the public and private keys, the file tag t, and the description of a machine implementing the prover's role in the proof-of-retrievability protocol: for example, the description of an interactive Turing machine, or of a circuit in an appropriately augmented model. The algorithm's output is the file $M \in \{0,1\}^*$. Note that Extr is given non–black-box access to $\mathcal{P}'$ and can, in particular, rewind it.

Consider the following setup game between an adversary $\mathcal{A}$ and an environment:

1. The environment generates a keypair (pk, sk) by running Kg, and provides pk to $\mathcal{A}$.
2. The adversary can now interact with the environment. It can make queries to a store oracle, providing, for each query, some file M. The environment computes $(M^*, t) \xleftarrow{\text{R}} \mathsf{St}(sk, M)$ and returns both M^* and t to the adversary.
3. For any M on which it previously made a store query, the adversary can undertake executions of the proof-of-retrievability protocol, by specifying the corresponding tag t. In these protocol executions, the environment plays the part of the verifier and the adversary plays the part of the prover: $\mathcal{V}(pk, sk, t) \rightleftharpoons \mathcal{A}$. When a protocol execution completes, the adversary is provided with the output of $\mathcal{V}$. These protocol executions can be arbitrarily interleaved with each other and with the store queries described above.
4. Finally, the adversary outputs a challenge tag t returned from some store query, and the description of a prover $\mathcal{P}'$.

The cheating prover $\mathcal{P}'$ is *ϵ-admissible* if it convincingly answers an ϵ fraction of verification challenges, i.e., if $\Pr\big[\big(\mathcal{V}(pk, sk, t) \rightleftharpoons \mathcal{P}'\big) = 1\big] \geq \epsilon$. Here the probability is over the coins of the verifier and the prover. Let M be the message input to the store query that returned the challenge tag t (along with a processed version M^* of M).

Definition 1. *We say a proof-of-retrievability scheme is ϵ-sound if there exists an extraction algorithm* Extr *such that, for every adversary $\mathcal{A}$, whenever $\mathcal{A}$, playing the* setup *game, outputs an ϵ-admissible cheating prover $\mathcal{P}'$ for a file M, the extraction algorithm recovers M from $\mathcal{P}'$ – i.e.,* $\mathsf{Extr}(pk, sk, t, \mathcal{P}') = M$ *– except possibly with negligible probability.*

Note that it is okay for $\mathcal{A}$ to have engaged in the proof-of-retrievability protocol for M in its interaction with the environment. Note also that each run of the proof-of-retrievability protocol is independent: the verifier implemented by the environment is stateless.

Finally, note that we require that extraction succeed (with all but negligible probability) from an adversary that causes $\mathcal{V}$ to accept with any nonnegligible probability ϵ. An adversary that passes the verification even a very small but nonnegligible fraction of the time – say, once in a million interactions – is fair game. Intuitively, recovering enough blocks to reconstruct the original file from such an adversary should take $O(n/\epsilon)$ interactions; our proofs achieve essentially this bound.

Concrete or asymptotic formalization. A proof-of-retrievability scheme is secure if no efficient algorithm wins the game above except rarely, where the precise meaning of "efficient" and "rarely" depends on whether we employ a concrete of asymptotic formalization.

It is possible to formalize the notation above either concretely or asymptotically. In a concrete formalization, we require that each algorithm defining the proof-of-retrievability scheme run in at most some number of steps, and that for any algorithm $\mathcal{A}$ that runs in time t steps, that makes at most q_S store queries, and that undertakes at most q_P proof-of-retrievability protocol executions, extraction from an ϵ-admissible prover succeeds except with some small probability δ. In an asymptotic formalization, every algorithm is provided with an additional parameter 1^λ for security parameter λ, we require each algorithm to run in time polynomial in λ, and we require that extraction fail from an ϵ-admissible prover with only negligible probability in λ, provided ϵ is nonnegligible.

Public or private verification, public or private extraction. In the model above, the verifier and extractor are provided with a secret that is not known to the prover or other parties. This is a secret-verification, secret-extraction model model. If the verification algorithm does not use the secret key, any third party can check that a file is being stored, giving public verification. Similarly, if the extract algorithm does not use the secret key, any third party can extract the file from a server, giving public extraction.

3 Constructions

In this section we give formal descriptions for both our private and public verification systems. The systems here follow the constructions outlined in the introduction with a few added generalizations. First, we allow blocks to contain $s \geq 1$ elements of $\mathbb{Z}_p$. This allows for a tradeoff between storage overhead and communication overhead. Roughly the communication complexity grows as $s+1$ elements of $\mathbb{Z}_p$ and the ratio of authentication overhead to data stored (post encoding) is $1 : s$. Second, we describe our systems where the set of coefficients sampled from B can be smaller than all of $\mathbb{Z}_p$. This enables us to take advantage make more efficient systems in certain situations.

3.1 Common Notation

We will work in the group $\mathbb{Z}_p$. When we work in the bilinear setting, the group $\mathbb{Z}_p$ is the support of the bilinear group G, i.e., $\#G = p$. In queries, coefficients will come from a set $B \subseteq \mathbb{Z}_p$. For example, B could equal $\mathbb{Z}_p$, in which case query coefficients will be randomly chosen out of all of $\mathbb{Z}_p$.

After a file undergoes preliminary processing, the processed file is split into *blocks*, and each block is split into *sectors*. Each sector is one element of $\mathbb{Z}_p$, and there are s sectors per block. If the processed file is b bits long, then there are $n = \lceil b/s \lg p \rceil$ blocks. We will refer to individual file sectors as $\{m_{ij}\}$, with $1 \le i \le n$ and $1 \le j \le s$.

Queries. A query is an l-element set $Q = \{(i, \nu_i)\}$. Each entry $(i, \nu_i) \in Q$ is such that i is a block index in the range $[1, n]$, and ν_i is a multiplier in B. The size l of Q is a system parameter, as is the choice of the set B.

The verifier chooses a random query as follows. First, she chooses, uniformly at random, an l-element subset I of $[1, n]$. Then, for each element $i \in I$ she chooses, uniformly at random, an element $\nu_i \xleftarrow{R} B$. We observe that this procedure implies selection of l elements from $[1, n]$ *without* replacement but a selection of l elements from B *with* replacement.

Although the set notation $Q = \{(i, \nu_i)\}$ is space-efficient and convenient for implementation, we will also make use of a vector notation in the analysis. A query Q over indices $I \subset [1, n]$ is represented by a vector $\boldsymbol{q} \in (\mathbb{Z}_p)^n$ where $\boldsymbol{q}_i = \nu_i$ for $i \in I$ and $\boldsymbol{q}_i = 0$ for all $i \notin I$. Equivalently, letting $\boldsymbol{u}_1, \ldots, \boldsymbol{u}_n$ be the usual basis for $(\mathbb{Z}_p)^n$, we have $\boldsymbol{q} = \sum_{(i,\nu_i) \in Q} \nu_i \boldsymbol{u}_i$.[17]

If the set B does not contain 0 then a random query (according to the selection procedure defined above) is a random weight-l vector in $(\mathbb{Z}_p)^n$ with coefficients in B. If B does contain 0, then a similar argument can be made, but care must be taken to distinguish the case "$i \in I$ and $\nu_i = 0$" from the case "$i \notin I$."

Aggregation. For its response, the server responds to a query Q by computing, for each j, $1 \le j \le s$, the value

$$\mu_j \leftarrow \sum_{(i,\nu_i) \in Q} \nu_i m_{ij} \ .$$

That is, by combining sectorwise the blocks named in Q, each with its multiplier ν_i. Addition, of course, is modulo p. The response is $(\mu_1, \ldots, \mu_s) \in (\mathbb{Z}_p)^s$.

Suppose we view the message blocks on the server as an $n \times s$ element matrix $M = (m_{ij})$, then, using the vector notation for queries given above, the server's response is given by $\boldsymbol{q}M$.

[17] We are using subscripts to denote vector elements (for $\boldsymbol{q}$) and to choose a particular vector from a set (for $\boldsymbol{u}$); but no confusion should arise.

3.2 Construction for Private Verification

Let $f\colon \{0,1\}^* \times \mathcal{K}_{\text{prf}} \to \mathbb{Z}_p$ be a PRF.[18] The construction of the private verification scheme Priv is:

Priv.Kg(). Choose a random symmetric encryption key $k_{\text{enc}} \xleftarrow{R} \mathcal{K}_{\text{enc}}$ and a random MAC key $k_{\text{mac}} \xleftarrow{R} \mathcal{K}_{\text{mac}}$. The secret key is $sk = (k_{\text{enc}}, k_{\text{mac}})$; there is no public key.

Priv.St(sk, M)**.** Given the file M, first apply the erasure code to obtain M'; then split M' into n blocks (for some n), each s sectors long: $\{m_{ij}\}_{\substack{1\le i\le n\\ 1\le j\le s}}$. Now choose a PRF key $k_{\text{prf}} \xleftarrow{R} \mathcal{K}_{\text{prf}}$ and s random numbers $\alpha_1, \dots, \alpha_s \xleftarrow{R} \mathbb{Z}_p$. Let t_0 be $n \| \mathsf{Enc}_{k_{\text{enc}}}(k_{\text{prf}} \| \alpha_1 \| \cdots \| \alpha_s)$; the file tag is $t = t_0 \| \mathsf{MAC}_{k_{\text{mac}}}(t_0)$. Now, for each i, $1 \le i \le n$, compute

$$\sigma_i \leftarrow f_{k_{\text{prf}}}(i) + \sum_{j=1}^{s} \alpha_j m_{ij} \ .$$

The processed file M^* is $\{m_{ij}\}$, $1 \le i \le n$, $1 \le j \le s$ together with $\{\sigma_i\}$, $1 \le i \le n$.

Priv.$\mathcal{V}(pk, sk, t)$**.** Parse sk as $(k_{\text{enc}}, k_{\text{mac}})$. Use k_{mac} to verify the MAC on t; if the MAC is invalid, reject by emitting 0 and halting. Otherwise, parse t and use k_{enc} to decrypt the encrypted portions, recovering n, k_{prf}, and $\alpha_1, \dots, \alpha_s$. Now pick a random l-element subset I of the set $[1, n]$, and, for each $i \in I$, a random element $\nu_i \xleftarrow{R} B$. Let Q be the set $\{(i, \nu_i)\}$. Send Q to the prover. Parse the prover's response to obtain $\mu_1, \dots, \mu_s$ and σ, all in $\mathbb{Z}_p$. If parsing fails, fail by emitting 0 and halting. Otherwise, check whether

$$\sigma \stackrel{?}{=} \sum_{(i,\nu_i)\in Q} \nu_i f_{k_{\text{prf}}}(i) + \sum_{j=1}^{s} \alpha_j \mu_j \ ;$$

if so, output 1; otherwise, output 0.

Priv.$\mathcal{P}(pk, t, M^*)$**.** Parse the processed file M^* as $\{m_{ij}\}$, $1 \le i \le n$, $1 \le j \le s$, along with $\{\sigma_i\}$, $1 \le i \le n$. Parse the message sent by the verifier as Q, an l-element set $\{(i, \nu_i)\}$, with the i's distinct, each $i \in [1, n]$, and each $\nu_i \in B$. Compute

$$\mu_j \leftarrow \sum_{(i,\nu_i)\in Q} \nu_i m_{ij} \quad \text{for } 1 \le j \le s, \qquad \text{and} \qquad \sigma \leftarrow \sum_{(i,\nu_i)\in Q} \nu_i \sigma_i \ .$$

Send to the prover in response the values $\mu_1, \dots, \mu_s$ and σ.

---

[18] In fact, the domain need only be $\lceil \lg N \rceil$-bit strings, where N is a bound on the number of blocks in a file.

3.3 Construction for Public Verification

Let $e\colon G \times G \to G_T$ be a bilinear map, let g be a generator of G, and let $H\colon \{0,1\}^* \to G$ be the BLS hash, treated as a random oracle.[19] The construction of the public verification scheme Pub is:

Pub.Kg(). Generate a random signing keypair $(spk, ssk) \xleftarrow{R} \mathsf{SKg}$. Choose a random $\alpha \xleftarrow{R} \mathbb{Z}_p$ and compute $v \leftarrow g^\alpha$. The secret key is $sk = (\alpha, ssk)$; the public key is $pk = (v, spk)$.

Pub.St(sk, M). Given the file M, first apply the erasure code to obtain M'; then split M' into n blocks (for some n), each s sectors long: $\{m_{ij}\}_{\substack{1\le i\le n\\ 1\le j\le s}}$. Now parse sk as (α, ssk). Choose a random file name $name$ from some sufficiently large domain (e.g., $\mathbb{Z}_p$). Choose s random elements $u_1, \dots, u_s \xleftarrow{R} G$. Let t_0 be "$name\|n\|u_1\|\cdots\|u_s$"; the file tag t is t_0 together with a signature on t_0 under private key ssk: $t \leftarrow t_0\|\mathsf{SSig}_{ssk}(t_0)$. For each i, $1 \le i \le n$, compute

$$\sigma_i \leftarrow \Big(H(name\|i) \cdot \prod_{j=1}^{s} u_j^{m_{ij}}\Big)^\alpha .$$

The processed file M^* is $\{m_{ij}\}$, $1 \le i \le n$, $1 \le j \le s$ together with $\{\sigma_i\}$, $1 \le i \le n$.

Pub.$\mathcal{V}$ (pk, sk, t). Parse pk as (v, spk). Use spk to verify the signature on on t; if the signature is invalid, reject by emitting 0 and halting. Otherwise, parse t, recovering $name$, n, and $u_1, \dots, u_s$. Now pick a random l-element subset I of the set $[1, n]$, and, for each $i \in I$, a random element $\nu_i \xleftarrow{R} B$. Let Q be the set $\{(i, \nu_i)\}$. Send Q to the prover.
Parse the prover's response to obtain $(\mu_1, \dots, \mu_s) \in (\mathbb{Z}_p)^s$ and $\sigma \in G$. If parsing fails, fail by emitting 0 and halting. Otherwise, check whether

$$e(\sigma, g) \stackrel{?}{=} e\Big(\prod_{(i,\nu_i)\in Q} H(name\|i)^{\nu_i} \cdot \prod_{j=1}^{s} u_j^{\mu_j},\ v\Big) \ ;$$

if so, output 1; otherwise, output 0.

Pub.$\mathcal{P}$ (pk, t, M^*). Parse the processed file M^* as $\{m_{ij}\}$, $1 \le i \le n$, $1 \le j \le s$, along with $\{\sigma_i\}$, $1 \le i \le n$. Parse the message sent by the verifier as Q, an l-element set $\{(i, \nu_i)\}$, with the i's distinct, each $i \in [1, n]$, and each $\nu_i \in B$. Compute

$$\mu_j \leftarrow \sum_{(i,\nu_i)\in Q} \nu_i m_{ij} \in \mathbb{Z}_p \quad \text{for } 1 \le j \le s, \qquad \text{and} \qquad \sigma \leftarrow \prod_{(i,\nu_i)\in Q} \sigma_i^{\nu_i} \in G .$$

Send to the prover in response the values $\mu_1, \dots, \mu_s$ and σ.

[19] For notational simplicity, we present our scheme using a symmetric bilinear map, but efficient implementations will use an asymmetric map $e\colon G_1 \times G_2 \to G_T$. Translating our scheme to this setting is simple. User public keys v will live in G_2; file generators u_j will live in G_1, as will the output of H; and security will be reduced to co-CDH [5].

4 Security Proofs

In this section we prove that both of our systems are secure under the model we provided. Intutively, we break our proof into three parts. The first part shows that the attacker can never give a forged response back to the a verifier. The second part of the proof shows that from any adversary that passes the check a non-negligible amount of the time we will be able to extract a constant fraction of the encoded blocks. The second step uses the fact that (w.h.p.) all verified responses must be legitimate. Finally, we show that if this constant fraction of blocks is recovered we can use the erasure code to reconstruct the original file.

In this section we provide an outline of our proofs and state our main theorems and lemmas. We defer the proofs of these to the full paper [16]. The proof, for both schemes, is in three parts:

1. Prove that the verification algorithm will reject except when the prover's $\{\mu_j\}$ are correctly computed, i.e., are such that $\mu_j = \sum_{(i,\nu_i)\in Q} \nu_i m_{ij}$. This part of the proof uses cryptographic techniques.
2. Prove that the extraction procedure can efficiently reconstruct a ρ fraction of the file blocks when interacting with a prover that provides correctly-computed $\{\mu_j\}$ responses for a nonnegligible fraction of the query space. This part of the proof uses combinatorial techniques.
3. Prove that a ρ fraction of the blocks of the erasure-coded file suffice for reconstructing the original file. This part of the proof uses coding theory techniques.

Crucially, only the part-one proof is different for our two schemes; the other parts are *identical.*

4.1 Part-One Proofs

Scheme with Private Verifiability

Theorem 1. *If the MAC is unforgeable, the symmetric encryption scheme is semantically secure, and the PRF is secure, then (except with negligible probability) no adversary against the soundness of our private-verification scheme ever causes $\mathcal{V}$ to accept in a proof-of-retrievability protocol instance, except by responding with values $\{\mu_j\}$ and σ that are computed correctly, i.e., as they would be by* Priv.$\mathcal{P}$.

We prove the theorem in Appendix A.1 of the full paper [16].

Scheme with Public Verifiability

Theorem 2. *If the signature scheme used for file tags is existentially unforgeable and the computational Diffie-Hellman problem is hard in bilinear groups, then, in the random oracle model, except with negligible probability no adversary against the soundness of our public-verification scheme ever ever causes $\mathcal{V}$ to accept in a proof-of-retrievability protocol instance, except by responding with values $\{\mu_j\}$ and σ that are computed correctly, i.e., as they would be by* Pub.$\mathcal{P}$.

We prove the theorem in Appendix A.2 of the full paper [16].

4.2 Part-Two Proof

We say that a cheating prover $\mathcal{P}'$ is *well-behaved* if it never causes $\mathcal{V}$ to accept in a proof-of-retrievability protocol instance except by responding with values $\{\mu_j\}$ and σ that are computed correctly, i.e., as they would be by $\mathsf{Pub}.\mathcal{P}$. The part-one proofs above guarantee that all adversaries that win the soundness game with nonnegligible probability output cheating provers that are well-behaved, provided that the cryptographic primitives we employ are secure. The part-two theorem shows that extraction always succeeds against a well-behaved cheating prover:

Theorem 3. *Suppose a cheating prover $\mathcal{P}'$ on an n-block file M is well-behaved in the sense above, and that it is ϵ-admissible: i.e., convincingly answers and ϵ fraction of verification queries. Let $\omega = 1/\#B+(\rho n)^l/(n-l+1)^l$. Then, provided that $\epsilon - \omega$ is positive and nonnegligible, it is possible to recover a ρ fraction of the encoded file blocks in $O\big(n \;/\; (\epsilon - \omega)\big)$ interactions with $\mathcal{P}'$ and in $O\big(n^2 s + (1 + \epsilon n^2)(n) \;/\; (\epsilon - \omega)\big)$ time overall.*

We first make the following definition.

Definition 2. *Consider an adversary $\mathcal{B}$, implemented as a probabilistic polynomial-time Turing machine, that, given a query Q on its input tape, outputs either the correct response ($\mathbf{q}M$ in vector notation) or a special symbol $\perp$ to its output tape. Suppose $\mathcal{B}$ responds with probability ϵ, i.e., on an ϵ fraction of the query-and-randomness-tape space. We say that such an adversary is ϵ-polite.*

The proof of our theorem depends upon the following lemma that is proved in Appendix A.3 of the full paper [16].

Lemma 1. *Suppose that $\mathcal{B}$ is an ϵ-polite adversary as defined above. Let ω equal $1/\#B + (\rho n)^l/(n - l + 1)^l$. If $\epsilon > \omega$ then it is possible to recover a ρ fraction of the encoded file blocks in $O\big(n \;/\; (\epsilon - \omega)\big)$ interactions with $\mathcal{B}$ and in $O\big(n^2 s + (1 + \epsilon n^2)(n) \;/\; (\epsilon - \omega)\big)$ time overall.*

To apply Lemma 1, we need only show that a well-behaved ϵ-admissible cheating prover $\mathcal{P}'$, as output by a setup-game adversary $\mathcal{A}$, can be turned into an ϵ-polite adversary $\mathcal{B}$. But this is quite simple. Here is how $\mathcal{B}$ is implemented. We will use the $\mathcal{P}'$ to construct the ϵ-adversary $\mathcal{B}$. Given a query Q, interact with $\mathcal{P}'$ according to $\big(\mathcal{V}(pk, sk, t, sk) \rightleftharpoons \mathcal{P}'\big)$, playing the part of the verifier. If the output of the interaction is 1, write $(\mu_1, \ldots, \mu_s)$ to the output tape; otherwise, write $\perp$. Each time $\mathcal{B}$ runs $\mathcal{P}'$, it provides it with a clean scratch tape and a new randomness tape, effectively rewinding it. Since $\mathcal{P}'$ is well-behaved, a successful response will compute $(\mu_1, \ldots, \mu_s)$ as prescribed for an honest prover. Since $\mathcal{P}'$ is ϵ-admissible, on an ϵ fraction of interactions it answers correctly. Thus algorithm $\mathcal{B}$ that we have constructed is an ϵ-polite advesrary.

All that remains to to guarantee that $\omega = 1/\#B + (\rho n)^l/(n - l + 1)^l$ is such that $\epsilon - \omega$ is positive – indeed, nonnegligible. But this simply requires that each of $1/\#B$ and $(\rho n)^l/(n - l + 1)^l$ be negligible in the security parameter; see Section 1.1.

4.3 Part-Three Proof

Theorem 4. *Given a ρ fraction of the n blocks of an encoded file M^*, it is possible to recover the entire original file M with all but negligible probability.*

Proof. For rate-ρ Reed-Solomon codes this is trivially true, since any ρ fraction of encoded file blocks suffices for decoding; see Appendix B of the full paper [16]. For rate-ρ linear-time codes the additional measures described there guarantee that the ρ fraction of blocks retrieved will allow decoding with overwhelming probability.

Acknowledgements

We thank Dan Boneh, Moni Naor, and Guy Rothblum for helpful discussions regarding this work, and Eric Rescorla for detailed comments on the manuscript.

References

1. Ateniese, G., Burns, R., Curtmola, R., Herring, J., Kissner, L., Peterson, Z., Song, D.: Provable data possession at untrusted stores. In: De Capitani di Vimercati, S., Syverson, P. (eds.) Proceedings of CCS 2007, pp. 598–609. ACM Press, New York (2007)
2. Ateniese, G., Burns, R., Curtmola, R., Herring, J., Kissner, L., Peterson, Z., Song, D.: Provable data possession at untrusted stores. Cryptology ePrint Archive, Report 2007/202 (2007), `http://eprint.iacr.org/` (Version December 7, 2007) (visited February 10, 2008)
3. Ateniese, G., Di Pietro, R., Mancini, L., Tsudik, G.: Scalable and efficient provable data possession. In: Liu, P., Molva, R. (eds.) Proceedings of SecureComm 2008. ICST (September 2008) (to appear)
4. Barreto, P., Naehrig, M.: Pairing-friendly elliptic curves of prime order. In: Preneel, B., Tavares, S. (eds.) SAC 2005. LNCS, vol. 3897, pp. 319–331. Springer, Heidelberg (2006)
5. Boneh, D., Lynn, B., Shacham, H.: Short Signatures from the Weil Pairing. In: Boyd, C. (ed.) ASIACRYPT 2001. LNCS, vol. 2248, pp. 514–533. Springer, Heidelberg (2001)
6. Bowers, K.D., Juels, A., Oprea, A.: Proofs of retrievability: Theory and implementation. Cryptology ePrint Archive, Report 2008/175 (2008), `http://eprint.iacr.org/`
7. Cramer, R., Hanaoka, G., Hofheinz, D., Imai, H., Kiltz, E., Pass, R., Shelat, A., Vaikuntanathan, V.: Bounded CCA2-secure encryption. In: Kurosawa, K. (ed.) ASIACRYPT 2007. LNCS, vol. 4833, pp. 502–518. Springer, Heidelberg (2007)
8. Deswarte, Y., Quisquater, J.-J., Saïdane, A.: Remote integrity checking. In: Jajodia, S., Strous, L. (eds.) Proceedings of IICIS 2003. IFIP, vol. 140, pp. 1–11. Kluwer Academic, Dordrecht (2004)
9. Filho, D., Barreto, P.: Demonstrating data possession and uncheatable data transfer. Cryptology ePrint Archive, Report 2006/150 (2006), `http://eprint.iacr.org/`

10. Freeman, D., Scott, M., Teske, E.: A taxonomy of pairing-friendly elliptic curves. Cryptology ePrint Archive, Report 2006/372 (2006), http://eprint.iacr.org/
11. Heng, S.-H., Kurosawa, K.: k-resilient identity-based encryption in the standard model. CT-RSA 2004 E89-A.1(1), 39–46 (2006); Originally published at CT-RSA 2004
12. Juels, A., Kaliski, B.: PORs: Proofs of retrievability for large files. In: De Capitani di Vimercati, S., Syverson, P. (eds.) Proceedings of CCS 2007, pp. 584–597. ACM Press, New York (2007), http://www.rsa.com/rsalabs/staff/bios/ajuels/publications/pdfs/POR-preprint-August07.pdf
13. Lillibridge, M., Elnikety, S., Birrell, A., Burrows, M., Isard, M.: A cooperative Internet backup scheme. In: Noble, B. (ed.) Proceedings of USENIX Technical 2003. USENIX, pp. 29–41 (June 2003)
14. Naor, M., Rothblum, G.: The complexity of online memory checking. In: Tardos, E. (ed.) Proceedings of FOCS 2005, pp. 573–584. IEEE Computer Society, Los Alamitos (2005)
15. Schwarz, T., Miller, E.: Store, forget, and check: Using algebraic signatures to check remotely administered storage. In: Ahamad, M., Rodrigues, L. (eds.) Proceedings of ICDCS 2006. IEEE Computer Society, Los Alamitos (July 2006)
16. Shacham, H., Waters, B.: Compact proofs of retrievability. Cryptology ePrint Archive, Report 2008/073 (2008), http://eprint.iacr.org/
17. Shah, M., Swaminathan, R., Baker, M.: Privacy-preserving audit and extraction of digital contents. Cryptology ePrint Archive, Report 2008/186 (2008), http://eprint.iacr.org/

On the Security of HB$^\#$ against a Man-in-the-Middle Attack

Khaled Ouafi*, Raphael Overbeck**, and Serge Vaudenay

Ecole Polytechnique Fédérale de Lausanne (EPFL),
CH-1015 Lausanne, Switzerland

Abstract. At EuroCrypt '08, Gilbert, Robshaw and Seurin proposed HB$^\#$ to improve on HB$^+$ in terms of transmission cost and security against man-in-the-middle attacks. Although the security of HB$^\#$ is formally proven against a certain class of man-in-the-middle adversaries, it is only conjectured for the general case. In this paper, we present a general man-in-the-middle attack against HB$^\#$ and RANDOM-HB$^\#$, which can also be applied to all anterior HB-like protocols, that recovers the shared secret in 2^{25} or 2^{20} authentication rounds for HB$^\#$ and 2^{34} or 2^{28} for RANDOM-HB$^\#$, depending on the parameter set. We further show that the asymptotic complexity of our attack is polynomial under some conditions on the parameter set which are met on one of those proposed in [8].

Keywords: HB, authentication protocols, RFID.

1 Introduction

Designing secure cryptographic protocols using lightweight components is one of the main challenges of cryptography. Indeed, the emergence of new technology such as radio-frequency identification (RFIDs) with low computation and memory capabilities has stressed the need of such protocols.

These devices require protection from many threats. For example, for a company using RFIDs in inventories and supply-chain management, a RFID tag should be protected from cloning. Biometric passports also have a tight relation with RFIDs since they use contactless chips to communicate and authenticate the passport holder to some authorized authority. Using RFID tags as a replacement of barcodes by many merchant have also raised the issue of traceability and privacy protection. Thus, the need of authentication protocols providing efficiency, security and privacy protection has become a key factor for the future development of this technology. One of the most popular attempts to fulfill this need are the HB family of authentication protocol.

* Supported by a grant of the Swiss National Science Foundation, 200021-119847/1.
** Funded by DFG grant OV 102/1-1.

J. Pieprzyk (Ed.): ASIACRYPT 2008, LNCS 5350, pp. 108–124, 2008.
© International Association for Cryptologic Research 2008

The HB Family. Originally introduced by Hopper and Blum [11], the HB protocol aims at authenticating RFID tags to a reader using very lightweight operations while reducing its security to a well-known $\mathcal{NP}$-hard problem: the learning parity with noise (LPN) problem [1]. In fact, this protocol only requires a matrix multiplication and some basic XOR operations. But Juels and Weis [12] showed later that HB is insecure against adversaries able to interact with tags by impersonating readers and then proposed a new variant immune against this type of attacks: HB$^+$. As these two protocols were initially studied in a scenario of a sequential executions, Katz and Shin [13] extended both security proofs of HB and HB$^+$ to a more general concurrent and parallel setting. However, as Gilbert, Robshaw and Sibert noted in [6], the security of HB$^+$ is compromised if the adversary is given the ability to modify messages going from the reader to the tag. This model was later known as the GRS security model.

Since then, many HB-like protocols aiming security in the GRS model were proposed. Most notably, we mention the works of Bringer, Chabanne and Dottax on HB^{++} [3], Munilla and Peinado on HB-MP [15] and Duc and Kim on HB* [4]. But all these protocols were proven to be insecure in the GRS model, as all of them were successfully cryptanalyzed by Gilbert, Robshaw and Seurin in [7].

Tag (secret X, Y)		**Reader** (secret X, Y)
Choose $b \in_R \{0,1\}^{k_y}$	$\xrightarrow{\ b\ }$	
	$\xleftarrow{\ a\ }$	Choose $a \in_R \{0,1\}^{k_x}$
Choose $\nu \in_R \{0,1\}^m$ s.t. $\Pr[v_i = 1] = \eta$		
Compute $z = aX \oplus bY \oplus \nu$	$\xrightarrow{\ z\ }$	Accept iff: $\mathsf{wt}(aX \oplus bY \oplus z) \leq t$

Fig. 1. The RANDOM-HB$^{\#}$ and HB$^{\#}$ protocols. In RANDOM-HB$^{\#}$, $X \in \mathbb{F}_2^{k_x \times m}$ and $Y \in \mathbb{F}_2^{k_y \times m}$ are random matrices, in HB$^{\#}$ they are Toeplitz matrices. wt denotes the Hamming weight.

At EuroCrypt '08, Gilbert, Robshaw and Seurin [8,9], proposed a new variant of HB$^+$ named RANDOM-HB$^{\#}$ and its optimized version HB$^{\#}$. In these protocols, the tag and the reader share some secret matrices X and Y. During an authentication instance, both issue challenges of k_y-bit and k_x-bit length respectively and the final response of the tag is a m-bit message disturbed by a noise vector in which every bit has a probability η of being 1.

The details of the RANDOM-HB$^{\#}$ and HB$^{\#}$ protocols are outlined in Figure 1 and the proposed parameters (inspired from the results of [14]) in Table 1. The difference between these two versions lies in the structure of the secret matrices X and Y: while in RANDOM-HB$^{\#}$ these two are completely random, thus needing $(k_x + k_y)m$ bits of storage, HB$^{\#}$ reduces this amount to $k_x + k_y + 2m - 2$ by using Toeplitz matrices for X and Y.

Besides generating two random vectors ν and b, the operations performed by the tag to authenticate itself are very cheap: it only needs two matrix

Table 1. HB$^\#$ Parameter sets proposed in [8,9]. P_{FR} and P_{FA} denote the false rejection and false acceptance rates respectively. In the set III, the Hamming weight of the error vector ν generated by the tag is smaller than t.

Parameter set	k_x	k_y	m	η	t	P_{FR}	P_{FA}
I	80	512	1164	0.25	405	2^{-45}	2^{-83}
II	80	512	441	0.125	113	2^{-45}	2^{-83}
III	80	512	256	0.125	48	0	2^{-81}

multiplications to compute aX and bY which can be implemented using basic AND and XOR operations along with two bitwise XOR operations between two m-bit vectors. In some variant, the tag generates a random error vector ν until it has weight no larger than t requiring the tag to be able to compute a Hamming weight wt.

RANDOM-HB$^\#$ is also accompanied with a proof of security in the GRS security model if the parameters satisfy the condition $m\eta \leq t \leq m/2$. Under the conjecture that the Toeplitz-MHB puzzle is hard, HB$^\#$ is also secure in the same model. However, both protocols only provide "strong arguments" in favor of their resistance against man-in-the middle adversaries and formally proving their security in such a model was left as an open problem.

Our Contribution. In this paper, we present an attack against RANDOM-HB$^\#$ and HB$^\#$ in a general man-in-the-middle attack where the adversary is given the ability to modify all messages. The idea of our attack is to modify the messages of a session according to values obtained from a passive attack where the adversary eavesdrops on a protocol session between a reader and the tag.

Through this paper, we will denote b and z (resp. a) the values sent by the tag (resp. the reader) and $\hat{b}$ and $\hat{z}$ (resp. $\hat{a}$) the value received by the reader (resp. the tag) after corruption by the adversary. Thus the tag computes $z = \hat{a}X \oplus bY \oplus \nu$ while the reader checks that $\mathsf{wt}(aX \oplus \hat{b}Y \oplus \hat{z}) \leq t$.

Outline. Our paper is organized as follows. First, we show how it is possible to mount a man-in-the-middle attack against HB$^\#$ by proposing an algorithm able to compute the Hamming weight of the errors introduced by the tag in a session $(\bar{a}, \bar{b}, \bar{z})$. Then, we provide a complexity analysis of this initial attack needed by the man-in-the-middle to fully recover the secret matrices of RANDOM-HB$^\#$ and HB$^\#$. Afterwards, we present our optimized attack in Section 4 and give the complexity results applied to parameter sets I and II of HB$^\#$ of Table 1. After that, we investigate some open proposals to limit the Hamming weight of the error vector in HB-like protocols and present an attack against the parameter set III of HB$^\#$ shown in Table 1. At last, we show the lower bounds on the parameters for which our attack does not work.

2 Basic Attack

In this section, we show that, contrarily to what was conjectured in [8,9], both RANDOM-HB$^{\#}$ and HB$^{\#}$ are vulnerable against man-in-the-middle attacks by presenting a (non-optimized) attack.

2.1 Principle

The core of our attack is Algorithm 1 in which Φ denotes the cumulative distribution function of the normal distribution. It shows how an adversary able to modify messages going in both directions can compute the Hamming weight of the error vector $\bar{\nu} = \bar{a}X \oplus \bar{b}Y \oplus \bar{z}$ denoted $\bar{w} = \mathsf{wt}(\bar{\nu})$ introduced in a triplet $(\bar{a}, \bar{b}, \bar{z})$. The crucial observation is that since $z = \hat{a}X \oplus bY \oplus \nu$, at in each for-loop of Algorithm 1, the reader computes the Hamming weight $\mathsf{wt}(\nu \oplus \bar{\nu})$ of

$$aX \oplus \hat{b}Y \oplus \hat{z} = aX \oplus (\bar{b} \oplus b)Y \oplus (\bar{z} \oplus z) = (\hat{a}X \oplus bY \oplus z) \oplus (\bar{a}X \oplus \bar{b}Y \oplus \bar{z}) = \nu \oplus \bar{\nu}$$

and accepts iff $\mathsf{wt}(\nu \oplus \bar{\nu}) \leq t$.

Algorithm 1. Approximating $\bar{w}$

Input: $\bar{a}, \bar{b}, \bar{z}, n$
Output: $P^{-1}\left(\frac{c}{n}\right)$, an approximation of $\bar{w} = \mathsf{wt}(\bar{a}X \oplus \bar{b}Y \oplus \bar{z})$
where $P(\bar{w}) = \Pr[\mathsf{wt}(\nu \oplus \bar{\nu}) \leq t] = \Phi\left(\frac{t-(m-\bar{w})\eta - \bar{w}(1-\eta)}{\sqrt{m\eta(1-\eta)}}\right)$
Processing:
1: Initialize $c \leftarrow 0$
2: **for** $i = 1 \ldots n$ **do**
3: During a protocol, set $\hat{a} \leftarrow a \oplus \bar{a}$, $\hat{b} \leftarrow b \oplus \bar{b}$ and $\hat{z} \leftarrow z \oplus \bar{z}$
4: **if** reader accepts **then**
5: $c \leftarrow c + 1$
6: **end if**
7: **end for**

Correctness. We show, that the output of Algorithm 1 is indeed an estimation of $\mathsf{wt}(\nu \oplus \bar{\nu})$. The probability p that a bit of $(\nu \oplus \bar{\nu})$ is 1 is given by:

$$p = \Pr[(\nu \oplus \bar{\nu})_i = 1] = \begin{cases} \eta & \text{if } \bar{\nu}_i = 0 \\ 1 - \eta & \text{if } \bar{\nu}_i = 1. \end{cases}$$

Hence, $m - \bar{w}$ bits of $(\nu \oplus \bar{\nu})$ follow a Bernoulli distribution of parameter η and the other $\bar{w}$ bits follow a Bernoulli distribution of parameter $1 - \eta$, thus $\mathsf{wt}(\nu \oplus \bar{\nu})$ follows a binomial distribution. Because of the independence of all bits, the expected value and variance of $\mathsf{wt}(\nu \oplus \bar{\nu})$ are given by $\mu = (m - \bar{w})\eta + \bar{w}(1 - \eta)$ and $\sigma^2 = m\eta(1 - \eta)$ respectively.

We now define the function P as $P(\bar{w}) = \Pr[\mathsf{wt}(\nu \oplus \bar{\nu}) \leq t]$. By the definition of the standard normal cumulative distribution function Φ and the central limit theorem, we have that

$$P(\bar{w}) \approx \Phi(u), \text{ with } u = \frac{t-\mu}{\sigma} . \quad (1)$$

The random variable $\frac{c}{n}$ thus follows a normal distribution with expected value $P(\bar{w})$ and variance $\frac{1}{n}P(\bar{w})(1-P(\bar{w}))$. To decide whether $\mathsf{wt}(\bar{\nu}) = \bar{w}$ or not, the estimate $\frac{c}{n}$ for $P(\mathsf{wt}(\bar{\nu}))$ has to be good enough. The difference of the probabilities is at least $P(\bar{w}+1) - P(\bar{w}) \approx P'(\bar{w})$ which we can compute as

$$P'(\bar{w}) \approx -\frac{1-2\eta}{\sqrt{m\eta(1-\eta)}}\Phi'(u) = -\frac{1-2\eta}{\sqrt{m\eta(1-\eta)}} \times \frac{1}{\sqrt{2\pi}}e^{-\frac{u^2}{2}} .$$

By taking

$$n = \frac{\theta^2}{r^2}R(\bar{w}) \quad \text{with} \quad R(\bar{w}) = 2\frac{P(\bar{w})(1-P(\bar{w}))}{(P'(\bar{w}))^2} , \quad (2)$$

the probability that $|\frac{c}{n} - P(\bar{w})| > r|P'(\bar{w})|$ is $2\Phi(-\theta\sqrt{2}) = \mathsf{erfc}(\theta)$. With θ high enough, $\frac{c}{n}$ yields a estimate of $P(\bar{w})$ with precision $\pm rP'(\bar{w})$. Thus, Algorithm 1 is correct if n is chosen large enough.

Choice of Input. To determine a reasonable choice for the input n, we have to fix values for r and θ. If we can assume that $\bar{w} = \mathsf{wt}(\bar{\nu})$ is an integer close to some value w_0, we can call Algorithm 1 and $r = \frac{1}{2}$ to infer $\bar{w} = \lceil P^{-1}(\frac{c}{n})\rfloor$ with error probability $\mathsf{erfc}(\theta)$ (here, $\lceil\cdot\rfloor$ refers to normal rounding). On the other hand, if we know that $\bar{w} \in \{w_0 - 1, w_0 + 1\}$, we can choose $r = 1$ to infer $\bar{w}$ by the closest value to $P^{-1}(\frac{c}{n})$. The error probability is $\frac{1}{2}\mathsf{erfc}(\theta)$. In both cases, Algorithm 1 is an oracle of complexity $n = \frac{\theta^2}{r^2}R(w_0)$ that can be used to compute $\bar{w}$ given $\bar{a}, \bar{b}, \bar{z}$ and succeeding with an probability of error smaller than $\mathsf{erfc}(\theta)$.

Since we have to recover ℓ secret bits by Algorithm 1, $\mathsf{erfc}(\theta)$ should be less than the inverse of the number of secret bits ℓ. Using the approximation $\Phi(-x) \approx \varphi(x)/x$ when x is large (so $\Phi(-x)$ is small) we obtain

$$\theta = \sqrt{\ln \ell} \quad \Longrightarrow \quad \mathsf{erfc}(\theta) = 2\Phi(-\theta\sqrt{2}) \approx 2\frac{\varphi(\theta\sqrt{2})}{\theta\sqrt{2}} = \frac{e^{-\theta^2}}{\theta\sqrt{\pi}} < \frac{1}{\ell} ,$$

and thereby a reasonable choice for θ.

Recovering the whole secret key. Algorithm 2 shows how to recover the secret key by building a system of linear equations with the help of Algorithm 1.

Clearly the complexity of Algorithm 2 is $\theta^2(4R(\bar{w}) + mR(\bar{w}))$ and we have to call it ℓ/m times on independent $(\bar{a}, \bar{b})$ pairs to fully recover X and Y, where ℓ is the length of the secret key (Note that $\ell = (k_x + k_y)m$ in Random-HB$^\#$ and $\ell = k_x + k_y + 2m - 2$ in HB$^\#$). The expected number of errors in the equation system defining X and Y is $\ell \cdot \mathsf{erfc}(\theta)$. The probability that a passive attack gives an $(\bar{a}, \bar{b})$ linearly dependent from the i previous ones is $\frac{2^{i-1}}{2^{k_x+k_y}}$. The number of

Algorithm 2. Getting linear equations for X and Y

Input: $\bar{a}, \bar{b}, \bar{z}$ and $\bar{w}_{\mathsf{est}}$ the expected weight of $\bar{\nu} = \bar{a}X \oplus \bar{b}Y \oplus \bar{z}$
Output: A linear equation $\bar{a}X \oplus \bar{b}Y = \bar{c}$
Processing:
1: Initialize m-bit vector $\bar{c} \leftarrow \bar{z}$
2: Call Algorithm 1 on input $(\bar{a}, \bar{b}, \bar{z}, n = 4\theta^2 R(\bar{w}_{\mathsf{est}}))$ to get $\bar{w}$
3: **for** $i = 1 \ldots m$ **do**
4: Flip bit i of $\bar{z}$ to get $\bar{z}'$
5: Call Algorithm 1 on input $(\bar{a}, \bar{b}, \bar{z}', n = \theta^2 R(\bar{w}))$ to get $\bar{w}'$
6: **if** $\bar{w}' = \bar{w} - 1$ **then**
7: $\bar{c}_i \leftarrow \bar{c}_i \oplus 1$
8: **end if**
9: **end for**

passive attacks to get the inputs for Algorithm 2 is thus and can be neglected in comparison to the ℓ/m calls of Algorithm 2.

$$C = \sum_{i=1}^{\lceil \ell/m \rceil} \frac{1}{1 - \frac{2^{i-1}}{2^{k_x+k_y}}} < 2 + \frac{\ell}{m} \tag{3}$$

Computational complexity. The computational complexity of the given attack is quite low in comparison to the number of authentications needed: For each call of Algorithm 1 we have at most n incrementation of a counter and one evaluation of P^{-1}. For RANDOM-HB$^\#$, after running Algorithm 2 we have m linear binary equation systems in $k_x + k_y$ variables (one for each row of the matrix $[X^\top | Y^\top]$), which can thus be solved in $O(m(k_x + k_y)^3)$ operations. This number is negligible in comparison to the number of authentications needed to perform Algorithm 2 and is even lower for HB$^\#$. Throughout the paper we thus measure the complexity of our attack in terms of (intercepted) authentications between the tag and the reader.

2.2 Asymptotic Complexity Analysis

The complexity of the attack is related to the complexity of Algorithm 2 which is in its turn related to the complexity of Algorithm 1. Thus, the main component of the attack affecting the overall complexity is the input n in Algorithm 1. Equation (2) yields that $n = O((\theta^2 e^{\frac{u^2}{2}})/(1-2\eta)^2)$ so the complexity of our attack is exponential in u^2 as we can use a θ logarithmic in ℓ.

Parameters with optimal complexity. The minimal value of n is reached when $u = 0$ which happens when the estimated value $\bar{w}_{\mathsf{est}}$ of $\mathsf{wt}(\bar{\nu})$ is

$$\bar{w}_{\mathsf{est}} = \bar{w}_{\mathsf{opt}} = \frac{t - m\eta}{1 - 2\eta} \, .$$

In this case we obtain

$$P(\bar{w}_{\mathsf{opt}}) = \frac{1}{2} \ ,$$
$$P'(\bar{w}_{\mathsf{opt}}) = -\frac{1-2\eta}{\sqrt{2\pi m\eta(1-\eta)}},$$
$$R(\bar{w}_{\mathsf{opt}}) = \frac{\pi m}{4}\left(\frac{1}{(1-2\eta)^2} - 1\right).$$

Obviously, our attack has optimal complexity if we can call Algorithm 2 on input of valid triplets $(\bar{a}, \bar{b}, \bar{z})$ with $\mathsf{wt}(\bar{\nu}) = \bar{w}_{\mathsf{opt}}$, only. As clearly, for most parameter sets the latter is not true for random triplets obtained by passive attacks, we would like to manipulate errors in $\bar{z}$ to reach an expected value of $\bar{w}_{\mathsf{opt}}$. Unfortunately, due to the hardness of the LPN problem, we cannot *remove* errors from $\bar{z}$ if $\bar{w} > \bar{w}_{\mathsf{opt}}$. However, if $\bar{w} \leq \bar{w}_{\mathsf{opt}}$ then we can *inject* errors in $\bar{z}$ so that the resulting vector has an expected weight of $\bar{w}_{\mathsf{opt}}$ and the attack remains polynomial. This case happens when:

$$m\eta \leq \frac{t - m\eta}{1 - 2\eta} \quad \Longleftrightarrow \quad t \geq 2m\eta(1-\eta) \ ,$$

using the approximation $\bar{w}_{\mathsf{est}} \approx m\eta$ when a valid triplet $(\bar{a}, \bar{b}, \bar{z})$ is obtained by a passive attack and the false rejection rate of the HB$^{\#}$ protocol is negligible. Thus in this case, our attack remains optimal.

Categorization of parameter sets. We have seen, that for $u = 0$, our attack has subquadratic running time. However, even if $u = O(\sqrt{\ln \ell})$, we obtain a polynomial time attack. Thus, from Formula (2) we distinguish three cases:

1. *Subquadratic complexity:* If $t \geq 2m\eta(1-\eta)$ the attack has a complexity of $O(\frac{\ell \ln \ell}{(1-2\eta)^2})$ since Algorithm 1 is called $O(\ell)$ times.
2. *Polynomial complexity:* $t = 2m\eta(1-\eta) - c\sqrt{m\eta(1-\eta)}, c = O(\sqrt{\ln \ell})$: the above complexity is multiplied by an e^{c^2} factor. Thus, Algorithm 1 is still polynomial.
3. *Exponential complexity:* All other cases.

Depending on the category of the parameter set, there are different strategies to find the triplets $(\bar{a}, \bar{b}, \bar{z})$ which serve as input for Algorithm 2 (and thus Algorithm 1). We present those strategies in the following and give numbers for the according parameter sets.

2.3 Strategy for the Case $t \geq 2m\eta(1-\eta)$

Thanks to the hypothesis $t \geq 2m\eta(1-\eta)$, we have that $\bar{w}_{\mathsf{opt}} \geq \bar{w} = m\eta$. Thus, the best strategy is to optimize the complexity of Algorithm 1 by having a triplet $(\bar{a}, \bar{b}, \bar{z})$ with an error vector of expected Hamming weight $\bar{w}_{\mathsf{opt}}$. Using a triplet $(\bar{a}, \bar{b}, \bar{z})$ obtained from a passive attack, we can flip the last $(\bar{w}_{\mathsf{opt}} - m\eta)/(1-2\eta)$ bits of $\bar{z}$ to get $\bar{\nu}$ of expected Hamming weight $\bar{w}_{\mathsf{opt}}$ and then use the attack described previously.

Application to parameter vector II. As these parameters are in the case $t \geq 2m\eta(1-\eta)$, we can use Algorithm 2 in its optimum complexity to attack both RANDOM-HB# and HB#. After computing $\bar{w}_{\mathsf{opt}} = 77.167$, $P'(\bar{w}_{\mathsf{opt}}) = 0.0431$, $R(\bar{w}_{\mathsf{opt}}) = 269.39$ and the expected value of $\bar{w} = m\eta = 55$, we have to flip $f = 29$ bits to get an expected value close to $\bar{w}_{\mathsf{opt}}$. For RANDOM-HB# the number of bits to retrieve is $\ell = (k_x + k_y)m = 261\,072$ for which we can use $\theta = 3.164$. The total complexity is $\ell\theta^2 R(\bar{w}_{\mathsf{opt}}) = 2^{29.4}$. In the case of HB# the number of secret bits is $\ell = k_x + k_y + 2m - 2 = 1\,472$ for which we use $\theta = 2.265$ and end up with complexity of $\ell\theta^2 R(\bar{w}_{\mathsf{opt}}) = 2^{21}$.

2.4 Strategy for t Close to $2m\eta(1-\eta)$

The case $t < 2m\eta(1-\eta)$ is trickier to address since the expected value of $\bar{w}$ becomes *greater* than w_{opt}. To achieve the same complexity as the previous case we would have to reduce the Hamming weight of $\bar{\nu}$ which is infeasible in polynomial time due to the hardness of the LPN problem.

However, if t is a only a little less than $2m\eta(1-\eta)$ then the expected value of $\bar{w}$ is not far from w_{opt}. So, we can use Algorithm 2 without flipping any bit of $\bar{z}$ and the complexity is still polynomial. To further speed up the attack, we can remove errors from $\bar{z}$ in step 9 of Algorithm 2 until we reach $\bar{w} = w_{\mathsf{opt}}$ which we can expect to happen at iteration $i = \left\lceil \frac{\bar{w}_{\mathsf{est}} - \bar{w}_{\mathsf{opt}}}{\bar{w}_{\mathsf{est}}} \right\rceil$.

Application to parameter set I. For parameter set I we have $t < 2m\eta(1-\eta)$. We first compute $\bar{w}_{\mathsf{est}} = m\eta = 291$, $\bar{w}_{\mathsf{opt}} = 228$, $P'(\bar{w}_{\mathsf{opt}}) = 0.0135$, $R(\bar{w}_{\mathsf{est}}) = 15\,532$ and $R(\bar{w}_{\mathsf{opt}}) = 2742.6$. For RANDOM-HB#, the number of key bits is $\ell = (k_x + k_y)m = 689\,088$ and $\theta = 3.308$ is enough to guarantee that $\mathsf{erfc}(\theta) \leq \frac{1}{689\,088}$. We obtain a total complexity of $\ell\theta^2(\frac{\bar{w}_0 - \bar{w}_{\mathsf{opt}}}{\bar{w}_{\mathsf{est}}} R(\bar{w}_{\mathsf{est}}) + \frac{\bar{w}_{\mathsf{opt}}}{\bar{w}_{\mathsf{est}}} R(\bar{w}_{\mathsf{opt}})) = 2^{35.4}$. For HB#, we have $\ell = k_x + k_y + 2m - 2 = 2\,918$ secret bits to retrieve, so $\theta_2 = 2.401$ is enough and we get a total complexity of $\ell\theta^2(\frac{\bar{w}_0 - \bar{w}_{\mathsf{opt}}}{\bar{w}_{\mathsf{est}}} R(\bar{w}_{\mathsf{est}}) + \frac{\bar{w}_{\mathsf{opt}}}{\bar{w}_{\mathsf{est}}} R(\bar{w}_{\mathsf{opt}})) = 2^{26.6}$.

2.5 Strategy for Lower t

The case of lower t, the false acceptance rate will be very low but the false rejection rate of HB# becomes high (e.g. 0.5 for $t = m\eta$; Please remember that for $t < m\eta$, HB# is no longer provable secure in the GRS security model.) so that it would require more than one authentication in average for the tag to authenticate itself. The main advantage of this approach is that the complexity of Algorithm 1 becomes exponential. Here, we present a better strategy than calling Algorithm 2 with an triplet $(\bar{a}, \bar{b}, \bar{z})$ obtained by a simple passive attack.

Our goal is to call Algorithm 2 with a $\bar{w}_{\mathsf{est}}$ as low as possible. During the protocol, we can set $(\hat{a}, \hat{b}, \hat{z})$ to $(a, b, z \oplus \bar{\nu})$ with $\bar{\nu}$ of weight $\bar{w}$ until the reader accepts $\hat{z}$. Then, we launch our attack with $(\bar{a}, \bar{b}, \bar{z}) = (a, b, z)$. A detailed description is shown in Algorithm 3.

Algorithm 3. Getting (a, b, z) with low Hamming weight

Input: $\bar{w}$
Output: (a, b, z) such that $(aX \oplus bY \oplus z)$ has low weight.
Processing:
1: Pick random vector $\bar{\nu}$ of Hamming weight $\bar{w}$
2: **repeat**
3: During a protocol with messages (a, b, z), set $\hat{z} = z \oplus \bar{\nu}$
4: **until** reader accepts

The probability that $\hat{z}$ gets accepted by the verifier is $P(\bar{w})$ which can be written in an equivalent way to Equation (1) as:

$$P(\bar{w}) = \sum_{j=0}^{t} \left(\binom{m-\bar{w}}{j} \eta^j (1-\eta)^{m-\bar{w}-j} \cdot \sum_{i=0}^{t-j} \binom{\bar{w}}{i} \eta^{\bar{w}-i} (1-\eta)^i \right) \quad (4)$$

For an accepted $\hat{z}$, the $m - \bar{w}$ positions not in the support of $\bar{\nu}$ are erroneous with probability

$$\eta_{\bar{w}} = \frac{\sum_{j=0}^{t} \left(j \binom{m-\bar{w}}{j} \eta^j (1-\eta)^{m-\bar{w}-j} \sum_{i=0}^{t-j} \binom{\bar{w}}{i} \eta^{\bar{w}-i} (1-\eta)^i \right)}{(m-\bar{w}) P(\bar{w})}. \quad (5)$$

On the other hand, the other positions of $\hat{z}$ in the support of $\bar{\nu}$ are non-zero with probability

$$\eta^{\circ}_{\bar{w}} = \frac{\sum_{j=0}^{t} \left(\binom{m-\bar{w}}{j} \eta^j (1-\eta)^{m-\bar{w}-j} \cdot \sum_{i=0}^{t-j} i \binom{\bar{w}}{i} \eta^{\bar{w}-i} (1-\eta)^i \right)}{\bar{w} P(\bar{w})}. \quad (6)$$

Thus, because of the high false rejection rate, if $\hat{z}$ gets accepted in our MIM-Attack with $(\bar{a}, \bar{b}, \bar{z}) = (0, 0, \bar{\nu})$, we can expect that the error vector ν, introduced in (a, b, z) the output of Algorithm 3, has weight $\bar{w}_{\text{est}} = (m - \bar{w})\eta_{\bar{w}} + \bar{w}(1 - \eta^{\circ}_{\bar{w}})$.

Application to parameter set II with $t = 55$. Assume that for the parameter set II we set $t = m\eta \approx 55$. Then, an accepted vector obtained by a passive attack will most likely have weight $\bar{w}_{\text{est}} = (m - \bar{w})\eta_0 + \bar{w}(1 - \eta^{\circ}_0) \approx 50$ and it will take $4\theta^2 R(\bar{w}_{\text{est}}) = 2^{30}$ operations to determine its correct weight. Calling Algorithm 3, e.g., with $\bar{w} = 41$, we get (a, b, z) with error vector ν of weight $\bar{w}_{\text{est}} = (m - \bar{w})\eta_{41} + \bar{w}(1 - \eta^{\circ}_{41}) \approx 33$ in $\frac{1}{P(\bar{w})} = 2^{20}$ authentications and can recover the weight of ν in another $4\theta^2 R(33) = 2^{20}$ operations with Algorithm 1. We determined the optimal input $\bar{w}$ by exhaustive search minimizing the sum of the complexity of the consecutive execution of Algorithms 3 and Algorithm 1.

The following table we consider parameter sets I and II with modified t. It shows the costs to learn one bit about the secret key, i.e. calling Algorithm 1 with a random vector obtained by a passive attack in comparison to calling Algorithm 3 first and then Algorithm 1 with its output. Note, that recovering successive bits is always cheaper.

Table 2. Attack cost for the initial bit of the shared key for HB# applied to $t = \lceil \eta m \rfloor$

Parameter set	Algorithm 1	Algorithms 3 + 1
I	2^{78}	$2^{58.5}$
II	2^{30}	2^{21}

3 Optimizing the Attack

In this Section, we present our best attack on RANDOM-HB# and HB#. First, we optimize Algorithm 2. Using an adaptive solution to the weighing problem [5] we show how to efficiently recover the error vector. Then, we present our full attack.

3.1 Optimizing Algorithm 2

The problem we are solving in Algorithm 2 can be formulated as follows: given a m-bit vector ν of Hamming weight w and an oracle measuring the sum of some selected bits (Algorithm 1), what is the minimal number of measurements to fully recover ν?

The naïve solution to this problem employed in Algorithm 2 takes m measurements. A more sophisticated solution to to fully recover a vector ν of arbitary weight was already given by Erdős and Rényi in [5]. They show that the minimal number of measurements required is upper-bounded by $(m \log_2 9)/\log_2 m$. To recover ν in the given complexity, they define a fixed series of measurements for each m. However, in our case, the vector ν is known to be of small weight ($\leq m\eta$), which allows us to improve on the solution by Erdős and Rényi. Our proposal, Algorithm 4, does not use a fixed series of measurements but takes into account the partial information obtained by all previous measurements.

To determine the error positions in a k-bit window by measuring the weight, Algorithm 4 uses a divide-and-conquer strategy: it splits the vector into two windows of the same length then measures each of them. For those parts which do not have full or zero weight it then applies this strategy recursively leading to a lower number of measurements comparing to measuring a k-bit window bit by bit as Algorithm 2 does.

The number of invocations of Algorithm 1, $C_w(k)$, to fully recover a k-bit window with known Hamming weight w by Algorithm 4 is

$$C_w(k) = \begin{cases} 0 & \text{if } w = 0 \text{ or } w = k \\ 1 + \sum_{i=0}^{k/2} \frac{\binom{\lfloor k/2 \rfloor}{i}\binom{\lceil k/2 \rceil}{w-i}}{\binom{k}{w}} \left(C_i(\lfloor k/2 \rfloor) + C_{w-i}(\lceil k/2 \rceil)\right) & \text{otherwise} \end{cases}$$

Let $C(k)$ be the average number of invocations of Algorithm 1 to first determine the number of errors in a k-bit window and then recover their positions using Algorithm 4:

$$C(k) = 1 + \sum_{w=0}^{k} C_w(k) \binom{k}{w} \eta^w (1-\eta)^{k-w}$$

Algorithm 4. Finding errors in $|J|$-bit windows

Input: $\bar{a}, \bar{b}, \bar{z}$, $\bar{w} = \mathsf{wt}(\bar{a}X \oplus \bar{b}Y \oplus \bar{z})$, a set $J \subseteq \{0, 1, \cdots m\}$ and w_J the number of non-zero $(\bar{a}X \oplus \bar{b}Y \oplus \bar{z})_j$, $j \in J$
Output: $I \subseteq J$ containing the j with non-zero $(\bar{a}X \oplus \bar{b}Y \oplus \bar{z})_j$, $j \in J$.
Processing:
1: **if** $w_J = 0$ **then**
2: $\quad I \leftarrow \emptyset$
3: **else if** $w_J = |J|$ **then**
4: $\quad I \leftarrow J$
5: **else**
6: $\quad$ Choose $J_1 \subseteq J$ such that $|J_1| = \lceil |J|/2 \rceil$.
7: $\quad$ Set ν' the m-vector with $\nu'_j = 1$ iff $j \in J_1$
8: $\quad$ Call Algorithm 1 on input $(\bar{a}, \bar{b}, \bar{z} \oplus \nu', n = 4\theta^2 R(\bar{w}))$ to get w'.
9: $\quad$ Call Algorithm 4 with $(\bar{a}, \bar{b}, \bar{z}, \bar{w}, J_1, w_{J_1} = (\bar{w} + |J_1| - w')/2)$ to get I_1
10: $\quad$ Call Algorithm 4 with $(\bar{a}, \bar{b}, \bar{z}, \bar{w}, J \setminus J_1, w_J - w_{J_1})$ to get I_2
11: $\quad I \leftarrow I_1 \cup I_2$
12: **end if**

Table 3. Complexity of measuring a 16-bit window for parameter set II

	Parameter Set I		Parameter Set II	
k	$C(k)\frac{16}{k}$	Cost measurement	$C(k)\frac{16}{k}$	Cost measurement
2	11	$2^{15.95}$	9.75	$2^{12.43}$
4	9.72	$2^{15.96}$	7.404	$2^{12.49}$
8	9.51	$2^{15.99}$	6.71	$2^{12.75}$
16	9.51	$2^{16.11}$	6.69	$2^{13.90}$

We note that $C(k)/k$ is minimal when k is a power of 2. Although, it is clear from Table 3 that the number of measurements decreases when k increases, the cost of measuring the weight of a k-bit window also increases faster with k, so a good tradeoff is to use $k = 8$.

Now that we have an efficient algorithm to find error positions in fixed size windows, we introduce Algorithm 5 which takes benefit from Algorithm 4 to optimize the number of measurements needed to localize the introduced errors and output m linear equations. Algorithm 5 splits the error vector introduced in a triplet $(\bar{a}, \bar{b}, \bar{z})$ to m/k k-bit windows, each one of these is then recovered using Algorithm 4. Additionally, using the learned bits, it adjusts $\bar{z}$ so that the next measurements cost less. The number of calls to Algorithm 4 we need before we reach $\bar{w} = \bar{w}_{\mathsf{opt}}$, is then

$$i = \begin{cases} \frac{\bar{w}_{\mathsf{opt}} - \bar{w}_{\mathsf{est}}}{k(m - \bar{w}_{\mathsf{est}})} m & \text{if } \bar{w}_{\mathsf{opt}} \geq \bar{w}_{\mathsf{est}} \\ \frac{\bar{w}_{\mathsf{est}} - \bar{w}_{\mathsf{opt}}}{k \cdot \bar{w}_{\mathsf{est}}} m & \text{if } \bar{w}_{\mathsf{opt}} \leq \bar{w}_{\mathsf{est}} \end{cases}$$

Algorithm 5. Optimizing Algorithm 2

Input: $\bar{a}, \bar{b}, \bar{z}$ and $\bar{w}_{\mathsf{est}}$ the expected value of $\bar{\nu} = \bar{a}X \oplus \bar{b}Y \oplus \bar{z}$, k
Output: A linear equation $\bar{a}X \oplus \bar{b}Y = \bar{c}$
Processing:
1: Initialize m-bit vector $\bar{c} \leftarrow \bar{z}$
2: Initialize $M \leftarrow \emptyset$
3: Call Algorithm 1 on input $(\bar{a}, \bar{b}, \bar{z}, n = 4\theta^2 R(\bar{w}_{\mathsf{est}}))$ to get $\bar{w}$
4: Define a set $\mathcal{S}$ of $J_i = \{ik+1, \ldots, \min((i+1)k, m)\}$, $i = 1 \ldots \lceil \frac{m}{k} \rceil$
5: **repeat**
6: Choose $J \in \mathcal{S}$
7: Call Algorithm 1 on input $(\bar{a}, \bar{b}, \bar{z} \oplus J, n = \theta^2 R(\bar{w}))$ to get $\bar{w}' = \mathsf{wt}(\bar{\nu}\ \mathsf{AND}\ J)$
8: Call Algorithm 4 with $(\bar{a}, \bar{b}, \bar{z}, \bar{w}, J, w_J = (\bar{w} + |J| - \bar{w}')/2)$ to get I
9: Set $\bar{c}_i \leftarrow \bar{c}_i \oplus 1$ for all $i \in I$
10: $M \leftarrow M \cup I$
11: Remove J from $\mathcal{S}$
12: **if** $\bar{w} > \bar{w}_{\mathsf{opt}}$ **then**
13: Flip $\min(|I|, \bar{w} - \bar{w}_{\mathsf{opt}})$ bits $\bar{z}_i$ for which $i \in I$
14: $\bar{w} \leftarrow \bar{w} - \min(|I|, \bar{w} - \bar{w}_{\mathsf{opt}})$
15: **else if** $\bar{w} < \bar{w}_{\mathsf{opt}}$ **then**
16: Flip $\min(|J \setminus I|, \bar{w}_{\mathsf{opt}} - \bar{w})$ bits $\bar{z}_i$ for which $i \in J \setminus I$
17: $\bar{w} \leftarrow \bar{w} + \min(|J \setminus I|, \bar{w}_{\mathsf{opt}} - \bar{w})$
18: **end if**
19: **until** $\mathcal{S}\emptyset$

So the full complexity of Algorithm 5 is given by

$$N = \theta^2 \left(iR(\bar{w}_{\mathsf{est}}) + \left\lceil \frac{m}{k} - i \right\rceil R(\bar{w}_{\mathsf{opt}}) \right) C(k) \ .$$

3.2 Final Algorithm

The final attack is described in Algorithm 6. The idea is to get a vector with low expected weight using Algorithm 3 and then find all the erroneous positions inserted by the tag to obtain m linear equations and iterate this until we get enough equations to solve and find the secrets X and Y. To get the lower complexity, we can flip the last bits of $\bar{z}$ so that we end up with an expected weight of $\bar{w}_{\mathsf{opt}}$. We note that introducing errors in a full segment as defined by Step 4 of Algorithm 5 does not increase the needed number of measurements as $C_w(k) = C_{k-w}(k)$. Using Formula (3), we deduce the full complexity in terms of intercepted authentications as

$$\left\lceil \frac{\ell}{m} \right\rceil \theta^2 \left(iR(\bar{w}_{\mathsf{est}}) + \left\lceil \frac{m}{k} - i \right\rceil R(\bar{w}_{\mathsf{opt}}) \right) C(k) \ + \ (2 + \frac{\ell}{m}) \frac{1}{P(w)} \ . \tag{7}$$

Application to parameter set I. With input $k = 8$ and $w = 300$ we obtain $P(w) = 2^{-7}$, $\bar{w}_{\mathsf{est}} = 273$ and $\bar{w}_{\mathsf{opt}} = 228$, $i = 24$, $R(\bar{w}_{\mathsf{opt}}) = 2742.6$, $R(\bar{w}_{\mathsf{est}}) = 7\,026.4$. So the full complexity of the attack is then given by Equation

Algorithm 6. Final attack on RANDOM-HB$^{\#}$ and HB$^{\#}$

Input: k, w
Output: X, Y the secrets of the tag
Processing:
1: Initialize $\mathcal{S} \leftarrow \emptyset$
2: **for** $i = 1 \dots 2 + \lceil \frac{\ell}{m} \rceil$ **do**
3: Call algorithm 3 on input w to get $\bar{a}, \bar{b}, \bar{z}$ with an error vector of expected weight $\bar{w}_{\text{est}} = (m - w)\eta_w + w(1 - \eta_w^\circ)$
4: **if** $\bar{w}_{\text{opt}} > \bar{w}_{\text{est}}$ **then**
5: Flip the last $(\bar{w}_{\text{opt}} - m\eta)/(1 - 2\eta)$ bits of $\bar{z}$
6: Set $\bar{w}_{\text{est}} \leftarrow \bar{w}_{\text{opt}}$
7: **end if**
8: Call Algorithm 5 on input $(\bar{a}, \bar{b}, \bar{z}, \bar{w}_{\text{est}}, k)$ to get m linear equations
9: Insert linear equations in $\mathcal{S}$
10: **end for**
11: Solve $\mathcal{S}$

(7) with θ and ℓ as in Section 2.4. This is 2^{25} sessions for HB$^{\#}$ and $2^{33.8}$ for RANDOM-HB$^{\#}$.

Application to parameter set II. In this case, we have $k = 8$, $w = 0$ and $\bar{w}_{\text{est}} = 55$. We flip 29 bits to obtain an error vector of expected weight $\bar{w}_{\text{opt}} = 77$, which yields $R(\bar{w}_{\text{opt}}) = 269.39$ and $i = 0$. The complexity is $2^{19.7}$ sessions for HB$^{\#}$ and $2^{28.1}$ for RANDOM-HB$^{\#}$.

4 Attacking Parameter Vectors without False Rejections

To thwart the previous attacks without taking parameter sets with huge m or high false rejection rate, we could change the protocol so that the prover generates a vector ν of constant or bounded Hamming weight like it was proposed for parameter set III. In this section we will show that this leads to different attacks.

Assume that the prover accepts (a, b, z) iff $w = \mathsf{wt}(aX \oplus bY \oplus z) = t$, then from this triplet the attacker learns

$$\bigoplus_{i=1}^{m} (aX \oplus bY)_i = \bigoplus_{i=1}^{m} z_i \oplus \begin{cases} 1 & \text{if } t \text{ odd} \\ 0 & \text{if } t \text{ even} \end{cases}$$

It is possible to recover the matrices X and Y by sending $z \oplus \bar{\nu}$ instead of the Tag's response z to the Reader, where $\bar{\nu}$ is a m-bit vector of Hamming weight 2. Doing so, the attacker learns

$$(aX \oplus bY)\bar{\nu}^\top = z\bar{\nu}^\top \oplus \begin{cases} 1 & \text{if } \hat{z} \text{ accepted} \\ 0 & \text{if } \hat{z} \text{ rejected} \end{cases}$$

since the verifier accepts $\hat{z}$ on challenges a, b if there was exactly one error in the flipped positions, which is the case with probability $\binom{m-w}{1}\binom{w}{1}/\binom{m}{2}$.

The above approach may be generalized to the case where the Hamming weight of ν is bounded in the original protocol, i.e. if the verifier accepts if $w \leq t$ and the prover discards error vectors which are going to be rejected. This was suggested for the parameter vector III. Again, the attacker can replace the Tag's answer z by $z \oplus \bar{\nu}$ where $\bar{\nu}$ is of weight 2. Now, the attackers response $z \oplus \bar{\nu}$ gets rejected iff $w \in \{t-1, t\}$ and the attacker flipped two non-erroneous positions. Thus, in the case of a rejection, the attacker learns

$$(aX \oplus bY)_i = z_i, \quad \bar{\nu}_i \neq 0$$

which happens with probability

$$q = \frac{\sum_{i=0}^{1} \binom{m}{t-i} \eta^{t-i} (1-\eta)^{m-t+i} \frac{\binom{m-t+i}{2}}{\binom{m}{2}}}{\sum_{i=0}^{t} \binom{m}{i} \eta^{i} (1-\eta)^{m-i}}$$

Application to parameter set III. For the parameter vector III, the attacker learns two bits about the secret key every $1/q = 2^{9.02} \approx 512$ iterations. This is 16 times faster than an attack by Algorithm 1 and needs only $\ell \cdot 2/q = 2^{26}$ authentications to recover a RANDOM-HB# secret key (2^{19} for HB#).

5 Lower Bounds on Secure Parameters

In this section, we investigate the lower bounds on the parameter sets for which our attack is not effective. We say that HB# is secure if recovering one bit of information about the secret key requires an attack with complexity (in terms of protocol sessions) within an order of magnitude of at least 2^s and time complexity "reasonably comparable".

Let us assume that Algorithm 3 succeeds with a total error weight of $t = \mathsf{wt}(\nu \oplus \bar{\nu})$ when the added error vector has weight $\bar{w}$. To obtain this vector, the attacker limited to 2^{80} operations can choose the input $\bar{w}$ in any way, such that $1/P(\bar{w}) = 1/\Phi(\frac{t-\mu}{\sigma}) \leq 2^{80}$. Since $\Phi(-10.2) \approx 2^{-80}$ we can be sure, that the $\bar{w}$ chosen by the attacker satisfies that

$$\begin{aligned}
\tfrac{t-\mu}{\sigma} = \quad & \tfrac{t-(m-\bar{w})\eta-\bar{w}(1-\eta)}{\sqrt{m\eta(1-\eta)}} \geq -10.2 \\
\Leftrightarrow (m-\bar{w})\eta + \bar{w}(1-\eta) &\leq 10.2\sqrt{m\eta(1-\eta)} + t \\
\Leftrightarrow \quad -\bar{w}\eta + \bar{w}(1-\eta) &\leq 10.2\sqrt{m\eta(1-\eta)} + t - m\eta \\
\Leftrightarrow \quad \bar{w}(1-2\eta) &\leq -m\eta + t + 10.2\sqrt{m\eta(1-\eta)} \\
\Leftrightarrow \quad \bar{w} &\leq \tfrac{1}{1-2\eta}(10.2\sqrt{m\eta(1-\eta)} + t - m\eta) \,.
\end{aligned} \tag{8}$$

Fixing $t = \lfloor m\eta \rceil$ for which our attack has the maximal complexity, we get the lowest value for a secure m, thus $\bar{w} = \frac{10.2\sqrt{m\eta(1-\eta)}}{1-2\eta}$.

We can now calculate the value $\bar{w}_{\mathsf{est}}$ by using equations (4), (5) and (6) and then by using Formula (2) with $r = 1/2$ and $\theta = 1/2$ (which leads to $\mathsf{erfc}(\theta) = 0.4795$) we can estimate the total cost of the attack. By using an exhaustive

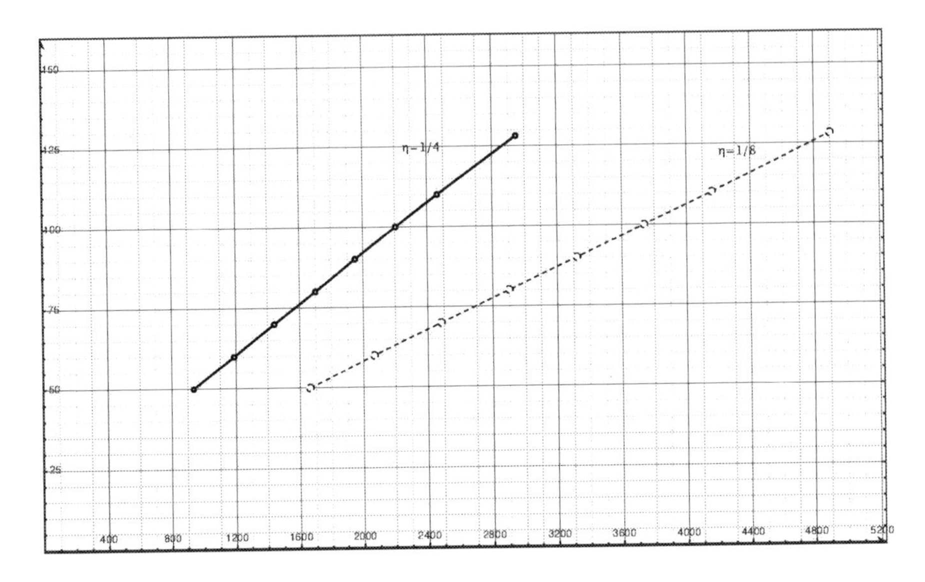

Fig. 2. Security level in logarithmic scale in comparison to m when $t = m\eta$

search on m we obtain that $m = 1\,697$ for $\eta = 1/4$ and $m = 2\,903$ for $\eta = 1/8$ is the lowest choice achieving 2^{80}-security and 50% of false rejection rate. The full results with the intermediates values are summarized in Table 4.

Table 4. Lowest values of m and $t = \lfloor m\eta \rceil$ for which our attack is not effective

η	m	t	$\bar{w}$	$\bar{w}_{\text{est}}$	$\eta_{\bar{w}}$	$\eta^{\circ}_{\bar{w}}$	$1/P(\bar{w})$	n
0.25	1697	424	364	340	$2^{-2.73}$	$2^{-0.7}$	2^{80}	2^{80}
0.125	2903	363	242	229	$2^{-3.93}$	$2^{-0.36}$	2^{80}	2^{80}

Following this method we obtain the graphs of Fig. 2 showing how the security scales with growing m. To reach this security with a more acceptable false rejection rate (ideally negligible), it requires m to be higher.

6 Conclusion

In this article, we proved that the conjecture about the security of RANDOM-HB$^{\#}$ and HB$^{\#}$ is wrong. We presented a basic attack against these protocols that allows to retrieve the shared secret between a reader and a tag. We showed a lower bound on the parameter set for which our attack is not effective but such parameters are unpractical to use in RFID tags.

Although it may not be the most effective for all versions, our attack is valid against all anterior protocols of the HB family.

Table 5. Summary of the complexity of our attacks

Parameter Set	k_X	k_Y	m	η	t	Random-HB#	HB#
I	80	512	1164	0.25	405	2^{34}	2^{25}
II	80	512	441	0.125	113	2^{28}	2^{20}
III(w bounded)	80	512	256	0.125	48	2^{26}	2^{19}

There are still new versions in the HB family. PUF-HB, proposed by Hammouri and Sunar [10] uses a physical unclonable function but does not carry any proof of security against man-in-the-middle attacks within. Indeed, a closer look reveals several possible points of attack for a man in the middle like flipping the last bit in the challenge vector a. On the other side, Trusted-HB, proposed by Bringer and Chabanne [2], is proved secure against general man-in-the-middle attacks. However, this comes at the cost of adding a check on the integrity of the error vector using a secure cryptographic hash function which on its own would be sufficient to allow authentication by shared secrets.

Acknowledgment

We would like to thank Henri Gilbert, Matt Robshaw and Yannick Seurin for fruitful discussions.

References

1. Berlekamp, E.R., McEliece, R., van Tilborg, H.C.A.: On the inherent intractability of certain coding problems (corresp.). IEEE Transactions on Information Theory 24(3), 384–386 (1978)
2. Bringer, J., Chabanne, H.: Trusted-HB: a low-cost version of HB$^+$ secure against man-in-the-middle attacks. CoRR, abs/0802.0603 (2008)
3. Bringer, J., Chabanne, H., Dottax, E.: HB^{++}: a lightweight authentication protocol secure against some attacks. In: Second International Workshop on Security, Privacy and Trust in Pervasive and Ubiquitous Computing (SecPerU 2006), Lyon, France, June 29, pp. 28–33. IEEE Computer Society, Los Alamitos (2006)
4. Duc, D.N., Kim, K.: Securing HB$^+$ against GRS man-in-the-middle attack. In: Institute of Electronics, Information and Communication Engineers, Symposium on Cryptography and Information Security, Sasebo, Japan, January 23-26, p. 123 (2007)
5. Erdős, P., Rényi, A.: On two problems of information theory. Publ. Math. Inst. Hung. Acad. Sci. 8(21), 229–243 (1963)
6. Gilbert, H., Robshaw, M., Sibert, H.: Active attack against HB$^+$: a provably secure lightweight authentication protocol. IEEE Electronics Letters 41(21), 1169–1170 (2005)
7. Gilbert, H., Robshaw, M.J.B., Seurin, Y.: Good variants of HB$^+$ are hard to find. In: Tsudik, G. (ed.) FC 2008. LNCS, vol. 5143, pp. 156–170. Springer, Heidelberg (2008)

8. Gilbert, H., Robshaw, M.J.B., Seurin, Y.: $HB^{\#}$: Increasing the security and efficiency of HB^{+}. In: Smart, N.P. (ed.) EUROCRYPT 2008. LNCS, vol. 4965, pp. 361–378. Springer, Heidelberg (2008)
9. Gilbert, H., Robshaw, M.J.B., Seurin, Y.: $HB^{\#}$: Increasing the security and efficiency of HB^{+}, full version. Cryptology ePrint Archive, Report 2008/028 (2008)
10. Hammouri, G., Sunar, B.: PUF-HB: A tamper-resilient HB based authentication protocol. In: Bellovin, S.M., Gennaro, R., Keromytis, A.D., Yung, M. (eds.) ACNS 2008. LNCS, vol. 5037, pp. 346–365. Springer, Heidelberg (2008)
11. Hopper, N.J., Blum, M.: Secure human identification protocols. In: Boyd, C. (ed.) ASIACRYPT 2001. LNCS, vol. 2248, pp. 52–66. Springer, Heidelberg (2001)
12. Juels, A., Weis, S.A.: Authenticating pervasive devices with human protocols. In: Shoup, V. (ed.) CRYPTO 2005. LNCS, vol. 3621, pp. 293–308. Springer, Heidelberg (2005)
13. Katz, J., Shin, J.S.: Parallel and concurrent security of the HB and HB^{+} protocols. In: Vaudenay, S. (ed.) EUROCRYPT 2006. LNCS, vol. 4004, pp. 73–87. Springer, Heidelberg (2006)
14. Levieil, É., Fouque, P.-A.: An improved LPN algorithm. In: De Prisco, R., Yung, M. (eds.) SCN 2006. LNCS, vol. 4116, pp. 348–359. Springer, Heidelberg (2006)
15. Munilla, J., Peinado, A.: HB-MP: A further step in the HB-family of lightweight authentication protocols. Computer Networks 51(9), 2262–2267 (2007)

Hash Functions from Sigma Protocols and Improvements to VSH

Mihir Bellare and Todor Ristov

Department of Computer Science and Engineering, University of California San Diego, 9500 Gilman Drive, La Jolla, CA 92093-0404, USA
www-cse.ucsd.edu/users/mihir, www-cse.ucsd.edu/users/tristov

Abstract. We present a general way to get a provably collision-resistant hash function from any (suitable) Σ-protocol. This enables us to both get new designs and to unify and improve previous work. In the first category, we obtain, via a modified version of the Fiat-Shamir protocol, the fastest known hash function that is provably collision-resistant based on the *standard* factoring assumption. In the second category, we provide a modified version VSH* of VSH which is faster when hashing short messages. (Most Internet packets are short.) We also show that Σ-hash functions are chameleon, thereby obtaining several new and efficient chameleon hash functions with applications to on-line/off-line signing, chameleon signatures and designated-verifier signatures.

1 Introduction

The failure of popular hash functions MD5 and SHA-1 [42, 43] lends an impetus to the search for new ones. The contention of our paper is that there will be a "niche" market for proven-secure even if not-so-fast hash functions. Towards this we provide a general paradigm that yields hash functions provably secure under number-theoretic assumptions, and also unifies, clarifies and improves previous constructs. Our hash functions have extra features such as being chameleon [25]. Let us now look at all this in more detail.

THE NEED FOR PROVEN-SECURE HASHING. Suppose an important document has been signed with a typical hash-then-sign scheme much as PKCS#1 [24]. If collisions are found in the underlying hash function the public key needs to be revoked and the signature can no longer be accepted. Yet there are instances in which we want a public key and signatures under it to survive for twenty or more years. This might be the case for a central and highly disseminated certificate or an important contract. Revocation of a widely disseminated public key is simply too costly and error-prone. In such a case, we want to be able to trust that collisions in our hash function will not be found even twenty years down the line.

Given the failure of MD5 and SHA-1, it would be understandable, from this twenty-year perspective, to feel uncertain about any hash function designed by "similar" methods. On the other hand, we may be very willing to pay a (reasonable!) computational price for security because documents or certificates of the

J. Pieprzyk (Ed.): ASIACRYPT 2008, LNCS 5350, pp. 125–142, 2008.
© International Association for Cryptologic Research 2008

ultra-importance we are considering may not need to be signed often. In this case, hash functions with *proven* security are interesting, and the faster they are the better. Our contribution is a general transform that yields a plurality of such hash functions, not only providing new ones but "explaining" or improving old ones.

FROM Σ TO HASH. We show how to construct a collision-resistant hash function from any (suitable) Σ-protocol. Recall that Σ-protocols are a class of popular 3-move identification schemes. Canonical examples are the Schnorr [37], Fiat-Shamir [17] and GQ [20] protocols, but there are many others as well [8, 21, 29, 31, 32, 33, 36]. Briefly, our hash function is defined using the simulator underlying a strong form of the usual honest-verifier zero-knowledge property of Σ-protocols. (We stress that the computation of the hash is deterministic even though the simulator is randomized!) The collision-resistance stems from strong special soundness [7], a well-studied property of Σ-protocols. The advantage of our approach is that there is a rich history in constructing proven-secure Σ-protocols and we can now leverage this to get collision-resistant hash functions. For future reference let us refer to a hash function derived from our approach as a Σ-hash function.

Damgard [16] and Cramer, Damgard and Mckenzie [13] have previously shown that it is possible to design commitment schemes based on Σ-protocols, but prior to our work it has not been observed that one can design collision-resistant hash functions from Σ-protocols. Note that secure commitment is not known to imply collision-resistant hashing and in fact is unlikely to do so because the former can be based on one-way functions [30] and the latter probably not [39]. Perhaps as a consequence, our construction requires slightly stronger properties from the Σ-protocols than do the constructions of [13, 16].

SPECIFIC DESIGNS. The Schnorr [37] and GQ [20] schemes are easily shown to meet our conditions, yielding collision resistant Σ-hash functions $\mathcal{H}\text{-}\mathcal{S}ch$ and $\mathcal{H}\text{-}\mathcal{GQ}$ based, respectively, on discrete log and RSA. More interesting is the Fiat-Shamir protocol $\mathcal{FS}$ [17]. It doesn't satisfy strong special soundness but we modify it to a protocol $\mathcal{SFS}$ (strong $\mathcal{FS}$) that we prove does under the factoring assumption, thereby obtaining a Σ-hash function $\mathcal{H}\text{-}\mathcal{SFS}$. From a modified version of the Micali-Shamir protocol [29] we obtain a Σ-hash function $\mathcal{H}\text{-}\mathcal{SMS}$ with security based on the SRPP (Square Roots of Prime Products) assumption of [29]. We also obtain a Σ-hash $\mathcal{H}\text{-}\mathcal{O}ka$ from Okamoto's protocol [32] and a pairing-based Σ-hash $\mathcal{H}\text{-}\mathcal{HS}$ from an identification protocol of [3] derived from the identity-based signature scheme of Hess [21].

HOW FAST? One question we consider interesting is, how fast can one hash while maintaining a proof of security under the *standard* factoring assumption? Figure 1 compares $\mathcal{H}\text{-}\mathcal{SFS}$ to the fastest known factoring-based functions and shows that the former emerges as the winner. (VSH is faster than all these, but is based on a non-standard assumption related to the difficulty of extracting modular square roots of products of small primes. We will discuss VSH, and our improvement to it, in a bit.) In Figure 1, $\mathcal{H}\text{-}\mathcal{D}a$ is the most efficient

Pre	$\mathcal{H}$-$\mathcal{D}a$	$\mathcal{H}$-$\mathcal{ST}$	$\mathcal{H}$-$\mathcal{SFS}$
0	1	0.22	2
2048	1	0.33	4
16384	1	2	8

Fig. 1. Performance of factoring-based hash functions. The modulus and output size are 1024 bits and the block size is 512 bits. "Pre" is the amount of pre-computation, in number of group elements stored. The table entry is the rate, defined as the average number of bits of data hashed per modular multiplication.

factoring-based instantiation known of Damgård's claw free permutation-based hash function [14, 19, 25]. $\mathcal{H}$-$\mathcal{ST}$ is the hash function of Shamir and Tauman [38]. The table entries are the rate, defined as the average number of bits of data hashed per modular multiplication in MD mode with a block size of 512 bits and a modulus and output size of 1024 bits. The figure shows that without pre-computation, $\mathcal{H}$-$\mathcal{SFS}$ is twice as fast as $\mathcal{H}$-$\mathcal{D}a$ and 9 times as fast as $\mathcal{H}$-$\mathcal{ST}$. But $\mathcal{H}$-$\mathcal{SFS}$ is amenable to pre-computation based speedup and $\mathcal{H}$-$\mathcal{D}a$ is not, so the gap in their rates increases swiftly with storage. $\mathcal{H}$-$\mathcal{ST}$ is also amenable to pre-computation based speedup but $\mathcal{H}$-$\mathcal{SFS}$ remains a factor 4 faster for any given amount of storage. We also remark that additionally $\mathcal{H}$-$\mathcal{SFS}$ is amenable to parallelization, unlike the other functions. We remark that $\mathcal{H}$-$\mathcal{SMS}$ is faster than $\mathcal{H}$-$\mathcal{SFS}$ but based on a stronger assumption. In Section 4 we recall $\mathcal{H}$-$\mathcal{D}a$ and $\mathcal{H}$-$\mathcal{ST}$ and justify the numbers in Figure 1. We also discuss implementation results.

Features of Σ-hash functions. Krawczyk and Rabin [25] introduced chameleon hashing. The functions they show have this property are that of [10] —$\mathcal{H}$-$\mathcal{S}ch$ in our taxonomy— and $\mathcal{H}$-$\mathcal{D}a$. Shamir and Tauman [38] add one more example, namely $\mathcal{H}$-$\mathcal{ST}$. We add five more examples, namely $\mathcal{H}$-$\mathcal{GQ}$, $\mathcal{H}$-$\mathcal{SFS}$, $\mathcal{H}$-$\mathcal{SMS}$, $\mathcal{H}$-$\mathcal{O}ka$, and $\mathcal{H}$-$\mathcal{HS}$. We obtain this as a consequence of a general result (Theorem 2) showing that any Σ-hash is chameleon.

Chameleon hashing has numerous applications. One of these is Shamir and Tauman's [38] chameleon hash based method for on-line, off-line signing. This means that when one uses a Σ-hash one can completely eliminate the on-line cost of signing. (This cost is shifted entirely to the off-line phase.) This compensates to some extent for the reduced efficiency of Σ-hash functions compared to conventional ones. (MD5 and SHA-1 are not chameleon and do not allow one to use the Shamir-Tauman construction.) Another application is chameleon signatures [25], which provides a recipient with a non-repudiable signature of a message without allowing it to prove to a third party that the signer signed this message. As explained in [25] this is an important tool for privacy-respecting authenticity in the signing of contracts and agreements. Finally, chameleon hash functions are used in designated-verifier signatures to achieve privacy [23, 40]. By adding new and more efficient chameleon hash functions to the pool of existing ones we enable new and more efficient ways to implement all these applications.

Another attribute of Σ-hash functions is that they are keyed. While one can, of course, simply hardwire into the code a particular key to get an unkeyed function in the style of MD5 or SHA-1, it is advantageous, as explained in [5], to allow each user to choose their own key. The reason is that damage from a collision is now limited to the user whose key is involved, and the attacker must re-invest resources to attack another key. This slows down the rate of attacks and gives users time to get patches in place or revoke keys.

Finally, the reductions underlying the security proofs of Σ-hash functions are tight, so that the proven security guarantees hold with normal values of the security parameters.

A REVERSE CONNECTION. As indicated above, Theorem 2 shows that Σ-hash functions are chameleon. Theorem 1 shows that the converse is true as well. Namely, all chameleon-hash functions are Σ-hash functions. We prove this by associating to any chameleon hash function $\mathcal{H}$ a Σ-protocol $\mathcal{SP}$ such that applying our Σ2H (Σ-to-hash) transform to $\mathcal{SP}$ returns $\mathcal{H}$. We thereby have a characterization of chameleon hash functions as Σ-hash functions, which we consider theoretically interesting. We also obtain numerous new Σ-protocols, and thus identification protocols and, via [13, 16], commitment schemes, from existing chameleon hash functions such as $\mathcal{H}$-$\mathcal{D}a$ [14] and $\mathcal{H}$-$\mathcal{ST}$ [38]. However, we are not aware of any practical benefit of these constructs over known ones.

UNIFYING PREVIOUS WORK. $\mathcal{H}$-$\mathcal{S}ch$ turns out to be exactly the classical hash function of Chaum, Van Heijst and Pfitzmann [10], and $\mathcal{H}$-$\mathcal{O}ka$ an extension thereof [10]. (Our other hash functions $\mathcal{H}$-$\mathcal{GQ}$, $\mathcal{H}$-$\mathcal{SFS}$, $\mathcal{H}$-$\mathcal{SMS}$ and $\mathcal{H}$-$\mathcal{HS}$ are new.) The re-derivation of these two hash functions as Σ-hashes sheds new light on the designs and shows how the Σ paradigm explains and unifies previous constructs.

But the most interesting connection in this regard is one we make between VSH [11] and $\mathcal{H}$-$\mathcal{SMS}$, the Σ-hash function emanating from the protocol of Micali and Shamir [29]. The latter is a more efficient version of the Fiat-Shamir protocol in which the public key, rather than consisting of random quadratic residues, consists of small primes. Interestingly $\mathcal{H}$-$\mathcal{SMS}$ turns out to be the VSH compression function [11] modulo some details. We suggest that this provides some intuition for the VSH design. It turns out that we can exploit this connection to get some improvements to VSH.

VSH*. In number-theoretic hashing there is (as elsewhere) a trade-off between speed and assumptions. We saw above that $\mathcal{H}$-$\mathcal{SFS}$ is the fastest known hash function under the standard factoring assumption. We now turn to non-standard factoring-related assumptions. Here the record-holder is VSH [11] with a proof based on the VSSR assumption of [11]. Our contribution is a modification VSH* of VSH that is faster for short messages. (Our implementations show that VSH* is up to 5 times faster than VSH on short messages. On long messages they have the same performance.) This is important because short messages are an important case in practice. (For example, most Internet packets are short.) VSH* remains provably collision-resistant under the same VSSR assumption as VSH.

We provide analogous improvements for the Fast-VSH variant of VSH provided by [11]. Again we can provide Fast-VSH* whose underlying compression function (unlike that of Fast-VSH) is proven collision-resistant, leading to speedups in hashing short messages. However, the speed gains are smaller than in the previous case.

Overall we believe that, even putting performance aside, having a collision resistant compression function underlying a hash function is a plus since it can be used directly and makes the hash function more misuse-resistant.

WHAT Σ-HASH FUNCTIONS AREN'T. Some recent work [1, 4, 12] suggests that general-purpose hash functions should have extra properties like pseudorandomness. Σ-hash functions are merely collision-resistant and chameleon; they do not offer these extra attributes. But as indicated above, Σ-hash functions are not intended to be general purpose. The envisaged applications are chameleon hashing and proven-secure, reasonable cost (purely) collision-resistant hashing.

RELATED WORK. Damgård [14] presents a construction of collision-resistant hash functions from claw-free permutation pairs [19]. As noted above, his factoring-based instantiation, based on [19] and also considered in [25, 38], is slower than our $\mathcal{H}$-$\mathcal{SFS}$.

Ishai, Kushilevitz and Ostrovsky [22] show how to transform homomorphic encryption (or commitment) schemes into collision-resistant hash functions. This is an interesting theoretical connection between the primitives. As far as we can tell, however, the approach is not yet practical. Specifically, their quadratic-residuosity (QR) based instantiation has a rate of 1/40 (that is, 40 modular multiplications per bit) with a 1024 bit modulus. (Their matrix needs 80 rows to get the 80-bit security corresponding to a 1024-bit modulus.) Hence their function is much slower than the constructs of Figure 1 in addition to being based on a stronger assumption (QR as opposed to factoring). Additionally it has a $80 \cdot 1024$ bit output so in a practical sense is not really hashing. Other instantiations of their construction that we know (El Gamal under DDH, Paillier [34] under DCRA) are also both slower than known ones and based on stronger assumptions.

Charles, Goren and Lauter [9] present a construct based on the assumed hardness of some problems related to elliptic curves. Their constructs are slower than ours and additionally are based on assumptions that are non-standard and should be treated with care [41]. Lyubashevsky, Micciancio, Peikert and Rosen [27] present a fast hash function SWIFFT with an asymptotic security proof based on assumptions about the hardness of lattice problems [26, 35], but the proof would not seem to yield guarantees for the parameter sizes proposed in [27]. In contrast, our reductions are tight and the proofs provide guaranties for standard values of the security parameters. Bellare and Micciancio's construction [2] (whose goal was to achieve incrementality) uses random oracles, but these can be eliminated by using a small block size, such as one bit. In this case their MuHASH is provably collision-resistant based only on the discrete-log assumption, and runs at 0.33 bits per group operation in MD mode. In comparison, $\mathcal{H}$-$\mathcal{S}ch$ (also discrete log based) is faster, at 0.57 bits per group operation in MD mode.

2 Definitions

NOTATION AND CONVENTIONS. We denote by $a_1\|\cdots\|a_n$ a string encoding of $a_1,\ldots,a_n$ from which the constituent objects are uniquely recoverable. We denote the empty string by ε. Unless otherwise indicated, an algorithm may be randomized. If A is a randomized algorithm then $y \leftarrow_{\$} A(x_1,\ldots)$ denotes the operation of running A with fresh coins on inputs $x_1,\ldots$ and letting y denote the output. We denote by $[A(x_1,\ldots)]$ the set of all y that have positive probability of being output by A on input $x_1,\ldots$. If S is a (finite) set then $s \leftarrow_{\$} S$ denotes the operation of picking s uniformly at random from S. If $X = x_1\|x_2\|\ldots\|x_n$, then $x_1\|x_2\|\ldots x_n \leftarrow X$ denotes the operation of parsing X into its constituents. Similarly, if $X = (x_1,x_2,\ldots,x_n)$ is an n-tuple, then $(x_1,x_2,\ldots,x_n) \leftarrow X$ denotes the operation of parsing X into its elements. We denote the security parameter by k, and by 1^k its unary encoding. Vectors are denoted in boldface, for example $\mathbf{u}$. If $\mathbf{u}$ is a vector then $|\mathbf{u}|$ is the number of its components and $\mathbf{u}[i]$ is its i-th component. "PT" stands for polynomial time.

Σ-PROTOCOLS. A Σ-protocol is a three-move interactive protocol conducted by a prover and a verifier. Formally, it is a tuple $\mathcal{SP} = (\mathsf{K},\mathsf{P},\mathsf{V},\mathrm{CmSet},\mathrm{ChSet},\mathrm{RpSet})$, where K,P are PT algorithms and V is a deterministic boolean algorithm. The key-generation algorithm K takes input 1^k and returns a pair (pk, sk) consisting of a public and secret key for the prover. The latter is initialized with pk, sk while the verifier is initialized with pk. The parties interact as depicted in Figure 2. The prover begins by applying P to pk, sk to yield his first move $\mathrm{Y} \in \mathrm{CmSet}(pk)$, called the commitment, together with state information y, called the ephemeral secret key. The commitment is sent to the verifier, who responds with a challenge c drawn at random from $\mathrm{ChSet}(pk)$. The prover computes its response z by applying P to pk, sk, the challenge and the ephemeral secret key y. (This computation may use fresh coins although in the bulk of protocols it is deterministic.) Upon receiving c the verifier applies V to the public key and transcript $\mathrm{Y}\|\mathrm{c}\|\mathrm{z}$ of the conversation to decide whether to accept or reject. We require *completeness*, which means that an interaction between the honest prover and verifier is always accepting. Formally, for all $k \in \mathbb{N}$ we have $\mathrm{d} = 1$ with probability 1 in the experiment

$$(pk, sk) \leftarrow_{\$} \mathsf{K}(1^k);\ (\mathrm{Y},\mathrm{y}) \leftarrow_{\$} \mathsf{P}(pk, sk);\ \mathrm{c} \leftarrow_{\$} \mathrm{ChSet}(pk);$$

$$\mathrm{z} \leftarrow_{\$} \mathsf{P}(pk, sk, \mathrm{c}, \mathrm{y});\ \mathrm{d} \leftarrow \mathsf{V}(pk, \mathrm{Y}\|\mathrm{c}\|\mathrm{z}).$$

The verifier given $pk, \mathrm{Y}\|\mathrm{c}\|\mathrm{z}$ should always check that $\mathrm{Y} \in \mathrm{CmSet}(\mathrm{Y})$ and $\mathrm{c} \in \mathrm{ChSet}(pk)$ and $\mathrm{z} \in \mathrm{RpSet}(pk)$ and reject otherwise. We implicitly assume this is done throughout.

SECURITY NOTIONS. We provide formal definitions of strong special soundness (sss) and strong honest verifier zero-knowledge (StHVZK). Strong special soundness of Σ-protocol $\mathcal{SP} = (\mathsf{K},\mathsf{P},\mathsf{V},\mathrm{CmSet},\mathrm{ChSet},\mathrm{RpSet})$ [7] asks that it be computationally infeasible, given only the public key, to produce a pair of accepting transcripts that are commitment-agreeing but challenge-response-disagreeing.

Prover		**Verifier**
Input: pk, sk		Input: pk
$(\mathrm{Y}, \mathrm{y}) \leftarrow_{\$} \mathsf{P}(pk, sk)$	$\xrightarrow{\mathrm{Y}}$	
	$\xleftarrow{\mathrm{c}}$	$\mathrm{c} \leftarrow_{\$} \mathrm{ChSet}(pk)$
$\mathrm{z} \leftarrow_{\$} \mathsf{P}(pk, sk, \mathrm{y}, \mathrm{c})$	$\xrightarrow{\mathrm{z}}$	$\mathrm{d} \leftarrow \mathsf{V}(pk, \mathrm{Y} \| \mathrm{c} \| \mathrm{z})$

Fig. 2. Σ-protocol. Keys pk and sk are produced using key-generation algorithm K.

Formally an sss-adversary, on input pk, returns a tuple $(\mathrm{Y}, \mathrm{c}_1, \mathrm{z}_1, \mathrm{c}_2, \mathrm{z}_2)$ such that $\mathrm{Y} \in \mathrm{CmSet}(pk)$; $\mathrm{c}_1, \mathrm{c}_2 \in \mathrm{ChSet}(pk)$; $\mathrm{z}_1, \mathrm{z}_2 \in \mathrm{RpSet}(pk)$ and $(\mathrm{c}_1, \mathrm{z}_1) \neq (\mathrm{c}_2, \mathrm{z}_2)$. The advantage $\mathbf{Adv}^{\mathrm{sss}}_{\mathcal{SP},A}(k)$ of such an adversary is defined for all $k \in \mathbb{N}$ as the probability that $\mathsf{V}(pk, \mathrm{Y} \| \mathrm{c}_1 \| \mathrm{z}_1) = 1$ and $\mathsf{V}(pk, \mathrm{Y} \| \mathrm{c}_2 \| \mathrm{z}_2) = 1$ in the experiment where $\mathsf{K}(1^k)$ is first executed to get (pk, sk) and then $A(pk)$ is executed to get $(\mathrm{Y}, \mathrm{c}_1, \mathrm{z}_1, \mathrm{c}_2, \mathrm{z}_2)$. We say that $\mathcal{SP}$ has strong special soundness if $\mathbf{Adv}^{\mathrm{sss}}_{\mathcal{SP},A}(\cdot)$ is negligible for all PT sss-adversaries A. To define StHVZK, let $\mathcal{Tr}_{\mathcal{SP}}$ be the algorithm that on input pk, sk executes P and V as per Figure 2 and returns the transcript $\mathrm{Y} \| \mathrm{c} \| \mathrm{z}$. Recall that a PT algorithm Sim is a HVZK simulator for $\mathcal{SP}$ if the outputs of the processes

$$(pk, sk) \leftarrow_{\$} \mathsf{K}(1^k); \quad \textbf{Return } (pk, \mathsf{Sim}(pk))$$

and

$$(pk, sk) \leftarrow_{\$} \mathsf{K}(1^k); \quad \textbf{Return } (pk, \mathcal{Tr}_{\mathcal{SP}}(pk, sk))$$

are identically distributed. We say that a PT algorithm StSim is a strong HVZK (StHVZK) simulator for $\mathcal{SP}$ if StSim is deterministic and the algorithm Sim defined on input pk by

$$\mathrm{c} \leftarrow_{\$} \mathrm{ChSet}(pk); \; \mathrm{z} \leftarrow_{\$} \mathrm{RpSet}(pk); \; \mathrm{Y} \leftarrow \mathsf{StSim}(pk, \mathrm{c}, \mathrm{z}); \; \textbf{Return } \mathrm{Y} \| \mathrm{c} \| \mathrm{z}$$

is a HVZK simulator for $\mathcal{SP}$. We say that $\mathcal{SP}$ is StHVZK if it has a PT StHVZK simulator. We denote by $\Sigma(\mathrm{sss})$ the set of all Σ-protocols that satisfy strong special soundness and by $\Sigma(\mathrm{StHVZK})$ the set of all Σ-protocols that are strong HVZK.

DISCUSSION. While the basic format of Σ-protocols as 3-move protocols of the type above is agreed upon, when it comes to security properties, there are different choices and variations in the literature. Our formalization of strong special soundness is from [7]. Strong HVZK seems to be new, but is natural since we will find many protocols that posses it.

COLLISION-RESISTANT HASH FUNCTIONS. A family of n-input hash functions (where $n \geq 1$ is a constant) is a tuple $\mathcal{H} = (\mathsf{KG}, \mathsf{H}, \mathrm{D}_1, \ldots, \mathrm{D}_n, \mathrm{R})$. The key-generation algorithm KG takes input 1^k and returns a key K describing a particular function $\mathsf{H}_K : \mathrm{D}_1(K) \times \ldots \mathrm{D}_n(K) \rightarrow \mathrm{R}(K)$. As this indicates, $\mathrm{D}_1, \ldots, \mathrm{D}_n, \mathrm{R}$ are functions that given K return sets. A cr-adversary, on input K returns distinct tuples $(x_1, \ldots, x_n), (y_1, \ldots, y_n)$ such that $x_i, y_i \in \mathrm{D}_i(K)$ for all $1 \leq i \leq n$. The advantage $\mathbf{Adv}^{\mathrm{cr}}_{\mathcal{H},B}(k)$ of such an adversary B is defined for all

$k \in \mathbb{N}$ as the probability that $\mathsf{H}(K, x_1, \ldots, x_n) = \mathsf{H}(K, y_1, \ldots, y_n)$ in the experiment where $\mathsf{KG}(1^k)$ is first executed to get K and then $B(K)$ is executed to get $((x_1, \ldots, x_n), (y_1, \ldots, y_n))$. We say that $\mathcal{H}$ is collision resistant if the cr-advantage of any PT adversary B is negligible.

3 Σ-Hash Theory

This section covers the theory of Σ-hash functions. We present and justify the $\Sigma 2\mathsf{H}$ transform that turns a Σ-protocol $\mathcal{SP} \in \Sigma(\text{sss}) \cap \Sigma(\text{StHVZK})$ into a collision-resistant hash function $\mathcal{H}\text{-}\mathcal{SP}$. Then we find Σ-protocols which we can prove have the required properties and derive specific Σ-hash functions. Finally we relate Σ and chameleon hash functions. In Section 4 we discuss the practical and performance aspects of our Σ-hash functions.

THE TRANSFORM. We show how to build a collision-resistant hash function from any Σ-protocol $\mathcal{SP} = (\mathsf{K}, \mathsf{P}, \mathsf{V}, \text{CmSet}, \text{ChSet}, \text{RpSet}) \in \Sigma(\text{sss}) \cap \Sigma(\text{StHVZK})$ that satisfies strong special soundness and strong HVZK. Let StSim be a strong HVZK simulator for $\mathcal{SP}$. We define the 2-input family of hash functions $\mathcal{H} = (\mathsf{KG}, \mathsf{H}, \text{ChSet}, \text{CmSet}, \text{RpSet})$ by $\mathsf{KG} = \mathsf{K}^{(1)}$ and $\mathsf{H}_{pk}(\mathrm{C}, \mathrm{Z}) = \mathsf{StSim}(pk, \mathrm{C}, \mathrm{Z})$, where $\mathsf{K}^{(1)}$ is the algorithm that on input 1^k lets $(pk, sk) \leftarrow_{\$} \mathsf{K}(1^k)$ and returns pk. In other words, the key is the prover's public key. (The secret key is discarded.) The inputs to the hash function are regarded as the challenge and response in the Σ-protocol. The output is the corresponding commitment. The existence of a StHVZK simulator is exploited to *deterministically* compute this output. We refer to a family of functions defined in this way as a Σ-hash. We write $\mathcal{H} = \Sigma 2\mathsf{H}(\mathcal{SP})$ to indicate that $\mathcal{H}$ has been derived as above from Σ-protocol $\mathcal{SP}$. The following theorem says that a Σ-hash family is collision-resistant.

Theorem 1. *Let* $\mathcal{SP} = (\mathsf{K}, \mathsf{P}, \mathsf{V}, \text{CmSet}, \text{ChSet}, \text{RpSet}) \in \Sigma(sss) \cap \Sigma(\text{StHVZK})$ *be a Σ-protocol. Let* $\mathcal{H} = (\mathsf{KG}, \mathsf{H}, \text{ChSet}, \text{RpSet}, \text{CmSet}) = \Sigma 2\mathsf{H}(\mathcal{SP})$ *be the family of hash functions associated to* $\mathcal{SP}$ *as above. For every cr adversary* B *against* $\mathcal{H}$ *there exists an sss-adversary* A *against* $\mathcal{SP}$ *such that for all* k *we have* $\mathbf{Adv}^{\mathrm{cr}}_{H,B}(k) \leq \mathbf{Adv}^{\mathrm{sss\text{-}na}}_{\mathcal{SP},A}(k)$*, and the running time of* B *is that of* A*.* ∎

The proof of this theorem, given in [6], is simple, but we note some subtleties, which is the way it relies on the (strong) HVZK and completeness of the Σ-protocol in addition to the strong special soundness. To construct Σ-hash functions we now seek Σ-protocols which we can show are in $\Sigma(\text{sss}) \cap \Sigma(\text{StHVZK})$.

OVERVIEW OF CONSTRUCTIONS. We begin, as illustrative examples, with the Schnorr [37] and GQ [20] Σ-protocols, which we can easily show to have the desired properties. The discrete log based Σ-hash $\mathcal{H}\text{-}\mathcal{S}ch$ obtained in the first case is that of [10] and its re-derivation as a Σ-hash sheds new light on its design and also shows how the Σ-hash paradigm unifies and explains existing work. The RSA based Σ-hash $\mathcal{H}\text{-}\mathcal{GQ}$ obtained in the second case is new. More interesting is the Fiat-Shamir [17] Σ-protocol. It doesn't satisfy strong special soundness, but we modify it to a Σ-protocol $\mathcal{SFS}$ that we prove is in $\Sigma(\text{sss}) \cap \Sigma(\text{StHVZK})$

Algorithm $\mathsf{K}(1^k)$	Prover		Verifier	$\mathsf{H}_{pk} : \mathbb{Z}_p \times \mathbb{Z}_p \rightarrow G$
$(\langle G\rangle, p, g) \leftarrow_{\$} \mathcal{G}(1^k)$	$\mathrm{Y} \leftarrow_{\$} \mathbb{Z}_p;\ \mathrm{Y} \leftarrow g^{\mathrm{y}}$	$\xrightarrow{\mathrm{Y}}$		$\mathsf{H}_{pk}(\mathrm{c}, \mathrm{z}) = X^{\mathrm{c}} g^{\mathrm{z}}$
$x \leftarrow_{\$} \mathbb{Z}_p$		$\xleftarrow{\mathrm{c}}$	$\mathrm{c} \leftarrow_{\$} \mathbb{Z}_p$	
$X \leftarrow g^{-x};\ sk \leftarrow x$	$\mathrm{z} \leftarrow \mathrm{y} + x \cdot \mathrm{c} \bmod p$	$\xrightarrow{\mathrm{z}}$	$\mathrm{d} \leftarrow (X^{\mathrm{c}} g^{\mathrm{z}} = \mathrm{Y})$	
$pk \leftarrow ((\langle G\rangle, p, g, X)$				
Return (pk, sk)				

Fig. 3. $\mathcal{Sch}$ Σ-protocol and the derived Σ-hash family, where $\mathcal{G}$ is a prime-order group generator

under the standard factoring assumption. With non-standard factoring-related assumptions (that it is hard to extract modular square roots of products of small primes) we get a faster Σ-hash $\mathcal{H}$-$\mathcal{SMS}$ from a modification of the Micali-Shamir Σ-protocol [29]. In [6] we show how to get another discrete-log based Σ-hash from Okamoto's protocol [32] and a pairing based one from the $\mathcal{HS}$ protocol [3, 21]. We proceed to the details.

$\mathcal{Sch}$. We fix a prime-order group generator, by which we mean a PT algorithm $\mathcal{G}$ that on input 1^k returns the description $\langle G\rangle$ of a group G of prime order $p \in \{2^{k-1}, \ldots, 2^k - 1\}$ together with p and a generator g of G. The key-generation process and protocol underlying the $\mathcal{Sch}$ Σ-protocol of [37] are then as shown in Figure 3. The algorithm that on input $pk = ((\langle G\rangle, p, g, X)$ picks $\mathrm{c}, \mathrm{z} \leftarrow_{\$} \mathbb{Z}_p$ and returns $X^c g^z \| \mathrm{c} \| \mathrm{z}$ is a HVZK simulator for $\mathcal{Sch}$, so $\mathcal{Sch} \in \Sigma(\mathrm{StHVZK})$ and the derived Σ-hash $\mathcal{H}$-$\mathcal{Sch}$ is as shown in Figure 3. The key observation for strong special soundness is that if $X^{\mathrm{c}_1} g^{\mathrm{z}_1} = X^{\mathrm{c}_2} g^{\mathrm{z}_2}$ and $(\mathrm{c}_1, \mathrm{z}_1) \neq (\mathrm{c}_2, \mathrm{z}_2)$ then it must be that $\mathrm{c}_1 \neq \mathrm{c}_2$. To sss-adversary A, this leads us to associate the discrete log finder D that on input $\langle G\rangle, p, g, X$ runs A on the same input to get $(\mathrm{Y}, \mathrm{c}_1, \mathrm{z}_1, \mathrm{c}_2, \mathrm{z}_2)$ and returns $(\mathrm{z}_2 - \mathrm{z}_1)(\mathrm{c}_1 - \mathrm{c}_2)^{-1} \bmod p$. Then for all k we have $\mathbf{Adv}^{\mathrm{sss}}_{\mathcal{Sch},A}(k) \leq \mathbf{Adv}^{\mathrm{dl}}_{\mathcal{G},D}(k)$, where the latter is defined as the probability that $x' = x$ in the experiment where we let $(\langle G\rangle, p, g) \leftarrow_{\$} \mathcal{G}(1^k)$ and $x \leftarrow_{\$} \mathbb{Z}_p$ and then let $x' \leftarrow_{\$} D((\langle G\rangle, p, g, g^x)$. This shows that $\mathcal{Sch}$ has strong special soundness as long as the discrete log problem is hard relative to $\mathcal{G}$. By Theorem 1 $\mathcal{H}$-$\mathcal{Sch}$ is collision-resistant under the same assumption.

$\mathcal{GQ}$. We fix a prime-exponent RSA generator with associated challenge length $L(\cdot)$, by which we mean a PT algorithm $\mathcal{G}_{\mathrm{rsa}}$ that on input 1^k returns an RSA modulus $N \in \{2^{k-1}, \ldots, 2^k - 1\}$ and an RSA encryption exponent $e > 2^{L(k)}$ that is a prime. The key-generation process and protocol underlying Σ-protocol $\mathcal{GQ}$ of [20] are then as shown in Figure 4. The algorithm that on input $pk = (N, e, l, X)$ picks $\mathrm{c} \leftarrow_{\$} \{0,1\}^l; \mathrm{z} \leftarrow_{\$} Z_N^*$ and returns $\mathrm{Y} \| \mathrm{c} \| \mathrm{z}$, where $\mathrm{Y} = \mathrm{z}^{\mathrm{c}} \mathrm{z}^2 \bmod N$, is a HVZK simulator for $\mathcal{GQ}$, so $\mathcal{GQ} \in \Sigma(\mathrm{StHVZK})$ and the derived Σ-hash $\mathcal{H}$-$\mathcal{GQ}$ is as shown in Figure 4. Again observe that if $X^{\mathrm{c}_1} \mathrm{z}_1^e = X^{\mathrm{c}_2} \mathrm{z}_2^e$ and $(\mathrm{c}_1, \mathrm{z}_1) \neq (\mathrm{c}_2, \mathrm{z}_2)$ then $\mathrm{c}_1 \neq \mathrm{c}_2$. To adversary A attacking the strong special soundness, this leads us to associate the inverter I that on input N, e, X runs A on input N, e, l, X where $l = L(\lfloor \log_2(N) \rfloor + 1)$ to get $(\mathrm{Y}, \mathrm{c}_1, \mathrm{z}_1, \mathrm{c}_2, \mathrm{z}_2)$ and returns $(\mathrm{z}_2 \mathrm{z}_1^{-1})^b X^a \bmod N$ where a, b satisfy $ae + b(\mathrm{c}_1 - \mathrm{c}_2) = 1$ and are

Algorithm $\mathsf{K}(1^k)$	Prover		Verifier
$(N,e) \leftarrow_\$ \mathcal{G}_{\mathrm{rsa}}(1^k)$	$\mathrm{y} \leftarrow_\$ \mathbb{Z}_N^*$;		
$x \leftarrow_\$ \mathbb{Z}_N^*$	$\mathrm{Y} \leftarrow \mathrm{y}^e \bmod N$	$\xrightarrow{\mathrm{Y}}$	
$X \leftarrow x^{-e} \bmod N$		$\xleftarrow{\mathrm{C}}$	$\mathrm{C} \leftarrow_\$ \{0,\ldots,2^l-1\}$
$l \leftarrow L(k)$; $sk \leftarrow x$	$\mathrm{Z} \leftarrow x^{\mathrm{C}} \cdot \mathrm{y} \bmod N$	$\xrightarrow{\mathrm{Z}}$	$\mathrm{d} \leftarrow (\mathrm{Y} = X^{\mathrm{C}} \cdot \mathrm{Z}^e \bmod N)$
$pk \leftarrow (N,e,l,X)$	$\mathsf{H}_{pk} : \{0,\ldots,2^l-1\} \times \mathbb{Z}_N^* \rightarrow \mathbb{Z}_N^*$		
Return (sk,pk)	$\mathsf{H}_{pk}(\mathrm{C},\mathrm{Z}) = X^{\mathrm{C}}\mathrm{Z}^e \bmod N$		

Fig. 4. $\mathcal{GQ}$ Σ-protocol and the derived Σ-hash family, where $\mathcal{G}_{\mathrm{rsa}}$ is a prime exponent RSA generator with associated challenge length L

found via the extended gcd algorithm. (This is where we use the fact that e is prime.) Then for all k we have $\mathbf{Adv}^{\mathrm{sss}}_{\mathcal{GQ},A}(k) \leq \mathbf{Adv}^{\mathrm{rsa}}_{\mathcal{G}_{\mathrm{rsa}},I}(k)$, where the latter is defined as the probability that $x' = x$ in the experiment where we let $(N,e) \leftarrow_\$ \mathcal{G}_{\mathrm{rsa}}(1^k)$ and $x \leftarrow_\$ \mathbb{Z}_N^*$ and then let $x' \leftarrow_\$ I(N,e,x^e \bmod N)$. This shows that $\mathcal{GQ}$ has strong special soundness if RSA is one-way relative to $\mathcal{G}_{\mathrm{rsa}}$. By Theorem 1, $\mathcal{H}$-$\mathcal{GQ}$ is collision-resistant under the same assumption.

$\mathcal{FS}$ AND $\mathcal{SFS}$. We fix a modulus generator, namely a PT algorithm $\mathcal{G}_{\mathrm{mod}}$ that on input 1^k returns a modulus $N \in \{2^{k-1},\ldots,2^k-1\}$ and distinct primes p,q such that $N = pq$. We also fix a challenge length $L(\cdot)$. If C is a l-bit string and $\mathbf{u} \in (Z_N^*)^l$ then we let $\mathbf{u}^{\mathrm{C}} = \prod \mathbf{u}[i]^{\mathrm{C}[i]}$ where the product is over $1 \leq i \leq l$ and $\mathrm{C}[i]$ denotes the i-th bit of C. The key-generation algorithm and protocol underlying the $\mathcal{FS}$ Σ-protocol are then as shown in Figure 5. However this protocol does not satisfy strong special soundness because if $\mathrm{Y}\|\mathrm{C}\|\mathrm{Z}$ is an accepting transcript relative to $pk = (N,l,\mathbf{u})$ then so is $\mathrm{Y}\|\mathrm{C}\|\mathrm{Z}'$ where $\mathrm{Z}' = N - \mathrm{Z}$. We now show how to modify $\mathcal{FS}$ so that it has strong special soundness. First, some notation. For $w \in \mathbb{Z}_N$ we let $[w]_N$ equal w if $w \leq N/2$ and $N - w$ otherwise. Let $\mathbb{Z}_N^+ = \mathbb{Z}_N^* \cap \{1,\ldots,N/2\}$. The modified protocol $\mathcal{SFS}$ (Strong $\mathcal{FS}$) is shown in Figure 5. Here CmSet, ChSet are as in $\mathcal{FS}$ but RpSet$((N,l,\mathbf{u}))$ is now equal to $\mathbb{Z}_N^+$ rather than $\mathbb{Z}_N^*$ as before. In [6] we show how to associate to any PT sss-adversary A a PT factoring adversary B such that for all $k \in \mathbb{N}$ we have $\mathbf{Adv}^{\mathrm{sss}}_{\mathcal{SFS},A} \leq 2 \cdot \mathbf{Adv}^{\mathrm{fac}}_{\mathcal{G}_{\mathrm{mod}},B}(k)$, where the latter is defined as the probability that $r \in \{p,q\}$ in the experiment where we let $(N,p,q) \leftarrow_\$ \mathcal{G}_{\mathrm{mod}}(1^k)$ and $r \leftarrow_\$ B(N)$. (Briefly, if $\mathrm{Y}\|\mathrm{C}_1\|\mathrm{Z}_1$ and $\mathrm{Y}\|\mathrm{C}_2\|\mathrm{Z}_2$ are accepting transcripts then if $\mathrm{C}_1 \neq \mathrm{C}_2$ we obtain the square root of some component of the public key and if $\mathrm{C}_1 = \mathrm{C}_2$ but $\mathrm{Z}_1 \neq \mathrm{Z}_2$ then $\mathrm{Z}_1, \mathrm{Z}_2$ are non-trivial square roots of the same square and we can factor N.) This shows that $\mathcal{SFS}$ has strong special soundness under the *standard* hardness of factoring assumption. Now, the algorithm that on input $pk = (N,l,\mathbf{u})$ lets $\mathrm{C} \leftarrow_\$ \{0,1\}^l$; $\mathrm{Z} \leftarrow_\$ Z_N^+$; $\mathrm{Y} \leftarrow \mathbf{u}^{\mathrm{C}} \cdot \mathrm{Z}^2 \bmod N$ and returns $\mathrm{Y}\|\mathrm{C}\|\mathrm{Z}$ is a HVZK simulator for $\mathcal{SFS}$. Accordingly $\mathcal{SFS} \in \Sigma(\mathrm{StHVZK})$ and we derive from $\mathcal{SFS}$ the Σ-hash family $\mathcal{H}$-$\mathcal{SFS}$ shown in Figure 5. Theorem 1 implies that $\mathcal{H}$-$\mathcal{SFS}$ is collision resistant under the standard factoring assumption.

$\mathcal{MS}$ AND $\mathcal{SMS}$. The Micali-Shamir protocol [29] is a variant of $\mathcal{FS}$ in which verification time is reduced by choosing the coordinates of $\mathbf{u}$ to be small primes.

Algorithm $\mathsf{K}(1^k)$	Prover		Verifier
$(N,p,q) \leftarrow_{\$} \mathcal{G}_{\rm mod}(1^k)$;	$\mathrm{y} \leftarrow_{\$} \mathbb{Z}_N^*$;		
$l \leftarrow L(k)$;	$\mathrm{Y} \leftarrow \mathrm{y}^2$	$\xrightarrow{\mathrm{Y}}$	
For $i = 1, \ldots, l$ do		$\xleftarrow{\mathrm{C}}$	$\mathrm{C} \leftarrow_{\$} \{0,1\}^l$
$\quad \mathbf{s}[i] \leftarrow_{\$} \mathbb{Z}_N^*$; $\mathbf{u}[i] \leftarrow \mathbf{s}[i]^{-2}$	$\mathrm{Z} \leftarrow \mathrm{y} \cdot \mathbf{s}^{\mathrm{C}}$	$\xrightarrow{\mathrm{Z}}$	$\mathrm{d} \leftarrow (\mathrm{Y} = \mathbf{u}^{\mathrm{C}} \cdot \mathrm{Z}^2)$
$sk \leftarrow \mathbf{s}$; $pk \leftarrow (N, l, \mathbf{u})$			
Return pk, sk			
Algorithm $\mathsf{K}(1^k)$	Prover		Verifier
$l \leftarrow L(k)$	$\mathrm{y} \leftarrow_{\$} \mathbb{Z}_N^*$;		
$(N,p,q,\mathbf{u}) \leftarrow_{\$} \mathcal{G}_{\mathcal{SP}}(1^k)$	$\mathrm{Y} \leftarrow \mathrm{y}^2$	$\xrightarrow{\mathrm{Y}}$	
For $i = 1, \ldots, l$ do		$\xleftarrow{\mathrm{C}}$	$\mathrm{C} \leftarrow_{\$} \{0,1\}^l$
$\quad \mathbf{s}[i] \leftarrow_{\$} \mathsf{SQR}(\mathbf{u}[i]^{-1}, p, q)$	$\mathrm{Z} \leftarrow [\mathrm{y} \cdot \mathbf{s}^{\mathrm{C}}]_N$	$\xrightarrow{\mathrm{Z}}$	$\mathrm{d} \leftarrow (\mathrm{Y} = \mathbf{u}^{\mathrm{C}} \cdot \mathrm{Z}^2)$
$pk \leftarrow (N, l, \mathbf{u})$; $sk \leftarrow \mathbf{s}$			
Return pk, sk			

$$\mathsf{H}_{pk} : \{0,1\}^l \times \mathbb{Z}_N^+ \rightarrow \mathbb{Z}_N^*$$
$$\mathsf{H}_{pk}(\mathrm{C}, \mathrm{Z}) = \mathbf{u}^{\mathrm{C}} \cdot \mathrm{Z}^2$$

Fig. 5. $\mathcal{FS}$, $\mathcal{SFS}$, $\mathcal{MS}$ and $\mathcal{SMS}$ protocols and the Σ-hash derived from $\mathcal{SFS}$, $\mathcal{SMS}$. The upper left key-generation algorithm is that of $\mathcal{FS}$ and $\mathcal{SFS}$, while the lower left one is that of $\mathcal{MS}$ and $\mathcal{SMS}$. The upper protocol is that of $\mathcal{FS}$ and $\mathcal{MS}$ while the lower protocol is that of $\mathcal{SFS}$ and $\mathcal{SMS}$. Here $\mathcal{G}_{\rm mod}$ is a modulus generator and $\mathcal{G}_{\rm SP}$ is a small prime modulus generator. The computations are in $\mathbb{Z}_N^*$, meaning modulo N.

As with $\mathcal{FS}$ it does not satisfy sss, but we can modify it to do so and thereby obtain a collision-resistant hash function $\mathcal{H}\text{-}\mathcal{SMS}$ that is faster than $\mathcal{H}\text{-}\mathcal{SFS}$ at the cost of a stronger assumption for security. To detail all this, let $\mathcal{G}_{\mathcal{SP}}$ be a small prime modulus generator with challenge length $L(\cdot)$, by which we mean a PT algorithm that on input 1^k returns a modulus $N \in \{2^{k-1}, \ldots, 2^k - 1\}$, distinct primes p, q such that $N = pq$, and an $L(k)$-vector $\mathbf{u}$ each of whose coordinates is a prime in $\mathrm{QR}(N) = \{x^2 \bmod N : x \in Z_N^*\}$. For efficiency we would choose these primes to be as small as possible. (For example $\mathbf{u}[i]$ is the i-th prime in $\mathrm{QR}(N)$.) An spr-adversary B against $\mathcal{G}_{\rm SP}, L$ takes input N and $\mathbf{u} \in (\mathbb{Z}_N^*)^{L(k)}$ and returns (x, S) where $x \in \mathbb{Z}_N^*$ and S is a non-empty subset of $\{0,1\}^l$. Its spr-advantage is defined for all k by

$$\mathbf{Adv}^{\rm spr}_{\mathcal{G}_{\rm SP},L,B}(k) = \Pr\left[x^2 \equiv \textstyle\prod_{i \in S} \mathbf{u}[i] \pmod N : \begin{array}{l} (N,p,q,\mathbf{u}) \leftarrow_{\$} \mathcal{G}_{\rm SP}(1^k) \;; \\ (x,S) \leftarrow_{\$} B(N,\mathbf{u}) \end{array} \right].$$

The SRPP (Square Root of Prime Products) assumption [29] says that the spr-advantage of any PT B is negligible. Now, Figure 5 shows our modified version $\mathcal{SMS}$ of the Micali-Shamir protocol. It is in Σ2H(StHVZK) for the same reason as $\mathcal{SFS}$ and hence the derived hash function is again as shown, where $\mathsf{SQR}(\cdot, p, q)$ takes input $w \in \mathrm{QR}(N)$ and returns at random one of the four square roots of w modulo $N = pq$, computed using the primes p, q. Strong special soundness of $\mathcal{SMS}$ is proven in [6] under the SRPP assumption. Theorem 1 now implies that $\mathcal{H}\text{-}\mathcal{SMS}$ is collision-resistant under the SRPP assumption.

Σ = CHAMELEON. We move from examples of Σ-hash functions to a general property of the class, namely that any Σ-hash function is chameleon. This is captured by the following.

Theorem 2. *Let* $\mathcal{SP} = (\mathsf{K}, \mathsf{P}, \mathsf{V}, \mathrm{CmSet}, \mathrm{ChSet}, \mathrm{RpSet}) \in \Sigma(\mathrm{StHVZK}) \cap \Sigma(\mathrm{sss}) \cap \Sigma(\mathrm{sc})$ *be a* Σ-protocol. *Then the* Σ-hash *family* $\mathcal{H}\text{-}\mathcal{SP} = \Sigma 2\mathsf{H}(\mathcal{SP}) = (\mathsf{KG}, \mathsf{H}, \mathrm{ChSet}, \mathrm{CmSet}, \mathrm{RpSet})$ *is chameleon.*

Refer to [6] for the proof of the above and the relevant definitions. As a consequence, we obtain the following new chameleon hash functions: $\mathcal{H}\text{-}\mathcal{GQ}$, $\mathcal{H}\text{-}\mathcal{SFS}$, $\mathcal{H}\text{-}\mathcal{SMS}$, $\mathcal{H}\text{-}\mathcal{O}k\alpha$, $\mathcal{H}\text{-}\mathcal{HS}$. ($\mathcal{H}\text{-}\mathcal{S}ch$ was already known to be chameleon [25].) This yields numerous new and more efficient instantiations of on-line/off-line signatures [38], chameleon signatures [25] and designated-verifier signatures [23, 40].

Even more interestingly, we prove the converse. The following theorem says that any chameleon hash family is a Σ-hash family, meaning the result of applying our $\Sigma 2\mathsf{H}$ transform to some Σ-protocol.

Theorem 3. *Let* $\mathcal{H} = (\mathsf{KG}, \mathsf{H}, \mathrm{ChSet}, \mathrm{CmSet}, \mathrm{RpSet})$ *be a family of chameleon hash functions. Then there is a* Σ-protocol $\mathcal{SP} = (\mathsf{K}, \mathsf{P}, \mathsf{V}, \mathrm{CmSet}, \mathrm{ChSet}, \mathrm{RpSet}) \in \Sigma(\mathrm{StHVZK}) \cap \Sigma(\mathrm{sss}) \cap \Sigma(\mathrm{sc})$ *such that* $\mathcal{H} = \Sigma 2\mathsf{H}(\mathcal{SP})$ *is the* Σ-hash *family corresponding to* $\mathcal{SP}$.

The proof is in [6]. Applying this to known chameleon-hash functions like $\mathcal{H}\text{-}\mathcal{D}a$ [14, 25] and $\mathcal{H}\text{-}\mathcal{ST}$ [38] yields new Σ-protocols and hence new identification schemes and, via [13, 16], new commitment schemes.

4 Σ-Hash Practice and Performance

In this section we cover practical issues related to Σ-hash functions, including performance, performance comparison with existing constructions and implementation results.

EXTENDING THE DOMAIN. A Σ-hash family $\mathcal{H}$ as defined above is actually a (keyed) compression function since the domain is relatively small. In practice however we need to hash messages of long and variable length. This would not at first appear to be much of a problem since we should be able to do MD iteration [15, 28]. In fact this is essentially true but one has to be careful about a few things. What one would naturally like to do is use the second argument to H_{pk} as the chaining variable. But this requires that outputs of the compression function can be regarded as chaining values, meaning $\mathrm{CmSet}(pk)$ be a subset of $\mathrm{RpSet}(pk)$. Sometimes this is true, as for $\mathcal{H}\text{-}\mathcal{GQ}$, which in this way lends itself easily and naturally to MD iteration. But in the case of $\mathcal{SFS}$ and $\mathcal{SMS}$ we have $\mathrm{CmSet}((N, l, \mathbf{u})) = \mathbb{Z}_N^* \subsetneq \mathbb{Z}_N^+ = \mathrm{RpSet}((N, l, \mathbf{u}))$. In [6] we show how to resolve these problems by appropriate "embeddings" that effectively allow the second input of the compression function to be used as a chaining variable at the cost of 1 bit in throughput and in particular allows us to run any of our Σ-hash functions in MD mode. We won't detail the general transform here,

Σ-hash	w	KB/s	space
$\mathcal{H}$-$\mathcal{SFS}$	0	30.85	n/a
$\mathcal{H}$-$\mathcal{SFS}$	4	67.41	2048
$\mathcal{H}$-$\mathcal{SFS}$	8	118.1	16384
$\mathcal{H}$-$\mathcal{SMS}$	0	914.3	n/a

Table 1. Implementation results. Here w is the "width" parameter determining pre-computation and the space is the number of group elements that need to be stored.

but it is instructive to describe the modified compression function. The public key has the form $(N, l, \mathbf{u}, v)$ where $N, l, \mathbf{u}$ are as before and $v \in \mathrm{QR}(N)$, and $\mathsf{H}_{pk} : \{0,1\}^l \times \mathbb{Z}_N^* \to \mathbb{Z}_N^*$ is defined by

$$\mathsf{H}_{pk}(\mathrm{C}, \mathrm{Z}) = \mathbf{u}^{\mathrm{C}} \cdot \mathrm{Z}^2 \cdot v^{f_N(\mathrm{Z})} \bmod N, \tag{1}$$

where $f_N(\mathrm{Z}) = 0$ if $\mathrm{Z} \in \mathbb{Z}_N^+$ and 1 otherwise. It can be shown that this modified function is also a Σ-hash, meaning the result of applying $\Sigma 2\mathsf{H}$ to a suitably modified version of the original Σ-protocol that retains the sss, StHVZK and sc properties of the original. But now $\mathrm{CmSet}((N, l, \mathbf{u}, v)) = \mathbb{Z}_N^* = \mathrm{RpSet}((N, l, \mathbf{u}, v))$ so MD-iteration is possible.

METRICS. We measure performance of a hash function in terms of rate, which we define as the average number of bits hashed per group operations. (By "average" we mean when the data is random.) In this measure, an exponentiation $a \mapsto A^a$ costs $1.5n$ group operations and a two-fold multi-exponentiation $a, b \mapsto A^a B^b$ costs $1.75n$ group operations where n is the length of a and also of b. We will use these estimates extensively below. We can consider two modes of operation of a given Σ-hash function $\mathcal{H}$-$\mathcal{SP}$, namely compression and MD. In the first case the data to be hashed by H_{pk} is the full input C, Z, while in the second case it is only C. (The second input is the chaining variable which is not part of the data.) The rate in MD mode is lower than in compression mode for most hash functions. ($\mathcal{SFS}$ is an interesting exception.) Compression mode is relevant when the function is being used as a chameleon hash, since the data can then be compressed with a standard (merely collision-resistant) hash function such as SHA-1 before applying the Σ-hash [25, Lemma 1]. MD mode is relevant when one wants to avoid conventional hash functions and get the full provable guarantees of the Σ-hash by using it alone. Our performance evaluations will consider MD mode.

PERFORMANCE OF Σ-hash FUNCTIONS. $\mathcal{H}$-$\mathcal{S}ch$ and $\mathcal{H}$-$\mathcal{GQ}$ can be computed with one two-fold multi-exponentiation so that they use 1.75 group operations per bit of data (in MD mode). We now turn to $\mathcal{H}$-$\mathcal{SFS}$. Since we are considering MD mode performance we refer to the MD-compatible version of the function from Equation (1). (But in fact performance is hardly affected by the modification.) On the average about half the bits of C are 1 so $\mathcal{H}$-$\mathcal{SFS}$ comes in at about 0.5 modular multiplications per bit. This explains the claim of Figure 1 in regard to $\mathcal{H}$-$\mathcal{SFS}$ without pre-computation. Now we look at how pre-computation speeds it up, using a block size of $l = 512$ (the same as MD5 and SHA-1) for illustration. The method is obvious. Pick a "width" w that divides l and let $t = l/w$. Letting $pk = (N, l, \mathbf{u}, v)$ denote the public key, pre-compute and store the table T with entries

$$T[i,x] = \prod_{j=1}^{w} \mathbf{u}[(i-1)w+j]^{x} \bmod N \quad (1 \leq i \leq t,\ x \in \{0, \ldots 2^{w}-1\})$$

The size of the table is $t2^{w} = l2^{w}/w$ group elements. Now computing $\mathcal{H}\text{-}\mathcal{SFS}$ takes $t + 2 = 2 + l/w$ multiplications since

$$\mathsf{H}_{pk}(\mathrm{C}, \mathrm{Z}) = \left(\prod_{i=1}^{t} T[i, x_i]\right) \cdot \mathrm{Z}^{2} \cdot v^{f_N(\mathrm{Z})} \bmod N,$$

where x_i is the integer with binary representation $\mathrm{C}[(i-1)w+1] \ldots c[iw]$ $(1 \leq i \leq t)$. The number of group operations per bit is thus $[2 + l/w]/l \approx 1/w$, meaning the rate is w. Figure 1 showed the storage and this rate for $w = 4$ and $w = 8$.

Analytical assessment of the performance of $\mathcal{H}\text{-}\mathcal{SMS}$ is difficult, but we have implemented both it and (for comparison) $\mathcal{H}\text{-}\mathcal{SFS}$. The implementation used a 1024 bit modulus and (for MD mode) a 512 bit block size. Table 1 shows that $\mathcal{H}\text{-}\mathcal{SMS}$ is about 30 times faster than the basic (no pre-computation) version of $\mathcal{H}\text{-}\mathcal{SFS}$. The gap drops to a factor of 15 and 7.5 when compared with the $w = 4$ and $w = 8$ pre-computation levels of $\mathcal{H}\text{-}\mathcal{SFS}$, respectively. Note that $\mathcal{H}\text{-}\mathcal{SMS}$ here is without pre-computation. (The latter does not seem to help it much.) These implementation results are on a Dual Pentium IV, 3.2 GHz machine, running Linux 2.6 kernel and using the gmp library [18].

COMPARISONS. We now assess performance of previous schemes, justifying claims in Section 1. Damgård [14] shows how to construct collision-resistant hash functions from claw-free permutations [19]. Of various factoring-based instantiations of his construction, the one of [19, 25], which we denote $\mathcal{H}\text{-}\mathcal{D}a$, seems to be the most efficient. The key is a modulus N product of two primes, one congruent to 3 mod 8 and the other to 7 mod 8, and the hash function $\mathsf{H}_N : \{0,1\}^l \times \mathbb{Z}_N^* \rightarrow \mathbb{Z}_N^*$ is defined by $\mathsf{H}_N(m, r) = 4^m \cdot r^s \bmod N$ where $s = 2^l$. Since multiplying by 4 is cheap, we view it as free and the cost is then one multiplication per bit, meaning $\mathcal{H}\text{-}\mathcal{SFS}$ is twice as fast. But pre-computation does not help $\mathcal{H}\text{-}\mathcal{D}a$ since r is not fixed, and the gap in rates increases as we allow pre-computation for $\mathcal{H}\text{-}\mathcal{SFS}$ as shown in Figure 1.

The key of Shamir and Tauman's [38] hash function is a modulus N and an $a \in \mathbb{Z}_N^*$. With a 1024 bit modulus the chaining variable needs to be 1024 bits as well, so that with a 512 bit block size the function would take a $512 + 1024$ bit input, regard it as an integer s, and return $a^s \bmod N$. The computation takes 1.5 multiplications per bit of the full input, which is $1.5 \cdot (1024 + 512)/512 = 4.5$ per bit of data, meaning the rate is $1/4.5 \approx 0.22$ as claimed in Figure 1. Since a is fixed, one can use the standard pre-computation methods for exponentiation. For any v dividing $1024 + 512 = 1536$, the computation takes $1536/v$ multiplications with a table of $2^v \cdot 1536/v$ group elements. Note that per data bit the rate is $512/(1536/v) = v/3$. To compare to $\mathcal{H}\text{-}\mathcal{SFS}$ we need to choose parameters so that the storage for the two is about the same, meaning $2^w(512/w) \approx 2^v(1536/v)$.

This yields $v = 1$ for $w = 4$ and $v = 6$ for $w = 8$. This explains the rates shown in Figure 1.

5 Improvements to VSH

The performance of a hash function on short inputs is important in practice. (For example, a significant fraction of Internet traffic consists of short packets.) We present a variant VSH* of VSH that is up to 5 times faster in this context while remaining proven-secure under the same assumption as VSH. The improvement stems from VSH*, unlike VSH, having a collision-resistant compression function.

BACKGROUND. The key of Contini, Lenstra and Steinfeld's VSH function [11] is a modulus N product of two primes. The VSH compression function $\mathsf{vsh}_N : \{0,1\}^l \times \mathbb{Z}_N^* \to \mathbb{Z}_N^*$ is defined by

$$\mathsf{vsh}_N(\mathrm{C}, \mathrm{Z}) = \mathrm{Z}^2 \cdot \prod_{i=1}^{l} p_i^{\mathrm{C}[i]} \bmod N,$$

where p_i is the i-th prime and $\mathrm{C}[i]$ is the i-th bit of C. The hash function VSH is obtained by MD-iteration of vsh with initial vector 1. A curious feature of VSH is that the compression function is *not* collision-resistant. Indeed, $\mathsf{vsh}_N(c, z) = \mathsf{vsh}_N(c, N - z)$ for any $c \in \{0,1\}^l$ and $z \in \mathbb{Z}_N^*$. Nonetheless, it is shown in [11] that the hash function VSH is collision-resistant based on the VSSR assumption. The latter states that given N, l it is hard to find $x \in \mathbb{Z}_N^*$ and integers $e_1, \ldots, e_l$, not all even, such that $x^2 \equiv p_1^{e_1} \cdot \ldots \cdot p_l^{e_l} \pmod N$. The proof makes crucial use of the fact that the initial vector is set to 1.

VSH*. We alter the compression function of VSH so that it becomes (provably) collision-resistant and then define VSH* by MD iteration with the initial vector being part of the data to be hashed. The first application of the compression function thus consumes much more (1024 bits more for a 1024 bit modulus, for example) of the input, resulting in significantly improved rate for the important practical case of hashing short messages. For example, the implementation results of Table 2 show speed increases of a factor of 5 over VSH when hashing 1024 bit messages. Performance for long messages is the same as for VSH. VSH* and its compression function vsh* are provably collision-resistant under the same VSSR assumption as VSH.

The inspiration comes from $\mathcal{H}$-$\mathcal{SMS}$ which we notice is very similar to vsh but, unlike the latter, is collision-resistant. The difference is that in $\mathcal{H}$-$\mathcal{SMS}$ the primes $\mathbf{u}[1], \ldots, \mathbf{u}[l], v$ —referring to the MD-compatible version of the function from Equation (1)— are quadratic residues. But this turns out to be important for the completeness of the Σ-protocol rather than for collision-resistance. This leads to the compression function $\mathsf{vsh}_N^* : \{0,1\}^l \times \mathbb{Z}_N^* \to \mathbb{Z}_N^*$ defined by

$$\mathsf{vsh}_N^*(\mathrm{C}, \mathrm{Z}) = \left(\prod_{i=1}^{l} p_i^{\mathrm{C}[i]}\right) \cdot p_{l+1}^{f_N(\mathrm{Z})} \cdot \mathrm{Z}^2 \bmod N,$$

where p_i is the i-th prime and $\mathrm{C}[i]$ is the i-th bit of C. As a check notice that $\mathsf{vsh}_N^*(\mathrm{C}, \mathrm{Z})$ is unlikely to equal $\mathsf{vsh}_N^*(\mathrm{C}, N - \mathrm{Z})$ because $f_N(\mathrm{Z}) \neq f_N(N - \mathrm{Z})$, meaning the attack showing vsh is not collision-resistant does not apply. Of course

Table 2. The size of the modulus used here is 1024. The block and the input size are given in bits.

Hash Function	block size	input size	Iterations	Avg. time
VSH	128	8×128	9	$140\mu s$
VSH*	128	8×128	1	$25\mu s$

this is not the only possible attack, but the proof of strong special soundness of $\mathcal{SMS}$ [6] can be adapted to show that vsh^* is collision-resistant under the VSSR assumption. Finally VSH^* is obtained by MD iteration of vsh^* but with the initial vector being the first $k-1$ bits of the input. For MD-strengthening, the standard padding method of SHA-1 is used.

The implementation results given in Table 2 were again obtained on a Dual Pentium IV, 3.2 GHz machine running Linux kernel 2.6 and using the gmp library [18]. We set the block size to 128 for both functions and considered hashing a 1024 bit input. In this case (even taking into account the increase in length due to MD strengthening) VSH^* needs 1 application of its compression function. On the other hand VSH (with their own form of strengthening) needs 9. The implementation shows that VSH^* is 5.6 times faster. We need to add that our implementations (unlike those of [11]) are not optimized, but our goal was more to assess the comparative than the absolute performance of these hash functions, and this is achieved because both are tested on the same platform.

References

1. Andreeva, E., Neven, G., Preneel, B., Shrimpton, T.: Seven-property-preserving iterated hashing: ROX. In: Kurosawa, K. (ed.) ASIACRYPT 2007. LNCS, vol. 4833, pp. 130–146. Springer, Heidelberg (2007)
2. Bellare, M., Micciancio, D.: A new paradigm for collision-free hashing: Incrementality at reduced cost. In: Fumy, W. (ed.) EUROCRYPT 1997. LNCS, vol. 1233, pp. 163–192. Springer, Heidelberg (1997)
3. Bellare, M., Namprempre, C., Neven, G.: Security proofs for identity-based identification and signature schemes. In: Cachin, C., Camenisch, J.L. (eds.) EUROCRYPT 2004. LNCS, vol. 3027, pp. 268–286. Springer, Heidelberg (2004)
4. Bellare, M., Ristenpart, T.: Multi-property-preserving hash domain extension and the EMD transform. In: Lai, X., Chen, K. (eds.) ASIACRYPT 2006. LNCS, vol. 4284, pp. 299–314. Springer, Heidelberg (2006)
5. Bellare, M., Ristenpart, T.: Hash functions in the dedicated-key setting: Design choices and MPP transforms. In: Arge, L., Cachin, C., Jurdziński, T., Tarlecki, A. (eds.) ICALP 2007. LNCS, vol. 4596, pp. 399–410. Springer, Heidelberg (2007)
6. Bellare, M., Ristov, T.: Hash Functions from Sigma Protocols and Improvements to VSH. Full Version of this paper. IACR eprint archive (2008)
7. Bellare, M., Shoup, S.: Two-tier signatures, strongly unforgeable signatures, and Fiat-Shamir without random oracles. In: Okamoto, T., Wang, X. (eds.) PKC 2007. LNCS, vol. 4450, pp. 201–216. Springer, Heidelberg (2007)

8. Beth, T.: Efficient zero-knowledge identification scheme for smart cards. In: Günther, C.G. (ed.) EUROCRYPT 1988. LNCS, vol. 330, pp. 77–84. Springer, Heidelberg (1988)
9. Charles, D., Goren, E., Lauter, K.: Cryptographic hash functions from expander graphs. In: Second NIST Hash Function Workshop (2006)
10. Chaum, D., Heijst, E.V., Pfitzmann, B.: Cryptographically strong undeniable signatures, unconditionally secure for the signer. In: Feigenbaum, J. (ed.) CRYPTO 1991. LNCS, vol. 576, pp. 470–484. Springer, Heidelberg (1992)
11. Contini, S., Lenstra, A.K., Steinfeld, R.: VSH, an efficient and provable collision-resistant hash function. In: Vaudenay, S. (ed.) EUROCRYPT 2006. LNCS, vol. 4004, pp. 165–182. Springer, Heidelberg (2006)
12. Coron, J.-S., Dodis, Y., Malinaud, C., Puniya, P.: Merkle-Damgård revisited: How to construct a hash function. In: Shoup, V. (ed.) CRYPTO 2005. LNCS, vol. 3621, pp. 430–448. Springer, Heidelberg (2005)
13. Cramer, R., Damgrd, I., MacKenzie, P.: Efficient zero-knowledge proofs of knowledge without intractability assumptions. In: Naccache, D., Paillier, P. (eds.) PKC 2002. LNCS, vol. 2274. Springer, Heidelberg (2002)
14. Damgård, I.: Collision free hash functions and public key signature schemes. In: Günther, C.G. (ed.) EUROCRYPT 1988. LNCS, vol. 330. Springer, Heidelberg (1988)
15. Damgård, I.: A design principle for hash functions. In: Brassard, G. (ed.) CRYPTO 1989. LNCS, vol. 435, pp. 416–427. Springer, Heidelberg (1990)
16. Damgård, I.: On the existence of bit commitment schemes and zero-knowledge proofs. In: Brassard, G. (ed.) CRYPTO 1989. LNCS, vol. 435, pp. 17–27. Springer, Heidelberg (1990)
17. Fiat, A., Shamir, A.: How to prove yourself: Practical solutions to identification and signature problems. In: Odlyzko, A.M. (ed.) CRYPTO 1986. LNCS, vol. 263, pp. 186–194. Springer, Heidelberg (1987)
18. The GNU MP bignum library, http://gmplib.org/
19. Goldwasser, S., Micali, S., Rivest, R.: A Digital Signature Scheme Secure Against Adaptive Chosen-Message Attacks. SIAM J. on Computing 17 (1988)
20. Guillou, L.C., Quisquater, J.-J.: A "paradoxical" indentity-based signature scheme resulting from zero-knowledge. In: Goldwasser, S. (ed.) CRYPTO 1988. LNCS, vol. 403, pp. 216–231. Springer, Heidelberg (1990)
21. Hess, F.: Efficient identity based signature schemes based on pairings. In: Nyberg, K., Heys, H.M. (eds.) SAC 2002. LNCS, vol. 2595, pp. 310–324. Springer, Heidelberg (2003)
22. Ishai, Y., Kushilevitz, E., Ostrovsky, R.: Sufficient conditions for collision-resistant hashing. In: Kilian, J. (ed.) TCC 2005. LNCS, vol. 3378, pp. 445–456. Springer, Heidelberg (2005)
23. Jakobsson, M., Sako, K., Impagliazzo, R.: Designated verifier proofs and their applications. In: Maurer, U.M. (ed.) EUROCRYPT 1996. LNCS, vol. 1070, pp. 143–154. Springer, Heidelberg (1996)
24. Jonsson, J., Kaliski, B.: Public-key cryptography standards (PKCS) #1: RSA cryptography, specifications version 2.1. Internet RFC 3447 (2003)
25. Krawczyk, H., Rabin, T.: Chameleon hashing and signatures. In: ISOC Network and Distributed System Security Symposium (NDSS 2000) (2000)
26. Lyubashevsky, V., Micciancio, D.: Generalized compact knapsacks are collision resistant. In: Bugliesi, M., Preneel, B., Sassone, V., Wegener, I. (eds.) ICALP 2006. LNCS, vol. 4052, pp. 144–155. Springer, Heidelberg (2006)

27. Lyubashevsky, V., Micciancio, D., Peikert, C., Rosen, A.: SWIFFT: a modest proposal for FFT hashing. In: Nyberg, K. (ed.) FSE 2008. LNCS, vol. 5086, pp. 54–72. Springer, Heidelberg (2008)
28. Merkle, R.C.: A certified digital signature. In: Brassard, G. (ed.) CRYPTO 1989. LNCS, vol. 435, pp. 218–238. Springer, Heidelberg (1990)
29. Micali, S., Shamir, A.: An improvement of the Fiat-Shamir identification and signature scheme. In: Goldwasser, S. (ed.) CRYPTO 1988. LNCS, vol. 403, pp. 244–247. Springer, Heidelberg (1990)
30. Naor, M.: Bit commitment using pseudorandomness. Journal of Cryptology 4(2), 151–158 (1991)
31. Ohta, K., Okamoto, T.: A modification of the Fiat-Shamir scheme. In: Goldwasser, S. (ed.) CRYPTO 1988. LNCS, vol. 403, pp. 232–243. Springer, Heidelberg (1990)
32. Okamoto, T.: Provably secure and practical identification schemes and corresponding signature schemes. In: Stinson, D.R. (ed.) CRYPTO 1993. LNCS, vol. 773. Springer, Heidelberg (1994)
33. Ong, H., Schnorr, C.-P.: Fast signature generation with a Fiat-Shamir like scheme. In: Damgård, I.B. (ed.) EUROCRYPT 1990. LNCS, vol. 473, pp. 432–440. Springer, Heidelberg (1991)
34. Paillier, P.: Public-key cryptosystems based on composite degree residuosity classes. In: Stern, J. (ed.) EUROCRYPT 1999. LNCS, vol. 1592, p. 223. Springer, Heidelberg (1999)
35. Peikert, C., Rosen, A.: Efficient Collision-Resistant Hashing from Worst-Case Assumptions on Cyclic Lattices. In: Halevi, S., Rabin, T. (eds.) TCC 2006. LNCS, vol. 3876, pp. 145–166. Springer, Heidelberg (2006)
36. Sakai, R., Ohgishi, K., Kasahara, M.: Cryptosystems based on pairing. In: SCIS 2000 (2000)
37. Schnorr, C.-P.: Efficient signature generation by smart cards. Journal of Cryptology 4(3), 161–174 (1991)
38. Shamir, A., Tauman, Y.: Improved online/offline signature schemes. In: Kilian, J. (ed.) CRYPTO 2001. LNCS, vol. 2139, p. 355. Springer, Heidelberg (2001)
39. Simon, D.R.: Finding collisions on a one-way street: Can secure hash functions be based on general assumptions? In: Nyberg, K. (ed.) EUROCRYPT 1998. LNCS, vol. 1403, pp. 334–345. Springer, Heidelberg (1998)
40. Steinfeld, R., Wang, H., Pieprzyk, J.: Efficient extension of standard Schnorr/RSA signatures into universal designated-verifier signatures. In: Bao, F., Deng, R., Zhou, J. (eds.) PKC 2004. LNCS, vol. 2947, pp. 86–100. Springer, Heidelberg (2004)
41. Tillich, J.-P., Zemor, G.: Collisions for the LPS expander graph hash function. In: Smart, N.P. (ed.) EUROCRYPT 2008. LNCS, vol. 4965, pp. 254–269. Springer, Heidelberg (2008)
42. Wang, X., Yin, Y.L., Yu, H.: Finding collisions in the full SHA-1. In: Shoup, V. (ed.) CRYPTO 2005. LNCS, vol. 3621, pp. 17–36. Springer, Heidelberg (2005)
43. Wang, X., Yin, Y.L., Yu, H.: How to break MD5 and other hash functions. In: Cramer, R. (ed.) EUROCRYPT 2005. LNCS, vol. 3494, pp. 19–35. Springer, Heidelberg (2005)

Slide Attacks on a Class of Hash Functions

Michael Gorski[1], Stefan Lucks[1], and Thomas Peyrin[2]

[1] Bauhaus-University Weimar
{Michael.Gorski,Stefan.Lucks}@uni-weimar.de
[2] Orange Labs and University of Versailles
Thomas.Peyrin@gmail.com

Abstract. This paper studies the application of slide attacks to hash functions. Slide attacks have mostly been used for block cipher cryptanalysis. But, as shown in the current paper, they also form a potential threat for hash functions, namely for sponge-function like structures. As it turns out, certain constructions for hash-function-based MACs can be vulnerable to forgery and even to key recovery attacks. In other cases, we can at least distinguish a given hash function from a random oracle.

To illustrate our results, we describe attacks against the GRINDAHL-256 and GRINDAHL-512 hash functions. To the best of our knowledge, this is the first cryptanalytic result on GRINDAHL-512. Furthermore, we point out a slide-based distinguisher attack on a slightly modified version of RADIOGATÚN. We finally discuss simple countermeasures as a defense against slide attacks.

Keywords: slide attacks, hash function, GRINDAHL, RADIOGATÚN, MAC, sponge function.

1 Introduction

A hash function $H : \{0,1\}^* \rightarrow \{0,1\}^n$ is used to compute an n-bit fingerprint from an arbitrarily-sized input. Established security requirements for cryptographic hash functions are collision resistance, preimage and 2nd preimage resistance – but ideally, cryptographers expect a good hash function to somehow *behave like a random oracle*.

Current practical hash functions, such as SHA-1 or SHA-2 [25, 26], are *iterated* hash functions, using a *compression function* with a fixed-length input, say $h : \{0,1\}^{n+l} \rightarrow \{0,1\}^n$, and the Merkle-Damgård (MD) transformation [14, 24] for the full hash function H with arbitrary input sizes. The core idea is to split the message M into l-bit blocks $M_1, \ldots, M_m \in \{0,1\}^l$ (with some padding, to ensure all the blocks are of size l-bit), to define an *initial value* X_0, and to apply the recurrence $X_i = h(X_{i-1}, M_i)$. The final *chaining variable* X_i is used as the hash output. The main benefit of the MD transformation is that it preserves collision resistance: if the compression function is collision resistant, then so is the hash function. Recent results, however, highlight some intrinsic limitations of the MD approach. This includes being vulnerable to multicollision attacks [16], long second-preimages attacks [19], and herding [18]. Even though the practical

J. Pieprzyk (Ed.): ASIACRYPT 2008, LNCS 5350, pp. 143–160, 2008.
© International Association for Cryptologic Research 2008

relevance of these attacks is unclear, they highlight some security issues, which designers of new hash functions should avoid.

In general, and due to certain structural weaknesses, MD-based hash functions do not *behave like a random oracles*. Consider, e.g., a secret key K, a message M and define a Message Authentication Code $\mathrm{MAC}(K, M) = H(K||M)$. If we model H as a random oracle, this construction meets the expected security requirements for a MAC. But for an MD-based hash function H, one can easily forge authentication codes: given $\mathrm{MAC}(K, M) = H(K||M)$, compute a valid $\mathrm{MAC}(K, M||Y) = H(K||M||Y)$ without knowing the secret key K. Coron et al. [11] recently discussed a formal model to prove hash functions being free from such structural weaknesses (but still weak against multicollision attacks).

Our contribution. Newly proposed hash function designs should not suffer from length extension. So for a new and well-designed hash function, the $\mathrm{MAC}(K, M) = H(K||M)$ *should* be a secure MAC. We will show that this is not the case for some recently proposed hash functions. In contrast to the case of MD-based hash functions, where one can forge messages but cannot recover K, our attacks allow, in general, the adversary to find K (much faster than by exhaustively searching for it).

Our attacks are an application of *slide attacks*. These are a classical tool for block ciphers cryptanalysis, but have so far not been used for hash function cryptanalysis.

The Targets for Our Attacks. A natural idea for thwarting the MD limitations is to increase the size of the internal chaining variables in the iterated process, see, e.g., [23]. Using a similar patch, sponge functions [3] followed the idea to employ a huge internal state (to hold a huge chaining variable) and to claim a *capacity* c, typically $c \gg n$. This defends against attackers even if these can perform $\gg 2^{n/2}$ operations (but are still restricted to $\ll 2^{c/2}$ units of work). Here n is considered a typical hash function output size (sponge functions may also provide for arbitrary output sizes, rather than for a fixed n).

Several recent hash functions follow this approach, including GRINDAHL [22] and RADIOGATÚN [2]. As far as we know, there are no cryptanalytic attack on either RADIOGATÚN or the 512-bit version of GRINDAHL while some collision attacks for the 256-bit version of GRINDAHL have already been published [20, 27].

In the current paper, we study the applicability of slide attacks for sponge functions. Our results indicate that slide attacks can be a serious threat for hash functions fitting into the sponge framework. On the other hand, if the hash function designer is aware of slide attacks, we believe it is easy to defend against such attacks. We give concrete examples by providing attacks against GRINDAHL [22] and two slightly tweaked versions of RADIOGATÚN [2]. Our attack applies for both published flavours of Grindahl, the 256-bit version and the 512-bit version. As far as we know, this is the first cryptanalytic result for the 512-bit version.

Outline: in Section 2 we recall the slide attacks basics, study the case of hash functions and focus on the case of sponge functions. Then, in Section 3 we give an example by applying our results to the GRINDAHL hash function and discuss the vulnerability of RADIOGATÚN to slide attacks in Section 4. Finally, we describe cheap and simple defenses against slide attacks and conclude in Section 5.

2 Slide Attacks

Block ciphers are often designed as a keyed permutation which is applied many rounds. It is a common belief that increasing the number of rounds makes the cipher stronger, but this is just true for statistical attacks such as differential or linear cryptanalysis. Some attacks can be applied even for block cipher variants with an arbitrary number of rounds. This is true for certain related key attacks, and for slide attacks. The usual defense is to strengthen the key schedule and the keyed permutation itself. Related key attacks have been introduced by Biham [5] and independently by Knudsen [21]. Slide attacks [8] utilize the self-similarity of the cipher, typically caused by a periodic key schedule. An r round block cipher with the same keyed permutation F^i in each round can be attacked by slide attacks if F^i is a weak permutation, i.e. the key used in F^i can be found with a slid plaintext-ciphertext pair.

2.1 Slide Attacks on Block Ciphers

Slide attacks on block ciphers have been applied to some ciphers with a weak key schedule (see [6, 8, 9, 12, 15, 17, 28, 29]). The original slide attack [8] works as follows. An n-bit block cipher E with r rounds is split into b identical rounds of the same keyed permutation F^i for $i = \{1, \ldots, b\}$. In the simplest case we have $b = r$ where the key schedule produces the same key in each round[1]. Thus we write the cipher as $E = F^1 \circ F^2 \circ \cdots \circ F^b = F \circ F \circ \cdots \circ F$. A plaintext P_j is then encrypted as

$$P_j \xrightarrow{F} X^{(1)} \xrightarrow{F} X^{(2)} \xrightarrow{F} \cdots \xrightarrow{F} C_j$$

where $X^{(i)}$ represents the intermediate encryption value after application of F^i and $X^{(b)} = C_j$ is the corresponding ciphertext. To mount a slide attack one has to find a slid pair of plaintexts (P_j, P_k), such that

$$P_k = F(P_j) \quad \text{and} \quad C_k = F(C_j) \tag{1}$$

hold, see also Figure 1.

Slide attacks can only be applied to a small class of ciphers with weak permutations periodic key schedules. A permutation is weak if, given the two equations in (1), it is easy to extract a non negligible part of the secret key. With $2^{n/2}$ known plaintext/ciphertext pairs (P_i, C_i) we expect at least one pair satisfying $P_k = F^i(P_j)$ among these texts by the birthday paradox. This gives us a slid

[1] Note that F^i might include more than one rounds of the cipher. If the key schedule produces identical keys with period p then F^i includes p rounds of the original cipher.

$$P_j \xrightarrow{F} X^{(1)} \xrightarrow{F} X^{(2)} \xrightarrow{F} X^{(3)} \xrightarrow{F} \cdots \xrightarrow{F} C_j$$
$$P_k \xrightarrow{F} X^{(1)} \xrightarrow{F} X^{(2)} \xrightarrow{F} \cdots \xrightarrow{F} X^{(b-1)} \xrightarrow{F} C_k$$

Fig. 1. A slide attack on block ciphers

pair. Thus, the classical slide attack allows to recover the unknown key of an n-bit block cipher using $O(2^{n/2})$ known plaintexts.

Advanced sliding techniques like *complementation slide* and *sliding with a twist* were introduced by Biryukov and Wagner [9]. These techniques allow to attack ciphers with a more complex key schedule. The basic concept of complementation slide is to slide two encryptions against each other where the inputs to the rounds may have a difference which is canceled out by a difference in the keys, while an encryption is slid against a decryption using a sliding with a twist technique. The *realigning slide attack* [28] allows to slide encryptions with unslid rounds in the middle of the slide. Biham et al. [6] improved the slide attack to detect a large amount of slid pairs very efficiently by using the relation between the cycle structure of the entire cipher and that of the keyed permutation.

2.2 Slide Attacks on Hash Functions

Slide attacks in a hash function setting have attracted very few consideration in the literature. To our knowledge, the only paper considers an attack on the internal block cipher from SHA-1 [31]. However, yet no direct way to transform it into a practical attack on the hash function has been found.

Slide attacks for block ciphers are different in some aspects from those applied on hash functions. By definition, block cipher computations depend on a secret key, and slide attacks are typically employed to distinguish a block cipher from a random permutation – and often for a key recovery attack to follow.

In the hash function case, there is no secret key to recover, just the message to be hashed, and the adversary is allowed to know this message – or even to choose it. Typical attacks on hash functions are about finding collisions or preimages – and it is hard to see how slide attacks could be employed in that context. But even for hash functions, a slide property that (or which) can be detected with some significant probability will allow us to differentiate the scheme from a random oracle. Indeed, with such a property, one can show a non random behavior of the hash function. This is already an issue, since hash functions are often utilized to simulate a random oracle as they are considered to be the closest practical primitive to this theoretical object. Going further, when secret data is used as a part of the input of the hash function, one can try to recover some information from it. The natural primitive where hash functions handle secret data are of course the Message Authentication Codes (MAC), that permit to authenticate a message M with a symmetric secret key K. For example, constructions such as HMAC [1] are implemented in a lot of different applications and make only two calls to a hash function. HMAC has the advantage to only require the internal function to be *weakly collision resistant* and also to provide secure MACs with MD-based hash functions. Note that a HMAC-based patch is one of the new domain extension algorithm proposed by Coron et al. [11] to thwart the simple MD-based MACs attacks. Those attacks are no more than a slide attack on the MD domain extension algorithm.

Generally, a good hash function H should provide a good MAC with the following computations: $\mathrm{MAC}(K, M) = H(K||M)$ or $\mathrm{MAC}(K, M) = H(K||M||K)$. Just like for block ciphers, if the hash function considered is not protected, one may be able to recover some non negligible part of the secret key K with a slide property that can be detected with a good probability. One has to note a work from Sasaki et al. [32] that attacks prefix, suffix and hybrid approaches for MAC constructions by using inner collisions for MD4, and a work from Preneel and Van Oorschot [30] that studies the envelope approache instantiated with MD5.

2.3 Slide Attacks on "Extended" Sponge Constructions

We analyze in this section how one can apply slide attacks to sponge-based hash functions, a newly introduced framework for building hash functions [2, 3]. More precisely, we use the "extended" sponge functions, a more general framework.

The "extended" sponge framework. Assume that H is an iterative hash function with an internal state of c words of p-bit each and a final output size of n bits. Let $M = M^1||M^2||\cdots||M^l$ be the $m \times p$-bit blocks of the message to hash with $M^l \neq 0^{m\times p}$ (the message is padded before split into blocks). Let M^i be the message block hashed at each round i and X^i the internal state after proceeding M^i, with $X^0 = IV$. We then have $X^i = F(S(X^{i-1}, M^i))$, where F is the round function and S defines how the message is incorporated in the internal state. Once all the l message blocks have been processed, r blank rounds (rounds with no message input) are applied $X^i = F(X^{i-1})$ and $A := X^{l+r}$ is the final internal state. Finally, we derivate n output bits by using the final output function $T(X^{l+r})$. Such a hash function can be written as

$$H(M) = X^0 \overset{F(S(X^0,M^1))}{\longrightarrow} \cdots \overset{F(S(X^{l-1},M^l))}{\longrightarrow} X^l \overbrace{\overset{F(X^l)}{\longrightarrow} \cdots \overset{F(X^{l+r-1})}{\longrightarrow}}^{\text{r times}} \overbrace{X_i^{l+r}}^{A} \overset{T(A)}{\longrightarrow} T_A,$$

where T_A represents the hash output. One has to note that for efficiency reasons and since the internal state will be big in practice, F is usually a quite light and fast round permutation.

This framework is really general and especially more general than the original sponge function one. More precisely, in the original model, S introduces the message blocks by XORing them to particular positions of the internal state. However, in our situation, we can also consider a function S that replaces some bits of the internal state by the message bits. We call the former *XOR sponge* and the latter *overwrite sponge*. Moreover, in the original model, the final output function T continues to apply some blank rounds and extract some bits from the internal state at the end of each application, until n bits have been received. In our framework we can also consider the case where the output bits come from a direct truncation of the final internal state A, and we call it *truncated sponge*.

There are two security issues, related to the general design of sponge functions. One issue is *invertibility*: one can run the function F into both directions. The second issue is *self-similarity*: all the blank rounds behave identically, and even a normal round can behave as a blank round if we have $X^{i-1} = S(X^{i-1}, M^i)$ (the effect of adding the message block is void). In the case of a XOR sponge we need $M^i = 0$ and in the case of an overwrite sponge we require that M_i is equal to the overwritten part of the internal state.

We will exploit self-similarity for our slide attacks. The idea is that if one message $M_1 = M^1||\ldots||M^l$ is the prefix of message $M_2 = M^1||\ldots||M^l||M^{l+1}$, the extended state after processing the first l blocks is the same. Now, if $X^{l+1} = S(X^l, M^{l+1})$, processing the next message block M^{l+1} for the longer message is the same as the first blank round when hashing the shorter message – the extended states remain identical. We call these two messages a *slid* pair : the two final internal states are just one permutation away $B := X_j^{l+r+1} = F(X_i^{l+r})$. The slide attack is shown in Table 1. Once we were able to generate a slid pair, we need to detect it. This fully depends on the output function T. When T is defined as in the original sponge framework, it is very easy to detect a slid pair : most of the output bits will be equal, just shifted by one round. If T is a truncation, we need to do a case by case analysis depending on the strength of the round function F and the number of bits thrown away. Yet finding and detecting a slid pair already allows us to differentiate the hash function from a random oracle.

We can try to go further, by attacking a MAC with prefix key, i.e. MAC(K, M). Note that such a construction makes sense as using HMAC based on a sponge hash function will turn out to be very inefficient. This is due to the fact that hashing very short messages (required in HMAC by the second hash function call) is quite

Table 1. A slide attack on hash functions

	X_i^0		X_j^0	
$F(S(X_i^0, M^0))$	$\downarrow$	$\leftarrow M^0 \rightarrow$	$\downarrow$	$F(S((X_j^0, M^0))$
	$\vdots$		$\vdots$	
$F(S(X_i^{l-1}, M^l))$	$\downarrow$	$\leftarrow M^l \rightarrow$	$\downarrow$	$F(S((X_j^{l-1}, M^l))$
	X_i^l		X_j^l	
$F(X_i^l)$	$\downarrow$	$M^{l+1} \rightarrow$	$\downarrow$	$F(S((X_j^l, M^{l+1}))$
	$\vdots$		X_j^{l+1}	
$F(X_i^{l+r-1})$	$\downarrow$		$\downarrow$	$F(X_j^{l+1})$
	$X_i^{l+r} = A$		$\vdots$	
$T(A)$	$\downarrow$		$\downarrow$	$F(X_j^{l+r})$
	T_A		$X_j^{l+r+1} = B$	
			$\downarrow$	$T(B)$
			T_B	

slow because of the blank rounds. Therefore, Bertoni et al. [4] proposed to use prefix-MAC instead of HMAC.

Consider a secret key K. For simplicity and without loss of generality, we assume some K to be a uniformly distributed $(k \times m \times p)$-bit random value (i.e. k message words long), for some public integer constant k. We will write $K = (K^1, \ldots, K^m) \in (\{0,1\}^{m \times p})^k$. The adversary is allowed to choose message challenges C_i, while the oracle replies $\mathrm{MAC}(K, C_i) = H(K||C_i)$. Ideally, finding K in such a scenario would require the adversary to exhaustively search over the set of all possible $K \in \{0,1\}^{k \times m \times p}$, thus taking $2^{k \times m \times p - 1}$ units of time on average. Forging a valid MAC depends on the size of the hash output and the size of the key, with a generic attack it requires $\min\{2^{k \times m \times p-1}, 2^n\}$ units of time. A pair of challenges (C_i, C_j), with $C_i = C_i^1||C_i^2||\cdots||C_i^l$ and $C_j = C_i||C_j^l$ is called a slid pair for K if their final internal state are slid by one application of the blank round function as:

$$X_j^{k+l+r+1} = F(X_i^{k+l+r})$$

Provided that one can generate slid pairs and detect them, one can also try to retrieve the internal state X_i^{k+l+r} thanks to this information. Again, a case by case analysis is required here. When X_i^{k+l+r} is known, one can invert all the blank rounds and get X_i^{k+l}. Note that with this information, an attacker can directly forge valid MACs for any message that contains M as prefix (exactly like the extension attacks against MD-based hash functions). If the round function with the message is also invertible, we can continue to invert all the challenge rounds and get X_i^k. This will allow us to recover some non trivial information on the secret key K.

A general outline of the attack is as follows:

1. Find and detect slid pairs of messages
2. Recover the internal state
3. Uncover some part of the secret key or forge valid MACs

The padding is very important here : for the XOR sponge functions, an appropriate padding can avoid slide attacks. Indeed, in that case, we require $M^l = 0^{m \times p}$ to get a slid pair. This gives an explanation why the condition $M^l \neq 0^{m \times p}$ is needed for the indifferentiability proofs of XOR sponge functions. However, for the truncated sponge function, a padding is ineffective to avoid slide attacks.

3 Applications

3.1 The Grindahl Design

Grindahl is a new hash function introduced by Knudsen et al. in [22], that fits our extended sponge framework. More precisely, it is an overwrite sponge function. There are two concrete instantiations of the Grindahl hash function family: a 256-bit and a 512-bit hash function proposed in the original Grindahl paper [22]. The parameters of these instantiations in our framework are defined as follows:

Grindahl-256 [22]. Grindahl-256 is a 256-bit hash function with $N_r = 4$ and $N_c = 12$. The rotation amounts are $(\rho_0, \ldots, \rho_3) = (1, 2, 4, 10)$.
Grindahl-512 [22]. Grindahl-512 is a 512-bit hash function with $N_r = 8$ and $N_c = 12$. The rotation amounts are $(\rho_0, \ldots, \rho_7) = (1, 2, \ldots, 8)$.

Note that the internal state of GRINDAHL can also be viewed as a matrix. Therefore, we define N_r and N_c to be the number of rows and columns of p-bit word respectively: we have $N_r \times N_c = c$. For each instance of GRINDAHL we have $p = 8$. The message chunk entering at each round can then be viewed as one column, thus $m = N_r$.

For GRINDAHL the padding consists of 10- and length-padding:

1. *10-padding* appends a "1"-bit to the message, followed by as many "0"-bits as needed to complete the last message block.
2. *Length-padding* then appends the number of message blocks (not bits!) for the entire padded message as a 64-bit value.

One effect of the 10-padding is that the last message block before the Length-padding can be any value, except for the all-zero block. (Or equivalently, any nonzero block B can be split up into an incomplete block R plus 10-padding: $B = R + P^{``10''}$. Note that R is 0 bit long if $B = 1000\ldots0$.)

A message $M = M^1||\ldots||M^l$ of 32-bit blocks M^i in the case of GRINDAHL-256, and an incomplete block M^l, will be padded to $Pad(M) = M^1||\ldots||M^l + P_1^{``10''}||M^{l+1}||M^{l+2}$, where $P_1^{``10''}$ is the 10-padding. This padded message has the following properties:

1. The last-but-two message block is not zero: $M^l + P_1^{``10''} \neq 0^{32}$.
2. The final two message blocks contain the 64-bit integer l: $(M^{l+1}||M^{l+2}) = l$. (From the GRINDAHL sample implementation, we conclude that the 32 least significant bits of the 64-bit value are stored in M^{l+2}, while the high-significant bits go into M^{l+1}.)

Similarly for GRINDAHL-512, a message $M = M^1||\ldots||M^l$ of 64-bit blocks M^i, where M^l is also incomplete, is padded to $Pad(M) = M^1||\ldots||M^l + P_1^{``10''}||M^{l+1}$ has the following properties after padding:

1. The last-but-one message block is not zero: $M^l + P_1^{``10''} \neq 0^{64}$.
2. The last message block contains the 64-bit integer l: $M^{l+1} = l$.

Most hash functions for variably-sized inputs iterate an underlying compression function for fixed-size inputs. GRINDAHL is no exception. At the end, the output will be the first $n/(p \times N_r)$ columns of of the final internal state. I.e., GRINDAHL is a truncated sponge. Internally, GRINDAHL uses a state of $(N_r \times N_c)$ words of p bit each. The compression function takes one m-word message block and an $(N_r \times N_c)$-word internal state as its input and generates new internal state (again of the size $(N_r \times N_c)$ words, of course), as its output.

Regarding this compression function, GRINDAHL follows a general three-step design strategy. Assume a m-word message block, which we write as M^i and a $(N_r \times N_c)$-word internal state, which we write as a N_c tuple of N_r-words: $(X^1, \ldots, X^{N_c}) \in (\{0,1\}^{p \times N_r})^{N_c}$. The incorporation step which concatenates a message block to the internal state is straightforward:

$$S : \{0,1\}^{p \times N_r} \times \{0,1\}^{p \times N_r \times N_c} \rightarrow \{0,1\}^{p \times N_r + p \times N_r \times N_c},$$
$$S(M^i, (X^1, \ldots, X^{N_c}) = (M^i, X^1, \ldots, X^{N_c}).$$

The $(p \times N_r + p \times N_r \times N_c)$-bit output of the incorporating S is the *extended state* $(X^0, \ldots, X^{N_c})$. The second step is a permutation over the extended state:

$$F : \{0,1\}^{p \times N_r + p \times N_r \times N_c} \rightarrow \{0,1\}^{p \times N_r + p \times N_r \times N_c},$$
$$F(X^0, \ldots, X^{N_c}) = (Y^0, \ldots, Y^{N_c}).$$

F is a permutation based on RIJNDAEL [13] primitives:

$$\begin{aligned} &\mathrm{F}(X^0, \ldots, X^{N_c}) \\ &\quad = \mathrm{MixColumns} \circ \mathrm{ShiftRows} \circ \mathrm{SubBytes} \circ \mathrm{AddConstant}(X^0, \ldots, X^{N_c}). \end{aligned}$$

MixColumns. Is a linear matrix multiplication of each state matrix column with a constant vector. This transformation is defined as in the RIJNDAEL specifications for the 256-bit version of GRINDAHL.

ShiftRows. This transformation cyclically shifts bytes a number of positions along each row. Thus, the i-th row is rotated by ρ_i positions to the right.

SubBytes. The only non-linear part of the permutation, exactly defined as the SubBytes function of RIJNDAEL.

AddConstant. This function is a simple XORing of the state matrix with a constant matrix M of the same size, where all bytes are zero except for one.

See [22] for a detailed description of GRINDAHL. The third operation is as straightforward as the first one – the first $p \times N_r$-bits of the $(p \times N_r + p \times N_r \times N_c)$-bit extended state are truncated away, to get a new $p \times N_r \times N_c$-bit internal state $(Y^1, \ldots, Y^{N_c})$:

$$\mathrm{R}\colon \{0,1\}^{p \times N_r + p \times N_r \times N_c} \rightarrow \{0,1\}^{p \times N_r \times N_c},\ \mathrm{R}(Y^0, \ldots, Y^{N_c}) = (Y^1, \ldots, Y^{N_c}).$$

See Figure 2 for a visual illustration of this design strategy. Note that the final truncation in one iteration and the initial concatenation of the b-bit message block in the next iteration together are tantamount to simply overwriting the corresponding column of the extended internal state. The final truncation is specified as

$$\mathrm{T}\colon \{0,1\}^{p \times N_r + p \times N_r \times N_c} \rightarrow \{0,1\}^n,\ \mathrm{T}(Y^0, \ldots, Y^{N_c}) = (Y^1, \ldots, Y^{n/(p \times N_r)}).$$

Let α be the internal state matrix with N_c columns and N_r rows, while $\hat{\alpha}$ represents the extended internal state with $N_c + 1$ columns and N_r rows. For

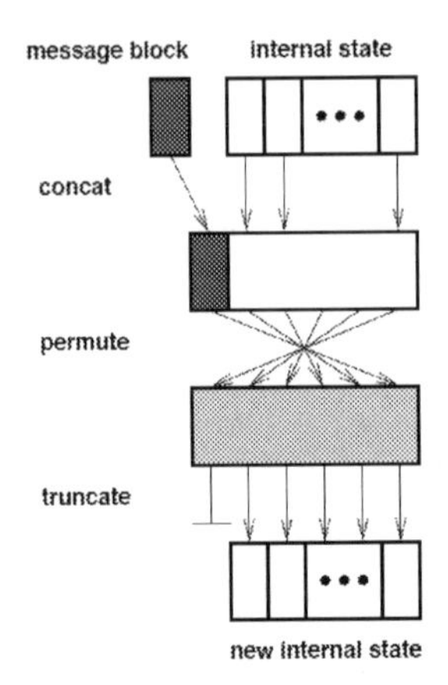

Fig. 2. The general design of the GRINDAHL compression function

a padded message $M = M^1||\ldots||M^d$ the GRINDAHL hash function does for $0 < i < d$:

$$\alpha \leftarrow R(P(S(M^i, \alpha)))$$

For the last message input M^d GRINDAHL performs $\hat{\alpha} \leftarrow P(S(M^d, \alpha))$. The truncation R is omitted after the last message input and finally 8 blank rounds with no message input are performed. These rounds only consists of the P operation on $\hat{\alpha}$. The final output remains after performing the output truncation T, which leaves the n-bit output.

3.2 Slide Attacks on GRINDAHL-512

Find slid pairs of messages. Building the challenge that generates a slid pair works as follows. We choose a message $M_1 = M_1^0||M_1^1||\ldots||M_1^{l-1}||M_1^l$, where M_1^l is a non complete block which will be padded. The MAC therefore processes

$$Pad(K||M_1) = K||M_1^0||M_1^1||\ldots||M_1^{l-1}||M_1^l + P_1^{\text{``10''}}||P_1^L$$

where $P_1^{\text{``10''}}$ is the 10-padding to $M_1^l$ and $P_1^L$ is the one-block of the message length. The value of $P_1^L$ can be chosen by the attacker while modifying the message length. For each $M_1$ we build the message $M_2 = M_1^0||M_1^1||\ldots||M_1^{l-1}||M_1^l + P_1^{\text{``10''}}||R$, where R is a random incomplete block. The MAC proceeds

$$Pad(K||M_2) = K||M_1^0||M_1^1||\ldots||M_1^{l-1}||M_1^l + P_1^{\text{``10''}}||R + P_2^{\text{``10''}}||P_2^L$$

and in some cases we have

$$Pad(K||M_2) = K||M_1^0||M_1^1||\ldots||M_1^{l-1}||M_1^l + P_1^{\text{``10''}}||P_1^L||P_2^L.$$

The messages M_1 and M_2 only differ in one additional block at the end. A pair (M_1, M_2) will be a slid pair with probability 2^{-64}. Detecting a slid pair is quite

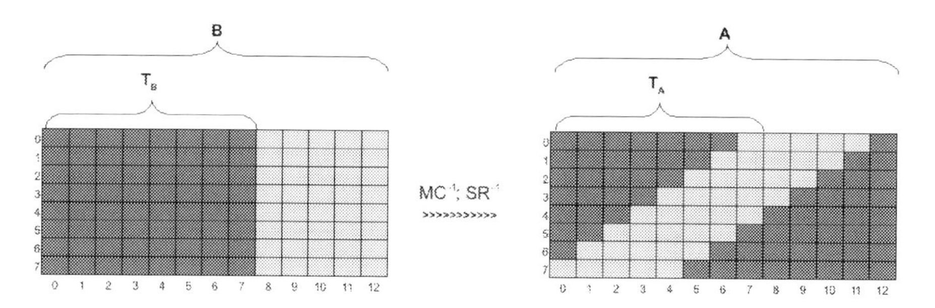

Fig. 3. Detecting a slid pair of messages for GRINDAHL-512. Cells in dark gray mark known bytes while cells in light gray mark unknown bytes. The inverse MixColumns (MC^{-1}) and the inverse ShiftRows (SR^{-1}) are the only two operations which are important for our analysis: AddConstant and SubBytes functions leave a known (respectively unknown) bytes known respectively unknown). Therefore we prevent the other operations.

simple. Let $T_A = A^0, \ldots, A^7$ and $T_B = B^0, \ldots, B^7$ be the query output (the truncated final internal states A and B). Then the condition $B = P(A)$ holds for a slid pair only. We could not directly apply another blank round to A since we only know T_A and not A. However, T_A and T_B leave enough information for detecting a slid pair. We can invert T_B one blank round and compare the resulting bytes with the bytes known from T_A. Thus, we can compare 34 bytes of T_A with the known bytes obtained from inverting T_B. In this way we can detect a slide pair since one occurs among 2^{64} pairs. Using the computation described above we can filter $2^{8 \cdot 34} = 2^{272}$ false pairs. Figure 3 shows the backward computation of one blank round.

Recover the internal state. A challenge (M_1, M_2) which produces a slid pair (T_A, T_B) can be used to recover the final internal state A (corresponding to the computation of M_1) just before the final truncation. Since the columns A^8 to A^{12} are unknown, we have to recover 40 bytes. As shown in Figure 3, we can directly recover 30 bytes from A by computing T_B one blank round backward, exactly as when we tried to detect slid pairs: we can fully invert the MixColumns transformation for the eight first columns (where all the bytes are known), then it is also very easy to invert ShiftRows, SubBytes and AddConstant transformations. So, when looking at Figure 3, it is clear that the attacker can directly get 30 unknown bytes from A. The remaining 10 unknown bytes can be recovered in a different way. For each possibility among those bytes ($2^{8 \cdot 10} = 2^{80}$ possibilities), we invert all the blank rounds and check if the last added word (the first encountered when computing backward) is P_1^L. Indeed, when inverting the real internal state A, we surely come to the insertion of P_1^L and this can be easily detected since we know this message block and since the message insertion overwrite the first column of the internal state. Now we are dealing with $2^{80-64} = 2^{16}$ possibilities only and we have to be careful, since some bytes become undetermined, if we continue the backward computation. The undetermined bytes are those which are replaced by the inserted message word during the message input step (due to the overwriting). However, we don't need them to discriminate among the 2^{16} lasting possibilities and we can compute one more round backward to check if we finally obtain the message word $M_1^l + P_1^{“10”}$ inserted. This leaves us the complete internal state A.

Uncover some parts of the secret key or forge valid MACs. By knowing the whole internal state A it is straightforward to invert the blank rounds. With this information, we can directly generate new valid MACs for messages which contain M_1 as prefix: we just have to continue the computation of the hash function by ourselves.

We can also try to invert the rounds where known message words are inserted. Some parts of the internal state are undetermined because of the truncation when adding message words as mentioned in the previous section. We can guess those undetermined columns by only keeping those which lead to the good inserted message words in the first column. This is equal to what we did above to recover the final internal state. By trying all the possible values of the truncated column, we can continue going backward and check which one leads to the known correct values of the message blocks inserted a few rounds before. Some trials will lead to wrong message blocks inserted and can be discarded. The one leading to the good values have a good chance to be the real erased bytes. Thus, we can go backward for all the known message words and recover the erased columns until we have to stop this procedure when we reach the unknown secret key word. The last unknown column which can be recovered is the column before inserting M_1^2. Now, with all those informations we can recover 4 bytes from 8 of the last unknown message block we encounter (the first when computing backward), which is part of the secret key. The rest of the secret can be then computed exhaustively (at a lower cost than brute force without slide attacks) or we can use a trick[2]. Indeed, we know that the initial internal state is equal to zero and one can accelerate the secret recovery with a meet-in-the-middle attack: we compute forward from the known initial internal state and we compute backward as we described before.

3.3 Slide Attacks on Grindahl-256

Applying the slide attack on Grindahl-256 is a little bit more difficult than on the 512 bit version, since the message block size is of 32 bit an the padding adds two additional blocks to the message. This makes it harder to control the message words and to find a slid pair. We describe the slide attack on Grindahl-256 in Appendix A.

4 Slide Attacks on Modified Versions of RadioGatún

We are able to use the presented technique to attack slightly modified versions of RadioGatún [2]. There are two possible modifications. Either we change the

[2] If the size of the key is not too big, we don't even require to do any exhausive search.

padding rule such that the last message block can also be an all zero input block. Or we change the message input step such that the input block enters the state via a replacement of the current state column. I.e., we turn RADIOGATÚN from an XOR sponge into an overwrite sponge. This modification is inspired by the message input step of GRINDAHL.

Consider the first case. The padding rule requires the final message block always to be non-zero, e.g., by applying the usual 10-padding. For an application where the message length always happens to be a multiple of the block size, this padding may appear to be moot. So consider an implementation without padding. Now the final message block might be all-zero. This gives an easy way to generate slid pairs (M_i, M_j) of messages – just take any M_i and set $M_j := (M_i||0)$ (M_i, concatenated by an all-zero message block). In this case, slide attacks are straightforward. Given for example a MAC such as

$$H(K||M_i) = Z_i^1, Z_i^2, Z_i^3, \ldots, Z_i^k \quad \text{and}$$
$$H(K||M_i||M^{zero}||M^{zero}) = Z_i^3, \ldots, Z_i^k, Z_i^{k+1}, Z_i^{k+2},$$

where Z_i^r represents the r-th output stream, one can easily forge the MAC $Z_i^2, \ldots, Z_i^{k+1}$, for the message $M_i||M^{zero}$.

For the second case (turning RADIOGATÚN into an overwrite sponge), consider a pair of messages $M_i = M_i^1||\ldots||M_i^d$ and $M_j = M_i||M_j^{d+1}$, with M_i being a prefix of M_j and M_j being one block longer. Both final blocks M_i^d and M_j^{d+1} being non-zero are slid with a probability of $2^{-p\times m}$. It is easy to detect slid pairs by comparing $k-1$ of the output blocks. If the pair (M_i, M_j) is slid, then we obtain:

$$H(K||M_i) = Z_i^1, Z_i^2, Z_i^3, \ldots, Z_i^k \quad \text{and}$$
$$H(K||M_i||M_j^{d+1}) = Z_i^2, Z_i^3, \ldots, Z_i^k, Z_i^{k+1}$$

This shows that our slide attack can be used to distinguish some hash functions, e.g. sponge-based one, from a random oracle if the designer do not take care to avoid sliding properties of their hash functions.

Slide-like distinguishing attacks are also applicable for other schemes, i.e. a modified version of PANAMA even leaves more non-trivial information of the internal state than our attack on modified RADIOGATÚN.

5 Possible Countermeasures and Conclusion

It only takes a negligible effort to defend hash functions from against slide attacks. Hash function designers, like block cipher designers, must be aware of possible slide attacks and be on guard for too much self-similarity in their constructions. For sponge-based hash functions, a simple patch would be to just add a nonzero constant just before running the blank rounds and extracting the hash value. Another option would be to marginally change the blank rounds. E.g., Grindahl could be changed such that the blank rounds use different rotation amounts (while maintaining the old rotation amounts for all the other rounds). Well-chosen padding rules also help. In the case of xor sponges, a good padding even seems to suffice as a defense against slide attacks.

We have studied the applicability of slide attacks for sponge functions. These are a classical tool for block cipher cryptanalysis, but have not been used for hash function cryptanalysis so far. Our results indicate that slide attacks can be a serious threat for sponge-based hash functions. If the hash function designer is aware of slide attacks, we believe that it is easy to defend against slide attacks. In our slide attacks on Grindahl and modified version of RadioGatún we demonstrated the power of these attacks. Our attacks apply for both published flavours of Grindahl, the 256-bit version and the 512-bit version. As far as we know, this is the first cryptanalytic result for the 512-bit version.

Acknowledgements

The authors wish to thank the anonymous reviewers for helpful comments.

References

[1] Bellare, M., Canetti, R., Krawczyk, H.: Keying hash functions for message authentication (1996)
[2] Bertoni, G., Daemen, J., Peeters, M., Van Assche, G.: Radiogatun, a belt-and-mill hash function. In: Second Cryptographic Hash Workshop, Santa Barbara, August 24-25 (2006), `http://radiogatun.noekeon.org/`
[3] Bertoni, G., Daemen, J., Peeters, M., Van Assche, G.: On the Indifferentiability of the Sponge Construction. In: Smart, N.P. (ed.) EUROCRYPT 2008. LNCS, vol. 4965, pp. 181–197. Springer, Heidelberg (2008)
[4] Bertoni, G., Daemen, J., Peeters, M., Van Assche, G.: Sponge Functions. In: ECRYPT Hash Workshop (2007)
[5] Biham, E.: New Types of Cryptanalytic Attacks Using Related Keys. J. Cryptology 7(4), 229–246 (1994)
[6] Biham, E., Dunkelman, O., Keller, N.: Improved Slide Attacks. In: Biryukov [7], pp. 153–166
[7] Biryukov, A. (ed.) FSE 2007. LNCS, vol. 4593. Springer, Heidelberg (2007)
[8] Biryukov, A., Wagner, D.: Slide Attacks. In: Knudsen, L.R. (ed.) FSE 1999. LNCS, vol. 1636, pp. 245–259. Springer, Heidelberg (1999)
[9] Biryukov, A., Wagner, D.: Advanced Slide Attacks. In: Preneel, B. (ed.) EUROCRYPT 2000. LNCS, vol. 1807, pp. 589–606. Springer, Heidelberg (2000)
[10] Brassard, G.(eds.) CRYPTO 1989. LNCS, vol. 435. Springer, Heidelberg (1990)
[11] Coron, J.-S., Dodis, Y., Malinaud, C., Puniya, P.: Merkle-Damgård Revisited: How to Construct a Hash Function. In: Shoup, V. (ed.) CRYPTO 2005. LNCS, vol. 3621, pp. 430–448. Springer, Heidelberg (2005)
[12] Courtois, N.T., Bard, G.V., Wagner, D.: Algebraic and Slide Attacks on KeeLoq. Cryptology ePrint Archive, Report 2007/062 (2007), `http://eprint.iacr.org/`
[13] Daemen, J., Rijmen, V.: The Design of Rijndael: AES - The Advanced Encryption Standard. Springer, Heidelberg (2002)
[14] Damgård, I.: A Design Principle for Hash Functions. In: Brassard [10], pp. 416–427
[15] Furuya, S.: Slide Attacks with a Known-Plaintext Cryptanalysis. In: Kim, K.-c. (ed.) ICISC 2001. LNCS, vol. 2288, pp. 214–225. Springer, Heidelberg (2002)

[16] Joux, A.: Multicollisions in Iterated Hash Functions. In: Franklin, M. (ed.) CRYPTO 2004. LNCS, vol. 3152, pp. 306–316. Springer, Heidelberg (2004)
[17] Kavut, S., Yücel, M.D.: Slide Attack on Spectr-H64. In: Menezes, A., Sarkar, P. (eds.) INDOCRYPT 2002. LNCS, vol. 2551, pp. 34–47. Springer, Heidelberg (2002)
[18] Kelsey, J., Kohno, T.: Herding Hash Functions and the Nostradamus Attack. In: Vaudenay, S. (ed.) EUROCRYPT 2006. LNCS, vol. 4004, pp. 183–200. Springer, Heidelberg (2006)
[19] Kelsey, J., Schneier, B.: Second Preimages on n-Bit Hash Functions for Much Less than 2^n Work. In: Cramer, R. (ed.) EUROCRYPT 2005. LNCS, vol. 3494, pp. 474–490. Springer, Heidelberg (2005)
[20] Khovratovich, D.: Cryptanalysis of hash functions with structures. In: ECRYPT Hash Workshop (2008)
[21] Knudsen, L.R.: Cryptanalysis of LOKI91. In: Zheng, Y., Seberry, J. (eds.) AUSCRYPT 1992. LNCS, vol. 718, pp. 196–208. Springer, Heidelberg (1993)
[22] Knudsen, L.R., Rechberger, C., Thomsen, S.S.: The Grindahl Hash Functions. In: Biryukov [7], pp. 39–57
[23] Lucks, S.: A Failure-Friendly Design Principle for Hash Functions. In: Roy, B. (ed.) ASIACRYPT 2005. LNCS, vol. 3788, pp. 474–494. Springer, Heidelberg (2005)
[24] Merkle, R.C.: One Way Hash Functions and DES. In: Brassard [10], pp. 428–446
[25] National Institute of Standards and Technology. FIPS 180-1: Secure Hash Standard (April 1995), `http://csrc.nist.gov`
[26] National Institute of Standards and Technology. FIPS 180-2: Secure Hash Standard (August 2002), `http://csrc.nist.gov`
[27] Peyrin, T.: Cryptanalysis of Grindahl. In: Kurosawa, K. (ed.) ASIACRYPT 2007. LNCS, vol. 4833, pp. 551–567. Springer, Heidelberg (2007)
[28] Phan, R.C.-W.: Advanced Slide Attacks Revisited: Realigning Slide on DES. In: Dawson, E., Vaudenay, S. (eds.) Mycrypt 2005. LNCS, vol. 3715, pp. 263–276. Springer, Heidelberg (2005)
[29] Phan, R.C.-W., Furuya, S.: Sliding Properties of the DES Key Schedule and Potential Extensions to the Slide Attacks. In: Lee, P.J., Lim, C.H. (eds.) ICISC 2002. LNCS, vol. 2587, pp. 138–148. Springer, Heidelberg (2003)
[30] Preneel, B., van Oorschot, P.C.: On the security of two mac algorithms. In: Maurer, U.M. (ed.) EUROCRYPT 1996. LNCS, vol. 1070, pp. 19–32. Springer, Heidelberg (1996)
[31] Saarinen, M.-J.O.: Cryptanalysis of Block Ciphers Based on SHA-1 and MD5. In: Johansson, T. (ed.) FSE 2003. LNCS, vol. 2887, pp. 36–44. Springer, Heidelberg (2003)
[32] Sasaki, Y., Wang, L., Ohta, K., Kunihiro, N.: Password Recovery on Challenge and Response: Impossible Differential Attack on Hash Function. In: Vaudenay, S. (ed.) AFRICACRYPT 2008. LNCS, vol. 5023, pp. 290–307. Springer, Heidelberg (2008)

A A Slide Attack on GRINDAHL-256

A.1 Find Slid Pairs of Messages

Building the challenge that generates a slid pair works as follows. We choose a message $M_1 = M_1^0||M_1^1||\dots||M_1^{l-1}||M_1^l$, where M_1^l is a non complete block which will be padded. The MAC therefore processes the hash input

$$Pad(K||M_1) = K||M_1^0||M_1^1||\dots||M_1^{l-1}||M_1^l + P_1^{\text{“10”}}||P_1^{L1}||P_1^{L2},$$

where $P_1^{\text{“10”}}$ is the 10-padding to M_1^l and $P_1^{L1}||P_1^{L2}$ is the two-block of the message length. Before building the second message, we want the condition

$$0^n \neq P_1^{L1} = P_1^{L2}$$

to always hold for M_1. Then, for each M_1 we build the message $M_2 = M_1^0||M_1^1||$ $M_i^2||\dots||M_1^{l-1}||M_1^l+P_1^{\text{“10”}}||R$, where R is an incomplete block which, after 10-padding, is the same as P_1^{L1}. As P_1^{L1} is nonzero, such an R exists. In this case, the hash input is

$$\begin{aligned} Pad(K||M_2) &= K||M_1^0||M_1^1||\dots||M_1^{l-1}||M_1^l + P_1^{\text{“10”}}||\ R + P_2^{\text{“10”}}\ ||P_2^{L1}||P_2^{L2} \\ &= K||M_1^0||M_1^1||\dots||M_1^{l-1}||M_1^l + P_1^{\text{“10”}}||\quad P_1^{L1}\quad ||P_1^{L2}||P_2^{L2} \end{aligned}$$

This holds because of the conditions fulfilled by P_1^{L1} and P_1^{L2}. In other words, M_1 and M_2 only differ in an additional block at the end. Such a pair (M_1, M_2) is slid with a probability of 2^{-32}. Detecting a slid pair is as simple as in the case of GRINDAHL-512. Here also the condition $B = P(A)$ holds for a slid pair only. T_A leaves enough information to compute column B^4 by performing one blank round on T_A. In this way the output (T_A, T_B) of a challenge (M_1, M_2) can be checked for a value of B^4 what we will expect for a slid pair. We can further check by using other columns than B^4, even if for them only a subspace of the potential solutions are determined by T_A. On the average, we need 2^{31} pairs until we find a slid one. Thus, we need to make about 2^{32} function calls to obtain and detect a slid pair. Figure 4 shows the backward computation of one blank round.

A.2 Recover the Internal State

A challenge (M_1, M_2) which produces a slid pair (T_A, T_B) can be used to recover the final internal state A (corresponding to the computation of M_1) just before the final truncation. Since the columns A^8 to A^{12} are unknown we have to recover 20 bytes. We can directly recover 10 bytes from A by computing T_B one blank round backward, exactly as when we tried to detect slid pairs: we can fully invert the MixColumns transformation for the eight first columns (where all the bytes are known), then it is also very easy to invert ShiftRows, SubBytes and AddConstant transformations. So, when looking at Figure 4, it is clear than the attacker can directly get 10 unknown bytes from A. The remaining 10 unknown bytes can be recovered in a different way. For each possibility among those bytes ($2^{8\cdot 10} = 2^{80}$ possibilities), we invert all the blank rounds and check if the last added word (the first encountered when computing backward) is P_1^{L2}. Indeed, when inverting the real internal state A, we surely come to the insertion of the block P_1^{L2} and this can be easily detected since we know this message block and since the message insertion overwrite the first column of the internal

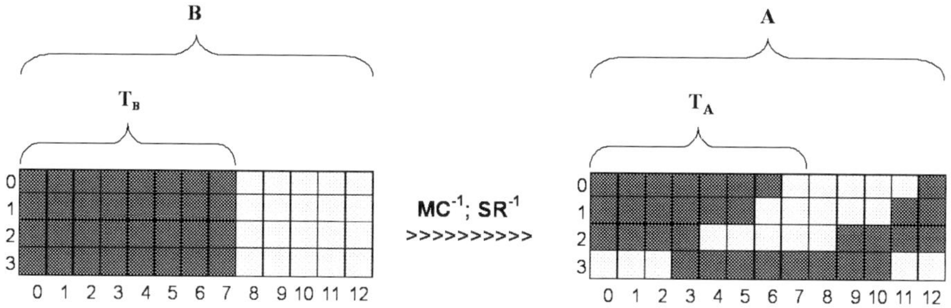

Fig. 4. Detecting a slide pair of messages for GRINDAHL-256. Cells in dark gray mark known bytes while cells in light gray mark unknown bytes. The inverse MixColumns (MC^{-1}) and the inverse ShiftRows (SR^{-1}) are the only two operations which are important for our analysis: AddConstant and SubBytes functions leave a known (respectively unknown) bytes known respectively unknown). Therefore we prevent the other operations.

state. We can continue to compute backward with the word P_1^{L1} even if some parts of the internal state at this point becomes undetermined due to the truncation when inserting the message words and thus we only have $2^{48-32} = 2^{16}$ possibilities. Finally, we can continue to the message word $M_1^l + P_1^{\text{"10"}}$ which leads to a recovery of the full internal state A.

A.3 Using Only Short Messages

Note that the above attack required $0^n \neq P_1^{L1} = P_1^{L2}$, i.e., the most significant and the least significant word of the length field of $(K||M_1)$ must the same – and nonzero. Thus, the smallest possible choice for $P_1^{L1} = P_1^{L2}$ is $P_1^{L1} = P_1^{L2} = 1$, implying a message length (for $(K||M)$, i.e., including the key) of $1 + 2^{32}$ blocks. If dealing with such long messages is an issue, we can modify the attack so use short messages. The modified attack goes as follows.

We choose a message $M_1 = M_1^0||M_1^1||\dots||M_1^{l-1}||M_1^l + P_1^{\text{"10"}}$, where the final block M_1^l is incomplete. The MAC processes the hash input

$$Pad(K||M_1) = K||M_1^0||M_1^1||\dots||M_1^{l-1}||M_1^l||P_1^{L1}||P_1^{L2},$$

with a length-field $P_1^{L1}||P_1^{L2}$. Note that P_1^{L2} holds the 32 least significant bits, while P_1^{L1} holds the 32 most significant bits. We assume short messages, thus $P_1^{L1} = 0^n$. This time, we want the MAC to process the hash input

$$\begin{aligned}
&Pad(K||M_2)\\
&= K||M_1^0||M_1^1||\dots||M_1^{l-1}||M_1^l + P_1^{\text{"10"}}||P_1^{L1}||S + P_2^{\text{"10"}}||P_2^{L1}||P_2^{L2}\\
&= K||M_1^0||M_1^1||\dots||M_1^{l-1}||M_1^l + P_1^{\text{"10"}}||P_1^{L1}||P_1^{L2}||P_2^{L1}||P_2^{L2},
\end{aligned}$$

Thus, M_1 and M_2 only differ in *two* additional blocks at the end. Accordingly, we choose

$$M_2 = M_1^0||M_1^1||\dots||M_1^{l-1}||M_1^l + P_1^{\text{"10"}}||P_1^{L1}||S.$$

As P_1^{L2} is nonzero, an incomplete block S with $S + P_2^{\text{"}10''} = P_1^{L2}$ does exist.

Now we define M_1 and M_2 as a slid-by-two pair, if, when processing the shorter message M_1, the first two empty rounds behave exactly the last two nonempty rounds when processing M_2. This happens with a probability of $(2^{-32})^2$, and on the average, we need 2^{63} pairs to find slid-by-two pair.

A pair of messages is slid-by-two, if and only if the two corresponding states A and B satisfy $B = P(P(A))$. Detecting slid-by-two pairs from $T(A)$ and $T(B)$ and then recovering the internal state A is slightly more complicated, compared to "ordinarily" slid-by-one pairs, but still feasible.

A.4 Uncover Some Parts of the Secret Key or Forge Valid MACs

By knowing the whole internal state A it is straightforward to invert the blank rounds. With this information, we can directly generate new valid MACs for messages which contain M_1 as prefix: we just have to continue the computation of the hash function by ourselves.

We can also try to invert the rounds where known message words are inserted. Some parts of the internal state are undetermined because of the truncation when adding message words. We do not known what was in the first column before erasing it with a message word, except for the first undetermined column which is equal to P_1^{L2} as described above. But we can guess those undetermined columns by only keeping those which lead to the good inserted message words in the first column. This is equal to what we did above to recover the final internal state. By trying all the possibles values the truncated column, we can continue going backward and check which one leads to the known correct values of the message blocks inserted a few rounds before. Some tries will lead to wrong message blocks inserted and can be discarded. The one leading to the good values have a good chance to be the real erased bytes. Thus, we can go backward for all the known message words and recover the erased columns until we have to stop this procedure when we reach the unknown secret key word. The last unknown column which can be recovered is the column before inserting M_1^3. Now, with all those informations we can recover 1 bytes from 4 of the last unknown message block we encounter (the first when computing backward), which is part of the secret key. The rest of the secret can be then computed exhaustively (at a lower cost than brute force without slide attacks) or we can use a trick[3]. Indeed, we know that the initial internal state is equal to zero and one can accelerate the secret recovery with a meet-in-the-middle attack: we compute forward from the known initial internal state and we compute backward as we described before.

[3] If the size of the key is not too big, we don't even require to do any exhaustive search.

Basing PRFs on Constant-Query Weak PRFs: Minimizing Assumptions for Efficient Symmetric Cryptography*

Ueli Maurer and Stefano Tessaro

Department of Computer Science, ETH Zurich, 8092 Zurich, Switzerland
{maurer,tessaros}@inf.ethz.ch

Abstract. Although it is well known that all basic private-key cryptographic primitives can be built from one-way functions, finding weak assumptions from which practical implementations of such primitives exist remains a challenging task. Towards this goal, this paper introduces the notion of a *constant-query weak PRF*, a function with a secret key which is computationally indistinguishable from a truly random function when evaluated at a *constant* number s of known random inputs, where s can be as small as two.

We provide iterated constructions of (arbitrary-input-length) PRFs from constant-query weak PRFs that even improve the efficiency of previous constructions based on the stronger assumption of a weak PRF (where polynomially many evaluations are allowed).

One of our constructions directly provides a new mode of operation using a constant-query weak PRF for IND-CPA symmetric encryption which is essentially as efficient as conventional PRF-based counter-mode encryption. Furthermore, our constructions yield efficient modes of operation for keying hash functions (such as MD5 and SHA-1) to obtain iterated PRFs (and hence MACs) which rely solely on the assumption that the underlying compression function is a constant-query weak PRF, which is the weakest assumption ever considered in this context.

1 Introduction

1.1 Minimizing Assumptions: Constant-Query Weak PRFs

Most cryptographic security proofs are *reductions*: Under the *assumption* that a primitive P exists, the existence of a second primitive P' is shown by means of a concrete construction that uses an implementation of P (usually in a black-box manner) to implement P'. For example, P' could be a *pseudorandom function* (PRF), i.e. a function with a secret key which is computationally indistinguishable from a truly random function under arbitrary (adaptive) access. These functions are central primitives as they provide a direct solution to the problems of provably secure symmetric encryption and message authentication.

* This research was partially supported by the Swiss National Science Foundation (SNF), project no. 200020-113700/1.

J. Pieprzyk (Ed.): ASIACRYPT 2008, LNCS 5350, pp. 161–178, 2008.
© International Association for Cryptologic Research 2008

Ideally, one would like the underlying primitive P to be as *weak* as possible, as in practice it is more likely that an efficient and secure candidate is successfully designed. Also, it is a safe practice to assume that already existing cryptographic functions (such as block ciphers or compression functions of hash functions) only fulfill weaker properties than what they have been originally designed for. Sometimes, however, reductions to weak assumptions turn out to be inefficient and involve large security losses (cf. [14] for a typical example), and hence designers of cryptographic systems are frequently confronted with a *trade-off* between the strength of the underlying assumption and the complexity of the resulting construction.

With the aim of proposing new weak assumptions for the purpose of building symmetric-key primitives, this paper introduces the notion of *constant-query weak pseudorandom functions*: Informally, for some constant s, a function $F : \{0,1\}^{\kappa} \times \{0,1\}^m \to \{0,1\}^n$ with $\kappa < s \cdot n$ is an *s-query weak PRF* (s-WPRF) if $F(K, \cdot)$ (under a secret key K) is indistinguishable from a random function when evaluated at s independent *known* random inputs.[1] This notion weakens significantly the regular concept of a *weak pseudorandom function* (WPRF) [19], where indistinguishability for polynomially many random inputs is required. We point out that a WPRF is by itself already much weaker than a PRF, as it possibly exhibits several non-random properties (such as having weak inputs or being commutative, i.e. $F(k, F(k', x)) = F(k', F(k, x))$). On top of this, an s-WPRF allows for even more structure: For instance, any $s+1$ distinct inputs $x_1, \ldots, x_{s+1}$ and the corresponding outputs $F(k, x_1), \ldots, F(k, x_{s+1})$ under a secret key k may satisfy an easily verifiable relation with no impact on the pseudorandomness of the function.

In this work, we address the problem of using s-WPRFs to construct PRFs. Since s-WPRFs imply the existence of one-way functions, a straightforward construction can be obtained using the results of [13, 14]. However, the inefficiency and the security loss of the resulting reduction make this approach unsuitable for any practical use, even if the underlying s-WPRF is both highly efficient and secure. For this reason, this paper deals with the question of finding *efficient* constructions of PRFs from s-WPRFs: Surprisingly, we are able to provide constructions which are more efficient than existing reductions of PRFs to WPRFs, while only requiring the underlying function to be an s-WPRF, for s as low as two. Furthermore, our constructions are iterated and can process inputs of arbitrary input length. This structure makes them well suited to be derived from properly keyed hash functions with very weak compression functions.

The next two sections are devoted to discussing previous work in the contexts of building PRFs from WPRFs and of iterated PRFs and MACs, respectively, and to relating it to our results.

---

[1] The assumption that s-WPRFs exist implies the existence of one-way functions, since the mapping $(k, r) \mapsto F(k, r)$ is easily verified to be one-way as long as $\kappa < s \cdot n$. For $\kappa \geq s \cdot n$, such functions can be constructed unconditionally, e.g. using s-wise independent functions. (However, optimal unconditional constructions with $\kappa = s \cdot n$ are not known for all parameters m).

1.2 Construction of PRFs from Weak PRFs

The first construction of a PRF from a WPRF is due to Naor and Reingold [19], and a further construction was later proposed by Maurer and Sjödin [17]. Both assume[2] a *length-preserving* underlying function $F : \{0,1\}^n \times \{0,1\}^n \to \{0,1\}^n$ (which can be obtained e.g. from a block cipher) and realize a keyed function mapping ℓ-bit strings to n-bit strings (for a fixed input length ℓ).

THE NAOR-REINGOLD CONSTRUCTION [19]. The construction NR_ℓ takes an ℓ-bit input (with ℓ being a power of two) and its secret key consists of 2ℓ n-bit strings $k_{1,0}, k_{1,1}, \ldots, k_{\ell,0}, k_{\ell,1}$. The computation on input $x = (x_1, \ldots, x_\ell)$ proceeds as follows: First, we define $y_i^{(\log \ell + 1)} := k_{i,x_i}$ for all $i = 1, \ldots, \ell$. Then, for all $j = \log \ell, \ldots, 1$ we compute $y_i^{(j)} := F(y_{2i-1}^{(j+1)}, y_{2i}^{(j+1)})$ for all $i = 1, \ldots, 2^{j-1}$ and finally output $y_1^{(1)}$. In other words, the elements of the key corresponding to the individual input bits are chosen as the values of the ℓ leaves of a complete binary tree which is evaluated in a bottom-up fashion by computing the value of each inner vertex as $F(y_l, y_r)$, where y_l and y_r are the values of its children, and finally outputting the value of the root. Hence, one evaluation of the construction needs $\frac{\ell}{2} + \frac{\ell}{4} + \cdots + \frac{\ell}{\ell} = \ell - 1$ calls to the underlying function F. A more involved construction (which we call $\overline{\mathsf{NR}}_{s,\ell}$) by the same authors uses a key consisting of s n-bit values and improves the total number of calls to roughly $\ell / \log s$ per evaluation, but only accepts ℓ and $\log s$ to have the form $2^j + 2$ for some $j \geq 0$. (For both constructions, other input lengths can be achieved through appropriate paddings.)

THE IC-CONSTRUCTION [17]. The construction IC_ℓ takes a $(\kappa + 2n)$-bit key consisting of three values $k_1 \in \{0,1\}^\kappa$ and $r, r' \in \{0,1\}^n$. (The value r' can even be made public.) It first precomputes the values $k_i := F(k_{i-1}, r')$ for all $i = 2, \ldots, \ell$. Furthermore, on an ℓ-bit input $x = (x_1, \ldots, x_\ell)$, it sets $y_0 := r$, and for all $j = 1, \ldots, \ell$, computes $y_j := F(k_j, y_{j-1})$ if $x_j = 1$, and $y_j := y_{j-1}$ else. Finally, it outputs y_ℓ. The construction IC_ℓ requires $w(x)$ calls to F when evaluated on input x, where $w(x) \leq \ell$ is the hamming weight of x. If memory restrictions do not allow storage of the keys $k_2, \ldots, k_\ell$, their values have to be computed at each evaluation and thus the construction requires $(\ell - 1) + w(x)$ calls to F per evaluation, which can be as high as $2\ell - 1$.

A central remark is that in order for all the aforementioned constructions to be secure PRFs for adversaries issuing q queries, the underlying WPRF must also be secure when evaluated at q random inputs. (The concrete security bounds for these constructions are discussed in the full version.) Moreover, in this paper we will focus on iterated constructions of PRFs and MACs where candidates for WPRFs may arise from (keyed) compression functions of hash functions, which have the form $F : \{0,1\}^\kappa \times \{0,1\}^n \to \{0,1\}^\kappa$ (where e.g. $\kappa = 160$ and $n = 512$ for SHA-1). The above constructions can all be extended in a straightforward

[2] In fact, the construction of [19] relies on an intermediate primitive, called a *synthesizer*, but a WPRF $F : \{0,1\}^n \times \{0,1\}^n \to \{0,1\}^n$ is in fact a synthesizer.

way[3] to handle such functions as well, but for the same input length ℓ the number of calls would increase considerably if $n > \kappa$ (roughly, by a factor of $\lceil \frac{n}{\kappa} \rceil$ with respect to the case $n = \kappa$, which is e.g. 4 for SHA-1). This holds even if we just want κ-bit outputs. Hence, this calls for a construction for which the condition $n > \kappa$ does not have a negative impact on the efficiency of the construction.

1.3 Assumptions in Iterated MACs and PRFs

Bellare et al. [2] proposed two efficient message authentication codes called HMAC and NMAC, obtained by appropriately keying an iterated[4] hash functions $H : \{0,1\}^\kappa \times \{0,1\}^* \to \{0,1\}^\kappa$ (where the first input is the initialization value) as $\mathsf{HMAC}(k_1 \| k_2, x) := H(IV, k_2 \| H(IV, k_1 \| x))$ (for a fixed known IV and $|k_1|, |k_2|$ both equal to the block length of H) and as $\mathsf{NMAC}(k_1 \| k_2, x) := H(k_2, H(k_1, x))$, respectively.[5] (Note that HMAC only requires black-box usage of H.) Even though alternative designs of MACs exist (such as CBC-MAC [5] and UMAC [8] to name a few), these constructions have enjoyed widespread usage due to the large availability of hash function implementations (both in hardware and in software). From the theoretical standpoint, security of HMAC/NMAC has been first proved [2] under the assumption that the compression function of H is a PRF (when keyed through the chaining value), and that H is *weakly collision resistant*, i.e. it is hard to find two distinct messages x, x' with $H(K, x) = H(K, x')$ for a secret key K (given oracle access to $H(K, \cdot)$). Bellare [1] subsequently proved HMAC/NMAC to be an arbitrary-input-length PRF under the sole assumption of the compression function being a PRF. We point out that the *cascade construction* by Bellare et al. [3] can also be seen as a way to key a hash function with a single key to obtain a PRF under the same assumption, at the expense of using a prefix-free encoding of the inputs. More recently, Fischlin [12] presented security proofs for HMAC/NMAC (when used as a MAC rather than as a PRF) relying on non-malleability properties of the underlying compression function. A further recent line of research [15, 22] has been concerned with increasing the efficiency of the HMAC/NMAC constructions by imposing slightly stronger requirements on the underlying compression function (i.e. pseudorandomness under mild types of related-key attacks).

The bottom line is that in order to deploy one of these constructions in practice, it is relevant to assess the level of confidence one is willing to put in the given compression function, but in view of continuous cryptanalytic achievements this is far from being a simple task. This issue motivates us to take steps in the

---

[3] One can simply base the above constructions on the function $F' : (k_1 \| \ldots \| k_c, r) \mapsto F(k_1, r) \| \ldots \| F(k_c, r)$ (possibly chopping some bits) where $c = \lceil n/\kappa \rceil$ (the function F' can be shown to be a WPRF). Note that more involved range-extension techniques (such as those from [11, 17, 20]) do not work here, as they require a length-preserving function beforehand.

[4] i.e. based on the Merkle-Damgård construction [10, 18], cf. also Section 2.

[5] Practical implementations usually consider single-keyed versions which, for simplicity, are not discussed here.

opposite direction: We raise the question of constructing iterated MACs (and PRFs) with *very low* requirements on the given compression function, while guaranteeing limited impact on the performance when compared with constructions with stronger underlying security assumptions. In particular, we consider constructions which only require the underlying compression function to be an s-WPRF (for s as small as two).

1.4 Contributions and Outline of This Paper

This paper initiates the study of constant-query WPRFs, and in particular investigates the problem of constructing efficient PRFs from these primitives.

- In Section 3, we present our first construction (called the RC-construction) of an arbitrary-input-length PRF from any s-WPRF $F : \{0,1\}^\kappa \times \{0,1\}^n \to \{0,1\}^\kappa$ (for some constant $s \geq 2$). As a special case of our construction, one obtains a fixed-input-length PRF which, for input length ℓ, requires $\approx \frac{\ell}{\log s}$ calls to F per evaluation, hence improving on earlier constructions despite the weaker underlying assumption of an s-WPRF.
- Careful instantiation of the RC-construction yields efficient counter-mode symmetric encryption relying on the sole assumption of an s-WPRF (for some $s \geq 2$), while requiring (on average) only $1 + \frac{1}{s-1}$ calls to F per κ-bit block of encrypted data and minimal storage overhead. Furthermore, the RC-construction directly yields constructions of efficient PRGs from s-WPRFs.
- Section 4 presents a further construction, called the *nested* RC-*construction*, which improves the throughput of the RC-construction for long messages making a novel use of pairwise independence, while still solely relying on the underlying function being an s-WPRF.
- Finally, Section 5 addresses the problem of deriving our constructions by keying iterated hash functions (such as SHA-1 or MD5) whose compression function is an s-WPRF: If minimal (and natural) regularity properties are additionally guaranteed by the compression function, the keying can be done in an entirely black-box way. Furthermore, this is the weakest assumption on the compression function for which modes of operations leading to secure PRFs and MACs have ever been considered.

The basic tools needed in the rest of the paper are reviewed in Section 2.

2 Preliminaries

2.1 Notational Conventions

Throughout this paper, for a set $\mathcal{U}$, we denote as $\mathcal{U}^n$, $\mathcal{U}^*$, and $\mathcal{U}^+$ the sets of *sequences* $s = (u_1, u_2, \ldots, u_{|s|})$ of elements of $\mathcal{U}$ of length $|s| = n$, of arbitrary length with the empty sequence ϵ, and of arbitrary length $|s|$ without the empty sequence ϵ, respectively. (For the case $\mathcal{U} = \{0,1\}$ we usually talk of *strings*.) The notation $s \| s'$ stands for the concatenation of sequences s and s', and u^r is the sequence $(u, u, \ldots, u)$ consisting of r repetitions of the symbol $u \in \mathcal{U}$. Given a two-argument function $F : \mathcal{U} \times \mathcal{V} \to \mathcal{Y}$ we denote by $F(u, \cdot)$ the function $\mathcal{V} \to \mathcal{Y}$ obtained by fixing the first input to u. Finally, $A^{\mathcal{O}}(r)$ denotes the (oracle) algorithm A which runs on input r with access to the oracle $\mathcal{O}$. Algorithms are in general randomized, and throughout this paper we fix some RAM model of computation for these algorithms. In particular, an algorithm A is said to have *running time* t if the sum of its description length and the worst-case number of steps it takes (counting oracle queries as single steps), taken over all randomness values, all inputs, and all compatible oracles, is at most t.

2.2 Cryptographic Functions

PSEUDORANDOM FUNCTIONS (PRFS). For some set $\mathcal{X}$ (generally $\mathcal{X} = \{0,1\}^{\ell}$ or $\mathcal{X} = \{0,1\}^{*}$) we consider *keyed* functions of the form $F : \{0,1\}^{\kappa} \times \{0,1\}^{\rho} \times \mathcal{X} \to \{0,1\}^{n}$, where the first and the second parameters are called the *public* and the *private* part of the *key*,[6] respectively. The third parameter is the *input* of F. We define the *PRF advantage* of D in distinguishing F from random as the quantity

$$\mathbf{Adv}_F^{\mathsf{PRF}}(D) := \left| \mathsf{P}\left[D^{F(K,R,\cdot)}(R) = 1\right] - \mathsf{P}\left[D^{\mathbf{R}_{\mathcal{X},n}}(R) = 1\right] \right|,$$

where K and R are independent and uniformly chosen from $\{0,1\}^{\kappa}$ and $\{0,1\}^{\rho}$, respectively, whereas $\mathbf{R}_{\mathcal{X},n}$ is a *random function* mapping elements of $\mathcal{X}$ to n-bit strings, i.e. an oracle which associates with each $x \in \mathcal{X}$ a uniformly-distributed independent n-bit string. (Whenever $\mathcal{X}$ is finite, this is equivalent to a randomly chosen function $\mathcal{X} \to \{0,1\}^{n}$.) For notational convenience we introduce the shorthand $\mathbf{Adv}_F^{\mathsf{PRF}}(t, q)$ to indicate the best advantage taken over all distinguishers with running time t and making at most q queries. Informally, F is a PRF if $\mathbf{Adv}_F^{\mathsf{PRF}}(t, q)$ is "negligible" for all t and q polynomial in some (understood) security parameter.[7] We often consider the case $\mathcal{X} = \{0,1\}^{*}$: Such a PRF is called an *arbitrary-input-length PRF* (AIL-PRF), and for this case we define $\mathbf{Adv}_F^{\mathsf{PRF}}(t, q, \ell)$ as the maximal advantage taken over all distinguishers with running time t making at most q queries each of length at most ℓ.

MESSAGE AUTHENTICATION CODES (MACS). A keyed function $F : \{0,1\}^{\kappa} \times \{0,1\}^{\rho} \times \{0,1\}^{*} \to \{0,1\}^{n}$ is a MAC if it is "unpredictable" under a secret key. Formally, for an adversary A, we define its *MAC advantage* as

$$\mathbf{Adv}_F^{\mathsf{MAC}}(A) := \mathsf{P}[A^{F(K,R,\cdot)}(R) = (x, y) \wedge F(K, R, x) = y \wedge x \text{ new}],$$

where K and R are random independent κ- and ρ-bit strings, respectively, and "x new" means that x was not queried by A to the given oracle. We define $\mathbf{Adv}_F^{\mathsf{MAC}}(t, q, \ell)$ to be the best advantage of an adversary with running time t

[6] We take this unconventional point of view as the constructions of this paper will allow part of the key to be publicly revealed with no harm to their security, and there are settings where this is a useful feature.

[7] If one considers both parts of the key as a single secret key, this implies that F is a PRF according to the usual definition considered in the literature.

issuing at most $q-1$ queries to $F(K,R,\cdot)$, each of length at most ℓ (and the message x output has also length at most ℓ). It is a well-known fact that a secure AIL-PRF is also a good MAC, namely $\mathbf{Adv}_F^{\mathsf{MAC}}(t,q,\ell) \le \mathbf{Adv}_F^{\mathsf{PRF}}(t',q,\ell)+\frac{1}{2^n}$, where $t \approx t'$.

Weak Pseudorandom Functions (WPRFs). This notion weakens a PRF to only withstand attacks where the function is queried on independent random *known* inputs. (Sometimes, this is called a *known-plaintext attack* (KPA) in the literature.) Formally, for some function g, we let $\mathcal{S}^g$ be the oracle that returns an ordered pair $(r, g(r))$ for a fresh random r each time it is invoked. Then, for a keyed function $F:\{0,1\}^\kappa \times \{0,1\}^m \to \{0,1\}^n$ we define the *WPRF advantage* of the distinguisher D in distinguishing F from random as

$$\mathbf{Adv}_F^{\mathsf{WPRF}}(D) := \left|\mathsf{P}[D^{\mathcal{S}^{F(K,\cdot)}}=1] - \mathsf{P}[D^{\mathcal{S}^{\mathbf{R}_{m,n}}}=1]\right|,$$

where $\mathbf{R}_{m,n}$ is a random function mapping m-bit strings to n-bit strings and K is a random κ-bit secret key.[8] Additionally $\mathbf{Adv}_F^{\mathsf{WPRF}}(t,q)$ stands for the best advantage taken over all distinguishers with running time t making at most q queries. For a constant s, we call a function $F:\{0,1\}^\kappa \times \{0,1\}^m \to \{0,1\}^n$ with $\kappa < s \cdot n$ an *s-weak pseudorandom function* (s-WPRF) if $\mathbf{Adv}_F^{\mathsf{WPRF}}(t,s)$ is negligible for all polynomial running times t, and we simply call it a *weak pseudorandom function* (WPRF) if $\mathbf{Adv}_F^{\mathsf{WPRF}}(t,q)$ is negligible for all polynomially bounded t and q.

Cascade and Iterated Hash Functions. For $F:\{0,1\}^\kappa \times \{0,1\}^n \to \{0,1\}^\kappa$, it is convenient to define its *cascade* $F^*:\{0,1\}^\kappa \times (\{0,1\}^n)^+ \to \{0,1\}^\kappa$ as the function which, on input $k \in \{0,1\}^\kappa$ and $(x_1,\ldots,x_\lambda) \in (\{0,1\}^n)^+$ (with $x_1,\ldots,x_\lambda \in \{0,1\}^n$) first computes $y_0 := k$ and $y_i = F(y_{i-1}, m_i)$ for all $i=1,\ldots,\lambda$, and subsequently outputs y_λ. In this work we also consider *iterated hash functions* [10, 18] $H:\{0,1\}^* \to \{0,1\}^\kappa$ with underlying *compression function* $F:\{0,1\}^\kappa \times \{0,1\}^n \to \{0,1\}^\kappa$ (n is generally called the *block length*) and *initialization value* $IV \in \{0,1\}^\kappa$ which are defined such that every input $x \in \{0,1\}^*$ is first padded as $(x_1,\ldots,x_\lambda) \in (\{0,1\}^n)^+$ and subsequently the value $F^*(IV,(x_1,\ldots,x_\lambda))$ is output. In general, the last block x_λ contains some padding bits as well as the length of the message (the so-called *MD-strengthening*) to preserve collision resistance of the compression function. Examples of such functions are those from the MD and the SHA families.

Universal Hashing. Let $H:\{0,1\}^\kappa \times \{0,1\}^* \to \{0,1\}^n$, and let $\delta:\mathbb{N}\to\mathbb{R}^+$. We say that H is *δ-almost universal* (δ-AU) if

$$\mathsf{p}_H^{\mathsf{COLL}}(x,x') := \mathsf{P}[H(K,x)=H(K,x')] \le \delta(\max\{|x|,|x'|\})$$

for all distinct $x,x' \in \{0,1\}^*$, where K is a randomly chosen κ-bit key. We stress that we extend the standard notion [9, 21] to deal with arbitrary input lengths

[8] In contrast to the definitions of PRFs and MACs, here we only consider a fully-secret key.

by letting δ be a function of the message length. The following lemma extends to the arbitrary-input-length case the well-known fact that δ-AU hash functions can be used to extend the domain of PRFs. (We omit its proof which follows the lines of the fixed-input-length case.)

Lemma 1. *Let* $H : \{0,1\}^\kappa \times \{0,1\}^* \to \{0,1\}^m$ *be* δ*-AU, and let* $F : \{0,1\}^\kappa \times \{0,1\}^\rho \times \{0,1\}^m \to \{0,1\}^n$ *be a keyed function. Define* $HF : \{0,1\}^{\kappa+\kappa'} \times \{0,1\}^\rho \times \{0,1\}^* \to \{0,1\}^n$ *such that* $HF(k\|k', r, x) := F(k', r, H(k, x))$*. Then we have*

$$\mathbf{Adv}_{HF}^{\mathsf{PRF}}(t, q, \ell) \leq \mathbf{Adv}_{F}^{\mathsf{PRF}}(t', q) + \frac{1}{2} \cdot q^2 \cdot \delta(\ell),$$

where $t' = t + q \cdot t_H(\ell)$*, with* $t_H(\ell)$ *being the time needed to evaluate* H *on inputs of length at most* ℓ*.*

3 The Randomized Cascade Construction

3.1 Description and Security of the Construction

In this section, we present the first iterated construction of this paper. It is reminiscent of the cascade construction of Bellare et al. [3], but only requires the underlying function $F : \{0,1\}^\kappa \times \{0,1\}^n \to \{0,1\}^\kappa$ to be an s-WPRF with $s \geq 2$ being a parameter of the construction. As in [3], the construction relies on the concept of a prefix-free encoding, which we briefly introduce.

PREFIX-FREE ENCODINGS. For a set $\mathcal{X}$, the efficiently computable function $\mathsf{ENC} : \mathcal{X} \to \{1, \ldots, s\}^+$ (i.e. outputting a non-empty sequence of elements of $\{1, \ldots, s\}$) is a *prefix-free encoding scheme* if for all distinct $x, x' \in \mathcal{X}$ the sequence $\mathsf{ENC}(x)$ is not a prefix of the sequence $\mathsf{ENC}(x')$. (In particular, ENC must be injective.) If $\mathcal{X} = \{0,1\}^*$, a prefix-free encoding scheme is e.g. obtained by encoding canonically the input as a sequence in $\{1, \ldots, s-1\}^*$, and then appending the symbol s to the sequence. Other variants exist, but it is generally desirable that ENC operates *on-line*, i.e. the encoding is progressively output while the input bits are provided, without the need to know the entire input before starting the encoding process. If $\mathcal{X} = \{0,1\}^\ell$ for some fixed ℓ, then prefix-freeness is achieved "for free" by encoding all inputs as sequences in $\{1, \ldots, s\}^*$ of equal length $\lceil \frac{\ell}{\log_2 s} \rceil$.

CONSTRUCTION. The *randomized cascade construction* with parameter s and input set $\mathcal{X}$ (where usually either $\mathcal{X} = \{0,1\}^*$ or $\mathcal{X} = \{0,1\}^\ell$ for a fixed ℓ) for the function F and prefix-free encoding scheme ENC, denoted $\mathsf{RC}^F_{s,\mathcal{X},\mathsf{ENC}}$, is a mapping $\{0,1\}^\kappa \times \{0,1\}^{sn} \times \mathcal{X} \to \{0,1\}^\kappa$: It takes a key consisting of a κ-bit private part k and an sn-bit long public part, which is interpreted as the concatenation of s n-bit strings $r_1, \ldots, r_s$. On input $x \in \mathcal{X}$, the κ-bit output is computed through the following two steps:

1. Compute $\mathsf{ENC}(x) = (m_1, \ldots, m_\lambda) \in \{1, \ldots, s\}^+$;
2. Output $F^*(k, (r_{m_1}, \ldots, r_{m_\lambda}))$.

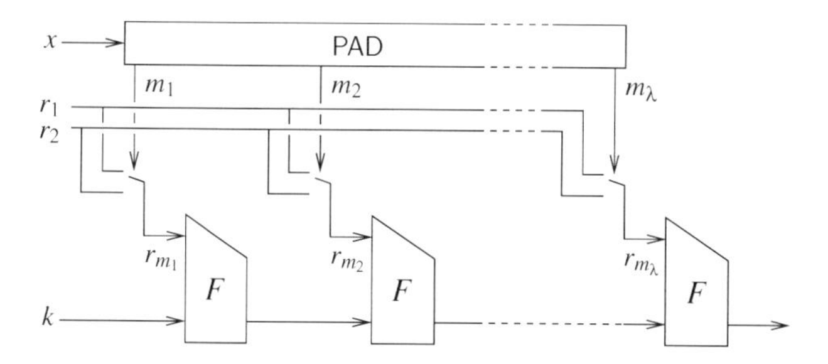

Fig. 1. The construction $\mathsf{RC}^F_{2,\mathsf{ENC}}$

As an example, the construction is depicted in Figure 1 for the special case $s = 2$. For notational convenience, we use the shorthands $\mathsf{RC}^F_{s,\mathsf{ENC}}$ for $\mathcal{X} = \{0,1\}^*$ (and omit the prefix-free encoding when it is generally understood from the context), as well as $\mathsf{RC}^F_{s,\ell}$ for $\mathcal{X} = \{0,1\}^\ell$ (where the canonical encoding described above is used). We also generically refer to the construction as the RC-construction.

Efficiency Comparisons. A fair comparison between the RC-construction and previous results can be undertaken for the fixed-input-length construction $\mathsf{RC}_{s,\ell}$ only. In the length-preserving case ($\kappa = n$), the construction $\mathsf{RC}_{\ell,s}$ is comparable to (for the case $s = 2$) the NR- and the IC-constructions in terms of calls to F, and outperforms them for $s > 2$. Furthermore, we obtain the same space-time trade-off of the $\overline{\mathsf{NR}}_{s,\ell}$-construction, but we allow for all possible values of s. Our construction also limits the effects of possibly very long input paddings in the NR- and $\overline{\mathsf{NR}}$-constructions. The efficiency improvement of our construction is however more evident in the case where $n > \kappa$, as even if $s = 2$, the number of calls to F of (the extended versions of) all other constructions is larger at least by a factor $\lceil \frac{n}{\kappa} \rceil$ (the factor is e.g. 4 when instantiating F with the compression function of SHA-1). Finally, because of the iterated structure, efficient sequential evaluation of $\mathsf{RC}_{s,\ell}$ requires (beside sufficient storage for the key material) κ bits only to store the "chaining value".

Security. In order to give precise security bounds for the RC-construction, it is convenient to think of the prefix-free encoding ENC in terms of a (possibly infinite) directed tree $\mathcal{T} = (\mathcal{V}, \mathcal{E})$ with vertex set $\mathcal{V}$ consisting of all sequences $(m_1, \ldots, m_j)$ which are a prefix of $\mathsf{ENC}(x)$ for some input x (in particular, including the encodings themselves and the empty sequence ϵ). Furthermore, for each $(m_1, \ldots, m_j) \in \mathcal{V}$ there exists a directed edge to $(m_1, \ldots, m_j, m_{j+1})$ for all $m_{j+1} \in \{1, \ldots, s\}$ such that $(m_1, \ldots, m_{j+1}) \in \mathcal{V}$. Hence, it is easy to see that ϵ is the root of the directed tree and its leaves are exactly the encodings of the inputs. We provide two examples of such trees in Figure 2.

Every sequence of queries to the RC-construction defines a subtree of $\mathcal{T}$ consisting of the paths from the root to the encodings of the queries: For notational convenience, we define the shorthand $L(x_1, \ldots, x_q)$, for q inputs $x_1, \ldots, x_q$, to be the amount of inner vertices (i.e. vertices which are not leaves) of the sub-tree induced by the evaluations of $x_1, \ldots, x_q$. It is easy to verify that for $\mathsf{RC}_{s,\ell}$ we have $L(x_1, \ldots, x_q) \leq 1 + q(\lceil \frac{\ell}{\log s} \rceil - 1)$. Also, for the case where the inputs are strings with arbitrary length, we define (always with respect to the understood encoding) $L(q, \ell) := \max_{x_1,\ldots,x_q : |x_i| \leq \ell} L(x_1, \ldots, x_q)$.

Consequently, one can see an interaction with the RC-construction as a process where the tree $\mathcal{T} = (\mathcal{V}, \mathcal{E})$ defined by ENC is traversed and κ-bit values are assigned to all visited vertices: While the root ϵ is assigned a random κ-bit value, the value of each visited vertex $(m_1, \ldots, m_j)$ is set to $F(z, r_{m_j})$, with z being the value of the parent vertex $(m_1, \ldots, m_{j-1})$. A query with input x is answered with the value at the corresponding leaf $\mathsf{ENC}(x)$. By the definition of an s-WPRF, it is easy to see that evaluating F under some given (pseudo-)random secret key at s independent random inputs produces s pseudorandom outputs,[9] and hence intuitively the above process sets the values of all visited vertices to pseudorandom values (and in particular this holds for the leaves). However, to formalize this intuition, we have to show that it is indeed possible to recycle the same values $r_1, \ldots, r_s$ for each invocation of F.

The following theorem formally captures the main security statement for the RC-construction (for a general input set $\mathcal{X}$).

Theorem 1. *Let $s \geq 2$, let $\mathcal{X}$ be a set, and let $\mathsf{ENC} : \mathcal{X} \to \{1, \ldots, s\}^+$ be a prefix-free encoding scheme. Furthermore, let $F : \{0,1\}^\kappa \times \{0,1\}^n \to \{0,1\}^\kappa$. For all L and all distinguishers D with running time t and with $L(x_1, x_2, \ldots) \leq L$ for all possible query sequences $x_1, x_2, \ldots \in \mathcal{X}$, there exists a distinguisher $D' = D'(D)$ such that*

$$\mathbf{Adv}^{\mathsf{PRF}}_{\mathsf{RC}^F_{s,\mathcal{X},\mathsf{ENC}}}(D) \leq L \cdot \left[\mathbf{Adv}^{\mathsf{WPRF}}_F(D') + s^2 \cdot 2^{-(n+1)}\right],$$

where D' makes exactly s queries and has running time $t' = t + \mathcal{O}(L \cdot t_F)$, with t_F being the time needed to evaluate F.

In Appendix A we provide a precise description of the distinguisher D', and refer the reader to the full version of this paper for the complete proof.

We remark that the term $s^2 2^{-(n+1)}$ is negligible, as s is assumed to be constant. Combined with the above observations on L, the theorem directly yields the following security bounds for the specialized variants of the RC-construction:

$$\mathbf{Adv}^{\mathsf{PRF}}_{\mathsf{RC}^F_s}(t, q, \ell) \leq L(q, \ell) \cdot \left[\mathbf{Adv}^{\mathsf{WPRF}}_F(t', s) + s^2 \cdot 2^{-(n+1)}\right],$$

$$\mathbf{Adv}^{\mathsf{PRF}}_{\mathsf{RC}^F_{s,\ell}}(t, q) \leq \left[1 + q\left(\left\lceil \frac{\ell}{\log s} \right\rceil - 1\right)\right] \cdot \left[\mathbf{Adv}^{\mathsf{WPRF}}_F(t'', s) + s^2 \cdot 2^{-(n+1)}\right],$$

with $t' = t + \mathcal{O}(L(q, \ell) \cdot t_F)$ and $t'' = t + \mathcal{O}((1 + q(\lceil \ell / \log s \rceil - 1)) \cdot t_F)$.

The most important observation is that all variants of the RC-construction require F to be only an s-WPRF. A minor positive aspect of the randomized cascade construction (if compared with other constructions) is the absence of any q-dependent birthday-like term in the above inequalities. Furthermore, if

[9] Except in the case where two of the random inputs $r_1, \ldots, r_s$ collide, which happens with small probability only.

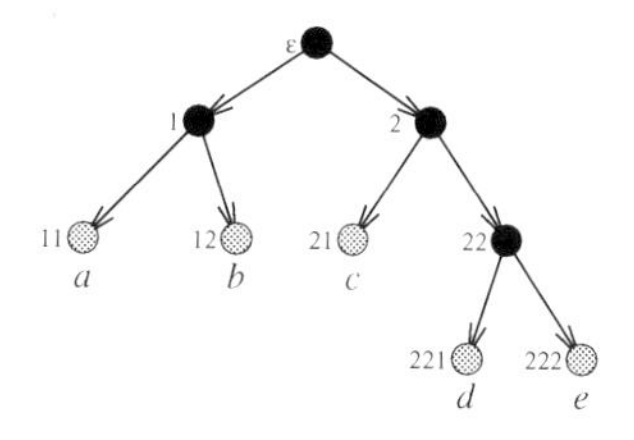

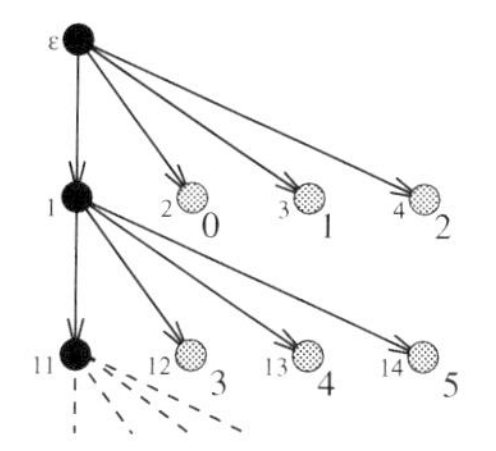

Fig. 2. Example trees associated with prefix-free encodings. Left: Encoding mapping inputs a, b, c, d, and e to sequences $(1,1)$, $(1,2)$, $(2,1)$, $(2,2,1)$, and $(2,2,2)$, respectively. Right: Encoding CTRENC used for efficient counter-mode evaluation.

we assume that F is indeed secure against q queries, the security of the $\mathsf{RC}_{s,\ell}$-construction is comparable to the one of the IC_ℓ-construction if we assume (in fact, very optimistically) that the best WPRF-distinguishing advantage grows linearly in the number of queries, i.e. $\mathbf{Adv}_F^{\mathsf{WPRF}}(t,q) = \Theta(q \cdot \mathbf{Adv}_F^{\mathsf{WPRF}}(t,s))$.

Larger Output Sizes. It is easy to increase the output size of the RC-construction (if needed) with the addition of a minor number of invocations of F per evaluation, which is independent of the input length: To obtain a construction $\overline{\mathsf{RC}}^F : \{0,1\}^\kappa \times \{0,1\}^{ns} \times \mathcal{X} \to \{0,1\}^{\phi\kappa}$ with output size $\phi \cdot \kappa$, we fix ϕ distinct strings $a_1, \ldots, a_\phi \in \mathcal{X}$ such that $L(a_1, \ldots, a_\phi)$ is minimal. Then, given key with private part k and public part $r_1, \ldots, r_s$, on input $x \in \mathcal{X}$, to compute $\overline{\mathsf{RC}}^F(k, r_1\|\ldots\|r_s, x)$ we first compute $k' := \mathsf{RC}^F(k, r_1\|\ldots\|r_s, x)$ and finally output $\mathsf{RC}^F(k', r_1\|\ldots\|r_s, a_1)\|\ldots\|\mathsf{RC}^F(k', r_1\|\ldots\|r_s, a_\phi)$. Security of this construction can be inferred by the fact that evaluating it at input x accounts to evaluating at inputs $(x, a_1), \ldots, (x, a_\phi)$ a variant of the RC-construction with input set $\mathcal{X} \times \{a_1, \ldots, a_\phi\}$ and prefix-free encoding $\mathsf{ENC}'(x, a) := \mathsf{ENC}(x)\|\mathsf{ENC}(a)$.

3.2 Efficient Encryption and PRGs from the RC-Construction

This section addresses two important applications of the RC-construction. For lack of space, we omit the proofs of the technical claims (which are mostly corollaries of Theorem 1 or are based on standard techniques).

Symmetric Encryption from the RC-Construction. Given a PRF $F : \{0,1\}^\kappa \times \{0,1\}^m \to \{0,1\}^n$ (in practice usually realized by a block cipher) one obtains an efficient stateful IND-CPA[10] encryption scheme for arbitrary-length messages by using F in so-called *counter-mode*, i.e. given a secret key k, we keep a counter ctr (initially 0), and the plaintext x (padded such that $|x|$ is a multiple of n) is encrypted as $[\mathsf{ctr}, x \oplus (F(k,\mathsf{ctr})\|F(k,\mathsf{ctr}+1)\|\ldots\|F(k,\mathsf{ctr}+|x|/n-1))]$

[10] Informally, a (stateful or randomized) encryption scheme (E, D) is *IND-CPA secure* [4, 16] if for a secret key K no polynomial-time adversary can distinguish the encryptions $E(K, x_0)$ and $E(K, x_1)$ for any two equally long messages x_0, x_1 of its choice even if it can obtain adaptively chosen encryptions $E(K, x)$ for arbitrary x's.

(and ctr is increased by $|x|/n$), where integers are canonically mapped to m-bit strings. Note in particular that we need one call to F for each n-bit block of encrypted data. Variants of *randomized stateless* counter-mode encryption (where one chooses a fresh random counter at every encryption instead of keeping a state) based on any WPRF $F : \{0,1\}^n \times \{0,1\}^n \to \{0,1\}^n$ were presented in [11, 17]. As with a full PRF, these schemes only require one call per n-bit block of encrypted data, but the underlying WPRF must be secure against as many queries as the amount of encrypted message blocks.

One can substantially weaken the assumption to an s-WPRF by using the RC-construction in stateful counter mode (with any encoding scheme). However, a dramatic increase of efficiency is achieved using a prefix-free encoding scheme $\mathsf{CTRENC} : \mathbb{N} \to \{1,\ldots,s\}^+$ tailored at this mode of operation, defined as

$$\mathsf{CTRENC}(i) := 1^{i \operatorname{div} s-1} \| (2 + (i \bmod s - 1)).$$

The tree arising from this encoding scheme is illustrated in Figure 2: In particular, it is clear that the sequence of values $\mathsf{RC}^F_{s,\mathsf{CTRENC}}(0), \mathsf{RC}^F_{s,\mathsf{CTRENC}}(1), \ldots$ can be computed very efficiently in an iterated way using only $\kappa + sn$ bits of memory and needing approximately $1 + \frac{1}{s-1}$ calls to F per κ-bit block of encrypted data. Furthermore, the values $r_1, \ldots, r_s$ can be chosen publicly by one communicating party (provided an authenticated channel is available), hence reducing the cost of key establishment to the generation of the κ-bit private part of the key. Security against (adaptive) chosen-ciphertext attacks based on any s-WPRF can be then obtained by standard techniques appending a MAC of the ciphertext [7] (e.g. using any of the PRF constructions presented in this paper).

Pseudorandom Generators from s-WPRFs. Recall that a *pseudorandom generator* (PRG) is a length-expanding function $G : \{0,1\}^\kappa \to \{0,1\}^m$ such that $G(K)$ is computationally indistinguishable from a random m-bit string under a random K. Surprisingly, constructing a good PRG from a WPRF (or an s-WPRF) turns out not to be a straightforward task: In contrast to PRFs, a WPRF F does not generally allow to find few "good" inputs $x_1, \ldots, x_t$ such that the mapping $k \mapsto F(k, x_1) \| \ldots \| F(k, x_t)$ is a PRG. However, one can use this approach employing the RC-construction as the underlying PRF: For any t *fixed* inputs $x_1, \ldots, x_t$ ($t > 2$) the mapping $\mathsf{G}^F : \{0,1\}^{sn+\kappa} \to \{0,1\}^{sn+t\kappa}$ such that $\mathsf{G}^F(r_1, \ldots, r_s, k)$ equals

$$r_1 \| \cdots \| r_s \| \mathsf{RC}^F_s(k, r_1 \| \ldots \| r_s, x_1) \| \cdots \| \mathsf{RC}^F_s(k, r_1 \| \ldots \| r_s, x_t)$$

is a PRG if F is an s-WPRF. (The order of the strings in the concatenation is irrelevant.) Note that an important advantage is that the strings $r_1, \ldots, r_s$ can be output as well. For example, given a 2-WPRF $F : \{0,1\}^n \times \{0,1\}^n \to \{0,1\}^n$, the mapping $\overline{\mathsf{G}}^F : \{0,1\}^{3n} \to \{0,1\}^{6n}$ such that $\overline{\mathsf{G}}^F(k, r_0, r_1)$ is set to

$$r_0 \| F(F(k, r_0), r_0) \| F(F(k, r_0), r_1) \| F(F(k, r_1), r_0) \| F(F(k, r_1), r_1) \| r_1 \quad (1)$$

is a length-doubling PRG which requires 6 calls to F. In particular, 3 calls are necessary in order to input only one both halves of the output. This improves a construction given in [17], which needed 3 and 4 calls, respectively.

An alternative approach to building a PRF from an s-WPRF F would consist of first constructing a length-doubling PRG G from F, and subsequently using the well-known GGM-construction [13] to build a PRF with a κ-bit key and ℓ-bit inputs by outputting, on input $x = (x_1, \ldots, x_{\ell-1}, x_\ell) \in \{0,1\}^\ell$ and key k, the κ-bit value $G_{x_\ell}(G_{x_{\ell-1}}(\cdots G_{x_1}(k)\cdots))$, where $G_i(k)$ for $i = 0, 1$ gives the first and the second half of the output of G, respectively. However, it is not hard to see that all constructions following this approach turn out to be less efficient than using the RC-construction directly (e.g. using the PRG of Equation 1 one needs 3 calls of F per input bit).

4 The Nested Randomized Cascade Construction

Even though the RC-construction can be practically efficient in special instantiation scenarios discussed earlier, its throughput is a major bottleneck in the case where the construction is used as a PRF (or a MAC) which is invoked at arbitrary inputs with variable lengths. Furthermore, the prefix-free encoding can be a limiting factor in the arbitrary-input-length case. This section presents a construction with better efficiency for long messages (i.e. longer than κ bits) and with no prefix-freeness requirements. Its core ingredient is a novel use of pairwise independence.

PAIRWISE-INDEPENDENT MAPPINGS. Recall that a mapping[11] $M : \{0,1\}^\kappa \times \{0,1\}^m \to \{0,1\}^n$ is *pairwise independent* if the values $M(K, x)$ and $M(K, x')$ are independent and uniformly distributed for all distinct $x, x' \in \{0,1\}^m$ under a random κ-bit key K. Most pairwise-independent mappings satisfy the following property, which will be central in our construction.

Definition 1. *A pairwise-independent mapping $M : \{0,1\}^\kappa \times \{0,1\}^m \to \{0,1\}^n$ is* key programmable *if there exists a (possibly randomized) algorithm SAMPLE which on input (x, x', y, y') (where possibly $x = x'$, $y = y'$) returns a uniformly chosen element from the set $\{k \mid M(k, x) = y, M(k, x') = y'\}$.*

If M is key programmable, the following two random experiments are equivalent to sampling a random κ-bit key K: (i) For some m-bit string x, sample Y as a uniform random n-bit string and $K := \mathsf{SAMPLE}(x, x, Y, Y)$; and (ii) For n-bit strings $x \neq x'$, sample Y, Y' as independent random n-bit strings and $K := \mathsf{SAMPLE}(x, x', Y, Y')$. Both the last two sampling strategies are used to ensure that $M(K, x) = Y$ (and possibly $M(K, x') = Y'$) for values $Y, Y' \in \{0,1\}^n$ which, although uniform and independent, are provided externally.

We provide two examples of key-programmable pairwise-independent mappings.

Example 1. Let M be such that given $k_1, k_2 \in \{0,1\}^n$ and the input $x \in \{0,1\}^n$, the output $M(k_1 \| k_2, x)$ equals $k_1 \oplus (k_2 \odot x)$, where $\oplus$ and $\odot$ are addition and multiplication of n-bit strings interpreted as elements of the extension field $GF(2^n)$.

[11] We use the word mapping, rather than hash function, to stress the fact that $m = n$ may also hold.

The unique $k_1\|k_2$ such that $M(k_1\|k_2, x) = y$ and $M(k_1\|k_2, x') = y'$ (with $x \neq x'$) can efficiently be found solving the corresponding system of two equalities. Is only a single constraint $M(k_1\|k_2, x) = y$ given, one chooses a random n-bit string k_2 and sets $k_1 := (k_2 \odot x) \oplus y$.

Example 2. An alternative is the mapping M' whose $(nm + n)$-bit key consists of an $(m \times n)$-binary matrix $\mathbf{A}$ and of a n-dimensional binary column vector $\mathbf{b}$, and on input x the output is $\mathbf{A}x + \mathbf{b}$, where x is interpreted as an m-dimensional column vector, and addition and multiplications are modulo 2. The function M' needs a larger key than M described above, but avoids finite-field multiplications.

Construction. The main idea of the nested RC-construction (called NRC, for short) is to combine an iterated phase where blocks are processed at a higher rate (but which satisfies a property weaker than pseudorandomness) with a second phase where the $\mathsf{RC}_{s,\kappa}$-construction (for fixed input length κ and a parameter s) is invoked on the output of the first phase (with independent key material).

More precisely, let $M : \{0,1\}^{\kappa'} \times \{0,1\}^m \to \{0,1\}^n$ be a key-programmable pairwise-independent mapping and let $F : \{0,1\}^{\kappa} \times \{0,1\}^n \to \{0,1\}^{\kappa}$ be the given compression function. The construction $\mathsf{PI}_M^F : \{0,1\}^{\kappa+\kappa'} \times \{0,1\}^* \to \{0,1\}^{\kappa}$ takes a key $k\|k'$, where $k \in \{0,1\}^{\kappa}$ and $k' \in \{0,1\}^{\kappa'}$. On input $x \in \{0,1\}^*$, it pads[12] x as $(x_1, \ldots, x_\lambda)$, where $x_1, \ldots, x_\lambda \in \{0,1\}^m$, and outputs $F^*(k, (M(k', x_1), \ldots, M(k', x_\lambda)))$.

Moreover, given the additional parameter s, we define the nested construction $\mathsf{NRC}_{M,s}^F : \{0,1\}^{2\kappa+\kappa'} \times \{0,1\}^{sn} \times \{0,1\}^* \to \{0,1\}^{\kappa}$ such that

$$\mathsf{NRC}_{M,s}^F(k_1\|k_2\|k', r_1\|\ldots\|r_s, x) := \mathsf{RC}_{s,\kappa}^F(k_1, r_1\|\ldots\|r_s, \mathsf{PI}_M^F(k_2\|k', x)).$$

It is easy to verify that in order to process a message x, the construction needs totally $\left\lceil \frac{|x|+1}{m} \right\rceil + \lceil \frac{\kappa}{\log s} \rceil$ calls to the underlying function F.

It is tempting to increase the throughput of the construction by choosing a mapping M with m much larger than n. However, all known constructions of pairwise-independent hash functions (in particular key-programmable ones) require keys twice as long as the *input* (rather than the output), and hence such an approach would entail a much longer key. In fact, we believe the length-preserving mapping M presented above to be a viable practically efficient solution: This special case of the construction is depicted in Figure 3.

Security. The following theorem precisely quantifies the security of the NRC-construction. We give only a compact statement, as well as an overview of the proof. The complete proof and the concrete reduction arising from it are given in the full version.

[12] According to the canonical padding which pads a string x to have length being a multiple of m by appending a 1 and sufficiently many 0's: The resulting padded string consists hence of $\left\lceil \frac{|x|+1}{m} \right\rceil$ m-bit blocks.

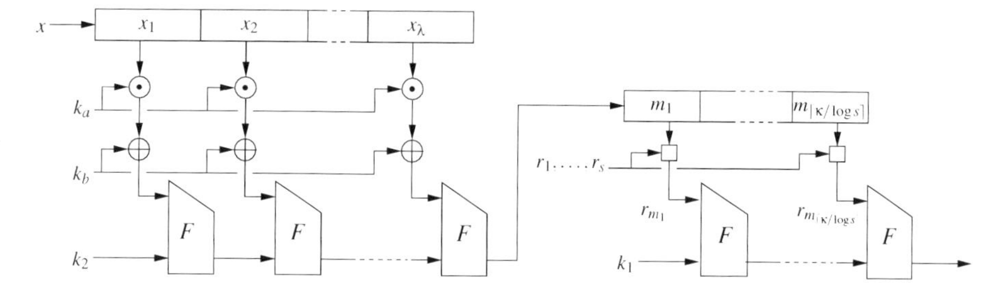

Fig. 3. The construction $\mathsf{NRC}^F_{M,s}$ for the special case $M(k_a \| k_b, x) = (k_a \odot x) \oplus k_b$

Theorem 2. *Let $M : \{0,1\}^{\kappa'} \times \{0,1\}^m \to \{0,1\}^n$ be a key-programmable pairwise-independent mapping, and $F : \{0,1\}^\kappa \times \{0,1\}^n \to \{0,1\}^\kappa$. For all $s \geq 2$ and for all t, q, and ℓ we have*

$$\mathbf{Adv}^{\mathsf{PRF}}_{\mathsf{NRC}^F_{M,s}}(t,q,\ell) \leq \left(1 + q\left(\left\lceil \frac{\kappa}{\log s} \right\rceil - 1\right)\right) \cdot \left(\mathbf{Adv}^{\mathsf{WPRF}}_F(t',s) + s^2 \cdot 2^{-(n+1)}\right) + \left\lceil \frac{\ell+1}{m} \right\rceil \cdot q^2 \cdot \left(\mathbf{Adv}^{\mathsf{WPRF}}_F(t'',2) + 2^{-n}\right) + q^2 \cdot 2^{-(\kappa+1)},$$

where $t' = t + \mathcal{O}(q(\frac{\ell}{m} + \frac{\kappa}{\log s}) \cdot t_F)$ and $t'' = \mathcal{O}\left(\frac{2\ell}{m} \cdot t_F\right)$, with t_F being the time needed for an evaluation of F.

The core of the proof consists of showing that whenever F is a WPRF for two-query adversaries, the PI-construction is δ-AU for a suitable function δ to be computed below. In the following, given two inputs x, x' with corresponding padded strings $(x_1, \ldots, x_\lambda)$ and $(x'_1, \ldots, x'_{\lambda'})$ (where without loss of generality $\lambda < \lambda'$), let λ^* be maximal with the property that $x_1 = x'_1, \ldots, x_{\lambda^*} = x'_{\lambda^*}$ (in particular, $\lambda^* := 0$ if $x_1 \neq x'_1$), and define the quantity $\Lambda(x,x')$ as $\lambda + \lambda' - \lambda^* - 1$ if $(x_1, \ldots, x_\lambda)$ is not a prefix of $(x'_1, \ldots, x'_{\lambda'})$, and as $\lambda + 1$ otherwise. In particular, note that $\Lambda(x,x') \leq \lambda + \lambda' \leq 2\max\{\lambda, \lambda'\} \leq 2\lceil \frac{\ell+1}{m} \rceil$ if $|x|, |x'| \leq \ell$.

The following lemma provides a precise upper bound on the collision probability of the PI-construction in terms of the WPRF distinguishing advantage of a distinguisher $D_{x,x'}$ (which in particular only depends on x and x') for F. We refer the reader to the full version of this paper for its proof.

Lemma 2. *For all distinct inputs $x, x' \in \{0,1\}^*$, there exists a two-query distinguisher $D_{x,x'}$ such that*

$$\mathsf{p}^{\mathsf{COLL}}_{\mathsf{PI}^F_M}(x,x') \leq \Lambda(x,x') \cdot \left(\mathbf{Adv}^{\mathsf{WPRF}}_F(D_{x,x'}) + 2^{-n}\right) + 2^{-\kappa},$$

where $D_{x,x'}$ has running time $\mathcal{O}\left(\Lambda(x,x') \cdot t_F\right)$.

In particular, given some ℓ, let $t'' = \mathcal{O}\left(\frac{2\ell}{m} \cdot t_F\right)$ be the maximal running time of the distinguisher $D_{x,x'}$ taken over all x, x' with $|x|, |x'| \leq \ell$. We define $\delta(\ell) := 2\left\lceil \frac{\ell+1}{m} \right\rceil \cdot (\mathbf{Adv}^{\mathsf{WPRF}}_F(t'',2) + 2^{-n}) + 2^{-\kappa}$. The function PI^F_M is δ-universal by Lemma 2, and this implies Theorem 2 using Lemma 1 and Theorem 1.

5 Black-Box Keying of Iterated Hash Functions

The iterated structure of the RC- and the NRC-constructions makes compression functions ideal candidates for instantiating the underlying s-WPRF. In general, however, we may be constrained to only have black-box access to an implementation of an iterated hash function $H : \{0,1\}^* \to \{0,1\}^\kappa$ (cf. Section 2) with direct access neither to the initialization value IV nor to the underlying compression function $F : \{0,1\}^\kappa \times \{0,1\}^n \to \{0,1\}^\kappa$. To overcome this obstacle, we encode (as in HMAC) an n-bit key as the first block of the input to the hash function H. More precisely, given the prefix-free encoding scheme $\mathsf{ENC} : \{0,1\}^* \to \{1,\ldots,s\}^+$, we consider the construction $\mathsf{HRC}^F_{s,\mathsf{ENC}}$ which takes a key with private part $k \in \{0,1\}^n$ and public part $r_1,\ldots,r_s \in \{0,1\}^n$, and on input x with $\mathsf{ENC}(x) = (m_1,\ldots,m_\lambda)$ outputs the value

$$\mathsf{HRC}^H_{s,\mathsf{ENC}}(k, r_1\|\ldots\|r_s, x) := H(k\|r_{m_1}\|\ldots\|r_{m_\lambda}),$$

and analogously we define $\mathsf{HRC}_{s,\ell}$ for inputs of fixed-length ℓ (using the canonical encoding to the base s). Furthermore, with $M : \{0,1\}^{\kappa'} \times \{0,1\}^m \to \{0,1\}^n$ being a key-programmable pairwise-independent mapping, we consider the construction $\mathsf{HNRC}^H_{M,s}$ which takes a key with private part $k_1, k_2 \in \{0,1\}^n$, $k' \in \{0,1\}^{\kappa'}$ and public parts $r_1,\ldots,r_s$. On input input x (padded as $(x_1,\ldots,x_\lambda)$) it outputs

$$\mathsf{HNRC}^H_{M,s}(k_1\|k_2\|k', r_1\|\ldots\|r_s, x) :=$$
$$\mathsf{HRC}^H_{s,\kappa}(k_1, r_1\|\ldots\|r_s, H(k_2\|M(k',x_1)\|\ldots\|M(k',x_\lambda))).$$

In order to lift the security statements of the RC- and the NRC-constructions to both the HRC- and HNRC-constructions, the assumption that F is an s-WPRF is not sufficient: First, it is necessary that the κ-bit output $F(IV, K)$ is computationally indistinguishable from a uniformly-distributed random string of length κ (under a secret random K); This guarantees that the chaining value obtained after the first evaluation of F is pseudorandom and can be used as the "key" for the RC- or the PI-construction. A further problem is due to the fact that we generally cannot enforce the last n-bit block processed by F to be random because of the padding introduced by H, and this issue should not destroy the pseudorandomness of the outputs. To our rescue, however, comes the fact that each such block is processed keying F with a *fresh* pseudorandom value: It is hence enough to additionally guarantee that for an arbitrary *fixed* n-bit string x and a random secret κ-bit string K, the string $F(K,x)$ is computationally indistinguishable from a random κ-bit string.

We stress that both these extra properties are very weak requirements: In fact, a good compression function should satisfy them even unconditionally. It is sufficient, for example, that $F(IV,\cdot)$ and $F(\cdot,x)$ (for all $x \in \{0,1\}^n$) are all (nearly-)regular functions. (We refer the reader to [6] for a discussion on regularity-properties of hash functions.). With these two additional assumptions on the compression function F of H, the security bounds of the RC and the NRC-construction can be lifted to their black-box counterparts. For lack of space, we omit the proofs, which are very similar to the ones of the original constructions.

6 Conclusions and Open Problems

We have shown that efficient arbitrary-input-length PRFs (and consequently MACs and encryption schemes) can be constructed under very weak assumptions, i.e. weak PRFs where security holds only for a limited number of queries. Our results provide new insights into the property of weak pseudorandomness.

A natural open question is whether there exist constructions of PRFs from WPRFs which take explicit advantage of more secure WPRFs (i.e. tolerating many queries) to achieve more efficient constructions than what we propose and what was considered in the literature (e.g. processing linearly-many bits per invocation even for short inputs). We conjecture, however, that this is not possible. A further direction arising from our work consists of finding further examples of cryptographic primitives where restricting adversaries in terms of queries leads to interesting phenomena such as those observed in this paper for weak pseudorandomness.

References

1. Bellare, M.: New proofs for NMAC and HMAC: Security without collision-resistance. In: Dwork, C. (ed.) CRYPTO 2006. LNCS, vol. 4117, pp. 602–619. Springer, Heidelberg (2006)
2. Bellare, M., Canetti, R., Krawczyk, H.: Keying hash functions for message authentication. In: Koblitz, N. (ed.) CRYPTO 1996. LNCS, vol. 1109, pp. 1–15. Springer, Heidelberg (1996)
3. Bellare, M., Canetti, R., Krawczyk, H.: Pseudorandom functions revisited: The cascade construction and its concrete security. In: FOCS 1996, pp. 514–523 (1996)
4. Bellare, M., Desai, A., Jokipii, E., Rogaway, P.: A concrete security treatment of symmetric encryption. In: FOCS 1997, pp. 394–403 (1997)
5. Bellare, M., Kilian, J., Rogaway, P.: The security of the cipher block chaining message authentication code. Journal of Computer and System Sciences 61(3), 362–399 (2000)
6. Bellare, M., Kohno, T.: Hash function balance and its impact on birthday attacks. In: Cachin, C., Camenisch, J.L. (eds.) EUROCRYPT 2004. LNCS, vol. 3027, pp. 401–418. Springer, Heidelberg (2004)
7. Bellare, M., Namprempre, C.: Authenticated encryption: Relations among notions and analysis of the generic composition paradigm. In: Okamoto, T. (ed.) ASIACRYPT 2000. LNCS, vol. 1976, pp. 531–545. Springer, Heidelberg (2000)
8. Black, J., Halevi, S., Krawczyk, H., Krovetz, T., Rogaway, P.: UMAC: Fast and secure message authentication. In: Wiener, M. (ed.) CRYPTO 1999. LNCS, vol. 1666, pp. 216–233. Springer, Heidelberg (1999)
9. Carter, J.L., Wegman, M.N.: Universal classes of hash functions. Journal of Computer and System Sciences 18(2), 143–154 (1979)
10. Damgård, I.B.: A design principle for hash functions. In: Brassard, G. (ed.) CRYPTO 1989. LNCS, vol. 435, pp. 416–427. Springer, Heidelberg (1990)
11. Damgård, I.B., Nielsen, J.B.: Expanding pseudorandom functions; or: From known-plaintext security to chosen-plaintext security. In: Yung, M. (ed.) CRYPTO 2002. LNCS, vol. 2442, pp. 449–464. Springer, Heidelberg (2002)

12. Fischlin, M.: Security of NMAC and HMAC based on non-malleability. In: Malkin, T.G. (ed.) CT-RSA 2008. LNCS, vol. 4964, pp. 138–154. Springer, Heidelberg (2008)
13. Goldreich, O., Goldwasser, S., Micali, S.: How to construct random functions. In: FOCS 1984, pp. 464–479 (1984)
14. Håstad, J., Impagliazzo, R., Levin, L.A., Luby, M.: A pseudorandom generator from any one-way function. SIAM Journal on Computing 28(4), 1364–1396 (1999)
15. Hirose, S., Park, J.H., Yun, A.: A simple variant of the Merkle-Damgård scheme with a permutation. In: Kurosawa, K. (ed.) ASIACRYPT 2007. LNCS, vol. 4833, pp. 113–129. Springer, Heidelberg (2007)
16. Katz, J., Yung, M.: Complete characterization of security notions for probabilistic private-key encryption. In: STOC 2000, pp. 245–254 (2000)
17. Maurer, U., Sjödin, J.: A fast and key-efficient reduction of chosen-ciphertext to known-plaintext security. In: Naor, M. (ed.) EUROCRYPT 2007. LNCS, vol. 4515, pp. 498–516. Springer, Heidelberg (2007)
18. Merkle, R.C.: A certified digital signature. In: Brassard, G. (ed.) CRYPTO 1989. LNCS, vol. 435, pp. 218–238. Springer, Heidelberg (1990)
19. Naor, M., Reingold, O.: Synthesizers and their application to the parallel construction of pseudo-random functions. Journal of Computer and System Sciences 58(2), 336–375 (1999)
20. Pietrzak, K., Sjödin, J.: Range extension for weak PRFs; the good, the bad, and the ugly. In: Naor, M. (ed.) EUROCRYPT 2007. LNCS, vol. 4515, pp. 517–533. Springer, Heidelberg (2007)
21. Stinson, D.R.: Universal hashing and authentication codes. In: Feigenbaum, J. (ed.) CRYPTO 1991. LNCS, vol. 576, pp. 74–85. Springer, Heidelberg (1992)
22. Yasuda, K.: Boosting Merkle-Damgård hashing for message authentication. In: Kurosawa, K. (ed.) ASIACRYPT 2007. LNCS, vol. 4833, pp. 216–231. Springer, Heidelberg (2007)

A Description of D' in the Proof of Theorem 1

We define $L+1$ hybrid experiments $H_0, H_1, \ldots, H_L$ where D is given random inputs $r_1, \ldots, r_s$ and interacts which a (randomized) oracle $\mathcal{X} \to \{0,1\}^\kappa$ that keeps track of all vertices of the subtree of $\mathcal{T}$ induced by the queries of D. In particular, it assigns to all *internal* vertices v of this subtree increasing integer values $l(v)$ according to the order in which they are visited for the first time, with $l(\epsilon) := 0$. Furthermore, it associates κ-bit values $z(v)$ with all visited vertices: Initially only $z(\epsilon)$ is defined and set to a random value. In H_i an oracle query $x \in \mathcal{X}$ (with $\mathsf{ENC}(x) = (m_1, \ldots, m_\lambda)$) by D is answered by looking for the highest λ^* such that $z(m_1, \ldots, m_{\lambda^*})$ is defined and for all $j = \lambda^* + 1, \ldots, \lambda$ assigning to $z(m_1, \ldots, m_j)$ a fresh random value if $l(m_1, \ldots, m_{j-1}) < i$ and $F(z(m_1, \ldots, m_{j-1}), r_{m_j})$ otherwise. Finally, $z(m_1, \ldots, m_\lambda)$ is returned to D as the oracle's output. Clearly, H_0 behaves as the experiment where D interacts with the RC-construction, whereas H_L answers all queries of D randomly.

For all $i = 0, \ldots, L-1$ one then constructs a distinguisher D_i for $\mathcal{S}^{F(k,\cdot)}$ and $\mathcal{S}^{\mathbf{R}_{n,\kappa}}$ which first issues s queries to the given oracle, obtaining s pairs $(r_1, y_1), \ldots, (r_s, y_s)$ and subsequently simulates the interaction of D with H_i, except that $z(m_1, \ldots, m_j)$ is set to y_{m_j} whenever $l(m_1, \ldots, m_{j-1}) = i$. Finally, the distinguisher $D'(D)$ chooses a random $i \in \{0, \ldots, L-1\}$ and runs D_i.

We refer the reader to the full version for the concrete analysis of the distinguishing advantage of D'.

Universally Composable Adaptive Oblivious Transfer*

Matthew Green and Susan Hohenberger

The Johns Hopkins University
Information Security Institute
3400 N. Charles Street; Baltimore, MD 21218, USA
{mgreen,susan}@cs.jhu.edu

Abstract. In an oblivious transfer (OT) protocol, a Sender with messages $M_1, \ldots, M_N$ and a Receiver with indices $\sigma_1, \ldots, \sigma_k \in [1, N]$ interact in such a way that at the end the Receiver obtains $M_{\sigma_1}, \ldots, M_{\sigma_k}$ without learning anything about the other messages and the Sender does not learn anything about $\sigma_1, \ldots, \sigma_k$. In an *adaptive* protocol, the Receiver may obtain $M_{\sigma_{i-1}}$ before deciding on σ_i. Efficient adaptive OT protocols are interesting as a building block for secure multiparty computation and for enabling oblivious searches on medical and patent databases.

Historically, adaptive OT protocols were analyzed with respect to a "half-simulation" definition which Naor and Pinkas showed to be flawed. In 2007, Camenisch, Neven, and shelat, and subsequent other works, demonstrated efficient adaptive protocols in the full-simulation model. These protocols, however, all use standard rewinding techniques in their proofs of security and thus are not universally composable. Recently, Peikert, Vaikuntanathan and Waters presented universally composable (UC) *non-adaptive* OT protocols for the 1-out-of-2 variant, in the static corruption model using certain trusted setup assumptions. However, it is not clear how to preserve UC security while extending these protocols to the adaptive k-out-of-N setting. Further, any such attempt would seem to require $O(N)$ computation per transfer for a database of size N. In this work, we present an efficient and UC-secure *adaptive* k-out-of-N OT protocol in the same model as Peikert *et al.*, where after an initial commitment to the database, the cost of each transfer is *constant*. Our construction is secure under bilinear assumptions in the standard model.

1 Introduction

Oblivious transfer (OT) was introduced by Rabin [31] and generalized by Even, Goldreich and Lempel [19] and Brassard, Crépeau and Robert [8]. It is a two-party protocol, where a Sender with messages $M_1, \ldots, M_N$ and a Receiver with indices $\sigma_1, \ldots, \sigma_k \in [1, N]$ interact in such a way that at the end the Receiver obtains $M_{\sigma_1}, \ldots, M_{\sigma_k}$ without learning anything about the other messages and

* This work was supported by the NSF under grant CT-0716142 and Hohenberger's Microsoft New Faculty Fellowship.

J. Pieprzyk (Ed.): ASIACRYPT 2008, LNCS 5350, pp. 179–197, 2008.
© International Association for Cryptologic Research 2008

the Sender does not learn anything about $\sigma_1, \ldots, \sigma_k$. Naor and Pinkas were the first to consider an *adaptive* setting, $\mathsf{OT}_{k\times 1}^N$, where the Receiver may obtain $M_{\sigma_{i-1}}$ before deciding on σ_i [28]. Efficient OT schemes are very important. OT_1^4 is a key building block for secure multi-party computation [21, 25, 34]. $\mathsf{OT}_{k\times 1}^N$ is a useful and interesting tool in its own right, enabling oblivious databases for applications such as medical record storage and patent searches [29].

Developing efficient *adaptive* protocols appears to be a more difficult and involved process than the non-adaptive protocols. Indeed, even finding the right security definition has proven challenging. Historically, many OT constructions were analyzed under a "half-simulation" definition, where the Sender and Receiver's security are described by a combination of simulation and game-based definitions. Naor and Pinkas [28] showed that schemes analyzed under this definition may admit practical attacks on the Receiver's privacy. To address this, Camenisch, Neven and shelat [10] and subsequently Green and Hohenberger [22] proposed efficient and fully-simulatable $\mathsf{OT}_{k\times 1}^N$ protocols under bilinear assumptions. Each of these protocols achieve the optimal total communication cost of $O(N+k)$ with reasonable constants. Unfortunately, their security proofs use adversarial rewinding, and thus do not imply security under concurrent execution.

Recently, Lindell [26] showed how to achieve efficient and fully-simulatable *non-adaptive* OT_1^2 under the DDH, Nth residuosity and quadratic residuosity assumptions, as well as the assumption that homomorphic encryption exists. Simultaneously, Peikert, Vaikuntanathan and Waters [30] proposed several non-adaptive, but universally composable OT_1^2 protocols based on DDH, quadratic residuosity and lattice assumptions. While both of these works add to our collective knowledge for non-adaptive OT, they do not shed much light on how to achieve efficient *adaptive* protocols. Indeed, Lindell points out that the adaptive case is considerably harder [26].

The general framework used in [26, 30] (where the Receiver chooses the encryption keys) seems inherently at odds with allowing efficient adaptive schemes. Each transfer requires $O(N)$ work for the Sender, whereas this can be *constant* in our protocols. Even more alarming, it isn't clear how (without killing the efficiency and perhaps the UC security of [30]) a Sender could convince the Receiver that he is not changing the database values with each request. This problem of ensuring a *consistent* database gets even worse when multiple Receivers are considered, as we do in Section 5.

Our Results. In this work, we take a different approach to constructing OT protocols, which allows them to be simultaneously efficient, adaptive, universally composable and globally consistent. We summarize what is known about $\mathsf{OT}_{k\times 1}^N$ protocols in Figure 1. Let us describe some highlights.

1. *Universal Composability:* The Universal Composability framework [13] allows for the design of concurrent and composable cryptographic protocols, which are important properties in any practical deployment of an oblivious database. Canetti and Fischlin showed that OT cannot be UC-realized without trusted setup assumptions such as the existence of a Common Reference

Protocol	Rounds	Communication	Assumption
Half Simulation:			
NP99 [28]	$\ell k\ log\ N + 1/2$	–	Sum Consistent Synthesizers + ℓ-round OT^2_1
CT05 [18]	$O(k) + 1/2$	$O(N)$	Decisional DH (in ROM)
Full Simulation:			
CNS07 [10]	$4k + 1/2$	$O(N)$	y-Power Decisional DH + q-Strong DH
CNS07 [10]	$O(k) + 1/2$	$O(N)$	Unique blind signature (in ROM)
GH07 [22]	$k + 1/2$	$O(N)$	Decisional Bilinear DH (in ROM)
UC ($\mathcal{F}_{CRS}$-hybrid):			
This work (§4)	$k + 1/2$	$O(N)$	SXDH + DLIN + q-Hidden LRSW

Fig. 1. Survey of efficient, adaptive k-out-of-N Oblivious Transfer protocols

String (CRS) [15]. This is formally referred to as the $\mathcal{F}_{CRS}$-hybrid model, and is assumed by the constructions of Peikert *et al.* [30] as well as those in this work. As in [30], we work in a static corruption model.

2. *Efficiency:* Our protocol is practical. For a database of N objects, the initialization phase requires $O(N)$ communication cost, and each transfer phase requires only constant cost, for reasonable constants. In contrast, simply repeating a OT^N_1 scheme (such as [30]) k times would require $O(N)$ communication cost for *each* transfer plus the additional work required for the Sender to convince the Receiver that he isn't changing the database values dynamically. Moreover, the message space of our protocol is a group element (so at least 160 bits), whereas the quadratic residuosity and lattice-based schemes of [30] have *one*-bit message spaces. We note, however, that the DDH-based scheme of [30] allows for multiple bit messages.
3. *Model and Assumptions:* We focus on protocols secure in the standard model. Our construction can be implemented assuming SXDH [2, 5, 24, 32], Decision Linear [5], and q-Hidden LRSW (a non-interactive variant of the LRSW assumption [27], for which we give a generic group proof in the full version of this work [23].) We note that our decisional assumptions, SXDH and Decision Linear, are much more simple than the q-Power Decisional Diffie-Hellman assumption used in the (non-UC) adaptive OT of Camenisch *et al.* [10]. In the full version, we also provide a second construction that is secure in symmetric groups (i.e., where SXDH does not hold) under an alternative set of hardness assumptions. See Figure 1 for more.

Intuition behind the Construction. Oblivious Transfer protocols can be roughly divided into two categories. Let's restrict our attention to non-adaptive OT^N_1 for the moment. In approach (1), which is used by [19, 26, 30, 31], the Receiver transmits a collection of specially-formed encryption keys to the Sender, who encrypts each message and returns the N ciphertexts to the Receiver. The protocol is secure provided that the encryption keys are formed such that a Receiver is able to decrypt at most *one* of the resulting ciphertexts. In approach (2), which is used by [10, 18, 22] and this work, the Sender encrypts the message collection under keys of her own choosing, and— in some interactive protocol with the Receiver— helps to decrypt *one* ciphertext.

While both approaches can be used to implement adaptive OT in theory, the first approach requires that the Sender generate a new set of ciphertexts at *each* transfer stage (for *each* receiver), requiring at least $O(N \cdot k)$ cost. Even worse, the Sender might be able to maliciously change the database between transfers and present different versions of the database to different receivers.

The latter approach is much better suited for the adaptive case. A single database can be committed to and then each decryption can be performed in constant computational and communication cost, for a total $O(N + k)$ cost. This approach is taken by the fully-simulatable protocols of [10], which both use rewinding in their simulations to (1) simulate proofs and (2) extract knowledge.[1]

An appealing naive approach to realizing UC-secure adaptive OT would be to modify the efficient standard-model protocol of Camenisch *et al.* [10] by simply replacing rewinding-based proofs with the non-interactive proof techniques of Groth and Sahai [24]. Unfortunately, this is non-trivial for two reasons. First, the Groth-Sahai techniques provide broad support for non-interactive, *witness indistinguishable* proofs of algebraic assertions in bilinear groups, but only provide non-interactive, *zero-knowledge* proofs for a restricted class of algebraic assertions. Unfortunately, the proof statements required by [10] fall outside of this class, and it does not seem easy to rectify this problem. Secondly, the protocol of [10] requires some form of extraction (e.g., extracting the chosen index from the adversarial Receiver or extracting the secret encryption keys from the adversarial Sender) for proofs containing elements of $\mathbb{Z}_p$; unfortunately, Groth-Sahai proofs of knowledge are f-extractable (but not fully extractable), where only some one-way function of the witness, $f(w)$, can be extracted (e.g., g^w) and not the witness w itself. Dealing with this limitation would necessitate substantial changes to the CNS protocol.

Instead, our construction starts from scratch. While we follow the "assisted decryption" framework of the CNS protocol, we are able to do so without the need for strong q-based decisional assumptions. We instead base the security of the ciphertexts in our scheme on the Decision Linear assumption [5]. Finally, since the Groth-Sahai proofs have not yet been shown to be either simulation-sound or UC in general, we develop techniques that permit UC simulation (even in the advanced case where multiple receivers interact with a single sender).

2 Definitions

Notation. By OT_k^N (resp., $\mathsf{OT}_{k\times 1}^N$), we denote a non-adaptive (resp., adaptive) k-out-of-N oblivious transfer protocol. Let $\stackrel{c}{\approx}$ denote computational indistinguishability, as defined in [13].

Adaptive k-out-of-N Oblivious Transfer. $\mathsf{OT}_{k\times 1}^N$ protocols consist of two phases: Initialization and Transfer. In the Initialization phase, the Sender commits to the input database $M_1, \ldots, M_N$. Subsequently, the Sender and Receiver

[1] Along the same lines, the half-simulation protocols of [20, 28] use a form of oblivious pseudorandom function evaluation (OPRF) to encrypt and obliviously decrypt the message database. Unfortunately, the evaluation protocols described in those works appear vulnerable to selective-failure attacks, and the modifications necessary to achieve UC security (or full simulation) seem substantial.

engage in up to k Transfers. During the ith Transfer, the Receiver adaptively selects a message index $\sigma_i \in [1, N]$ and engages in a protocol such that it obtains M_{σ_i} (or $\perp$ if the protocol fails) and nothing else, while the Sender learns nothing about σ_i. The simulation-based nature of the security definition we use ensures that protocol failures must occur independently of the message index σ_i chosen by the Receiver (capturing the strong selective-failure blindness property [10].)

Universally Composable Security. As in [30], we work in the standard UC framework with static corruptions, where all parties are modeled as p.p.t. interactive Turing machines. Security of protocols is defined by comparing the protocol execution to an *ideal process* for carrying out the desired task. More formally, there is an *environment* $\mathcal{Z}$ whose task is to distinguish between two worlds: ideal and real. In the ideal world, "dummy parties" (some of whom may be corrupted by the *ideal adversary* $\mathcal{S}$) interact with an *ideal functionality* $\mathcal{F}$. In the real world, parties (some of whom may be corrupted by the *real world adversary* $\mathcal{A}$) interact with each other according to some protocol π. We refer to Canetti [13, 14] for a fuller description, as well as a definition of the ideal world ensemble $\mathsf{IDEAL}_{\mathcal{F},\mathcal{S},\mathcal{Z}}$ and the real world ensemble $\mathsf{EXEC}_{\pi,\mathcal{A},\mathcal{Z}}$. We use the established notion of a protocol π *securely realizing* an ideal functionality $\mathcal{F}$ as:

Definition 1. *Let $\mathcal{F}$ be a functionality. A protocol π* UC-realizes *$\mathcal{F}$ if for any adversary $\mathcal{A}$, there exists a simulator $\mathcal{S}$ such that for all environments $\mathcal{Z}$,*

$$\mathsf{IDEAL}_{\mathcal{F},\mathcal{S},\mathcal{Z}} \stackrel{c}{\approx} \mathsf{EXEC}_{\pi,\mathcal{A},\mathcal{Z}}.$$

Canetti and Fischlin showed that OT cannot be UC-realized without a trusted setup assumption [15]. Thus, as in [16, 30], we assume the existence of an honestly-generated Common Reference String (crs), and work in the so-called $\mathcal{F}_{CRS}$-hybrid model. The functionality is parameterized by a distribution D and a set $\mathcal{P}$ of recipients. For our purposes, $\mathcal{P}$ will include the OT Sender and Receiver only. Here the environment learns about the reference string from the adversary, and thus the simulator can set up a string with "trapdoor information", etc.

Figure 2 describes the $\mathcal{F}_{CRS}$ functionality and Figure 3 describes the $\mathcal{F}_{OT}^{N\times 1}$ functionality.

We briefly mention that there are techniques for designing and analyzing multiple OT protocols which use a single reference string; i.e., a multi-session extension. One might worry that if multiple protocols now share some joint state, then they can no longer be analyzed separately and then composed later. Fortunately, this is addressed by *universal composition with joint state* (JUC) [17] and could be done in our case. A second issue with sharing the reference string is that we make no guarantee about the security of protocols which use the same reference string in ways other than those specified by the OT protocol, and here we explicitly assume that the crs is only available to certain parties. This is at odds with the notion that the crs is a "global" entity, however, there are strong impossibility results for UC-realizing OT in a setting where the crs is available to everyone (including the environment) and can no longer be crafted by the simulator. There are models, such as the *augmented CRS* functionality

Functionality $\mathcal{F}_{CRS}^{\mathcal{D},P}$

Upon receiving input (sid, crs) from party P, first verify that $p \in \mathcal{P}$; else ignore the input. If there is no value r recorded, then choose and record $r \leftarrow D$. Finally send output (sid, crs, r) to P.

Fig. 2. Ideal functionality for the common reference string [14]

Functionality $\mathcal{F}_{OT}^{N \times 1}$

$\mathcal{F}_{OT}^{N \times 1}$ proceeds as follows, parameterized with integers N, ℓ and running with an oblivious transfer Sender **S**, a receiver **R** and an adversary $\mathcal{S}$.

- Upon receiving a message $(\mathsf{sid}, \mathsf{sender}, m_1, \ldots, m_N)$ from **S**, where each $m_i \in \{0,1\}^\ell$, store $(m_1, \ldots, m_N)$.
- Upon receiving a message $(\mathsf{sid}, \mathsf{receiver}, \sigma)$ from **R**, check if a $(\mathsf{sid}, \mathsf{sender}, \ldots)$ message was previously received. If no such message was received, send nothing to **R**. Otherwise, send (sid, request) to **S** and receive the tuple $(\mathsf{sid}, b \in \{0,1\})$ in response. Pass (sid, b) to the adversary, and: If $b = 0$, send $(\mathsf{sid}, \perp)$ to **R**. If $b = 1$, send (sid, m_σ) to **R**.

Fig. 3. Functionality for adaptive Oblivious Transfer, based on the OT_1^2 definition from [16]

$\mathcal{F}_{\text{ACRS}}$ [12], which overcome these impossibility results, but we do not explore these advanced UC issues with respect to our OT construction in this work.

3 Preliminaries

Bilinear Groups. Let BMsetup be an algorithm that, on input 1^κ, outputs the parameters for a bilinear mapping as $\gamma = (p, \mathbb{G}_1, \mathbb{G}_2, \mathbb{G}_T, e, g \in \mathbb{G}_1, \tilde{g} \in \mathbb{G}_2)$, where g generates $\mathbb{G}_1$ and $\tilde{g}$ generates $\mathbb{G}_2$, the groups $\mathbb{G}_1, \mathbb{G}_2, \mathbb{G}_T$ each have prime order p, and $e : \mathbb{G}_1 \times \mathbb{G}_2 \rightarrow \mathbb{G}_T$.

Symmetric External Diffie-Hellman Assumption (SXDH) [2, 5, 24, 32]**:** Let $\mathsf{BMsetup}(1^\kappa) \rightarrow \gamma = (p, \mathbb{G}_1, \mathbb{G}_2, \mathbb{G}_T, e, g, \tilde{g})$. The SXDH assumption states that the Decisional Diffie-Hellman problem is hard within both $\mathbb{G}_1$ and $\mathbb{G}_2$.

Groups where SXDH holds is one of the three settings for Groth-Sahai proofs [24].

Decision Linear Assumption (DLIN) [5]**:** Let $\mathsf{BMsetup}(1^\kappa) \rightarrow (p, \mathbb{G}_1, \mathbb{G}_2, \mathbb{G}_T, e, g, \tilde{g})$. For all p.p.t. adversaries Adv, the following probability is strictly less than $1/2 + 1/\text{poly}(\kappa)$:

$$\Pr[a, b, c, d \xleftarrow{\$} \mathbb{Z}_p; f \leftarrow g^c; \tilde{f} \leftarrow \tilde{g}^c; h \leftarrow g^d; \tilde{h} \leftarrow \tilde{g}^d;$$
$$z_0 \leftarrow h^{a+b}; z_1 \xleftarrow{\$} \mathbb{G}_1; d \leftarrow \{0,1\} : \mathsf{Adv}(\gamma, g, \tilde{g}, f, \tilde{f}, h, \tilde{h}, g^a, f^b, z_d) = d].$$

Note that this is a weaker asymmetric version of the original DLIN assumption of Boneh, Boyen and Shacham [5], which was set in symmetric groups.

q-Hidden LRSW Assumption: Let $\mathsf{BMsetup}(1^\kappa) \to \gamma = (p, \mathbb{G}_1, \mathbb{G}_2, \mathbb{G}_T, e, g, \tilde{g})$. For all p.p.t. adversaries Adv, the following probability is strictly less than $1/\text{poly}(\kappa)$:

$$\begin{aligned} &\Pr[s,t \xleftarrow{\$} \mathbb{Z}_p; \tilde{S} \leftarrow \tilde{g}^s, \tilde{T} \leftarrow \tilde{g}^t; \forall i \in [1 \dots q], x_i, y_i \xleftarrow{\$} \mathbb{Z}_p, b_i \leftarrow g^{y_i}, \tilde{b}_i \leftarrow \tilde{g}^{y_i}; \\ &A \leftarrow \mathsf{Adv}(\gamma, \tilde{S}, \tilde{T}, \{b_1, b_1^{s+x_1 st}, b_1^{x_1}, b_1^{x_1 t}, g^{x_1}, \tilde{b}_1\}, \dots, \{b_q, b_q^{s+x_q st}, b_q^{x_q}, b_q^{x_q t}, g^{x_q}, \tilde{b}_q\}) : \\ &\qquad A = (a_1, a_2, a_3, a_4, a_5, a_6) \wedge x \notin \{x_1, \dots, x_q\} \wedge x \in \mathbb{Z}_p^* \wedge a_1 \in \mathbb{G}_1 \wedge \\ &\qquad\qquad a_2 = a_1^{s+xst} \wedge a_3 = a_1^x \wedge a_4 = a_1^{xt} \wedge a_5 = g^x \wedge e(a_1, \tilde{g}) = e(g, a_6)]. \end{aligned}$$

Related formulations of the above assumption in an oracle-setting, where the x_i values are chosen dynamically by Adv, are the LRSW assumption which was introduced by Lysyanskaya *et al.* [27] and the Strong LRSW assumption of Ateniese *et al.* [1]. We eliminate the oracle and instead give q random tuples, which are also slightly changed. In the full version of this work [23], we show that the above assumption admits a proof in Shoup's generic group model [33].

3.1 Groth-Sahai Proofs

The Groth-Sahai proof system [24] permits a variety of efficient non-interactive proofs of the satisfiability of one or more pairing product equations. For variables $\{\mathcal{X}\}_{1\dots m} \in \mathbb{G}_1, \{\mathcal{Y}\}_{1\dots n} \in \mathbb{G}_2$ and constants $\{\mathcal{A}\}_{1\dots n} \in \mathbb{G}_1, \{\mathcal{B}\}_{1\dots m} \in \mathbb{G}_2, a_{i,j} \in \mathbb{Z}_p$, and $t_T \in \mathbb{G}_T$, these equations have the form:

$$\prod_{i=1}^{n} e(\mathcal{A}_i, \mathcal{Y}_i) \prod_{i=1}^{m} e(\mathcal{X}_i, \mathcal{B}_i) \prod_{i=1}^{m}\prod_{j=1}^{n} e(\mathcal{X}_i, \mathcal{Y}_j)^{a_{i,j}} = t_T$$

Groth and Sahai show how to construct Witness Indistinguishable proof-of-knowledge of a satisfying witness to such an equation, in prime-order groups where the SXDH or Decision Linear assumptions hold. The proof system they describe can be composed over multiple equations involving the same variables. They point out that in some special cases, their techniques can be strengthened to provide Zero Knowledge. Unlike the interactive proofs used in [10, 22], the Groth-Sahai proofs do not use adversarial rewinding in their security analysis.

Groth-Sahai Commitments [24]. At the core of the Groth-Sahai system is a homomorphic commitment scheme to elements of $\mathbb{G}_1$ or $\mathbb{G}_2$.[2] The public parameters for the commitment scheme can be generated in two ways. Method (1) leads to a perfectly-binding commitment scheme, while method (2) leads to a perfectly-*hiding* scheme. Note that the two parameter distributions are computationally indistinguishable under the SXDH assumption. When the GS commitment parameters are configured according to method (1), they are equivalent

[2] As noted in [3, 24] commitment scheme can also be used to commit to elements of $\mathbb{Z}_p$, though we use this only in the context of simulating proofs.

to an Elgamal encryption of a group element, and can be decrypted by a party that knows a trapdoor to the commitment parameters. When commitments are configured according to method (2), a "simulation" trapdoor can be used on random commitments to open them to any value g^x (or $\tilde{g}^x$) for known x.

The Proof System. We now describe the proof system at a high level, adopting some notation and exposition from [3]. For this description we will conceal many of the underlying details, though the reader can refer to [3, 24] for a more detailed explanation. The proof system contains the following (possibly probabilistic) polynomial time algorithms:

$\mathsf{GSSetup}(\gamma)$. On input $\gamma \in \mathsf{BMsetup}(1^\kappa)$, outputs a string GS containing parameters for the proof system. This string embeds binding parameters for the G-S commitment scheme.

$\mathsf{GSProve}(GS, S, W)$. On input a statement S describing the equation, and a satisfying witness $W \in \langle \{\mathcal{X}\}_{1 \ldots m}, \{\mathcal{Y}\}_{1 \ldots n} \rangle$, outputs a proof π. To formulate this proof, a commitment $\hat{C}_i$ is generated for each element in W. The proof embeds openings to the commitments in such a way that a prover can ascertain that S is verifiably satisfied, and yet the elements of W remain hidden.

$\mathsf{GSVerify}(GS, \pi)$. Verifies the proof π (using the commitments and opening values) and outputs ACCEPT if π is valid, REJECT otherwise. (For compactness of notation, we will specify that π embeds the statement S).

Above we describe the proof system in normal operation. In our security proofs we will additionally use:

$\mathsf{GSExtractSetup}(\gamma)$. Outputs GS (distributed identically to the output of $\mathsf{GSSetup}(\gamma)$) and an extraction trapdoor td_{ext} containing a trapdoor for the commitment scheme. This trapdoor permits an extraction of a valid witness from the commitments embedded within a proof.

$\mathsf{GSExtract}(GS, td_{ext}, \pi)$. Given a proof π and the extraction trapdoor, extracts $\mathcal{X}_i$ or $\mathcal{Y}_i$ from each commitment $\hat{C}_i$, and outputs the witness $W = \langle \{\mathcal{X}\}_{1 \ldots M}, \{\mathcal{Y}\}_{1 \ldots N} \rangle$ that satisfies the equations.

$\mathsf{GSSimulateSetup}(\gamma)$. Outputs parameters GS' that are computationally indistinguishable from the output of $\mathsf{GSSetup}(\gamma)$, as well as a simulation trapdoor td_{sim} which consists of a simulation trapdoor for the commitment scheme.

$\mathsf{GSSimProve}(GS', td_{sim}, S)$. Given simulation parameters GS' and trapdoor td_{sim}, outputs a proof π of statement S that such that $\mathsf{GSVerify}(GS', \pi) =$ ACCEPT. Note that this algorithm operates on certain restricted classes of statements (see below).

GS proofs can be defined over multiple pairing product equations. In this case, satisfiability implies knowledge of a witness for the full set of equations. In our constructions, we will denote a GS proof statement using the notation of Camenisch and Stadler [11]. For instance, $NIWI_{GS}\{(a_1, a_2) : e(a_1, a_2)e(g, h^{-1}) = 1 \wedge e(a_2, g_2)e(d_2^{-1}, a_3) = 1\}$ represents a non-interactive Witness Indistinguishable proof of knowledge, formed under parameters GS, of a witness $W = \langle a_1, a_2 \rangle$

that simultaneously satisfies both listed equations. All values not in enclosed within the initial ()'s are assumed to be known to the verifier.

Witness Indistinguishability and Zero Knowledge. In general, Groth-Sahai proofs satisfy a strong definition of Witness Indistinguishability in groups where the SXDH assumption holds (complete security definitions can be found in the full version of this work [23]). However, for certain restricted classes of statements, the proof system can also be used to construct non-interactive Zero Knowledge (NIZK) proofs. For certain trivial statements, this is simply a matter of using a WI proof for which a witness can easily be found. E.g., in the special case where $t_T = 1$ for a pairing product equation, a simulator can always compute a satisfying witness by selecting each $\mathcal{X}_i$ or $\mathcal{Y}_i$ to be g^0 or $\tilde{g}^0$ respectively.

More practically, Groth and Sahai observe that some non-trivial statements can be proven in Zero Knowledge by applying the simulation trapdoor for the Groth-Sahai commitment scheme. This trapdoor allows the simulator to open a random commitment to any g^x or $\tilde{g}^x$ (for known x), and can be applied such that the same commitment is opened *differently* for each equation within the statement. In some cases, we may need to re-write a statement in order to construct a ZK proof. For example, consider a proof of the statement $e(a, d) = e(g, h)$ made on variable a and constants d, g, h. By adding a second variable b and a further equation, we obtain an equivalent statement which can be proven using the following zero knowledge proof:

$$NIZK_{GS}\{(a,b) : e(a,d)e(b,h^{-1}) = 1 \ \wedge \ e(b,g)e(g^{-1},g) = 1\}$$

Note that the equivalence holds by the property that $b = g$ is the only valid solution to the revised equation. However, using the simulation trapdoor we can open the appropriate commitments such that $a = b = g^0$ in the first equation, while in the second equation $b = g$. We will use similar techniques to simulate the Zero-Knowledge proofs in our constructions.

3.2 Additional Tools

Modified CL Signatures. Our constructions use a variant of the Camenisch-Lysanskyaya signature scheme [9], altered to operate on messages in $\mathbb{G}_1$. Whereas CL signatures rely on the interactive LRSW assumption to achieve security against adaptive chosen-message attacks, in the context of our construction we will require only a non-interactive q-Hidden LRSW assumption to achieve a weaker property (unforgeability given a set of signatures on *random* messages).

$\mathsf{CLKeyGen}(\gamma, g, \tilde{g})$. On input $\gamma = (p, \mathbb{G}_1, \mathbb{G}_2, \mathbb{G}_T, e, \dots)$ and generators $(g, \tilde{g})$, select $s, t \stackrel{\$}{\leftarrow} \mathbb{Z}_p$ and set $\tilde{S} \leftarrow \tilde{g}^s, \tilde{T} \leftarrow \tilde{g}^t$. Output $vk = (\gamma, g, \tilde{g}, \tilde{S}, \tilde{T})$, and $sk = (vk, s, t)$.

$\mathsf{CLSign}_{sk}(m)$. On input a message $m \in \mathbb{G}_1$, select $w \stackrel{\$}{\leftarrow} \mathbb{Z}_p$ and output the signature $\mathsf{sig} = (g^w, m^w, g^{ws}m^{wst}, m^{wt}, \tilde{g}^w) \in \mathbb{G}_1^4 \times \mathbb{G}_2$.

$\mathsf{CLVerify}_{vk}(\mathsf{sig}, m)$. On input the value $m \in \mathbb{G}_1$ and $\mathsf{sig} = (a_1, a_2, a_3, a_4, \tilde{a}_5)$, verify that $e(g, \tilde{a}_5) = e(a_1, \tilde{g}) \wedge e(m, \tilde{a}_5) = e(a_2, \tilde{g}) \wedge e(a_2, \tilde{T}) = e(a_4, \tilde{g}) \wedge e(a_3, \tilde{g}) = e(a_1 a_4, \tilde{S})$.

Note that the verification algorithm can be represented as a set of pairing product equations, and thus it is possible to prove knowledge of a pair (m, sig) using the GS proof system. To prove knowledge of m, sig, first select $y \xleftarrow{\$} \mathbb{Z}_p$, compute $\mathsf{sig}' = \langle a_1', a_2', a_3', a_4', \tilde{a}_5' \rangle = \langle a_1^y, a_2^y, a_3^y, a_4^y, \tilde{a}_5^y \rangle$ and release the pair $a_1', \tilde{a}_5'$ along with the following witness indistinguishable proof:

$$\pi = NIWI_{GS}\{(m, a_2', a_3', a_4') :$$
$$e(m, \tilde{a}_5')e(a_2', \tilde{g}^{-1}) = 1 \wedge e(a_2', \tilde{T})e(a_4', \tilde{g}^{-1}) = 1 \wedge e(a_3', \tilde{g})e(a_4'^{-1}, \tilde{S}) = e(a_1', \tilde{S})\}$$

The verifier checks both the proof and the fact that $e(a_1', \tilde{g}) = e(g, \tilde{a}_5')$.

Selective-message Secure Boneh-Boyen Signatures. Our constructions also make use of a weak signature scheme built from the Boneh-Boyen selective-ID IBE scheme [4] (§4).

$\mathsf{BBKeyGen}(\gamma, g_1, \tilde{g}_1)$. On input $\gamma = (p, \mathbb{G}_1, \mathbb{G}_2, \mathbb{G}_T, e, \ldots)$ and bases $(g_1, \tilde{g}_1)$, select $\alpha, z \xleftarrow{\$} \mathbb{Z}_p$, $g \leftarrow g_1^{1/\alpha}$, $\tilde{g} \leftarrow \tilde{g}_1^{1/\alpha}$, $g_2 \leftarrow g^z$, $\tilde{g}_2 \leftarrow \tilde{g}^z$, $h \xleftarrow{\$} \mathbb{G}_1$. Output $vk = (\gamma, g, \tilde{g}, g_1, g_2, h, \tilde{g}_2)$, and $sk = (vk, g_2^\alpha)$.

$\mathsf{BBSign}_{sk}(m)$. On input a message $m \in \mathbb{G}_1$, select $r \xleftarrow{\$} \mathbb{Z}_p$ and output the signature $\mathsf{sig} = ((mh)^r g_2^\alpha, \tilde{g}^r, g^r) \in \mathbb{G}_1^2 \times \mathbb{G}_2$.

$\mathsf{BBVerify}_{vk}(\mathsf{sig}, m)$. On input $m \in \mathbb{G}_1$ and $\mathsf{sig} = (s_1, \tilde{s}_2, s_3)$, verify that $e(s_1, \tilde{g}) \;/\; e(mh, \tilde{s}_2) = e(g_1, \tilde{g_2})$ and $e(g, \tilde{s}_2) = e(s_3, \tilde{g})$.

We can prove knowledge of a pair (m, sig) as follows. Select $y \xleftarrow{\$} \mathbb{Z}_p$ and set $\mathsf{sig}' = (s_1', \tilde{s}_2', s_3') = (s_1(mh)^y, \tilde{s}_2\tilde{g}^y, s_3 g^y)$. Output $\tilde{s}_2', s_3'$ and the WI proof:

$$\pi = NIWI_{GS}\{(m, s_1') : e(s_1', \tilde{g})e(m, \tilde{s}_2'^{-1}) = e(h, \tilde{s}_2')e(g_1, \tilde{g_2})\}$$

The verifier checks the proof and the fact that $e(g, \tilde{s}_2') = e(s_3', \tilde{g})$.

Double-Trapdoor BBS Encryption. Our OT constructions employ an encryption scheme with a "double-trapdoor" (so that both the simulator in charge of the crs and the sender in charge of the pk can extract the messages of the ciphertext.) It is crucial that the holder of either secret key can verify the consistency of the ciphertext with respect to the other secret key (*i.e.*, that decryption using the other key would reveal the same plaintext.) We use a variant of Boneh-Boyen-Shacham encryption [5], which has a public consistency check.

Let $\mathsf{BMsetup}(1^\kappa) \to \gamma = (p, \mathbb{G}_1, \mathbb{G}_2, \mathbb{G}_T, e, g, \tilde{g})$. Publish global parameters $\gamma, h, \tilde{h}$ such that $e(g, \tilde{h}) = e(\tilde{g}, h)$, and for $i \in [1, 2]$ select $sk_i \leftarrow (x_i, y_i \in_R \mathbb{Z}_p)$ and $pk_i = (u_i, v_i, \tilde{u}_i, \tilde{v}_i) \leftarrow (h^{1/x_i}, h^{1/y_i}, \tilde{h}^{1/x_i}, \tilde{h}^{1/y_i})$. To encrypt a message $m \in \mathbb{G}_1$ under pk_1/pk_2, first select random values $r, s \in \mathbb{Z}_p$ and output the ciphertext $(u_1^r, v_1^s, u_2^r, v_2^s, h^{r+s}m)$. To decrypt a message $(c_1, \ldots, c_5)$ under $sk_1 = (x_1, y_1)$, output $c_5/(c_1^{x_1} \cdot c_2^{y_1})$. To decrypt under $sk_2 = (x_2, y_2)$, output $c_5/(c_3^{x_2} \cdot c_4^{y_2})$. Note that the structure of a ciphertext can be verified using the bilinear map, by checking that $e(c_1, \tilde{u}_2) = e(c_3, \tilde{u}_1) \wedge e(c_2, \tilde{v_2}) = e(c_4, \tilde{v}_1)$ In the full version [23] we show that scheme above is semantically-secure under the DLIN assumption.

Protocol OTA

OTA is parameterized by the algorithms (OTGenCRS, OTInitialize, OTRequest, OTRespond, OTComplete).

When **S** is activated with (sid, sender, $\langle M_1, \ldots, M_N \in \{0,1\}^\ell \rangle$):

1. **S** queries $\mathcal{F}_{CRS}$ with (sid, **S**, **R**) and receives (sid, crs). **R** then queries $\mathcal{F}_{CRS}$ with (sid, **S**, **R**) and receives (sid, crs).[a]
2. **S** computes $(T, sk) \leftarrow$ OTInitialize(crs, $M_1, \ldots, M_N$), sends (sid, T) to **R** and stores (sid, T, sk).

When **R** is activated with (sid, receiver, σ), and **R** has previously received (sid, T) and (sid, crs):

1. **R** runs $(Q, Q_{priv}) \leftarrow$ OTRequest(crs, T, σ), sends (sid, Q) to **S** and stores (sid, Q_{priv}).
2. **S** gets (sid, Q) from **R**, runs $R \leftarrow$ OTRespond(crs, T, sk, Q), and sends (sid, R) to **R**.
3. **R** receives (sid, R) from **S**, and outputs (sid, OTComplete(crs, T, R, Q_{priv})).

[a] $\mathcal{F}_{CRS}$ computes computes crs $\leftarrow$ OTGenCRS(1^κ).

Fig. 4. A high-level outline of the $\mathsf{OT}_{k\times 1}^N$ protocol, with details of each algorithm described in Section 4. We make no explicit mention of the value k, the total transfers permitted by the Sender, because our protocol does not depend on it. The Sender may choose to stop answering the Receiver's queries at any point, in which case OTRespond outputs "reject" and OTComplete accepts this as the message $\perp$.

4 A UC-Secure Adaptive OT Construction

Our adaptive oblivious transfer protocol, $\mathsf{OT}_{k\times 1}^N$ follows the framework described in Figure 4. We now describe one instantiation of the algorithms (OTGenCRS, OTInitialize, OTRequest, OTRespond, OTComplete). In the full version [23], we provide a second instantiation, under different assumptions.

OTGenCRS(1^κ). Given security parameter κ, generate parameters for a bilinear mapping $\gamma = (p, \mathbb{G}_1, \mathbb{G}_2, \mathbb{G}_T, e, g, \tilde{g}) \leftarrow$ BMsetup(1^κ). Compute $GS_S \leftarrow$ GSSetup(γ) and $GS_R \leftarrow$ GSSetup(γ). Choose $a, b, c \xleftarrow{\$} \mathbb{Z}_p$, and set $(g_1, g_2, h, \tilde{g}_1, \tilde{g}_2, \tilde{h}) \leftarrow (g^a, g^b, g^c, \tilde{g}^a, \tilde{g}^b, \tilde{g}^c)$. Output crs $= (\gamma, GS_S, GS_R, g_1, g_2, h, \tilde{g}_1, \tilde{g}_2, \tilde{h})$. (In the full version [23], we describe how this common reference string can be replaced by a common random string.)

OTInitialize(crs, $m_1, \ldots, m_N$). This algorithm is executed by the Sender. On input a collection of N messages and the crs, it outputs a commitment to the database, T, for publication to the Receiver, as well as a Sender secret key, sk. We treat messages as elements of $\mathbb{G}_1$, since there exist efficient mappings between strings in $\{0,1\}^\ell$ and elements in $\mathbb{G}_1$ (e.g., [1, 6]).

1. Parse crs to obtain $GS_S, g_1, g_2, h, \tilde{g}_1, \tilde{g}_2, \tilde{h}$ and γ.
2. Choose random values $x_1, x_2 \in \mathbb{Z}_p$.

3. Set $(u_1, u_2) \leftarrow (h^{1/x_1}, h^{1/x_2})$, $(\tilde{u}_1, \tilde{u}_2) \leftarrow (\tilde{h}^{1/x_1}, \tilde{h}^{1/x_2})$.
4. Set $(vk_1, sk_1) \leftarrow \mathsf{CLKeyGen}(\gamma, u_1, \tilde{u}_1)$, $(vk_2, sk_2) \leftarrow \mathsf{CLKeyGen}(\gamma, u_2, \tilde{u}_2)$ and $(vk_3, sk_3) \leftarrow \mathsf{BBKeyGen}(\gamma, u_1, \tilde{u}_1)$.
5. Set $pk \leftarrow (u_1, u_2, \tilde{u}_1, \tilde{u}_2, vk_1, vk_2, vk_3)$.
6. For $j = 1, \ldots, N$ encrypt each message m_j as:
   (a) Select random $r, s, t \in \mathbb{Z}_p$.
   (b) Compute $\mathsf{sig}_1 \leftarrow \mathsf{CLSign}_{sk_1}(u_1^r)$, $\mathsf{sig}_2 \leftarrow \mathsf{CLSign}_{sk_2}(u_2^s)$, and $\mathsf{sig}_3 \leftarrow \mathsf{BBSign}_{sk_3}(u_1^r u_2^s)$.
   (c) Set $C_j \leftarrow (u_1^r,\ u_2^s,\ g_1^r,\ g_2^s,\ m_j \cdot h^{r+s},\ \mathsf{sig}_1,\ \mathsf{sig}_2,\ \mathsf{sig}_3)$.
7. Set $T \leftarrow (pk, C_1, \ldots, C_N)$ and $sk \leftarrow (x_1, x_2)$. Output (T, sk).

Each ciphertext C_j above can be thought of as a signcryption where it is the *randomness* for each ciphertext that is signed, rather than the plaintext itself. Each plaintext m_j is encrypted under **S**'s public key u_1, u_2, as well as a "key" g_1, g_2 drawn from crs. This "double-trapdoor" encryption is necessary for the security proof of the OT scheme.

To verify the format of each ciphertext $C_j = (c_1, \ldots, c_5,\ \mathsf{sig}_1,\ \mathsf{sig}_2,\ \mathsf{sig}_3)$ in T, anyone can check that $\mathsf{CLVerify}_{vk_1}(c_1, \mathsf{sig}_1)$, $\mathsf{CLVerify}_{vk_2}(c_2, \mathsf{sig}_2)$, and $\mathsf{BBVerify}_{vk_3}(c_1 c_2, \mathsf{sig}_3)$ each succeed, and that $e(c_1, \tilde{g}_1) = e(c_3, \tilde{u}_1) \wedge e(c_2, \tilde{g}_2) = e(c_4, \tilde{u}_2)$.

$\mathsf{OTRequest}(\mathsf{crs}, T, \sigma)$. This algorithm is executed by a Receiver. On input T generated by the Sender, along with an item index σ, generates a query Q for transmission to the Sender.

1. Parse T as $(pk, C_1, \ldots, C_N)$, and ensure that it is correctly formed (see above). If T is not correctly formed, abort the protocol. (This is only necessary on the first transfer.)
2. Parse crs to obtain $(GS_R, \tilde{h})$, and parse pk as $(u_1, u_2, \tilde{u}_1, \tilde{u}_2, vk_1, vk_2, vk_3)$. Parse the σth ciphertext C_σ as $(c_1, \ldots, c_5, \mathsf{sig}_1, \mathsf{sig}_2, \mathsf{sig}_3)$.
3. Select random $v_1, v_2 \in \mathbb{Z}_p$.
4. Set $d_1 \leftarrow (c_1 \cdot u_1^{v_1})$, $d_2 \leftarrow (c_2 \cdot u_2^{v_2})$, $t_1 \leftarrow h^{v_1}$, $t_2 \leftarrow h^{v_2}$.
5. Use the Groth-Sahai techniques and reference string GS_R to compute a Witness Indistinguishable proof π that the values d_1, d_2 pertaining to the ciphertext C_σ (which the Receiver wishes to have the Sender help him open) have the correct structure:

$$\begin{aligned}\pi = NIWI_{GS_R}\{(c_1, c_2, t_1, t_2, \mathsf{sig}_1, \mathsf{sig}_2, \mathsf{sig}_3) :\\ e(c_1, \tilde{h})e(t_1, \tilde{u}_1) = e(d_1, \tilde{h}) \ \wedge\ e(c_2, \tilde{h})e(t_2, \tilde{u}_2) = e(d_2, \tilde{h}) \ \wedge\\ \mathsf{CLVerify}_{vk_1}(c_1, \mathsf{sig}_1) = 1 \ \wedge\ \mathsf{CLVerify}_{vk_2}(c_2, \mathsf{sig}_2) = 1 \ \wedge\\ \mathsf{BBVerify}_{vk_3}(c_1 c_2, \mathsf{sig}_3) = 1\}\end{aligned}$$

6. Set request $Q \leftarrow (d_1, d_2, \pi)$, and private state $Q_{priv} \leftarrow (Q, \sigma, v_1, v_2)$. Output (Q, Q_{priv}).

To explain what is happening in the statement of step (5), first observe that the signature proofs of knowledge ensure that the values c_1, c_2 and the product $(c_1 c_2)$ each correspond to a valid signature held by the Receiver. The remaining equations ensure that the values d_1, d_2 correspond to "blinded" versions of the elements c_1, c_2. These checks guarantee that the witness used by the Receiver, and thus the decryption request being made, corresponds to one of the N ciphertexts published by the Sender.

$\mathsf{OTRespond}(\mathsf{crs}, T, sk, Q)$. This algorithm is executed by the Sender. If the Sender does not wish to answer any more requests for the Receiver, then the Sender outputs the message "reject". Otherwise, the Sender processes the Receiver's request Q as:

1. Parse crs to obtain $(GS_R, \tilde{g}, \tilde{h})$, and parse T as $(pk, C_1, \ldots, C_N)$, and sk as (x_1, x_2).
2. Parse pk (from T) as $(u_1, u_2, \tilde{u}_1, \tilde{u}_2, vk_1, vk_2, vk_3)$.
3. Parse Q as (d_1, d_2, π) and verify proof π using GS_R. Abort if check fails.
4. Set $a_1 \leftarrow d_1^{x_1}$, $a_2 \leftarrow d_2^{x_2}$, and $s \leftarrow a_1 \cdot a_2$.
5. Use the Groth-Sahai techniques and reference string GS_S to formulate a zero-knowledge proof[3] that the decryption value s is properly computed:

$$\delta = NIZK_{GS_S}\{(a_1, a_2) : e(a_1, \tilde{u}_1)e(d_1^{-1}, \tilde{h}) = 1 \\ \wedge\ e(a_2, \tilde{u}_2)e(d_2^{-1}, \tilde{h}) = 1 \ \wedge\ e(a_1 a_2, \tilde{h})e(s^{-1}, \tilde{h}) = 1\}$$

   The third equation ensures that $s = a_1 \cdot a_2$, while the first two, since the values $(u_1, d_1, u_2, d_2, \tilde{h})$ are known to both parties, ensure that $a_1 = d_1^{x_1}$ and $a_2 = d_2^{x_2}$.
6. Output $R \leftarrow (s, \delta)$.

$\mathsf{OTComplete}(\mathsf{crs}, T, R, Q_{priv})$. This algorithm is executed by the Receiver. On input R generated by the Sender in response to a request Q, along with state Q_{priv}, outputs a message m or $\perp$. If R is the message "reject", then the Receiver outputs $\perp$. Otherwise, the Receiver does:

1. Parse crs to obtain (GS_S, h). Parse T as $(pk, C_1, \ldots, C_N)$, R as (s, δ), and Q_{priv} as (Q, σ, v_1, v_2).
2. Verify proof δ using GS_S. If verification fails, output $\perp$.
3. Parse C_σ to obtain the first five elements $(c_1, \ldots, c_5)$ and output $m = c_5/(s \cdot h^{-v_1} \cdot h^{-v_2})$. Map this element to a value in $\{0,1\}^\ell$ [1].

4.1 Efficiency Analysis

When the protocol in Figure 4 is implemented using the algorithms described above, we obtain a $(k+1/2)$-round protocol with communications cost $O(N+k)$, where $k \leq N$. More concretely, the crs is comprised of 7 elements in $\mathbb{G}_1$ and 7 elements of $\mathbb{G}_2$, the Sender's public key contains 5 elements in $\mathbb{G}_1$ and 6 elements in $\mathbb{G}_2$. Each of the N ciphertexts in T requires 15 elements in $\mathbb{G}_1$ and 3 elements in $\mathbb{G}_2$. Moreover, each item transfer involves transmission of 68 elements of $\mathbb{G}_1$

[3] We present a simplified version of this proof above. However, to permit simulation, we must add a third variable $\tilde{a}_3 = \tilde{h}$ and re-write the proof as $NIZK_{GS_S}\{(a_1, a_2, \tilde{a}_3) : e(a_1, \tilde{u}_1)e(d_1^{-1}, \tilde{a}_3) = 1 \wedge e(a_2, \tilde{u}_2)e(d_2^{-1}, \tilde{a}_3) = 1 \wedge e(a_1 a_2, \tilde{a}_3)e(s^{-1}, \tilde{a}_3) = 1 \wedge e(u_1, \tilde{a}_3) = e(u_1, \tilde{h})\}$. See the full version for details.

and 38 elements of $\mathbb{G}_2$ from Receiver to Sender, and then 20 elements of $\mathbb{G}_1$ and 18 elements of $\mathbb{G}_2$ from Sender to Receiver. The message space of our OT protocol is elements in $\mathbb{G}_1$, which will be sufficient for transferring a symmetric encryption key to unlock a file of arbitrary size.

4.2 Security Analysis

Theorem 1. *Instantiated with the above algorithms,* OTA *securely realizes the functionality* $\mathcal{F}_{OT}^{N\times 1}$ *in the* $\mathcal{F}_{CRS}$-hybrid *model under the SXDH, DLIN, and q-Hidden LRSW assumptions.*

Due to space considerations, we provide only a sketch of Theorem 1 below (the complete proof can be found in the full version of this work [23]). When either the Sender or the Receiver is corrupted, we wish to describe a simulator $\mathcal{S}$ such that it can interact with the ideal functionality $\mathcal{F}_{OT}^{N\times 1}$ (which we'll denote simply as $\mathcal{F}$) and the environment $\mathcal{Z}$ appropriately; i.e., $\mathsf{IDEAL}_{\mathcal{F},\mathcal{S},\mathcal{Z}} \stackrel{c}{\approx} \mathsf{EXEC}_{\mathsf{OTA},\mathcal{A},\mathcal{Z}}$.

Simulating the case where only S is corrupted. We first consider the case where the real-world adversary $\mathcal{A}$ corrupts the Sender, and thus $\mathcal{S}$ must interact with $\mathcal{F}$ as the ideal Sender and with (an internal copy of) $\mathcal{A}$ as a real-world Receiver. Here $\mathcal{S}$ does the following:

1. Ask $\mathcal{A}$ to begin an OT protocol, and set the crs *for these two parties* by running $\gamma = (p, \mathbb{G}_1, \mathbb{G}_2, \mathbb{G}_T, e, g \in \mathbb{G}_1, \tilde{g} \in \mathbb{G}_2) \leftarrow \mathsf{BMsetup}(1^\kappa)$, $GS_S \leftarrow \mathsf{GSSetup}(\gamma)$, $GS_R \leftarrow \mathsf{GSSetup}(\gamma)$, selecting random elements $a_1, a_2 \in \mathbb{Z}_p$, and setting $g_1^{a_1} = g_2^{a_2} = h$ (and a corresponding relationship for $\tilde{g}_1, \tilde{g}_2, \tilde{h}$). Set $\mathsf{crs} = (\gamma, GS_S, GS_R, g_1, g_2, h, \tilde{g}_1, \tilde{g}_2, \tilde{h})$. When the parties query $\mathcal{F}_{CRS}$, return $(\mathsf{sid}, \mathsf{crs})$.
2. Obtain the database commitment T from $\mathcal{A}$. Verify that T is well-formed, abort if not. Otherwise, $\forall i \in [1, N]$ use a_1, a_2 to decrypt each ciphertext $C_i = (c_1, \ldots, c_5, \ldots)$ as $m_i = c_5/(c_3^{a_1} c_4^{a_2})$. Map each element $m_i \in \mathbb{G}_1$ to a string in $\{0,1\}^\ell$ [1]. Send $(\mathsf{sid}, \mathbf{S}, m_1, \ldots, m_N)$ to $\mathcal{F}$.
3. Upon receiving $(\mathsf{sid}, \mathsf{request})$ from $\mathcal{F}$, return $\mathsf{OTRequest}(\mathsf{crs}, T, 1)$ to $\mathcal{A}$. This response includes two random values d_1, d_2 and a non-interactive witness indistinguishable proof π with respect to $GS_R \in \mathsf{crs}$ that d_1, d_2 are "blinded" values corresponding to ciphertext C_1. This proof can be performed honestly and without rewinding.
4. If $\mathcal{A}$ issues a "reject" message or responds with anything other than a value in $\mathbb{G}_1$ and a valid NIZK proof, then $\mathcal{S}$ tells $\mathcal{F}$ to fail the request by sending message $(\mathsf{sid}, 0)$. Otherwise, $\mathcal{S}$ sends the message $(\mathsf{sid}, 1)$ to $\mathcal{F}$.

The indistinguishability argument here follows from the indistinguishability of the crs (which is identically distributed to a real crs), the perfect extraction of the messages in step (2),[4] and the Witness Indistinguishability of the

[4] Note that a ciphertext that passes the validity check can be represented as $C = (u_1^r, u_2^s, g_1^r, g_2^s, h^{r+s}m, \ldots)$ for some $r, s \in \mathbb{Z}_p$, and when (g_1, g_2, h) have the relationship described above, decryption using a_1, a_2 always produces m.

GS proof π issued during each request phase, which guarantees that $\mathcal{A}$ (the corrupt Sender) cannot distinguish a request to decrypt C_1 from a request to decrypt any other valid ciphertext. Thus, $\mathcal{S}$ can adequately mimic its response pattern.

Simulating the case where only R is corrupted. Next, we consider the case where the real world adversary $\mathcal{A}$ corrupts the Receiver, and thus $\mathcal{S}$ must interact with $\mathcal{F}$ as the ideal Receiver and with (and internal copy of) $\mathcal{A}$ as real-world Receiver. This case requires that the $q = N$ for the *q-Hidden LRSW* assumption. Here $\mathcal{S}$ does the following:

1. Ask $\mathcal{A}$ to begin an OT protocol, and set the crs *for these two parties* by running $\gamma = (p, \mathbb{G}_1, \mathbb{G}_2, \mathbb{G}_T, e, g \in \mathbb{G}_1, \tilde{g} \in \mathbb{G}_2) \leftarrow \mathsf{BMsetup}(1^\kappa)$, $(GS_S, td_{sim}) \leftarrow \mathsf{GSSimulateSetup}(\gamma)$ and $(GS_R, td_{ext}) \leftarrow \mathsf{GSExtractSetup}(\gamma)$. Select random elements for $g_1, g_2, h, \tilde{g}_1, \tilde{g}_2, \tilde{h}$. Set $\mathsf{crs} \leftarrow (\gamma, GS_S, GS_R, g_1, g_2, h, \tilde{g}_1, \tilde{g}_2, \tilde{h})$. When the parties query $\mathcal{F}_{CRS}$, return $(\mathsf{sid}, \mathsf{crs})$.
2. $\mathcal{S}$ must commit to a database of messages for $\mathcal{A}$ *without* knowing the messages $m_1, \ldots, m_N$. Thus, $\mathcal{S}$ simply commits to random junk messages, and sends the corresponding T to $\mathcal{A}$.
3. When $\mathcal{A}$ makes a transfer request, $\mathcal{S}$ uses td_{ext} to extract the witness W corresponding to $\mathcal{A}$'s decryption request from the NIWI proof. (This extraction is done via opening perfectly-binding commitments which are included in the WI proof and does not require any rewinding.) This witness includes the first two elements (c_1, c_2) of the ciphertext that $\mathcal{A}$ is requesting to decrypt, and from these it is possible to determine the index σ' of the ciphertext that $\mathcal{A}$ has requested to open.
4. $\mathcal{S}$ now sends $(\mathsf{sid}, \mathbf{R}, \sigma')$ to $\mathcal{F}$ to obtain the real $m_{\sigma'}$ message.
5. Finally, $\mathcal{S}$ returns a response to $\mathcal{A}$ which opens $C_{\sigma'}$ to $m_{\sigma'}$ and then uses td_{sim} to simulate an NIZK proof that this opening is correct. The NIZK proof here is designed in such a way that simulation is always possible and no rewinding is necessary.

The indistinguishability argument here follows from the indistinguishability of the crs (from a real crs), the indistinguishability of the "fake" database T, the ability to extract witnesses from the NIWI proofs, and the zero-knowledge property of "fake" NIZK proofs. In particular, note that the *N-Hidden LRSW* assumption ensures that any decryption request made by the receiver corresponds to a valid ciphertext from the database T (if $\mathcal{A}$ produces a proof π embedding invalid ciphertext values, we can use $\mathcal{A}$ to solve *N-Hidden LRSW* or the co-CDH problem [7], which is implied by *N-Hidden LRSW*).[5] Unlike the protocol of [10]

[5] Note that we are using both an existentially unforgeable signature scheme, as well as a selective-ID IBE scheme that has been "retasked" as signature scheme. The latter leads to a signature that is only secure for a polynomial-sized, fixed message space. In the full version, we show that this limitation is acceptable given that we are signing the product of other messages which have been signed using the stronger signature scheme. Since there are at most a polynomial number of such products, the construction is secure.

we are able to base the semantic security of the ciphertexts on a standard decisional assumption (the Decision Linear assumption). This is possible because the full ciphertext can be constructed using only the DLIN input (see the note on Ciphertext security below). Notice that $\mathcal{S}$ is never *both* simulating and extracting via the same (subsection of the) common reference string; indeed, we do not require that the proofs be simulation-sound.

Simulating the remaining cases. When both the Receiver and Sender are corrupted, $\mathcal{S}$ knows the inputs to $\mathbf{S}$ and $\mathbf{R}$ and can simulate a protocol execution by generating the real messages exchanged between the two parties. In the case where neither party is corrupted, then: when $\mathcal{S}$ receives messages of the form (sid, b_i) indicating that transfers have occurred, $\mathcal{S}$ generates a simulated transcript between the honest $\mathbf{S}$ and $\mathbf{R}$. In this case, $\mathcal{S}$ runs the protocol as specified, using as $\mathbf{S}$'s input a random database $(\hat{m}_1, \ldots, \hat{m}_N)$, and (for each transfer), $\mathbf{R}$'s input $\sigma' = 1$. If in the ith transfer $b_i = 0$ then $\mathbf{S}$'s responds with an invalid R (the empty string). Else, $\mathbf{S}$ returns a valid response as in the protocol.

Ciphertext security. We briefly elaborate on the security of the ciphertexts in our scheme. To prove security when Receiver is corrupted, we must show that a ciphertext vector encrypting random messages is indistinguishable from a vector encrypting the real message database. We argue that this is the case under the Decision Linear assumption. Let $D = (g, \tilde{g}, f, \tilde{f}, h, \tilde{h}, g^a, f^b, z_d)$ be a candidate Decision Linear tuple. We consider a simulation that behaves as follows:

1. Set $u_1 = g, u_2 = f, \tilde{u}_1 = \tilde{g}, \tilde{u}_2 = \tilde{f}$. Select random $y_1, y_2 \in \mathbb{Z}_p$, and set $g_1 = u_1^{y_1}, g_2 = u_2^{y_2}$ (and similarly for $\tilde{g}_1, \tilde{g}_2$). Fix $\mathsf{crs} \leftarrow (\gamma,\ GS'_S,\ GS'_R,\ g_1,\ g_2,\ h,\ \tilde{g}_1,\ \tilde{g}_2,\ \tilde{h})$.
2. Generate $(vk_1, sk_1), (vk_2, sk_2), (vk_3, sk_3)$ as in normal operation. Set $pk = (u_1, u_2, \tilde{u}_1, \tilde{u}_2, vk_1, vk_2, vk_3)$.
3. For $i = 1$ to N, choose fresh random $s, t_1, t_2 \in \mathbb{Z}_p$ and set $c_1 = g^{as} g^{st_1}, c_2 = f^{bs} f^{st_2}$. Set C_i:
$$C_i = (c_1, c_2, c_1^{y_1}, c_2^{y_2}, z_d^s h^{s(t_1+t_2)} m_j, \mathsf{sig}_1, \mathsf{sig}_2, \mathsf{sig}_3)$$
where $\mathsf{sig}_1, \mathsf{sig}_2, \mathsf{sig}_3$ are generated normally using the proper secret keys.
4. Set $T \leftarrow (pk, C_1, \ldots, C_N)$.
5. The simulation answers requests from the malicious Receiver by extracting from its proof and simulating correct responses (as described above.)

Note that in the above, if $z_d = h^{a+b}$, then the above simulation perfectly encrypts $(m_1, \ldots, m_N)$. However, when z_d is a random element of $\mathbb{G}_1$, then the ciphertexts correspond to encryptions of random elements in $\mathbb{G}_1$. Now, suppose for the sake of contradiction, that there exists an environment $\mathcal{Z}$ who can distinguish case one from case two with non-negligible probability ϵ. Then, it is easy to see that we can use $\mathcal{Z}$ to decide Decision Linear.

5 On Multiple Receivers

OT is traditionally described as a two-party protocol between a Sender and Receiver. We presented our main construction in this setting. However, since we are motivated by the application of OT to database systems, we would also like to support applications where multiple users share a single database. Naively this can be accomplished by requiring the database to run separate OT protocol instances with each user. However, this approach can be quite inefficient, and moreover does not ensure *consistency* in the database viewed by individual Receivers. Consider a strengthening of the security definition of $\mathcal{F}_{OT}^{N\times 1}$ (in Figure 3) to include the additional requirement that all Receivers "view" the same database, i.e., the database owner cannot selectively alter the messages in the database when interacting with different receivers – on query σ from *any* receiver, he must return a value in $\{m_\sigma, \perp\}$. In the full version of this work [23] we discuss extensions to our protocol designed to achieve this property.

References

1. Ateniese, G., Camenisch, J., de Medeiros, B.: Untraceable RFID tags via insubvertible encryption. In: CCS 2005, pp. 92–101. ACM Press, New York (2005)
2. Ballard, L., Green, M., de Medeiros, B., Monrose, F.: Correlation-resistant storage from keyword searchable encryption. Cryptology ePrint Archive, Report 2005/417 (2005)
3. Belenkiy, M., Chase, M., Kolweiss, M., Lysyanskaya, A.: Non-interactive anonymous credentials. In: Canetti, R. (ed.) TCC 2008. LNCS, vol. 4948, pp. 356–374. Springer, Heidelberg (2008)
4. Boneh, D., Boyen, X.: Efficient selective-ID secure Identity-Based Encryption without random oracles. In: Cachin, C., Camenisch, J.L. (eds.) EUROCRYPT 2004. LNCS, vol. 3027, pp. 223–238. Springer, Heidelberg (2004)
5. Boneh, D., Boyen, X., Shacham, H.: Short group signatures. In: Franklin, M. (ed.) CRYPTO 2004. LNCS, vol. 3152, pp. 45–55. Springer, Heidelberg (2004)
6. Boneh, D., Franklin, M.K.: Identity-based encryption from the Weil Pairing. In: Kilian, J. (ed.) CRYPTO 2001. LNCS, vol. 2139, pp. 213–229. Springer, Heidelberg (2001)
7. Boneh, D., Lynn, B., Shacham, H.: Short signatures from the Weil Pairing. In: Boyd, C. (ed.) ASIACRYPT 2001. LNCS, vol. 2248, pp. 514–532. Springer, Heidelberg (2001)
8. Brassard, G., Crépeau, C., Robert, J.-M.: All-or-nothing disclosure of secrets. In: Odlyzko, A.M. (ed.) CRYPTO 1986. LNCS, vol. 263, pp. 234–238. Springer, Heidelberg (1987)
9. Camenisch, J., Lysyanskaya, A.: Signature schemes and anonymous credentials from bilinear maps. In: Franklin, M. (ed.) CRYPTO 2004. LNCS, vol. 3152, pp. 56–72. Springer, Heidelberg (2004)
10. Camenisch, J., Neven, G., Shelat, A.: Simulatable adaptive oblivious transfer. In: Naor, M. (ed.) EUROCRYPT 2007. LNCS, vol. 4515, pp. 573–590. Springer, Heidelberg (2007)
11. Camenisch, J., Stadler, M.: Efficient group signature schemes for large groups. In: Sommer, G., Daniilidis, K., Pauli, J. (eds.) CAIP 1997. LNCS, vol. 1296, pp. 410–424. Springer, Heidelberg (1997)

12. Canetti, R., Dodis, Y., Pass, R., Walfish, S.: Universally composable security with pre-existing setup. In: Vadhan, S.P. (ed.) TCC 2007. LNCS, vol. 4392, pp. 61–85. Springer, Heidelberg (2007)
13. Canetti, R.: Universally Composable Security: A new paradigm for cryptographic protocols. In: FOCS 2001, pp. 136–145. IEEE Computer Society, Los Alamitos (2001), http://eprint.iacr.org/2000/067
14. Canetti, R.: Universally composable security: Towards the bare bones of trust. In: Kurosawa, K. (ed.) ASIACRYPT 2007. LNCS, vol. 4833, pp. 88–112. Springer, Heidelberg (2007)
15. Canetti, R., Fischlin, M.: Universally composable commitments. In: Kilian, J. (ed.) CRYPTO 2001. LNCS, vol. 2139, pp. 19–40. Springer, Heidelberg (2001)
16. Canetti, R., Lindell, Y., Ostrovsky, R., Sahai, A.: Universally composable two-party and multi-party secure computation. In: STOC 2002, pp. 494–503. ACM Press, New York (2002)
17. Canetti, R., Rabin, T.: Universal composition with joint state. In: Boneh, D. (ed.) CRYPTO 2003. LNCS, vol. 2729, pp. 265–281. Springer, Heidelberg (2003)
18. Chu, C.-K., Tzeng, W.-G.: Efficient k-out-of-n oblivious transfer schemes with adaptive and non-adaptive queries. In: Vaudenay, S. (ed.) PKC 2005. LNCS, vol. 3386, pp. 172–183. Springer, Heidelberg (2005)
19. Even, S., Goldreich, O., Lempel, A.: A randomized protocol for signing contracts. In: CRYPTO 1982, pp. 205–210 (1982)
20. Freedman, M.J., Ishai, Y., Pinkas, B., Reingold, O.: Keyword search and oblivious pseudorandom functions. In: Kilian, J. (ed.) TCC 2005. LNCS, vol. 3378, pp. 205–210. Springer, Heidelberg (2005)
21. Goldreich, O., Micali, S., Wigderson, A.: How to play any mental game or a completeness theorem for protocols with honest majority. In: STOC 1987, pp. 218–229 (1987)
22. Green, M., Hohenberger, S.: Blind identity-based encryption and simulatable oblivious transfer. In: Kurosawa, K. (ed.) ASIACRYPT 2007. LNCS, vol. 4833, pp. 265–282. Springer, Heidelberg (2007)
23. Green, M., Hohenberger, S.: Universally composable adaptive oblivious transfer. Cryptology ePrint Archive, Report 2008/163 (2008), http://eprint.iacr.org/2008/163
24. Groth, J., Sahai, A.: Efficient non-interactive proof systems for bilinear groups. In: Smart, N.P. (ed.) EUROCRYPT 2008. LNCS, vol. 4965, pp. 415–432. Springer, Heidelberg (2008)
25. Kilian, J.: Founding cryptography on oblivious transfer. In: STOC 1988, pp. 20–31 (1988)
26. Lindell, Y.: Efficient fully-simulatable oblivious transfer. In: Malkin, T.G. (ed.) CT-RSA 2008. LNCS, vol. 4964, pp. 52–70. Springer, Heidelberg (2008)
27. Lysyanskaya, A., Rivest, R.L., Sahai, A., Wolf, S.: Pseudonym systems. In: Heys, H.M., Adams, C.M. (eds.) SAC 1999. LNCS, vol. 1758, pp. 184–199. Springer, Heidelberg (2000)
28. Naor, M., Pinkas, B.: Oblivious transfer with adaptive queries. In: Wiener, M. (ed.) CRYPTO 1999. LNCS, vol. 1666, pp. 573–590. Springer, Heidelberg (1999)
29. Ogata, W., Kurosawa, K.: Oblivious keyword search. Special issue on coding and cryptography Special issue on coding and cryptography Journal of Complexity 20(2-3), 356–371 (2004)
30. Peikert, C., Vaikuntanathan, V., Waters, B.: A framework for efficient and composable oblivious transfer. In: Wagner, D. (ed.) CRYPTO 2008. LNCS, vol. 5157, pp. 554–571. Springer, Heidelberg (2008)

31. Rabin, M.: How to exchange secrets by oblivious transfer. Technical Report TR-81, Aiken Computation Laboratory, Harvard University (1981)
32. Scott, M.: Authenticated id-based key exchange and remote log-in with simple token and pin number (2002), http://eprint.iacr.org/2002/164
33. Shoup, V.: Lower bounds for discrete logarithms and related problems. In: Fumy, W. (ed.) EUROCRYPT 1997. LNCS, vol. 1233, pp. 256–266. Springer, Heidelberg (1997)
34. Yao, A.: How to generate and exchange secrets. In: FOCS, pp. 162–167 (1986)

A Linked-List Approach to Cryptographically Secure Elections Using Instant Runoff Voting*

Jason Keller[1] and Joe Kilian[2]

[1] Department of Computer Science, Rutgers University, Piscataway, NJ 08854 USA
jakeller@eden.rutgers.edu
[2] PNYLAB, LLC
joe@pnylab.com

Abstract. Numerous methods have been proposed to conduct cryptographically secure elections. Most of these protocols focus on 1-out-of-n voting schemes. Few protocols have been devised for preferential voting systems, in which voters provide a list of rankings of the candidates, and many of those treat ballots as if they were ballots in a 1-out-of-n voting scheme. We propose a linked-list-based scheme that provides improved privacy over current schemes, hiding voter preferences that should not be revealed. For large lists of candidates we achieve improved asymptotic performance.

Keywords: Electronic Voting, Secure Computation.

1 Introduction

Electronic voting is by far the most mature area of secure computation, with a vast literature (c.f. [17]). Most electronic voting protocols may be viewed as attempts to emulate the following physical metaphor: Voters cast ballots into a large box, at the conclusion of which the box is shaken and opened.

Much work has gone into efficiently and securely approximating this physical paradigm. However, this type of balloting represents merely one way of specifying and aggregating preferences. Numerous ways of aggregating preferences have been proposed, and indeed, are used in major political elections. We consider one such system, known as *instant runoff voting.*

1.1 Instant Runoff Voting

Ballots in a single transferable vote (STV) system are submitted as a list of ordinal preferences. The voters' first choices are counted, and any candidate receiving a certain quota of votes is declared a winner. One such example is the Hare-Clark quota, used in Australian elections:

$$\frac{\text{number of eligible votes}}{\text{number of open seats} + 1} + 1.$$

* Supported in part by NSF grant CCF-0728937. Work done in part while at Rutgers University.

J. Pieprzyk (Ed.): ASIACRYPT 2008, LNCS 5350, pp. 198–215, 2008.
© International Association for Cryptologic Research 2008

Votes in excess of the quota are proportionally "returned" to the voters, and applied to the next viable choice on their list. If not enough candidates reach their quota in this fashion, the candidate with the fewest number of votes is eliminated, and the process continues until all of the open seats are filled.

Although Arrow's theorem guarantees that there will be some cases for which Hare-Clark voting induces some pathology, it is attractive in practice for its ability to avoid "wasted" votes. One has comparatively less incentive (though some still exists) for strategically not supporting ones favorite candidate because the candidate is either assured to win or very likely to lose. Beyond its aesthetic appeal, the fact that it is in actual use for an important election motivates our attention.

We focus on the special case of Hare-Clark in which there is one open seat, and thus a candidate needs to win a majority of the votes in order to win the election. This is a special case known as Instant Runoff Voting (IRV), which is used in certain local jurisdictions in the United States, including elections in San Francisco [18] and Cambridge, Massachusetts [16]. In this scheme, if a candidate has a majority of votes, then he is elected. Otherwise, the candidate with the fewest votes is eliminated; counters look at the next choices of each ballot that had a vote for the recent loser. We note that for this special case, there is no need to redistribute excess winning votes; however, it remains necessary to eliminate candidates and redistribute these votes.

1.2 Difficulties with the Physical Paradigm

In simple voting an ideal physical ballot box with paper ballots is the "gold standard" against which electronic protocols are judged; indeed, there have been perhaps over-nostalgic calls for its use in practice. However, with instant runoff voting, merely severing the identification between voters and their preference list gives insufficient privacy. Particularly in the case where there is a large number of candidates, a full preference order may conceivably be used to identify a voter and thus leak information far beyond that revealed by the final vote counts, with obvious implications for privacy and coercibility. We note that this problem is not specific to a protocol implementation, but to the nature of what is to be revealed.

As a result, in actual physical elections, one has the choice of either revealing extra information or placing a great deal of trust in the discretion and trustworthiness of the election officials.

The secure multi-party computation paradigm [5, 14] is arguably a superior gold standard than any physical ballot box. One endeavors to simulate trusted election officials, who compute the correct results, but then only reveal that which is supposed to be revealed.

Thus, an intriguing aspect of this type of voting is that a cryptographic protocol may potentially offer a solution that is qualitatively superior to current best practices.

1.3 Related Work

Electronic voting has been a model problem of secure multi-party computation since it was proposed by Chaum [7]. Many protocols have been proposed for single-vote, first-past-the-post-style elections, leveraging homomorphic encryption or mix-network technologies; see, for example, [2, 4, 8, 9, 12, 22, 23, 24, 26]).

Without leaving the realm of simple elections, variations are possible in the security and privacy guarantees of the voting protocol. For example, receipt-free and incoercible voting schemes aim to prevent voter intimidation and vote selling by preventing the voter from being able to prove how they voted; see, for example, [3, 20, 24]. One may view this property as a closer approximation to the physical paradigm, in which the voter cannot prove which ballot is theirs. It should be noted that incoercibility does not follow from the generic multi-party solutions (though incoercibility can be generalized to this setting [6]).

Hevia and Kiwi [15] consider the problem of revealing the winner of the election, but keeping secret the vote tally. As with the problem we consider, the "ideal" physical implementation of voting does not guarantee as strong privacy conditions.

The techniques of "standard" electronic voting also yield solutions to simple preference voting, in which a voter may cast either zero or one votes for each candidate. For example, one can implement a k-candidate preference voting election by k simple 2-candidate elections in which the ith election is used to count votes for the ith candidate.

Protocols for preferential voting schemes, such as IRV, adopt a similar approach. Aditya et al. consider elections for the Australian Senate and House of Representatives [1]. They examine the efficiency of balloting using a naive balloting representation and straight mix-network and homomorphic encryption schemes. For an election with k candidates, their scheme using homomorphic encryptions requires posting a ballot of size $O(k!)$ bits. Their basic mix-network based scheme requires a voter to post a number between 1 and $k!$, corresponding to each set of preferences. In their most efficient scheme, they leverage Australia's voting machine structure, and adapt it to the vector-ballot approach introduced by Kiayias and Yung [21] to handle elections with write-in ballots. Each vote is a 3-vector. The first position contains a homomorphically-encrypted vote, corresponding to one of twenty preset choices. The other two positions are used to represent "write-in" votes (in which voters list their preferences rather than choosing from a preset list). The write-in votes are submitted in blocks with some preset preferential votes to a shrink-and-mix network, while blocks with no write-in votes are tabulated.

1.4 Our Contribution

We contribute a new protocol for instant runoff voting that has superior asymptotic performance when there are a large number of candidates and superior privacy guarantees.

The protocols of Aditya et al. may be applied to the case we consider, as it is a special case of their own. We thus compare our protocol to this solution, noting that the comparison is somewhat unfair due to their greater generality.

Although the work required of the voter in the protocol of [1] was small in other respects, the message length scales super-exponentially in the number of candidates. In our solution, the work per ballot construction is roughly quadratic in the number of candidates.

An arguably more important improvement is in our privacy guarantees. The protocol of [1] essentially attempts to mirror the privacy properties of existing systems. Thus, it is acceptable in their framework to reveal individual preference lists once the direct linkage with voters has been eliminated. Hence, this protocol necessarily suffers from the weaknesses of the physical solution with respect to privacy and coercion.

In our protocol, we first reveal the counts of the first-choice preferences each candidate obtained. Whenever a candidate is eliminated and their votes recast (using the next viable preference on the preference list), the new counts are also revealed. However, only these intermediate results are revealed.

One could, of course, strive for even stronger privacy guarantees, such as revealing only the winner(s), or only revealing the order of elimination. One might argue that our protocol necessarily reveals statistics, such as the second-choice preference statistics of those voters whose first choice candidate is the first to be eliminated.

However, revealing such intermediate counts seems to be reasonable and indeed often necessary from a procedural point of view. For most elections, the electorate wishes to know the final counts, not merely the winner. It would likely be considered unreasonable to declare that a candidate is eliminated without giving the actual vote count that was the basis of their elimination.

Furthermore, one can imagine using our protocols on a precinct by precinct basis, with intermediate counts reported to a conventional voting authority that decides who next to eliminate. Such regional counts can be useful in detecting vote fraud. Thus, it may be essential that the tallies from each round be revealed, and that elimination decisions can be made externally and in principle independently of a the results within an individual precinct.

1.5 Techniques Used

We make original use of standard electronic voting techniques, particularly the use of re-encryption mix networks (c.f. [7]) and group cryptography (c.f. [10]) and efficient proofs on committed values (c.f. [8]). On a very high level, voters generate linked lists of encrypted votes that specify their preferences. The encryptions are done with respect to a key that is held in aggregate by the election committee, who can decrypt elements using group cryptography. The head of the list corresponds to the highest ranked viable candidate. By using group decryption to decrypt these heads, the first round vote counts may be computed.

When a candidate is eliminated, we must efficiently search out the next element in the list. However, we must be very careful about leaking extraneous information. For example, it cannot be revealed what was the original ranking of the current head of a list. Nor can we reveal for any list the history of which elements are moved to the head (or we will reveal the list). For this reason, we keep all but the (current) head elements in a separate table of elements that is constantly remixed. This separation complicates the problem of finding the next element of a list. We use a system of random ID tags to allow us to use group decryption to find the next elements in the set.

An important technical problem we must deal with is that it would reveal too much to follow a link from an eliminated top-choice vote only to find another eliminated candidate. We must therefore perform surgery on our linked lists, deleting eliminated candidates from interiors of lists so we will never arrive at them.

To perform all of these list manipulations, we use three mix networks in different ways. Pieces of the ballots are proved consistent before being distributed among the mix networks. The consistency proofs are done using standard proofs of equality on committed values. We use standard witness-hiding techniques and heuristically replace the honest-verifiers with hash function using Gennaro's variant [13] of the Fiat-Shamir heuristic [11] (designed to avoid vote duplication attacks).

Summarizing, we present a scheme that uses a linked-list structure to represent a ballot, treats all ballots equally using three mix-networks, and also improves privacy by hiding preferences.

ROAD MAP: In Section 2, we present the basic cryptographic elements of the protocol: mix-networks, group decryption, and plaintext equality proofs. We discuss the ballot design and voting procedure in Section 3. We briefly discuss efficiency and security in Section 4. We discuss other possible research directions in Section 5.

2 Preliminaries

We use a number of basic cryptographic primitives, which we review for self-containment of the exposition.

RE-ENCRYPTION MIX-NETWORKS: Mix-networks (or mixnets), which are used to create communication channels that are difficult to trace, consist of a series of servers that take a series of texts $M_1, \ldots, M_n$ and output a permutation $\pi(M_1), \ldots, \pi(M_n)$ of these texts. In re-encryption mixnets, each mix server takes in a series of encrypted messages and applies a re-randomization to each cipher text. In the case of an El Gamal cipher text this re-encryption corresponds to a selecting a random group element and applying a small number of group operations. Neff [22] describes a protocol for the shuffling of sequences of El Gamal pairs. We use a variant of Neff's protocol in which blocks of encryptions are mixed - the block are re-encrypted in random order, but the (plaintext) values within each block are preserved in their original order.

SECRET SHARING AND GROUP DECRYPTION: We proceed with secret sharing as in [9]. To generate a private El Gamal key to distribute to counters, we use the (t, n) threshold protocol of Shamir [25]. Namely, for the secret exponent s, we announce shares $s_1, \ldots s_n$ for the counters, such that for any set Γ of t shares, we can recover the secret.

Using group cryptography, the authorities can simulate a single entity that alone has access to the decryption key. Decryptions of encrypted values by the group is comparatively straightforward and efficient. In our analysis, we will treat such decryptions as basic operations.

PLAINTEXT EQUALITY PROOFS AND PROOFS OF KNOWLEDGE: Given El Gamal encryptions of M_1 and M_2, $(\alpha_1, \beta_1) = (g^r, M_1 h^r)$ and $(\alpha_2, \beta_2) = (g^s, M_2 h^s)$, we can execute an efficient plaintext equality proof protocol, that proves that M_1 and M_2 are the same. Also, given an encryption of M and a known value of r, we must be able to produce (with proof) an encryption of $M' = M + r$. For most homomorphic encryption systems, one can compute the encryption of $M + r$ from an encryption of M.

It is also crucial that we can perform σ proofs of knowledge of encrypted values (i.e., proofs in which the prover sends an honest verifier a message, the honest verifier sends a random challenge to the prover, and the prover sends a reply). In practice, we "compress" such proofs using Gennaro's variant of the Fiat-Shamir heuristic in which the verifier's challenge is computed as a hash of the first message and the prover's identity (so as to avoid replaying other player's proofs). This heuristic results in a single message "certificate" that the player knows the values being committed to. We heuristically analyze our protocol as if the actual proofs were invoked.

The use of proofs of knowledge is crucial to both the correctness and privacy of our protocol. Intuitively, proving knowledge of a committed value prevents malleability attacks in which one commits to values that one doesn't know, but which are somehow related to other committed values.

3 Voting Scheme

3.1 Preliminary Setup

The protocol uses three mix networks. The pool of first place votes is sent to mix network 1, subsequent choices of each voter are sent to mix network 2, and elimination links are sent to mix network 3. At the start of each election, the authorities announce the public key used for all encryptions. Shares of the corresponding private key are distributed to the counters using the secret-sharing scheme described in the previous section.

We also assume the existence of a public "bulletin board" that is used as a staging area for the mix networks. As we describe below, the encrypted values sent through the mix networks are subject to various constraints that must be verified. The encrypted values and their consistency proofs are posted to the bulletin board and checked before being routed through the mix networks.

3.2 Counter Initialization

The voting authorities collectively set up an El Gamal based public-key group encryption scheme. The public key is made public and is used for the re-encryption mixer. The private key is held in a distributed fashion by the group.

3.3 Ballot Design: Constructing the Linked List

On a high level, a ballot is composed of a set of preference elements, each of which consists of preference data and additional keys used to link the preference element. In the following discussion, i will denote the preference in the list. We will have multiple elimination rounds, index by j, each requiring separate links.

To establish a link, each preference element has a set of incoming keys (thought of as a large random number) $\text{in}_{i,j}$, used to establish a connection with the preceding element in the list, and a set of outgoing keys, $\text{out}_{i,j}$, used to establish links with following elements. To establish that $x_{i'}$ follows x_i in the linked list we set $\text{out}_{i,j} = \text{in}_{i',j}$. We similarly set up random tags $\text{lose}_{i,j}$ that will aid in the removal of x_i if it corresponds to a candidate being eliminated.

For an election with k candidates, a (proper) voter does the following to construct a ballot (see Figure 1 in the appendix):

1. Determine the order of preferences, $x_1, \ldots, x_k$, where each x_i is a name (or number) representing each candidate.
2. For $i = 1, \ldots, k+1$ and $j = 1, \ldots, k$
   - Select the keys $\text{in}_{i,j}$ for $i = 1, \ldots, k+1$ and $j = 1, \ldots, k$ independently at random (in fact, we require a further step, to ensure that keys are distinct; see Section 3.6). .
   - If $i \neq k+1$, let $\text{out}_{i,j} = \text{in}_{i+1,j}$. This operation creates the links between choices.
   - Otherwise, select $\text{out}_{k+1,j}$ independently at random. This operation ends the list at the terminal choice.
   - Select the keys $\text{lose}_{i,j}$ independently at random.
3. Post $(\widehat{x}_1, \widehat{\text{in}}_{1,j}, \widehat{\text{out}}_{1,j}, \widehat{\text{lose}}_{1,j})$, encryptions of $(x_1, \text{in}_{1,j}, \text{out}_{1,j})$, for $j=1, \ldots, k$ to mix network 1.
4. For $i = 2, \ldots, k+1$ and $j = 1, \ldots, k$, post the tuple $(\widehat{x}_i, \widehat{\text{in}}_{i,j}, \widehat{\text{out}}_{i,j}, \widehat{\text{lose}}_{i,j})$ to mix network 2.
5. For $i = 1, \ldots, k+1$ and $j = 1, \ldots, k$, post the tuple $(\widehat{x}_i, \widehat{\text{lose}}_{i,j})$ to mix network 3.

To complete the ballot, the voter posts plaintext equality proofs [19] made non-interactive by Gennaro's modification to the Fiat-Shamir heuristic [13] to verify that the linked list is composed properly, namely that $\text{in}_{i+1,j} = \text{out}_{i,j}$. To verify that the removal links point to the proper candidate to be removed, the voter must also prove that x_i and $\text{lose}_{i,j}$ are equal across mix networks. Similarly, the voter posts proofs of knowledge of the encrypted values. All such proofs are posted to the public bulletin board, and may be verified by all interested parties.

Remark. For our analysis, it is useful to enforce other constraints on the ballot. For example, there is no real point in having a duplicated a name on ones list, and we may optionally wish to restrict the names to a specific list of candidates. The former may be accomplished using proofs of inequality. The latter may be accomplished used standard mix-net proofs - one writes down a list of encrypted names and proves that it is a permutation of the allowed list.

Figure 1 shows an example of each component: a portion of a vote and a removal tag, for an election with 3 candidates. A concrete example and diagram showing a full voter's posting are included in the next subsection.

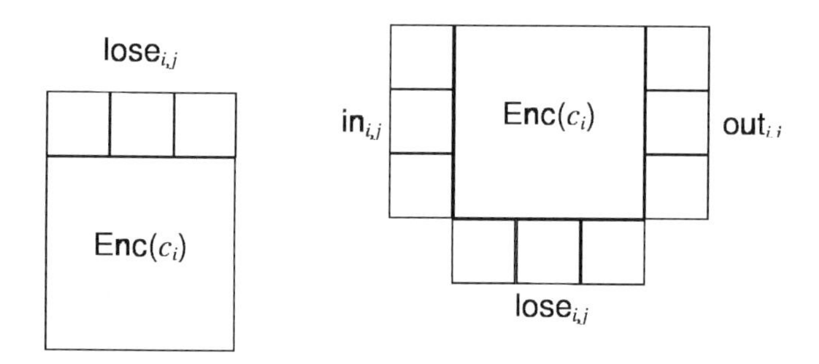

Fig. 1. A visualization of the components of a voter's ballot. A choice posted to mix networks 1 or 2 is on the left. A removal tag posted to mix network 3 is on the right. See figure 2, in the appendix, for an example of a complete ballot posted by a voter.

3.4 An Example

Consider an election with three candidates: A. Smith, B. Jones, and C. Johnson, in which a voter wants to post a vote of (Johnson, Smith, Jones) in that order. His ballot will be constructed as follows (we give a graphical example of a three candidate ballot in Figure 2):

x_1, C. Johnson
- Encrypt x_1.
- For $j = 1, 2, 3$
  - Select $\text{in}_{1,j}$ independently (indeed, select all keys $\text{in}_{\cdot,j}$ at random).
  - Set $\text{out}_{1,j} = \text{in}_{2,j}$ after $\text{in}_{2,j}$ has been selected.
  - Select $\text{lose}_{1,j}$ independently.
- Encrypt $\text{in}_{1,j}$, $\text{out}_{1,j}$, and $\text{lose}_{1,j}$.
- Create copies of $\widehat{x}_1$ and $\widehat{\text{lose}_{1,j}}$ by re-randomizing the encryption. As a tuple, these copies are the removal tag that gets posted to mix network 3.
- Post the tuple $(\widehat{x}_1, \widehat{\text{in}_{1,j}}, \widehat{\text{out}_{1,j}}, \widehat{\text{lose}_{1,j}})$ to mix network 1.

x_2, A. Smith and x_3, B. Jones
- Proceed as with x_1. Compute the tuples $(\widehat{x}_2, \widehat{\text{in}_{2,j}}, \widehat{\text{out}_{2,j}}, \widehat{\text{lose}_{2,j}})$ and $(\widehat{x}_3, \widehat{\text{in}_{3,j}}, \widehat{\text{out}_{3,j}}, \widehat{\text{lose}_{3,j}})$ as above.
- Post those tuples to mix network 2.
- Post the (re-encrypted) removal tags $(\widehat{x}_2, \widehat{\text{lose}_{2,j}})$ and $(\widehat{x}_3, \widehat{\text{lose}_{3,j}})$ to mix network 3.

x_4, the terminal choice
- Encrypt x_4.
- For $j = 1, 2, 3$

  - Select $\text{in}_{4,j}$ randomly and encrypt.
  - Select $\text{out}_{4,j}$ randomly and encrypt.
  - Select $\text{lose}_{4,j}$ randomly and encrypt.
- Post $(\widehat{x}_4, \widehat{\text{in}}_{4,j}, \widehat{\text{out}}_{4,j}, \widehat{\text{lose}}_{4,j})$ to mix network 2.

In order to prove that a vote is valid, the voter must prove the following using plaintext equality proofs:

- Given $\widehat{\text{in}}_{2,j}$ and $\widehat{\text{out}}_{1,j}$, show that $\text{in}_{2,j} = \text{out}_{1,j}$ (i.e., that $\widehat{\text{in}}_{2,j}$ and $\widehat{\text{out}}_{1,j}$ encrypt the same value)
- Given $\widehat{\text{in}}_{3,j}$ and $\widehat{\text{out}}_{2,j}$, show that $\text{in}_{3,j} = \text{out}_{2,j}$.
- Given $\widehat{\text{in}}_{4,j}$ and $\widehat{\text{out}}_{3,j}$, show that $\text{in}_{4,j} = \text{out}_{3,j}$.

Similarly, show that

- x_1 in network 1 $= x_1$ in network 3.
- x_i in network 2 $= x_i$ in network 3 (for $i > 1$).
- $\text{lose}_{1,j}$ in network 1 $= \text{lose}_{1,j}$ in network 3.
- $\text{lose}_{i,j}$ in network 2 $= \text{lose}_{i,j}$ in network 3 (for $i > 1$).

3.5 Counting and Elimination

COUNTING: After polls close, counters begin tallying votes:

1. The counters verify the posted proofs of plaintext equality, and accept those votes whose proofs pass.
2. The mix networks shuffle the pools of votes. The removal tags are mixed in round 1 only.
3. The counters leave the output of mix network 2, the voters' subsequent choices, encrypted.
4. The counters decrypt the first slots, representing the choice of candidate, of the first-place votes (from mix network 1) and of the removal tags.
5. Counters discard terminal choices or votes for eliminated candidates that show up in the primary vote pool.
6. Actual counting is trivial. The counters read the decrypted names of the first-place votes. A candidate is declared the winner if he has enough votes. Otherwise, a candidate is eliminated.

ELIMINATION: When a candidate L is eliminated, the counters act accordingly:

1. They announce the candidate L to be eliminated in round r, and locate the removal tags corresponding to L in mix network 3. Recall that this network contains pairs consisting of encrypted names and encrypted lose values. The counters can collectively decrypt all of the names, and then for all entries corresponding to L, decrypt the corresponding lose values. These values may then be efficiently matched to their corresponding entries in mix net 2, as discussed below.
2. For each choice c in the pools of votes, the counters decrypt $\widehat{\text{lose}}_{c,r}$ and $\widehat{\text{in}}_{c,r}$.

3. For each removal tag, the counters decrypt $\widehat{\text{lose}}_{L,r}$, and search for $\text{lose}_{L,r}$ in the pools of votes.
4. When a matching lose key is found, the counters check that the choice slot encrypts L, to ensure that they are eliminating the proper vote.
5. Link forwarding is now performed; see Figure 3. The counters decrypt $\widehat{\text{out}}_{L,r}$ and search for an incoming key $\text{in}_{c,r}$. The counters use a plaintext equality test to ensure that the correct link is being followed.
6. The counters set $\widehat{\text{in}}_{c,j} = \widehat{\text{in}}_{L,j}$, for $j = r, \ldots, k$. This redirects the links from the eliminated choice to a choice that is still competing in the election.
7. If a vote for L was in the primary choice pool, the counters promote the choice found by following the link.
8. At the end of round r, the counters discard $\text{in}_{c,r}$, $\text{out}_{c,r}$, and $\text{lose}_{c,r}$ are discarded for each candidate c. All keys corresponding to round r are now discarded, and counters will use keys corresponding to round $r+1$ for the next elimination.
9. Counters remix the votes using mix networks 1 and 2.

Remark. Eliminating a candidate and forwarding links illustrates the need for a terminal choice. If a voter's last choice is eliminated, the previous choice will now link to the terminal choice, instead of having hanging links. The terminal choice serves as an "anchor" that will always be among the pool of candidates.

3.6 Ensuring Distinctness and Unrelatedness of Keys

Recall that a link is created by generating a random tag that appears in multiple places in the mix net. The correctness of the protocol requires that the tags be distinct and the privacy of the protocol depends on the the inability of an

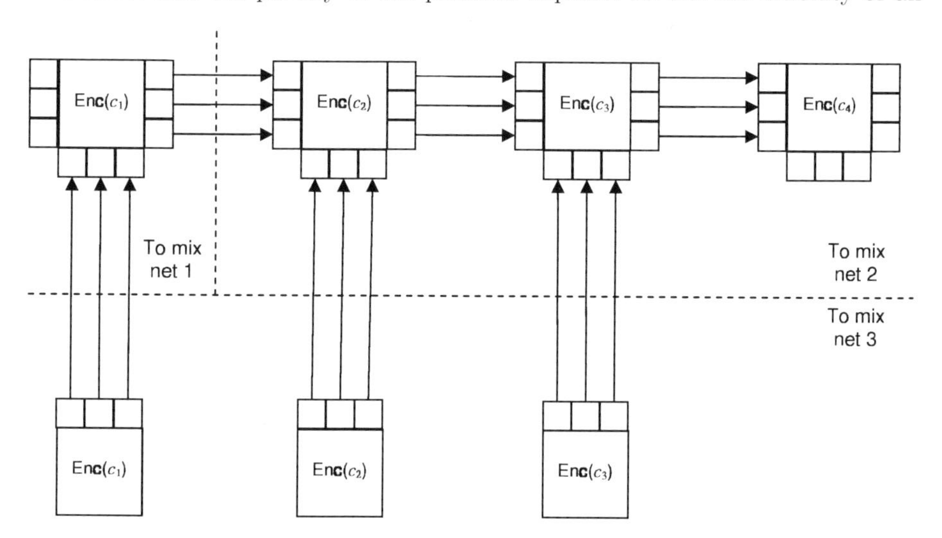

Fig. 2. A sample ballot for an election with three candidates

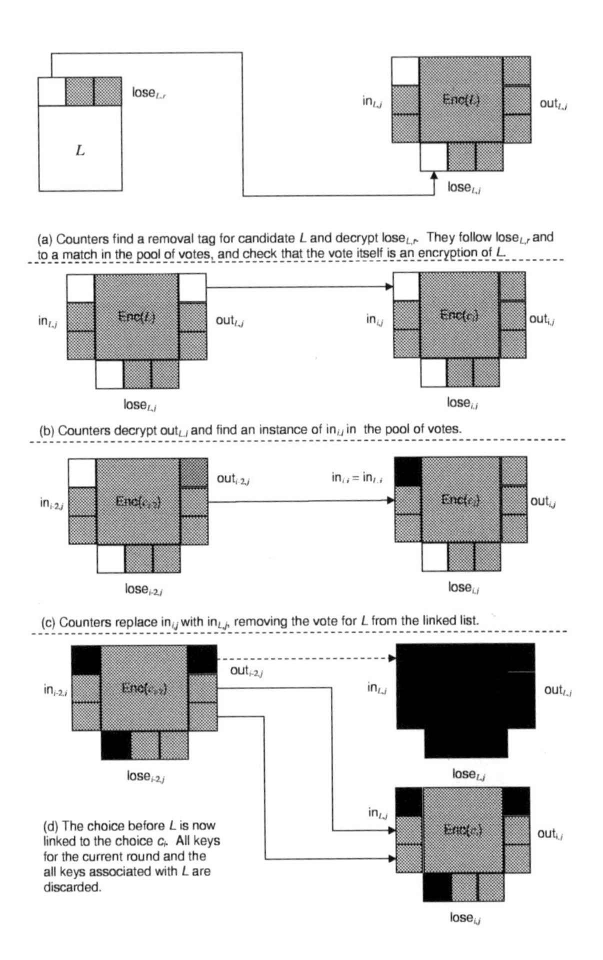

Fig. 3. An example of link forwarding. Encrypted items are in gray, decrypted items are in white, and discarded items are in black.

adversarial coalition to create nontrivial relations between their tags and those of good voters.

The latter problem is implicitly dealt with in the full privacy analysis, and follows from the fact that all of the tags come with proofs of knowledge (here we assume the idealized version of the protocol, where the proofs of knowledge are carried out). The values of the tags chosen by the adversarial players must be decided upon, and known to the adversarial players (via the extractor for the proof), given only the encryptions of these tags and zero-knowledge proofs based on these encryptions. If any nontrivial polynomial-time relation R held (with probability greater than chance) between the values chosen by the good voters and the values known to the adversaries, this could be used to obtain a distinguisher that breaks the underlying probabilistic encryption scheme.

However, nothing stops colluding voters (or even a single voter) from making two tags equal when they should not be. We solve this problem by using a standard coin-flipping in the well protocol. The interactive form of this protocol is as follows:

1. The tag creator generates a random tag T, and encrypts it, generating C.
2. A randomizer generates a random r.
3. The tag creator generates an encryption C' of $T' = T + r$. Note that for most homomorphic encryption systems, C' can be generated from C and r.

In this ideal interactive scenario, the value of T' is random. Following Gennaro, we heuristically choose r as a hash of C, the identity of the tag creator, and a representation of the "place" of this tag in the protocol as a whole (we simply ask that this representation never appear twice in the same election).

Of course, if a tag is prescribed to be equal to an earlier generated value, we simply create the commitment with this earlier value (and prove equality).

It can be shown that if T and C are chosen correctly (a random value and a random encryption), then the distribution of T' is indistinguishable from random. This is not true if T is chosen adversarially. However, by a standard argument, T' cannot be chosen to collide with any other tag value, except with negligible probability, if one replaces the hash function with a random oracle. We heuristically assume the same holds true for a suitable cryptographic hash function.

We note that the tags are homomorphically encrypted for use in the mix-net; one can achieve greater efficiency (at some loss of clarity) by putting a randomization step in at this point. Even further efficiency can be obtained by limiting the range of r, say to 192 bits even if the range of the tags is much larger.

4 Analysis

4.1 The Framework and Limits of Our Analysis

Aside from the analysis of efficiency, we cannot formally analyze our protocol in its recommended usage, which makes use of variants of the Fiat-Shamir heuristic. We instead, following a long tradition, analyze the "idealized" protocol, in which the parties engage in true proofs of knowledge and coin-flipping protocols with a trusted external party.

We also assume that while some of the counters may be corrupt, sufficiently many are honest so that the mix-net and group decryption protocols are secure and serially composable.

We also assume that the (essentially external) decisions as to which candidate is eliminated in any phase are independent of the "internals" of the protocol (i.e., based on the encrypted , though they may of course depend on the tallies of who has how many votes. We note that any sensible decision procedure will not look any deeper than the precincts vote sub-totals. This limitation may be relaxed, particularly if k is small - essentially giving the adversary full choice over the elimination sequence requires a $k!$ increase in the computational hardness of breaking the probabilistic encryptions and subverting the mix-net, coin-flipping and group decryption protocols.[1]

[1] We suspect that with some care, the $k!$ factor may be reduced to $k^{O(1)}$. However, a slightly more intricate analysis is required.

Thus, we view and analyze our protocol, and the attacks on it, as follows.

1. The voters, both good and malicious, prepare their encrypted lists, and perform the requisite proofs and coin-flipping protocols with an honest party. The malicious voters may see the encryptions generated by the good voters, and the transcripts of these protocols, but must engage in the proofs and coin flipping protocols anew (this is why we use Gennaro's trick to prevent the reuse of the Fiat-Shamir proofs). It is in the creation of these encrypted ballots that we allow the adversary the most freedom of operation.
2. For each phase of the counting process, the counters engage in various secure computations (mix net operations and group decryptions) on the encrypted values. As we assume that the adversary is unable to corrupt these protocols (sufficiently), we assume that
   - The operations proceed correctly.
   - The adversary is able to see the inputs and output of these operations, but not the actual operation of the protocol.

   These two assumptions are justified based on the correctness and simulatability of the underlying sub-protocols. Given the inputs and outputs, anyone can simulate the set of messages comprising the execution of the secure computation.

After some of these secure computations, tallies of votes for each surviving candidate are generated. We call these tallies *ideal snapshots.* We call the output of the secure computations *protocol snapshots.*

Thus, we can view the attack on the protocol as comprising the (mis)generation of ballots followed by the observation of a series of protocol snapshots. We compare such an attack with an ideal attack, which works as follows:

1. The voters, adversarial or not, create ordered lists of candidates.
2. Initially, or after a candidate has been eliminated, the tallies of current first choice votes for candidate are revealed, corresponding to the ideal snapshot defined above.

To analyze correctness, we observe that our protocol (at least in its idealized form) ensures that the ballots correspond to well-defined lists of candidates, and that the resulting "ideal snapshots" are what they should be given given this list. To analyze privacy, we go on to show that given the information that may be extracted from the adversarial voters and the ideal snapshots, one may generate simulated protocol snapshots that are computationally indistinguishable from the actual protocol snapshots.

4.2 Efficiency

In a correct vote, each choice consists of a name slot and $O(k)$ keys. The complete construction of the linked list requires $O(k^2)$ key values. Because El Gamal encryption and the plaintext equality proof take a constant number of exponentiations, a quadratic number of exponentiations is needed to cast a vote. Each ballot will also require $O(k^2)$ encryptions. The centers must perform shuffles on $O(nk^2)$ encrypted values per elimination round. Group decryptions must be performed on $O(nk)$ encrypted values per elimination round.

4.3 Correctness

To show that this protocol is correct, we show that accepted ballots correspond to independent, well-formed lists of names, and that the protocol performs the correct operations on these lists.

Lemma 1 summarizes the result of the zero-knowledge proofs of knowledge and coin-flipping protocols.

Lemma 1. *Suppose we have a collection of submitted ballots that have passed the zero-knowledge proofs of knowledge given in the Section 3.3. Then, assuming that all the ballot creators run in probabilistic polynomial time and that the probabilistic encryptions are secure, the following will hold almost always:*

1. *All accepted ballots can be mapped to a well-formed list of names and well formed tag values; all such values may be extracted from the entity submitting the ballot (and hence performing the proofs of knowledge).*
2. *All tag values that are specified by the protocol to be equal will be equal; any two tag values that are not specified to be equal will not be equal.*

One important consequence of the proofs of knowledge is that vote duplication or other forms of mauling are impossible. Suppose that the good voters have vote lists $\{L\}$ and generate the (essentially) random tags $\{t\}$ used for the linked lists. We consider two types of adversary. The *ideal model* adversary, A', chooses vote lists $\{L\}$ and tags $\{t'\}$, without seeing $\{L\}$ and $\{t\}$. The *real model* adversary, A^* sees a transcript consisting of the actual ballots generated by the good voters, and is allowed to generate ballots for itself. However, it must perform the specified proofs of equality and knowledge on these ballots; let $\{L^*\}$ and $\{t^*\}$ be the lists and tags obtained by the extractor for these proofs (by Lemma 1, these lists are well defined with all but negligible probability). Lemma 2 asserts that A^* cannot use its extra information to any better effect than A'.

Lemma 2. *For any probabilistic polynomial time adversary, A^*, there is a probabilistic polynomial time adversary A' such that $(\{L\}, \{t\}, \{L'\}, \{t'\})$ is computationally indistinguishable from $(\{L\}, \{t\}, \{L^*\}, \{t^*\})$.*

Proof. (Sketch) We use a standard hybrid argument. Given A^*, we create a hybrid adversary, A_1, that runs A^* given the encryptions, but with simulated proofs instead of actual proofs. The output of this adversary must be computationally indistinguishable from that of A^*, or we have a violation of the zero-knowledge property. We define A' as the adversary that generates random encrypted values and runs A_1. The output of A' must be computationally indistinguishable from that of A_1, or there would be a violation of the semantic security of the encryption.

We pause to reflect on the meaning of Lemma 1 and Lemma 2 for the types of attacks that can be staged during the ballot reconstruction phase. The adversary must create ballots that correspond to well formed lists and tags, such that the set of tags have no spurious duplications. The lists and tag values had might as well be chosen independently of the honest voters. In short, the adversary acts no differently than an adversary that chooses its lists and tags and engages in the protocol.

It remains to consider the remainder of the protocol. Recall, we assume that the adversary is assumed not to be able to corrupt enough counters to interfere with the mix-net and group decryption operations.

We observe that the details of the ballots (other than the fact that they are valid) are essentially irrelevant to the rest of the protocol. The proofs are essentially dropped once they are verified, leaving only the choice of encryptions. Recall that a re-encrypting mix-net replaces the encryption of some value x with a random encryption of x. Thus, the precise encryptions chosen by the adversary almost immediately become irrelevant, as summarized in Lemma 3.

Lemma 3. *The result of the first re-encrypting mix-net operation depends only on the values of the lists and tags encrypted in the ballots, not on the ballots themselves.*

Thus, the only effective difference between a general adversary that chooses its ballots and a comparatively ideal adversary that chooses its list of candidates and then participates in the protocol is that the general adversary can specify its tags arbitrarily (but not to collide spuriously). By a straightforward but tedious argument, one can show the following:

Lemma 4. *Given a set of well-formed ballots, corresponding to a set of lists of candidates, with no spurious tag collisions, and sequence of candidate eliminations, the vote counts produced at each round will be the same as that produced by the ideal vote-counting algorithm on these lists of candidates.*

Hence, the (partial) freedom to choose the tag values is irrelevant to the intermediate counts of the protocol.

The above Lemmas imply the correctness of our (idealized) protocol.

Privacy. The methodology of the previous section can be extended to simultaneously establish privacy as well. Consider the view of the adversary attempting to corrupt the election. At the time it selects its ballots, it has only seen probabilistic encryptions of the good voters' lists and tags, and zero-knowledge proofs on these values. As with the proof of Lemma 2, we can simulate this view with simulated proofs on random committed values. It remains to simulate the views of the later parts of the protocols. As before, we use the extraction property of the proofs to extract the lists $\{L'\}$ and tags $\{t'\}$ specified by the adversary. By the previous section (particularly Lemma 3), once the ballots have been constructed and tested, these values are the only aspects that are relevant to future steps of the protocol.

We consider the view of the adversary in the ideal and actual settings. In the ideal setting, the adversary sees $\{L'\}$ and $\{t'\}$ and then sees the sequence of intermediate vote counts (one initial, and one for each elimination phase). In reality, the adversary sees a sequence of "snapshots" consisting of encrypted values output by the mix net, of which some subset are revealed at each stage, as specified by the protocol and which candidates are eliminated. Additionally, there is the adversaries view of the actual secure computations we are invoking, but these are assumed to be simulatable. Lemma 5 states that one can simulate the snapshots given the information available in the ideal model.

Lemma 5. *Given the vote lists $\{L'\}$ and tags $\{t'\}$ given by the adversary, and the sequence of vote totals generated in each elimination phase, and the identities of each eliminated candidate, one can in probabilistic polynomial time generate simulations of the output of each secure computation operation that are computationally indistinguishable from the outputs of the protocol.*

The proof is a tedious but straightforward hybrid argument.

5 Discussion

RECEIPT FREENESS: One of the more obvious deficiencies of this protocol is its lack of receipt-freeness. It seems likely that, at the cost of modestly greater complexity, one can make a receipt-free version of this protocol using standard techniques (though we do not claim such a result). The natural approach would be for voters to interact with a voting entity to securely compute a ballot; the voter inputs its preferences, but has no more knowledge of the proofs and encryptions than if another voter had cast a ballot with the same preference list. While general secure computation is impractical, the operations required for constructing a ballot, namely creating randomized encryptions for the candidate names, random tags and proofs of equality of these tags, are quite amenable to this approach.

PRACTICALITIES: It should be pointed out that we have ignored an entire space of trust and security issues, assuming for example that voters have completely trustworthy implementations of their part of the protocol. We view this work as an early step towards efficient preference-based voting.

EXTENSION TO MULTIPLE WINNERS: This protocol only covers the case of an election with a single victor. If the election is for multiple seats, winners get "eliminated." They keep a quota's worth of first-choice votes, with the surplus getting redistributed with a fractional weight. From this protocol, a STV protocol, which modifies this protocol by preserving preference hiding and using the same ideas for link forwarding, but taking the fractional redistribution of votes into account, may arise.

HANDLING MULTIPLE LOSERS AND WRITE-IN VOTES: It may be foreseeable that a number of candidates with relatively small tallies of votes will not be able to garner enough votes to win the election. In this protocol, the votes have to be reshuffled after each elimination, or authorities may reveal significant link information. We would like to modify this protocol so that multiple losing candidates can be removed efficiently. This would also allow for the inclusion of write-in candidates. Write-in candidates with a significant number of votes will stay in the vote pool, while the occasional sporadic write-in vote will be eliminated promptly.

Incomplete Voting: A voter may not need to fill out a complete ballot, instead opting for ranking t-out-of-k candidates. In San Francisco elections, for example, voters select only three out of k candidates when voting. This scheme is adaptable to such an incomplete vote, so long as voters post one key per candidate. Each vote listing t candidates will take $O(tk)$ bits. Schemes that encode a full list of choices in one ballot will now require at least $\log(k!) + 1$ bits. If t is sufficiently small, then this system also improves on the space efficiency of previous schemes. On the other hand, the privacy of some ballots will be compromised, as terminal choices will appear in the primary pool of votes; counters may be able to reconstruct ballots consisting of only eliminated candidates. One potential solution to this is to have a voter post dummy choices to fill out the ballot.

Acknowledgments

We thank the anonymous reviewers for many useful comments.

References

1. Aditya, R., Boyd, C., Dawson, E., Viswanathan, K.: Secure e-voting for preferential elections. In: Traunmüller, R. (ed.) EGOV 2003. LNCS, vol. 2739, pp. 246–249. Springer, Heidelberg (2003)
2. Baudron, O., Fouque, P.-A., Pointcheval, D., Stern, J., Poupard, G.: Practical multi-candidate election system. In: PODC 2001: Proceedings of the twentieth annual ACM symposium on Principles of distributed computing, pp. 274–283. ACM Press, New York (2001)
3. Benaloh, J., Tuinstra, D.: Receipt-free secret-ballot elections (extended abstract). In: STOC 1994: Proceedings of the twenty-sixth annual ACM symposium on Theory of computing, pp. 544–553. ACM, New York (1994)
4. Benaloh, J.C., Yung, M.: Distributing the power of a government to enhance the privacy of voters. In: PODC 1986: Proceedings of the fifth annual ACM symposium on Principles of distributed computing, pp. 52–62. ACM, New York (1986)
5. Yao, A.C.: How to generate and exchange secrets. In: IEEE Symposium on Foundations of Computer Science, pp. 162–167 (1986)
6. Canetti, R., Gennaro, R.: Incoercible multiparty computation (extended abstract). In: IEEE Symposium on Foundations of Computer Science, pp. 504–513 (1996)
7. Chaum, D.: Untraceable electronic mail, return addresses, and digital pseudonyms. Communications of the ACM 24, 84–88 (1981)
8. Cramer, R., Damgård, I., Schoenmakers, B.: Proofs of partial knowledge and simplified design of witness hiding protocols. In: Desmedt, Y.G. (ed.) CRYPTO 1994. LNCS, vol. 839, pp. 174–187. Springer, Heidelberg (1994)
9. Cramer, R., Gennaro, R., Schoenmakers, B.: A secure and optimally efficient multi-authority election scheme. In: Fumy, W. (ed.) EUROCRYPT 1997. LNCS, vol. 1233, pp. 103–118. Springer, Heidelberg (1997)
10. Desmedt, Y.G., Frankel, Y.: Threshold cryptosystems. In: Brassard, G. (ed.) CRYPTO 1989. LNCS, vol. 435, pp. 307–315. Springer, Heidelberg (1990)
11. Fiat, A., Shamir, A.: How to prove yourself: Practical solutions to identification and signature problems. In: Odlyzko, A.M. (ed.) CRYPTO 1986. LNCS, vol. 263, pp. 186–194. Springer, Heidelberg (1987)

12. Furukawa, J., Sako, K.: An efficient scheme for proving a shuffle. In: Kilian, J. (ed.) CRYPTO 2001. LNCS, vol. 2139, pp. 368–387. Springer, Heidelberg (2001)
13. Gennaro, R.: Achieving independence efficiently and securely. In: PODC 1995: Proceedings of the fourteenth annual ACM symposium on Principles of distributed computing, pp. 130–136. ACM Press, New York (1995)
14. Goldreich, O., Micali, S., Wigderson, A.: How to play any mental game. In: STOC 1987: Proceedings of the nineteenth annual ACM conference on Theory of computing, pp. 218–229. ACM, New York (1987)
15. Hevia, A., Kiwi, M.: Electronic jury voting protocols. Theor. Comput. Sci. 321(1), 73–94 (2004)
16. http://www.cambridgema.gov/Election/proprep.html
17. http://www.cs.st-andrews.ac.uk/tws/research/bibliography.pdf
18. http://www.sfgov.org/site/election_index.asp
19. Jakobsson, M., Juels, A.: Addition of elgamal plaintexts. In: Okamoto, T. (ed.) ASIACRYPT 2000. LNCS, vol. 1976, pp. 346–358. Springer, Heidelberg (2000)
20. Juels, A., Catalano, D., Jakobsson, M.: Coercion-resistant electronic elections. In: WPES 2005: Proceedings of the 2005 ACM workshop on Privacy in the electronic society, pp. 61–70. ACM, New York (2005)
21. Kiayias, A., Yung, M.: The vector-ballot e-voting approach. In: Juels, A. (ed.) FC 2004. LNCS, vol. 3110, pp. 72–89. Springer, Heidelberg (2004)
22. Neff, C.A.: A verifiable secret shuffle and its application to e-voting. In: CCS 2001: Proceedings of the 8th ACM conference on Computer and Communications Security, pp. 116–125. ACM Press, New York (2001)
23. Peng, K., Boyd, C., Dawson, E.: Simple and efficient shuffling with provable correctness and zk privacy. In: Shoup, V. (ed.) CRYPTO 2005. LNCS, vol. 3621, pp. 188–204. Springer, Heidelberg (2005)
24. Sako, K., Kilian, J.: Receipt-free mix-typc voting scheme-a practical solution to the implementation of a voting booth. In: Guillou, L.C., Quisquater, J.-J. (eds.) EUROCRYPT 1995. LNCS, vol. 921, pp. 393–403. Springer, Heidelberg (1995)
25. Shamir, A.: How to share a secret. Commun. ACM 22(11), 612–613 (1979)
26. Wikstrom, D.: A universally composable mix-net (2004)

Towards Robust Computation on Encrypted Data*

Manoj Prabhakaran and Mike Rosulek

University of Illinois, Urbana-Champaign
{mmp,rosulek}@uiuc.edu

Abstract. Encryption schemes that support computation on encrypted data are useful in constructing efficient and intuitively simple cryptographic protocols. However, the approach was previously limited to stand-alone and/or honest-but-curious security. In this work, we apply recent results on "non-malleable homomorphic encryption" to construct new protocols with Universally Composable security against active corruption, for certain interesting tasks. Also, we use our techniques to develop non-malleable homomorphic encryption that can handle homomorphic operations involving more than one ciphertext.

1 Introduction

Computation on encrypted data is one of the most intriguing problems in cryptography today. There is a long history of works investigating this problem in various general settings [1, 2, 3, 5, 11, 12, 13, 17, 22, 23], as well as in relation to specific computational tasks (e.g., searching on encrypted inputs [4, 8, 10, 13, 14, 15, 18, 19, 24]). As demonstrated by these works, being able to compute on encrypted inputs leads to simple intuitive protocols for many cryptographic tasks.

However, compared to some of the core areas in cryptography like encryption, authentication and secure multi-party computation, the state of the art for computation on encrypted inputs remains quite limited. The majority of encryption schemes that allow computations on encrypted data are only known to achieve security against chosen-plaintext attacks. As such, protocols that manipulate encrypted data often have to employ complicated machinery of zero-knowledge proofs and/or distributed key management to provide protection against malicious participants. Similarly, issues like composability of protocols have hardly been explored for this problem.

In this work we take a closer look at the composability and non-malleability aspects of computation on encrypted data. Our goal is to construct protocols that are secure in the demanding setting of Universally Composable (UC) security [7]. The main challenge is in forbidding a malicious party from manipulating encrypted data in unwanted ways. The traditional solution to this problem is to use zero-knowledge proofs to enforce honest behavior. However, general zero-knowledge proofs are not possible in the UC framework.

* Partially supported by NSF grant CNS 07-47027.

J. Pieprzyk (Ed.): ASIACRYPT 2008, LNCS 5350, pp. 216–233, 2008.
© International Association for Cryptologic Research 2008

Instead, our approach is to restrict malicious parties' capabilities via strong non-malleable guarantees on the encryption scheme itself. This approach has the additional benefit that shifting some of the security burden to the encryption scheme allows us to construct conceptually simple protocols that still achieve strong security against malicious parties.

Requiring "non-malleability" for an encryption scheme may seem counter-productive to the goal of computing on its encrypted data. Indeed, a scheme must necessarily be malleable in some way for its encrypted data to be manipulated. However, a security notion called *Homomorphic-CCA (HCCA)* security has recently been defined in [20], meaningfully combining homomorphic computational features and non-malleability. Briefly, a scheme that achieves HCCA security is homomorphic with respect to certain operations, but explicitly forbids all other manipulations to the underlying plaintext.

The HCCA security requirement is strong enough to be meaningful in the UC framework, but unlike general-purpose UC zero-knowledge proofs, can be achieved in the plain model. Indeed, such a scheme has been constructed in [20], under a standard assumption. However, that construction only supports a very limited class of homomorphic operations. In particular, it does not support operations which combine multiple encrypted inputs, which are relevant in the context of computation on encrypted data. Our contribution in this work is to show that when used with appropriately encoded data, the relatively unexpressive scheme from [20] can be used to robustly implement more sophisticated computations on data encrypted in multiple ciphertexts.

1.1 Overview of Our Results

Background: Non-Malleable Homomorphic Encryptions. Computation on encrypted data necessitates having an encryption scheme that supports some homomorphic operations. However, when considering security against malicious parties, a non-malleability requirement is also generally needed.

A key component in our constructions is a public-key encryption scheme that meaningfully combines both non-malleability and homomorphic operations. Such schemes were introduced in [20]. We review the relevant security definitions for these schemes in Section 2. For the purposes of this overview, the reader may consider a "non-malleable (unary) homomorphic encryption scheme" to be one in which the *only* ways to construct a valid ciphertext are: (1) encrypting a known message, or (2) applying a homomorphic operation to some $\mathsf{Enc}(m)$ to obtain $\mathsf{Enc}(T(m))$, for any function T in a set of *allowed transformations*. The set of allowed transformations is a fixed parameter of the encryption scheme, and it is infeasible for an adversary to generate a ciphertext whose value depends on other ciphertexts in any other way. Furthermore, ciphertexts derived via the homomorphic operation are completely indistinguishable (even to the recipient) from ciphertexts generated by the standard encryption operation. In [20], a construction was given for a family of encryption schemes that support these requirements for a range of allowed transformation operations related to cyclic group operations. Our results do not rely on any additional properties of that construction, but uses the primitive in a black-box manner, and as such, can be instantiated with the construction in [20] or any future construction satisfying the appropriate security requirements.

The common technique in our constructions is to exploit the power of this encryption scheme as follows: We encode the input data with some special randomized "integrity" information into a vector of several ciphertexts. The integrity information is intended to correlate the vector of ciphertexts together into one "bundle." The homomorphic property of the scheme ensures that the integrity information and data can be manipulated in certain ways. For instance, in both of our main results, the integrity information can be "re-randomized" using the scheme's homomorphic operations.

When using a homomorphic non-malleable encryption scheme in a protocol, already by the non-malleability property of the encryption scheme, ciphertexts can only be derived from others using a certain limited class of operations. By employing an appropriate integrity encoding, we further enforce that among the small set of allowed operations, the only ones which preserve/maintain the integrity information are the legitimate operations prescribed by the protocol. In other words, the integrity encoding provides a means to give and verify an implicit zero-knowledge proof that the protocol is being honestly implemented.

Opinion Polling. Our first result is an "opinion poll" protocol that elegantly illustrates the power of the combination of non-malleability, unlinkability and homomorphism in a single encryption scheme. The protocol is motivated by the following scenario: A pollster wishes to collect information from many respondents. However, the respondents are concerned about the anonymity of their responses. Indeed, it is in the interest of the pollster to set things up so that the respondents are guaranteed anonymity, especially if the subject of the poll is sensitive personal information.

To help collect responses anonymously, the pollster can enlist the help of an external tabulator. The respondents require that the external tabulator too does not see their responses, and that if the tabulator is honest, then responses are anonymized for the pollster (i.e., so that he cannot link responses to respondents). The pollster, on the other hand, does not want to trust the tabulator at all: if the tabulator tries to modify any responses, the pollster should be able to detect this so that the poll can be invalidated.

A relevant view of this problem is as an instance of a model that we call *crypto-computing on third-party inputs* — a model that extends the "crypto-computing" model from [23]. In this new model, the inputs to the computation are owned by a set of parties other than the client (who receives the output — the pollster in our case) and the server (who does the actual computation on encrypted data — the tabulator in our case). This separation of roles introduces new security requirements: (1) Privacy for the input parties: the client should not learn anything other than the intended output value. The server should not learn anything either. (The input providers are not necessarily interested in the correctness of the computation.) (2) Robustness: a malicious server cannot make the client accept an output that is inconsistent with the parties' inputs.

The opinion poll scenario is similar to the classic setting for mix-nets [9], where a group of servers accepts a list of ciphertexts and outputs a random permutation of their decrypted values. However, in many mix-net protocols it can be quite complicated to enforce the correctness of outputs against a malicious (i.e., actively corrupt) server (in our case, the tabulator in particular). Often zero-knowledge proofs [16], or distributed decryption via verifiable secret sharing are used to enforce the integrity of operations performed on the ciphertexts. In contrast, our use of non-malleable homomorphic encryption leads to a simple and elegant UC-secure protocol.

The main idea in our protocol is to use an encryption scheme whose *only* homomorphic operation is $\mathsf{Enc}(\alpha, \beta) \mapsto \mathsf{Enc}(\alpha, t\beta)$, where t, α, β are elements of some cyclic group. In other words, plaintexts consist of a pair of group elements. Anyone can multiply (apply the group operation to) the second plaintext component with a known value t, but the first component is completely non-malleable, and the two components remain "tied together." Now, to implement the opinion poll protocol, the pollster generates a (multiplicative) secret sharing $r_1, \dots, r_n$ of a random secret group element R, then sends to the ith respondent a share r_i. Each respondent sends $\mathsf{Enc}(m_i, r_i)$ to the tabulator, where m_i is his response to the poll. Now the tabulator can blindly re-randomize the shares (multiply the ith share by a random s_i, such that $\prod_i s_i = 1$), shuffle the resulting ciphertexts, and send them to the pollster. The pollster will ensure that the shares encode the secret R and accept the results.

Informally, security is argued as follows. The pollster only sees a random permutation of the responses, and since the multiplicative sharing of R is re-randomized, there is no way to link any responses to the r_i shares he originally dealt to the respondents. The tabulator sees only encrypted data, and in particular has no information about the secret R or any individual shares r_i. The only way the tabulator could successfully (with non-negligible probability) generate ciphertexts whose second components are a multiplicative share of R is by making exactly one of his ciphertexts be derived from each respondent's ciphertext. By the non-malleability of the encryption scheme, each response m_i is inextricably "tied to" the corresponding share r_i and cannot be modified, so each respondent's response should be represented exactly once in the tabulator's output. Finally, observe that the responses of malicious respondents must be independent of honest parties' responses – by "copying" an honest respondent's ciphertext to the tabulator, a malicious respondent also "copies" the corresponding r_i. The resulting shares would be inconsistent with overwhelming probability.

We also show a similar protocol where the computation performed is a boolean-OR of the respondents' boolean inputs (where the tabulator also provides an input). Again, the non-triviality in these constructions is not in the complexity of the computation performed, but in ensuring (using only the properties of the encryption scheme, and in particular no zero-knowledge proofs) that a malicious server cannot do anything unwanted without detection.

Binary Homomorphic Encryption. Our second contribution is an extension of the non-malleable homomorphic encryption scheme of [20]. The scheme of [20] is homomorphic in an inherently *unary* way; it prohibits operations that combine multiple ciphertexts together in a homomorphic way. However, many existing applications of (plain) homomorphic encryption schemes rely on combining multiple ciphertexts together. Unfortunately, in [20], it was shown that it is impossible to achieve the natural extension of the security definitions to the setting where the homomorphic operations act on multiple ciphertexts. The complication arose from the tension between the non-malleability requirement and the *unlinkability* requirement (namely, that a ciphertext not leak whether it was derived as a normal encryption or via one of the homomorphic operations).

In this work, we show that a meaningful relaxation of these definitions can be achieved. Instead of settling for absolute unlinkability, we consider a relaxation similar to that used in [23], in which ciphertexts grow in size after applying the operations. Thus, a ciphertext will reveal no more than (an upper bound on) the number of homomorphic operations that have been applied to derive it. However, unlike in [23], our goal is to achieve non-malleability and robustness against malicious adversaries.

We construct an encryption scheme that supports the binary group operation in a cyclic group; i.e., anyone can transform $\mathsf{Enc}^*(\alpha)$ and $\mathsf{Enc}^*(\beta)$ into $\mathsf{Enc}^*(\alpha\beta)$, but the scheme is otherwise non-malleable. Lacking a "standard" security definition for such an encryption scheme, we prove that our construction is a UC-secure realization of a natural ideal functionality, whose details are motivated by extending the UC functionality considered in [20].

The main idea in our construction is to encode a message m as a vector $\mathsf{Enc}(m_1), \ldots, \mathsf{Enc}(m_k)$, where the m_i's are a random multiplicative sharing of m in the group. and Enc is a non-malleable homomorphic encryption scheme that supports (unary) group operations (from [20]). To "multiply" two such encrypted encodings, we can simply concatenate the two vectors of ciphertexts together, and rerandomize the new set of shares (multiply each component by s_i, where $\prod_i s_i = 1$, as in the opinion poll protocol) to bind the sets together.

The above approach captures the main intuition, but our actual construction uses a slightly different approach to ensure UC security. In the scheme described above, anyone can split the vector $\mathsf{Enc}(m_1), \ldots, \mathsf{Enc}(m_k)$ into two smaller vectors that encode two (random) elements whose product is m. We interpret this as a violation of our desired properties, since it is a way to make two encodings whose values are related to a *longer* encoding. To get around this problem of "breaking apart" these ciphertexts, we encode m as $\mathsf{Enc}(\alpha_1, \beta_1), \ldots, \mathsf{Enc}(\alpha_k, \beta_k)$, where the α_i's and β_i's form two *independently random* secret sharings of m. Rerandomizing these encodings is possible when we use a scheme that is homomorphic with respect to the operations $(\alpha, \beta) \mapsto (t\alpha, s\beta)$. Now these encodings cannot be split up in such a way that the first components and second components are shares of the same value. Note that it is crucial here that because of the non-malleability properties of the scheme, the (α_i, β_i) pairs cannot themselves be "broken apart."

2 Preliminaries

Homomorphic Encryption Syntax and Security. Our constructions use homomorphic encryption schemes that have *unary* homomorphic operations on the plaintext messages. That is, we suppose there is a procedure CTrans, which takes a ciphertext and a (description) of a function T on plaintexts, such that $\mathsf{Dec}_{SK}(\mathsf{CTrans}(\zeta, T)) = T(\mathsf{Dec}_{SK}(\zeta))$ is satisfied.

Prabhakaran and Rosulek [20] introduced security definitions for homomorphic encryptions that combine non-malleability as well as robust homomorphic features. Schemes satisfying these definitions are vital for achieving UC security in our constructions. We present a high-level overview of their security definitions below; we refer the reader to Appendix A for the complete formal definitions.

Informally, a homomorphic encryption scheme achieves *Homomorphic-CCA (HCCA) security* with respect to a set of functions $\mathcal{T}$ if the scheme is non-malleable except for the possibility of changing an encryption of m into an encryption of $T(m)$, for $T \in \mathcal{T}$ (i.e., no other operations are possible in the scheme). We also consider the complementary requirement: Informally, a homomorphic scheme is *unlinkable* with respect to $\mathcal{T}$ if it is indeed possible to change encryptions of m into encryptions of $T(m)$ for $T \in \mathcal{T}$ as a feature (using the CTrans operation), in such a way that ciphertexts do not reveal whether they were generated via Enc or via CTrans.

Formalizing the intuitive HCCA requirement in a general way is non-trivial. It is achieved in [20] by requiring that there be an additional procedure RigEnc_{PK} (used only in the analysis) which outputs a special "rigged" ciphertext ζ and some auxiliary information S, such that ζ is indistinguishable from a normal ciphertext. The rigged ciphertext does not necessarily encode a message; however, there is a corresponding procedure $\mathsf{RigExtract}_{SK}$ which, when given another ciphertext ζ' and the auxiliary information S, determines whether ζ' was obtained by applying a transformation to ζ, and if so, outputs that transformation. The formal HCCA security experiment enforces the indistinguishability of rigged and normal ciphertexts, as well as the correctness of $\mathsf{RigExtract}$'s output. Intuitively, if $\mathsf{RigExtract}$ only outputs transformations in $\mathcal{T}$, then ciphertexts can only depend on the values of other ciphertexts according to transformations in $\mathcal{T}$.

The unlinkability requirement is formalized via a more straight-forward security experiment. At a high level, the experiment enforces that for all adversarially generated ciphertexts ζ such that $\mathsf{Dec}_{SK}(\zeta) \neq \bot$, the two distributions $\mathsf{Enc}_{PK}(T(\mathsf{Dec}_{SK}(\zeta)))$ and $\mathsf{CTrans}(\zeta, T)$ are indistinguishable, even in the presence of a decryption oracle.

Concrete constructions. Prabhakaran and Rosulek [20] give a construction achieving the desired properties for various kinds of homomorphic operations, under the Decisional Diffie-Hellman assumption.

Let $\mathbb{G}$ be a cyclic group, and let $\mathbb{G}^n$ denote the product group, where we extend the group operation in $\mathbb{G}$ component-wise. For $\sigma \in \mathbb{G}^n$, define the function $T_\sigma : \mathbb{G}^n \to \mathbb{G}^n$ as the "multiplication by σ" operation: $T_\sigma(\alpha) = \sigma\alpha$. Finally, for any $\mathbb{H} \subseteq \mathbb{G}^n$, define $\mathcal{T}_{\mathbb{H}} = \{T_\sigma \mid \sigma \in \mathbb{H}\}$.

Theorem 1 ([20]). *For any $n \geq 1$ and any subgroup $\mathbb{H}$ of $\mathbb{G}^n$, there is an encryption scheme with message space $\mathbb{G}^n$ that is simultaneously HCCA-secure and unlinkable, with $\mathcal{T}_{\mathbb{H}}$ as the set of allowed operations, provided that the Decisional Diffie-Hellman (DDH) assumption holds in $\mathbb{G}$ and any subgroup of $\mathbb{Z}^*_{|\mathbb{G}|}$.*

Our two main results use instantiations of the above construction with $n = 2$, and $\mathbb{H} = \{1\} \times \mathbb{G}$ and $\mathbb{H} = \mathbb{G}^2$, respectively.

3 Opinion Polling

We describe an intuitively simple yet robust protocol for the opinion polling application described in Section 1.1, using HCCA encryption as a component.

Formally, we give a secure protocol for the UC ideal functionality $\mathcal{F}_{\mathsf{poll}}$, described in Figure 1. For the opinion polling application, we associate the pollster with party P_{client}, the tabulator with P_{server}, and the respondents with the input parties $P_1, \ldots, P_n$. Note that in $\mathcal{F}_{\mathsf{poll}}$, P_{client} learns only a random permutation of the parties' inputs, while P_{server} learns nothing about their inputs (except the knowledge of who has submitted inputs). Also, P_{server} and each input party can cause the process to abort without P_{client} accepting any output.

On input $[\textsc{setup}, P_{\mathsf{client}}, P_{\mathsf{server}}, P_1, \ldots, P_n]$ from party P_{client}:

- Send $[\textsc{setup}, P_{\mathsf{client}}, P_{\mathsf{server}}]$ to each party P_i.
- Send $[\textsc{setup}, P_{\mathsf{client}}, P_1, \ldots, P_n]$ to P_{server}.

On input $[\textsc{input}, x_i]$ from input party P_i:

- Send $[\textsc{inputfrom}, P_i]$ to P_{server}, and remember x_i.

On input "OK" from P_{server}:

- If P_{server} is corrupt, expect to receive from P_{server} a permutation σ on $\{1, \ldots, n\}$. If P_{server} is honest, choose σ at random.
- If not all $P_1, \ldots, P_n$ parties have supplied an input, or if some $x_i = \perp$, then send $\perp$ to P_{client}.
- Otherwise, give $(x_{\sigma(1)}, \ldots, x_{\sigma(n)})$ to P_{client}.

On input "CANCEL" from a corrupt P_{server}, send $\perp$ to P_{client}.

Fig. 1. UC ideal functionality $\mathcal{F}_{\mathsf{poll}}$

The Protocol. We present our protocol for $\mathcal{F}_{\mathsf{poll}}$ following the high-level overview given in Section 1.1. We then prove that the protocol is a UC-secure realization of $\mathcal{F}_{\mathsf{poll}}$, provided that at least one of $\{P_{\mathsf{client}}, P_{\mathsf{server}}\}$ are honest.

Let $\mathcal{E}$ = (KeyGen, Enc, Dec, CTrans) be an unlinkable HCCA-secure scheme, whose message space is $\mathbb{G}^2$ for a cyclic group $\mathbb{G}$, and whose allowed (unary) transformations are $(\alpha, \beta) \mapsto (\alpha, t\beta)$ for all $t \in \mathbb{G}$. We suppose the CTrans operation accepts arguments as $\mathsf{CTrans}(C,t)$, where $t \in \mathbb{G}$ specifies the transformation $(\alpha,\beta) \mapsto (\alpha,t\beta)$. We abbreviate the $\mathsf{CTrans}(C,t)$ operation as "$t * C$". Thus $t * \mathsf{Enc}_{PK}(\alpha,\beta)$ is indistinguishable from $\mathsf{Enc}_{PK}(\alpha,t\beta)$, in the sense of the unlinkability definition.

The protocol proceeds as follows:

1. P_{client} generates a key pair $(SK, PK) \leftarrow \mathsf{KeyGen}$ and chooses random elements $r_1,\ldots,r_n \leftarrow \mathbb{G}$, remembering $R = \prod_i r_i$. She then sends $(PK, r_i, P_{\mathsf{server}})$ to each party P_i, and sends $(P_{\mathsf{client}}, P_1,\ldots,P_n)$ to P_{server}.
2. Input party P_i holds input x_i. He receives $(PK, r_i, P_{\mathsf{server}})$ from P_{client}, then sends $\mathsf{Enc}_{PK}(x_i, r_i)$ to P_{server} through a secure channel.
3. P_{server} collects ciphertext C_i from each input party P_i, then chooses a random permutation σ on $[n]$ and random $s_1,\ldots,s_n \leftarrow \mathbb{G}$ subject to $\prod_i s_i = 1$. He computes $C'_i = s_{\sigma(i)} * C_{\sigma(i)}$ and sends $(C'_1,\ldots,C'_n)$ to P_{client}.
4. P_{client} decrypts each C'_i as $(x'_i, r'_i) \leftarrow \mathsf{Dec}_{SK}(C'_i)$. If any decryptions fail, or if $\prod_i r'_i \neq R$, she aborts. Otherwise, she outputs $(x'_1,\ldots,x'_n)=(x_{\sigma(1)},\ldots,x_{\sigma(n)})$.

Theorem 2. *If $\mathcal{E}$ is unlinkable and HCCA-secure with message space $\mathbb{G}^2$, and allowed transformations as described above, where $|\mathbb{G}|$ is superpolynomial in the security parameter, then our protocol is a secure realization (with respect to static corruptions) of $\mathcal{F}_{\mathsf{poll}}$, against adversaries who corrupt at most one of $\{P_{\mathsf{server}}, P_{\mathsf{client}}\}$.*

Proof. Given a real-world adversary $\mathcal{A}$, we construct a simulator $\mathcal{S}$. We break the proof down into 3 cases according to which parties $\mathcal{A}$ corrupts:

Case 1: If $\mathcal{A}$ corrupts neither P_{server} nor P_{client}, then suppose by symmetry that $\mathcal{A}$ corrupts some input parties $P_1,\ldots,P_k$. Then the main task for $\mathcal{S}$ is to extract the inputs of each corrupt P_i and send them to $\mathcal{F}_{\mathsf{poll}}$. $\mathcal{S}$ simply does the following:

- On receiving [SETUP, $P_{\mathsf{client}}, P_{\mathsf{server}}, P_1,\ldots,P_n$] from $\mathcal{F}_{\mathsf{poll}}$, generate $(PK, SK) \leftarrow \mathsf{KeyGen}$. Choose random $r_1,\ldots,r_k \leftarrow \mathbb{G}$ and simulate that P_{client} sent $(PK, r_i, P_{\mathsf{server}})$ to each corrupt input party P_i.
- If not all corrupt parties P_i send a ciphertext C_i to P_{server}, then abort. Otherwise, set $(x_i, r'_i) \leftarrow \mathsf{Dec}_{SK}(C_i)$.
- If any of the above decryption fails, or if $\prod_i r'_i \neq \prod_i r_i$, then send [INPUT, $\perp$] to $\mathcal{F}_{\mathsf{poll}}$ on behalf of *each* corrupt input party P_i.
- Otherwise send [INPUT, x_i] to $\mathcal{F}_{\mathsf{poll}}$ on behalf of each corrupt input party P_i.

It is straight-forward to see that in the cases where $\mathcal{S}$ sends [INPUT, $\perp$], then by the honest behavior of P_{server} and P_{client}, the protocol would have mandated that P_{client} refuse the output.

Case 2: If $\mathcal{A}$ corrupts P_{client} and (without loss of generality) input parties $P_1,\ldots,P_k$, then $\mathcal{S}$ does the following:

- When corrupt P_{client} sends $(PK, r_i, P_{\mathsf{server}})$ to each honest input party P_i, send [SETUP, $P_{\mathsf{client}}, P_{\mathsf{server}}, P_1,\ldots,P_n$] to $\mathcal{F}_{\mathsf{poll}}$ on behalf of P_{client}.

- When a corrupt input party P_i sends a ciphertext C_i to honest P_{server}, send $[\text{INPUT}, 1]$ to $\mathcal{F}_{\mathsf{poll}}$ on behalf of P_i.
- When $\mathcal{F}_{\mathsf{poll}}$ gives the final output to $\mathcal{S}$, remove as many 1's from the output list as there are corrupt input parties. Call the remaining outputs x_{k+1}, $\ldots, x_n$. Honestly simulate the remainder of the protocol on behalf of the honest input parties, using x_i as the input for honest party P_i.

Since P_{client} is corrupt, $\mathcal{S}$ can legally obtain the set of honest input parties' inputs. The only difference therefore between the view of $\mathcal{A}$ in the real world and our simulation is that the honest parties are simulated with inputs that may be permuted. However, since P_{server} is honest, P_{client}'s view in the protocol is independent of any permutation on the honest parties' inputs.

Case 3: If $\mathcal{A}$ corrupts P_{server} and input parties $P_1, \ldots, P_k$, then $\mathcal{S}$ does the following:

- When $\mathcal{F}_{\mathsf{poll}}$ gives $[\text{SETUP}, P_{\mathsf{client}}, P_1, \ldots, P_n]$ to $\mathcal{S}$, generate $(PK, SK) \leftarrow$ KeyGen. Pick random $r_1, \ldots, r_n \leftarrow \mathbb{G}$ and simulate that P_{client} sent $(PK, r_i,$ $P_{\mathsf{server}})$ to each corrupt P_i.
- When $\mathcal{F}_{\mathsf{poll}}$ gives $[\text{INPUTFROM}, P_i]$ to $\mathcal{S}$ for an honest party $(i > k)$, generate $(C_i, S_i) \leftarrow \mathsf{RigEnc}_{PK}$ and simulate that P_i sent C_i to P_{server}. Remember S_i.
- When P_{server} sends P_{client} a list of ciphertexts $(C'_1, \ldots, C'_n)$, do the following for each i:
  - If $\mathsf{Dec}_{SK}(C'_i) \neq \bot$, then set $(x_i, r'_i) \leftarrow \mathsf{Dec}_{SK}(C'_i)$.
  - Else, if $\mathsf{RigExtract}_{SK}(C'_i, S_j) \neq \bot$ for some j, set $r'_i := r_i \cdot \mathsf{RigExtract}_{SK}$ (C'_i, S_j).
  - If both these operations fail, send CANCEL to $\mathcal{F}_{\mathsf{poll}}$ on behalf of P_{server}.

  If $\prod_i r'_i \neq \prod_i r_i$ or for some $j > k$, there is more than one i such that $\mathsf{RigExtract}_{SK}(C'_i, S_j) \neq \bot$, then send CANCEL to $\mathcal{F}_{\mathsf{poll}}$ on behalf of P_{server}. Otherwise, let σ be any permutation on $[n]$ that maps each $j > k$ to the unique i such that $\mathsf{RigExtract}_{SK}(C'_i, S_j) \neq \bot$. Send $[\text{INPUT}, x_\sigma(i)]$ to $\mathcal{F}_{\mathsf{poll}}$ on behalf of corrupt P_i $(i \leq k)$, and then send OK to $\mathcal{F}_{\mathsf{poll}}$ on behalf of P_{server}, with σ as the permutation that $\mathcal{F}_{\mathsf{poll}}$ expects.

In this case, the primary task of $\mathcal{S}$ is to determine whether the corrupt P_{server} gives a valid list of ciphertexts to P_{client}. Applying the HCCA definition in a sequence of hybrid interactions, we see that the behavior of the real world interaction versus this simulation interaction is preserved when appropriately replacing $\mathsf{Enc}/\mathsf{Dec}$ with $\mathsf{RigEnc}/\mathsf{RigExtract}$.

Note that the adversary's view is independent of $r_{k+1}, \ldots, r_n$. If $\mathsf{Dec}_{SK}(C'_i) \neq \bot$, then the corresponding r'_i value computed by the simulator is also independent of $r_{k+1}, \ldots, r_n$. Thus the only way $\prod_i r_i = \prod_i r'_i$ can be satisfied with non-negligible probability is if for each honest party P_j, exactly one i satisfies $\mathsf{RigExtract}_{SK}(C'_i, S_j) \neq \bot$. In this case, there will be exactly as many x_i's as corrupt players, and the simulator can legitimately send these to $\mathcal{F}_{\mathsf{poll}}$ as instructed (with the appropriate permutation).

Boolean OR on Encrypted Data. Using a similar technique, we can obtain a UC-secure protocol for a boolean-OR functionality. This functionality is identical to $\mathcal{F}_{\mathsf{poll}}$ except that P_{server} also gets to provide an input (say we identify P_{server} with P_0), and instead of giving $(x_{\sigma(0)}, \ldots, x_{\sigma(n)})$, it gives $\bigvee_i x_i$ as the output to P_{client}.

We can achieve this new functionality with a similar protocol — this time, using an encryption scheme that is unlinkable HCCA-secure with respect to all group operations in $\mathbb{G}^2$. P_{client} sends shares r_i to the input parties as before. The input parties send $\mathsf{Enc}_{PK}(x_i, r_i)$ to P_{server}, where $x_i = 1$ if P_i's input is 0, and x_i is randomly chosen in $\mathbb{G}$ otherwise. Then, P_{server} rerandomizes the r_i shares as before, and also randomizes the x_i's in the following way: P_{server} multiplies each x_i by s_i such that $\prod_i s_i = 1$ if P_{server}'s input is 0, and $\prod_i s_i$ is random otherwise (P_{server} can randomize both sets of shares simultaneously using the homomorphic operation). P_{client} receives the processed ciphertexts and ensures that $\prod_i r'_i = 1$. Then if $\prod_i x'_i = 1$, it outputs 0, else it outputs 1.

We note that this approach to evaluating a boolean OR (where the induced distribution is a fixed element if the result is 0, and is random if the result is 1) has previously appeared elsewhere, e.g., [5, 6].

Relation to Voting. Our opinion polling protocol falls short of a solution for the classic election scenario in several aspects. First, in our scheme, respondents can cause the entire protocol to abort. Second, the respondents have no stake in the correctness of the results; if the pollster publishes the entire set of responses, there is no way for respondents to verify its correctness. Respondents may submit their vote accompanied by a randomly chosen nonce — this would allow a respondent to verify that his own response was included, but not that the entire set of responses is valid. Adding a publicly published nonce also allows trivial vote-selling. We finally note that an election protocol (in which all participants receive guaranteed correct results) is not possible in the plain UC model, given the impossibility results of [21].

4 Non-malleable Homomorphic Encryption for Binary Operations

In [20], it was shown that no homomorphic encryption can be completely unlinkable and also allow a group operation over the message space as a *binary* homomorphic operation — that is, an operation that multiplies two encrypted group elements. Still, the impossibility result left open the possibility of achieving a relaxation of these requirements. We consider a relaxation similar to [23]; namely, we allow the ciphertext to leak the number of operations applied to it (i.e., the depth of the circuit applied), but ideally no additional information.

Informally, we associate a *length* parameter with each ciphertext. If a length-ℓ and a length-ℓ' ciphertext are combined, then the result is a length $\ell + \ell'$ ciphertext.

Security Definition. Our formal definition is in the form of an ideal functionality in the UC framework. It is a generalization of the "homomorphic message posting" functionality presented in [20], to the case where multiple messages can be

The functionality keeps track of a database of records of the form $(\mathsf{handle}, \ell, m)$. Let $\mathsf{GetHandle}(args)$ be a subroutine which sends $[\text{HANDLE-REQ}, args]$ to the adversary and expects in return a string handle. If handle is previously recorded in the database, abort; otherwise, return handle.

Setup: On receiving a command [SETUP] from a party P: If a previous SETUP command has been processed, abort. Else, send $[\text{ID-REQ}, P]$ to the adversary, and expect in response a string id. Broadcast $[\text{ID-ANNOUNCE}, P, \mathsf{id}]$ to all other parties.

Dummy handles: On receiving a command $[\text{DUMMY}, \ell, \mathsf{handle}]$ from a *corrupt party only*, internally record $(\mathsf{handle}, \ell, \perp)$ and broadcast $[\text{HANDLE-ANNOUNCE}, \mathsf{handle}]$ to all parties.

Posting messages: On receiving a command $[\text{POST}, \ell, m_0, \mathsf{handle}_1, \ldots, \mathsf{handle}_k]$ from a party sender: If any handle_i is not recorded internally, or $m_0 \notin \mathbb{G}$, ignore the request. Otherwise, suppose $(\mathsf{handle}_i, \ell_i, \mathsf{msg}_i)$ is recorded for each i. If $\ell < \sum_i \ell_i$, ignore the request. Let $D = \{i \mid m_i = \perp\} \subseteq [k]$, the indices of the dummy handles. Set $m^* = m_0 * \prod_{i \notin D} m_i$, the product of known plaintexts involved.

- If $D = \emptyset$ (no dummy handles involved): If P is corrupt, set $\mathsf{handle}^* \leftarrow \mathsf{GetHandle}(\mathsf{sender}, \ell, m^*)$; otherwise let $\mathsf{handle}^* \leftarrow \mathsf{GetHandle}(\mathsf{sender}, \ell)$. Internally record $(\mathsf{handle}^*, \ell, m^*)$ and broadcast $[\text{HANDLE-ANNOUNCE}, \mathsf{handle}^*]$ to all parties.
- If $\ell > \sum_{i \in D} \ell_i$ (not entirely derived from dummy handles): If P is corrupt, set $\mathsf{handle}' \leftarrow \mathsf{GetHandle}(\mathsf{sender}, \ell', m^*)$, else set $\mathsf{handle}' \leftarrow \mathsf{GetHandle}(\mathsf{sender}, \ell')$. Internally record $(\mathsf{handle}', \ell', m^*)$.
  Set $\mathsf{handle}^* \leftarrow \mathsf{GetHandle}(\mathsf{sender}, \ell, \{\mathsf{handle}'\} \cup \{\mathsf{handle}_i \mid i \in D\})$. Internally record $(\mathsf{handle}^*, \ell, \perp)$ and send $[\text{HANDLE-ANNOUNCE}, \mathsf{handle}^*]$ to all parties.
- Otherwise (dummy handles only), Set $\mathsf{handle}^* \leftarrow \mathsf{GetHandle}(\mathsf{sender}, \ell, m_0, \{\mathsf{handle}_i \mid i \in D\})$. Internally record $(\mathsf{handle}^*, \ell, \perp)$ and send $[\text{HANDLE-ANNOUNCE}, \mathsf{handle}^*]$ to all parties.

Message reading: On receiving a command $[\text{GET}, \mathsf{handle}]$ from party P (who gave the first SETUP command): If $(\mathsf{handle}, \ell, \mathsf{msg})$ is recorded internally, send msg to P; else send $\perp$.

Fig. 2. UC ideal functionality $\mathcal{F}_{\mathbb{G}}$, parametrized by a cyclic group $\mathbb{G}$

combined. The functionality, called $\mathcal{F}_{\mathbb{G}}$, is given in full detail in Figure 2. Below we explain and motivate the details of the definition.

The $\mathcal{F}_{\mathbb{G}}$ functionality allows users to post messages to each other, as on a bulletin board. The messages are stored in the functionality's memory, and are not given out except to the designated recipient. Instead, messages can be referred to using abstract *handles*, which reveal no information about the message.

Following our desired intuition, users can only generate new messages in two ways (for uniformity, all handled in the same part of the functionality's code). A user can simply post a message by supplying a group element m (this is the case where $k = 0$ in the user's POST command). Alternatively, a user can provide a list of existing handles along with a group element m. If all these handles correspond to honestly-generated posts, then this has the same effect as if the user posted the product of all the corresponding messages (though note that the user does not have to know what these messages are to do this). We model the fact that handles reveal nothing about the message by letting the adversary choose the actual handle string, without knowledge of the message. The designated recipient can obtain the message by providing a handle to the functionality. Note that there is no way (even for corrupt parties) to generate a handle derived from existing handles in a non-approved way.

However, (as in [20]) adversaries can also post *dummy handles*, which contain no message. When a user posts a derived message using such a handle, the resulting handle also contains no message. However, the adversary is also told that the handle was used in a derived POST command. The adversary also gets access to an "intermediate" handle corresponding to all the non-DUMMY handles that were combined in the POST request. Still, the adversary learns nothing about the messages corresponding to these handles. This weakness is slight and natural, since the adversary could output a ciphertext encrypted under some key unknown to the other participants. The ciphertext would be meaningless to the other parties, but the adversary could also be able to detect when someone has derived another message using it.

One may of course consider *interactive* protocols for $\mathcal{F}_{\mathbb{G}}$. However, we restrict attention to non-interactive protocols obtained via *encryption schemes* — where KeyGen implements the SETUP command, Enc and CTrans implement the POST command, and Dec implements the GET command, all in the natural ways.

The Construction. Let $\mathcal{E} = (\mathsf{KeyGen}, \mathsf{Enc}, \mathsf{Dec}, \mathsf{CTrans})$ be an unlinkable HCCA-secure scheme, whose message space is $\mathbb{G}^2$ for a cyclic group $\mathbb{G}$, and whose allowed (unary) transformations are all group operations in $\mathbb{G}^2$. We suppose the CTrans operation accepts arguments as $\mathsf{CTrans}(C, (r, s))$, where $r, s \in \mathbb{G}$ specify the transformation $(\alpha, \beta) \mapsto (r\alpha, s\beta)$. We abbreviate the $\mathsf{CTrans}(C, (r, s))$ operation as "$(r, s) * C$". Thus $(r, s) * \mathsf{Enc}_{PK}(\alpha, \beta)$ is indistinguishable from $\mathsf{Enc}_{PK}(r\alpha, s\beta)$, in the sense of the unlinkability definition.

The new scheme $\mathcal{E}^*$ is given by the following algorithms:

Key generation (KeyGen^*)**.** Same as KeyGen.

Encryption (Enc^*)**.** To encrypt an element $m \in \mathbb{G}$ in a length-ℓ ciphertext, output

$$C = \Big(\mathsf{Enc}_{PK}(\alpha_1, \beta_1), \ldots, \mathsf{Enc}_{PK}(\alpha_\ell, \beta_\ell)\Big)$$

where α_i, β_i are randomly chosen in $\mathbb{G}$ subject to the constraint $\prod_i \alpha_i = \prod_i \beta_i = m$.

Decryption (Dec^*)**.** To decrypt a ciphertext $C = (C_1, \ldots, C_\ell)$, decrypt each C_i to get (α_i, β_i). If any decryption returns $\perp$, or if $\prod_i \alpha_i \neq \prod_i \beta_i$, output $\perp$. Else output $\prod_i \alpha_i$.

Transformation operation (CTrans^*)**.** To "multiply" two given ciphertexts $C = (C_1, \ldots, C_\ell)$ and $C' = (C_1, \ldots, C_{\ell'})$, output a random permutation of:

$$\Big((r_1, s_1) * C_1, \ldots, (r_\ell, s_\ell) * C_\ell, (r_{\ell+1}, s_{\ell+1}) * C'_1, \ldots, (r_{\ell+\ell'}, s_{\ell+\ell'}) * C'_{\ell'}\Big)$$

where r_i, s_i are randomly chosen in $\mathbb{G}$ subject to $\prod_i r_i = \prod_i s_i = 1$
To "multiply" a single given ciphertext $C = (C_1, \ldots, C_\ell)$ by a given known group element $R \in \mathbb{G}$ (without increasing the ciphertext length), output:

$$\Big((r_1, s_1) * C_1, \ldots, (r_\ell, s_\ell) * C_\ell\Big)$$

where r_i, s_i are randomly chosen in $\mathbb{G}$ subject to $\prod_i r_i = \prod_i s_i = R$.

We note that the syntax of CTrans^* can be naturally extended to support multiplying several ciphertexts and/or a known group element at once, simply by composing the operations described above.

Theorem 3. *If $\mathcal{E}$ is unlinkable and HCCA-secure with respect to $\mathbb{G}^2$, where $|\mathbb{G}|$ is superpolynomial in the security parameter, then $\mathcal{E}^*$ (as described above) is a secure realization of $\mathcal{F}_{\mathbb{G}}$, with respect to static corruptions.*

Proof. Let $\mathcal{E} = (\mathsf{KeyGen}, \mathsf{Enc}, \mathsf{Dec}, \mathsf{CTrans})$ be the unlinkable HCCA-secure scheme used as the main component in our construction, and let RigEnc and $\mathsf{RigExtract}$ be the procedures guaranteed by HCCA security.

We proceed by constructing an ideal-world simulator for any arbitrary real-world adversary $\mathcal{A}$. The simulator $\mathcal{S}$ is constructed by considering a sequence of hybrid functionalities that culminate in $\mathcal{F}_{\mathbb{G}}$. These hybrids differ from $\mathcal{F}_{\mathbb{G}}$ only in how much they reveal in their HANDLE-REQ requests to the adversary.

Correctness. Note that $\mathcal{F}_{\mathbb{G}}$ only makes two kinds of HANDLE-REQ requests: those containing a lone message, and those containing a list of handles.

Let $\mathcal{F}_1$ be the functionality that behaves exactly as $\mathcal{F}_{\mathbb{G}}$, except that every time it sends a HANDLE-REQ to the simulator, it also includes the entire party's input that triggered the HANDLE-REQ. Define $\mathcal{S}_1$ to be the simulator that internally runs the adversary $\mathcal{A}$, and does the following:

- When $\mathcal{F}_1$ gives $(\text{ID-REQ}, P)$ to $\mathcal{S}_1$, it generates a key pair $(PK, SK) \leftarrow \mathsf{KeyGen}$ and responds with PK. It simulates to $\mathcal{A}$ that party P broadcast PK.
- When $\mathcal{F}_1$ gives a HANDLE-REQ to $\mathcal{S}_1$, it generates the handle appropriately — with either Enc^*_{PK} or CTrans^* on an existing handle, depending on the party's original command which is included in the HANDLE-REQ. It simulates to $\mathcal{A}$ that the appropriate party output the handle.
- When $\mathcal{A}$ broadcasts a length-ℓ ciphertext C, $\mathcal{S}_1$ tries to decrypt it with Dec^*_{SK}. If it decrypts (say, to m), then $\mathcal{S}_1$ sends a (POST, ℓ, m) command to $\mathcal{F}_1$ and later gives C as the handle; else it sends (DUMMY, ℓ, C).

$\mathcal{S}_1$ exactly simulates the honest parties' behavior in the real world interaction. By the correctness properties of $\mathcal{E}^*$, the outputs of the honest ideal-world parties match that of the real world, except with negligible probability; thus, $\text{REAL}^{\mathcal{E}^*}_{\mathcal{Z},\mathcal{A}} \approx \text{IDEAL}^{\mathcal{F}_1}_{\mathcal{Z},\mathcal{S}_1}$ for all environments $\mathcal{Z}$.

Unlinkability. Let $\mathcal{F}_2$ be exactly like $\mathcal{F}_1$, except for the following change: For requests of the form $[\text{HANDLE-REQ}, \mathsf{sender}, \ell, m]$, $\mathcal{F}_2$ does not send the handles that caused this request. That is, whereas $\mathcal{F}_1$ would tell the simulator that the handle is being requested for a POST command combining some non-dummy handles, $\mathcal{F}_2$ would instead act like sender had sent $[\text{POST}, \ell, m]$ (that this is closer to what $\mathcal{F}_{\mathbb{G}}$ does; internally behaving identically for such requests). Let $\mathcal{S}_2 = \mathcal{S}_1$, since $\mathcal{F}_1$ is only sending one fewer type of HANDLE-REQ to the simulator.

By a standard hybrid argument, we can see that $\text{IDEAL}^{\mathcal{F}_1}_{\mathcal{Z},\mathcal{S}_1} \approx \text{IDEAL}^{\mathcal{F}_2}_{\mathcal{Z},\mathcal{S}_2}$ for all environments $\mathcal{Z}$. The hybrids are over the number of POST requests affected by this change. Consecutive hybrids differ by whether a single handle was generated by Enc^* or by CTrans^*. The only handles that are affected here are non-DUMMY handles, and thus ciphertexts which decrypt successfully under SK. Thus distinguishing between consecutive hybrids can be reduced to succeeding in the unlinkability experiment (by further hybridizing over the individual Enc ciphertext components).

HCCA. If the owner P of the functionality is corrupt, then $\mathcal{S}_2$ is already a suitable simulator for $\mathcal{F}_{\mathbb{G}}$, and we can stop at this point.

Otherwise, the difference between $\mathcal{F}_{\mathbb{G}}$ and $\mathcal{F}_2$ is that $\mathcal{F}_{\mathbb{G}}$ does not reveal the message in certain HANDLE-REQ requests. Namely, those in which the simulator receives $[\text{HANDLE-REQ}, \mathsf{sender}, \ell]$.

Let $\mathcal{S}_3$ be exactly like $\mathcal{S}_2$, except for the following changes: Each time $\mathcal{S}_2$ would generate a ciphertext component via $\mathsf{Enc}_{PK}(\alpha, \beta)$, $\mathcal{S}_3$ instead generates it with RigEnc_{PK}. It keeps track of the auxiliary information S and records (S, α, β) internally. Also, whenever $\mathcal{S}_2$ would decrypt a ciphertext component using Dec_{SK}, $\mathcal{S}_3$ instead decrypts it via:

$$D(C) = \begin{cases} (r\alpha, s\beta) & \text{if any } (S, \alpha, \beta) \text{ is recorded such that } (r, s) \leftarrow \mathsf{RigExtract}_{SK}(C, S) \\ \mathsf{Dec}_{SK}(C) & \text{otherwise} \end{cases}$$

By a straight-forward hybrid argument (where distinguishing between consecutive hybrids reduces to distinguishing in one execution of the HCCA experiment), we have that $\text{IDEAL}^{\mathcal{F}_2}_{\mathcal{Z},\mathcal{S}_2} \approx \text{IDEAL}^{\mathcal{F}_2}_{\mathcal{Z},\mathcal{S}_3}$ for all environments $\mathcal{Z}$.

Suppose the internal records (S, α, β) are labeled as (S_j, α_j, β_j) for $j \geq 1$. Now for each HANDLE-REQ request q sent to $\mathcal{S}_3$, we define J_q to be the set of indices j such that (S_j, α_j, β_j) was generated as a result of servicing request q.

Each α, β is chosen randomly in $\mathbb{G}$, subject to a constraint on some of their products, as prescribed by Enc^* and CTrans^*. However, the ciphertexts given to the adversary are generated by RigEnc_{PK}, and thus independent of these random choices. In fact, the entire adversary's view is (essentially) independent of the random choices of α, β, subject to $\prod_{j \in J_q} \alpha_j/\beta_j$ being fixed (we pessimistically assume that $\mathcal{A}$ knows this fixed value for each q). Put another way, $\prod_{j \in J'}(\alpha_j/\beta_j)$ is uniformly distributed for a multiset J' if and only if for all q, all elements of J_q have the same multiplicity in J'.

We now examine when a ciphertext given by the adversary is successfully decrypted by the simulator (and thus given to the functionality as a POST instead of as a DUMMY handle).

Given a ciphertext (sequence of HCCA ciphertexts) $C = (C_1, \ldots, C_\ell)$, $\mathcal{S}_3$ first decrypts each C_i to obtain $(\alpha_i, \beta_i) = D(C_i)$. The overall decryption succeeds if $\prod_i(\alpha_i/\beta_i) = 1$. Let J' be the multiset of indices j such that $\bot \neq \mathsf{RigExtract}_{SK}(C_i, S_j)$, with multiplicity for each i where this holds. The decryption constraint above is uniformly distributed (and thus equality holds only with negligible probability) unless all elements of J_q have the same multiplicity in J'. However, when all elements of J_q have the same multiplicity in J', we may cancel all the α_j/β_j terms in the constraint. What remains are terms of the form α_i/β_i, where $(\alpha_i, \beta_i) \leftarrow \mathsf{Dec}_{SK}(C_i)$, and terms of r_i/s_i, where $(r_i, s_i) \leftarrow \mathsf{RigExtract}_{SK}(C_i, S_j)$. The ciphertext then decrypts successfully if and only if the constraint holds with respect to these remaining terms.

Thus, we can consider a simulator $\mathcal{S}_4$ which behaves just like $\mathcal{S}_3$, except that when $\mathcal{A}$ outputs a ciphertext $C = (C_1, \ldots, C_\ell)$, it processes it as follows:

- If some C_i is such that $D(C_i) = \bot$, the ciphertext is invalid; send $[\text{DUMMY}, C]$ to the functionality.
- Define J' as above. If for some q, the elements of J_q do not all have the same multiplicity in J', the ciphertext is invalid; send $[\text{DUMMY}, C]$ to the functionality.
- Let I be the set of indices such that $\bot \neq (\alpha_i, \beta_i) \leftarrow \mathsf{Dec}_{SK}(C_i)$. If $\prod_{i \in I} (\alpha_i/\beta_i) \neq 1$, then the ciphertext is invalid; send $[\text{DUMMY}, C]$ to the functionality.
- Let $(r_i, s_i) \leftarrow \mathsf{RigExtract}_{SK}(C_i, S_j)$ for each $i \notin I$, If $\prod_{i \notin I}(r_i/s_i) \neq 1$, then the ciphertext is invalid; send $[\text{DUMMY}, C]$ to the functionality.
- Otherwise, send $[\text{POST}, \ell, m_0, \{\mathsf{handle}_j \mid j \in J'\}]$ to the functionality, where $m_0 = \prod_{i \in I} \alpha_i \prod_{i \notin I} r_i$.

Except with negligible probability, $\mathcal{S}_4$ interacts identically with the functionality as $\mathcal{S}_3$. However, note that $\mathcal{S}_4$ does not actually look at the α_j, β_j values that are recorded for each call to RigEnc. Thus $\mathcal{S}_4$ can be successfully implemented even if the functionality does not reveal m in messages of the form $[\text{HANDLE-REQ}, \mathsf{sender}, \ell, m]$. Therefore $\mathcal{S}_4$ is a suitable simulator for $\mathcal{F}_{\mathbb{G}}$ itself, and $\text{IDEAL}^{\mathcal{F}_2}_{\mathcal{Z},\mathcal{S}_3} \approx \text{IDEAL}^{\mathcal{F}_{\mathbb{G}}}_{\mathcal{Z},\mathcal{S}_4}$ for all environments $\mathcal{Z}$.

Acknowledgments

We would like to thank Josh Benaloh and the anonymous referees for suggesting helpful improvements.

References

1. Abadi, M., Feigenbaum, J.: Secure circuit evaluation. J. Cryptology 2(1), 1–12 (1990)
2. Abadi, M., Feigenbaum, J., Kilian, J.: On hiding information from an oracle. J. Comput. Syst. Sci. 39(1), 21–50 (1989)
3. Aiello, W., Ishai, Y., Reingold, O.: Priced oblivious transfer: How to sell digital goods. In: Pfitzmann, B. (ed.) EUROCRYPT 2001. LNCS, vol. 2045, pp. 119–135. Springer, Heidelberg (2001)

4. Boneh, D., Crescenzo, G.D., Ostrovsky, R., Persiano, G.: Public key encryption with keyword search. In: Cachin, C., Camenisch, J.L. (eds.) EUROCRYPT 2004. LNCS, vol. 3027, pp. 506–522. Springer, Heidelberg (2004)
5. Boneh, D., Goh, E.-J., Nissim, K.: Evaluating 2-DNF formulas on ciphertexts. In: Kilian, J. (ed.) TCC 2005. LNCS, vol. 3378, pp. 325–341. Springer, Heidelberg (2005)
6. Broadbent, A., Tapp, A.: Information-theoretic security without an honest majority. In: Kurosawa, K. (ed.) ASIACRYPT 2007. LNCS, vol. 4833, pp. 410–426. Springer, Heidelberg (2007)
7. Canetti, R.: Universally composable security: A new paradigm for cryptographic protocols. Cryptology ePrint Archive, Report 2000/067 (2005)
8. Chang, Y.-C., Mitzenmacher, M.: Privacy preserving keyword searches on remote encrypted data. In: Ioannidis, J., Keromytis, A.D., Yung, M. (eds.) ACNS 2005. LNCS, vol. 3531, pp. 442–455. Springer, Heidelberg (2005)
9. Chaum, D.: Untraceable electronic mail, return addresses, and digital pseudonyms. Commun. ACM 4(2) (February 1981)
10. Chor, B., Gilboa, N., Naor, M.: Private information retrieval by keywords. TR CS0917, Department of Computer Science, Technion (1997)
11. Desmedt, Y.: Computer security by redefining what a computer is. In: NSPW 1992-1993: Proceedings on the 1992-1993 workshop on New security paradigms, pp. 160–166. ACM Press, New York (1993)
12. Feigenbaum, J.: Encrypting problem instances: Or.., can you take advantage of someone without having to trust him? In: Williams, H.C. (ed.) CRYPTO 1985. LNCS, vol. 218, pp. 477–488. Springer, Heidelberg (1986)
13. Freedman, M.J., Ishai, Y., Pinkas, B., Reingold, O.: Keyword search and oblivious pseudorandom functions. In: Kilian, J. (ed.) TCC 2005. LNCS, vol. 3378, pp. 303–324. Springer, Heidelberg (2005)
14. Goh, E.-J.: Secure indexes. Cryptology ePrint Archive, Report 2003/216 (2003), http://eprint.iacr.org/2003/216/
15. Golle, P., Staddon, J., Waters, B.R.: Secure conjunctive keyword search over encrypted data. In: Jakobsson, M., Yung, M., Zhou, J. (eds.) ACNS 2004. LNCS, vol. 3089, pp. 31–45. Springer, Heidelberg (2004)
16. Groth, J.: A verifiable secret shuffle of homomorphic encryptions. In: Desmedt, Y.G. (ed.) PKC 2003. LNCS, vol. 2567, pp. 145–160. Springer, Heidelberg (2002)
17. Ishai, Y., Paskin, A.: Evaluating branching programs on encrypted data. In: Vadhan, S.P. (ed.) TCC 2007. LNCS, vol. 4392, pp. 575–594. Springer, Heidelberg (2007)
18. Ogata, W., Kurosawa, K.: Oblivious keyword search. J. Complexity 20(2-3), 356–371 (2004)
19. Park, D.J., Kim, K., Lee, P.J.: Public key encryption with conjunctive field keyword search. In: Lim, C.H., Yung, M. (eds.) WISA 2004. LNCS, vol. 3325, pp. 73–86. Springer, Heidelberg (2005)
20. Prabhakaran, M., Rosulek, M.: Homomorphic encryption with CCA security. In: Aceto, L., Damgård, I., Goldberg, L.A., Halldórsson, M.M., Ingólfsdóttir, A., Walukiewicz, I. (eds.) ICALP 2008, Part II. LNCS, vol. 5126, pp. 667–678. Springer, Heidelberg (2008), http://eprint.iacr.org/2008/079
21. Prabhakaran, M., Rosulek, M.: Homomorphic encryption with CCA security. In: Aceto, L., Damgård, I., Goldberg, L.A., Halldórsson, M.M., Ingólfsdóttir, A., Walukiewicz, I. (eds.) ICALP 2008, Part II. LNCS, vol. 5126, pp. 667–678. Springer, Heidelberg (2008), http://eprint.iacr.org/2008/079

22. Rivest, R.L., Adleman, L., Dertouzos, M.L.: On data banks and privacy homomorphisms. In: Foundations of secure computation (Workshop, Georgia Inst. Tech., Atlanta, Ga, pp. 169–179. Academic, New York (1978)
23. Sander, T., Young, A., Yung, M.: Non-interactive cryptocomputing for NC^1. In: FOCS, pp. 554–567 (1999)
24. Song, D.X., Wagner, D., Perrig, A.: Practical techniques for searches on encrypted data. In: IEEE Symposium on Security and Privacy, pp. 44–55 (2000)

A Security Definitions for Non-Malleable Homomorphic Encryption

The formal definitions in this section are summarized from [20] for reference:

HCCA Security. The main security definition, called *Homomorphic-CCA (HCCA)* security, formalizes the intuition that a homomorphic encryption scheme is "non-malleable except for a certain set of operations." The complete security experiment is given in Figure 3, and we give an overview and motivation below.

Definition 1. *A homomorphic encryption scheme is* Homomorphic-CCA (HCCA) secure *with respect to $\mathcal{T}$ if there are PPT algorithms* RigEnc *and* RigExtract, *where the range of* RigExtract *is $\mathcal{T} \cup \{\bot\}$, and such that for all PPT adversaries $\mathcal{A}$, the advantage of $\mathcal{A}$ in the IND-HCCA experiment (Figure 3) is negligible.*

When $b = 0$ in the experiment, the adversary simply receives an encryption of his chosen plaintext msg^*, and gets access to an unrestricted decryption oracle. However, when $b = 1$ in the experiment, instead of an encryption of msg^*, the adversary receives a "rigged" ciphertext generated by RigEnc, without knowledge of msg^*. Such a rigged ciphertext need not encode any actual message, so if the adversary asks for it (or any of its derivatives via the homomorphic operations) to be decrypted, the decryption oracle's response must be compensated in some way, or else it would be easy to distinguish the $b = 0$ from $b = 1$ scenarios. For this purpose, the RigEnc procedure also produces some (secret) extra state information, which makes it possible to identify (via the RigExtract procedure) all ciphertexts derived from that particular rigged ciphertext, as well as *how* they were derived. So in the $b = 1$ scenario, the decryption oracle first uses RigExtract to check whether the given ciphertext was derived via a homomorphic operation of the scheme, and if so, compensates in its response. For example, if the query ciphertext was derived by applying the T transformation, then the decryption oracle should respond with $T(\mathsf{msg}^*)$, to mimic the $b = 0$ case.

It is easily seen that if it is feasible for an adversary to modify an encryption of $\mathsf{Enc}(\mathsf{msg})$ into a related encryption $\mathsf{Enc}(T(\mathsf{msg}))$, but RigExtract never outputs T, then there is a way for an adversary to distinguish between $b = 0$ and $b = 1$ in the experiment. Thus by restricting the range of the RigExtract procedure in the security definition, we limit the feasible malleability of the scheme.

Finally, because RigExtract uses the private key, as well as secret auxiliary information from RigEnc, we should provide an oracle for these procedures. We do so in a "guarded" way that keeps the auxiliary shared information hidden from the adversary in the experiment.

Unlinkability. The second security definition, called *unlinkability*, formalizes of the natural requirement that a ciphertext hides not only its plaintext, but also its "history"

Setup: Pick $(PK, SK) \leftarrow \mathsf{KeyGen}$ and give PK to $\mathcal{A}$.
Phase I: $\mathcal{A}$ gets access to the $\mathsf{Dec}_{SK}(\cdot)$ oracle and the following two "guarded" RigEnc and RigExtract oracles:

$$\mathsf{GRigEnc}_{PK}() = \zeta_i, \text{ where } (\zeta_i, S_i) \leftarrow \mathsf{RigEnc}_{PK}, \text{ when called for the } i\text{th time}$$
$$\mathsf{GRigExtract}_{SK}(\zeta, i) = \mathsf{RigExtract}_{SK}(\zeta, S_i)$$

Challenge: $\mathcal{A}$ outputs a plaintext msg^*. We privately flip a coin $b \leftarrow \{0,1\}$. If $b = 0$, we compute $\zeta^* \leftarrow \mathsf{Enc}_{PK}(\mathsf{msg}^*)$. If $b = 1$, we compute $(\zeta^*, S^*) \leftarrow \mathsf{RigEnc}_{PK}$. In both cases, we give ζ^* to $\mathcal{A}$.
Phase II: $\mathcal{A}$ gets access to the same GRigEnc and GRigExtract oracles as in Phase I, as well as a "rigged" version of the decryption oracle RigDec. When $b = 0$, RigDec is simply the normal decryption oracle $\mathsf{Dec}_{SK}(\cdot)$. When $b = 1$, RigDec is implemented as follows:

$$\mathsf{RigDec}_{SK}(\zeta) = \begin{cases} T(\mathsf{msg}^*) & \text{if } \bot \neq T \leftarrow \mathsf{RigExtract}_{SK}(\zeta, S^*) \\ \mathsf{Dec}_{SK}(\zeta) & \text{otherwise} \end{cases}.$$

Output: $\mathcal{A}$ outputs a bit b'. The *advantage* of $\mathcal{A}$ is $\Pr[b' = b] - \frac{1}{2}$.

Fig. 3. IND-HCCA security experiment, parametrized by $\mathcal{T}$

Setup: Pick $(PK, SK) \leftarrow \mathsf{KeyGen}$ and give PK to $\mathcal{A}$.
Phase I: $\mathcal{A}$ is given access to the decryption oracle $\mathsf{Dec}_{SK}(\cdot)$.
Challenge: Flip a coin $b \leftarrow \{0,1\}$. $\mathcal{A}$ outputs a ciphertext ζ and a transformation $T \in \mathcal{T}$. If $\mathsf{Dec}_{SK}(\zeta) = \bot$, do nothing. Else give ζ^* to $\mathcal{A}$ where

$$\zeta^* \leftarrow \begin{cases} \mathsf{Enc}_{PK}(T(\mathsf{Dec}_{SK}(\zeta))) & \text{if } b = 0 \\ \mathsf{CTrans}(\zeta, T) & \text{if } b = 1 \end{cases}.$$

Phase II: $\mathcal{A}$ is given access to the decryption oracle $\mathsf{Dec}_{SK}(\cdot)$.
Output: $\mathcal{A}$ outputs a bit b'. The *advantage* of $\mathcal{A}$ is $\Pr[b' = b] - \frac{1}{2}$.

Fig. 4. Unlinkability security experiment, parametrized by $\mathcal{T}$

— i.e., whether it was generated as a normal Enc, or by applying the homomorphic operations to some other ciphertext.

We note that the definition is more than just a correctness property, as it involves the behavior of the scheme's algorithms on maliciously-crafted ciphertexts. The security experiment also includes a decryption oracle, making it applicable even to adversaries with chosen-ciphertext attack capabilities.

Definition 2. *A homomorphic encryption scheme is* unlinkably homomorphic *with respect to $\mathcal{T}$ if for all PPT adversaries $\mathcal{A}$, the advantage of $\mathcal{A}$ in the unlinkability experiment (Figure 4) is negligible.*

Efficient Protocols for Set Membership and Range Proofs

Jan Camenisch[1], Rafik Chaabouni[1,2], and abhi shelat[3]

[1] IBM Research
[2] EPFL
[3] U. of Virginia

Abstract. We consider the following problem: Given a commitment to a value σ, prove in zero-knowledge that σ belongs to some discrete set Φ. The set Φ can perhaps be a list of cities or clubs; often Φ can be a numerical range such as $[1, 2^{20}]$. This problem arises in e-cash systems, anonymous credential systems, and various other practical uses of zero-knowledge protocols.

When using commitment schemes relying on RSA-like assumptions, there are solutions to this problem which require only a constant number of RSA-group elements to be exchanged between the prover and verifier [5, 15, 16]. However, for many commitment schemes based on bilinear group assumptions, these techniques do not work, and the best known protocols require $O(k)$ group elements to be exchanged where k is a security parameter.

In this paper, we present two new approaches to building set-membership proofs. The first is based on bilinear group assumptions. When applied to the case where Φ is a range of integers, our protocols require $O(\frac{k}{\log k - \log\log k})$ group elements to be exchanged. Not only is this result asymptotically better, but the constants are small enough to provide significant improvements even for small ranges. Indeed, for a discrete logarithm based setting, our new protocol is an order of magnitude more efficient than previously known ones.

We also discuss alternative implementations of our membership proof based on the strong RSA assumption. Depending on the application, e.g., when Φ is a published set of values such a frequent flyer clubs, cities, or other ad hoc collections, these alternative also outperform prior solutions.

Keywords: Range proofs, set membership proofs, proofs of knowledge, bi-linear maps.

1 Introduction

In this paper we consider zero-knowledge protocols which allow a prover to convince a verifier that a digitally committed value is a member of a given public set. A special case of this problem is when to show that the committed value lies in a specified integer range.

The first problem, which we denote the *set membership* proof, occurs for instance in the context of anonymous credentials. Consider a user who is issued

J. Pieprzyk (Ed.): ASIACRYPT 2008, LNCS 5350, pp. 234–252, 2008.
© International Association for Cryptologic Research 2008

a credential containing a number of attributes such as address. Further assume the user needs to prove that she lives in a European capital. Thus, we are given a list of all such cities and the user has to show that she possesses a credential containing one of those cities as address (without of course, leaking the city the user lives in). Or, consider a user who has a subscription to a journal (e.g., the news and the sports section). Further assume that some general sections are to all subscribers of a list of sections. Thus, using our protocol, the user can efficiently show that she is a subscriber to one of the required kinds.

The second problem, which we denote the *range proof*, also occurs often in anonymous credential and e-cash scenarios. For example, a user with passport credential might wish to prove that her age is within some range, e.g. greater than 18, or say between 13 and 18 in the case of a teen-community website. This problem is a special case of the set membership proof. Since the elements of the set occur in consecutive order, special techniques can be applied.

1.1 Our Results

Given a set $\Phi = \{\phi_1, \phi_2, \ldots, \phi_n\}$ and a commitment[1] C, a typical approach to the set membership problem is to use a zero-knowledge proof of the form

> "C is a commitment to the element ϕ_1 OR it is a commitment to ϕ_2 OR it is a commitment to ϕ_3 $\cdots$ OR it is a commitment to ϕ_n."

Even though there exist efficient algebraic Σ (Sigma) protocols for handling a single such OR clause, such a proof still has length which is proportional to n. One might argue that such proofs necessarily have length proportional to n since the task of describing the set Φ itself requires space n.

However, in many practical situations, the set Φ is often specified in advance by the verifying party. In other words, Φ can be considered a common input to both Prover and Verifier, and thus we might ask whether it is possible to prove a commitment is a commitment to an element of Φ *without* having to explicitly list Φ in the proof.

To the best of our knowledge, we are the first to propose such a scheme for general, unstructured sets. Our approach is incredibly simple. We provide a way to "encode" the set Φ in a way that allows for $O(1)$-sized proofs that a committed element belongs to Φ. Specifically, we let the verifier specify Φ by providing "digital signatures" on the elements of Φ under a new verification key vk. Now if we consider this set of digital signatures as a common input, the proof becomes a statement of the form:

[1] One might wonder what it means to say "the element committed to in C" when the commitment scheme is not a perfectly-binding one. In such a case, technically, the proof is only *computationally sound*—often called an *argument* instead of a proof. In other words, we assume that a computationally-bounded prover knows only one way to open the commitment C and cannot deduce other ways. Indeed, such protocols are technically called *arguments* instead of proofs. Since prior work refers to the problem as a "proof," we continue to use that term.

"The prover knows a signature under vk for the element committed to in C."

We provide two types of protocols that are instantiations of this idea. The first one is based on a bilinear-group signature scheme which enables an efficient way to make this proof. The second way is based on the Strong RSA assumption and uses the idea of cryptographic accumulators. In both cases, the actual proof of the statement requires $O(1)$ group elements to be exchanged between prover and verifier.

The special case of Range proofs. A popular special case of the set membership problem occurs when the set Φ consists of a range $[a, a+1, a+2, \ldots, b]$—which we denote $[a, b]$. This problem has been well-studied because it occurs so often in practice. Indeed, under the Strong RSA assumption, there are very efficient proofs for this problem as we discuss in the prior work section below. However, in cases when the range is small or the same range is used in many protocol instantiations, our protocol will be more efficient (by a factor of about 8-10, depending on the group employed).

If one is not willing to rely on the Strong RSA assumption, the folklore method to the problem of range proofs is to have the Prover commit to all k bits of his secret, prove that these commitments all encode either a 0 or a 1 and prove that the commitments indeed commit to all the bits of s. The verifier is then convinced that the secret lies in $[0, 2^{k+1} - 1]$ since there were only k commitments. The method can be generalized to any range. The size of such a proof is thus $O(k)$ group elements.

Using the simple idea of the set membership proof, we are able to reduce this size both asymptotically and in practice for many often-occurring ranges. Our simple idea is as follows: Instead of committing to the individual bits of the committed value, we write the secret value in base-u (for some optimally chosen u) and commit to these u-ary digits. If we only provide ℓ such commitments and prove that the secret can be written in u-ary notation, then we implicitly prove that the secret is in the range $[0, u^{\ell}]$. A generalization of this technique can be used to prove that the secret is in $[a, b]$ for arbitrary integers a and b. The key technique is to use the set-membership protocol in order to prove that each committed digit is indeed a digit in base-u. Writing the secret in base-u (instead of base 2) is indeed an obvious step. However, with prior methods, doing so does not reduce the proof size. With prior methods, proving that a committed digit is a u-ary digit requires a u-wise OR proof of size $O(u)$; since this u-wise OR proof must be done ℓ times independently, prior methods require communication $O(u \cdot \ell)$.

The key insight in our scheme is to design a scheme which can reuse part of one u-ary digit proof in all ℓ proof instances. Specifically, the verifier can send *one* list of u signatures representing u-ary digits, and the prover can use this *same* list to prove that all ℓ digits are indeed u-ary digits. Thus, the total communication complexity of our approach is $O(u + \ell)$. With well-selected values for u and ℓ, we show that this approach yields a proof of size $O(\frac{k}{\log k - \log \log k})$ which is both asymptotically and practically better than the only other known method.

Note that if the range is small or the same range is used for many protocols, then it is more efficient to employ the set membership protocol directly.

1.2 Prior and Related Work

Assume for concreteness the Pedersen commitment scheme over a prime order group. Let g, h be elements of a group G of prime order q. Let $C = g^s h^r$ be the commitment that the prover has sent to the verifier, where s is the secret of which the prover want to show that it lies in a specific range and r is a randomly chosen element from $\mathbb{Z}_q$.

There are a number of known ways that a prover can convince a verifier that the secret committed in C lies in a given range assuming the hardness of the Strong (or sometimes called flexible) RSA problem. Let us review them here.

The most frequent method used in practice is the following. First, the verifier picks a safe prime product $\mathfrak{n} = (2\mathfrak{p}+1)(2\mathfrak{q}+1)$ and two random quadratic residues $\mathfrak{g}, \mathfrak{h}$ modulo $\mathfrak{n}$, and proves to the prover that $\mathfrak{g} \in \langle \mathfrak{h} \rangle$ is true. Next, the verifier computes $\mathfrak{c} = \mathfrak{g}^s \mathfrak{h}^{r'} \bmod \mathfrak{n}$, sends this value to the prover and then runs the following protocol with him:

$$PK\{(s, r, r') : \mathfrak{c} = \mathfrak{g}^s \mathfrak{h}^{r'} \pmod{\mathfrak{n}} \ \wedge \ C = g^s h^r \ \wedge \ s \in [-A, A]\}$$

The protocol is basically a generalized Schnorr proof (in a group of unknown order), where the verifier in addition to accepting the basic proof also verifies whether the answer corresponding to the secret s lies in $[-A/2, A/2]$. If it does so, then the verifier can conclude that the secret must lie in the range $[-A, A]$ (this becomes apparent when one considers the knowledge extractor for the protocol). The drawback of this proof is that it in fact works only if the secret lies in the smaller range $[-A2^{-(k'+k'')}, A2^{-(k'+k'')}]$, with k' being the number of bits of the challenge sent by the verifier and k'' determining the statistical zero-knowledge property, i.e., the secret must be $k' + k''$ bits smaller. Therefore the protocol cannot be used for situations where one has to show that a secret lies exactly in a given range.

Boudot [5] provided an efficient proof that did not have this drawback. He used the observation that any positive number can be composed as the sum of four squares. Thus, to show that a secret s lies in $[A, B]$, one just needs to show that the values $s_1 = s - A$ and $s_2 = B - s$ are positive. So basically, what the prover has to do is to give commitments to s_1 and s_2 and to the numbers $s_{(1,1)}, \ldots, s_{(1,4)}$ and $s_{(2,1)}, \ldots, s_{(2,4)}$, the sum of whose squares are equal to s_1 and s_2 respectively. Of course, if these commitments were, e.g., Pedersen commitments in a group of prime order q, them all we could conclude is that s_1 and s_2 are the sum of four square modulo q, which is not very helpful. Luckily, Okamoto and Fujisaki [13] have shown that when the commitments and the proof is done in a group where the order is not known to the prover, then these relations hold over the integers and thus one can really assert that s_1 and s_2 are positive.

Thus, we get the following protocol: First the prover computes the following commitments $\mathfrak{c}_{(i,j)} = \mathfrak{g}^{s_{(i,j)}} \mathfrak{h}^{r_{(i,j)}} \bmod \mathfrak{n}$ for some randomly chosen $r_{(i,j)}$, sends these to the verifier and then engages in the following proof with him :

$$
\begin{aligned}
PK\{(s, r, r', s_1^{(1)}, \ldots, s_1^{(4)}, s_2^{(1)}, \ldots, s_2^{(4)}, r'', r^*) :& \\
\mathfrak{c}_{(1,1)} = \mathfrak{g}^{s_{(1,1)}} \mathfrak{h}^{r_{(1,1)}} \wedge \ldots \wedge \mathfrak{c}_{(1,4)} = \mathfrak{g}^{s_{(1,4)}} \mathfrak{h}^{r_{(1,4)}} \wedge& \\
\mathfrak{c}_{(2,1)} = \mathfrak{g}^{s_{(2,1)}} \mathfrak{h}^{r_{(2,1)}} \wedge \ldots \wedge \mathfrak{c}_{(2,4)} = \mathfrak{g}^{s_{(2,4)}} \mathfrak{h}^{r_{(2,4)}} \wedge& \\
\mathfrak{c}/\mathfrak{g}^A = {\mathfrak{c}_{(1,1)}}^{s_{(1,1)}} \cdots {\mathfrak{c}_{(1,4)}}^{s_{(1,4)}} \mathfrak{h}^{r''} \wedge \mathfrak{g}^B/\mathfrak{c} = {\mathfrak{c}_{(2,1)}}^{s_{(2,1)}} \cdots {\mathfrak{c}_{(2,4)}}^{s_{(2,4)}} \mathfrak{h}^{r^*} \wedge& \\
\mathfrak{c} = \mathfrak{g}^s \mathfrak{h}^{r'} \pmod{\mathfrak{n}} \wedge C = g^s h^r\}&
\end{aligned}
$$

We see that this protocol requires the prover to compute 22 modular exponentiations (including the computations of the commitments) and the verifier to compute 12 modular exponentiations. The communication complexity is in about 35 group elements. Groth [15] optimizes this protocol by exploiting the fact that special integers can be written as the sum of 3 squares instead of 4 squares. The major drawback of these approaches is that the Rabin and Shallit algorithm typically used to find the 4 (or 3) squares which sum to the secret takes time $O(k^4)$ where k is the size of the interval. Lipmaa [16] provides another algorithm to find this squares that improves somewhat on the Rabin-Shallit one. However, in practice, these algorithms running times quickly make this approach preventive.

Independently to our work, Teranishi and Sako [20] presented a k-Times Anonymous Authentication in which they present a range proof using Boneh-Boyen signature scheme [4], that can be obtained from our generalized set membership. However their range proof does not compete with ours as our verifier publishes significantly less signatures.

Schoenmakers [18, 19] studied and discussed several recursive relations which can be used to reduce the number of basic Schnorr proofs when committing to the individual bits of the secret. In particular, he writes the upper bound L of the positive range $[0, L)$ as either the product or the sum of two numbers. By doing this scheme recursively he decreased the amount of work needed. However the overall communication load in his protocols is still $O(k)$, where $2^{k-1} < L \leqslant 2^k$. We note that some of his techniques for reducing certain ranges to other more convenient ranges can be used with any range proof technique.

Micali, Kilian, and Rabin [17] considered a more general problem in which a polynomial-time prover wants to commit to a finite set Φ of strings so that, later on, he can, for any string x, reveal with a proof whether $x \in \Phi$ or $x \notin \Phi$ without leaking any knowledge beyond the membership assertions. In particular, the proofs do not even reveal the size of Φ—much less the actual elements. Thus, these protocols are not directly comparable to ours.

1.3 Organization

In section 2, we recall zero-knowledge proofs, Σ-protocols and define proofs of set membership and range proofs. In section 3, we describe our new signature-based set membership together with its corresponding proof. In section 4, we explain how to apply our new signature-based set membership for efficient range proof. We also emphasis on the communication complexity and show how our new range proof is asymptotically better. To have a better insight of our state of the art, we provide a concrete example together with some comparison of previous work. In section 5, we recall cryptographic accumulators together with their proofs, and we describe our new accumulator-based set membership.

2 Definitions

Zero-knowledge proofs and Σ-protocols. We use definitions from [2, 11]. A pair of interacting algorithms (P, V) is a proof of knowledge (PK) for a relation $R = \{(\alpha, \beta)\} \subseteq \{0,1\}^* \times \{0,1\}^*$ with knowledge error $\kappa \in [0,1]$ if (1) for all $(\alpha, \beta) \in R$, $\mathsf{V}(\alpha)$ accepts a conversation with $\mathsf{P}(\beta)$ with probability 1; and (2) there exists an expected polynomial-time algorithm E, called the *knowledge extractor*, such that if a cheating prover P^* has probability ϵ of convincing V to accept α, then E, when given rewindable black-box access to P^*, outputs a witness β for α with probability $\epsilon - \kappa$.

A proof system (P, V) is *honest-verifier zero-knowledge* if there exists a p.p.t. algorithm Sim, called the *simulator*, such that for any $(\alpha, \beta) \in R$, the outputs of $V(\alpha)$ after interacting with $\mathsf{P}(\beta)$ and that of $\mathsf{Sim}(\alpha)$ are computationally indistinguishable.

Note that standard techniques can be used to transform an honest-verifier zero-knowledge proof system into a general zero-knowledge one [11]. This is especially true of special Σ-protocols that will be presented later in the paper. Thus, for the remainder of the paper, our proofs will be honest-verifier zero-knowledge. (This also allows us to make more accurate comparisons with the other proof techniques since they are usually also presented as honest-verifier protocols).

A Σ-protocol is a proof system (P, V) where the conversation is of the form (a, c, z), where a and z are computed by P, and c is a challenge chosen at random by V. The verifier accepts if $\phi(\alpha, a, c, z) = 1$ for some efficiently computable predicate ϕ. Given two accepting conversations (a, c, z) and (a, c', z') for $c \neq c'$, one can efficiently compute a witness β. Moreover, there exists a polynomial-time simulator Sim that on input α and a random string c outputs an accepting conversation (a, c, z) for α that is perfectly indistinguishable from a real conversation between $\mathsf{P}(\beta)$ and $\mathsf{V}(\alpha)$.

We use notation introduced by Camenisch and Stadler [9] for the various zero-knowledge proofs of knowledge of discrete logarithms and proofs of the validity of statements about discrete logarithms. For instance,

$$PK\{(\alpha, \beta, \gamma) : y = g^\alpha h^\beta \ \wedge \ \mathfrak{y} = \mathfrak{g}^\alpha \mathfrak{h}^\gamma \ \wedge \ (u \le \alpha \le v)\}$$

denotes a "*zero-knowledge* Proof *of* Knowledge *of integers* α, β, *and* γ *such that* $y = g^\alpha h^\beta$ *and* $\mathfrak{y} = \mathfrak{g}^\alpha \mathfrak{h}^\gamma$ *holds, where* $v \le \alpha \le u$," where $y, g, h, \mathfrak{y}, \mathfrak{g}$, and $\mathfrak{h}$ are elements of some groups $G = \langle g \rangle = \langle h \rangle$ and $\mathfrak{G} = \langle \mathfrak{g} \rangle = \langle \mathfrak{h} \rangle$. The convention is Greek letters denote quantities the knowledge of which is being proved, while all other parameters are known to the verifier. Using this notation, a proof-protocol can be described by just pointing out its aim while hiding all details. We note that all of the protocols we present in this notation can be easily instantiated as Σ-protocols.

Definition 1 (Proof of Set Membership). *Let $C = (\text{Gen}, \text{Com}, \text{Open})$ be the generation, the commit and the open algorithm of a string commitment scheme. For an instance c, a proof of set membership with respect to commitment scheme C and set Φ is a proof of knowledge for the following statement:*

$$PK\{(\sigma, \rho) : c \leftarrow \text{Com}(\sigma;\ \rho) \wedge \sigma \in \Phi\}$$

Remark: The proof system is defined with respect to *any* commitment scheme. Thus, in particular, if Com is a perfectly-hiding scheme, then the language Γ_S consists of all commitments (assuming that S is non-empty). Thus for soundness, it is important that the protocol is a proof of knowledge.

Definition 2 (Range Proof). *A range proof with respect to a commitment scheme C is a special case of a proof of set membership in which the set Φ is a continuous sequence of integers $\Phi = [a, b]$ for $a, b \in \mathbb{N}$.*

3 Signature-Based Set Membership

Here we present a new set membership protocol that is inspired by the oblivious transfer protocol presented by Camenisch, Neven, and shelat [8]. The basic idea is that the verifier first sends the prover a signature of every element in the set Φ. Thus, the prover receives a signature on the particular element σ to which C is a commitment. The prover then "blinds" this received signature and performs a proof of knowledge that she possesses a signature on the committed element. Notice that the communication complexity of this proof depends on the cardinality of Φ—in particular because the verifier's first message contains a signature of every element in Φ. The rest of the protocol, however, requires only a constant number of group elements to be sent. The novelty of this approach is that the first verifier message can be re-used in other proofs of membership; indeed, we use this property to achieve our results for range proofs.

Computational Assumptions. Our protocols in this section require bilinear groups and associated hardness assumptions. Let PG be a pairing group generator that on input 1^k outputs descriptions of multiplicative groups $\mathbb{G}_1$ and $\mathbb{G}_\mathrm{T}$ of prime order p where $|p| = k$. Let $\mathbb{G}_1^* = \mathbb{G}_1 \setminus \{1\}$ and let $g \in \mathbb{G}_1^*$. The generated groups are such that there exists an admissible bilinear map $e : \mathbb{G}_1 \times \mathbb{G}_1 \rightarrow \mathbb{G}_\mathrm{T}$, meaning that (1) for all $a, b \in \mathbb{Z}_p$ it holds that $e(g^a, g^b) = e(g, g)^{ab}$; (2) $e(g, g) \neq 1$; and (3) the bilinear map is efficiently computable.

Definition 3 (Strong Diffie-Hellman Assumption [4]). *We say that the q-SDH assumption associated to a pairing generator PG holds if for all p.p.t. adversaries A, the probability that $A(g, g^x, \ldots, g^{x^q})$ where $(\mathbb{G}_1, \mathbb{G}_\mathrm{T}) \leftarrow \mathsf{PG}(1^k)$, $g \leftarrow \mathbb{G}_1^*$ and $x \leftarrow \mathbb{Z}_p$, outputs a pair $(c, g^{1/(x+c)})$ where $c \in \mathbb{Z}_p$ is negligible in k.*

A recent work by Cheon [10] shows a "weakness" in the q-SDH assumption. However, this "weakness" is not so relevant when q is a very small number like 50 as it is in our paper.

Boneh-Boyen Signatures. Our scheme relies on the elegant Boneh-Boyen short signature scheme [4] which we briefly summarize. The signer's secret key is $x \leftarrow \mathbb{Z}_p$, the corresponding public key is $y = g^x$. The signature on a message m is $\sigma \leftarrow g^{1/(x+m)}$; verification is done by checking that $e(\sigma, y \cdot g^m) = e(g, g)$. This scheme is similar to the Dodis and Yampolskiy verifiable random function [12].

Security under weak chosen-message attack is defined through the following game. The adversary begins by outputting ℓ messages $m_1, \ldots, m_\ell$. The challenger generates a fresh key pair and gives the public key to the adversary, together with signatures $\sigma_1, \ldots, \sigma_\ell$ on $m_1, \ldots, m_\ell$. The adversary wins if it succeeds in outputting a valid signature σ on a message $m \notin \{m_1, \ldots, m_\ell\}$. The scheme is said to be unforgeable under a chosen-message attack if no p.p.t. adversary A has non-negligible probability of winning this game. Our scheme relies on the following property of the Boneh-Boyen short signature [4] which we paraphrase below:

Lemma 1 ([4](Lemma 3.2)). *Suppose the q-Strong Diffie Hellman assumption holds in* $(\mathbb{G}_1, \mathbb{G}_T)$*. Then the basic Boneh-Boyen signature scheme is q-secure against an existential forgery under a weak chosen message attack.*

A Note on Protocol Clarity. In order to make our protocols more readable in this version, we do not specifically mention standard checks such as verifying that a received number is a prime, verifying that an element is a proper generator and in the correct group, and, specifically related to our protocols, whether all of the received verifier values are signatures, etc. Again, many of these checks only apply when compiling from honest-verifier zero-knowledge to full zero-knowledge; as we mentioned above, we only consider the honest case.

Common Input: g, h, a commitment C, and a set Φ
Prover Input: σ, r such that $C = g^\sigma h^r$ and $\sigma \in \Phi$.

$P \xleftarrow{y, \{A_i\}} V$ Verifier picks $x \in_R \mathbb{Z}_p$ and sends $y \leftarrow g^x$ and $A_i \leftarrow g^{\frac{1}{x+i}}$ for every $i \in \Phi$.

$P \xrightarrow{V} V$ Prover picks $v \in_R \mathbb{Z}_p$ and sends $V \leftarrow A_\sigma^v$.

Prover and Verifier run $\mathrm{PK}\{(\sigma, r, v) : C = g^\sigma h^r \ \wedge \ V = g^{\frac{v}{x+\sigma}}\}$

$P \xrightarrow{a, D} V$ Prover picks $s, t, m \in_R \mathbb{Z}_p$ and sends $a \leftarrow e(V, g)^{-s} e(g, g)^t$ and $D \leftarrow g^s h^m$.

$P \xleftarrow{c} V$ Verifier sends a random challenge $c \in_R \mathbb{Z}_p$.

$P \xrightarrow{z_\sigma, z_v, z_r} V$ Prover sends $z_\sigma \leftarrow s - \sigma c$, $z_v \leftarrow t - vc$, and $z_r \leftarrow m - rc$.

Verifier checks that $D \stackrel{?}{=} C^c h^{z_r} g^{z_\sigma}$ and that $a \stackrel{?}{=} e(V, y)^c \cdot e(V, g)^{-z_\sigma} \cdot e(g, g)^{z_v}$

Fig. 1. Set membership protocol for set Φ

Theorem 1. *If the $|\Phi|$-Strong Diffie-Hellman assumption associated with a pairing generator* PG *holds, then protocol in Fig. 1 is a zero-knowledge argument of set membership for a set Φ.*

Proof. The completeness of the protocol follows by inspection. The soundness follows from the extraction property of the proof of knowledge and the unforgeability of the random function. In particular, the extraction property implies that for any prover P^* that convinces V with probability ϵ, there exists an extractor which interacts with P^* and outputs a witness (σ, r, v) with probability $poly(\epsilon)$. Moreover, if we assume that the extractor input consists of two transcripts, i.e.,

$$\{y, \{A_i\}, V, a, D, c, c', z_\sigma, z'_\sigma, z_v, z'_v, z_r, z'_r\},$$

the witness can be obtained by computing:

$$\sigma = \frac{z_\sigma - z'_\sigma}{c' - c}; \quad r = \frac{z_r - z'_r}{c' - c}; \quad v = \frac{z_v - z'_v}{c' - c}$$

The extractor succeeds when $(c' - c)$ is invertible in $\mathbb{Z}_p$. If $\sigma \notin \Phi$, then P^* can be (almost) directly be used to mount a weak chosen-message attack against the Boneh-Boyen signature scheme with probability $poly(\epsilon)$ of succeeding. Thus, ϵ must be negligible.

Finally, to prove honest-verifier zero-knowledge, we construct a simulator Sim that will simulate all interactions with any honest verifier V^*, see Fig. 2.

1. Sim retrieves $y, \{A_i\}$ from V^*.
2. Sim chooses $\sigma \in_R \Phi$, $v \in_R \mathbb{Z}_p$ and sends $V \leftarrow A_\sigma^v$ to V^*.
3. Sim chooses $s, t, m \in_R \mathbb{Z}_p$ and sends $a \leftarrow e(V, g)^{-s} e(g, g)^t$ and $D \leftarrow g^s h^m$ to V^*.
4. Sim receives c from V^*
5. Finally Sim computes and sends $z_\sigma \leftarrow s - \sigma c$, $z_v \leftarrow t - vc$, and $z_r \leftarrow m - rc$ to V^*.

Fig. 2. Simulator for the set membership protocol

Since $\mathbb{G}_1$ is a prime-order group, then the blinding is perfect in the first two steps; thus the zero-knowledge property follows from the zero-knowledge property of the Σ-protocol (Steps 3 to 5).

4 Range Proofs

We now turn our attention to the range proofs.

First note that the protocol for set membership can be directly applied to the problem of range proofs. This will not be efficient for ranges spanning more than a few hundred elements. However, if the particular range is fixed over many protocols as it might often be (as is for instance the case when one needs to prove that one is between 13 and 18 years old), then the verifier can publish the signatures once and for all. Thus, the proofs become just the second phase which amounts to one pairing and two exponentiation for the prover and the verifier. This will be about a factor of 8-10 times more efficient than employing Boudot's method.

For the remainder assume, however, that the range is large or that the cost of publishing/sending the signatures on the set elements cannot be amortized.

Instead, our approach is to write the secret σ in u-ary notation, i.e., $\sigma = \sum_j^\ell \sigma_j \cdot u^j$. We may now easily prove that $\sigma \in [0, u^\ell)$ by simply providing (and proving) commitments to the u-ary digits of σ. This problem, however, can be solved by repeating the basic set-membership protocol from above on the set $[0, u-1]$. Moreover, the first verifier message, which requires the most communication, can be re-used for each of the ℓ digits. Assuming that $\sigma \in [0, B)$, the goal is thus to minimize the communication load under the constraint $u^\ell \geqslant B$.

4.1 Range Proofs From Our Signature-Based Set-Membership Protocol

We first present how to prove that our secret σ lies in $[0, u^\ell)$ (see Figure 3). Write σ in the base u to obtain ℓ elements as such: $\sigma = \sum_j \left(\sigma_j u^j\right)$.

Common Input: g, h, u, ℓ, and a commitment C
Prover Input: σ, r such that $C = g^\sigma h^r$ and $\sigma \in [0, u^\ell)$.

$P \xleftarrow{\;y, \{A_i\}\;} V$ Verifier picks $x \in_R \mathbb{Z}_p$ and sends $y \leftarrow g^x$ and $A_i \leftarrow g^{\frac{1}{x+i}}$ for every $i \in \mathbb{Z}_u$.

$P \xrightarrow{\;\{V_j\}\;} V$ Prover picks $v_j \in_R \mathbb{Z}_p$ and sends $V_j \leftarrow A_{\sigma_j}^{v_j}$ for every $j \in \mathbb{Z}_l$, s.t. $\sigma = \sum_j \left(\sigma_j u^j\right)$

Prover and Verifier run $\mathrm{PK}\{(\sigma_j, r, v_j) : C = h^r \prod_j (g^{u^j})^{\sigma_j} \wedge V_j = g^{\frac{v_j}{x+\sigma_j}}\}$

$P \xrightarrow{\;\{a_j\}, D\;} V$ Prover picks $s_j, t_j, m_j \in_R \mathbb{Z}_p$ for every $j \in \mathbb{Z}_l$ and sends $a_j \leftarrow e(V_j, g)^{-s_j} e(g, g)^{t_j}$ and $D \leftarrow \prod_j \left(g^{u^j s_j}\right) h^{m_j}$.

$P \xleftarrow{\;c\;} V$ Verifier sends a random challenge $c \in_R \mathbb{Z}_p$.

$P \xrightarrow{\;\{z_{\sigma_j}\}, \{z_{v_j}\}, z_r\;} V$ Prover sends $z_{\sigma_j} \leftarrow s_j - \sigma_j c$, $z_{v_j} \leftarrow t_j - v_j c$ for every $j \in \mathbb{Z}_\ell$, and $z_r = m - rc$.

Verifier checks that $D \stackrel{?}{=} C^c h^{z_r} \prod_j \left(g^{u^j z_{\sigma_j}}\right)$ and that $a_j \stackrel{?}{=} e(V_j, y)^c \cdot e(V_j, g)^{-z_{\sigma_j}} \cdot e(g, g)^{z_{v_j}}$ for every $j \in \mathbb{Z}_l$

Fig. 3. Range proof protocol for range $[0, u^\ell)$

Lemma 2. *If the* $(\log k)$*-Strong Diffie Hellman assumption associated to a pairing generator* $\mathsf{PG}(1^k)$ *holds, there exists a zero-knowledge range argument for the range* $[0, u^\ell)$ *where* $u^\ell < \{0,1\}^{k-1}$.

Proof. (Sketch)
Completeness follows from inspection. As before, the soundness follows from the unforgeability of the Boneh-Boyen signature and the extraction property of the proof of knowledge protocol. The honest-verifier zero-knowledge property follows from the perfect blinding of the signatures in the first phase, and the corresponding honest-verifier zero-knowledge property of the Σ-protocol.

Remark: The prover will have to compute 5ℓ exponentiations.

4.2 Communication Complexity

The first message consisting of u signatures and a verification key sent by the verifier to the prover, is not counted as part of the protocol $((u+1)\cdot|\mathbb{G}_1|)$. The prover then sends ℓ blinded values back. Thus, the first phase requires $Init_l(u,\ell) = \ell \cdot |\mathbb{G}_1|$ communication. The second phase of the protocol involves a proof of knowledge. The prover sends $\ell+1$ first-messages of a Σ-protocol. The verifier sends a single challenge, and the prover responds with $2\ell+1$ elements. Thus the overall communication load according to the parameters u and ℓ is:

$$Com(u,\ell) = \ell \cdot (|\mathbb{G}_1| + |\mathbb{G}_T| + 2\cdot|\mathbb{Z}_p|) + (|\mathbb{G}_1| + 2\cdot|\mathbb{Z}_p|) \tag{1}$$

Finding the optimal u and ℓ thus involves solving

$$\min\ c_1 u + c_2 \ell + c_3 \ \text{ s.t. } u^\ell \geqslant B$$

Notice that the bit-committing protocol corresponds to a setting where $u = 2$ and $\ell = k$ which leads to a total communication complexity $O(k)$. Since our protocol allows us to choose more suitable u, we first show that the asymptotic complexity of our approach is smaller than the prior protocols.

Asymptotic Analysis. For the asymptotic analysis, we may ignore the constants c_1, c_2 and c_3. Moreover, we can take $B \approx p/2$ as this is sufficient for showing that a committed value is "positive," i.e., in the range $[0,(p-1/2)]$. Since $p/2 \approx 2^k$, the constraint becomes $u^\ell \geqslant 2^{k-1}$.

By taking logs and dividing, we have that $\ell \approx \frac{k}{\log u}$. Setting $u = \frac{k}{\log k}$ then we get that

$$u = O\left(\frac{k}{\log k}\right), \quad \ell = O\left(\frac{k}{\log k - \log\log k}\right)$$

resulting in a total communication complexity of

$$Com(u,\ell) = O\left(\frac{k}{\log k - \log\log k}\right)$$

which is asymptotically smaller than $O(k)$.

Concrete Optimization. Not only is our solution asymptotically better, but it also performs well for realistic concrete parameters. In order to perform the optimization for concrete parameters we substitute the constraint that $u^\ell \approx B$ into the equation $u + \ell$ above. To minimize, we set the derivative with respect to u to 0 and attempt to solve the equation:

$$c_1 - \frac{c_2 \log B}{u \log^2 u} = 0$$

which simplifies to

$$u \log^2 u = \frac{c_2 \log B}{c_1}. \tag{2}$$

where $\frac{c_2}{c_1} \approx 10$ when standard bilinear groups are used [14]. This equation cannot be solved analytically. However, given B, c_1 and c_2, we can use numerical methods to find a good u as described in [3].

4.3 Handling Arbitrary Ranges $[a, b]$

The above protocol works for the range $[0, u^\ell)$. In order to handle an arbitrary range $[a, b]$, we use an improvement of a folklore reduction described by Schoenmakers in [18] and [19]. Suppose that $u^{\ell-1} < b < u^\ell$. To show the $\sigma \in [a, b]$, it suffices to show that

$$\sigma \in [a, a + u^\ell] \text{ AND } \sigma \in [b - u^\ell, b]$$

Proving that our secret lies in both subsets can be derived from our general

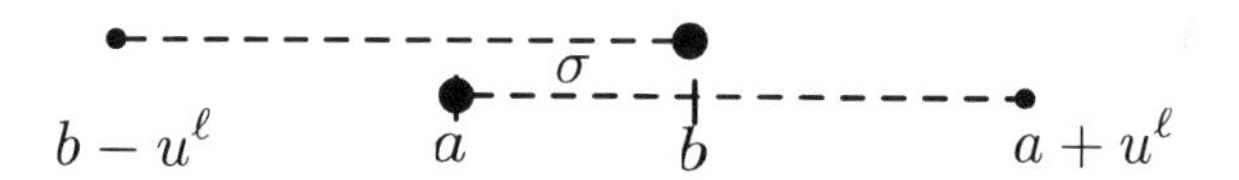

proof that $\sigma \in [0, u^\ell)$ as illustrated in the figure:

$$\sigma \in [b - u^\ell, b) \Longleftrightarrow \sigma - b + u^\ell \in [0, u^\ell)$$
$$\sigma \in [a, a + u^\ell) \Longleftrightarrow \sigma - a \in [0, u^\ell).$$

Note that the u signatures and the verification key need to be sent only once for both subsets. Since both a, b are public, the only modification necessary is the verifier's check, which should now be:

$$D \stackrel{?}{=} C^c g^{-B+u^\ell} h^{z_r} \prod_j (g^{z_{\sigma_j}}), \quad D \stackrel{?}{=} C^c g^{-A} h^{z_r} \prod_j (g^{z_{\sigma_j}}).$$

Thus, essentially 3ℓ extra elements are sent in the protocol, and the prover will have to compute in overall 7ℓ exponentiations.

This scheme can be further optimized when $A+u^{\ell-1} < B$ with an OR-composition. Indeed, the decomposition becomes:

$$[A, B) = [B - u^{\ell-1}, B) \cup [A, A + u^{\ell-1}).$$

The needed modifications are similar to the previous case; the efficiency arises from the fact that we are now working with $\mathbb{Z}_{\ell-1}$. The length of the range set can also be optimized. Indeed if $B - A = u^{\ell}$ then the proof reduces to proving that $\sigma - A \in [0, u^{\ell})$.

Combining this analysis with Lemma 2 yields the following theorem.

Theorem 2. *If the* $\log k$*-Strong Diffie Hellman assumption associated to a pairing generator* $\mathsf{PG}(1^k)$ *holds, there exists a zero-knowledge range argument for the range* $[a, b]$ *where* $0 < a < b < \{0, 1\}^{k-1}$ *whose communication complexity is* $O(\frac{k}{\log k - \log\log k})$.

4.4 Concrete Example and Discussion

Let us discuss our protocol and compare it with other available solutions. The bottom line is the performance of the different methods depend on the application at hand as well as for the assumptions one is willing to make. Assume for a while, all assumptions are fine. Then, for very small intervals (a couple of bits), the standard bit-by-bit method and Schoenmaker's method will probably be the most efficient one. For very large intervals, the method by Boudot will probably be the one of choice as it is mostly independent of the size of the interval. More precisely, it is independent for the verifier but not for the prover as the prover needs to run the Rabin-Shallit algorithm to represent numbers as the sum of four squares and this algorithm has complexity $\mathcal{O}(n^4)$ where n is the bit-length of the number to be decomposed.

Having said that, our methods will typically be the most efficient one when the signatures can be made part of the system parameters, which is probably the case in many scenarios. Of course, at some point it will no longer possible to publish signature of all elements in the range and thats where one will have to restrict these signatures and employ the protocol in this section. When this becomes necessary, one will in practice to make a choice whether it is more efficient to use our algorithm or Boudot's one, the other two will definitely be less efficient.

If one is not restricted by the assumptions one is willing to make, the case is not so clear cut. Let us give a concrete example to shed some light on this. If we pick $B = 599644800$ (which will represent people born before 1989, with their birth date encoded using the Unix Epoch system), we can find the optimal values of u and ℓ by either computing them numerically or by following [3]. Both methods will lead us to $u = 57$ and $\ell = 5$, which minimize the overall communication load:

$$Com_l(57, 5) = 6 \cdot |\mathbb{G}_1| + 5 \cdot |\mathbb{G}_T| + 12 \cdot |\mathbb{Z}_p| \quad (3)$$

Let us illustrate this optimization case with a concrete example. We will assume that an airline company wants to provide special offers to its young clients from a third party. However the exact age of clients should not be divulged to the third party. This offer targets those who are born between 1981 and 1989 (not included). Following the previous example, the birth date will be a secret number between $[347184000, 599644800)$. Here the best option will be to use the OR-composition as $A + u^{\ell-1} < B$ (we know from the previous example that $u = 57$ and $\ell = 5$). Using parameters from Galbraith, Paterson, and Smart [14], we estimate that the size of $\mathbb{G}_1$ is 256 bits, the size of $\mathbb{G}_T$ is 3072 bits and the size of $\mathbb{Z}_p$ is upper-bounded by 256 bits. This leads to an overall communication load of:

$$Com_l(u = 57, \ell = 5) = \ell \cdot |\mathbb{G}_1| + (2\ell - 2) \cdot |\mathbb{G}_T| + 4\ell \cdot |\mathbb{Z}_p| = 30976 \; bits \quad (4)$$

In order to have a better appreciation of this result, let us compare it to previous protocols:

Scheme	*Communication Complexity*
Our new range proof	30976 bits
Boudot's method	48946 bits
Standard bit-by-bit method	96768 bits
Schoenmaker's method	50176 bits

Fig. 4. Communication load comparison for range proof $[347184000, 599644800)$

Let us also discuss the computational complexities. For the verifier, the figure are about similar to the communication complexities as basically the verifier needs to do some computation with the elements received. For the prover it is about the same with the exception that for Boudot's method where the prover needs to run the Rabin-Shallit algorithms. Experiments show that the later algorithm dominates by far the other operations the prover needs to do.

Now, when one does not want to resort to the (strong) RSA assumption, our methods is the only one that provides an efficient proof except when the interval is only a couple of bits.

5 Alternative Set Membership Proofs

The protocol in the previous section employed a set-membership proof as a building block. The set-membership proof protocol we presented in Section 3 has the verifier to produce signatures on the set elements, send them to the prover and then has the prover to show that he knows a signature (by the verifier) and the element he holds. Concretely, we employed the weak signature scheme by Boneh and Boyen in that section. We now discuss alternative solutions to the set membership protocol, i.e., essentially so that the whole protocol could be based on different assumptions. Due to space restriction we do not give all the details here but only in the full version of this paper. However, the solution presented previously is the most efficient one, the alternatives discussed in this section are of similar efficiency.

5.1 Using Alternative Signature Schemes

The protocol that we presented in Section 3 required the prover to be able to prove the knowledge of a signature on a value that he has committed to, where we used Pedersen commitment scheme. Apart for the weak Boneh-Boyen signature scheme, there are other signature schemes that could be employed. In terms of assumptions, one notable alternative would be the one by Camenisch and Lysyanskaya [7] that is based on the strong RSA assumption. It is not hard to adapt the protocol given in Section 3 to that signature scheme, in particular, as Camenisch and Lysyanskaya give protocols to prove knowledge of a committed value in their paper [7].

5.2 Alternative Protocol Using Cryptographic Accumulators

The reasons why we employed a signature scheme in our set-membership protocol is that the prover needed to show that he committed to a value for which he knows an authenticator without revealing that value or authenticator. Now it turns out that one can achieve exactly the same goal with cryptographic accumulators with similar complexities.

Recall cryptographic accumulators. A cryptographic accumulator is an algorithm that allows one to compress a list of elements into a single accumulator value. For each element there exists a witness attesting to the fact that the element is indeed contained in the accumulator value. For some cryptographic accumulator, there exists efficient proof protocols that allow a prover holding the element and the witness to prove to a verifier in zero knowledge that he indeed is privy to an element that is contained in the accumulator. Camenisch and Lysyanskaya have given an implementation of such an accumulator and a protocol that a committed value is indeed contained in the accumulator based on the strong RSA assumption[6].

Now the idea to build an efficient set-membership proof with dynamic accumulator is very similar to the signature based one: The verifier add each element in the set into the accumulator and sends the accumulator value to the prover together with the witness for each element. The prover then proves to the verifier that the value he has committed to is indeed contained in the accumulator produced by the verifier using the witness obtained for the verifier. This protocol is depicted in Appendix A for the SRSA-based accumulator.

Acknowledgements

The research leading to these results has received funding from the European Community's Seventh Framework Programme (FP7/2007-2013) under grant agreement n° 216483.

References

1. Bangerter, E., Camenisch, J., Maurer, U.M.: Efficient proofs of knowledge of discrete logarithms and representations in groups with hidden order. In: Vaudenay, S. (ed.) PKC 2005. LNCS, vol. 3386, pp. 154–171. Springer, Heidelberg (2005)
2. Bellare, M., Goldreich, O.: On defining proofs of knowledge. In: Brickell, E.F. (ed.) CRYPTO 1992. LNCS, vol. 740, pp. 390–420. Springer, Heidelberg (1993)
3. Black, K.: Classroom note: Putting constraints in optimization for first-year calculus students. SIAM Rev. 39(2), 310–312 (1997)
4. Boneh, D., Boyen, X.: Short signatures without random oracles. In: Cachin, C., Camenisch, J.L. (eds.) EUROCRYPT 2004. LNCS, vol. 3027, pp. 56–73. Springer, Heidelberg (2004)
5. Boudot, F.: Efficient proofs that a committed number lies in an interval. In: Preneel, B. (ed.) EUROCRYPT 2000. LNCS, vol. 1807, pp. 431–444. Springer, Heidelberg (2000)
6. Camenisch, J., Lysyanskaya, A.: Dynamic accumulators and application to efficient revocation of anonymous credentials. In: Yung, M. (ed.) CRYPTO 2002. LNCS, vol. 2442, pp. 61–76. Springer, Heidelberg (2002)
7. Camenisch, J., Lysyanskaya, A.: A signature scheme with efficient protocols. In: Cimato, S., Galdi, C., Persiano, G. (eds.) SCN 2002. LNCS, vol. 2576, pp. 268–289. Springer, Heidelberg (2003)
8. Camenisch, J., Neven, G., shelat, a.: Simulatable adaptive oblivious transfer. In: Naor, M. (ed.) EUROCRYPT 2007. LNCS, vol. 4515, pp. 573–590. Springer, Heidelberg (2007)
9. Camenisch, J., Stadler, M.: Efficient group signature schemes for large groups. In: Kaliski Jr., B.S. (ed.) CRYPTO 1997. LNCS, vol. 1294, pp. 410–424. Springer, Heidelberg (1997)
10. Cheon, J.H.: Security analysis of the strong diffie-hellman problem. In: Vaudenay, S. (ed.) EUROCRYPT 2006. LNCS, vol. 4004, pp. 1–11. Springer, Heidelberg (2006)
11. Cramer, R., Damgård, I., MacKenzie, P.D.: Efficient zero-knowledge proofs of knowledge without intractability assumptions. In: Imai, H., Zheng, Y. (eds.) PKC 2000. LNCS, vol. 1751, pp. 354–373. Springer, Heidelberg (2000)
12. Dodis, Y., Yampolskiy, A.: A verifiable random function with short proofs and keys. In: Public Key Cryptography, pp. 416–431 (2005)
13. Fujisaki, E., Okamoto, T.: Statistical zero knowledge protocols to prove modular polynomial relations. In: Kaliski Jr., B.S. (ed.) CRYPTO 1997. LNCS, vol. 1294, pp. 16–30. Springer, Heidelberg (1997)
14. Galbraith, S.D., Paterson, K.G., Smart, N.P.: Pairings for cryptographers. Cryptology ePrint Archive, Report 2006/165 (2006)
15. Groth, J.: Non-interactive zero-knowledge arguments for voting. In: Ioannidis, J., Keromytis, A.D., Yung, M. (eds.) ACNS 2005. LNCS, vol. 3531, pp. 467–482. Springer, Heidelberg (2005)
16. Lipmaa, H.: On diophantine complexity and statistical zero-knowledge arguments. In: Laih, C.-S. (ed.) ASIACRYPT 2003. LNCS, vol. 2894, pp. 398–415. Springer, Heidelberg (2003)
17. Micali, S., Rabin, M., Kilian, J.: Zero-knowledge sets. In: FOCS 2003: Proceedings of the 44th Annual IEEE Symposium on Foundations of Computer Science, Washington, DC, USA. IEEE Computer Society Press, Los Alamitos (2003)

18. Schoenmakers, B.: Some efficient zeroknowledge proof techniques. In: International Workshop on Cryptographic Protocols, Monte Verita, Switzerland (March 2001)
19. Schoenmakers, B.: Interval proofs revisited. In: International Workshop on Frontiers in Electronic Elections, Milan, Italy (September 2005)
20. Teranishi, I., Sako, K.: K-times anonymous authentication with a constant proving cost. In: Yung, M., Dodis, Y., Kiayias, A., Malkin, T.G. (eds.) PKC 2006. LNCS, vol. 3958, pp. 525–542. Springer, Heidelberg (2006)

A Accumulator Based Membership Proof

A.1 Cryptographic Accumulators and Proofs for Them

Definition 4. *[6] A secure accumulator for a family of inputs $\{\mathcal{X}_k\}$ is a family of families of functions $\mathcal{G} = \{\mathcal{F}_k\}$ with the following properties:*

Efficient generation: *There is an efficient probabilistic algorithm G that on input 1^k produces a random element f of $\mathcal{F}_k$. Moreover, along with f, G also outputs some auxiliary information about f, denoted t_f.*

Efficient evaluation: *$f \in \mathcal{F}_k$ is a polynomial-size circuit that, on input $(u, x) \in \mathcal{U}_f \times \mathcal{X}_k$, outputs a value $v \in \mathcal{U}_f$, where $\mathcal{U}_f$ is an efficiently-samplable input domain for the function f; and $\mathcal{X}_k$ is the intended input domain whose elements are to be accumulated.*

Quasi-commutative: *For all k, for all $f \in \mathcal{F}_k$, for all $u \in \mathcal{U}_f$, for all $x_1, x_2 \in \mathcal{X}_k$, $f(f(u, x_1), x_2) = f(f(u, x_2), x_1)$. If $X = \{x_1, \ldots, x_m\} \subset \mathcal{X}_k$, then by $f(u, X)$ we denote $f(f(\ldots(u, x_1), \ldots), x_m)$.*

Witnesses: *Let $v \in \mathcal{U}_f$ and $x \in \mathcal{X}_k$. A value $w \in U_f$ is called a* witness *for x in v under f if $v = f(w, x)$.*

Security: *Let $\mathcal{U}'_f \times \mathcal{X}'_k$ denote the domains for which the computational procedure for function $f \in \mathcal{F}_k$ is defined (thus $U_f \subseteq \mathcal{U}'_f$, $\mathcal{X}_k \subseteq \mathcal{X}'_k$). For all probabilistic polynomial-time adversaries $\mathcal{A}_k$,*

$$\Pr[f \leftarrow G(1^k); u \leftarrow \mathcal{U}_f; (x, w, X) \leftarrow \mathcal{A}_k(f, U_f, u) : \\ X \subset \mathcal{X}_k; w \in \mathcal{U}'_f; x \in \mathcal{X}'_k; x \notin X; f(w, x) = f(u, X)] = \mathrm{neg}(k)$$

Note that only the legitimate accumulated values, $(x_1, \ldots, x_m)$, must belong to $\mathcal{X}_k$; the forged value x can belong to a possibly larger set $\mathcal{X}'_k$.

Implementation Based on the Strong RSA Assumption. Here we recall the Camenisch-Lysyanskaya accumulator [6].

- $\mathcal{F}_k$ is the family of functions that correspond to exponentiating modulo safe-prime products drawn from the integers of length k. Choosing $f \in \mathcal{F}_k$ amounts to choosing a random modulus $n = pq$ of length k, where $p = 2p' + 1$, $q = 2q' + 1$, and p,p',q,q' are all prime. We will denote f corresponding to modulus n and domain $\mathcal{X}_{A,B}$ by $f_{n,A,B}$. We denote $f_{n,A,B}$ by f_n or by f when it does not cause confusion.
- $\mathcal{X}_{A,B}$ is the $\{e \in \texttt{primes} : e \neq p', q' \;\wedge\; A \leq e \leq B\}$, where A and B can be chosen with arbitrary polynomial dependence on the security parameter k, as long as $2 < A$ and $B < A^2$. $\mathcal{X}'_{A,B}$ is (any subset of) of the set of integer from $[2, A^2 - 1]$ such that $\mathcal{X}_{A,B} \subseteq \mathcal{X}'_{A,B}$.

Common Input: g, h, a commitment C, and a set $\S$
Prover Input: s_j, r such that $C = g^{s_j} h^r$ and $s_j \in \S$.

$P \xleftarrow{\mathfrak{n}, \S_{ew}} V$ Verifier picks a safe prime product $\mathfrak{n} = (2\mathfrak{p}+1)(2\mathfrak{q}+1)$ and
a random quadratic residues $\mathfrak{u}, \mathfrak{g}, \mathfrak{h}$ modulo $\mathfrak{n}$,
picks random $u_i \in \{0,1\}^{k'}$ such that $e_i = s_i 2^k + u_i$ are prime.
computes $\mathfrak{v} \leftarrow \mathfrak{u}^{2\prod e_i} \bmod \mathfrak{n}; \mathfrak{w}_i \leftarrow \mathfrak{v}^{1/e_i} \bmod \mathfrak{n}$,
sends $\mathfrak{n}$, $\mathfrak{u}$, $\mathfrak{v}$, $\mathfrak{g}$, $\mathfrak{h}$, and $\S_{ew} \leftarrow \{(s_1, e_1, \mathfrak{w}_1).....(s_n, e_n, \mathfrak{w}_n)\}$
convinces the prover that $\mathfrak{g} \in \langle \mathfrak{h} \rangle$
(we will discuss the details separately below).

$P \xrightarrow{\mathfrak{W}, \mathfrak{R}, \mathfrak{C}} V$ Prover picks $r_1, r_2, r_3 \in \{0, \ldots, n2^\ell\}$,
where ℓ is a security parameter and
sends $\mathfrak{W} \leftarrow \mathfrak{w}_j \mathfrak{u}^{r_1} \bmod n$, $\mathfrak{R} \leftarrow \mathfrak{g}^{r_1} \mathfrak{h}^{r_2} \bmod n$
and $\mathfrak{C} \leftarrow \mathfrak{g}^{e_j} \mathfrak{h}^{r_3} \bmod n$
Prover and Verifier run

$$\begin{aligned} \mathrm{PK}\{(\alpha, \beta, \gamma, \delta, \epsilon, \rho, \rho_1, \rho_2, \rho_3, \phi, \xi, \nu) :\ & C = g^\sigma h^\rho \ \wedge \\ & \mathfrak{C} = (\mathfrak{g}^{2^k})^\sigma \mathfrak{g}^\mu \mathfrak{h}^{\rho_3} \pmod n \ \wedge\ \mathfrak{R} = \mathfrak{g}^{\rho_1} \mathfrak{h}^{\rho_2} \pmod n \ \wedge \\ & \mathfrak{v} = \mathfrak{W}^\epsilon (\tfrac{1}{\mathfrak{u}})^\delta \pmod n \ \wedge\ 1 = \mathfrak{R}^\epsilon \mathfrak{g}^\delta \mathfrak{h}^\phi \pmod n \\ & \wedge\ \mu \in [-2^{k-1}, 2^{k-1}]\} \end{aligned}$$

Fig. 5. Set membership protocol for set $\S$

- For $f = f_n$, the auxiliary information t_f is the factorization of n.
- For $f = f_n$, $\mathcal{U}_f = \{u \in QR_n : u \neq 1\}$ and $\mathcal{U}'_f = \mathbb{Z}_n^*$.
- For $f = f_n$, $f(u, x) = u^x \bmod n$.
  Note that $f(f(u, x_1), x_2) = f(u, \{x_1, x_2\}) = u^{x_1 x_2} \bmod n$

A.2 Membership Proof with Cryptographic Accumulators

We are now ready to employ the accumulator for the membership proof which can be used as an alternative building block for our range proof presented in Section 4.

One complication that we have to deal with here is that the accumulator allows one to accumulator prime number only whereas our set is arbitrary bits strings. We thus need to encode a mapping. This can be done as follows. Let $\{s_1, \ldots, s_n\}$ be our set, where we assume that the s_i are integers. We let $e_i = s_i 2^k + u_i$ where $u_i < 2^{k'} < 2^k$ is selected so that e_i is prime as k and k' are security parameters (we discuss them below). With this encoding, the verifier can accumulate all the e_i's and send the accumulator value, the e_i, and the corresponding witnesses to the prover. Now the prover has to prove that e_i that corresponds to the s_i in his commitment is contained in the accumulators. The resulting protocol is given in Figure A.1, where we adapt the accumulator proof given by Camenisch and Lysyanskaya [6] to our setting. That is, we have to additionally prove that the correspondence between the e_i and the committed

s_i holds. For this to work, the prover need that show he knows some u_i such that $e_i = s_i 2^k + u_i$ holds. Here it is of course important that this u_i be at most 2^{k-1} bits. This can be enforced efficiently provided that in reality u_i is a couple of bits smaller, i.e., k' bits, where in practice the difference should be about 300 bits for this to work. More precisely, we employ the first range proof discussed in Section 1.2.

Remarks: 1) We need to discuss how the verifier can convince the prover that $\mathfrak{g} \in \langle \mathfrak{h} \rangle$ holds. One way to achieve this, is that the prover runs with the verifier the protocol $PK\{(\alpha) : \mathfrak{g} = \mathfrak{h}^{\alpha} \pmod{\mathfrak{n}}\}$ using binary challenges. Another, more efficient, way is described by Bangerter et al.[1].

2) We note also, that for many applications, the parameters $\mathfrak{n}$, $\mathfrak{u}$, $\mathfrak{v}$, $\mathfrak{g}$, $\mathfrak{h}$, and $\S_{ew} \leftarrow \{(s_1, e_1, \mathfrak{w}_1).....(s_n, e_n, \mathfrak{w}_n)\}$ can be computed and published once and for all (possibly a trusted third party). In this case the computational complexity of our protocols becomes independent of the number of members in the set.

Preimage Attacks on 3, 4, and 5-Pass HAVAL

Yu Sasaki and Kazumaro Aoki

NTT, 3-9-11 Midoricho, Musashino-shi, Tokyo, 180-8585 Japan

Abstract. This paper proposes preimage attacks on hash function HAVAL whose output length is 256 bits. This paper has three main contributions; a preimage attack on 3-pass HAVAL at the complexity of 2^{225}, a preimage attack on 4-pass HAVAL at the complexity of 2^{241}, and a preimage attack on 5-pass HAVAL reduced to 151 steps at the complexity of 2^{241}. Moreover, we optimize the computational order for brute-force attack on full 5-pass HAVAL and its complexity is $2^{254.89}$. As far as we know, the proposed attack on 3-pass HAVAL is the best attack and there is no preimage attack so far on 4-pass and 5-pass HAVAL. Note that the complexity of the previous best attack on 3-pass HAVAL is 2^{230}. Technically, our attacks find pseudo-preimages of HAVAL by combining the meet-in-the-middle and local-collision approaches, then convert pseudo-preimages to a preimage by using a generic algorithm.

Keywords: HAVAL, splice-and-cut, meet-in-the-middle, local collision, hash function, one-way, preimage.

1 Introduction

Cryptographic hash functions are important primitives to build secure schemes. A hash function takes arbitrarily long bit string and outputs a hash value with a fixed length. A hash function is required to satisfy the security properties such as collision resistance, 2nd preimage resistance, and preimage resistance. When the length of the hash value is n bits, a collision, a 2nd preimage, and a preimage should not be computed faster than $2^{n/2}$, 2^n, and 2^n operations, respectively.

HAVAL [18] is one of the dedicated hash functions and has relatively long history. HAVAL is based on Merkle-Damgård construction, and its compression function is similar to MD5 [10]. The basic operation of HAVAL is done in 32 bits that is the same as MD5. Therefore, 32-bit values are called *words*. However, the interface of the HAVAL compression function is doubled compared to MD5, that is, the number of chaining variables and the message length of the compression function are 8 words and 32 words respectively. The nonlinear function of HAVAL takes seven words as input and outputs a word. So, one step of HAVAL only changes one word out of 8 words of the internal state. To satisfy several security requirements, HAVAL has three variants called x-pass HAVAL ($x = 3, 4, 5$). x-pass HAVAL consists of $32x$ steps.

Due to the simple structure, there are several cryptoanalytic results on HAVAL as shown in the next paragraph. However, regarding the preimage attack, there is only one result on 3-pass HAVAL [2]. In this paper, we propose preimage attacks

J. Pieprzyk (Ed.): ASIACRYPT 2008, LNCS 5350, pp. 253–271, 2008.
© International Association for Cryptologic Research 2008

Table 1. Comparison of preimage attacks on HAVAL

Attack target	Number of steps	Attack type	Previous work [2]	**Our attack**	
				strategy 1*	strategy 2
3-pass	96 (Full)	Pseudo-preimage	2^{224}	2^{253}	2^{192}
		Preimage	2^{230}		2^{225}
4-pass	128 (Full)	Pseudo-preimage	-	$2^{254.43}$	2^{224}
		Preimage	-		2^{241}
5-pass	151 (Steps 0-150)	Pseudo-preimage	-	Not evaluated	2^{224}
		Preimage	-		2^{241}
	160 (Full)	Pseudo-preimage	-	$2^{254.89}$	$2^{253.81}$*
		Preimage	-		—

* This attack is a kind of brute force attack, but the computation is optimized.

on HAVAL: the best attack on 3-pass HAVAL so far, the first attack on 4-pass HAVAL and 5-pass HAVAL.

Known previous results except for the preimage attack are as follows: collision attacks on 3-pass HAVAL are discussed in Ref. [12,11,13,14], and those on 4-pass HAVAL are discussed in Ref. [15,17]. Note that a real collision has been generated up to 4-pass HAVAL. Theoretically, a collision of 5-pass HAVAL can be generated in 2^{123} compression function evaluations [17] that is faster than the birthday paradox for 256-bit output. (Hereafter, we omit the unit of complexity whenever it is obvious and it is the number of "compression function evaluation.") Non-randomness of 4-pass and 5-pass of HAVAL in the encryption mode is analyzed by Ref. [6,16]. The security of the HMAC-HAVAL is analyzed by Ref. [5]. A 2nd preimage attack on 3-pass HAVAL and its application to HMAC-3-pass HAVAL are proposed by Ref. [7]. However, this 2nd preimage attack is different from the one usually considered. In Ref. [8] a useful statement to clarify the difference of these two types of 2nd preimage attacks is shown. The attack of Ref. [7] can generate a 2nd preimage at the complexity of one compression function with a probability of 2^{-114} for a given random message, and it requires the complexity of 2^{128} with a probability of $1-2^{-114}$. Therefore, the average complexity is very close to 2^{128}. Consequently, no result that produces a 2nd preimage of any given message is known. Moreover, no result is known on preimage attack on HAVAL, except for the recent result on 3-pass HAVAL [2].

1.1 Related Work Regarding Preimage Attack

In 2008, a preimage attack on MD4 was proposed by Leurent [8]. The attack first generates pseudo-preimages based on the Dobbertin's pioneering work [4], and converts a pseudo-preimage attack to a preimage attack by using the generic approach [9, Fact9.99][1]. Preimage attacks on step-reduced MD5 and full 3-pass

[1] The following works that compute a preimage from partial-pseudo-preimages also use this kind of conversion. The method of the conversion from partial-pseudo-preimages to a preimage is improved by using hash-tree [8] and P^3graph [3].

HAVAL are proposed by Aumasson et al. [2], whose approach is based on the meet-in-the-middle technique. Preimage attacks on full MD4 and 63-step MD5 are proposed by [1], whose approach is also based on the meet-in-the-middle technique. Note, both of [2,1] use the conversion algorithm of [9, Fact9.99].

In the meet-in-the-middle attack of Aumasson et al. [2], a compression function is divided into the first half and the last half, and both computation results are compared in the middle. They also use new techniques that make the attack efficient by using the absorption properties of Boolean functions. On the other hand, Aoki and Sasaki propose new techniques to apply the meet-in-the-middle attack to not only the first half and the last half but also any two consecutive parts of a compression function [1]. This paper combines the techniques of Ref. [1,2], and attacks more passes of HAVAL.

1.2 Our Contributions

In this paper, we propose preimage attacks on 3-, 4-, and 5-pass HAVAL whose output length is 256 bits. First, we consider a strategy to find preimages of 3-, 4-, and 5-pass HAVAL faster than the brute force attack by a few bits (strategy 1). Second, we consider another strategy that can find a preimages of 3-, 4-, and 5-pass HAVAL much more efficiently by combining techniques of [1] and [2] (strategy 2). As a result of applying strategy 2 to each pass of HAVAL, we find the best preimage attack so far on 3-pass HAVAL by using the techniques of [1], the first preimage attack on 4-pass HAVAL by combining techniques of [1] and [2], and the first preimage attack on step-reduced 5-pass HAVAL by combining techniques of [1] and [2] and further improving a technique of [2]. We summarize the results of the previous work and ours in Table 1.

Organization of this paper is as follows. Section 2 introduces the specification of HAVAL and techniques of existing attacks. Section 3 gives two strategies of the preimage attack that can be applied to HAVAL and *other hash functions* whose message expansion is similar to HAVAL. Regarding the technique in Ref. [2] as an application of a local collision, we can compute preimages of a hash function that has more rounds. Section 4 describes attacks on HAVAL following the strategy 1. Section 5 describes attacks on HAVAL following the strategy 2. Finally, we conclude this paper in Section 6.

2 Previous Works: Specification and Techniques for Preimage Attacks

2.1 Description of HAVAL

HAVAL is a hash function proposed by Zheng et al. in 1992, which compresses a message up to $(2^{64}-1)$ bits into either 128, 160, 192, 224, or 256 bits. Since this paper only analyzes 256-bit version, we only describe the specification for 256 bits. HAVAL iteratively computes a step function to compute a hash value. The number of steps is chosen from either 96, 128, or 160, where corresponding HAVAL algorithms are called 3-pass HAVAL, 4-pass HAVAL, and 5-pass

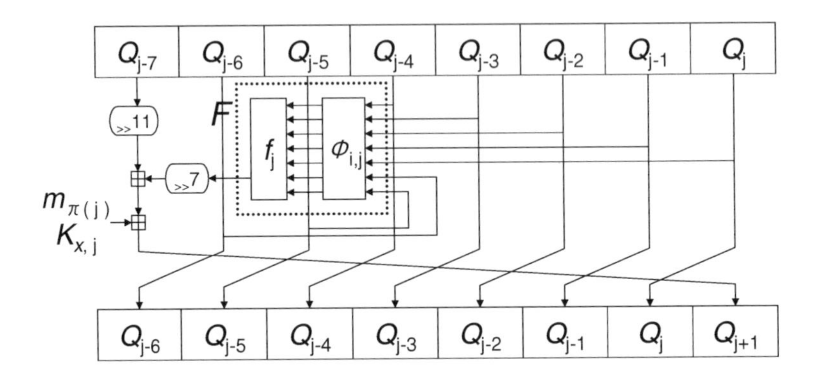

Fig. 1. Step function of HAVAL

Table 2. HAVAL message expansion

0	1	2	3	4	5	6	7	8	9	10	11	12	13	14	15	16	17	18	19	20	21	22	23	24	25	26	27	28	29	30	31
5	14	26	18	11	28	7	16	0	23	20	22	1	10	4	8	30	3	21	9	17	24	29	6	19	12	15	13	2	25	31	27
19	9	4	20	28	17	8	22	29	14	25	12	24	30	16	26	31	15	7	3	1	0	18	27	13	6	21	10	23	11	5	2
24	4	0	14	2	7	28	23	26	6	30	20	18	25	19	3	22	11	31	21	8	27	12	9	1	29	5	15	17	10	16	13
27	3	21	26	17	11	20	29	19	0	12	7	13	8	31	10	5	9	14	30	18	6	28	24	2	23	16	22	4	1	25	15

HAVAL, respectively. HAVAL has the Merkle-Damgård structure, which uses 256-bit (8-word) chaining variables and a 1024-bit (32-word) message block to compute a compression function.

An input message M is processed to be a multiple of 1024 bits by the padding procedure. A single bit '1' is appended followed by '0's until the length becomes 944 modulo 1024. At the last, 3-bit field representing a version number of HAVAL, 3-bit field representing the number of the pass used, 10-bit field representing the output length, and 64-bit field representing an unpadded message length are appended.

Padded message M^* is separated into 1024-bit message blocks ($M_0, M_1, \ldots, M_{n-1}$). Let $CF : \{0,1\}^{256} \times \{0,1\}^{1024} \rightarrow \{0,1\}^{256}$ be the compression function of HAVAL. A hash value is computed as follows.

1. $H_0 \leftarrow IV$,
2. $H_{i+1} \leftarrow CF(H_i, M_i)$ for $i = 0, 1, \ldots, n-1$,

where H_i is a 256-bit value and IV is the initial value defined in the specification. Finally, H_n is output as a hash value of M.

Compression Function. The compression function $H_{i+1} \leftarrow CF(H_i, M_i)$ is computed as follows.

1. M_i is divided into 32-bit message words m_j ($j = 0, 1, \ldots, 31$).
2. $p_0 \leftarrow H_i$.

$$
\begin{aligned}
0 \leq j \leq 31 : f_j(x_6, x_5, \ldots, x_0) &= x_1x_4 \oplus x_2x_5 \oplus x_3x_6 \oplus x_0x_1 \oplus x_0 \\
32 \leq j \leq 63 : f_j(x_6, x_5, \ldots, x_0) &= x_1x_2x_3 \oplus x_2x_4x_5 \oplus x_1x_2 \oplus x_1x_4 \oplus \\
&\quad x_2x_6 \oplus x_3x_5 \oplus x_4x_5 \oplus x_0x_2 \oplus x_0 \\
64 \leq j \leq 95 : f_j(x_6, x_5, \ldots, x_0) &= x_1x_2x_3 \oplus x_1x_4 \oplus x_2x_5 \oplus x_3x_6 \oplus x_0x_3 \oplus x_0 \\
96 \leq j \leq 127 : f_j(x_6, x_5, \ldots, x_0) &= x_1x_2x_3 \oplus x_2x_4x_5 \oplus x_3x_4x_6 \oplus \\
&\quad x_1x_4 \oplus x_2x_6 \oplus x_3x_4 \oplus x_3x_5 \oplus \\
&\quad x_3x_6 \oplus x_4x_5 \oplus x_4x_6 \oplus x_0x_4 \oplus x_0 \\
128 \leq j \leq 159 : f_j(x_6, x_5, \ldots, x_0) &= x_1x_4 \oplus x_2x_5 \oplus x_3x_6 \oplus x_0x_1x_2x_3 \oplus x_0x_5 \oplus x_0
\end{aligned}
$$

$x_a x_b$ represents bitwise AND operation.

Fig. 2. Boolean Functions of HAVAL

Table 3. Wordwise rotation of HAVAL

	$x_6\ x_5\ x_4\ x_3\ x_2\ x_1\ x_0$ $\downarrow\ \downarrow\ \downarrow\ \downarrow\ \downarrow\ \downarrow\ \downarrow$		$x_6\ x_5\ x_4\ x_3\ x_2\ x_1\ x_0$ $\downarrow\ \downarrow\ \downarrow\ \downarrow\ \downarrow\ \downarrow\ \downarrow$		$x_6\ x_5\ x_4\ x_3\ x_2\ x_1\ x_0$ $\downarrow\ \downarrow\ \downarrow\ \downarrow\ \downarrow\ \downarrow\ \downarrow$
$\phi_{3,1}$	$x_1\ x_0\ x_3\ x_5\ x_6\ x_2\ x_4$	$\phi_{4,1}$	$x_2\ x_6\ x_1\ x_4\ x_5\ x_3\ x_0$	$\phi_{5,1}$	$x_3\ x_4\ x_1\ x_0\ x_5\ x_2\ x_6$
$\phi_{3,2}$	$x_4\ x_2\ x_1\ x_0\ x_5\ x_3\ x_6$	$\phi_{4,2}$	$x_3\ x_5\ x_2\ x_0\ x_1\ x_6\ x_4$	$\phi_{5,2}$	$x_6\ x_2\ x_1\ x_0\ x_3\ x_4\ x_5$
$\phi_{3,3}$	$x_6\ x_1\ x_2\ x_3\ x_4\ x_5\ x_0$	$\phi_{4,3}$	$x_1\ x_4\ x_3\ x_6\ x_0\ x_2\ x_5$	$\phi_{5,3}$	$x_2\ x_6\ x_0\ x_4\ x_3\ x_1\ x_5$
-	-	$\phi_{4,4}$	$x_6\ x_4\ x_0\ x_5\ x_2\ x_1\ x_3$	$\phi_{5,4}$	$x_1\ x_5\ x_3\ x_2\ x_0\ x_4\ x_6$
-	-	-	-	$\phi_{5,5}$	$x_2\ x_5\ x_0\ x_6\ x_4\ x_3\ x_1$

3. $p_{j+1} \leftarrow R_j(p_j, m_{\pi(j)})$ for $j = 0, 1, \ldots, k$, where $k = 32x - 1$ for x-pass.
4. Output $H_{i+1}(= p_k + H_i)$, where "+" denotes a 32-bit word-wize addition. In this paper, we similarly use "−" to denote a 32-bit word-wize subtraction.

R_j is the step function for Step j. Let Q_j be a 32-bit value that satisfies $p_j = (Q_{j-7} \| Q_{j-6} \| Q_{j-5} \| Q_{j-4} \| Q_{j-3} \| Q_{j-2} \| Q_{j-1} \| Q_j)$. R_j for x-pass HAVAL ($x \in \{3, 4, 5\}$) is defined as follows:

$$
\begin{cases}
T = f_j \circ \phi_{x,j}(Q_{j-6}, Q_{j-5}, Q_{j-4}, Q_{j-3}, Q_{j-2}, Q_{j-1}, Q_j) \\
R_j(p_j, m_{\pi(j)}) = (Q_{j-7} \ggg 11) + (T \ggg 7) + m_{\pi(j)} + K_{x,j}
\end{cases}
$$

where f_j is a bitwise Boolean function defined in Fig. 2, $\phi_{i,j}$ is a word-wize permutation defined in Table 3, π_j is a message expansion function defined in Table 2, $\ggg n$ is n-bit right rotation, and $K_{x,j}$ is a constant defined in the specification. We show a graph of the step function in Fig. 1. Note that $R_j^{-1}(\cdot, m_{\pi(j)})$ can be computed in almost the same complexity as that of R_j.

2.2 Converting Pseudo-preimages to a Preimage

For a given hash value y, a pseudo-preimage is a pair of (x, M) such that $CF(x, M) = y$, where x may not equal to IV and CF is a compression function of a Merkle-Damgård hash function. There is a generic algorithm that converts a pseudo-preimage attack to a preimage attack [9, Fact9.99]. Let the complexity of a pseudo-preimage attack be 2^k. The procedure of this attack when the hash value is n-bit long is as follows.

1. Generate $2^{(n-k)/2}$ pseudo-preimages at the complexity of $2^k \cdot 2^{(n-k)/2}$.
2. Generate $2^{(n+k)/2}$ messages that start from the IV, and compute their hash values.

One of these hash values is expected to be matched. The complexity of this attack is $2^k \cdot 2^{(n-k)/2} + 2^{(n+k)/2} = 2^{1+(n+k)/2}$.

This algorithm has been used in previous preimage attacks [8,2,1].

2.3 Preimage Attacks on 3-Pass HAVAL

Aumasson et al. proposed two attacks that find a preimage of 3-pass HAVAL at the complexity of 2^{230}, and the attacks require 16×2^{64} words of memory [2]. Both attacks take an approach of the meet-in-the-middle attack. In this paper, we are particularly interested in the Attack A of their paper [2, Algorithm 4].

In the Attack A of [2, Algorithm 4], the authors focused attention on the location of the message words m_5 and m_6, where m_5 appears at Step 5, 32, and 94 and m_6 appears at Step 6, 55, and 89 as shown in Table 2[2]. First, chaining variables p_0 to p_6, where p_0 is IV and p_{i+1} is the 256-bit output of the i-th step, are fixed so that the change of m_6 in Step 6 is guaranteed to be absorbed by changing Q_{-1}, which is the seventh word of the IV. Similarly, chaining variables p_{95} and p_{96} are fixed so that the change of m_5 in Step 94 is guaranteed to be absorbed by changing Q_{95}, which is the seventh word of p_{96}. Due to this effort, computation for Step 0 to 47 becomes independent of m_6, and computation for Step 95 to 48 becomes independent of m_5. The authors of [2] and we call these independent words *neutral words*.

Finally, the authors apply the meet-in-the-middle attack to find a pseudo-preimage of a given hash value $H_n = (H^a \| H^b \| H^c \| H^d \| H^e \| H^f \| H^g \| H^h)$. The rough sketch of the procedure is as follows. Refer to [2] for details.

Algorithm

1. Fix $m_x, x \notin \{5, 6\}$ and $p_y, y \in \{0, \ldots, 6, 95, 96\}$ so that changes of m_6 in Step 6 and of m_5 in Step 94 are absorbed and $p_0 + p_{96} = H_n$ is satisfied except for $Q_{-1} + Q_{95} = H^g$.
2. For all 64 bits of (m_5, Q_{-1}), compute $p_{j+1} \leftarrow R_j(p_j, m_{\pi(j)})$ for $j = 0, 1, \ldots, 47$, and store them in a table.
3. For all 64 bits of (m_6, Q_{95}), compute $p_j \leftarrow R_j^{-1}(p_{j+1}, m_{\pi(j)})$ for $j = 95, 94, \ldots, 48$. Then, check if resulting p_{48} are matched with p_{48}s in the table.
4. For all matched pairs, check if $Q_{-1} + Q_{95} = H^g$ is satisfied.

In the above procedure, the meet-in-the-middle attack saves the complexity of 64 bits but step 4 of the procedure succeeds with a probability of 2^{-32}. Thus, this attack is faster than the brute force attack by the factor of 2^{32}.

[2] We number the first step as 0.

2.4 Preimage Attacks on MD4 and MD5

Preimage attacks on MD4 and MD5 are proposed by Aoki and Sasaki [1]. They proposed a new technique called the splice-and-cut technique.
Splice-and-Cut: Splice the last and the first step and divide the attack target into two chunks of steps so that each chunk includes at least one message word that is independent of the other chunk. Then, pseudo-preimages are computed by the meet-in-the-middle approach.

Different from Aumasson et al. [2], Aoki and Sasaki focused attention on the property that chaining variables in the first and last steps can be considered to be consecutive by the equation $p_0 = H_n - p_{\text{last}}$. This idea enables them to start the meet-in-the-middle attack from *any* step.

Aoki and Sasaki also proposes another technique named *partial matching*. This technique enables attackers to skip several steps when they search for good chunks in the attack target. Assume that one of divided chunks provides the value of p_i, where $p_i = (Q_{i-7}\|Q_{i-6}\|Q_{i-5}\|Q_{i-4}\|Q_{i-3}\|Q_{i-2}\|Q_{i-1}\|Q_i)$ and the other chunk provides the value of p_{i-4}, where $p_{i-4} = (Q_{i-11}\|Q_{i-10}\|Q_{i-9}\|Q_{i-8}\|Q_{i-7}\|Q_{i-6}\|Q_{i-5}\|Q_{i-4})$. p_i and p_{i-4} cannot be directly compared, however, a part of these values, that is, $Q_{i-7}, Q_{i-6}, Q_{i-5}$, and Q_{i-4} can be compared immediately. In such a case, we can ignore the value of $m_{\pi(i-1)}, m_{\pi(i-2)}, m_{\pi(i-3)}$, and $m_{\pi(i-4)}$ when we perform the meet-in-the-middle attack.

3 General Strategies of Our Preimage Attack

3.1 Strategy 1: Speed Up the Brute-Force Attack

This is a technique that enables us to quickly search for a message which connects a given initial value IV and a given hash value H_n. The idea is to reuse an intermediate value of computation of a message when we compute different messages. Assume m_a and m_b form a local collision in the first round, that is, any change of m_a can be offset by changing m_b, and these messages appear at Steps $s_1, s_2, (s_1 < s_2)$ in the second round. In this case, the computation result until Step s_1 can be reused with all m_a and corresponding m_b.

Moreover, since IV and H_n are fixed, the values of chaining variables in the last round can also be reused. Let steps at which m_a and m_b are used be $s_3, s_4, (s_3 < s_4)$. In this case, the computation result from Step s_4 to the last can be reused.

Notice, this technique can also be achieved by inserting local collision in the last round.

3.2 Strategy 2: Finding Pseudo-preimages by the Meet-in-the-Middle Attack

Combining the splice-and-cut and local-collision. The technique proposed by Aumasson et al. [2] is for finding a pseudo-preimage by applying the meet-in-the-middle attack that starts from the first step and the last step. On the other

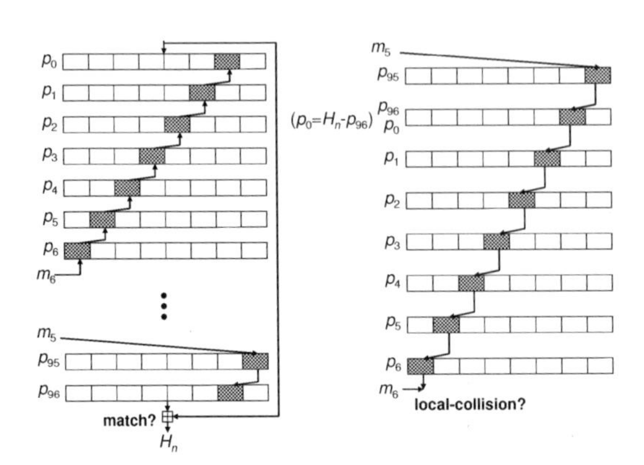

Fig. 3. A local collision formed by the neutral words used by Aumasson et al. [2]

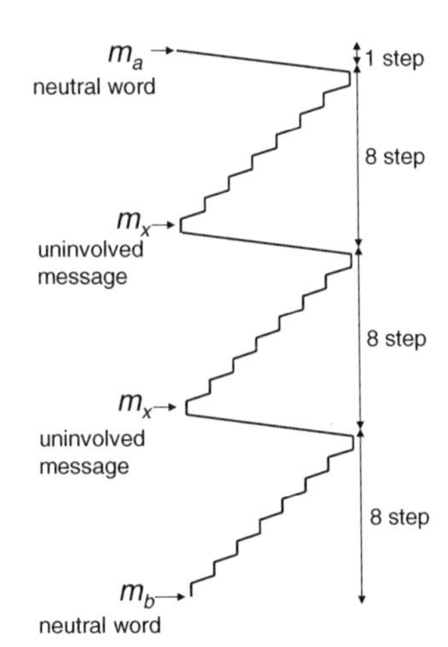

Fig. 4. A long collision pass used in the splice-and-cut technique

hand, the splice-and-cut and partial-matching techniques proposed by Aoki and Sasaki [1] are for finding a pseudo-preimage by applying the meet-in-the-middle attack that starts from an intermediate step. We found that these two techniques can be combined together, and more steps might be attacked.

Aumasson et al. use the fact that m_6 is used near the first step, m_5 is used near the last step, and corresponding chaining variables appear in the same equation for the computing hash value. We found that their technique can be used at not only the first and last several steps but also intermediate steps.

Observation: The key idea of the attack is searching for message words that can form a local collision. In fact, their selection of message words can be considered as a local collision that starts with Step 94 and ends with Step 6.

The graphical explanation is shown in Fig. 3. Cells denote 32-bit chaining variables and highlighted cells denote chaining variables whose values are changed according to the selection of values of neutral words (m_5, m_6). The left diagram explains the attack procedure of Aumasson et al., and the right diagram describes it in a different step order to show (m_5, m_6) forms a local collision. Note, in the splice-and-cut technique, the first and last steps are considered to be consecutive by the equation $p_0 = H_n - p_{96}$, which can be ignored when we analyze the dependency of message words.

As you can see in Fig. 3, the technique of Aumasson et al. [2] can be inserted in any part of an attack target. Therefore, this can be combined with the splice-and-cut technique. For convenience, we call this technique *local-collision technique*, and we summarize the property of the local-collision technique.

New technique 1. Local-Collision: When we search for chunks in an attack target, neutral words forming a local collision can be ignored. This occurs when neutral words appear $(L+1)$ steps away and other chaining variables can be guaranteed

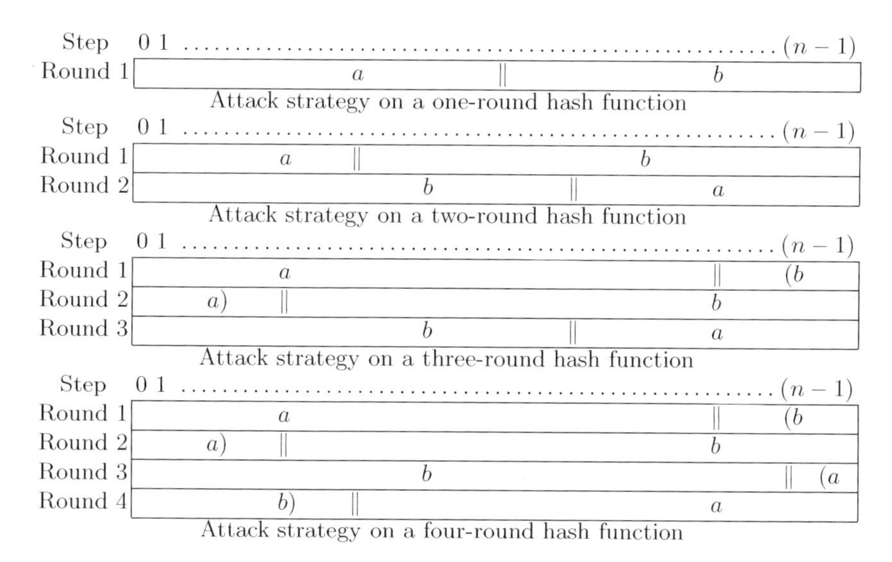

Fig. 5. Attack strategies on a hash function with up to 4 rounds

not to be affected by the local collision, where L represents the number of chaining variables (e.g. $L = 4$ for MD5*, $L = 8$ for* HAVAL*).*

Extension to use long collision paths. The local-collision technique described above can be extended to use a long collision path as shown in Fig. 4.

In HAVAL, the influence of changing $m_{\pi(i)}$ can be offset by changing $m_{\pi(i+8n)}$, $n \geq 1$. In this case, $m_{\pi(i+8k)}$, $1 \leq k < n$ can be any message word. We call $m_{\pi(i+8k)}$ *uninvolved messages*. As long as the meet-in-the-middle attack with a local collision such as the attack approach of Aumasson et al. is taken, neutral words can also be used as uninvolved messages. On the other hand, in our approach explained in Section 5.3, we use "meet-in-the-middle attack" which uses two tables but does not get the gain of the time-to-memory conversion. Thus, neutral words require to increase the complexity of about $n/$(number of all steps), since we need to fix all variables within local collision steps before we perform the "meet-in-the-middle attack". We also note that the changes of a 32-bit chaining variable corresponding to neutral words must be absorbed in the Boolean functions so that other chaining variables are not changed. Achieving this tends to be difficult if several message words appear twice or messages used as padding string appear in a local collision path.

Number of rounds that can be attacked. The meet-in-the-middle attack works very efficiently if the message expansion consists of a permutation of message word order in each round like MD5 or HAVAL. In this section, we formalize how many rounds can be attacked. Attack strategies are also drawn in Fig. 5.

We explain how to attack a hash function that has only one-round. Let us divide the attack target into the first half and the last half steps. In a round, each message appears only once. Therefore, any pair of message words used in the first and second chunks are independent each other, hence they can be used as the neutral words. Finally, we perform the meet-in-the-middle attack between the first chunk including m_a and the second chunk including m_b.

To attack a two-round hash function, we use the property that chaining variables in the first and last steps can be considered to be consecutive. Let a pair of message words (m_a, m_b) appear in the first round in this order. In the second round, if m_b is used in an earlier step than m_a, the attack target can be divided into two chunks so that one chunk includes a neutral word m_a and the other chunk includes m_b. Therefore, a pseudo-preimage attack can be achieved by the splice-and-cut technique.

A three-round hash function can be attacked by combining the splice-and-cut technique and one of the partial-matching or local collision techniques. Assume (m_a, m_b) is a pair of message words that can be skipped by using the partial-matching or local-collision technique. In Fig. 5, skipped steps are indicated by parentheses. If the same strategy for the two-round attack can be applied in the rest of steps, a pseudo-preimage attack can be achieved.

To attack a four-round hash function, we need to use all techniques. At the beginning of two chunks, we skip several steps by the local-collision technique, and at the end of two chunks, we skip several steps by the partial-matching technique. Both skipped steps need to include both neutral words.

4 Preimage Attacks on HAVAL Following the Strategy 1

We apply the general strategy 1 explained in Section 3 to all passes of HAVAL. The memory requirement of the attack is negligible.

First, we consider a preimage attack on 3-pass HAVAL. According to the message expansion of HAVAL shown in Table 2, if we make a local collision from Steps 9 to 17, computation results for 77 steps out of 96 steps can be reused among different messages. The message word distribution for this attack is shown in Table 4.

Table 4. Message word distribution for fast brute-force attack on 3-pass HAVAL

Step	0	1	2	3	4	5	6	7	8	9	10	11	12	13	14	15	16	17	18	19	20	21	···	29	30	31
index	0	1	2	3	4	5	6	7	8	⑨	10	11	12	13	14	15	16	⑰	18	19	20	21	···	29	30	31
	reused									local collision									reused							
Step	32	33	34	35	36	37	38	39	40	41	42	43	44	45	46	47	48	49	50	51	52	53	···	61	62	63
index	5	14	26	18	11	28	7	16	0	23	20	22	1	10	4	8	30	3	21	⑨	⑰	24	···	25	31	27
	reused																									
Step	64	65	66	67	68	69	70	71	72	73	74	75	76	77	78	79	80	81	82	83	84	85	···	93	94	95
index	19	⑨	4	20	28	⑰	8	22	29	14	25	12	24	30	16	26	31	15	7	3	1	0	···	11	5	2
							reused																			

The attack procedure is as follows:

Attack procedure

1. Fix m_{29}, m_{30}, and m_{31} to satisfy the padding for a 1-block message.
2. Temporarily determine m_9 and m_{17}, and determine chaining variables and messages $m_i, i \notin \{9, 17, 29, 30, 31\}$ so that Steps 9-17 form a local collision[3].
3. Randomly determine other message words that are not specified yet.
4. Compute $R_j(p_j, m_{\pi(j)})$ for $j = 0, 1, \ldots, 50$ and compute $R_j^{-1}(p_{j+1}, m_{\pi(j)})$ for $j = 95, 94, \ldots, 70$, where $p_{96} = H_n - IV$. Store the values of p_{51} and p_{70} in a table, where $p_{70} = (Q_{63} \| Q_{64} \| Q_{65} \| Q_{66} \| Q_{67} \| Q_{68} \| Q_{69} \| Q_{70})$.
5. For all 32 bits of m_9, compute m_{17} so that the value of Q_{18} does not change. Then, compute $R_j(p_j, m_{\pi(j)})$ for $j = 51, 52, \ldots, 62$ and check whether computed Q_{63} is in the table or not. If it is in the table, compute Q_{63}, ..., Q_{70} and check all values are matched. Otherwise, choose other m_9 and repeat this process.

The complexity of the above procedure is $2^{29}(= 2^{32} \cdot \frac{12}{96})$ and success probability of step 5 is $2^{-224}(= 2^{-256} \cdot 2^{32})$. Therefore, by repeating the procedure 2^{224} times by changing the values of $m_i, 18 \le i \le 28$, a message that connects a given IV and H_n will be found at the complexity of $2^{253}(= 2^{29} \cdot 2^{224})$.

On 4-pass HAVAL, the attack procedure is similar to 3-pass HAVAL. Applying local collision in the last round between Steps 102-110, the complexity of the attack is $2^{256} \cdot \frac{128-(19+59+7)}{128} \approx 2^{254.43}$. On 5-pass HAVAL, applying local collision in the first round between Steps 19-27, the complexity of the attack is $2^{256} \cdot \frac{160-(56+23+7)}{160} \approx 2^{254.89}$.

5 Preimage Attacks on HAVAL Following the Strategy 2

Our general strategy 1 can work for all passes of HAVAL, however, the efficiency is not so high. This section further reduces the complexity of preimage attacks by using the general strategy 2, which uses the meet-in-the-middle approach.

5.1 A Preimage Attack on 3-Pass HAVAL

We propose a preimage attack on 3-pass HAVAL, which finds a pseudo-preimage of 3-pass HAVAL at the complexity of 2^{192}, and is converted to a preimage attack of the complexity of 2^{225}. Thus, the resulting preimage is 2-block long. This attack uses the splice-and-cut and partial-matching techniques as shown in Table 5.

The attack procedure for a hash value $H_n = (H^a \| H^b \| H^c \| H^d \| H^e \| H^f \| H^g \| H^h)$ is as follows.

[3] How to determine the chaining variables and messages to obtain a local collision is explained in Section 5.2. A local collision for this attack can be obtained in the similar method.

Table 5. Message word distribution for 3-pass HAVAL

Step	0	1	2	3	4	5	6	7	8	9	10	11	12	13	···	21	22	23	24	25	26	27	28	29	30	31
index	⓪	①	2	3	4	⑤	6	7	8	9	10	⑪	12	13	···	21	22	23	24	25	26	27	28	29	30	31
	skip		first chunk																							
Step	32	33	34	35	36	37	38	39	40	41	42	43	44	45	···	53	54	55	56	57	58	59	60	61	62	63
index	⑤	14	26	18	⑪	28	7	16	⓪	23	20	22	①	10	···	24	29	6	19	12	15	13	2	25	31	27
	first chunk								second chunk																	
Step	64	65	66	67	68	69	70	71	72	73	74	75	···	83	84	85	86	87	88	89	90	91	92	93	94	95
index	19	9	4	20	28	17	8	22	29	14	25	12	···	3	①	⓪	18	27	13	6	21	10	23	⑪	⑤	2
	second chunk																							skip		

Attack procedure

1. Fix m_{29}, m_{30}, and m_{31} to satisfy the padding for a 2-block message.
2. Fix m_i ($i \notin \{0, 1, 5, 11\}$) and p_{40} to randomly chosen values.
3. For all (m_0, m_1), do: $p_{j+1} \leftarrow R_j(p_j, m_{\pi(j)})$ for $j = 40, 41, \ldots, 92$.
4. Make a table of $(m_0, m_1, p_{93}, (H^e - Q_{93}, H^d - Q_{92}, H^c - Q_{91}))$s which are computed in the last step, where $p_{93} = (Q_{86} \| Q_{87} \| Q_{88} \| Q_{89} \| Q_{90} \| Q_{91} \| Q_{92} \| Q_{93})$.
5. For all (m_5, m_{11}),
   (a) do the following: $p_j \leftarrow R_j^{-1}(p_{j+1}, m_{\pi(j)})$ for $j = 39, 38, \ldots, 2$, where, $p_2 = (Q_{-5} \| Q_{-4} \| Q_{-3} \| Q_{-2} \| Q_{-1} \| Q_0 \| Q_1 \| Q_2)$.
   (b) Check whether Q_{-5}, Q_{-4}, and Q_{-3} are matched with $H^c - Q_{91}, H^d - Q_{92}$, and $H^e - Q_{93}$ in the table.
   (c) If they are matched, compute $p_{94}, p_{95}, p_{96}, p_0$, and p_1 by using the matched pairs, and check whether $H_n = p_0 + p_{96}$ are satisfied.
   (d) If satisfied, the pair of corresponding message and p_0 is a pseudo-preimage of H_n.

In the above procedure, the complexity of step 3 is $2^{64} \cdot \frac{53}{96}$ and the complexity of step 5a is $2^{64} \cdot \frac{38}{96}$. After step 5b, $2^{32} (= 2^{128} \cdot 2^{-96})$ pairs are expected to be remained. After step 5c, $2^{-128} (= 2^{-160} \cdot 2^{32})$ pair are expected to be remained. Therefore, by repeating the above procedure 2^{128} times, we expect to obtain a pseudo-preimage, where the complexity is $2^{192} (= 2^{64} \cdot 2^{128})$. Finally, this pseudo-primage attack is converted to a preimage attack of the complexity of 2^{225} by the generic approach explained in Section 2.2[4]. Step 4 requires 13×2^{64} words of memory and other steps require negligible amount of memory.

5.2 A Preimage Attack on 4-Pass HAVAL

We propose a preimage attack on 4-pass HAVAL, which finds a pseudo-preimage of 4-pass HAVAL at the complexity of 2^{224}, and is converted to a preimage attack of the complexity of 2^{241}. Thus, the resulting preimage is 2-block long. This

[4] Combination of the attack proposed by Aumasson et al. described in Section 2.3 and P^3graph proposed in [3] will be the preimage attack with a complexity of 2^{225}. Moreover, following [1, Appendix], the complexity is further improved to 2^{224}, but the length of the preimage message will be very long.

Table 6. Message word distribution for 4-pass HAVAL

Step	0	1	2	3	4	5	6	7	···	20	21	22	23	24	25	26	27	28	29	30	31
index	0	1	2	3	4	⑤	6	7	···	20	21	22	23	㉔	25	26	27	28	29	30	31
	second chunk													local collision (1-cycle)							

Step	32	33	34	···	48	49	50	51	52	53	54	55	56	57	58	59	60	61	62	63
index	⑤	14	26	···	30	3	21	9	17	㉔	29	6	19	12	15	13	2	25	31	27
		first chunk																		

Step	64	65	66	67	68	69	70	71	72	73	74	75	76	77	···	90	91	92	93	94	95
index	19	9	4	20	28	17	8	22	29	14	25	12	㉔	30	···	21	10	23	11	⑤	2
	first chunk																			skip	

Step	96	97	98	···	112	113	114	115	116	117	118	119	120	121	122	123	124	125	126	127
index	㉔	4	0	···	22	11	31	21	8	27	12	9	1	29	⑤	15	17	10	16	13
		second chunk																		

Table 7. Fixed values for preimage attack on 4-pass HAVAL

step j	$m_{\pi(j)}$	Q_{j-7}	Q_{j-6}	Q_{j-5}	Q_{j-4}	Q_{j-3}	Q_{j-2}	Q_{j-1}	Q_j
24	$\textcircled{m}_{24}$	$\textcircled{Q}_{17}$	$\mathbf{C}_1$	$\mathbf{C}_2$	$\mathbf{C}_3$	$\mathbf{C}_4$	**1**	**0**	**0**
25	m_{25}	$\mathbf{C}_1$	$\mathbf{C}_2$	$\mathbf{C}_3$	$\mathbf{C}_4$	**1**	**0**	**0**	$*$
26	m_{26}	$\mathbf{C}_2$	$\mathbf{C}_3$	$\mathbf{C}_4$	**1**	**0**	**0**	$*$	**0**
27	m_{27}	$\mathbf{C}_3$	$\mathbf{C}_4$	**1**	**0**	**0**	$*$	**0**	**0**
28	m_{28}	$\mathbf{C}_4$	**1**	**0**	**0**	$*$	**0**	**0**	**0**
29	$\underline{m_{29}}$	**1**	**0**	**0**	$*$	**0**	**0**	**0**	$\mathbf{C}_5$
30	$\underline{m_{30}}$	**0**	**0**	$*$	**0**	**0**	**0**	$\mathbf{C}_5$	$\mathbf{C}_6$
31	$\underline{m_{31}}$	**0**	$*$	**0**	**0**	**0**	$\mathbf{C}_5$	$\mathbf{C}_6$	$\mathbf{C}_7$
32	$\textcircled{m}_5$	$*$	**0**	**0**	**0**	$\mathbf{C}_5$	$\mathbf{C}_6$	$\mathbf{C}_7$	$\mathbf{C}_8$
33		**0**	**0**	**0**	$\mathbf{C}_5$	$\mathbf{C}_6$	$\mathbf{C}_7$	$\mathbf{C}_8$	$\textcircled{Q}_{33}$

Messages used for the padding string are underlined.
Variables which we try all possible values are circled.

attack uses the splice-and-cut, partial-matching, and local-collision techniques as shown in Table 6.

In this attack, we need to guarantee that the neutral words form a local-collision in Steps 24-32. This is achieved by fixing chaining variables so that the change of a chaining variable corresponding to both neutral words does not propagate through the Boolean functions. How chaining variables are fixed is shown in Table 7, where, **0**, **1**, $\mathbf{C}_i$, and $*$ denote `0x00000000`, `0xffffffff`, a fixed value, and a flexible value which depends on the value of neutral words, respectively.

The attack procedure for a hash value $H_n = (H^a \| H^b \| H^c \| H^d \| H^e \| H^f \| H^g \| H^h)$ is as follows.

Attack procedure

1. Randomly choose the values of $\mathbf{C}_1, \ldots, \mathbf{C}_5$, and fix the values of chaining variables denoted by $\mathbf{C}_1, \ldots, \mathbf{C}_5, \mathbf{0}$, and $\mathbf{1}$ in Table 7.
2. Compute m_i ($i \in \{25, 26, 27, 28\}$) by solving the step function.

3. Fix m_{29}, m_{30}, and m_{31} to satisfy the padding for a 2-block message.
4. Compute Q_{30}, Q_{31}, and Q_{32} by the step function.
5. Randomly determine other message words that are not specified yet.
6. For all (m_5, Q_{17}), do the following:

$$\begin{cases} p_j & \leftarrow R_j^{-1}(p_{j+1}, m_{\pi(j)}) \quad \text{for } j = 23, 22, \ldots, 0, \\ p_{128} & \leftarrow H_n - p_0, \\ p_j & \leftarrow R_j^{-1}(p_{j+1}, m_{\pi(j)}) \quad \text{for } j = 127, 126, \ldots, 97. \end{cases}$$

7. Make a table of (m_5, Q_{17}, p_{97})s which are computed in the last step, where $p_{97} = (Q_{90} \| Q_{91} \| Q_{92} \| Q_{93} \| Q_{94} \| Q_{95} \| Q_{96} \| Q_{97})$.
8. For all (m_{24}, Q_{33}),
   (a) do the following: $p_{j+1} \leftarrow R_j(p_j, m_{\pi(j)})$ for $j = 33, 34, \ldots, 93$, where, $p_{94} = (Q_{87} \| Q_{88} \| Q_{89} \| Q_{90} \| Q_{91} \| Q_{92} \| Q_{93} \| Q_{94})$.
   (b) Check whether $Q_{94}, Q_{93}, Q_{92}, Q_{91}$, and Q_{90} are matched with those stored in the table.
   (c) If they are matched, compute p_{95}, p_{96}, and p_{97} with the matched pairs, and check whether they are matched with those stored in the table.
   (d) If matched, compute Q_{25}, which is denoted by $*$ in Table 7, by the step function for Step 24 with matched (m_{24}, Q_{17}) and by the step function for Step 33 with matched (m_5, Q_{33}).
   (e) Check whether both results of Q_{25} are matched.
   (f) If matched, the pair of corresponding message and p_0 is a pseudo-preimage of H_n.

In the above procedure, the complexity of step 6 is $2^{64} \cdot \frac{55}{128}$ and the complexity of step 8a is $2^{64} \cdot \frac{61}{128}$. After step 8b, $2^{-32} (= 2^{128} \cdot 2^{-160})$ pair is expected to be remained. After step 8c, $2^{-128} (= 2^{-32} \cdot 2^{-96})$ pair is expected to be remained. After step 8e, $2^{-160} (= 2^{-32} \cdot 2^{-128})$ pair is expected to be remained. Therefore, by repeating the above procedure 2^{160} times, we expect to obtain a pseudo-preimage, where the complexity is $2^{224} (= 2^{64} \cdot 2^{160})$. Finally, this pseudo-primage attack is converted to a preimage attack of the complexity of 2^{241} by the generic approach explained in Section 2.2. Step 7 requires 10×2^{64} words of memory and other steps require negligible amount of memory.

5.3 Notes on Preimage Attack on 5-Pass HAVAL

A preimage attack on 5-pass HAVAL reduced to 151 steps

5-pass HAVAL reduced to 151 steps, which use the first 151 steps of 5-pass HAVAL, can be attacked by using the almost same approach as the attack on 4-pass HAVAL. In Table 6, Step 127 is a part of the second chunk that includes m_5 and is independent of m_{24}. According to the message expansion shown in Table 2, Steps 128-150 are independent from m_{24}. Therefore, the attack on 4-pass HAVAL in the last section can also be applied to the first 151 steps of 5-pass HAVAL. The complexity is almost the same, so we can find a pseudo-preimage at the complexity of 2^{224}, and this attack is converted to a preimage

Table 8. Message word distribution for 5-pass HAVAL (full)

Step	0	1	2	3	4	5	···	19	20	21	22	23	24	25	26	27	28	29	30	31
index	0	1	2	3	4	5	···	19	20	21	22	23	24	25	26	27	28	29	30	31
	second chunk														skip					
Step	32	33	34	35	36	37	38	39	40	41	42	43	44	45	46	47	48	49	···	63
index	5	14	26	18	11	28	7	16	0	23	20	22	1	10	4	8	30	3	···	27
	skip																			
Step	64	65	66	67	68	69	70	71	72	73	74	75	76	77	78	79	80	81	···	95
index	19	9	4	20	28	17	8	22	29	14	25	12	24	30	16	26	31	15	···	2
	skip				first chunk															
Step	96	···	103	104	105	106	107	108	109	110	111	112	113	114	115	116	117	···	121	···
index	24	···	23	26	6	30	20	18	25	19	3	22	11	31	21	8	27	···	29	···
	first chunk						local collision (3-cycle)													
Step	128	129	130	131	132	133	134	135	136	137	138	139	140	141	142	143	144	145	···	159
index	27	3	21	26	17	11	20	29	19	0	12	7	13	8	31	10	5	9	···	15
	local collision				second chunk															

attack of the complexity of 2^{241}, and requires 10×2^{64} words of memory. Note, we experimentally confirmed that there is no selection of chunks that can attack more than 151 steps at the better complexity.

A preimage attack on full 5-pass HAVAL

As mentioned in Section 3.2, our attack works efficiently on a hash function with less than or equal to 4 rounds, but does not work on the one with more than 4 rounds. However, by combining the exhaustive search, we can find a pseudo-preimage at $2^{253.81}$.

To attack full 5-pass HAVAL, we need to use all the techniques explained: splice-and-cut, partial-matching, and local-collision techniques. The selection of the chunks are shown in Table 8. We stress that our computer search program did not find a pair of chunks that can be attacked with a 9-step local collision. This problem was solved by using a long collision path introduced in Section 3.2.

To guarantee that the neutral words form a local-collision in Steps 107-131, we fix chaining variables as shown in Table 9.

1. Fix the value of chaining variables as shown in Table 9, and derive the corresponding messages by using the step function.
2. Fix the value of message words that are not used inside the local collision steps. Note there is enough message space to find a pseudo-preimage.
3. For all 2^{32} values of Q_{108}, compute a corresponding value of Q_{124}. Store the result in a table named Table A.
4. For all 2^{64} values of (m_{26}, Q_{100}), do the following:

$$p_j \leftarrow R_j^{-1}(p_{j+1}, m_{\pi(j)}) \qquad \text{for } j = 106, 105, \ldots, 68.$$

Store $(m_{26}, Q_{100}, p_{68})$ in a table named Table B.

Table 9. Fixed values for preimage attack on 5-pass HAVAL

Round	Step j	$m_{\pi(j)}$	Q_{j-7}	Q_{j-6}	Q_{j-5}	Q_{j-4}	Q_{j-3}	Q_{j-2}	Q_{j-1}	Q_j
4R	107	$\textcircled{m}_{20}$	$\textcircled{Q}_{100}$	C_1	C_2	C_3	1	0	1	1
	108	m_{18}	C_1	C_2	C_3	1	0	1	1	$*(Q_{108})$
	109	m_{25}	C_2	C_3	1	0	1	1	*	1
	110	m_{19}	C_3	1	0	1	1	*	1	1
	111	$\circ m_3$	1	0	1	1	*	1	1	1
	112	m_{22}	0	1	1	*	1	1	1	1
	113	m_{11}	1	1	*	1	1	1	1	0
	114	m_{31}	1	*	1	1	1	1	0	1
	115	$(\circ m_{21})$	*	1	1	1	1	0	1	1
	116	m_8	1	1	1	1	0	1	1	$*(Q_{116})$
	117	$\circ m_{27}$	1	1	1	0	1	1	*	1
	118	m_{12}	1	1	0	1	1	*	1	1
	119	m_9	1	0	1	1	*	1	1	1
	120	m_1	0	1	1	*	1	1	1	1
	121	m_{29}	1	1	*	1	1	1	1	0
	122	m_5	1	*	1	1	1	1	0	0
	123	(m_{15})	*	1	1	1	1	0	0	0
	124	m_{17}	1	1	1	1	0	0	0	$*(Q_{124})$
	125	m_{10}	1	1	1	0	0	0	*	0
	126	m_{16}	1	1	0	0	0	*	0	0
	127	m_{13}	1	0	0	0	*	0	0	C_4
5R	128	$\circ m_{27}$	0	0	0	*	0	0	C_4	0
	129	$\circ m_3$	0	0	*	0	0	C_4	0	C_5
	130	$\circ m_{21}$	0	*	0	0	C_4	0	C_5	C_6
	131	$\textcircled{m}_{26}$	*	0	0	C_4	0	C_5	C_6	C_7
	132		0	0	C_4	0	C_5	C_6	C_7	$\textcircled{Q}_{132}$

Messages that appear twice are stressed with $\circ$.
Uninvolved messages are written in parentheses.

5. For all 2^{64} values of (m_{20}, Q_{132}), do the followings:

$$\begin{cases} p_{j+1} \leftarrow R_j(p_j, m_{\pi(j)}) & \text{for } j = 132, 133, \ldots, 159, \\ p_0 \quad \leftarrow H_n - p_{160}, \\ p_{j+1} \leftarrow R_j(p_j, m_{\pi(j)}) & \text{for } j = 0, 1, \ldots, 25. \end{cases}$$

   Store $(m_{20}, Q_{132}, p_{26})$ in a table named Table C.
6. For all 2^{96} values of $(m_{26}, Q_{100}, m_{20})$, do the followings.
   (a) Compute a value of Q_{108} by using (m_{20}, Q_{100}).
   (b) Find a value of corresponding Q_{124} by looking up Table A.
   (c) Compute a value of corresponding Q_{132} by using Q_{124} and m_{26}.
   (d) Find values of corresponding p_{68} and p_{26} by looking up Tables B and C.
   (e) Compute skipped steps, which are Steps 26-67, by using $(m_{26}, p_{26}, m_{20}, p_{68})$.
   (f) If skipped steps are matched, output corresponding messages.

In the above procedure, steps 1 and 2 finish in negligible time. Step 3 takes the complexity of about $2^{32} \cdot \frac{3}{160}$. Step 4 takes the complexity of $2^{64} \cdot \frac{39}{160}$, and step 5 takes the complexity of $2^{64} \cdot \frac{54}{160}$. Steps 6a to 6d finishes in negligible time for each of $(m_{26}, Q_{100}, m_{20})$. Step 6e seems to take the complexity of $2^{96} \cdot \frac{42}{160}$, but this can be easily improved to $2^{96} \cdot \frac{35}{160}$ by the partial-matching technique. Furthermore, the equation for computing Step 26 can be written as follows:

$$Q_{27} \leftarrow m_{\pi(26)} + (\text{term independent from } m_{\pi(26)}).$$

Therefore, Step 26 can be computed in negligible cost compared to one step function, and thus, the complexity becomes $2^{96} \cdot \frac{34}{160}$. After Step 6e, the number of matched message is evaluated as $2^{-160} (= 2^{-256} \cdot 2^{96})$. Therefore, by repeating steps 2 to 6 of the above procedure 2^{160} times, a pseudo-preimage can be found at the complexity of $2^{160} \cdot 2^{96} \cdot \frac{34}{160} \approx 2^{253.81}$. Steps 4 and 5 require 20×2^{64} words of memory in total and other steps require negligible amount of memory. To apply the depth first search for steps 4-6, Table B or C can be removed and memory requirement becomes half.

Notes on local collision shown in Table 9. In the local collision shown in Table 9, m_3, m_{21}, and m_{27} appear twice. Therefore, we need to be careful so that all fixed values in Table 9 can be achieved. m_{21} is used in Steps 115 and 130. Since a message used in Step 115 is an uninvolved message, we can determine m_{21} so that Step 130 is satisfied. We can ignore the influence to Step 115. Regarding m_3 and m_{27}, since they are used in Steps 129 and 128 whose outputs can be any value ($\mathbf{C}_6$ and $\mathbf{C}_5$), m_3 and m_{27} can be fixed so that Steps 111 and 117 are satisfied. This local collision also includes m_{29}, which is involved to the message padding. Unfortunately, this local collision needs to fix m_{29} to a unique value, since all input and output values of Step 121 are fixed. As a result, this attack cannot satisfy the message padding of 5-pass HAVAL. It is interesting that the uniquely fixed m_{29} satisfies the message padding rules of MD5. Since the padding rules of HAVAL require to produce more information than those of MD5, for example output length and pass number, the fixed m_{29} does not satisfy the padding for HAVAL but satisfies the padding for MD5.

6 Conclusion

In this paper, we proposed preimage attacks on HAVAL. We considered two general strategies to find a preimage. The first approach is speeding up the brute-force attack. By this approach, we can reduce the complexity of preimage attacks by a few bits. The second approach is the meet-in-the-middle approach. We found that the techniques proposed by [1] and [2] can be combined to attack hash functions with more rounds than previous works. As a result, we found a pseudo-preimage attack and a preimage attack on 3-pass HAVAL whose complexities are 2^{192} and 2^{225}, a pseudo-preimage attack and a preimage attack on 4-pass HAVAL whose complexities are 2^{224} and 2^{241}, and a pseudo-preimage attack and a preimage attack on 151-step 5-pass HAVAL whose complexities are also 2^{224} and 2^{241}. Moreover, we optimized the computational order for brute force attack on 5-pass HAVAL and its complexity is $2^{254.89}$. As far as we know, the proposed attack on 3-pass HAVAL is the best attack and proposed attacks on 4-pass HAVAL and 5-pass HAVAL are the first attacks.

References

1. Aoki, K., Sasaki, Y.: Preimage attacks on one-block MD4, 63-step MD5 and more. In: Avanzi, R., Keliher, L., Sica, F. (eds.) Selected Areas in Cryptography — Workshop Records of 15th Annual International Workshop, SAC 2008, Sackville, New Brunswick, Canada, pp. 82–98 (2008)
2. Aumasson, J.-P., Meier, W., Mendel, F.: Preimage attacks on 3-pass HAVAL and step-reduced MD5. In: Avanzi, R., Keliher, L., Sica, F. (eds.) Selected Areas in Cryptography — Workshop Records of 15th Annual International Workshop, SAC 2008, Sackville, New Brunswick, Canada, pp. 99–114 (2008), (also appeared in IACR Cryptology ePrint Archive: Report `http://eprint.iacr.org/2008/183`)
3. De Cannière, C., Rechberger, C.: Preimages for reduced SHA-0 and SHA-1. In: Wagner, D. (ed.) CRYPTO 2008. LNCS, vol. 5157, pp. 179–202. Springer, Heidelberg (2008) (slides on preliminary results were appeared at ESC 2008 seminar, `http://wiki.uni.lu/esc/`)
4. Dobbertin, H.: The first two rounds of MD4 are not one-way. In: Vaudenay, S. (ed.) FSE 1998. LNCS, vol. 1372, pp. 284–292. Springer, Heidelberg (1998)
5. Kim, J., Biryukov, A., Preneel, B., Hong, S.: On the security of HMAC and NMAC based on HAVAL, MD4, MD5, SHA-0 and SHA-1. In: De Prisco, R., Yung, M. (eds.) SCN 2006. LNCS, vol. 4116, pp. 242–256. Springer, Heidelberg (2006)
6. Kim, J., Biryukov, A., Preneel, B., Lee, S.: On the security of encryption modes of MD4, MD5 and HAVAL. In: Qing, S., Mao, W., López, J., Wang, G. (eds.) ICICS 2005. LNCS, vol. 3783, pp. 147–158. Springer, Heidelberg (2005)
7. Lee, E., Kim, J., Chang, D., Sung, J., Hong, S.: Second preimage attack on 3-pass HAVAL and partial key-recovery attacks on NMAC/HMAC-3-pass HAVAL. In: Nyberg, K. (ed.) FSE 2008. LNCS, vol. 5086, pp. 189–206. Springer, Heidelberg (2008)
8. Leurent, G.: MD4 is not one-way. In: Nyberg, K. (ed.) FSE 2008. LNCS, vol. 5086, pp. 412–428. Springer, Heidelberg (2008)
9. Menezes, A.J., van Oorschot, P.C., Vanstone, S.A.: Handbook of applied cryptography. CRC Press, Boca Raton (1997)
10. Rivest, R.L.: Request for Comments 1321: The MD5 Message Digest Algorithm. The Internet Engineering Task Force (1992), `http://www.ietf.org/rfc/rfc1321.txt`
11. Suzuki, K., Kurosawa, K.: How to find many collisions of 3-pass HAVAL. In: Miyaji, A., Kikuchi, H., Rannenberg, K. (eds.) IWSEC 2007. LNCS, vol. 4752, pp. 428–443. Springer, Heidelberg (2007) (A preliminary version was appeared in IACR Cryptology ePrint Archive: Report 2007/079, `http://eprint.iacr.org/2007/079`)
12. van Rompay, B., Biryukov, A., Preneel, B., Vandewalle, J.: Cryptanalysis of 3-pass HAVAL. In: Laih, C.-S. (ed.) ASIACRYPT 2003. LNCS, vol. 2894, pp. 228–245. Springer, Heidelberg (2003)
13. Wang, X., Feng, D., Yu, X.: An attack on hash function HAVAL-128. Science in China (Information Sciences) 48(5), 545–556 (2005)
14. Wang, X., Yu, H.: How to break MD5 and other hash functions. In: Cramer, R. (ed.) EUROCRYPT 2005. LNCS, vol. 3494, pp. 19–35. Springer, Heidelberg (2005)

15. Wang, Z., Zhang, H., Qin, Z., Meng, Q.: Cryptanalysis of 4-pass HAVAL. IACR Cryptology ePrint Archive: Report 2006/161 (2006), `http://eprint.iacr.org/2006/161`
16. Yoshida, H., Biryukov, A., De Cannière, C., Lano, J., Preneel, B.: Non-randomness of the full 4 and 5-pass HAVAL. In: Blundo, C., Cimato, S. (eds.) SCN 2004. LNCS, vol. 3352, pp. 324–336. Springer, Heidelberg (2005)
17. Yu, H., Wang, X., Yun, A., Park, S.: Cryptanalysis of the full HAVAL with 4 and 5 passes. In: Robshaw, M.J.B. (ed.) FSE 2006. LNCS, vol. 4047, pp. 89–110. Springer, Heidelberg (2006)
18. Zheng, Y., Pieprzyk, J., Seberry, J.: HAVAL — one-way hashing algorithm with variable length of output. In: Zheng, Y., Seberry, J. (eds.) AUSCRYPT 1992. LNCS, vol. 718, pp. 83–104. Springer, Heidelberg (1993)

How to Fill Up Merkle-Damgård Hash Functions

Kan Yasuda

NTT Information Sharing Platform Laboratories, NTT Corporation
3-9-11 Midoricho Musashino-shi, Tokyo 180-8585 Japan
yasuda.kan@lab.ntt.co.jp

Abstract. Many of the popular Merkle-Damgård hash functions have turned out to be not collision-resistant (CR). The problem is that we no longer know if these hash functions are even second-preimage-resistant (SPR) or one-way (OW), without the underlying compression functions being CR. We remedy this situation by introducing the "split padding" into a current Merkle-Damgård hash function H. The patched hash function $\bar{H}$ resolves the problem in the following ways: (i) $\bar{H}$ is SPR if the underlying compression function h satisfies an "SPR-like" property, and (ii) $\bar{H}$ is OW if h satisfies an "OW-like" property. The assumptions we make about h are provided with simple definitions and clear relations to other security notions. In particular, they belong to the class whose existence is ensured by that of OW functions, revealing an evident separation from the strong CR requirement. Furthermore, we get the full benefit from the patch at almost no expense: The new scheme requires no change in the internals of a hash function, runs as efficiently as the original, and as usual inherits CR from h. Thus the patch has significant effects on systems and applications whose security relies heavily on the SPR or OW property of Merkle-Damgård hash functions.

Keywords: hash function, Merkle-Damgård, padding, second-preimage resistance, one-wayness.

1 Introduction

Most of the modern cryptographic hash functions follow a design principle called the Merkle-Damgård construction. A main feature of such a hash function is that its collision resistance (CR) is guaranteed by that of its underlying compression function [21,8], yet unfortunately popular hash functions MD5 [28] and SHA-1 [24] are now shown to be *not* CR [37,38], hence losing the CR of their respective compression functions. These attacks have a profound impact on current systems using hash functions, not to mention those applications whose security is entirely based on the CR property of their hash-function components.

The loss of CR also exerts a strong influence on schemes whose security depends on the second-preimage resistance (SPR) or one-wayness (OW) of Merkle-Damgård hash functions. This is due to the fact that the SPR or OW security of such a hash function is hitherto ensured only by its CR (Recall that SPR is immediately implied by CR [23] and that OW is also implied by CR as long as

J. Pieprzyk (Ed.): ASIACRYPT 2008, LNCS 5350, pp. 272–289, 2008.
© International Association for Cryptologic Research 2008

the hash function is "uniform" [36] or "sufficiently compressing" [31]). Now that the popular hash functions are not CR, we lose our proof-based assurance of the SPR and OW properties for these hash functions. To summarize:

> ***We have no guarantee whatsoever of the SPR or OW property for a Merkle-Damgård hash function without CR by its underlying compression function.***

This is the main problem we explore in the paper. We come up with a solution by first making a slight modification to the design of current hash functions. The change is fully compatible with a standard Merkle-Damgård interface. We then show that the patched hash functions indeed accomplish SPR and OW, assuming weaker-than-CR properties of the underlying compression functions.

Obtaining Upper-Bound Results for SPR and OW. A more direct way to overcome our problem in hand is to analyze exactly what sorts of properties the underlying compression function must possess in order to ensure the SPR and OW security of the Merkle-Damgård construction. [26,10] takes this approach and identifies complexity assumptions about the underlying compression function, which assure the SPR or OW property of the whole hash function. However, they do not consider the non-randomness involved in the padding or length-encoding bits.

Rather, we treat the problem of padding and length-encoding bits in detail.[1] The importance of these bits are already pointed out by [18,4]. We then come up with simple formulations of the complexity assumptions about the compression function, showing that these assumptions are indeed weaker than the CR requirement.

Need for SPR and OW Hash Functions. CR, SPR and OW are the three classical requirements for security of keyless hash functions (*e.g.*, [32,23]). We already know that the notion of CR plays an important role in designing cryptographic schemes. However, there are situations in which CR is not necessarily required but SPR or OW is essential to the security of systems. For example, adversaries might be unable to control input data to the hash function, say by protocol specification or by the fact that inputs are encrypted under a secret key before hashing.

A CR hash function may not be best suited to above scenarios due to its large hash size. Recall that for n-bit security the hash size of a CR hash function needs to be (at least) $2n$ bits. Suppose we want to use an SPR hash function with n-bit security. If the SPR security of hash functions were guaranteed only by their CR, then we would have needed to use a $2n$-bit hash function, whereas we could just use an n-bit hash function if the SPR security is directly guaranteed (not via its CR).

Our Results. We apply a patch to the Merkle-Damgård construction so that the SPR and OW properties are now guaranteed by certain reasonable and simple assumptions about the compression function.

---

[1] On the other hand, we assume that messages are distributed uniformly at random, which may not hold true in some of the practical applications, as pointed out by [3].

Split Padding. This is the patch. The new scheme works exactly the same as the original hash function except for the very end; the split-padding method alters processing of the last two blocks of a message. Message expansion is minimal, requiring at most one extra block (and such a case is rare). The new scheme is compatible with fixed-IV (initialization vector) usage and Merkle-Damgård strengthening. It can also handle a message in stream by delaying processing and buffering the two most recent blocks of the message.

CR Preservation. There is "nothing to lose" by applying the patch. Namely, we show that the CR preservation property of the original Merkle-Damgård construction is still in action with our new scheme. This motivates us to apply the patch to the systems whose hash functions are still CR (*e.g.*, SHA-256).

SPR Guarantee. This is one of the main features of our new mode of operation. The entire hash function can be proven SPR based on the assumption that the underlying compression function satisfies a property we call "cs-SPR" (chosen-suffix SPR).

OW Guarantee. This is the other main feature of the new construction. We show that the OW security of the entire hash function is ensured by the "cs-OW" (chosen-suffix OW) property of the underlying compression function.

Justification for cs-SPR and cs-OW. We give the rationale behind our choice of these assumptions by demonstrating their relationships with several known versions of SPR and OW properties. In particular, we show that these assumptions are strictly weaker than the strong CR requirement, by proving that they belong to the class whose existence is guaranteed by that of an OW function.

Without Random Oracles. We avoid use of random oracles in our proofs of security. The proofs of the SPR and OW properties are conducted in the standard model, following the concrete-security-reduction methodology. For the proof of CR preservation, we adopt the "human-ignorance" approach developed by [29].

Organization of the Paper. In Sect. 2 we review previous work related to the topic. Section 3 provides necessary definitions for the security of hash functions. In Sect. 4 we give a description of our patch "split padding." The proofs for the CR, SPR and OW properties of our scheme are given in Sect. 5, 6 and 7, respectively. We analyze the assumption cs-SPR in Sect. 8, followed by a similar analysis of cs-OW in Sect. 9. Section 10 presents certain application of our scheme.

2 Related Work

Merkle-Damgård Construction. [2] points out that the SPR or OW assumption about a compression function alone is *not* sufficient for the SPR or OW property of its Merkle-Damgård iteration. [19] observes that the use of a fixed IV precludes the trivial "truncation" or "free-start" SPR attack and that Merkle-Damgård strengthening (*i.e.*, length encoding in the final block) appears to defeat "long-message" SPR attacks. Then later it is shown that birthday-type

SPR attacks are still possible on the Merkle-Damgård construction [9,17], disproving the effectiveness of Merkle-Damgård strengthening against long-message SPR attacks.

Keyed (Randomized) Hash Functions. Hash functions in theory are often in the dedicated-key setting [6]. [31] describes seven security notions in such a setting: Coll, Sec, eSec, aSec, Pre, ePre and aPre. ROX [2] is a powerful mode of operation that preserves all the seven properties of the underlying compression function. Since aSec and aPre correspond with SPR and OW in the keyless setting, ROX can be used with its keys fixed, thereby as a keyless hash function. Unfortunate aspects of ROX are that it requires major modifications to the design of current hash functions and that its security is based on the use of random oracles.

BCM [3] achieves Coll, Sec and Pre properties with its proof of Sec security conducted in the standard model (That of Pre is in the random oracle model). However, aSec property is not achieved, and hence BCM is not suited to our keyless setting (BCM construction yields a keyed, Sec-secure hash function, and the keys cannot be fixed because it does not assure aSec).

There exist a number of domain-extension constructions for eSec(=TCR, UOWHF). Prominent one is the Randomized Hashing [13]. The Randomized Hashing has a close connection with our split padding, *cf.* Sect. 10.

SPR and OW Attacks on Specific Hash Algorithms. MD4 [27] is now shown to be not OW [18]. The attack first finds "pseudo-preimages" for the compression function and then extends them to the entire hash function. [4] studies the SPR and OW properties of the Snefru, being already aware of the importance of padding scheme to these security notions.

Other Weaknesses of Merkle-Damgård Construction. It has been pointed out that there exist a number of properties that the Merkle-Damgård construction does not achieve. These include multi-collisions [15], herding attacks [16], indifferentiability [7] and balance [5]. It is certainly *not* the purpose of our construction to achieve these goals. Rather, we focus on the classical three notions of CR, SPR and OW.

3 Definitions

Notation. Given a finite bit string $x \in \{0,1\}^*$ we define $|x|$ as the bit length of x. The notation $x_1 \| x_2$ represents the concatenation of two strings $x_1, x_2 \in \{0,1\}^*$. We write $\lceil \cdot \rceil$ for the ceiling function. By $x \stackrel{\$}{\leftarrow} X$ we indicate the operation of selecting an element uniformly at random from the set X and assigning its value to variable x. We write 0^n for the bit string $00 \cdots 0$ (n times).

Given a function $f : X \to Y$, we say that a pair $(x, x') \in X \times X$ is *colliding* with respect to f if $f(x) = f(x')$ and $x \neq x'$. We write $x \bowtie_f x'$, or simply $x \bowtie x'$, to indicate the fact that x and x' are colliding. Similarly, given a keyed function $f_k : X \to Y$ we write $(k, x) \bowtie_f (k', x')$ when $f_k(x) = f_{k'}(x')$ and $(k, x) \neq (k', x')$.

Algorithm *MD-strengthening* (y)
101 Set $\eta \leftarrow \lceil (|y|+1)/(m-n) \rceil$
102 Divide $y = y[1] \parallel \cdots \parallel y[\eta-1] \parallel y[\eta]$
so that $|y[1]| = \cdots = |y[\eta-1]| = m-n$ and $0 \leq |y[\eta]| \leq m-n-1$
103 If $|y[\eta]| \leq m-n-\sigma-1$ then $z \leftarrow y \| 10^{m-n-\sigma-1-|y[\eta]|} \| \langle |y| \rangle$ EndIf
104 If $|y[\eta]| \geq m-n-\sigma$ then $z \leftarrow y \| 10^{m-n-1-|y[\eta]|} \| 0^{m-n-\sigma} \| \langle |y| \rangle$ EndIf
105 Output z

Algorithm *MD-iteration* (z) // Accepts only z with $|z|$ being a multiple of $m-n$
201 Set $\zeta \leftarrow |z|/(m-n)$
202 Divide $z = z[1] \parallel \cdots \parallel z[\zeta]$ so that $|z[1]| = \cdots = |z[\zeta]| = m-n$
203 Set $v[0] \leftarrow IV$; For $i = 1, \ldots, \zeta$ do $v[i] \leftarrow h\big(z[i] \parallel v[i-1]\big)$ EndFor; Output $v[\zeta]$

Fig. 1. Definitions of functions *MD-strengthening* and *MD-iteration*

An adversary A is a probabilistic algorithm that takes inputs. An adversary A may often be a pair of such algorithms, as $A = (A_1, A_2)$. We write $y \leftarrow A(x)$ to mean that adversary A outputs a value upon its input x, the output value being assigned to variable y.

Merkle-Damgård Construction. Throughout the paper we fix a compression function $h : \{0,1\}^m \to \{0,1\}^n$ with $m > n$. We also fix a value $IV \in \{0,1\}^n$. We choose a length-encoding function $\langle \cdot \rangle$, which takes an integer as its input and returns a σ-bit representation of the input value (A typical value of σ is 64). This restricts message lengths to a maximum of $2^\sigma - 1$ bits. Hence, the domain of the hash function should be written as $\{0,1\}^{\leq 2^\sigma - 1}$ formally, but for simplicity we write $\{0,1\}^*$ to indicate the message space. Now the Merkle-Damgård hash function $H : \{0,1\}^* \to \{0,1\}^n$ is defined as

$$H(x) \stackrel{\text{def}}{=} \mathit{MD\text{-}iteration}\big(\mathit{MD\text{-}strengthening}(x)\big),$$

where the functions *MD-strengthening* and *MD-iteration* are as described in Fig. 1.[2] We adopt the convention that on empty input (*i.e.*, null string) function *MD-iteration* returns the value *IV*.

Let z and z' be two strings whose lengths are multiples of $m-n$ bits. Divide them into $(m-n)$-bit blocks as $z = z[1] \parallel \cdots \parallel z[\zeta]$ and $z' = z'[1] \parallel \cdots \parallel z'[\zeta']$. Suppose $\mathit{MD\text{-}iteration}(z) = \mathit{MD\text{-}iteration}(z')$ and $z \neq z'$. We define

$$\mathit{index}(z, z') \stackrel{\text{def}}{=} \begin{cases} \zeta & \text{if } \zeta \neq \zeta', \\ i & \text{if } \zeta = \zeta', \end{cases}$$

where i is defined to be the largest integer in $\{1, 2, \ldots, \zeta\}$ such that following (i), (ii) and (iii) hold: (i) $\mathit{MD\text{-}iteration}\big(z[1] \parallel \cdots \parallel z[i]\big) = \mathit{MD\text{-}iteration}$

[2] For a technical reason we deliberately define the iteration as $h\big(z[i] \parallel v[i-1]\big)$ at line 203 rather than as $h\big(v[i-1] \parallel z[i]\big)$.

Table 1. Complexity assumptions about keyless compression function $h : \{0,1\}^m \to \{0,1\}^n$

	(Alias)	Game
fp-CR	(c-SPR [13])	$\tilde{x} \xleftarrow{\$} \{0,1\}^\mu$, $(a, x') \leftarrow A(\tilde{x})$, $\tilde{x}\|a \overset{?}{\bowtie} x'$
cs-SPR		$(a, St) \leftarrow A_1(\cdot)$, $\tilde{x} \xleftarrow{\$} \{0,1\}^\mu$, $x' \leftarrow A_2(\tilde{x}, St)$, $\tilde{x}\|a \overset{?}{\bowtie} x'$
SPR	(weak CR)	$x \xleftarrow{\$} \{0,1\}^m$, $x' \leftarrow A(x)$, $x \overset{?}{\bowtie} x'$
cs-OW		$(a, St) \leftarrow A_1(\cdot)$, $\tilde{x} \xleftarrow{\$} \{0,1\}^\mu$, $v \leftarrow h(\tilde{x}\|a)$, $x' \leftarrow A_2(v, St)$, $v \overset{?}{=} h(x')$
ks-OW	(partial OW [23])	$a \xleftarrow{\$} \{0,1\}^{m-\mu}$, $\tilde{x} \xleftarrow{\$} \{0,1\}^\mu$, $v \leftarrow h(\tilde{x}\|a)$, $x' \leftarrow A(a, v)$, $v \overset{?}{=} h(x')$
OW	(preimage resistance)	$x \xleftarrow{\$} \{0,1\}^m$, $v \leftarrow h(x)$, $x' \leftarrow A(v)$, $v \overset{?}{=} h(x')$

$(z'[1] \,\|\, \cdots \,\|\, z'[i])$, (ii) $z[j] = z'[j]$ for all $j \geq i+1$, (iii) Either $z[i] \neq z'[i]$ or $\mathit{MD\text{-}iteration}(z[1] \,\|\, \cdots \,\|\, z[i-1]) \neq \mathit{MD\text{-}iteration}(z'[1] \,\|\, \cdots \,\|\, z'[i-1])$.

Complexity Assumptions about Keyless Compression Functions. Table 1 lists six notions of security for the compression function $h : \{0,1\}^m \to \{0,1\}^n$. Here we have a fixed security parameter $1 \leq \mu \leq m$. A typical value of μ is $\mu = n/2$ or $\mu = n$. Of the six notions, the two most important ones in the current work are cs-SPR and cs-OW, because these are the assumptions that we make about the underlying compression function h. Others are fp-CR (forced-prefix CR, *cf.* [35]), SPR, ks-OW (known-suffix OW) and OW. These four appear in the list only for the purpose of analyzing the nature of cs-SPR and cs-OW in Sect. 8 and 9, respectively.

The notion of cs-SPR is a variant of SPR where a suffix is chosen by adversaries. Informally, the game of cs-SPR for the compression function $h : \{0,1\}^m \to \{0,1\}^n$ proceeds as follows: First a suffix a of a given length is chosen by an adversary, and then a challenge $\tilde{x}$ of a given length is randomly drawn and shown to the adversary. The goal of the adversary is to find a second preimage $x' \neq \tilde{x}\|a$ such that $h(\tilde{x}\|a) = h(x')$. Here we emphasize that the adversary is required to commit on the value a before observing the challenge $\tilde{x}$.

Security Goals for Keyless Hash Functions. Our SPR and OW goals for a hash function $H : \{0,1\}^* \to \{0,1\}^n$ is formalized in Table 2. Note that an adversary can choose the challenge length $\lambda \geq \mu$ at the beginning of each game. Also note that the adversary's response x' may be of length different from λ.

Security Notions for Keyed Function Family. We utilize four notions, Coll, eColl (enhanced Coll), TCR (target collision resistance) and eTCR (enhanced

Table 2. Security goals for keyless hash function $H : \{0,1\}^* \to \{0,1\}^n$

	Game
SPR	$(\lambda, St) \leftarrow A_1(\cdot)$, $x \xleftarrow{\$} \{0,1\}^\lambda$, $x' \leftarrow A_2(x, St)$, $x \overset{?}{\bowtie} x'$
OW	$(\lambda, St) \leftarrow A_1(\cdot)$, $x \xleftarrow{\$} \{0,1\}^\lambda$, $v \leftarrow H(x)$, $x' \leftarrow A_2(v, St)$, $v \overset{?}{=} H(x')$

Table 3. Security notions for keyed function family $\varphi_k : \{0,1\}^m \to \{0,1\}^n$

(Alias)	Game
Coll	$k \stackrel{\$}{\leftarrow} K$, $(x, x') \leftarrow A(k)$, $(k, x) \stackrel{?}{\bowtie} (k, x')$
eColl	$k \stackrel{\$}{\leftarrow} K$, $(x, k', x') \leftarrow A(k)$, $(k, x) \stackrel{?}{\bowtie} (k', x')$
TCR (eSec, UOWHF)	$(x, St) \leftarrow A_1(\cdot)$, $k \stackrel{\$}{\leftarrow} K$, $x' \leftarrow A_2(k, St)$, $(k, x) \stackrel{?}{\bowtie} (k, x')$
eTCR	$(x, St) \leftarrow A_1(\cdot)$, $k \stackrel{\$}{\leftarrow} K$, $(k', x') \leftarrow A_2(k, St)$, $(k, x) \stackrel{?}{\bowtie} (k', x')$

TCR) for analyzing cs-SPR in Sect. 8. See Table 3. The notion of eColl appears to be new.

Advantage Functions and Adversarial Resources. For a CR-like or SPR-like goal, we define the advantage function of an adversary A as $\mathbf{Adv}_f^{\mathrm{goal}}(A) \stackrel{\mathrm{def}}{=} \Pr\big[\stackrel{?}{\bowtie} \text{ holds}\big]$, where f is the target function (h, H, *etc.*). Similarly, we define $\mathbf{Adv}_f^{\mathrm{goal}}(A) \stackrel{\mathrm{def}}{=} \Pr\big[\stackrel{?}{=} \text{ holds}\big]$ for an OW-like goal. The probabilities are over all coins defined in game and used by A. We fix a model of computation and measure the time complexity of adversaries. The time complexity of an adversary is the time for execution of its overlying game plus its code size. We let $\mathsf{time}(h)$ denote the time complexity necessary for one computation of h. Define $\mathbf{Adv}_f^{\mathrm{goal}}(t, \ell) \stackrel{\mathrm{def}}{=} \max_A \mathbf{Adv}_f^{\mathrm{goal}}(A)$, where max runs over all adversaries A, with its time complexity being at most t, with λ, $|St|$ and $|x'|$ each being at most ℓ blocks. A "block" is $m - n$ bits. ℓ may be omitted from the notation if irrelevant in the context.

4 How to Insert Split Padding

Our new hash function $\bar{H}$ operates exactly the same as the original hash H except for the last two blocks of messages. More precisely, the new hash function is defined as

$$\bar{H}(x) \stackrel{\mathrm{def}}{=} H\big(\mathsf{split\text{-}padding}(x)\big),$$

with a plain Merkle-Damgård hash function $H : \{0,1\}^* \to \{0,1\}^n$. The definition of $\mathsf{split\text{-}padding}$ is given in Fig. 2 along with a pictorial representation in Fig. 3.

The basic idea of the "split padding" method is to make sure that every block input to the compression function h has at least μ bits of a message,[3] rather than being entirely padding bits or length-encoding bits. Indeed, when combined with $\mathsf{split\text{-}padding}$, the function $\mathsf{MD\text{-}strengthening}$ never invokes line 104 of Fig. 1.

For this mechanism to work, we need to impose a condition $\sigma + 1 + 2\mu \leq m - n$ on $h : \{0,1\}^m \to \{0,1\}^n$. As long as this condition is fulfilled, the algorithm $\mathsf{split\text{-}padding}$ is well-defined. Also observe that no message bits are shared across the blocks. We will come back to this issue after proving the following basic result.

Proposition 1. *The function* $\mathsf{split\text{-}padding}$ *is one-to-one (i.e., injective).*

[3] Of course, here we are assuming that the message is at least μ bits long to begin with.

Algorithm *split-padding*(x)

301 Set $\xi \leftarrow \lceil (|x|+1)/(m-n) \rceil$

302 Divide $x = x[1] \,\|\, \cdots \,\|\, x[\xi-1] \,\|\, x[\xi]$

so that $|x[1]| = \cdots = |x[\xi-1]| = m-n$ and $0 \le |x[\xi]| \le m-n-1$

303 If $\mu \le |x[\xi]| \le m-n-\sigma-2$ then *pad-plain* EndIf

304 If $|x[\xi]| \le \mu - 1$ then *pad-with-borrow* EndIf

305 If $|x[\xi]| \ge m-n-\sigma-1$ then *pad-with-carry* EndIf

306 Output y

310 **Subroutine** *pad-plain*

311 Put $y \leftarrow x \| 0$

320 **Subroutine** *pad-with-borrow*

321 If $\xi \ge 2$ then divide $x[\xi-1] = \tilde{x}[\xi-1] \,\|\, \mathit{brw}$

so that $|\mathit{brw}| = \mu$ and $|\tilde{x}[\xi-1]| = m-n-\mu$

322 Set $\tilde{x}[\xi] \leftarrow \mathit{brw} \| x[\xi]$

323 Put $y \leftarrow x[1] \,\|\, \cdots \,\|\, x[\xi-2] \,\|\, \tilde{x}[\xi-1] \,\|\, 10^{\mu-1} \,\|\, \tilde{x}[\xi] \,\|\, 1$ EndIf

324 If $\xi = 1$ then put $y \leftarrow x[1] \,\|\, 1$ EndIf

330 **Subroutine** *pad-with-carry*

331 Divide $x[\xi] = \tilde{x}[\xi] \,\|\, \mathit{cry}$ so that $|\mathit{cry}| = \mu$ and $|\tilde{x}[\xi]| = |x[\xi]| - \mu$

332 Set $\tilde{x}[\xi+1] \leftarrow \mathit{cry}$

333 Put $y \leftarrow x[1] \,\|\, \cdots \,\|\, x[\xi-1] \,\|\, \tilde{x}[\xi] \,\|\, 10^{m-n-|\tilde{x}[\xi]|-1} \,\|\, \tilde{x}[\xi+1] \,\|\, 1$

Fig. 2. Description of "split padding" algorithm

Proof. Let $x, x' \in \{0,1\}^*$. We want to prove that the equality *split-padding*(x) = *split-padding*(x') implies $x = x'$. So suppose we have x and x' such that the condition *split-padding*(x) = *split-padding*(x') holds. Set $y \leftarrow$ *split-padding*(x) and $y' \leftarrow$ *split-padding*(x'). Divide $y = w \| b$ and $y' = w' \| b'$ so that $|b| = |b'| = 1$. The equality $y = y'$ tells us that $b = b'$ and $w = w'$.

Case A: $b = b' = 0$. In this case we know that both y and y' come from *pad-plain*. Hence, we must have $x = w$ and $x' = w'$, which yields $x = x'$.

Case B: $b = b' = 1$. We observe that in this case both y and y' originate from either *pad-with-borrow* or *pad-with-carry*. We divide the case according to the size $|y| = |y'|$.

Case B1: $|y| = |y'| \le m - n$. Note that *pad-with-carry* always produces more than or equal to two blocks of output, which implies that both y and y' must have been processed through *pad-with-borrow* in this case. Consequently, we get $x = w$, $x' = w'$ and $x = x'$.

Case B2: $|y| = |y'| > m - n$. Put $\eta \leftarrow \lceil (|y|+1)/(m-n) \rceil$. We must have $\eta \ge 2$. Divide $y = y[1] \,\|\, \cdots \,\|\, y[\eta-1] \,\|\, y[\eta]$ and $y' = y'[1] \,\|\, \cdots \,\|\, y'[\eta-1] \,\|\, y'[\eta]$, so that $|y[1]| = \cdots = |y[\eta-1]| = |y'[1]| = \cdots = |y'[\eta-1]| = m-n$. Recall that *pad-with-borrow* sets the last block to

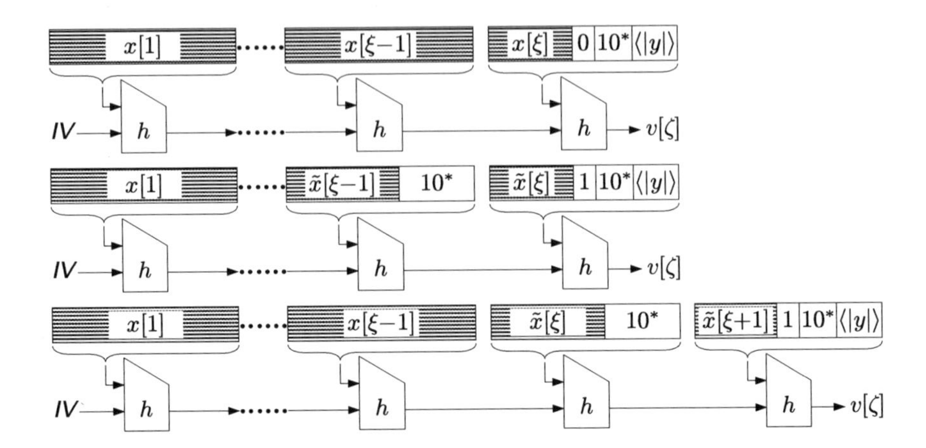

Fig. 3. Split padding: "plain" (top), "borrow" (middle) and "carry" (bottom), combined with *MD-strengthening*. Note that the last $10^* \| \langle|y|\rangle$ comes from *MD-strengthening*, not from *split-padding*. 10^* means $10\cdots0$ with an appropriate number of zeros.

a length between $\mu+1$ and 2μ bits, whereas *pad-with-carry* always sets the last block to a length of $\mu+1$ bits. Now we further divide the case according to the value $|y[\eta]|$.

Case B2a: $|y[\eta]| = |y'[\eta]| \geq \mu + 2$. This case assures that both y and y' originate from *pad-with-borrow*. It means that we can write $y[\eta-1] = \tilde{y}[\eta-1] \| 10^{\mu-1}$ and $y'[\eta-1] = \tilde{y}'[\eta-1] \| 10^{\mu-1}$. We can also write $y[\eta] = \tilde{y}[\eta] \| 1$ and $y'[\eta] = \tilde{y}'[\eta] \| 1$. Therefore, we obtain

$$x = y[1] \| \cdots \| y[\eta-2] \| \tilde{y}[\eta-1] \| \tilde{y}[\eta],$$
$$x' = y'[1] \| \cdots \| y'[\eta-2] \| \tilde{y}'[\eta-1] \| \tilde{y}'[\eta],$$

which immediately implies that $x = x'$.

Case B2b: $|y[\eta]| = |y'[\eta]| = \mu + 1$. There are multiple possibilities in this case: y and y' may come from either *pad-with-borrow* or *pad-with-carry*. To identify the case, we further divide $y[\eta-1] = \tilde{y}[\eta-1]\|10^{\alpha}$ and $y'[\eta-1] = \tilde{y}'[\eta-1]\|10^{\alpha}$ for some integer α. Observe that *pad-with-borrow* always sets $\alpha = \mu - 1$, whereas *pad-with-carry* sets $\alpha \geq \mu$. Hence, by looking at the value α we see that either (i) both y and y' are from *pad-with-borrow*, or (ii) both y and y' are from *pad-with-carry*. Write $y[\eta] = \tilde{y}[\eta] \| 1$ and $y'[\eta] = \tilde{y}'[\eta] \| 1$. We see that

$$x = y[1] \| \cdots \| y[\eta-2] \| \tilde{y}[\eta-1] \| \tilde{y}[\eta],$$
$$x' = y'[1] \| \cdots \| y'[\eta-2] \| \tilde{y}'[\eta-1] \| \tilde{y}'[\eta],$$

which gives us the desired equality $x = x'$.

Thus, we have shown that *split-padding* $(x) =$ *split-padding* (x') always implies $x = x'$. This proves the injectivity of the function *split-padding*. □

On the Constraint $\sigma + 1 + 2\mu \leq m - n$. We need to impose this constraint on the underlying compression function h. Thanks to the constraint, every block ends up containing at least μ bits of a message after the *split-padding* procedure.

The constraint is necessary for handling the last block properly. Recall that the subroutine *pad-with-borrow* produces the last block with a length *at most* $\mu + \mu - 1 + 1 = 2\mu$ bits. The subroutine *pad-with-carry* produces the last block with a length always equal to $\mu + 1$ bits. The constraint guarantees that the last block of either type, padded with *MD-strengthening*, fits neatly in a single block. The subroutine *pad-plain* by definition handles only the case when the last block fits in one block.

The constraint also guarantees that the second last block contains at least μ bits of the message. This holds true for *pad-plain*, *pad-with-borrow* and *pad-with-carry*.

The constraint is not problematic as long as we are dealing with MD5 or SHA-1 with $\mu = n/2$ or $n = \mu$. It puts an obstacle in the way of using SHA-256 with $\sigma = 64$ and $\mu = 256$. In such a case we are limited to setting the value of μ only up to $\mu \leq 223$.

5 CR of Merkle-Damgård with Split Padding

We follow the "human-ignorance" approach developed by [29] for formalizing the notion of CR.

Proposition 2. *Let $\bar{H} : \{0,1\}^* \to \{0,1\}^n$ be the Merkle-Damgård hash function with split padding constructed of a compression function $h : \{0,1\}^m \to \{0,1\}^n$. If there exists an explicitly-given adversary that finds a pair of colliding messages for $\bar{H}$ with a probability ε, spending time complexity at most t, each message being at most ℓ blocks, then there exists an explicitly-given adversary that finds a collision for h with a probability ε, spending time complexity at most $t' \approx t + 2\ell \cdot \mathsf{time}(h)$.*

Proof. This statement immediately follows from the injectivity of *split-padding* and the well-known CR reduction of the plain Merkle-Damgård iteration [21,8]. □

6 SPR of Merkle-Damgård with Split Padding

In this section we prove that the patched hash function $\bar{H}$ is SPR assuming that the underlying compression function h is cs-SPR. After proving our result, we also discuss the birthday bound implied by the reduction.

Adversary B

410 Run $A = (A_1, A_2)$ and obtain a bit length $(\lambda, St) \leftarrow A_1(\cdot)$ // $\lambda \geq \mu$

411 Generate a random message $x \stackrel{\$}{\leftarrow} \{0,1\}^\lambda$

412 Put $z \leftarrow \mathit{MD\text{-}strengthening}\big(\mathit{split\text{-}padding}(x)\big)$

413 Divide $z = z[1] \,\|\, z[2] \,\|\, \cdots \,\|\, z[\zeta]$ so that $|z[1]| = |z[2]| = \cdots = |z[\zeta]| = m - n$

414 Choose an index $i \stackrel{\$}{\leftarrow} \{1, 2, \ldots, \zeta\}$

415 Compute $v[i-1] \leftarrow \mathit{MD\text{-}iteration}\big(z[1] \,\|\, z[2] \,\|\, \cdots \,\|\, z[i-1]\big)$

416 Divide $z[i] = \alpha \| \beta$ so that $|\alpha| = \mu$ and $|\beta| = m - n - \mu$

420 Submit $\beta \| v[i-1]$ as a committed suffix and receive a challenge $\tilde{x} \in \{0,1\}^\mu$

421 Put $w \leftarrow (\mathit{MD\text{-}strengthening} \circ \mathit{split\text{-}padding})^{-1}\big(z[1] \,\|\, \cdots$
$\cdots \,\|\, z[i-1] \,\|\, \tilde{x} \,\|\, \beta \,\|\, z[i+1] \,\|\, \cdots \,\|\, z[\zeta]\big)$

430 Feed w and St to A_2 and obtain a second preimage $x' \leftarrow A_2(w, St)$

431 Put $z' \leftarrow \mathit{MD\text{-}strengthening}\big(\mathit{split\text{-}padding}(x')\big)$

432 Divide $z' = z'[1] \,\|\, \cdots \,\|\, z'[\zeta']$ so that $|z'[1]| = \cdots = |z'[\zeta']| = m - n$

433 If $\zeta \neq \zeta'$ then $v'[\zeta'-1] \leftarrow \mathit{MD\text{-}iteration}\big(z'[1] \,\|\, z'[2] \,\|\, \cdots \,\|\, z'[\zeta'-1]\big)$

434 $x^* \leftarrow z'[\zeta'] \,\|\, v'[\zeta'-1]$ EndIf

435 If $\zeta = \zeta'$ then $v'[i-1] \leftarrow \mathit{MD\text{-}iteration}\big(z'[1] \,\|\, z'[2] \,\|\, \cdots \,\|\, z'[i-1]\big)$

436 $x^* \leftarrow z'[i] \,\|\, v'[i-1]$ EndIf

440 Output x^*

Fig. 4. Definition of adversary B attacking $h : \{0,1\}^m \to \{0,1\}^n$ in the cs-SPR sense

Theorem 1. *Let $\bar{H} : \{0,1\}^* \to \{0,1\}^n$ be the Merkle-Damgård hash function with split padding constructed of a compression function $h : \{0,1\}^m \to \{0,1\}^n$. Then $\bar{H}$ is SPR if h is cs-SPR. More concretely, we have*

$$\mathbf{Adv}^{\mathrm{spr}}_{\bar{H}}(t, \ell) \leq (\ell + 1) \cdot \mathbf{Adv}^{\text{cs-spr}}_{h}(t'),$$

where $t' \approx t + 2\ell \cdot \mathsf{time}(h)$. Note that the security parameter μ is implicit in the statement.

Proof. Let $A = (A_1, A_2)$ be an adversary attacking the hash function $\bar{H} : \{0,1\}^* \to \{0,1\}^n$ in the SPR sense. Assume that A has time complexity at most t and only handles strings whose lengths are at most ℓ blocks. We shall construct an adversary B that uses A_1 and A_2 as black-boxes and that attacks the underlying compression function h in the cs-SPR sense. The definition of B is given in Fig. 4. The basic idea is that B simulates an SPR game for A with B's challenge embedded into a randomly chosen block. Then B "hopes" that A finds a second preimage colliding at that block.

Let us first check if B simulates an SPR game for A correctly. In order to do this, we only need to verify that the distribution of simulated challenges w at line 421 is uniformly random on the set $\{0,1\}^\lambda$. The only difference between this w and the $x \in \{0,1\}^\lambda$ at line 411 is that the α in the i-th block is replaced with the challenge $\tilde{x}$. Because of the split padding, all the bits of $\alpha \in \{0,1\}^\mu$

come from the random message x. Since $\tilde{x}$ is random and independent from x, we see that w is indeed drawn uniformly at random from the set $\{0,1\}^\lambda$. The point here is that B can choose any block, as every block contains at least μ bits of a message owing to the split padding.

We next evaluate the success probability of B. We see that B succeeds whenever A succeeds, provided that at the same time B correctly guesses the index i (Here we need (i) injectivity of *split-padding*, (ii) injectivity of *MD-strengthening*, and (iii) the length encoding in *MD-strengthening*). The choice of index i does not affect the overall distribution of $w \in \{0,1\}^\lambda$; the distribution is the same as $w \xleftarrow{\$} \{0,1\}^\lambda$, being independently random from i. Moreover, the value i is completely hidden from A. Therefore, the choice of i is independent from the transcript of A producing x'. Putting $u \leftarrow \textsf{MD-strengthening}(\textsf{split-padding}(w))$ we get:

$$\begin{aligned}\mathbf{Adv}_h^{\text{cs-spr}}(B) &\geq \Pr\big[w \bowtie_{\tilde{H}} x' \text{ and } i = \textsf{index}(u, z')\big] \\ &= \Pr\big[w \bowtie x'\big] \times \Pr\big[i = \textsf{index}(u, z') \mid w \bowtie x'\big] \\ &= \Pr\big[A \text{ succeeds}\big] \times (1/\zeta) \geq 1/(\ell+1) \cdot \mathbf{Adv}_{\tilde{H}}^{\text{spr}}(A).\end{aligned}$$

Lastly, we compute the time complexity of adversary B. It is about equal to the time complexity of A plus *two* executions of *MD-iteration* at lines 415, 433 and 435, each of which costs at most $\ell \cdot \textsf{time}(h)$. This proves the theorem. □

Remarks on the Birthday Bound. We note that our bound for SPR is of *quadratic* degradation in ℓ (a linear term in the coefficient of the advantage function and another one in time complexity). This means that the security guarantee becomes vacuous when $\ell \approx 2^{n/2}$ (with $\mu = n$; recall that with $t \approx 2^{n/2} \cdot \textsf{time}(h)$ the advantage increases to about $2^{-n/2}$, *cf.* [23]). In fact, the long-message SPR attacks described in [9,17] are still applicable to the patched construction. It also implies that our reduction essentially gives a tight bound.

This is neither regression to the plain Merkle-Damgård construction nor severe limitation in practice. In the original Merkle-Damgård construction, the SPR of a hash function is assured up to the birthday bound based on the strong CR assumption about the underlying compression function rather than cs-SPR. The birthday bound by CR is at $t \approx 2^{n/2} \cdot \textsf{time}(h)$ irrespective of the message length ℓ. Thus, our result provides a stronger bound than the one originally assured by CR. Also, in practice a typical value of σ restricts message lengths to less than $2^{n/2}$ blocks, so speaking of the security beyond $\ell \approx 2^{n/2}$ often becomes moot.

Moreover, many of the provably secure SPR (TCR) constructions, including Randomized Hashing [13] and Higher-Order UOWHF [14], are susceptible to the long-message SPR attacks, hence giving security only up to the birthday bound. It is true that there exist some constructions that accomplish SPR beyond the birthday bound, such as Wide-Pipe [20] and ROX [2], but the security of these constructions relies on the use of random oracles.

It seems a non-trivial task for us to construct a mode of operation that achieves the full SPR security without random oracles: The "dithering" and "checksum" require major modifications to the current Merkle-Damgård construction, and these techniques are shown to be not effective in precluding long-message attacks [1,11].

7 OW of Merkle-Damgård with Split Padding

In this section we prove that the Merkle-Damgård construction combined with split padding is OW provided that the underlying compression function is cs-OW. The result contrasts sharply with the one for SPR of the previous section in that we have a security reduction without the birthday bound.

Theorem 2. *Let $\bar{H} : \{0,1\}^* \to \{0,1\}^n$ be the Merkle-Damgård hash function with split padding constructed of a compression function $h : \{0,1\}^m \to \{0,1\}^n$. Then $\bar{H}$ is OW if h is cs-OW. More concretely, we have*

$$\mathbf{Adv}_{\bar{H}}^{\text{ow}}(t, \ell) \le \mathbf{Adv}_{h}^{\text{cs-ow}}(t'),$$

where $t' \approx t + 2\ell \cdot \mathsf{time}(h)$. Note that the security parameter μ is implicit in the statement.

Proof. Let $A = (A_1, A_2)$ be an adversary trying to invert the hash function $\bar{H} : \{0,1\}^* \to \{0,1\}^n$ in the OW sense. Assume that A has time complexity at most t and only handles strings whose lengths are at most ℓ blocks. We shall construct an adversary B that uses A_1 and A_2 as black-boxes and that tries to invert the underlying compression function h in the cs-OW sense. The definition of B is given in Fig. 5. A significant difference from the SPR case is that B simulates an OW game for A with its challenge embedded always into the last block.

We first check if B simulates an OW game for A correctly. For this, we need to verify that the distribution of the challenge v given at line 520 indeed coincides with the distribution $v \leftarrow \bar{H}(x)$, $x \xleftarrow{\$} \{0,1\}^\lambda$. By definition of cs-OW oracle, the value v at line 520 is computed as $v \leftarrow h(\tilde{x} \| \beta \| v[\zeta - 1])$ with $\tilde{x} \xleftarrow{\$} \{0,1\}^\mu$. Now note that all the bits of α at line 515 come from the random message $x \in \{0,1\}^\lambda$ and appear in no other blocks, owing to the split padding. Since the randomness of $\tilde{x}$ is independent from that of x, replacing α with $\tilde{x}$ does not affect the distribution of v. Thus we see that B indeed simulates the correct distribution of v.

It remains to evaluate the success probability of B. We observe that B succeeds whenever A succeeds in inversion, so $\mathbf{Adv}_{h}^{\text{cs-ow}}(B) \ge \mathbf{Adv}_{\bar{H}}^{\text{ow}}(A)$ holds. The time complexity of B is about that of A plus two executions of $\mathsf{MD\text{-}iteration}$ at lines 514 and 533. This proves the theorem. □

Tightness of the Bound. Unlike the case of SPR, this time the degradation is only linear in ℓ (*i.e.*, we do not have the coefficient $\ell+1$ in front of the advantage function).[4] This means that we still have some security left even when $\ell \approx 2^{n/2}$

[4] [26,10] obtains an OW result based on the OW assumption about h and the "output regularity" of h. The result, however, has a coefficient of ℓ in front of the advantage function (associated with the regularity).

Adversary B

510 Run $A = (A_1, A_2)$ and obtain a bit length $(\lambda, St) \leftarrow A_1(\cdot)$ // $\lambda \geq \mu$
511 Generate a random message $x \overset{\$}{\leftarrow} \{0,1\}^\lambda$
512 Put $z \leftarrow \mathit{MD\text{-}strengthening}(\mathit{split\text{-}padding}(x))$
513 Divide $z = z[1] \,\|\, z[2] \,\|\, \cdots \,\|\, z[\zeta]$ so that $|z[1]| = |z[2]| = \cdots = |z[\zeta]| = m - n$
514 Compute $v[\zeta - 1] \leftarrow \mathit{MD\text{-}iteration}(z[1] \,\|\, z[2] \,\|\, \cdots \,\|\, z[\zeta - 1])$
515 Divide $z[\zeta] = \alpha \| \beta$ so that $|\alpha| = \mu$ and $|\beta| = m - n - \mu$
520 Submit $\beta \| v[\zeta - 1]$ as a committed suffix and receive a challenge $v \in \{0,1\}^n$
530 Feed v to A_2 and obtain a preimage $x' \leftarrow A_2(v, St)$
531 Put $z' \leftarrow \mathit{MD\text{-}strengthening}(\mathit{split\text{-}padding}(x'))$
532 Divide $z' = z'[1] \,\|\, \cdots \,\|\, z'[\zeta']$ so that $|z'[1]| = \cdots = |z'[\zeta']| = m - n$
533 Compute $v'[\zeta' - 1] \leftarrow \mathit{MD\text{-}iteration}(z'[1] \,\|\, z'[2] \,\|\, \cdots \,\|\, z'[\zeta' - 1])$
534 Set $x^* \leftarrow z'[\zeta'] \,\|\, v'[\zeta' - 1]$
540 Output x^*

Fig. 5. Definition of adversary B attacking $h : \{0,1\}^m \rightarrow \{0,1\}^n$ in the cs-OW sense

(with $\mu = n$) and that long-message birthday attacks do not apply to the OW case.

Our bound for OW is "essentially" tight, except for the ℓ-degradation in time complexity. To see this, consider an inverter A (in the OW sense) attacking $\bar{H}$, who outputs $\lambda = \mu$ at the beginning of each game and receives a challenge $v \in \{0,1\}^n$. Then the challenge is computed as $v \leftarrow h(\tilde{x} \| a)$, $\tilde{x} \overset{\$}{\leftarrow} \{0,1\}^\mu$ with the suffix $a = 010^* \| \langle |\mu| \rangle \| \mathit{IV}$. This is "essentially" a cs-OW game on h, except that the suffix $a \in \{0,1\}^{m-\mu}$ is not completely "chosen" by A but rather "known" to A. We shall discuss more on the gap between cs-OW and ks-OW in Sect. 9.

8 Analysis of cs-SPR

The purpose of this section is to reveal the nature of cs-SPR. It is clear that CR implies cs-SPR[5] but not vice versa, so cs-SPR is a strictly weaker requirement than CR. Here we do want to say more; that is, we claim that cs-SPR is an assumption which is inherently weaker than CR, by showing:

A cs-SPR function exists if an OW function exists.

This is a complexity-theoretic result. It is known that the existence of Coll functions implies that of OW functions [12], but not vice versa—[34] shows that there exists no black-box construction of Coll functions from OW functions. This is a strong evidence of separation between the Coll property and the OW, and we show that cs-SPR belongs to the latter class.

[5] More formally, it should read "fp-CR implies cs-SPR."

Fig. 6. A ⇒ B indicates "A-secure implies B-secure," while A → B indicates "A-existence implies B-existence." Dotted boxes indicate black-box separation from the Coll requirement.

Our claim is based on the results [25,30] which prove that the existence of an OW function implies that of a TCR function family.[6] The existence of a TCR function family is equivalent to that of an SPR function [33], so we actually present explicit black-box construction of a cs-SPR function from an SPR function, thereby showing the existence of a cs-SPR function based on that of an OW function. For better understanding of cs-SPR, we also point out a symmetry between cs-SPR and eTCR; roughly speaking, cs-SPR can be regarded as an unkeyed version of eTCR. The diagram on the left in Fig. 6 summarizes the relationships among these various notions.

Proposition 3. *A cs-SPR function exists if an SPR function exists.*

Proof. Let $f : \{0,1\}^\mu \to \{0,1\}^\nu$ be an SPR function. Define $g : \{0,1\}^m \to \{0,1\}^{\nu+m-\mu}$ as $g(\tilde{x}\|a) \stackrel{\text{def}}{=} f(\tilde{x})\|a$ for $\tilde{x} \in \{0,1\}^\mu$ and $a \in \{0,1\}^{m-\mu}$. Then it can be directly verified that $\mathbf{Adv}_g^{\text{cs-spr}}(t) \le \mathbf{Adv}_f^{\text{spr}}(t')$ where $t' \approx t$. □

Proposition 4. *A cs-SPR function exists if and only if an eTCR function family exists.*

Proof. Let $f : \{0,1\}^m \to \{0,1\}^n$ be a cs-SPR function with a security parameter $\mu < m$. Define a family of functions $\varphi_k : \{0,1\}^{m-\mu} \to \{0,1\}^n$ with $k \in \{0,1\}^\mu$ as $\varphi_k(a) \stackrel{\text{def}}{=} f(k\|a)$. Then it is easy to see that $\mathbf{Adv}_\varphi^{\text{etcr}}(t) \le \mathbf{Adv}_f^{\text{cs-spr}}(t')$ where $t' \approx t$. Conversely, let $\varphi_k : \{0,1\}^{m-\mu} \to \{0,1\}^n$ be an eTCR function family with $k \in \{0,1\}^\mu$. Define $f : \{0,1\}^m \to \{0,1\}^n$ as $f(\tilde{x}\|a) \stackrel{\text{def}}{=} \varphi_{\tilde{x}}(a)$. Then we see that $\mathbf{Adv}_f^{\text{cs-spr}}(t) \le \mathbf{Adv}_\varphi^{\text{etcr}}(t')$ where $t' \approx t$. □

The notion of cs-SPR provides a bridge between SPR and CR, depending on the value μ. This can be viewed as an unkeyed version of the continuum developed in [22]. Also, the notion of cs-SPR contrasts sharply with that of fp-CR, as there is a clear distinction between the two: The notion of fp-CR is open to generic birthday attacks, whereas that of cs-SPR is not.

[6] These results are based on polynomially-bounded reductions. [25] proves the existence based on that of an OW *permutation*, while [30] on that of an OW function.

9 Analysis of cs-OW

There are obvious implications: cs-OW-secure ⇒ ks-OW-secure ⇒ OW-secure. Thus our assumption cs-OW is the strongest of these three notions. We show that cs-OW is, however, not "too far" from OW, by proving (see the diagram on the right in Fig. 6):

A cs-OW function exists if an OW function exists.

Unlike the case of cs-SPR, we have a direct black-box construction this time:

Proposition 5. *A cs-OW function exists if an OW function exists.*

Proof. Let $f : \{0,1\}^\mu \to \{0,1\}^\nu$ be an OW function. Define $g : \{0,1\}^m \to \{0,1\}^{\nu+m-\mu}$ as $g(\tilde{x}\|a) \stackrel{\text{def}}{=} f(\tilde{x})\|a$ for $\tilde{x} \in \{0,1\}^\mu$ and $a \in \{0,1\}^{m-\mu}$. Then it can be directly verified that $\mathbf{Adv}_g^{\text{cs-ow}}(t) \leq \mathbf{Adv}_f^{\text{ow}}(t')$ where $t' \approx t$. □

10 Application to Randomized Hashing

In closing the paper we point out that our split padding is compatible with the Randomized Hashing [13]. Recall that the Randomized Hashing first mixes a message with a randomly generated mask and then hashes the data using the Merkle-Damgård construction. A problem arises when line 104 of the algorithm *MD-strengthening* in Fig. 1 is invoked, because it would then result in an insufficient amount of randomness in the last block. The Randomized Hashing suggests using "double padding" for dealing with this problem. Our split padding offers an alternative to this method, assuring at least μ bits of randomness in the very last invocation to the compression function.

References

1. Andreeva, E., Bouillaguet, C., Fouque, P.A., Hoch, J.J., Kelsey, J., Shamir, A., Zimmer, S.: Second preimage attacks on dithered hash functions. In: Smart, N.P. (ed.) EUROCRYPT 2008. LNCS, vol. 4965, pp. 270–288. Springer, Heidelberg (2008)
2. Andreeva, E., Neven, G., Preneel, B., Shrimpton, T.: Seven-property-preserving iterated hashing: ROX. In: Kurosawa, K. (ed.) ASIACRYPT 2007. LNCS, vol. 4833, pp. 130–146. Springer, Heidelberg (2007)
3. Andreeva, E., Preneel, B.: A three-property-secure hash function. In: Avanzi, R., Keliher, L., Sica, F. (eds.) SAC 2008. Workshop Records, pp. 208–224 (2008)
4. Biham, E.: New techniques for cryptanalysis of hash functions and improved attacks on Snefru. In: Nyberg, K. (ed.) FSE 2008. LNCS, vol. 5086, pp. 444–461. Springer, Heidelberg (2008)
5. Bellare, M., Kohno, T.: Hash function balance and its impact on birthday attacks. In: Cachin, C., Camenisch, J.L. (eds.) EUROCRYPT 2004. LNCS, vol. 3027, pp. 401–418. Springer, Heidelberg (2004)

6. Bellare, M., Ristenpart, T.: Hash functions in the dedicated-key setting: Design choices and MPP transforms. In: Arge, L., Cachin, C., Jurdziński, T., Tarlecki, A. (eds.) ICALP 2007. LNCS, vol. 4596, pp. 399–410. Springer, Heidelberg (2007)
7. Coron, J.S., Dodis, Y., Malinaud, C., Puniya, P.: Merkle-Damgård revisited: How to construct a hash function. In: Shoup, V. (ed.) CRYPTO 2005. LNCS, vol. 3621, pp. 430–448. Springer, Heidelberg (2005)
8. Damgård, I.: A design principle for hash functions. In: Brassard, G. (ed.) CRYPTO 1989. LNCS, vol. 435, pp. 416–427. Springer, Heidelberg (1990)
9. Dean, R.D.: Formal Aspects of Mobile Code Security. PhD thesis, Princeton University (1999)
10. Dodis, Y., Puniya, P.: Getting the best out of existing hash functions; or what if we are stuck with SHA? In: Bellovin, S.M., Gennaro, R., Keromytis, A.D., Yung, M. (eds.) ACNS 2008. LNCS, vol. 5037, pp. 156–173. Springer, Heidelberg (2008)
11. Gauravaram, P., Kelsey, J.: Linear-XOR and additive checksums don't protect Damgård-Merkle hashes from generic attacks. In: Malkin, T.G. (ed.) CT-RSA 2008. LNCS, vol. 4964, pp. 36–51. Springer, Heidelberg (2008)
12. Goldreich, O.: The Foundations of Cryptography, vol. 2. Cambridge University Press, Cambridge (2004)
13. Halevi, S., Krawczyk, H.: Strengthening digital signatures via randomized hashing. In: Dwork, C. (ed.) CRYPTO 2006. LNCS, vol. 4117, pp. 41–59. Springer, Heidelberg (2006)
14. Hong, D., Preneel, B., Lee, S.: Higher order universal one-way hash functions. In: Lee, P.J. (ed.) ASIACRYPT 2004. LNCS, vol. 3329, pp. 201–213. Springer, Heidelberg (2004)
15. Joux, A.: Multicollisions in iterated hash functions. Application to cascaded constructions. In: Franklin, M. (ed.) CRYPTO 2004. LNCS, vol. 3152, pp. 306–316. Springer, Heidelberg (2004)
16. Kelsey, J., Kohno, T.: Herding hash functions and the Nostradamus attack. In: Vaudenay, S. (ed.) EUROCRYPT 2006. LNCS, vol. 4004, pp. 183–200. Springer, Heidelberg (2006)
17. Kelsey, J., Schneier, B.: Second preimages on n-bit hash functions for much less than 2^n work. In: Cramer, R. (ed.) EUROCRYPT 2005. LNCS, vol. 3494, pp. 474–490. Springer, Heidelberg (2005)
18. Leurent, G.: MD4 is not one-way. In: Nyberg, K. (ed.) FSE 2008. LNCS, vol. 5086, pp. 412–428. Springer, Heidelberg (2008)
19. Lai, X., Massey, J.L.: Hash function based on block ciphers. In: Rueppel, R.A. (ed.) EUROCRYPT 1992. LNCS, vol. 658, pp. 55–70. Springer, Heidelberg (1993)
20. Lucks, S.: A failure-friendly design principle for hash functions. In: Roy, B. (ed.) ASIACRYPT 2005. LNCS, vol. 3788, pp. 474–494. Springer, Heidelberg (2005)
21. Merkle, R.C.: One way hash functions and DES. In: Brassard, G. (ed.) CRYPTO 1989. LNCS, vol. 435, pp. 428–446. Springer, Heidelberg (1990)
22. Mironov, I.: Hash functions: From Merkle-Damgård to Shoup. In: Pfitzmann, B. (ed.) EUROCRYPT 2001. LNCS, vol. 2045, pp. 166–181. Springer, Heidelberg (2001)
23. Menezes, A.J., van Oorschot, P.C., Vanstone, S.A.: Handbook of Applied Cryptography. CRC Press, Boca Raton (1996)
24. NIST: Secure Hash Standard. FIPS 180-1 (1995)
25. Naor, M., Yung, M.: Universal one-way hash functions and their cryptographic applications. In: ACM 21st STOC, pp. 33–43. ACM, New York (1989)
26. Puniya, P.: New Design Criteria for Hash Functions and Block Ciphers. PhD thesis, New York University (2007)

27. Rivest, R.L.: The MD4 message digest algorithm. In: Menezes, A., Vanstone, S.A. (eds.) CRYPTO 1990. LNCS, vol. 537, pp. 303–311. Springer, Heidelberg (1991)
28. Rivest, R.L.: The MD5 Message-Digest Algorithm. RFC 1321. IETF (1992)
29. Rogaway, P.: Formalizing human ignorance: Collision-resistant hashing without the keys. In: Nguyên, P.Q. (ed.) VIETCRYPT 2006. LNCS, vol. 4341, pp. 211–228. Springer, Heidelberg (2006)
30. Rompel, J.: One-way functions are necessary and sufficient for secure signatures. In: ACM 22nd STOC, pp. 387–394. ACM, New York (1990)
31. Rogaway, P., Shrimpton, T.: Cryptographic hash-function basics: Definitions, implications, and separations for preimage resistance, second-preimage resistance, and collision resistance. In: Roy, B., Meier, W. (eds.) FSE 2004. LNCS, vol. 3017, pp. 371–388. Springer, Heidelberg (2004)
32. Schneier, B.: Applied Cryptography, 2nd edn. John Wiley & Sons, Chichester (1996)
33. Shoup, V.: A composition theorem for universal one-way hash functions. In: Preneel, B. (ed.) EUROCRYPT 2000. LNCS, vol. 1807, pp. 445–452. Springer, Heidelberg (2000)
34. Simon, D.R.: Finding collisions on a one-way street: Can secure hash functions be based on general assumptions? In: Nyberg, K. (ed.) EUROCRYPT 1998. LNCS, vol. 1403, pp. 334–345. Springer, Heidelberg (1998)
35. Stevens, M., Lenstra, A.K., de Weger, B.: Chosen-prefix collisions for MD5 and colliding X.509 certificates for different identities. In: Naor, M. (ed.) EUROCRYPT 2007. LNCS, vol. 4515, pp. 1–22. Springer, Heidelberg (2007)
36. Stinson, D.R.: Some observations on the theory of cryptographic hash functions. Des. Codes Cryptography 38(2), 259–277 (2006)
37. Wang, X., Yu, H.: How to break MD5 and other hash functions. In: Cramer, R. (ed.) EUROCRYPT 2005. LNCS, vol. 3494, pp. 19–35. Springer, Heidelberg (2005)
38. Wang, X., Yin, Y.L., Yu, H.: Finding collisions in the full SHA-1. In: Shoup, V. (ed.) CRYPTO 2005. LNCS, vol. 3621, pp. 17–36. Springer, Heidelberg (2005)

Limits of Constructive Security Proofs

Michael Backes[1,2] and Dominique Unruh[1]

[1] Saarland University, Saarbrücken, Germany
{backes,unruh}@cs.uni-sb.de
[2] Max-Planck-Institute for Software Systems, Saarbrücken, Germany
backes@mpi-sws.mpg.de

Abstract. The collision-resistance of hash functions is an important foundation of many cryptographic protocols. Formally, collision-resistance can only be expected if the hash function in fact constitutes a parametrized family of functions, since for a single function, the adversary could simply know a single hard-coded collision. In practical applications, however, unkeyed hash functions are a common choice, creating a gap between the practical application and the formal proof, and, even more importantly, the concise mathematical definitions.

A pragmatic way out of this dilemma was recently formalized by Rogaway: instead of requiring that no adversary exists that breaks the protocol (existential security), one requires that given an adversary that breaks the protocol, we can efficiently construct a collision of the hash function *using an explicitly given reduction* (constructive security).

In this paper, we show the limits of this approach: We give a protocol that is existentially secure, but that provably cannot be proven secure using a constructive security proof.

Consequently, constructive security—albeit constituting a useful improvement over the state of the art—is not comprehensive enough to encompass all protocols that can be dealt with using existential security proofs.

1 Introduction

The collision-resistance of hash functions is an important ingredient of many cryptographic protocols. Formally, collision-resistance can only be expected if the hash function in fact constitutes a parametrized family of functions, since for a single function, the adversary could simply have a collision hard-coded into its program. In practical applications, however, such unkeyed hash functions are often used (e.g., SHA-1), creating a gap between the practical application and the formal proof, and, even more importantly, the concise mathematical definitions.

A pragmatic way out of this dilemma was discussed by Stinson [10] and recently formalized by Rogaway [9]: instead of requiring that no adversary exists that breaks the protocol (existential security), one requires that given an adversary that breaks the protocol, one can efficiently construct a collision of the hash function *using an explicitly given reduction* (constructive security).

Slightly more formally, the dilemma can be described as follows: An existential security proof for a protocol π shows the following: If there exists a

J. Pieprzyk (Ed.): ASIACRYPT 2008, LNCS 5350, pp. 290–307, 2008.
© International Association for Cryptologic Research 2008

polynomial-time adversary A that has a non-negligible advantage in breaking the protocol, then there exists a polynomial-time adversary B that has a non-negligible advantage in breaking at least one of the assumptions of the protocol. Here, the exact meaning of the word *advantage* depends on the security notion under consideration; in a proof system for example, the advantage would be the probability to convince the verifier of a wrong fact. For collision-resistant hash functions, it would be the probability of finding a collision. Considering a protocol π whose security is based on the collision-resistance of an *unkeyed* hash function H, an existential security proof would show the following: If an adversary A has non-negligible advantage in breaking π, there is an adversary B that outputs a collision of H with non-negligible probability. However, this is vacuously true: There always exists an adversary that has a collision of H hard-coded into its program and outputs this collision with probability one. We, that is the totality of all human beings, might not know this adversary, but it exists nonetheless. To circumvent this problem, mathematical definitions and proofs usually make use of keyed hash functions. In this case, for every key K the collision might be different so that the assumption that no polynomial-time adversary can compute collisions for more than a small fraction of the keys is sensible.

But what if we are forced to use unkeyed hash functions, e.g., because of efficiency considerations or simply because industrial applications often rely on unkeyed hash functions? Do we lose all possibility to prove security, since we cannot expect an existential security proof in this case? Fortunately, this is not necessarily the case: we may ground security on the observation that although there always exists an adversary finding a collision of an unkeyed hash function, this adversary might not be explicitly known. This leads to the following approach that was recently formalized by Rogaway [9]: A constructive security proof for a protocol π that uses a hash function H is an efficient transformation C (that must be explicitly given) that, upon input an adversary A and the hash function H, outputs a collision of H. If someone finds a successful adversary A, he hence also knows a collision, thereby breaking the collision-resistance of the hash function.

Rogaway [9] stresses that most existential security proofs already constitute constructive security proofs and that all that must be done for concisely handling unkeyed hash functions is to rephrase those proofs in a constructive setting. Indeed, folklore has always believed that protocols with existential security proofs can be transformed into constructive ones. In some cases it may be as easy as rephrasing the theorem statement, in other cases it may be as hard as finding a different proof. E.g., [9] writes: "In general, it is well understood that one can rephrase provable-security results as assertions about explicitly given reductions". Although this folklore statement may hold true in many cases of practical interest, we show that it does not hold true in general. We construct a protocol (more exactly, a zero-knowledge argument of knowledge) that we show secure with an existential security proof, but for which we further show that there provably does not exist any constructive security proof.

Hence although constructive security proofs may constitute a useful improvement over the state of the art, there are applications where the use of unkeyed hash functions cannot be justified even with this technique.

1.1 Our Contribution

We show how hash functions can be used to construct protocols that can be shown secure using an existential security proof, but that cannot be proven secure using a constructive security proof.

The main idea underlying this separating example is to construct a protocol whose security is based on a *non-uniform* security reduction. Then, this reduction will only lead to a *non-uniform* collision-finding algorithm. Since an unkeyed hash function can only be secure against *uniform* adversaries, such a reduction does not lead to a contradiction when basing the protocol on an unkeyed hash function. Thus, in particular, a *non-uniform* reduction does not give rise to a constructive security proof. The main technical difficulty lies in actually proving that the security of the protocol can *only* be shown using a non-uniform reduction.

More specifically, we investigate argument systems (computationally sound proof systems) as our security notion of interest. The approach can be adapted to other notions as well, e.g., by constructing a protocol for another task that uses and depends on the argument system presented in this paper.

In more detail, we construct, depending on a hash function H, a proof system (P^H, V^H) of which we can show the following properties:

- Under two nonstandard but reasonable assumptions (discussed below in the paragraph on complexity assumptions and formalized in Assumption 1 in the body of the paper) and the assumption that H is a non-uniform collision-resistant hash function, we can give an *existential* security proof for (P^H, V^H).
- Using Assumption 1, we can prove that one cannot give a *constructive* security proof that reduces the security of (P^H, V^H) to the collision-resistance of H. This even holds independent of any additional assumptions we might use for the constructive security proof (as long as these assumptions are not false).

At a first glance, this separation may seem confusing because of the different layers of assumptions (in the proofs themselves and in the proofs about proofs). Thus the following view might help to improve the intuition underlying our result: In a world where Assumption 1 has been *proven* to hold, it will be possible to show *existentially* that (P^H, V^H) is secure if H is collision-resistant, but a *constructive* security proof for (P^H, V^H) reducing to the collision resistance of H will be impossible.

At this point, we consider it important to stress that our assumptions and in particular our proofs strongly rely on the careful distinction of non-uniform and uniform complexity. In particular, we use non-uniform techniques to prove results about uniform algorithms.

Basic Idea of the Construction. In order to construct a zero-knowledge argument of knowledge that has an existential proof of security but no constructive security proof, we use the following general approach. We take an existing zero-knowledge proof of knowledge $(P^\dagger, V^\dagger)$ and modify it as follows: Instead of directly showing that a given statement σ holds, the prover P^H shows (using $P^\dagger$) that one of the following two statements holds:

- he knows a witness for the statement σ, or
- he knows a ciphertext c that is the encryption of a collision of H.

The basic idea is that given an adversary that knows such a ciphertext c, one can break the argument. However, given an adversary with a hard-coded ciphertext, a constructive security proof should not be able to extract the collision contained in the ciphertext. We have to achieve the following two goals:

- If H is a collision-resistant keyed hash function, it is hard to find a ciphertext c that is the encryption of a collision of H. Otherwise the argument can be easily broken even if the hash function is secure, thus even defying the existential security proof.
- Given c, it is hard to extract a collision from c; in particular, the decryption key should be secret. Otherwise a constructive security proof can use a knowledge extractor to extract c from a successful prover and then extract a collision from c. Further, the decryption of c should not be part of the witness used for the proof system $(P^\dagger, V^\dagger)$ since this witness could then be extracted from the adversary.

We achieve the first goal as follows: To ensure that it is hard to find a ciphertext given a collision-resistant keyed hash function, we use an encryption scheme that can be broken by non-uniform adversaries, but that is secure against uniform adversaries. An adversary that breaks (P^H, V^H) entails an adversary that finds a ciphertext c that is the encryption of a collision of H. This again entails the existence of a *non-uniform* adversary decrypting these ciphertexts and thus finding collisions. Consequently, if we require H to be a keyed hash function that is collision-resistant against *non-uniform* adversaries, we obtain a contradiction. On the other hand, a constructive security proof cannot obtain the collisions in this way, since in such a proof the reduction would have to be explicitly given and thus in particular be a uniform algorithm.

The second goal is achieved as follows: We do not directly show (using $P^\dagger$) that c is the encryption of a collision of H, since this would necessitate to use the plaintext, i.e., the collision, as a witness, which in turn would allow to extract this witness. Instead, we introduce another proof system (P^*, V^*). This proof system is non-interactive (in the strong sense that it does not even use a common reference string), statistically sound (otherwise the overall scheme could be broken by non-uniform adversaries that know a single wrong proof) and it should hide the plaintext of the encryption c. The last condition roughly means that if some adversary can extract the plaintext of c given a proof N, then it could also extract the plaintext without knowledge of N with non-negligible probability. We call such a proof system a *content-hiding proof of content.* Given a content-hiding proof of content, we do not directly prove that c is an encryption of a collision, but that we know a non-interactive proof N that c is an encryption of a collision. Then in the constructive security proof, c and N might be extractable from an adversary, but this would not be of help: If one could extract a collision from c and N, one could extract one from c alone as well (since (P^*, V^*) is content-hiding). If the encryption scheme is IND-CPA secure, the encryption c alone is indistinguishable from a random encryption. Thus one could also find the collision without using c at all. A constructive security proof would hence imply the existence of an algorithm to find collisions.

Summary of the Construction. We now summarize our construction in a more detailed and a more concise manner. Let f be a one-way permutation that is secure against uniform adversaries, but can be inverted by non-uniform adversaries (Definition 2). From f we construct an encryption scheme $\mathcal{E}_f$ such that for each security parameter, there is a fixed public key, and such that the corresponding secret key can be found by a non-uniform adversary (Definition 3). The scheme $\mathcal{E}_f$ is shown to be IND-CPA secure (Lemma 2).

Let then (P^*, V^*) be a content-hiding proof of content for the encryption scheme $\mathcal{E}_f$ (Definitions 4 and 5). That is, using P^* we can show non-interactively that a given ciphertext c is the encryption of a cleartext m that fulfills a given property π. Since P^* is content-hiding, we know that if we can extract the plaintext from c given the non-interactive proof, we can also do so without access to the proof. Let $(P^\dagger, V^\dagger)$ be a computational zero-knowledge proof of knowledge. Let H be a hash function (keyed or unkeyed). Then we construct the argument system (P^H, V^H) as follows (Definition 6):

- The prover P^H takes as input a SAT-instance σ and a corresponding witness w. The verifier V^H expects a SAT-instance σ.
- To show his knowledge of w, the prover P^H invokes the prover $P^\dagger$ to show that either
  - he knows a witness w for σ, or
  - he knows a ciphertext c and a non-interactive proof N such that the proof N convinces the verifier V^* that the ciphertext c is an encryption of a collision of H.

  The prover can easily perform this proof since he knows the witness w.
- The verifier V^H uses $V^\dagger$ to verify the above proof.

Note that the prover P^* is never used in the above construction. The existence of P^* will however be used in the proofs.

On our Complexity Assumptions. Our proof is based on the existence of content-hiding proofs of content as well as on the existence of one-way permutations with non-uniform trapdoors, which constitute nonstandard complexity assumptions. To motivate these assumptions, we prove that relative to a random oracle these assumptions follow from standard ones.

At a first glance, it may seem that a result that needs such strong assumptions and involved constructions will not be of relevance for the provability of natural protocol constructions, i.e., construction which do not have the creation of a counterexample in mind. We would like to point out the following counter-arguments: First, one reason why we need such strong assumptions is that we do not only want a protocol that cannot be proven secure using constructive proofs, but that *provably* cannot be proven secure using constructive proofs. The reason for the complexity of our example may hence not follow from the fact that all natural protocols have constructive proofs, but rather from the fact that proving unprovability is in general a difficult task. Secondly, somewhat similar techniques have already been used in the literature: Barak [3] presents an argument system in which the prover proves that the statement under consideration is true or that he knows a short circuit describing (the data sent by) the verifier. This seemingly contrived construction then was shown to allow for argument systems that enjoy properties that where shown to be impossible for zero-knowledge argument systems that do not use the circuit of the adversary (i.e., black-box zero-knowledge argument systems). In that light it may well be possible that some useful protocol will have to use constructions similar to the ones presented in this work and therefore will have no constructive security proof.

1.2 Related Work

Hash functions where first formalised in [4]. In [9] the notion of a constructive security proof was made explicit, although the concept was already discussed or implicitly used in many other papers.

The idea of considering problems relative to oracles to analyze complexity assumptions was introduced by [2]. See also [6] for a survey and a discussion of such relativisation techniques.

An example of a non-constructive security proof can be found in [5, Section 8]. They give a resettable zero-knowledge proof in the timing-model, and the proof of soundness uses a non-constructive reduction. However, it is not shown that their protocol does not have a constructive proof. In contrast, the complexity of our constructions result from the necessity of creating a scheme where we can *prove* that no constructive security proof exists. We believe that the result of [5] and our result complement each other: [5] show that there are *natural* protocols where we *do not know* constructive security proofs, while we show that there are *contrived* protocols where constructive security proofs *do not exist* (under certain complexity assumptions).

2 Preliminaries and Notation

By $x \leftarrow A$ we mean assigning the output of the probabilistic algorithm A to x, and by $x \stackrel{\$}{\leftarrow} M$ assigning a uniformly randomly chosen element of M to x. By $\langle A, B\rangle$ we mean the output of B after an interaction of the interactive machines A and B. The variable k will always denote the security parameter.

An *unkeyed hash function* H is a function from $\{0,1\}^*$ to $\{0,1\}^n$ for some n that can be computed in deterministic polynomial time. A *keyed (family of) hash functions* consists of a family $\{H_K\}$ of functions together with an efficient key generation algorithm G_H such that the following holds: Given K and x, the image $H_K(x)$ can be computed in deterministic polynomial time. Further, for $K \leftarrow G_H(1^k)$, the function H_K maps $\{0,1\}^*$ to $\{0,1\}^{\ell(k)}$ for some polynomially bounded function ℓ.

Of central interest to this paper is the notion of a constructive security proof. In principle, a constructive security proof consists of two parts: an explicitly given reduction C from adversaries to collisions, and a proof that C is indeed such a reduction. Since we are only interested in negative results in this paper, it will be sufficient to show that no such reduction C exists. We therefore slightly abuse notation and define a constructive security proof to solely consist of this reduction C. That is, we do not even require that the reduction is proven to be correct.

Furthermore, we will confine ourselves to constructive security proofs that a given protocol is an argument system. This results in a less abstract definition, which is sufficient for our application. Examples of constructive security proofs for other properties are given in [9].

Let (P^H, V^H) be a proof system parametrized by an unkeyed hash function H that is assumed to be given as a circuit. For an adversary A (given as a circuit) and an unsatisfiable SAT-formula σ, we define

$$\mathrm{Adv}^{\mathrm{arg}}_{V^H,k}(A,\sigma) := \Pr[\langle A, V^H(1^k,\sigma)\rangle = 1].$$

Further, for an algorithm C, let

$$\mathrm{Adv}^{\mathrm{col}}_{H,k}(C,A,\sigma) := \Pr[(x,x') \leftarrow C(1^k,H,A,\sigma) : x \neq x' \text{ and } H(x) = H(x')].$$

Definition 1 (Constructive Security Proof). *Let (P^H, V^H) be a proof system parametrised by an unkeyed hash function H. We call an algorithm C a* constructive security proof that (P^H, V^H) is an argument *if C runs in uniform probabilistic polynomial-time and there exist some $c > 0$ and some negligible function μ such that for all circuits A, all unsatisfiable boolean formulas σ and all $k \in \mathbb{N}$ we have*

$$\mathrm{Adv}^{\mathrm{col}}_{H,k}(C,A,\sigma) \geq \left(\frac{\mathrm{Adv}^{\mathrm{arg}}_{V^H,k}(A,\sigma)}{k + |A| + |H| + |\sigma|}\right)^c - \mu(k).$$

Our notion of a constructive security proof slightly deviates from the notion put forward in [9]. The most obvious difference is that [9] does not contain any asymptotic definition of a constructive security proof. Instead, all results are given in terms of concrete security, i.e., the relation between the advantage to break the protocol and the advantage to find collisions is given explicitly. A negative statement, i.e., a claim that a given protocol has no constructive security proof, cannot rely on concrete security since one does not aim to show that a given relation between the two advantages does not hold, but that no (useful) lower bound for $\mathrm{Adv}^{\mathrm{col}}$ in terms of $\mathrm{Adv}^{\mathrm{arg}}$ exists. To characterize such useful lower bounds we have introduced the above asymptotic formulation. Since we are interested in a negative result, we have made the lower bound as weak as possible.

A notion of black-box constructive proofs has also been formalized in [9]. Since black-box is the stricter kind of reduction, our negative result encompasses this notion as well.

3 Assumptions Underlying Our Negative Result

In this section, we will present two cryptographic assumptions that are needed in our proof.

3.1 One-Way Permutations with Non-uniform Trapdoors

The first assumption roughly states that there are one-way permutations that are secure against uniform adversaries but that can be inverted by non-uniform ones.

Definition 2 (One-Way Permutations with Non-Uniform Trapdoors). *A function* $f : \{0,1\}^* \to \{0,1\}^*$ *is a* one-way permutation with non-uniform trapdoors, *if*

- *The function* f *is a length-preserving permutation that is computable in deterministic polynomial time.*
- *The function* f *is one-way against uniform adversaries.*
- *There exists a sequence* t_k *of polynomial-sized circuits, such that* $t_k(f(x)) = x$ *for all* $k \in \mathbb{N}$ *and all* $x \in \{0,1\}^k$.

The existence of one-way permutations with non-uniform trapdoors constitutes a nonstandard complexity assumption in cryptography. Although we did not succeed in reducing the existence of one-way permutations with non-uniform trapdoors to more common assumptions in general, we show that there is an oracle relative to which this is possible.

Lemma 1. *Assume that trapdoor one-way permutations with dense public keys*[1] *exist that are one-way against uniform probabilistic polynomial-time adversaries. Then there exists an oracle* $\mathcal{O}$ *relative to which one-way permutations with non-uniform trapdoors exist.*

The proof of this lemma is given in the full version [1].

The proof of Lemma 1 in fact shows a stronger statement: choosing a random oracle entails one-way permutations with non-uniform trapdoors with probability one. If we accept the random oracle heuristic, the following conjecture is thus made realistic by the proof of Lemma 1:

[1] We say a family of trapdoor permutations has dense public keys if the distribution of the public keys is near the uniform distribution on the set of strings of a given length. Intuitively, this means that we can choose the public key using only public coins.

Conjecture 1. Let R be a sufficiently unstructured, efficiently computable function. Then using R in the construction of the proof of Lemma 1 yields one-way permutations with non-uniform trapdoors.

Using one-way permutations with non-uniform trapdoors, we can use the standard construction for creating IND-CPA secure encryption schemes from one-way permutations. The result is an encryption scheme where for each security parameter there is a single public key, and where the corresponding secret keys can be recovered by non-uniform adversaries (but not by uniform ones).

Definition 3 (Singleton Encryption). *Let f be a one-way permutation with non-uniform trapdoors. We define the* singleton encryption scheme $\mathcal{E}_f, \mathcal{D}_f$ for *f as follows: Let $pk_k := 1^k$ and $sk_k := t_k$, where t_k denotes the trapdoors of the function f. For $x \in \{0,1\}$, we have $\mathcal{E}_f(pk, x) := (f(r_1), r_2, (r_1 \cdot r_2) \oplus x)$ where r_1, r_2 are uniformly random from $\{0,1\}^{|pk|}$. For $x \in \{0,1\}^*$, we have $\mathcal{E}_f(pk, x) := (\mathcal{E}_f(pk, x_1), \ldots, \mathcal{E}_f(pk, x_{|x|}))$.*

The corresponding (deterministic) decryption algorithm $\mathcal{D}_f$ proceeds as follows: Upon input $(pk, sk, (c_1, r_2, c_2))$ where sk is a circuit and (c_1, r_2, c_2) the encryption of a single bit, the decryption algorithm first verifies that $f(sk(c_1)) = c_1$ and that $|c_1| = |pk|$. If so, it outputs $(sk(c_1) \cdot r_2) \oplus c_2$. Otherwise, it outputs $\perp$. The encryption of multiple bits is handled by decrypting each bit individually (with output $\perp$ if one of the decryptions fails).

The set of valid public keys of $\mathcal{E}_f$ for security parameter k is hence $\{pk_k\}$; consequentely the public key generation algorithm is trivial. The corresponding secret-keys sk_k, i.e., the trapdoors of f, are guaranteed to exist, but they are not efficiently computable by a uniform adversary. We have $\mathcal{D}_f(pk_k, sk_k, \mathcal{E}_f(pk_k, m)) = m$ for all m by construction; moreover, $\mathcal{D}_f(pk_k, sk, c) = m \neq \perp$ for some (possibly invalid) secret key sk implies $\mathcal{D}_f(pk_k, sk_k, c) = m$ since the checks performed by $\mathcal{D}_f$ guarantee $sk(c_1) = sk_k(c_1)$.

The following lemma states that the construction given above indeed results in an IND-CPA secure encryption scheme, at least against uniform adversaries:

Lemma 2. *Let f be a one-way permutation with non-uniform trapdoors and let $\mathcal{E}_f$ be the singleton encryption scheme for f. Then $\mathcal{E}_f$ is IND-CPA secure against uniform adversaries in the following sense: For all uniform probabilistic polynomial-time algorithms A_1, A_2, we have that*

$$\Pr\Big[(m_0, m_1, z) \leftarrow A_1(1^k), b \overset{\$}{\leftarrow} \{0,1\}, c \leftarrow \mathcal{E}_f(pk_k, m_b) : \\ A_2(1^k, c, z) = b \ \wedge \ |m_0| = |m_1|\Big]$$

is negligible in k.

A proof of Lemma 2 can be found in [8, Section 5.3.4.1]. Although this proof applies to a slightly different definition of public-key encryption where the public and secret keys are chosen by an explicit key generation algorithm, the proof carries over, mainly because the secret keys are not used in the definition of IND-CPA security.

3.2 Proofs of Content

We now introduce the novel notion of a non-interactive proof of content. Intuitively, a proof of content is a non-interactive proof system that proves that a given ciphertext c is the encryption of some plaintext m that fulfills some predicate π.

We first introduce some additional notation: Given an encryption scheme $(\mathcal{E}, \mathcal{D})$ with deterministic decryption, a Boolean circuit π, a ciphertext c, a public key pk and a private key sk, let $\pi^{pk,sk}[c] := true$ if and only if $m := \mathcal{D}(pk, sk, c) \neq \perp$ and $\pi(m) = 1$, and let $\pi^{pk}[c] = true$ if there exists a secret key sk such that $\pi^{pk,sk}[c] = true$.

Definition 4 (Non-Interactive Proofs of Content). *A non-interactive proof of content for an encryption scheme $(\mathcal{E}, \mathcal{D})$ (where $\mathcal{D}$ is deterministic) consists of a polynomial-time prover P and a polynomial-time verifier V such that the following holds:*

- Polynomial length. *There exists a polynomial p such that for every π, c, pk, sk, and k, we have $|P(1^k, \pi, c, pk, sk)| \leq p(|(1^k, \pi, c, pk, sk)|)$.*
- Completeness. *There is a negligible function μ such that for every π, c, pk and sk satisfying $\pi^{pk,sk}[c] = true$ and for every k, we have*

$$\Pr\left[V(1^k, pk, \pi, c, P(1^k, \pi, c, pk, sk)) = 0\right] \leq \mu(k).$$

- Soundness. *There is a negligible function μ such that for every π, c, and pk satisfying $\pi^{pk}[c] = false$ and for every k and every string N, we have*

$$\Pr\left[V(1^k, pk, \pi, c, N) = 1\right] \leq \mu(k).$$

So far, a proof of content can be quite easily realized by revealing the secret key of the encryption scheme. This of course is not satisfying; hence we need an additional secrecy property. We cannot expect the proof system to be zero-knowledge (since it is non-interactive without a common reference string), but we can require that a proof will not help us to extract the plaintext from the ciphertext m (which would be clearly violated if we learned the secret key). We will call this property *content-hiding*.

We now define content-hiding proofs of content. This notion will crucially depend on the notion of a valid public key of a given encryption scheme, and of the notion of the corresponding secret key. The notion of a valid public key and corresponding secret key has a natural meaning for most public-key cryptosystems, but it may not be well-defined in general. However, in the remainder of the paper we will only consider the encryption scheme from Definition 3 where a public key is valid if and only if it has the form 1^k, and where the secret key corresponding to a given public key is uniquely determined as t_k. So for the sake of readability we abstain from formally specifying what a valid public key and the corresponding secret key are.

Definition 5 (Content-Hiding Proofs of Content). *A non-interactive proof of content (P, V) for an encryption scheme $(\mathcal{E}, \mathcal{D})$ is called* content-hiding *if the following holds:*

Let G be any polynomial-time algorithm that upon input 1^k outputs a valid public key pk for $\mathcal{E}$, a message $m \in \{0,1\}^*$, a circuit π and some auxiliary information $z \in \{0,1\}^*$. Let A be any polynomial-time algorithm such that

$$\Pr\Big[(pk, m, \pi, z) \leftarrow G(1^k),\ c \leftarrow \mathcal{E}(pk, m),\ N \leftarrow P(1^k, \pi, c, pk, sk),\\ m' \leftarrow A(1^k, pk, c, \pi, z, N)\ :\ m = m'\Big]$$

is not *negligible in k, where sk denotes the secret key corresponding to pk.*

Then there exists a polynomial-time algorithm S outputting a list of strings, such that

$$\Pr\Big[(pk, m, \pi, z) \leftarrow G(1^k),\ c \leftarrow \mathcal{E}(pk, m),\ M' \leftarrow S(1^k, pk, c, \pi, z) : m \in M'\Big].$$

is not *negligible in k.*

While the definition of content-hiding proof is similar to that of witness-hiding proofs, there is an important difference: Witness-hiding proofs guarantee that the witness cannot be guessed if the statement is chosen according to some fixed distribution, while we require that the content-hiding property holds for *any* efficiently sampleable distribution on the messages m. Furthermore, a witness-hiding proof only guarantees that the witness is not disclosed as a whole, while we only require that the *message* m is not disclosed as a whole; the latter requirement is weaker since a witness would consist of m and the randomness used for encryption.

The existence of content-hiding proofs of content constitutes a novel cryptographic assumption. We did not succeed in reducing it to existing assumptions, but we show that at least there is an oracle relative to which this is possible.

Lemma 3. *Assume that trapdoor one-way permutations with dense public keys exist that are secure against non-uniform probabilistic polynomial-time adversaries. Then there exists an oracle $\mathcal{O}$ relative to which content-hiding proofs of content with deterministic verifiers exist for any encryption scheme $(\mathcal{E}, \mathcal{D})$.*

The proof of Lemma 3 (which is given in the full version [1]) establishes the following slightly stronger statement: choosing a random oracle entails content-hiding proofs of content with probability one. Hence the following conjecture is again justified by the random oracle heuristic:

Conjecture 2. Let R be a sufficiently unstructured efficiently computable function. Then using R in the construction of the proof of Lemma 3 yields content-hiding proofs of content with deterministic verifiers.

In the next section we will need both the existence of one-way permutations with non-uniform trapdoors as well as of content-hiding proofs of content. We additionally use some standard complexity assumptions. All assumptions used are summarized in the following statement:

Assumption 1. *There exist a one-way function with non-uniform trapdoors f (Definition 2) and a content-hiding proof of content with a deterministic verifier*[2] *for the singleton encryption scheme $\mathcal{E}_f$ for f (Definition 3).*

Further, we assume the existence of one-way functions secure against non-uniform adversaries and the existence of a keyed family of hash functions that is collision-resistant against non-uniform adversaries.

4 Limits of Constructive Security Proofs

Based on the definitions and assumptions from the preceding sections, we are now ready to show the existence of an existentially secure argument system that does not have a constructive security proof.

In the following, let f be a length-regular one-way function with non-uniform trapdoors, let $\mathcal{E}_f$ be the singleton encryption scheme for f, and let (P^*, V^*) denote a content-hiding proof of content for $\mathcal{E}_f$. Let $(P^\dagger, V^\dagger)$ be a computational zero-knowledge proof of knowledge, which can be constructed from one-way functions secure against non-uniform polynomial-time adversaries (see e.g., [7, Section 4.7.3]). When passing an algorithm A as argument to a function or algorithm, we assume that A is encoded as a circuit in some canonical way. Let H be the description of a function from $\{0,1\}^*$ to $\{0,1\}^*$. When considering H as a circuit, we will always mean the circuit describing the function H restricted to the domain $\{0,1\}^k$.

Stating the construction in a concise manner necessitates a few auxiliary definitions:

- Let $\pi_H(x_1, x_2) := \mathit{true}$ if and only if $x_1, x_2 \in \{0,1\}^k$, $x_1 \neq x_2$ and $H(x_1) = H(x_2)$.
- Let $\gamma(H, c, N) := \mathit{true}$ if and only if $V^*(1^k, pk_k, \pi_H, c, N) = 1$.
- Let $\eta(H, \sigma, c, N, w) := \mathit{true}$ if and only if $\sigma(w) = 1$ or $\gamma(H, c, N) = \mathit{true}$.
- Let $l_c(k) := |\mathcal{E}_f(1^k, 1^{2k})|$ denote the length of an encryption of a $2k$-bit plaintext.
- Let l_P be a polynomial such that for all $k \in \mathbb{N}$ and $c \in \{0,1\}^{l_c(k)}$, the value $l_P(k+|H|)$ is an upper bound on $|P^*(1^k, \pi_H, c, t_k)|$ where $|H|$ denotes the size of the circuit H and t_k is the non-uniform trapdoor for f (cf. Definition 2). Such a polynomial l_P exists, since there are polynomial upper bounds on all arguments of P^*, and P^* satisfies the polynomial length property from Definition 4.
- Let L_η be the language consisting of all (H, σ) such that there exist a triple (c, N, w) with $|c| \leq l_c(k)$ and $|N| \leq l_P(k + |H|)$ that satisfies $\eta(H, \sigma, c, N, w) = \mathit{true}$. Obviously, $L_\eta \in \mathrm{NP}$. Note that if $\sigma(w) = 1$, then w is a witness for $(H, \sigma) \in L_\eta$.

[2] We could also weaken the assumption slightly by allowing a probabilistic verifier. While our results hold as well for probabilistic verifiers, we have chosen to use this slightly stronger formulation since it makes the separating example and the proof easier.

Using this notation, we can now describe the protocol that will have an existential security proof, but that will provably not have a constructive proof:

Definition 6 (The Separating Argument System). *The proof system (P^H, V^H) where H may be a keyed or unkeyed hash function, is defined as follows:*

- *The prover P^H is invoked with input $(1^k, \sigma, w)$ where σ is a Boolean circuit and w is an assignment such that $\sigma(w) = 1$. The verifier is invoked with input $(1^k, \sigma)$.*
- *The prover P^H invokes $P^\dagger$ on security parameter 1^k, L_η-instance (H, σ) and witness w; here H is treated as a circuit mapping $\{0,1\}^k$ to $\{0,1\}^*$.*
- *The verifier V^H invokes $V^\dagger(1^k, \sigma)$ to verify the proof given by the prover P^H.*

The notation introduced in front of Definitions 4 and 6 (e.g., $\pi^{pk}[c]$, γ, $P^\dagger$, etc.) will be used in the following proofs without explicit reference.

We have assumed in Assumption 1 that V^* is deterministic. If V^* was probabilistic, we would have to change the above proof system as follows: First, the prover commits to a witness (c, N, w). The prover and the verifier then perform a coin-toss to choose a random tape R for V^*. Finally the prover proves that $\sigma(w) = 1$ or that the verifier V^* accepts with random tape R. We have opted to consider the case of a deterministic verifier V^* to make the presentation more readable.

Theorem 1. *Under Assumption 1, if H_K is a keyed hash-function that is secure against non-uniform adversaries then the proof system (P^H, V^H) is a (non-uniformly secure) computational zero-knowledge argument of knowledge for SAT. (We assume the key K to be chosen by some key generation algorithm $\mathcal{K}(1^k)$.)*

Proof. Since $(P^\dagger, V^\dagger)$ is a computational zero-knowledge proof, the computational zero-knowledge property and the completeness of (P^H, V^H) follow from the construction.

We show that (P^H, V^H) is an argument of knowledge, i.e., we construct a knowledge extractor E such that there exists a polynomial q such that for any non-uniform polynomial-time prover $\tilde{P}$ and any sequence σ of SAT-instances of polynomial length, there is a negligible function μ such that the following holds for each $k \in \mathbb{N}$:

$$\begin{aligned} &\Pr[K \leftarrow \mathcal{K}(1^k) : \ E^{\tilde{P}(1^k, K)}(1^k, H_K, \sigma_k) \text{ is a SAT-witness for } \sigma_k] \\ \geq &\frac{1}{q(k)} \Pr[K \leftarrow \mathcal{K}(1^k) : \ \langle \tilde{P}(1^k, K), V^{H_K}(1^k, \sigma_k) \rangle = 1] - \mu(k). \end{aligned} \quad (1)$$

Here $E^{\tilde{P}(1^k, K)}(1^k, H_K, \sigma_k)$ denotes the extractor E with black-box access to $\tilde{P}(1^k, K)$ and that is given a description of H_K.

Let $E_\dagger$ be the knowledge-extractor of $(P^\dagger, V^\dagger)$. Then there is a polynomial q such that for every non-uniform polynomial-time prover $\hat{P}$ and every sequence of polynomial-sized L_η-instances (H_k, σ_k) there exists a negligible function ν such that for all k the following holds:

$$\Pr[E_\dagger^{\hat{P}(1^k)}(1^k, H_k, \sigma_k) \text{ is an } L_\eta\text{-witness for } (H_k, \sigma_k)]$$
$$\geq \frac{1}{q(k)} \Pr[\langle \hat{P}(1^k), V^\dagger(1^k, H_k, \sigma_k)\rangle = 1] - \nu(k). \tag{2}$$

Here $E^{\hat{P}(1^k)}$ denotes the extractor $E_\dagger$ with black-box access to H_K and $\hat{P}(1^k, K)$.

We construct the knowledge-extractor E as follows: When invoked with black-box access to $\tilde{P}$ and with input $(1^k, H, \sigma)$, it invokes $(c, N, w) \leftarrow E_\dagger^{\tilde{P}}(1^k, H, \sigma)$ and then returns w.

It is left to show that E satisfies (1). Let $\tilde{P}$ be a non-uniform polynomial-time prover as in (1) and σ a sequence of SAT-instances of polynomial length. Let K be a sequence of keys for the hash-function H. By (2) and by definition of L_η, there exists a negligible function ν such that

$$\Pr[(c, N, w) \leftarrow E_\dagger^{\tilde{P}(1^k, K_k)}(1^k, H_k, \sigma_k) : \eta(H_{K_k}, \sigma_k, c, N, w) = \mathit{true}]$$
$$\geq \frac{1}{q(k)} \Pr[\langle \tilde{P}(1^k, K_k), V^\dagger(1^k, H_{K_k}, \sigma_k)\rangle = 1] - \nu(k) \tag{3}$$

Since this holds for every sequence K of keys, we have for some negligible ν and all $k \in \mathbb{N}$:

$$\Pr[K \leftarrow \mathcal{K}(1^k),\ (c, N, w) \leftarrow E_\dagger^{\tilde{P}(1^k, K)}(1^k, H_K, \sigma_k) :$$
$$\eta(H_K, \sigma_k, c, N, w) = \mathit{true}]$$
$$\geq \frac{1}{q(k)} \Pr[K \leftarrow \mathcal{K}(1^k) : \langle \tilde{P}(1^k, K), V^\dagger(1^k, H_K, \sigma_k)\rangle = 1] - \nu(k). \tag{4}$$

(Otherwise we could simply use the worst-case sequence of keys to contradict (3).)

Let μ_1 be defined as follows:

$$\mu_1(k) := \Pr[K \leftarrow \mathcal{K}(1^k), (c, N, w) \leftarrow E_\dagger^{\tilde{P}(1^k, K)}(1^k, H_K, \sigma_k) : \gamma(H_K, c, N) = \mathit{true}].$$

By definition, $\gamma(H_K, c, N) = \mathit{true}$ is equivalent to $V^*(1^k, pk_k, \pi_{H_K}, c, N) = 1$ which in turn implies $\pi_{H_K}^{pk_k}[c] = \mathit{true}$. Hence there exists a secret key sk such that $\mathcal{D}_f(pk_k, sk, c) =: m \neq \bot$ and $\pi_{H_K}(m) = \mathit{true}$. Since $\mathcal{D}_f(pk_k, sk, c) = m \neq \bot$ implies $\mathcal{D}_f(pk_k, sk_k, c) = m$ by construction, it follows that $\pi_{H_K}(\mathcal{D}_f(pk_k, sk_k, c)) = \mathit{true}$. We therefore have

$$\mu_1(k) \leq \Pr[K \leftarrow \mathcal{K}(1^k),\ (c, N, w) \leftarrow E_\dagger^{\tilde{P}(1^k, K)}(1^k, H_K, \sigma_k),$$
$$m \leftarrow \mathcal{D}_f(pk_k, sk_k, c) :\ \pi_{H_K}(m) = \mathit{true}].$$

Since $(c, N, w) \leftarrow E_\dagger^{\tilde{P}(1^k, K)}(1^k, H_K, \sigma_k)$, $m \leftarrow \mathcal{D}_f(pk_k, sk_k, c)$ can be computed by a non-uniform polynomial-time algorithm (given 1^k and K), and since

$\pi_{H_K}(m) = \mathit{true}$ implies that m encodes a collision of H_K, we have constructed a non-uniform polynomial-time algorithm that finds collisions of H_K with probability at least μ_1. Since by assumption, H_K is collision-resistant against non-uniform polynomial-time adversaries, this implies that μ_1 is negligible.

By definition, we have $\eta(H_K, \sigma_k, c, N, w) = \mathit{true}$ if and only if $\sigma_k(w) = 1$ or $\gamma(H_K, c, N) = \mathit{true}$. So using the definition of E and V^H we get

$$\begin{aligned}
&\Pr[K \leftarrow \mathcal{K}(1^k),\ w \leftarrow E^{\tilde{P}(1^k,K)}(1^k, H_K, \sigma_k) : \sigma_k(w) = 1] \\
&= \Pr[K \leftarrow \mathcal{K}(1^k),\ (c, N, w) \leftarrow E_\dagger^{\tilde{P}(1^k,K)}(1^k, H_K, \sigma_k) : \sigma_k(w) = 1] \\
&\geq \Pr[K \leftarrow \mathcal{K}(1^k),\ (c, N, w) \leftarrow E_\dagger^{\tilde{P}(1^k,K)}(1^k, H_K, \sigma_k) : \\
&\qquad\qquad \eta(H_K, \sigma_k, c, N, w) = \mathit{true}] - \mu_1(k) \\
&\overset{(4)}{\geq} \frac{1}{q(k)} \Pr[K \leftarrow \mathcal{K}(1^k) : \langle \tilde{P}(1^k, K), V^\dagger(1^k, H_K, \sigma_k)\rangle = 1] - \nu(k) - \mu_1(k). \\
&= \frac{1}{q(k)} \Pr[K \leftarrow \mathcal{K}(1^k) : \langle \tilde{P}(1^k, K), V^{H_K}(1^k, \sigma_k)\rangle = 1] - \nu(k) - \mu_1(k). \qquad (5)
\end{aligned}$$

Setting $\mu := \nu + \mu_1$, this gives us (1) and thus shows that (P^H, V^H) is a (non-uniformly secure) computational zero-knowledge argument of knowledge. □

Theorem 2. *Under Assumption 1, there exists no constructive security proof C that (P^H, V^H) is an argument.*

In particular, the theorem implies that no constructive security proof exists that (P^H, V^H) is a computational zero-knowledge argument of knowledge.

Proof. Assume for contradiction that a constructive security proof C exists that (P^H, V^H) is an argument.

Let f be a one-way permutation with non-uniform trapdoors and let $\{\tilde{H}_K\}_{K \in \mathcal{K}}$ be a keyed family of hash functions that is one-way against non-uniform adversaries. Let $G_{\tilde{H}}$ be the key generation algorithm for $\tilde{H}_K$, and assume w.l.o.g. that for $K \leftarrow G_{\tilde{H}}(1^k)$ the function $\tilde{H}_K$ maps from $\{0,1\}^*$ to $\{0,1\}^k$.

We first construct a keyed family $\{H_{a,b,K}\}_{(a,b,K) \in Y \times \mathcal{K}}$ of hash functions $H_{a,b,K} : \{0,1\}^* \to \{0,1\}^{k+1}$ with $Y := \bigcup Y_k$ and $Y_k := \{(a,b) : a, b \in \{0,1\}^k, a \neq b\}$ as follows:

$$H_{a,b,K}(x) := \begin{cases} 0 \| \tilde{H}_K(x), & |x| \neq k, \\ 1 \| f(x), & |x| = k,\ f(x) \neq a, \\ 1 \| b, & |x| = k,\ f(x) = a. \end{cases} \qquad \text{for } a, b, x \in \{0,1\}^k.$$

It is easy to see that the only collision (x, x') of $H_{a,b,K}$ that satisfies $|x| = |x'| = k$ is $(f^{-1}(a), f^{-1}(b))$. Hence finding such a collision of $H_{a,b,K}$ for random (a,b) implies inverting f at a. Finding collisions (x, x') with $|x| \neq k$ or $|x'| \neq k$ breaks the collision-resistance of $\tilde{H}_K$. So $H_{a,b,K}$ is collision-resistant against uniform polynomial-time adversaries.

In the following, we write *k-collision* to denote a collision (x, x') with $|x| = |x'| = k$. Then there exists only a single k-collision (x, x') of $H_{a,b,K}$ (where $k = |a| = |b|$).

Let σ_{false} denote some fixed unsatisfiable circuit. Let $\tilde{P}$ be a prover that upon input $(1^k, H, c, N)$ invokes $P^\dagger$ on security parameter 1^k, L_η-instance (H, σ_{false}) and witness (c, N, w).

By construction of (P^H, V^H) and since $(P^\dagger, V^\dagger)$ is complete, there exists a negligible function μ_1 such that for all c, N with $|c| \le l_c(k)$ and $|N| \le l_P(k + |H|)$ such that N is a valid proof for $\pi_H^{pk_k}[c] = true$ (i.e., such that $V^*(1^k, pk_k, \pi_H, c, N) = 1$), we have

$$\Pr\Big[\langle \tilde{P}(1^k, H, c, N), V^H(1^k, \sigma_{false})\rangle = 1\Big] \ge 1 - \mu_1(k). \tag{6}$$

Consider the following game G_0:

$$(\tilde{a}, \tilde{b}) \xleftarrow{\$} Y_k,\ a := f(\tilde{a}),\ b := f(\tilde{b}),\ K \leftarrow G_{\tilde{H}}(1^k),\ H := H_{a,b,K}, \tag{7}$$

$$c \leftarrow \mathcal{E}_f(pk_k, (\tilde{a}, \tilde{b})),\ N \leftarrow P^*(1^k, \pi_H, c, pk_k, sk_k), \tag{8}$$

$$(\hat{a}, \hat{b}) \leftarrow C(1^k, H, \tilde{P}(1^k, H, c, N), \sigma_{false}). \tag{9}$$

That is, first, in (7) we construct a hash-function H such that we know the (only) k-collision $(\tilde{a}, \tilde{b})$. Then in (8) we construct an encryption c of that k-collision and a proof that c indeed contains a k-collision (i.e., that $\pi_H^{pk_k}[c] = true$). Finally, in (9) we invoke the generic security proof C with a description of the hash-function H, with a description of $\tilde{P}$ (instantiated with input $(1^k, H, c, N)$) and with the SAT-instance σ_{false}.

By the completeness of (P^*, V^*), there is a negligible function μ_2 such that in G_0 the following holds: $\Pr[V^*(1^k, pk_k, \pi_H, c, N) = 1] \ge 1 - \mu_2(k)$. Further, by definition of l_c and l_P it is $|c| \le l_c(k)$ and $|N| \le l_P(k + |H|)$. Then using (6) we get

$$\mathrm{Adv}_k^{\mathrm{arg}} := \Pr\Big[\langle \tilde{P}(1^k, H, c, N), V^H(1^k, \sigma_{false})\rangle = 1\Big] \ge 1 - \mu_1(k) - \mu_2(k)$$

when H, c and N are chosen as in game G_0.

Since σ_{false} is not satisfiable, this violates the soundness of the argument system (P^H, V^H). So by the definition of constructive security proofs, C should be able to extract a collision given 1^k, H, $\tilde{P}(1^k, H, c, N)$ and σ_{false}. More exactly, let p be a polynomial such that $p(k)$ bounds the length of $(1^k, H, \tilde{P}(1^k, H, c, N), \sigma_{false})$. Such a polynomial exists, since H is constructed by a polynomial-time algorithm and $\tilde{P}$ runs in polynomial time. Then there is a $c > 0$ and a negligible function μ_5 such that

$$\begin{aligned}\Pr\big[(\hat{a}, \hat{b}) \text{ is a collision of } H\big] &\ge \left(\frac{\mathrm{Adv}_k^{\mathrm{arg}}}{p(k)}\right)^c - \mu_5(k)\\ &\ge \left(\frac{1 - \mu_1(k) - \mu_2(k)}{p(k)}\right)^c - \mu_5(k) =: \nu(k).\end{aligned}$$

Then ν is not negligible. On the other hand, since $\tilde{H}_K$ is collision-resistant against non-uniform adversaries, and $(\hat{a},\hat{b})$ is computed by non-uniform polynomial-time algorithms in (7–9),[3] there is a negligible function μ_4 bounding the probability that $(\hat{a},\hat{b})$ is a collision of $\tilde{H}_K$. Since by construction of $H := H_{a,b,K}$, the only collision of H that is not a collision of $\tilde{H}_K$ is the k-collision $(f^{-1}(a), f^{-1}(b)) = (\hat{a},\hat{b})$, it follows that

$$\Pr\big[(\hat{a},\hat{b}) = (\tilde{a},\tilde{b})\big] \geq \nu(k) - \mu_4(k). \tag{10}$$

Let now $A(1^k, pk, c, \pi, H, N) := C(1^k, H, \tilde{P}(1^k, H, c, N), \sigma_{false})$. Since C and $\tilde{P}$ are polynomial-time algorithms, so is A. Further let $G(1^k)$ be an algorithm that chooses $m := (\tilde{a},\tilde{b})$ and H as in game G_0 and then outputs (pk_k, m, π_H, H). Then G runs in polynomial-time, too. Then the following game G_1 is just a rewriting of game G_0:

$$(pk, m, \pi, H) \leftarrow G(1^k),\ c \leftarrow \mathcal{E}_f(pk, m),$$
$$N \leftarrow P^*(1^k, \pi, c, pk, sk),\ m' \leftarrow A(1^k, pk, c, \pi, H, N)$$

with $(\hat{a},\hat{b}) := m'$ and with sk being the secret key corresponding to pk. So by (10) it follows that $\Pr[m = m'] \geq \nu(k) - \mu_4(k)$ in game G_0. This is not negligible. Since (P^*, V^*) is content-hiding, it follows that there is a polynomial-time simulator S such that

$$\nu_2(k) := \Pr\big[(pk, m, \pi, H) \leftarrow G(1^k),\ c \leftarrow \mathcal{E}_f(pk, m),$$
$$M' \leftarrow S(1^k, pk, c, \pi, H)\ :\ m \in M'\big] \tag{11}$$

is *not* negligible. Since $\mathcal{E}_f$ is IND-CPA by Lemma 2, and the algorithms in (11) are all uniform polynomial-time algorithms, we can replace $\mathcal{E}_f(pk, m)$ by $\mathcal{E}_f(pk, 0^{2k})$ (since $|m| = 2k$). (For this, note that G chooses $pk := pk_k$.) Then, for some negligible function μ_3, we have

$$\Pr[(pk, m, \pi) \leftarrow G(1^k),\ c \leftarrow \mathcal{E}_f(pk, 0^{2k}),$$
$$M' \leftarrow S(1^k, pk, c, \pi, H)\ :\ m \in M'] \geq \nu_2(k) - \mu_3(k)$$

Since given a description of $H_{a,b,K}$ with $a = f(\tilde{a})$ and $b = f(\tilde{b})$, we can efficiently verify whether for some m' we have $m' = (\tilde{a},\tilde{b})$, we can modify S so that it directly outputs $m = (\tilde{a},\tilde{b})$ if that m is in M'. Call the resulting algorithm S'. By substituting the definition of G we get

$$\Pr[(\tilde{a},\tilde{b}) \xleftarrow{\$} Y_k,\ a := f(\tilde{a}),\ b := f(\tilde{b}),\ K \leftarrow G_{\tilde{H}}(1^k),$$
$$(\hat{a},\hat{b}) \leftarrow S'(1^k, pk_k, \mathcal{E}_f(pk_k, 0^{2k}), \pi_{H_{a,b,K}}, H_{a,b,K})\ :$$
$$(\hat{a},\hat{b}) = (\tilde{a},\tilde{b})] \geq \nu_2(k) - \mu_3(k).$$

[3] The non-uniformity stems from the appearance of sk_k in game G_0.

Let the algorithm $T(1^k, a)$ perform as follows: First, it chooses b uniformly from $\{0,1\}^k \setminus \{a\}$ and K using $G_{\tilde{H}}(1^k)$. Then it executes $(\hat{a}, \hat{b}) \leftarrow S'(1^k, pk_k, \mathcal{E}_f(pk_k, 0^{2k}), \pi_{H_{a,b,K}}, H_{a,b,K})$ and outputs $\hat{a}$. Then the previous probability can be rewritten as

$$\Pr[\tilde{a} \stackrel{\$}{\leftarrow} \{0,1\}^k, \hat{a} := T(1^k, f(\tilde{a})) : \tilde{a} = \hat{a}] \geq \nu_2(k) - \mu_3(k).$$

Since $\nu_2 - \mu_3$ is not negligible and T is a uniform polynomial-time algorithm, this is a contradiction to f being one-way against uniform polynomial-time adversaries. Hence our assumption that C is a constructive security proof was wrong. □

References

1. Backes, M., Unruh, D.: Limits of constructive security proofs (2008), http://www.infsec.cs.uni-sb.de/~unruh/publications/backes08limits.html
2. Baker, T., Gill, J., Solovay, R.: Relativizations of the p $\stackrel{?}{=}$ NP question. SIAM Journal on Computing 4, 431–442 (1975)
3. Barak, B.: How to go beyond the black-box simulation barrier. In: 42th Annual Symposium on Foundations of Computer Science, Proceedings of FOCS 2001, pp. 106–115. IEEE Computer Society, Los Alamitos (2001), http://www.wisdom.weizmann.ac.il/~boaz/Papers/nonbb.ps
4. Damgård, I.: Collision free hash functions and public key signature schemes. In: Price, W.L., Chaum, D. (eds.) EUROCRYPT 1987. LNCS, vol. 304, pp. 203–216. Springer, Heidelberg (1988)
5. Dwork, C., Naor, M.: Zaps and their applications. ECCC TR02-001 (2002), http://eccc.hpi-web.de/eccc-reports/2002/TR02-001/index.html
6. Fortnow, L.: The role of relativization in complexity theory. Bulletin of the EATCS 52 (February 1994), http://people.cs.uchicago.edu/~fortnow/papers/relative.ps
7. Goldreich, O.: Foundations of Cryptography, vol. 1 (Basic Tools). Cambridge University Press, Cambridge (August 2001), http://www.wisdom.weizmann.ac.il/~oded/frag.html
8. Goldreich, O.: Foundations of Cryptography, vol. 2 (Basic Applications). Cambridge University Press, Cambridge (May 2004), http://www.wisdom.weizmann.ac.il/~oded/frag.html
9. Rogaway, P.: Formalizing human ignorance: Collision-resistant hashing without the keys. In: Nguyên, P.Q. (ed.) VIETCRYPT 2006. LNCS, vol. 4341, pp. 221–228. Springer, Heidelberg (2006), http://eprint.iacr.org/2006/281
10. Stinson, D.R.: Some observations on the theory of cryptographic hash functions. IACR ePrint Archive (March 2001), http://eprint.iacr.org/2001/020

Efficient Chosen Ciphertext Secure Public Key Encryption under the Computational Diffie-Hellman Assumption

Goichiro Hanaoka[1] and Kaoru Kurosawa[2]

[1] RCIS, AIST
[2] Ibaraki University

Abstract. Recently Cash, Kiltz, and Shoup [13] showed a variant of the Cramer-Shoup (CS) scheme [14] whose chosen-ciphertext (CCA) security relies on the *computational Diffie-Hellman* (CDH) assumption. The cost for this high security is that the size of ciphertexts is much longer than the CS scheme (which is based on the decisional Diffie-Hellman assumption). In this paper, we show how to achieve CCA-security under the CDH assumption without increasing the size of ciphertexts. We also show a more efficient scheme under the *hashed Diffie-Hellman* assumption.

Both of our schemes are based on a certain broadcast encryption (BE) scheme while the Cash-Kiltz-Shoup scheme is based on the Twin DH problem. Of independent interest, we also show a generic method of constructing CCA-secure PKE schemes from BE schemes.

1 Introduction

1.1 Background

Chosen-ciphertext security (CCA-security, for short) [35,16] is considered as a standard notion of security for public key encryption (PKE) in practice. Furthermore, this security also implies universally composable security [11]. So far, many CCA-secure PKE schemes have been proposed, both theoretical ones [31,16,36] and practical ones [14,38,12,26,10,1,25,22], and their security are proven under existence of enhanced trapdoor permutations (for theoretical schemes) or under various number theoretic assumptions (for practical schemes). Theoretical schemes pursue weaker assumptions and practical schemes pursue efficiency.

One of the most important research topics in this field is to design CCA-secure PKE schemes with weaker assumptions and better efficiency. Cramer and Shoup showed the first practical PKE scheme under the decisional Diffie-Hellman (DDH) assumption. Kurosawa and Desmedt showed a more efficient scheme under the DDH assumption [26].

However, there has been no (even theoretical) CCA-secure PKE scheme under the *computational Diffie-Hellman* (CDH) assumption except for a recent work by Cash, Kiltz, and Shoup [13].[1]

[1] We started our work independently of [13]. In fact, the authors of [13] kindly cited an earlier version of our paper as an independent work.

J. Pieprzyk (Ed.): ASIACRYPT 2008, LNCS 5350, pp. 308–325, 2008.
© International Association for Cryptologic Research 2008

1.2 Our Contribution

In this paper, we present a practical CCA-secure PKE scheme under the CDH assumption such that the size of a ciphertext is much smaller than that of the Cash-Kiltz-Shoup (CKS) scheme. Indeed, the ciphertext length of our scheme is the same as that of the Cramer-Shoup (CS) scheme (which is based on the DDH assumption). Specifically, ciphertext overhead of our CDH-based scheme is only three group elements for arbitrary plaintext length, while that of the CKS scheme is $k/\log k + 2$ group elements where k is the security parameter.

We also present a more efficient CCA-secure PKE scheme under the hashed Diffie-Hellman (HDH) assumption. This scheme is as efficient as the Kurosawa-Desmedt (KD) scheme [26] in terms of both computational costs and data sizes while the HDH assumption is weaker than the DDH assumption.[2]

Both of our schemes are based on the Naor-Pinkas broadcast encryption (BE) scheme while the CKS scheme is based on the Twin DH problem. Of independent interest, we show a generic method of transforming any selectively chosen-plaintext (CPA) secure *verifiable* BE scheme into a CCA-secure key encapsulation mechanism (KEM) with almost no cost, where we say that a BE scheme is verifiable if any receiver can tell whether all receivers decrypt a given ciphertext to the identical result or not.

Further, we show that almost all existing methods for achieving CCA-security, e.g. [16,14,12], can be explained by using verifiable BE schemes. It is also possible to construct a new PKE scheme based on this paradigm, for example, from the Boneh-Gentry-Waters (BGW) BE scheme [6]. Moreover, we can generically convert any CPA-secure verifiable BE into a CCA-secure BE with almost no cost. Our results imply that verifiable BE is a powerful tool to obtain CCA-security.

1.3 Related Works

Under Stronger Assumptions than CDH. After the KD scheme, several CCA-secure encryption schemes were constructed under stronger assumptions than the CDH assumption. The scheme of Boyen, Mei, and Waters [10] is based on the *bilinear Diffie-Hellman* (BDH) assumption. The scheme of Kiltz [25] is based on the *gap hashed Diffie-Hellman* (GHDH) assumption. The scheme of Hofheinz and Kiltz [22] is based on the *n-linear* DDH assumption.

KEM/DEM Framework. The KEM/DEM framework was formalized by Shoup [38] for the design of hybrid encryption schemes, and the CS hybrid encryption scheme was constructed. However, the KD scheme does not fit into this framework. To explain the KD scheme in a general framework, Abe, Gennaro, Kurosawa, and Shoup [1] established the Tag-KEM/DEM framework. Hofheinz and Kiltz [22] introduced the notion of Constrained CCA (CCCA) security of KEM.

[2] After an earlier version of this paper [21], in the latest full-version of [13], Cash, Kiltz, and Shoup pointed out that the Hofheinz-Kiltz scheme in [22] can be also proved to be secure under the HDH assumption.

How to Achieve CCA Security. Naor and Yung showed that a non-adaptively CCA-secure encryption scheme can be constructed from any semantically secure encryption [19] and *non-interactive zero knowledge* (NIZK) proof [4]. Dolev, Dwork, and Naor [16] and Sahai [36] improved this idea and presented adaptively CCA-secure constructions. However, it is not known if an NIZK proof can be constructed from any semantically secure encryption scheme. (A partial answer to this question is given in [32].)

Canetti, Halevi, and Katz [12] proposed another generic method such that a CCA-secure PKE scheme can be obtained from a selectively secure identity-based encryption (IBE) scheme [37,5]. Boneh and Katz [7] improved its efficiency. Kiltz [24] discussed a more relaxed condition for achieving CCA-security.

Broadcast Encryption. In the model of broadcast encryption (BE) schemes, there are multiple receivers. The sender broadcasts a ciphertext such that only privileged receivers can decrypt. Fiat and Naor [17] proposed the first non-trivial construction of BE. Naor, Naor, and Lotspiech [29] presented a significantly more efficient scheme. Naor and Pinkas [30] proposed a public key BE scheme by using ElGamal-like construction, and Dodis and Fazio [15] improved it to be secure against adaptive adversaries as well as chosen-ciphertext adversaries. Boneh, Gentry, and Waters [6] proposed the first fully collusion resistant (public key) BE scheme whose ciphertext and user decryption keys are of constant size.

1.4 Organization

Definitions are given in Sec. 2. Our main idea is described in Sec. 3. The proposed scheme under the CDH assumption is shown in Sec. 4. A more efficient scheme under the HDH assumption is presented in Sec. 5. A comparison with other PKE schemes is given in Sec. 6. Finally, we show a generic method to construct CCA-secure PKE schemes from verifiable BE in Sec. 7.

2 Definitions

2.1 Key Encapsulation Mechanisms

It is well-known that by combining a CCA-secure KEM and a CCA-secure data encryption mechanism (DEM), a CCA-secure PKE scheme is generically obtained [38], and furthermore, there exist some other flexible methods for hybrid encryption as well [1,22]. It is also known that a CCA-secure DEM can be generically constructed from any pseudorandom functions without redundancy [27,33].

A KEM consists of the following three algorithms: **Setup**(1^k) takes as input the security parameter 1^k and outputs a decryption key dk and a public key PK. **Encrypt**(PK) takes as input a public key PK and outputs a pair (ψ, K) where ψ is a ciphertext and $K \in \mathcal{K}$ is a data encryption key. **Decrypt**(dk, ψ, PK) takes as input the decryption key dk, a ciphertext ψ, and the public key PK, and outputs $K \in \mathcal{K}$ which will be used for decrypting the DEM part of hybrid encryption.

We require that if $(dk, PK) \stackrel{R}{\leftarrow}$ **Setup**(1^k) and $(\psi, K) \stackrel{R}{\leftarrow}$ **Encrypt**(PK) then **Decrypt**$(dk, \psi, PK) = K$.

CCA-security of a KEM is defined using the following game between an attack algorithm A and a challenger. Both the challenger and A are given 1^k as input.

Setup. The challenger runs **Setup**(1^k) to obtain a decryption key dk and a public key PK. The challenger also runs algorithm **Encrypt** to obtain $(\psi^\star, K^\star) \stackrel{R}{\leftarrow}$ **Encrypt**(PK) where $K^\star \in \mathcal{K}$. Next, the challenger picks a random $b \in \{0, 1\}$. It sets $K_0 = K^\star$ and picks a random $K_1 \in \mathcal{K}$. It then gives the public key PK and the challenge ciphertext $(\psi^\star, K_b)$ to algorithm A.

Query. Algorithm A adaptively issues decryption queries $\psi_1, ..., \psi_{q_D}$. For query $\psi_i (\neq \psi^\star)$, the challenger responds with **Decrypt**(dk, ψ_i, PK).

Guess. Algorithm A outputs its guess $b' \in \{0, 1\}$ for b and wins the game if $b = b'$.

Let $\mathsf{AdvKEM}_\mathsf{A}$ denote the probability that A wins the game.

Definition 1. We say that a KEM is (τ, ϵ, q_D) *CCA-secure* if for all τ-time algorithms A who make a total of q_D decryption queries, we have that $|\mathsf{AdvKEM}_\mathsf{A} - 1/2| < \epsilon$.

2.2 Number Theoretic Assumptions

The CDH, HDH, and DDH Assumptions. Let $\mathbb{G}$ be a multiplicative group with prime order p. Then, the CDH problem on $\mathbb{G}$ is stated as follows. Let A be an algorithm, and we say that A has advantage ϵ in solving the CDH problem on $\mathbb{G}$ if $\Pr[\mathsf{A}(g, g^\alpha, g^\beta) = g^{\alpha\beta}] \geq \epsilon$, where the probability is over the random choice of generator g in $\mathbb{G}$, the random choice of α and β in $\mathbb{Z}_p$, and the random bits consumed by A.

Definition 2. We say that the (τ, ϵ)-CDH assumption holds in $\mathbb{G}$ if no τ-time algorithm has advantage ϵ in solving the CDH problem on $\mathbb{G}$.

The *hashed Diffie-Hellman* (HDH) problem on $\mathbb{G}$ and function $h : \mathbb{G} \to \mathcal{D}$ is stated as follows. Let A be an algorithm, and we say that A has advantage ϵ in solving the HDH problem on $\mathbb{G}$ and h if

$$1/2 \cdot |\Pr[\mathsf{A}(g, g^\alpha, g^\beta, h(g^{\alpha\beta})) = 0] - \Pr[\mathsf{A}(g, g^\alpha, g^\beta, T) = 0]| \geq \epsilon,$$

where the probability is over the random choice of generator g in $\mathbb{G}$, the random choice of α and β in $\mathbb{Z}_p$, the random choice of $T \in \mathcal{D}$, and the random bits consumed by A.

Definition 3. We say that the (τ, ϵ)-HDH assumption holds in $\mathbb{G}$ and h if no τ-time algorithm has advantage ϵ in solving the HDH problem on $\mathbb{G}$ and h. Especially, we say that the (τ, ϵ)-DDH assumption holds in $\mathbb{G}$ if (τ, ϵ)-HDH assumption holds in $\mathbb{G}$ and h, where h is the identity function.

Important Implications. It is important to note that the HDH assumption is strictly weaker than the DDH assumption for appropriately chosen h. If h is a *key derivation function* [38], then the DDH assumption immediately implies the HDH assumption (but not vice versa). Furthermore, if h is a hardcore bit for the Diffie-Hellman key [18,9,8,23], then the CDH assumption is equivalent to the HDH assumption. Obviously, the CDH assumption is weaker than both the HDH and DDH assumptions.

Hardcore Bits for the Diffie-Hellman Key. Let A be a τ-time algorithm which has advantage ϵ in solving the HDH problem on $\mathbb{G}$ and $h : \mathbb{G} \to \{0, 1\}$.

Definition 4. We say that function $h : \mathbb{G} \to \{0, 1\}$ is a (p_1, p_2) hardcore bit function in $\mathbb{G}$ if there exists a $p_1(\tau)$-time algorithm B which for any given A, can solve the CDH problem with advantage $p_2(\epsilon)$ for some polynomials p_1 and p_2.

2.3 Public Key Broadcast Encryption Schemes

Model. Here, we review definitions for public key BE schemes. For simplicity, we define encryption schemes as key encapsulation mechanisms, and borrow the same notations as [6] with some slight modifications. A BE scheme consists of the following three algorithms: **Setup**$(1^k, n, t)$ takes as input the security parameter 1^k, the number of receivers n, and the maximum number of revoked users t $(t < n)$. It outputs n decryption keys $d_1, ..., d_n$ and a public key PK. **Encrypt**$(\mathcal{S}, PK)$ takes as input a subset $\mathcal{S} \subseteq \{1, ..., n\}$ with $|\mathcal{S}| \geq n - t$, and a public key PK. It outputs a pair (ψ, K) where ψ is called the header and $K \in \mathcal{K}$ is a message encryption key. Let M be a message to be broadcast to the set $\mathcal{S}$ and let C_M be the encryption of M under the symmetric key K. The broadcast to users in $\mathcal{S}$ consists of $(\mathcal{S}, \psi, C_M)$. The pair $(\mathcal{S}, \psi)$ is often called the full header and C_M is often called the broadcast body. **Decrypt**$(\mathcal{S}, i, d_i, \psi, PK)$ takes as input a subset $\mathcal{S} \subseteq \{1, ..., n\}$, a user index $i \in \{1, ..., n\}$ and the decryption key d_i for user i, a header ψ, and the public key PK. If $i \in \mathcal{S}$ and $|\mathcal{S}| \geq n - t$, then the algorithm outputs the message encryption key $K \in \mathcal{K}$. The key K can then be used to decrypt the broadcast body C_M and obtain the message body M.

As usual, we require that the scheme be correct, namely that for all $\mathcal{S} \subseteq \{1, ..., n\}$ and all $i \in \mathcal{S}$, if $((d_1, ..., d_n), PK) \stackrel{R}{\leftarrow}$ **Setup**$(1^k, n, t)$ and $(\psi, K) \stackrel{R}{\leftarrow}$ **Encrypt**$(\mathcal{S}, PK)$ then **Decrypt**$(\mathcal{S}, i, d_i, \psi, PK) = K$.

CCA Security. We define CCA-security of a BE scheme against a static adversary. Security is defined using the following game between an attack algorithm A and a challenger. Both the challenger and A are given 1^k, n and t, the total number of potential users and the maximum number of revoked users, respectively, as inputs.

Init. Algorithm A begins by outputting a set $\mathcal{S}^\star \subseteq \{1, ..., n\}$ of receivers that A wants to attack, where $|\mathcal{S}^\star| \geq n - t$.

Setup. The challenger runs **Setup**$(1^k, n, t)$ to obtain decryption keys $d_1, ..., d_n$ and a public key PK. The challenger also runs algorithm **Encrypt** to obtain

$(\psi^\star, K^\star) \xleftarrow{R}$ **Encrypt**$(\mathcal{S}^\star, PK)$ where $K^\star \in \mathcal{K}$. Next, the challenger picks a random $b \in \{0,1\}$. It sets $K_0 = K^\star$ and picks a random $K_1 \in \mathcal{K}$. It then gives $(\psi^\star, K_b)$ to algorithm A.

Query. Algorithm A adaptively issues decryption queries $q_1, ..., q_D$ where a decryption query consists of the triple $(u, \mathcal{S}, \psi)$ where $\psi \neq \psi^\star$, $\mathcal{S} \subseteq \mathcal{S}^\star$ and $u \in \mathcal{S}$. The challenger responds with K (or $\perp$) $=$ **Decrypt**$(\mathcal{S}, u, d_u, \psi, PK)$.

Guess. Algorithm A outputs its guess b' for b and wins the game if $b = b'$.

Let $\mathsf{AdvBr}_{\mathsf{A},n,t}$ denote the probability that A wins the game when the challenger is given n and t.

Definition 5. We say that a broadcast encryption scheme is $(\tau, \epsilon, n, t, q_D)$ *CCA-secure* if for all τ-time algorithms A who make a total of q_D decryption queries, we have that $|\mathsf{AdvBr}_{\mathsf{A},n,t} - 1/2| < \epsilon$. Especially, we say that a broadcast encryption scheme is (τ, ϵ, n, t) *semantically secure* if it is $(\tau, \epsilon, n, t, 0)$ CCA-secure.

Verifiability. For achieving CCA-security, we need an important property for underlying BE, which we call *verifiability.* Roughly speaking, we say that a BE scheme has verifiability if a valid receiver of a broadcasted message can verify if his decryption result is the same as that for any other receiver. We can define two flavors of verifiability: *public* verifiability and *private* verifiability. Their difference is that in a publicly verifiable BE scheme, a receiver can verify equality of keys without using his decryption key, and on the other hand, it is necessary in a privately verifiable scheme.

For public verifiability, we define adversary A's advantage $\mathsf{AdvVfy}_{\mathsf{A},n,t}$ as

$$\begin{aligned}&\mathsf{AdvVfy}_{\mathsf{A},n,t}\\&= \Pr[\exists i,j \in \mathcal{S}^\star,\ \mathbf{Decrypt}(\mathcal{S}^\star, i, d_i, \psi^\star, PK) \neq \mathbf{Decrypt}(\mathcal{S}^\star, j, d_j, \psi^\star, PK)|\\&\qquad ((d_1, ..., d_n), PK) \xleftarrow{R} \mathbf{Setup}(1^k, n, t);\ (\mathcal{S}^\star, \psi^\star) \xleftarrow{R} \mathsf{A}((d_1, ..., d_n), PK)].\end{aligned}$$

Definition 6. We say that a broadcast encryption scheme is (τ, ϵ, n, t) *publicly verifiable* if for all τ-time algorithms A, we have that $\mathsf{AdvVfy}_{\mathsf{A},n,t} < \epsilon$.

We can also define private verifiability in a similar manner, and its formal definition is given in the full version of this paper [21].

2.4 Other Cryptographic Tools

Target Collision Resistant Hash Functions. Let $\mathsf{TCR} : \mathcal{X} \rightarrow \mathcal{Y}$ be a hash function (we individually define the range and domain of TCR for each scheme), A be an algorithm, and A's advantage $\mathsf{AdvTCR}_\mathsf{A}$ be $\mathsf{AdvTCR}_\mathsf{A} = \Pr[\mathsf{TCR}(x') = \mathsf{TCR}(x) \in \mathcal{Y} \wedge x' \neq x|\ x \xleftarrow{R} \mathcal{X};\ x' \xleftarrow{R} \mathsf{A}(x)]$.

Definition 7. We say that TCR is a (τ, ϵ) *target collision resistant hash function* if for all τ-time algorithms A, we have that $\mathsf{AdvTCR}_\mathsf{A} < \epsilon$.

One-Time Signatures. A signature scheme consists of the following three algorithms: $\mathbf{Gen}(1^k)$ takes as input the security parameter 1^k, and outputs a verification key vk and a signing key sk. $\mathbf{Sign}(sk, m)$ takes as input a signing key sk and a message m, and outputs a signature σ. $\mathbf{Verify}(vk, m, \sigma)$ takes as input a verification key vk, a message m, and a signature σ, and outputs a bit $b \in \{0, 1\}$. We require that for all sk, all m in the message space, and all σ output by $\mathbf{Sign}(sk, m)$, we have $\mathbf{Verify}(vk, m, \sigma) = 1$.

Security is defined using the following game between an attack algorithm A and a challenger. Both the challenger and A are given 1^k as input.

Setup. The challenger runs $\mathbf{Gen}(1^k)$ to obtain vk and sk. It gives A the verification key vk.

Query. Algorithm A may issue at most one query m. The challenger responds with $\sigma \stackrel{R}{\leftarrow} \mathbf{Sign}(sk, m)$.

Forge. Algorithm A outputs $(m^\star, \sigma^\star)$ such that $(m^\star, \sigma^\star) \neq (m, \sigma)$.

Let $\mathsf{AdvOTS}_\mathsf{A}$ denote the probability that $\mathbf{Verify}(vk, m^\star, \sigma^\star) = 1$.

Definition 8. We say that a signature scheme is (τ, ϵ) *strongly unforgeable* if for all τ-time algorithms A, we have that $\mathsf{AdvOTS}_\mathsf{A} < \epsilon$.

3 Toward Efficient CCA-Secure Scheme under CDH

The Naor-Pinkas BE scheme [30] is one-way under the CDH assumption. In this section, we construct a verifiable BE scheme from the Naor-Pinkas BE scheme, where we say that a BE scheme is verifiable if any receiver can tell whether all receivers decrypt a given ciphertext to the identical result or not. The main difficulty in this paper is how to add verifiability to the Naor-Pinkas scheme.

Our CCA-secure PKE scheme under the CDH assumption is obtained from this variant of the Naor-Pinkas BE scheme. See Sec. 7 for details on this observation.

3.1 The Naor-Pinkas Broadcast Encryption Scheme

The Naor-Pinkas scheme [30], which was constructed based on [2], is as follows. Let $\mathbb{G}$ be a multiplicative group with prime order p, and $g \in \mathbb{G}$ be a generator. Suppose that there are at most t potential revoked users.

In the setup phase, the center chooses a polynomial $f(x) = \sum_{0 \le i \le t} a_i x^i$ over $GF(p)$ randomly, and computes $y_i = g^{a_i}$ for $0 \le i \le t$. The public key is $PK = (\mathbb{G}, g, y_0, ..., y_t)$. The center keeps $f(x)$ as the master key, and gives $d_i = f(i)$ to each user $i = 1, ..., p-1$ as his decryption key.

To revoke users $i_1, ..., i_t \in Z_p$, the sender generates a ciphertext $\psi = (g^r, (g^{f(i_1)})^r, ..., (g^{f(i_t)})^r)$ and a key $K = y_0^r$ where $r \stackrel{R}{\leftarrow} \mathbb{Z}_p$. Notice that $g^{f(i)}$ can be computed as $\prod_{0 \le j \le t} y_j^{i^j}$ for any $i \in \{1, ..., p-1\}$. On receiving $\psi = (C_0, ..., C_t)$, user $u \notin \{i_1, ..., i_t\}$ computes $C_u = C_0^{d_u}$ and recovers the key as $K = C_u^{\lambda(u)} \prod_{1 \le j \le t} C_j^{\lambda(i_j)}$ where $\lambda(x)$ is the Lagrange coefficient such that $\lambda(x) = \prod_{i' \in \{i, i_1, ..., i_t\} \setminus \{x\}} i' \cdot (i' - x)^{-1}$ over $\mathbb{Z}_p$.

3.2 Verifiability

As we mentioned, the main difficulty in this paper is how to add verifiability to the Naor-Pinkas scheme. Here we give a solution. Consider a modification of the Naor-Pinkas scheme such that user i is given $(f(i), f(rnd), rnd)$ as his decryption key, where $rnd \stackrel{R}{\leftarrow} \mathbb{Z}_p$. We note that a legitimate user i can decrypt a ciphertext in two different ways according to two different keys, i.e. $f(i)$ and $f(rnd)$. If these decryption results are not identical, then the user can detect that the ciphertext is in an invalid form. Notice that since rnd is random and not known to other users, it is difficult to generate an invalid ciphertext whose decryption results under $f(i)$ and $f(rnd)$ are identical.

Unfortunately, the above idea is faulty. Namely, even if user i is revoked and $f(i)$ does not work for decryption, he still has $f(rnd)$ and can decrypt a ciphertext by using it. Hence, the modified scheme is not secure any more. Therefore, we further modify the Naor-Pinkas scheme as follows: For at most t revoked users, in the setup phase, a polynomial $f(x) = \sum_{0 \le i \le 2t+1} a_i x^i$ is generated in the same manner as the original Naor-Pinkas scheme except that its degree is changed to be $2t+1$. The public key is $PK = (\mathbb{G}, g, y_0, ..., y_{2t+1})$. We assume that a user i has two unique identities $\mathbf{i}$ and i, where we denote $i = (\mathbf{i}, \mathsf{i}) \in \{1, ..., p-1\}^2$. The center keeps $f(x)$ as the master key, and for user $i = (\mathbf{i}, \mathsf{i}) \in \{1, ..., p-1\}^2$ he publishes $d_i = (f(\mathbf{i}), f(\mathsf{i}), f(rnd), rnd)$ as i's decryption key, where $rnd \stackrel{R}{\leftarrow} \mathbb{Z}_p$. Assuming that users $i_1 (= (\mathbf{i}_1, \mathsf{i}_1)), ..., i_t (= (\mathbf{i}_t, \mathsf{i}_t))$ are revoked, the sender generates $\psi = (g^r, (g^{f(\mathbf{i}_1)})^r, ..., (g^{f(\mathbf{i}_t)})^r, (g^{f(\mathsf{i}_1)})^r, ..., (g^{f(\mathsf{i}_t)})^r)$ and $K = y_0^r$ where $r \stackrel{R}{\leftarrow} \mathbb{Z}_p$.

On receiving $\psi = (C_0, ..., C_{2t})$, a user $i = (\mathbf{i}, \mathsf{i}) (\notin \{i_1, ..., i_t\})$ computes $C_{\mathbf{i}} = C_0^{f(\mathbf{i})}$, $C_{\mathsf{i}} = C_0^{f(\mathsf{i})}$, and $C_{rnd} = C_0^{f(rnd)}$. We notice that ψ can be decrypted by using any two of $C_{\mathbf{i}}$, C_{i}, and C_{rnd} with the Lagrange interpolation (for example, by using $(C_{\mathbf{i}}, C_{\mathsf{i}})$, the session key is recovered as $K = C_{\mathbf{i}}^{\lambda(\mathbf{i})} C_{\mathsf{i}}^{\lambda(\mathsf{i})} \prod_{1 \le j \le t} (C_j^{\lambda(\mathbf{i}_j)} C_{j+t}^{\lambda(\mathsf{i}_j)})$ where $\lambda(x)$ is the Lagrange coefficient such that $\lambda(x) = \prod_{i' \in \{\mathbf{i}, \mathsf{i}, \mathbf{i}_1, ..., \mathbf{i}_t, \mathsf{i}_1, ..., \mathsf{i}_t\} \setminus \{x\}} i' \cdot (i' - x)^{-1}$ over $\mathbb{Z}_p$). Then, user i carries out decryption in three different ways according to the three different choices of $(C_{\mathbf{i}}, C_{\mathsf{i}})$, $(C_{\mathbf{i}}, C_{rnd})$, and $(C_{\mathsf{i}}, C_{rnd})$. Then, user i can be convinced of the equality of decryption results for all legitimate subscribers if i's three decryption results are identical. Furthermore, when i is revoked, he cannot decrypt a ciphertext at all even though he still has $f(rnd)$. Now, we obtain a new verifiable BE scheme from Naor-Pinkas BE, and are ready to convert it into a CCA-secure PKE scheme.

4 Efficient CCA-Secure KEM from CDH

In this section, we show an efficient CCA-secure KEM under the CDH assumption such that the size of ciphertexts is the same as that of the CS scheme. Our

KEM is obtained from a verifiable BE scheme which was shown in Sec. 3. Let $\mathbb{G}$ be a multiplicative group with prime order p, and $g \in \mathbb{G}$ be a generator. Then, the construction of the scheme is as follows:

Setup(1^k)**:** Generate a random polynomial $f(x) = a_0 + a_1 x + \cdots + a_{k+2}x^{k+2}$ over $\mathbb{Z}_p$, and compute $y_i = g^{a_i}$ for $0 \leq i \leq k+2$. The decryption key is $f(x)$, and the public key is $PK = (\mathbb{G}, g, y_0, y_1, ..., y_{k+2}, \mathsf{TCR}_0, \mathsf{TCR}_1, h)$, where $\mathsf{TCR}_b : \mathbb{G} \rightarrow \mathcal{S}_b$ $(b = 0, 1)$ are target collision resistant hash functions such that $\mathcal{S}_0 \cup \mathcal{S}_1 \subseteq \mathbb{Z}_p^*$, $\mathcal{S}_0 \cap \mathcal{S}_1 = \emptyset$, and $h : \mathbb{G} \rightarrow \{0,1\}$ is a hardcore bit function for the Diffie-Hellman key in $\mathbb{G}$.[3]

Encrypt(PK)**:** Pick a random $r \stackrel{R}{\leftarrow} \mathbb{Z}_p$, and compute

$$\psi = (g^r, g^{r \cdot f(\mathbf{i})}, g^{r \cdot f(\mathrm{i})}), \qquad K = (h(y_0^r)||h(y_1^r)||...||h(y_{k-1}^r))$$

where $\mathbf{i} = \mathsf{TCR}_0(g^r)$ and $\mathrm{i} = \mathsf{TCR}_1(g^r)$. The final output is (ψ, K). (Notice that one can easily compute $g^{f(x)}$ as $g^{f(x)} = \prod_{0 \leq i \leq k+2} y_i^{x^i}$.)

Decrypt(dk, ψ, PK)**:** For a ciphertext $\psi = (C_0, C_1, C_2)$, check whether $(C_1, C_2) \stackrel{?}{=} (C_0^{f(\mathbf{i})}, C_0^{f(\mathrm{i})})$, where $\mathbf{i} = \mathsf{TCR}_0(C_0)$ and $\mathrm{i} = \mathsf{TCR}_1(C_0)$. If not, output $\perp$. Otherwise, output $K = (h(C_0^{a_0})||h(C_0^{a_1})||...||h(C_0^{a_{k-1}}))$.

Theorem 1. Let $\mathbb{G}$ be a multiplicative group with prime order p, TCR_0 and TCR_1 be (τ, ϵ_{tcr}) target collision resistant hash functions, and h be a (p_1, p_2) hardcore bit function for the Diffie-Hellman key in $\mathbb{G}$. Then, the above scheme is $(p_1^{-1}(\tau) - o(p_1^{-1}(\tau)), k \cdot p_2^{-1}(\epsilon_{cdh}) + 2\epsilon_{tcr} + q_D(2k/(p-3) + 1/(p-k-2)), q_D)$ CCA-secure under the (τ, ϵ_{cdh}) CDH assumption on $\mathbb{G}$.

Proof. Assume that for challenge ciphertext $(g^\beta, g^{\beta \cdot f(\mathbf{i}^\star)}, g^{\beta \cdot f(\mathrm{i}^\star)})$ such that $\mathbf{i}^\star = \mathsf{TCR}_0(g^\beta)$ and $\mathrm{i}^\star = \mathsf{TCR}_1(g^\beta)$, there exists an adversary A' which distinguishes $(h(y_0^\beta)||h(y_1^\beta)||...||h(y_{k-1}^\beta))$ from a random k-bit string. Then, by a standard hybrid argument, there also exists another adversary A which for some j such that $0 \leq j \leq k-1$ distinguishes

$$(h(y_0^\beta)||h(y_1^\beta)||...||h(y_j^\beta)||random_{k-j-1})$$

from

$$(h(y_0^\beta)||h(y_1^\beta)||...||h(y_{j-1}^\beta)||random_{k-j})$$

where $random_\ell$ denotes an ℓ-bit random string.

Now, assume we are given such an adversary A which distinguishes these two values with running time τ, advantage ϵ, and q_D decryption queries. We use A to construct another adversary B which for given (g, g^α, g^β) distinguishes $h(g^{\alpha\beta})$ from a random bit. Define adversary B as follows:

[3] h is a random string R if it is the Goldreich-Levin (GL) bit [18], where the size of R is equal to that of a group element. See also Appendix of [9] for the GL bit of the Diffie-Hellman keys.

1. For given (g, g^α, g^β), B picks target collision resistant hash functions TCR_0 and TCR_1, and computes $\mathbf{i}^\star = \mathsf{TCR}_0(g^\beta)$ and $\mathsf{i}^\star = \mathsf{TCR}_1(g^\beta)$.
2. B sets $y_j = g^\alpha$, and picks distinct randoms $rnd_j, ..., rnd_{k-1}$ from $\mathbb{Z}_p^* \backslash \{\mathbf{i}^\star, \mathsf{i}^\star\}$. B also picks randoms $u_{\mathbf{i}^\star}$, $u_{\mathsf{i}^\star}$, $a_0, ..., a_{j-1}$, and $u_j, ..., u_{k-1}$ from $\mathbb{Z}_p$.
3. B calculates $y_l = g^{a_l}$ for $0 \le l \le j-1$.
4. Let $f(x) = \sum_{i=0}^{k+2} a_i x^i$ be a polynomial over $\mathbb{Z}_p$ such that $a_j = \alpha$, $f(\mathbf{i}^\star) = u_{\mathbf{i}^\star}$, $f(\mathsf{i}^\star) = u_{\mathsf{i}^\star}$, and $f(rnd_j) = u_j, ..., f(rnd_{k-1}) = u_{k-1}$. Then, by using the Lagrange interpolation, B calculates $y_{j+1}, ..., y_{k+2}$ such that $g^{f(x)} = \prod_{0 \le j \le k+2} y_j^{x^j}$. Notice that $y_l = g^{a_l}$ holds for $0 \le l \le k+2$.
5. B inputs public key $PK = (\mathbb{G}, g, y_0, y_1, ..., y_{k+2}, \mathsf{TCR}_0, \mathsf{TCR}_1, h)$ and challenge ciphertext $\psi^\star = (g^\beta, (g^\beta)^{u_{\mathbf{i}^\star}}, (g^\beta)^{u_{\mathsf{i}^\star}})$ and
$$K^\star = (h((g^\beta)^{a_0}) || h((g^\beta)^{a_1}) || ... || h((g^\beta)^{a_{j-1}}) || \gamma || random_{k-j-1})$$
to A where γ is $h(g^{\alpha\beta})$ or a random bit.
6. When A makes decryption query $\psi = (C_0, C_1, C_2)$, B proceeds as follows:
   (a) If $C_0 = g^\beta$, then B responds $\perp$.
   (b) If $C_0 \neq g^\beta$ and $\mathsf{TCR}_b(C_0) \in \{\mathbf{i}^\star, \mathsf{i}^\star, rnd_j, ..., rnd_{k-2}, rnd_{k-1}\}$ for $b = 0$ or 1, then B aborts and outputs a random bit.
   (c) If $C_0 \neq g^\beta$ and $\mathsf{TCR}_b(C_0) \notin \{\mathbf{i}^\star, \mathsf{i}^\star, rnd_j, ..., rnd_{k-2}, rnd_{k-1}\}$ for both $b = 0$ and 1, B computes $C_0^{u_{\mathbf{i}^\star}}$, $C_0^{u_{\mathsf{i}^\star}}, C_0^{u_j}, ..., C_0^{u_{k-2}}$, and $C_0^{u_{k-1}}$. Let $\mathsf{TCR}_0(C_0) = \mathbf{i}$ and $\mathsf{TCR}_1(C_0) = \mathsf{i}$, and f_1, f_2, and f_3 be polynomials over $\mathbb{Z}_p$ with degree $k+2$ whose coefficient for x^l term is a_l for $0 \le l \le j-1$, such that
$$\begin{aligned}&(f_1(\mathbf{i}), f_1(\mathsf{i}), f_1(\mathbf{i}^\star), f_1(\mathsf{i}^\star), f_1(rnd_{j+1}), ..., f_1(rnd_{k-1}))\\ &\qquad = (\log_{C_0} C_1, \log_{C_0} C_2, u_{\mathbf{i}^\star}, u_{\mathsf{i}^\star}, u_{j+1}, ..., u_{k-1})\\ &(f_2(\mathbf{i}), f_2(\mathsf{i}), f_2(\mathbf{i}^\star), f_2(rnd_j), ..., f_2(rnd_{k-1}))\\ &\qquad = (\log_{C_0} C_1, \log_{C_0} C_2, u_{\mathbf{i}^\star}, u_j, ..., u_{k-1})\\ &(f_3(\mathbf{i}), f_3(\mathsf{i}), f_3(\mathsf{i}^\star), f_3(rnd_j), ..., f_3(rnd_{k-1}))\\ &\qquad = (\log_{C_0} C_1, \log_{C_0} C_2, u_{\mathsf{i}^\star}, u_j, ..., u_{k-1}).\end{aligned}$$
   Then, B calculates ${C_0}^{a_{1,l}}, {C_0}^{a_{2,l}}, {C_0}^{a_{3,l}}$ by using the Lagrange interpolation where $a_{1,l}$, $a_{2,l}$, and $a_{3,l}$ denote the coefficients of x^l term of f_1, f_2, and f_3 for $j \le l \le k-1$, respectively, and responds
$$K = (h(C_0^{a_0}) || ... || h(C_0^{a_{j-1}}) || h(C_0^{a_{1,j}}) || ... || h(C_0^{a_{1,k-1}}))$$
   if ${C_0}^{a_{1,j}} = {C_0}^{a_{2,j}} = {C_0}^{a_{3,j}}$, or "$\perp$" otherwise.
7. Finally, A outputs a bit b as his guess, and B outputs the same bit b as his own guess for $h(g^{\alpha\beta})$.

Let Win denote the event that A's guess is correct in the real world, Abort denote the event that A submits a ciphertext $\psi = (C_0, C_1, C_2)$ such that $C_0 \neq g^\beta$ and $\mathsf{TCR}_b(C_0) \in \{\mathbf{i}^\star, \mathsf{i}^\star, rnd_j, ..., rnd_{k-2}, rnd_{k-1}\}$ for $b = 0$ or 1, and $\mathsf{Invalid}$ denote the event that A submits a ciphertext $\psi = (C_0, C_1, C_2)$ such that B does not abort, ${C_0}^{a_{1,j}} = {C_0}^{a_{2,j}} = {C_0}^{a_{3,j}}$, but $(C_1, C_2) \neq (C_0^{f(\mathbf{i})}, C_0^{f(\mathsf{i})})$.

Then, B's advantage for guessing $h(g^{\alpha\beta})$ is estimated as follows:

$$\frac{1}{2} \cdot | \Pr[\mathsf{B}(g, g^\alpha, g^\beta, h(g^{\alpha\beta})) = 0] - \Pr[\mathsf{B}(g, g^\alpha, g^\beta, T) = 0]|$$

$$\geq | \Pr[\mathsf{Win}|\overline{\mathsf{Abort}} \wedge \overline{\mathsf{Invalid}}] \Pr[\overline{\mathsf{Abort}} \wedge \overline{\mathsf{Invalid}}] - \frac{1}{2}|$$

$$\geq | \Pr[\mathsf{Win}] - \Pr[\mathsf{Abort}] - \Pr[\mathsf{Invalid}] - \frac{1}{2}|.$$

Now, we prove following lemmas.

Lemma 1. $\Pr[\mathsf{Abort}] \leq 2\epsilon_{tcr} + \frac{2q_D k}{p-3}$.

Proof. Assume we are given an adversary A with $\Pr[\mathsf{Abort}] = p_A$. Then, we can construct another adversary B′ which for given $C \stackrel{R}{\leftarrow} \mathbb{G}$, finds $C'(\neq C) \in \mathbb{G}$ such that $\mathsf{TCR}_b(C') = \mathsf{TCR}_b(C)$ for $b = 0$ or 1 as follows: For given C, B′ generates decryption key $f(x)$ and public key $PK = (\mathbb{G}, g, y_0, y_1, ..., y_{k+2}, \mathsf{TCR}_0, \mathsf{TCR}_1, h)$, and computes challenge ciphertext $\psi^\star = (C, C^{u_{\mathbf{i}^\star}}, C^{u_{\mathsf{i}^\star}})$, where $u_{\mathbf{i}^\star} = f(\mathbf{i}^\star)$, $u_{\mathsf{i}^\star} = f(\mathsf{i}^\star)$, $\mathbf{i}^\star = \mathsf{TCR}_0(C)$, and $\mathsf{i}^\star = \mathsf{TCR}_1(C)$. B′ also picks distinct randoms $rnd_j, ..., rnd_{k-1}$ from $\mathbb{Z}_p^* \backslash \{\mathbf{i}^\star, \mathsf{i}^\star\}$, and gives PK and $(\psi^\star, K^\star)$ to A, where $K^\star$ is a correct key under $f(x)$ or a random element of $\mathbb{G}$ with probability 1/2.

Since $rnd_j, ..., rnd_{k-1}$ are information-theoretically hidden to A, for a query $\psi = (C_0, C_1, C_2)$, $\mathsf{TCR}_0(C_0)$ or $\mathsf{TCR}_1(C_0) \in \{rnd_j, ..., rnd_{k-2}, rnd_{k-1}\}$ happens with probability at most $2(k-j)/(p-3)$. Therefore, the probability that A submits a ciphertext $\psi = (C_0, C_1, C_2)$ $(C_0 \neq C)$ such that $\mathsf{TCR}_0(C_0) = \mathbf{i}^\star$ or $\mathsf{TCR}_1(C_0) = \mathsf{i}^\star$ is at least $p_A - 2q_D(k-j)/(p-3)$. B' outputs such C_0 as C'.

By using B′ as it is, we immediately have an algorithm B″ which for given $C \stackrel{R}{\leftarrow} \mathbb{G}$, finds $C''(\neq C) \in \mathbb{G}$ such that $\mathsf{TCR}_0(C'') = \mathsf{TCR}_0(C)$ with probability at least $p_A - 2q_D(k-j)/(p-3) - p_1$, where p_1 is the probability that B′ outputs C' such that $\mathsf{TCR}_1(C') = \mathsf{TCR}_1(C)$. Since $p_1 \leq \epsilon_{tcr}$, B″'s advantage is at least $p_A - 2q_D(k-j)/(p-3) - \epsilon_{tcr}$. Hence, $\epsilon_{tcr} \geq p_A - 2q_D(k-j)/(p-3) - \epsilon_{tcr}$, and therefore, we have $2\epsilon_{tcr} + 2q_D(k-j)/(p-3) \geq p_A$. □

Lemma 2. $\Pr[\mathsf{Invalid}] \leq \frac{q_D}{p-k-2}$.

Proof. Let $f_0(x) = \sum_{0 \leq l \leq j-1} a_l x^l$, and $f_1'(x), f_2'(x)$, and $f_3'(x)$ be polynomials such that $f_l(x) = f_0(x) + x^j \cdot f_l'(x)$ for $l = 1, 2, 3$. Let $f'(x)$ be a polynomial such that $f(x) = f_0(x) + x^j \cdot f'(x)$. Suppose $\psi = (C_0, C_1, C_2)$ is a ciphertext such that B does not abort, ${C_0}^{f_1'(0)} = {C_0}^{f_2'(0)} = {C_0}^{f_3'(0)}$, but $(C_1, C_2) \neq (C_0^{f(\mathbf{i})}, C_0^{f(\mathsf{i})})$. Then, we notice that f_1' and f_2' which are polynomials with degree $k-j+2$ have $k-j+3$ intersections, and consequently they have to be identical. Similarly, we have that $f_1' = f_2' = f_3'$. This implies that for [$\mathsf{Invalid} = \text{true}$], A has to choose C_1 and C_2 (without knowing $rnd_j, ..., rnd_{k-1}$) such that f_1' (with degree $k-j+2$) satisfies

1. $(f_1'(\mathbf{i}), f_1'(\mathsf{i}), f_1'(\mathbf{i}^\star), f_1'(\mathsf{i}^\star), f_1'(rnd_j), ..., f_1'(rnd_{k-1}))$
$= ((\log_{C_0} C_1 - f_0(\mathbf{i})) \cdot \mathbf{i}^{-j}, (\log_{C_0} C_2 - f_0(\mathsf{i})) \cdot \mathsf{i}^{-j}, f'(\mathbf{i}^\star), f'(\mathsf{i}^\star),$
$f'(rnd_j), ..., f'(rnd_{k-1})),$

2. $f'_1 \neq f'$.

Since f'_1 and f' have at most $k-j+2$ intersections and $k-j+1$ of them are $(\mathbf{i}^\star, f'(\mathbf{i}^\star))$, $(\mathsf{i}^\star, f'(\mathsf{i}^\star))$, $(rnd_{j+1}, f'(rnd_{j+1})), ..., (rnd_{k-1}, f'(rnd_{k-1}))$, there is only one remained intersection which must be $(rnd_j, f'(rnd_j))$. Therefore, [$\mathsf{Invalid}$ = true] happens only when A correctly guesses the value of rnd_j (even if A is given $rnd_{j+1}, ..., rnd_{k-1}$). Hence, for any invalid query ψ, the probability that B does not respond "$\perp$" is at most $1/(p-k+j-2)(\leq 1/(p-k-2))$. □

A's advantage is estimated as at least $1/k$ times A''s advantage due to the hybrid argument. □

5 Efficient CCCA-Secure KEM from HDH

In this section, based on the strategy in Sec. 3, we propose another KEM which is CCCA-secure [22] under the HDH assumption. This scheme is as efficient as the KD scheme [26] with a weaker assumption. As shown in [22], a CCA-secure PKE scheme can be constructed by combining any CCCA-secure KEM and authenticated symmetric key encryption [3] as a DEM. Let $\mathbb{G}$ be a multiplicative group with prime order p, and $g \in \mathbb{G}$ be a generator. Then, the construction of our CCCA-secure KEM is as follows:

Setup(1^k)**:** Generate a random polynomial $f(x) = a_0 + a_1 x + a_2 x^2$ over $\mathbb{Z}_p$, and compute $y_j = g^{a_j}$ for $0 \leq j \leq 2$. The decryption key is $f(x)$, and the public key is $PK = (\mathbb{G}, g, y_0, y_1, y_2, \mathsf{TCR}, h)$, where $\mathsf{TCR} : \mathbb{G} \to \mathbb{Z}_p^*$ is a target collision resistant hash function and $h : \mathbb{G} \to \{0,1\}^\nu$ is a hash function.

Encrypt(PK)**:** Pick a random $r \stackrel{R}{\leftarrow} \mathbb{Z}_p$, and compute $\psi = (g^r, g^{r \cdot f(i)})$ and $K = h(y_0^r)$, where $i = \mathsf{TCR}(g^r)$. The final output is (ψ, K). (Notice that one can easily compute $g^{f(x)}$ as $g^{f(x)} = \prod_{0 \leq j \leq 2} y_j^{x^j}$.)

Decrypt(dk, ψ, PK)**:** For a ciphertext $\psi = (C_0, C_1)$, check whether $C_1 \stackrel{?}{=} C_0^{f(i)}$, where $i = \mathsf{TCR}(C_0)$. If not, output $\perp$. Otherwise, output $K = h(C_0^{a_0})$.

The above scheme can be proved to be CCCA-secure, and its security is formally addressed in the full version of this paper [21].

6 Comparison

Table 1 shows a comparison of our schemes with other CCA-secure schemes, i.e. Cramer-Shoup (CS) [14,38], Kurosawa-Desmedt (KD) [26], Boyen-Mei-Waters (BMW) [10], Kiltz [25], Cash-Kiltz-Shoup (CKS) [13], and Hofheinz-Kiltz (HK) [22]. In the comparison, we utilize a redundancy-free CCA-secure DEM [20,33] for constructing a CCA-secure hybrid encryption scheme from a CCA-secure KEM.

As seen in Table 1, our proposed scheme in Sec. 4 yields both provable security under the CDH assumption and short ciphertext length which is comparable

Table 1. Efficiency comparison for CCA-secure PKE schemes. Some figures are borrowed from [10,25]. For efficiency, we count the number of pairings, multi(or sequential)-exponentiations [34], regular-exponentiations, and other group operations ("ops" denotes group operations) used for encryption and decryption. All symmetric operations (such as hash function/MAC/KDF) are ignored. Ciphertext overhead represents the difference between ciphertext and plaintext length, and $|g|$ and $|mac|$ are the length of a group element and an authentication tag, respectively. In the table, we let $k' = k/\log k$ where k is the security parameter, i.e. DEM-key length.

	Security Assumption	Ciphertext Overhead	Encryption	Decryption
			#pairings + #[multi,regular]-exp (+ #ops)	
CS [14]	DDH	$3\|g\|$	$0 + [1, 3]$	$0 + [1, 1]$
KD [26]	DDH	$2\|g\| + \|mac\|$	$0 + [1, 2]$	$0 + [1, 0]$
BMW [10]	BDH	$2\|g\|$	$0 + [1, 2]$	$1 + [0, 1]$
Kiltz [25]	GHDH	$2\|g\|$	$0 + [1, 2]$	$0 + [1, 0]$
CKS [13]	CDH	$(k' + 2)\|g\|$	$0 + [k' + 1, k' + 1]$	$0 + [1^{\ddagger}, 0]$
	HDH	$3\|g\|$	$0 + [2, 2]$	$0 + [1, 0]$
HK† [22]	HDH	$2\|g\| + \|mac\|$	$0 + [1, 2]$	$0 + [1, 0]$
Ours §4	CDH	$3\|g\|$	$0 + [2^{\ddagger}, k' + 1]$	$0 + [1^{\ddagger}, 0]$
Ours §5	HDH	$2\|g\| + \|mac\|$	$0 + [1, 2]$	$0 + [1, 0]$
Ours §7.3	2ℓ-BDHE	$2\|g\|$	$0 + [0, 3] + \ell$	$3 + [0, 0] + \ell$

† A slight modification by [13] is applied.

‡ Relatively more expensive computation is needed for one exponentiation.

to other practical schemes. Comparing with the CDH-based CKS scheme, our scheme in Sec. 4 is more efficient, and especially, the ciphertext overhead of our scheme, i.e. three group elements, is much shorter than that of the CKS scheme, i.e. $k/\log k + 2$ group elements, since $k/\log k \simeq 18$ for 128-bit security. In the comparison, we assume that $\log k$ hardcore bits can be extracted from a single DH key [18]. Furthermore, the ciphertext overhead of our scheme is the same as that of the CS scheme. Our scheme in Sec. 5 is as efficient as the KD scheme with a weaker underlying assumption. The Hofheinz-Kiltz scheme [22] (with a modification by [13]) has almost the same property as ours. (See also the footnote in Sec. 1.2.)

7 CCA-Security from BE with Verifiability

In this section, we observe that it is possible to construct a CCA-secure PKE scheme from an arbitrary verifiable BE scheme, and that security of many existing CCA-secure PKE schemes can also be explained from this viewpoint. This observation implies that one of promising approaches for achieving CCA-security is to concentrate on designing verifiable BE schemes. In fact, constructions of our proposed schemes are based on this approach.

7.1 The Generic Conversion

Given a verifiable BE scheme $\Pi' = (\mathbf{Setup}', \mathbf{Encrypt}', \mathbf{Decrypt}')$ which is CPA-secure against selective adversaries, we construct a CCA-secure KEM $\Pi = (\mathbf{Setup}, \mathbf{Encrypt}, \mathbf{Decrypt})$. In the construction, we use a strong one-time signature scheme Σ = (**Gen**, **Sign**, **Verify**) in which the verification key generated by $\mathbf{Gen}(1^k)$ has length k. We assume that the maximum number of potential users in Π' is n, and a sender can revoke t users where there exists an injective mapping (or a target collision resistant hash function) $\mathsf{INJ} : \{0,1\}^k \rightarrow \mathcal{P}$ and $\mathcal{P}$ is the set of all subsets $\mathcal{S} \subseteq \{1, ..., n\}$ with $|\mathcal{S}| = n - t$. Notice that for existence of such an injective mapping, it is necessary that ${}_nC_t \geq 2^k$ (for example, $(n, t) = (2k, k)$). The construction of Π is as follows:

Setup(1^k)**:** Choose n and t (which is a possible parameter choice for Π') such that ${}_nC_t \geq 2^k$. Run $\mathbf{Setup}'(1^k, n, t)$ to obtain $(d_1, ..., d_n, PK)$, and pick an injective mapping $\mathsf{INJ} : \{0,1\}^k \rightarrow \mathcal{P}$. The decryption key is $dk = (d_1, ..., d_n)$ and the public key is $\overline{PK} = (PK, \mathsf{INJ})$.

Encrypt$(\overline{PK})$**:** Run $\mathbf{Gen}(1^k)$ to obtain verification key vk and signing key sk (with $|vk| = k$), and compute $\mathcal{S}_{vk} = \mathsf{INJ}(vk)$, $(\psi, K) \leftarrow \mathbf{Encrypt}'(\mathcal{S}_{vk}, PK)$ and $\sigma \leftarrow \mathbf{Sign}(sk, \psi)$. The final output is $((\psi, vk, \sigma), K)$.

Decrypt$(dk, \psi, \overline{PK})$**:** For a ciphertext (ψ, vk, σ), check whether $\mathbf{Verify}(vk, \psi, \sigma) \stackrel{?}{=} 1$. If not, output $\perp$. Otherwise, compute $\mathcal{S}_{vk} = \mathsf{INJ}(vk)$ and output $K \leftarrow \mathbf{Decrypt}'(\mathcal{S}_{vk}, i, d_i, \psi, PK)$ where $i \in \mathcal{S}_{vk}$.

CCA-security of the above construction can be proven in a similar manner to [12]. We give an intuitive explanation for the security. Let A be an algorithm which can break CCA-security of Π. Then, it is possible to construct another algorithm B which can break Π' by using A as follows: B runs $(vk^\star, sk^\star) \leftarrow \mathbf{Gen}(1^k)$, and commits $\mathcal{S}^\star = \mathsf{INJ}(vk^\star)$ as the subset of users which will be attacked. For given public key PK of Π', B passes (PK, INJ) to A as a public key of Π. When A submits decryption query (ψ, vk, σ), B responds to it by simply decrypting the ciphertext with decryption key d_i such that $i \in \mathsf{INJ}(vk) \backslash \mathcal{S}^\star \subseteq \{1, ..., n\}$. We note that there always exists at least one such a decryption key unless $vk = vk^\star$, and $vk \neq vk^\star$ holds with an overwhelming probability if σ is a valid signature. Let $(\psi^\star, K^\star)$ be a challenge ciphertext of Π' from the challenger. Then, B gives $((\psi^\star, vk^\star, \sigma^\star), K^\star)$ to A as a challenge ciphertext of Π where $\sigma^\star \leftarrow \mathsf{Sign}(sk^\star, \psi^\star)$. A formal security proof is given in the full version of this paper [21].

Theorem 2. If Π' is a $(\tau, \epsilon_{cpa}, n, t)$ semantically secure and $(\tau, \epsilon_{vfy}, n, t)$ publicly verifiable broadcast encryption scheme such that ${}_nC_t \geq 2^k$, and Σ is a (τ, ϵ_{uf}) strongly unforgeable one-time signature scheme, then Π is a $(\tau - o(\tau), \epsilon_{cpa} + \epsilon_{vfy} + \frac{1}{2}\epsilon_{uf}, q_D)$ CCA-secure key encapsulation mechanism.

A similar result can also be obtained from *privately* verifiable BE schemes.

7.2 Remarks

We notice that the above generic conversion is identical to the Canetti-Halevi-Katz (CHK) paradigm [12] except that the underlying primitive of CHK, i.e. IBE, is replaced with verifiable BE in our construction. Kiltz [24] also showed that IBE is not always necessary for CHK and a weaker primitive which is called *tag-based encryption* (TBE) [28] is sufficient, and demonstrated to construct a

Table 2. Relation among broadcast encryption and public key encryption schemes. The column "(n,t)" denotes a possible and typical parameter setting for each underlying broadcast encryption scheme, and $\mathsf{poly}(k)$ and $\mathsf{exp}(k)$ denote polynomial and exponential functions for the security parameter k, respectively. For verifiability, related cryptographic tools are described, and $\surd$ means that the underlying broadcast encryption has verifiability as it is.

BE Scheme	(n,t)	Verifiability	$\Rightarrow$	PKE Scheme
Trivial BE	$(\mathsf{poly}(k), n/2)$	NIZK		DDN [16]
Naor-Pinkas [30]	$(\mathsf{exp}(k), 1)$	DDH		a variant of CS [14]
		GHDH	$\Rightarrow$	Kiltz [25]
		Sec. 3.2		Ours §4
IBE	$(\mathsf{exp}(k), n-1)$	$\surd$		CHK [12]
BGW [6]	$(\mathsf{poly}(k), n/2)$	$\surd$		Ours §7.3

concrete TBE scheme without using IBE-related techniques. There are also other CCA-secure schemes whose security can be explained via the TBE framework, e.g. [14,10,25]. Our proposed method is a generic construction of TBE from BE with verifiability.

Many existing CCA-secure PKE schemes can be explained via our observation in Sec. 7.1 with different underlying BE schemes, and relations among existing BE and CCA-secure PKE schemes are summarized in Table 2. We give more detailed explanations for this in the full version of this paper [21].

7.3 Another New CCA-Secure KEM from Boneh-Gentry-Waters

Based on the proposed methodology, we can construct yet another new practical CCA-secure KEM from the BGW BE scheme [6]. This can be a further evidence that BE with verifiability is a powerful tool for constructing CCA-secure PKE. The proposed scheme yields tight security reduction to the 2ℓ-BDHE problem [6] for relatively small ℓ, short ciphertexts and short decryption keys. The concrete construction of the scheme is as follows: Let $\mathbb{G}_1$ and $\mathbb{G}_2$ be multiplicative cyclic groups with prime order p, and $e : \mathbb{G}_1 \times \mathbb{G}_1 \rightarrow \mathbb{G}_2$ be a bilinear mapping [5]. **Setup**(1^k) chooses $\ell \in \mathbb{N}$ such that ${}_{2\ell}C_\ell \geq 2^k$, and picks a random generator $g \in \mathbb{G}_1$ and random $\alpha, \gamma \in \mathbb{Z}_p$. It also generates $g_1, ..., g_{4\ell}, v$, and Z where $g_i = g^{(\alpha^i)}$, $v = g^\gamma$, and $Z = e(g_{2\ell+1}, g)$. The decryption key is $dk = g^{\alpha^{2\ell+1}}$, and the public key is $PK = (g, g_1, ..., g_{2\ell}, g_{2\ell+2}, ..., g_{4\ell}, v, Z, \mathsf{TCR})$, where $\mathsf{TCR} : \mathbb{G}_1 \rightarrow \mathcal{P}$ is a target collision resistant hash function and $\mathcal{P} = \{\mathcal{S} | \mathcal{S} \subseteq \{1, ..., 2\ell\}, |\mathcal{S}| = \ell\}$. **Encrypt**$(PK)$ picks a random $r \in \mathbb{Z}_p$, sets $K = Z^r \in \mathbb{G}_2$, computes $\mathcal{S} = \mathsf{TCR}(g^r)$, and outputs (ψ, K) where $\psi = (g^r, (v \cdot \prod_{j \in \mathcal{S}} g_{2\ell+1-j})^r) \in \mathbb{G}_1^2$. For ciphertext $\psi = (C_0, C_1)$, **Decrypt**(dk, ψ, PK) computes $\mathcal{S} = \mathsf{TCR}(C_0)$, and checks whether $e(g, C_1) \stackrel{?}{=} e(v \cdot \prod_{j \in \mathcal{S}} g_{2\ell+1-j}, C_0)$. It outputs "$\perp$" if it is invalid, or $K = e(dk, C_0)$ otherwise. Security of this scheme can be proven by a straightforward combination of the proofs of Theorem 2 of this paper and Theorem 3.1 of [6]. Unfortunately, this scheme is not very advantageous to other schemes, but it is still comparably efficient to other practical schemes (see Table 1).

7.4 A Generic Construction of CCA-Secure Broadcast Encryption

By using our methodology, it is also generically possible to construct a CCA-secure BE scheme from CPA-secure one with public verifiability. The conversion is fairly simple, and the resulting CCA-secure scheme can be practical. When applying this to the BGW BE scheme, we can have a new CCA-secure BE scheme with verifiability whose computational cost is slightly better than the previous scheme [6]. More detailed explanation is given in the full version of this paper [21].

Acknowledgement

The authors would like to thank Nuttapong Attrapadung, David Cash, Eike Kiltz and Takahiro Matsuda for their helpful comments and suggestions. The authors also would like to thank anonymous reviewers of Asiacrypt'08 for their invaluable comments.

References

1. Abe, M., Gennaro, R., Kurosawa, K., Shoup, V.: Tag-KEM/DEM: a new framework for hybrid encryption and a new analysis of Kurosawa-Desmedt KEM. In: Cramer, R. (ed.) EUROCRYPT 2005. LNCS, vol. 3494, pp. 128–146. Springer, Heidelberg (2005)
2. Anzai, J., Matsuzaki, N., Matsumoto, T.: A quick group key distribution scheme with "entity revocation". In: Lam, K.-Y., Okamoto, E., Xing, C. (eds.) ASIACRYPT 1999. LNCS, vol. 1716, pp. 333–347. Springer, Heidelberg (1999)
3. Bellare, M., Namprempre, C.: Authenticated encryption: relations among notions and analysis of the generic composition paradigm. In: Okamoto, T. (ed.) ASIACRYPT 2000. LNCS, vol. 1976, pp. 531–545. Springer, Heidelberg (2000)
4. Blum, M., Feldman, P., Micali, S.: Non-interactive zero-knowledge and its applications. In: Proc. of STOC 1988, pp. 103–112 (1988)
5. Boneh, D., Franklin, M.K.: Identity-based encryption from the Weil pairing. In: Kilian, J. (ed.) CRYPTO 2001. LNCS, vol. 2139, pp. 213–229. Springer, Heidelberg (2001)
6. Boneh, D., Gentry, C., Waters, B.: Collusion resistant broadcast encryption with short ciphertexts and private keys. In: Shoup, V. (ed.) CRYPTO 2005. LNCS, vol. 3621, pp. 258–275. Springer, Heidelberg (2005)
7. Boneh, D., Katz, J.: Improved efficiency for CCA-secure cryptosystems built using identity-based encryption. In: Menezes, A. (ed.) CT-RSA 2005. LNCS, vol. 3376, pp. 87–103. Springer, Heidelberg (2005)
8. Boneh, D., Shparlinski, I.: On the unpredictability of bits of the elliptic curve Diffie-Hellman scheme. In: Kilian, J. (ed.) CRYPTO 2001. LNCS, vol. 2139, pp. 201–212. Springer, Heidelberg (2001)
9. Boneh, D., Venkatesan, R.: Hardness of computing the most significant bits of secret keys in Diffie-Hellman and related schemes. In: Koblitz, N. (ed.) CRYPTO 1996. LNCS, vol. 1109, pp. 129–142. Springer, Heidelberg (1996)
10. Boyen, X., Mei, Q., Waters, B.: Direct chosen ciphertext security from identity-based techniques. In: Proc. of CCS 2005, pp. 320–329 (2005)

11. Canetti, R.: Universally composable security: a new paradigm for cryptographic protocols. In: Proc. of FOCS 2001, pp. 136–145 (2001)
12. Canetti, R., Halevi, S., Katz, J.: Chosen-ciphertext security from identity-based encryption. In: Cachin, C., Camenisch, J.L. (eds.) EUROCRYPT 2004. LNCS, vol. 3027, pp. 207–222. Springer, Heidelberg (2004)
13. Cash, D., Kiltz, E., Shoup, V.: The twin Diffie-Hellman problem and applications. In: Smart, N.P. (ed.) EUROCRYPT 2008. LNCS, vol. 4965, pp. 127–145. Springer, Heidelberg (2008); The full version is IACR ePrint 2008/067
14. Cramer, R., Shoup, V.: A practical public key cryptosystem provably secure against adaptive chosen ciphertext attack. In: Krawczyk, H. (ed.) CRYPTO 1998. LNCS, vol. 1462, pp. 13–25. Springer, Heidelberg (1998)
15. Dodis, Y., Fazio, N.: Public key trace and revoke scheme secure against adaptive chosen ciphertext attack. In: Desmedt, Y.G. (ed.) PKC 2003. LNCS, vol. 2567, pp. 100–115. Springer, Heidelberg (2003)
16. Dolev, D., Dwork, C., Naor, M.: Non-malleable cryptography. In: Proc. of STOC 1991, pp. 542–552 (1991)
17. Fiat, A., Naor, M.: Broadcast encryption. In: Stinson, D.R. (ed.) CRYPTO 1993. LNCS, vol. 773, pp. 480–491. Springer, Heidelberg (1994)
18. Goldreich, O., Levin, L.A.: A hard-core predicate for all one-way functions. In: Proc. of STOC 1989, pp. 25–32 (1989)
19. Goldwasser, S., Micali, S.: Probabilistic encryption. J. Comput. Syst. Sci. 28(2), 270–299 (1984)
20. Halevi, S., Rogaway, P.: A tweakable enciphering mode. In: Boneh, D. (ed.) CRYPTO 2003. LNCS, vol. 2729, pp. 482–499. Springer, Heidelberg (2003)
21. Hanaoka, G., Kurosawa, K.: Efficient chosen ciphertext secure public key encryption under the computational Diffie-Hellman assumption, IACR ePrint 2008/211 (2008)
22. Hofheinz, D., Kiltz, E.: Secure hybrid encryption from weakened key encapsulation. In: Menezes, A. (ed.) CRYPTO 2007. LNCS, vol. 4622, pp. 553–571. Springer, Heidelberg (2007)
23. Kiltz, E.: A primitive for proving the security of every bit and about universal hash functions & hard core bits. In: Freivalds, R. (ed.) FCT 2001. LNCS, vol. 2138, pp. 388–391. Springer, Heidelberg (2001)
24. Kiltz, E.: Chosen-ciphertext security from tag-based encryption. In: Halevi, S., Rabin, T. (eds.) TCC 2006. LNCS, vol. 3876, pp. 581–600. Springer, Heidelberg (2006)
25. Kiltz, E.: Chosen-ciphertext secure key-encapsulation based on gap hashed Diffie-Hellman. In: Okamoto, T., Wang, X. (eds.) PKC 2007. LNCS, vol. 4450, pp. 282–297. Springer, Heidelberg (2007)
26. Kurosawa, K., Desmedt, Y.: A new paradigm of hybrid encryption scheme. In: Franklin, M. (ed.) CRYPTO 2004. LNCS, vol. 3152, pp. 426–442. Springer, Heidelberg (2004)
27. Luby, M., Rackoff, C.: How to construct pseudorandom permutations from pseudorandom functions. SIAM J. Comput. 17(2), 373–386 (1988)
28. MacKenzie, P.D., Reiter, M.K., Yang, K.: Alternatives to non-malleability: Definitions, constructions, and applications. In: Naor, M. (ed.) TCC 2004. LNCS, vol. 2951, pp. 171–190. Springer, Heidelberg (2004)
29. Naor, D., Naor, M., Lotspiech, J.: Revocation and tracing schemes for stateless receivers. In: Kilian, J. (ed.) CRYPTO 2001. LNCS, vol. 2139, pp. 41–62. Springer, Heidelberg (2001)

30. Naor, M., Pinkas, B.: Efficient trace and revoke schemes. In: Frankel, Y. (ed.) FC 2000. LNCS, vol. 1962, pp. 1–20. Springer, Heidelberg (2001)
31. Naor, M., Yung, M.: Public-key cryptosystems provably secure against chosen ciphertext attacks. In: Proc. of STOC 1990, pp. 427–437 (1990)
32. Pass, R., Shelat, A., Vaikuntanathan, V.: Construction of a non-malleable encryption scheme from any semantically secure one. In: Dwork, C. (ed.) CRYPTO 2006. LNCS, vol. 4117, pp. 271–289. Springer, Heidelberg (2006)
33. Phan, D.H., Pointcheval, D.: About the security of ciphers (semantic security and pseudo-random permutations). In: Handschuh, H., Hasan, M.A. (eds.) SAC 2004. LNCS, vol. 3357, pp. 182–197. Springer, Heidelberg (2004)
34. Pippenger, N.: On the evaluation of powers and related problems. In: Proc. of FOCS 1976, pp. 258–263 (1976)
35. Rackoff, C., Simon, D.R.: Non-interactive zero-knowledge proof of knowledge and chosen ciphertext attack. In: Feigenbaum, J. (ed.) CRYPTO 1991. LNCS, vol. 576, pp. 433–444. Springer, Heidelberg (1992)
36. Sahai, A.: Non-malleable non-interactive zero knowledge and adaptive chosen-ciphertext security. In: Proc. of FOCS 1999, pp. 543–553 (1999)
37. Shamir, A.: Identity-based cryptosystems and signature schemes. In: Blakely, G.R., Chaum, D. (eds.) CRYPTO 1984. LNCS, vol. 196, pp. 47–53. Springer, Heidelberg (1985)
38. Shoup, V.: Using hash functions as a hedge against chosen ciphertext attack. In: Preneel, B. (ed.) EUROCRYPT 2000. LNCS, vol. 1807, pp. 275–288. Springer, Heidelberg (2000)

Twisted Edwards Curves Revisited

Huseyin Hisil, Kenneth Koon-Ho Wong, Gary Carter, and Ed Dawson

Information Security Institute,
Queensland University of Technology, QLD, 4000, Australia
{h.hisil,kk.wong,g.carter,e.dawson}@qut.edu.au

Abstract. This paper introduces fast algorithms for performing group operations on twisted Edwards curves, pushing the recent speed limits of Elliptic Curve Cryptography (ECC) forward in a wide range of applications. Notably, the new addition algorithm uses[1] $8\mathbf{M}$ for suitably selected curve constants. In comparison, the fastest point addition algorithms for (twisted) Edwards curves stated in the literature use $9\mathbf{M}+1\mathbf{S}$. It is also shown that the new addition algorithm can be implemented with four processors dropping the effective cost to $2\mathbf{M}$. This implies an effective speed increase by the full factor of 4 over the sequential case. Our results allow faster implementation of elliptic curve scalar multiplication. In addition, the new point addition algorithm can be used to provide a natural protection from side channel attacks based on simple power analysis (SPA).

Keywords: Efficient elliptic curve arithmetic, unified addition, side channel attack, SPA.

1 Introduction

Edwards curves are drawing increasing attention with their low cost and memory friendly arithmetic in cryptographic applications. Recently, there has been a rapid development of Edwards curves and their use in cryptology. An outline of the previous work that closely relates to twisted Edwards curves is as follows.

- Building on the historical results of Euler and Gauss, Edwards introduced a normal form for elliptic curves and stated the addition law in [13]. These curves are defined by $x^2+y^2=c^2+c^2x^2y^2$.
- Bernstein and Lange introduced a more general version of these curves defined by $x^2+y^2=c^2(1+dx^2y^2)$ or simply $x^2+y^2=1+dx^2y^2$ together with the first algorithms for computing the group operations on projective coordinates in [5]. For instance, the addition costs $10\mathbf{M}+1\mathbf{S}+1\mathbf{D}$ with $c=1$. Here, and in the rest of this paper, multiplication by a curve constant is denoted by $\mathbf{D}$. With the definitions in [5], these curves are today known as the Edwards curves.
- Bernstein and Lange introduced the inverted Edwards coordinates in [6] which reduce the cost for the group operations on Edwards curves. For instance, the addition costs $9\mathbf{M}+1\mathbf{S}+1\mathbf{D}$.

[1] **M**: Field multiplication, **S**: Field squaring, **I**: Field inversion.

J. Pieprzyk (Ed.): ASIACRYPT 2008, LNCS 5350, pp. 326–343, 2008.
© International Association for Cryptologic Research 2008

- Bernstein, Birkner, Joye, Lange, and Peters introduced twisted Edwards curves $ax^2 + y^2 = 1 + dx^2y^2$ in [1], a generalization of Edwards curves.

In this paper, the speed of the arithmetic of twisted Edwards curves is increased by a suitable point representation. The new system is called *extended twisted Edwards coordinates* which adds an auxiliary coordinate to twisted Edwards coordinates. Despite the computational overhead of the additional coordinate, we develop faster ways of performing point addition since the new formulae are composed of polynomial expressions with lower total degrees. We show that the increase in the number of coordinates comes with an increase in the level of parallelism which is exploited for further improvements. We also provide optimizations for the scalar multiplication by mixing extended twisted Edwards coordinates with twisted Edwards coordinates.

The paper is organized as follows. A review of twisted Edwards curves together with some new results is given in Section 2. The new point representation is introduced in Section 3. Several applications of the new achievements are given in Section 4. We draw our conclusions in Section 5.

2 Twisted Edwards Curves

In what follows some terms related to the group law on elliptic curves will be extensively used. In particular, the term *unified* is used to emphasize that point addition formulae remain valid when two input points are identical, see [10, Section 29.1.2]. Therefore, unified addition formulae can be used for point doubling. The term *complete* is used to emphasize that addition formulae are defined for all inputs, see [5]. The term *readdition* is used to emphasize that a point addition has already taken place and some of the previously computed data is cached, see [5]. The term *mixed addition* refers to adding an affine point to a point in some projective representation, see [11]. We adapt the notation from [11], [5], and [1].

Let K be a field of odd characteristic. In [5], Bernstein and Lange introduce Edwards curves defined by $x^2 + y^2 = c^2(1 + dx^2y^2)$ where $c, d \in K$ with $cd(1 - dc^4) \neq 0$. In [1], this form is generalized to twisted Edwards form defined by

$$E_{E,a,d}\colon ax^2 + y^2 = 1 + dx^2y^2$$

where $a, d \in K$ with $ad(a - d) \neq 0$. Edwards curves are then a special case of twisted Edwards curve where a can be rescaled to 1. We next review some formulae regarding the group law on twisted Edwards curves which will be used with slight modifications in Section 3.

Affine addition formulae for twisted Edwards curves in [1] (also see [13], [5]):

$$(x_1, y_1) + (x_2, y_2) = \left(\frac{x_1y_2 + y_1x_2}{1 + dx_1y_1x_2y_2}, \frac{y_1y_2 - ax_1x_2}{1 - dx_1y_1x_2y_2}\right) = (x_3, y_3). \quad (1)$$

The point $(0, 1)$ is the identity element and the point $(0, -1)$ is of order 2. The negative of a point (x, y) is $(-x, y)$. For further facts such as the resolution of singularities or the points at infinity or the coverage of these curves or the group structure, we refer the reader to the original reference [1]. Also see [13], [5], [4], [6], and [3].

In [5] (where $a = 1$) and later in [1], it was proven that if d is not a square in K and a is a square in K then these formulae are complete. In Theorem 1, with reasonable assumptions, we show that it is possible to prevent exceptions in the addition formulae even if d is a square in K or a is not a square in K. We should note that this statement should not be considered as a recommendation for selecting d a square in K and/or a a non-square in K. The desired properties for a and d may change depending on the target application. We will recall Theorem 1 in Section 4.

Theorem 1. *Let K be a field of odd characteristic. Let $E_{E,a,d}$ be a twisted Edwards curve defined over K. Let $P = (x_1, y_1)$ and $Q = (x_2, y_2)$ be points on $E_{E,a,d}$. Assume that P and Q are of odd order. It follows that $1 - dx_1x_2y_1y_2 \neq 0$ and $1 + dx_1x_2y_1y_2 \neq 0$.*

Proof. In [5] (where $a = 1$) and later in [1], it is proven that the points at infinity (over the extension of K where they exist) are of even order. Assume that P and Q are of odd order. Thus, P, Q and $P + Q$ cannot be the points at infinity. Since the formulae (1) are complete (see [1]) provided that the points at infinity are not involved, the denominators of (1); $1 - dx_1x_2y_1y_2$ and $1 + dx_1x_2y_1y_2$ must be nonzero. □

Affine doubling formulae (independent of d) for twisted Edwards curves deduced from [1] (also see [5], [2], [3]):

$$2(x_1, y_1) = \left(\frac{2x_1y_1}{y_1^2 + ax_1^2}, \frac{y_1^2 - ax_1^2}{2 - y_1^2 - ax_1^2} \right) = (x_3, y_3). \tag{2}$$

The exceptional cases and how to prevent them are analogous to formulae (1).

Affine addition formulae (independent of d) for twisted Edwards curves adapted from our preprint [17]: Consider the relations obtained by the curve equation; $ax_1^2 + y_1^2 = 1 + dx_1^2y_1^2$, $ax_2^2 + y_2^2 = 1 + dx_2^2y_2^2$. After straight forward eliminations, we express a and d in terms of x_1, x_2, y_1, y_2 as follows,

$$a = \frac{(x_1^2y_1^2 - x_2^2y_2^2) - y_1^2y_2^2(x_1^2 - x_2^2)}{x_1^2x_2^2(y_1^2 - y_2^2)}, \qquad d = \frac{(x_1^2 - x_2^2) - (x_1^2y_2^2 - y_1^2x_2^2)}{x_1^2x_2^2(y_1^2 - y_2^2)}.$$

Ignoring any exceptions that can be introduced by these rational expressions, substitutions in the addition formulae (1) yield

$$\begin{aligned} x_3 &= \frac{x_1y_2 + y_1x_2}{1 + \frac{(x_1^2 - x_2^2) - (x_1^2y_2^2 - y_1^2x_2^2)}{x_1^2x_2^2(y_1^2 - y_2^2)} x_1y_1x_2y_2} = \frac{x_1x_2(y_1^2 - y_2^2)}{x_1y_1 - x_2y_2 - y_1y_2(x_1y_2 - y_1x_2)} \\ &= \frac{x_1y_1 + x_2y_2}{y_1y_2 + \frac{(x_1^2y_1^2 - x_2^2y_2^2) - y_1^2y_2^2(x_1^2 - x_2^2)}{x_1^2x_2^2(y_1^2 - y_2^2)} x_1x_2} = \frac{x_1y_1 + x_2y_2}{y_1y_2 + ax_1x_2}, \end{aligned}$$

$$y_3 = \frac{y_1y_2 - \frac{(x_1^2y_1^2 - x_2^2y_2^2) - y_1^2y_2^2(x_1^2 - x_2^2)}{x_1^2x_2^2(y_1^2 - y_2^2)} x_1x_2}{1 - \frac{(x_1^2 - x_2^2) - (x_1^2y_2^2 - y_1^2x_2^2)}{x_1^2x_2^2(y_1^2 - y_2^2)} x_1y_1x_2y_2} = \frac{x_1y_1 - x_2y_2}{x_1y_2 - y_1x_2}.$$

The addition formulae (independent of d) are then as follows,

$$(x_1, y_1) + (x_2, y_2) = \left(\frac{x_1 y_1 + x_2 y_2}{y_1 y_2 + a x_1 x_2}, \frac{x_1 y_1 - x_2 y_2}{x_1 y_2 - y_1 x_2} \right) = (x_3, y_3). \tag{3}$$

The formulae given by (3) produce the same outputs as the addition formulae (1). However, these formulae fail for point doubling. In addition, there are exceptional cases even if d is a not a square in K and a is a square in K. The following theorem states these points explicitly.

Theorem 2. *Let K be a field of odd characteristic. Let $E_{E,a,d}$ be a twisted Edwards curve defined over K. Let $P = (x_1, y_1)$ and $Q = (x_2, y_2)$ be points on $E_{E,a,d}$. Assume that P is fixed.*

If $x_1 = 0$ or $y_1 = 0$ then $y_1 y_2 + a x_1 x_2 = 0$ if and only if $Q \in S_x$ where $S_x = \{(y_1/\sqrt{a}, -x_1\sqrt{a}), (-y_1/\sqrt{a}, x_1\sqrt{a})\}$. Similarly, $x_1 y_2 - y_1 x_2 = 0$ if and only if $Q \in S_y$ where $S_y = \{(x_1, y_1), (-x_1, -y_1)\}$.

Otherwise (i.e. $x_1 \neq 0$ and $y_1 \neq 0$), S_x and S_y are given by

$$S_x = \left\{ \left(\frac{y_1}{\sqrt{a}}, -x_1\sqrt{a} \right), \left(-\frac{y_1}{\sqrt{a}}, x_1\sqrt{a} \right), \left(\frac{1}{x_1\sqrt{a\,d}}, -\frac{\sqrt{a}}{y_1\sqrt{d}} \right), \left(-\frac{1}{x_1\sqrt{a\,d}}, \frac{\sqrt{a}}{y_1\sqrt{d}} \right) \right\},$$

$$S_y = \left\{ (x_1, y_1), (-x_1, -y_1), \left(\frac{1}{y_1\sqrt{d}}, \frac{1}{x_1\sqrt{d}} \right), \left(-\frac{1}{y_1\sqrt{d}}, -\frac{1}{x_1\sqrt{d}} \right) \right\}.$$

Proof. $\Rightarrow$: The set of all solutions to the system of equations $y_1 y_2 + a x_1 x_2 = 0, a x_1^2 + y_1^2 = 1 + d x_1^2 y_1^2, a x_2^2 + y_2^2 = 1 + d x_2^2 y_2^2$ gives S_x. The set of all solutions to the system of equations $x_1 y_2 - y_1 x_2 = 0, a x_1^2 + y_1^2 = 1 + d x_1^2 y_1^2, a x_2^2 + y_2^2 = 1 + d x_2^2 y_2^2$ gives S_y. Clearly, all solutions are distinct since $(0, 0)$ is not on the curve.

$\Leftarrow$: Trivial, by substitution. □

Theorem 2 shows that suitable selection of a and d are not enough to eliminate all exceptional cases. Therefore the formulae given by (3) are not complete. Nevertheless, the exceptional inputs have a special property given by the following lemma.

Lemma 1. *Let $K, E_{E,a,d}, P, Q$ be defined as in Theorem 2. Assume that P is a fixed point of odd order. Assume that $Q \in S_x \cup S_y - \{P\}$. Then Q is of even order.*

Proof. The proof is given in Appendix-A. □

We now provide a practical solution to prevent exceptional cases. We will recall Corollary 1 in Section 4.

Corollary 1. *Let $E_{E,a,d}$ be a twisted Edwards curve defined over K. Let $P = (x_1, y_1)$ and $Q = (x_2, y_2)$ be points on $E_{E,a,d}$. Assume that P and Q are of odd order with $P \neq Q$. It follows that $y_1 y_2 + a x_1 x_2 \neq 0$ and $x_1 y_2 - y_1 x_2 \neq 0$.*

Proof. The proof follows from Theorem 2 and Lemma 1. □

Cryptographic applications involving elliptic curve scalar multiplication typically use points of prime order. If this is the case, Corollary 1 shows that the addition formulae given by (3) are exception-free for distinct input points. Furthermore, extending K cannot introduce any exception. Of course, one can still choose arbitrary points as the input at the expense of exception handling or leave the exceptions unhandled. However, this can lead active attackers to succeed in exceptional point attacks, see [19]. As a general solution, a suitable randomization technique can be used. For various randomization techniques, a comprehensive reference is [10, chapter 29].

The rest of the paper is about cryptographic applications. Therefore, we now further assume that K is finite. In some implementations the ratio $\mathbf{I}/\mathbf{M}$ is quite large. For this reason, a natural strategy is to prevent the frequent use of field inversions and a classical solution is using projective coordinates.

At this stage, consider the homogenous projective coordinates in [1]. In this system, each point (x, y) on $ax^2 + y^2 = 1 + dx^2y^2$ is represented as the triplet $(X : Y : Z)$ which corresponds to the affine point $(X/Z, Y/Z)$ with $Z \neq 0$. These triplets satisfy the homogenous projective equation

$$(aX^2 + Y^2)Z^2 = Z^4 + dX^2Y^2. \tag{4}$$

The curve defined by (4) is the projective closure of the curve $ax^2 + y^2 = 1 + dx^2y^2$. The identity element is represented by $(0 : 1 : 1)$. The negative of $(X : Y : Z)$ is $(-X : Y : Z)$. For all nonzero $\lambda \in K$, $(X : Y : Z) = (\lambda X : \lambda Y : \lambda Z)$. We denote this system by $\mathcal{E}$. The choice of $\mathcal{E}$ leads to inversion-free very efficient point addition algorithms recently proposed in [1, Section 6].

3 Extended Twisted Edwards Coordinates

To gain more speed, it is convenient to introduce an auxiliary coordinate $t = xy$ to represent a point (x, y) on $ax^2 + y^2 = 1 + dx^2y^2$ in extended affine coordinates (x, y, t). One can pass to the projective representation using the map $(x, y, t) \mapsto (x : y : t : 1)$. For all nonzero $\lambda \in K$, $(X : Y : T : Z) = (\lambda X : \lambda Y : \lambda T : \lambda Z)$ which satisfies (4) and corresponds to the extended affine point $(X/Z, Y/Z, T/Z)$ with $Z \neq 0$. The auxiliary coordinate T has the property $T = XY/Z$. This point representation is named *extended twisted Edwards coordinates* and is denoted by $\mathcal{E}^e$. The identity element is represented by $(0 : 1 : 0 : 1)$. The negative of $(X : Y : T : Z)$ is $(-X : Y : -T : Z)$. Given $(X : Y : Z)$ in $\mathcal{E}$ passing to $\mathcal{E}^e$ can be performed in $3\mathbf{M} + 1\mathbf{S}$ by computing (XZ, YZ, XY, Z^2). Given $(X : Y : T : Z)$ in $\mathcal{E}^e$ passing to $\mathcal{E}$ is cost-free by simply ignoring T.

3.1 Unified Addition in $\mathcal{E}^e$

Given $(X_1 : Y_1 : T_1 : Z_1)$ and $(X_2 : Y_2 : T_2 : Z_2)$ with $Z_1 \neq 0$ and $Z_2 \neq 0$, a unified addition can be performed as $(X_1 : Y_1 : T_1 : Z_1) + (X_2 : Y_2 : T_2 : Z_2) = (X_3 : Y_3 : T_3 : Z_3)$ where

$$\begin{aligned}X_3 &= (X_1Y_2 + Y_1X_2)(Z_1Z_2 - dT_1T_2),\\ Y_3 &= (Y_1Y_2 - aX_1X_2)(Z_1Z_2 + dT_1T_2),\\ T_3 &= (Y_1Y_2 - aX_1X_2)(X_1Y_2 + Y_1X_2),\\ Z_3 &= (Z_1Z_2 - dT_1T_2)(Z_1Z_2 + dT_1T_2).\end{aligned} \tag{5}$$

These unified formulae are derived from the addition formulae (1). We deduce from [5] and [1] that these formulae are also complete when d is not a square in K and a is a square in K. The operations can be performed with a $9\mathbf{M} + 2\mathbf{D}$ algorithm given by

$$\begin{gathered}A \leftarrow X_1 \cdot X_2, \quad B \leftarrow Y_1 \cdot Y_2, \quad C \leftarrow dT_1 \cdot T_2, \quad D \leftarrow Z_1 \cdot Z_2,\\ E \leftarrow (X_1 + Y_1) \cdot (X_2 + Y_2) - A - B, \quad F \leftarrow D - C, \quad G \leftarrow D + C,\\ H \leftarrow B - aA, \quad X_3 \leftarrow E \cdot F, \quad Y_3 \leftarrow G \cdot H, \quad T_3 \leftarrow E \cdot H, \quad Z_3 \leftarrow F \cdot G.\end{gathered}$$

An $8\mathbf{M} + 2\mathbf{D}$ mixed addition algorithm can then be derived by setting $Z_2 = 1$. This means that we are adding $(X_1 : Y_1 : T_1 : Z_1)$ and an extended affine point (x_2, y_2, x_2y_2) which is equally written as $(x_2 : y_2 : x_2y_2 : 1)$.

Choosing curve constants with extremely small sizes or extremely low (or high) hamming weight can be used to eliminate the computational overhead of a field multiplication. For instance see [9], [7], [12]. See also [1, Section 7] for an alternative strategy for the selection of constants. When using $\mathcal{E}^e$ the situation is even better if $a = -1$; we save $1\mathbf{M} + 1\mathbf{D}$ rather than just $1\mathbf{D}$. Consider a twisted Edwards curve given by

$$ax^2 + y^2 = 1 + dx^2y^2.$$

The map $(x, y) \mapsto (x/\sqrt{-a}, y)$ defines the curve,

$$-x^2 + y^2 = 1 + (-d/a)x^2y^2.$$

This map can be constructed if $-a$ is a square in K. It is worth pointing out here that the curve $-x^2 + y^2 = 1 + (-d/a)x^2y^2$ corresponds to the Edwards curve $x^2 + y^2 = 1 + (d/a)x^2y^2$ via the map $(x, y) \mapsto (ix, y)$ if $i \in K$ with $i^2 = -1$. For such curves a $10\mathbf{M} + 1\mathbf{S} + 1\mathbf{D}$ point addition algorithm is given in [4, add-2007-bl-4].

After a renaming of the constant $-d/a$ to d', the point addition on the twisted Edwards curve $-x^2 + y^2 = 1 + d'x^2y^2$ can now be performed with an $8\mathbf{M} + 1\mathbf{D}$ algorithm given by

$$\begin{gathered}A \leftarrow (Y_1 - X_1) \cdot (Y_2 - X_2), \quad B \leftarrow (Y_1 + X_1) \cdot (Y_2 + X_2), \quad C \leftarrow kT_1 \cdot T_2,\\ D \leftarrow 2Z_1 \cdot Z_2, \quad E \leftarrow B - A, \quad F \leftarrow D - C, \quad G \leftarrow D + C,\\ H \leftarrow B + A, \quad X_3 \leftarrow E \cdot F, \quad Y_3 \leftarrow G \cdot H, \quad T_3 \leftarrow E \cdot H, \quad Z_3 \leftarrow F \cdot G\end{gathered}$$

where $k = 2d'$. The optimization that leads to the removal of the extra multiplication is similar to the optimizations in [23] and [4, add-2007-bl-4]. A $7\mathbf{M} + 1\mathbf{D}$ mixed addition algorithm can be derived by setting $Z_2 = 1$.

In the case $a = -1$, we comment that it is possible to save two additions by further extending the coordinates to $(X: Y: T: Z: Y - X: Y + X)$. Alternatively, $(Y_2 - X_2), (Y_2 + X_2), 2Z_2$, and $k = 2d'$ can be cached to save two additions and two multiplications by 2 when performing readdition. We do not claim that these cachings are very useful in practice. On the other hand, a caching of kT_2 leads to readdition in **8M** rather than **8M**+**1D**. This can save time if **D** is large. As a consequence, readdition with $Z_2 = 1$ needs **7M** rather than **7M**+**1D**. Similar arguments can be easily extended over the other algorithms in Section 3 when appropriate.

3.2 Dedicated Addition in $\mathcal{E}^e$

Given the representations $(X_1: Y_1: T_1: Z_1)$ and $(X_2: Y_2: T_2: Z_2)$ of distinct points with $Z_1 \neq 0$ and $Z_2 \neq 0$, the point addition can be performed as $(X_1: Y_1: T_1: Z_1) + (X_2: Y_2: T_2: Z_2) = (X_3: Y_3: T_3: Z_3)$ where

$$\begin{aligned} X_3 &= (X_1Y_2 - Y_1X_2)(T_1Z_2 + Z_1T_2), \\ Y_3 &= (Y_1Y_2 + aX_1X_2)(T_1Z_2 - Z_1T_2), \\ T_3 &= (T_1Z_2 + Z_1T_2)(T_1Z_2 - Z_1T_2), \\ Z_3 &= (Y_1Y_2 + aX_1X_2)(X_1Y_2 - Y_1X_2). \end{aligned} \tag{6}$$

These formulae are independent of the curve constant d. These formulae are analogous to the addition formulae (3). The operations can be performed with a $9\mathbf{M} + 1\mathbf{D}$ algorithm given by

$$\begin{aligned} &A \leftarrow X_1 \cdot X_2, \quad B \leftarrow Y_1 \cdot Y_2, \quad C \leftarrow Z_1 \cdot T_2, \quad D \leftarrow T_1 \cdot Z_2, \\ &E \leftarrow D + C, \quad F \leftarrow (X_1 - Y_1) \cdot (X_2 + Y_2) + B - A, \quad G \leftarrow B + aA, \\ &H \leftarrow D - C, \quad X_3 \leftarrow E \cdot F, \quad Y_3 \leftarrow G \cdot H, \quad T_3 \leftarrow E \cdot H, \quad Z_3 \leftarrow F \cdot G. \end{aligned}$$

An $8\mathbf{M} + 1\mathbf{D}$ mixed addition algorithm can be derived by setting $Z_2 = 1$.

For the case $a = -1$, the operations can be performed with an **8M** algorithm given by

$$\begin{aligned} &A \leftarrow (Y_1 - X_1) \cdot (Y_2 + X_2), \quad B \leftarrow (Y_1 + X_1) \cdot (Y_2 - X_2), \quad C \leftarrow 2Z_1 \cdot T_2, \\ &D \leftarrow 2T_1 \cdot Z_2, \quad E \leftarrow D + C, \quad F \leftarrow B - A, \quad G \leftarrow B + A, \\ &H \leftarrow D - C, \quad X_3 \leftarrow E \cdot F, \quad Y_3 \leftarrow G \cdot H, \quad T_3 \leftarrow E \cdot H, \quad Z_3 \leftarrow F \cdot G. \end{aligned}$$

A **7M** mixed addition algorithm can be derived by setting $Z_2 = 1$. A parallel version of the dedicated addition algorithm is given in Section 4.4 for the case $a = -1$.

3.3 Dedicated Doubling in $\mathcal{E}^e$

Given $(X_1: Y_1: T_1: Z_1)$ with $Z_1 \neq 0$, point doubling can be performed as $2(X_1: Y_1: T_1: Z_1) = (X_3: Y_3: T_3: Z_3)$ where

$$\begin{aligned} X_3 &= 2X_1Y_1(2Z_1^2 - Y_1^2 - aX_1^2), \\ Y_3 &= (Y_1^2 + aX_1^2)(Y_1^2 - aX_1^2), \\ T_3 &= 2X_1Y_1(Y_1^2 - aX_1^2), \\ Z_3 &= (Y_1^2 + aX_1^2)(2Z_1^2 - Y_1^2 - aX_1^2). \end{aligned} \tag{7}$$

These formulae are independent of the curve constant d. These are essentially the same formulae from [1] plus the formula $T_3 = 2X_1Y_1(Y_1^2 - aX_1^2)$ which increases the number of multiplications needed to compute a point doubling by 1. The operations can be performed with a $4\mathbf{M} + 4\mathbf{S} + 1\mathbf{D}$ algorithm given by

$$A \leftarrow X_1^2, \quad B \leftarrow Y_1^2, \quad C \leftarrow 2Z_1^2, \quad D \leftarrow aA, \quad E \leftarrow (X_1 + Y_1)^2 - A - B,$$
$$G \leftarrow D + B, \quad F \leftarrow G - C, \quad H \leftarrow D - B, \quad X_3 \leftarrow E \cdot F, \quad Y_3 \leftarrow G \cdot H,$$
$$T_3 \leftarrow E \cdot H, \quad Z_3 \leftarrow F \cdot G.$$

This algorithm is similar to $3\mathbf{M} + 4\mathbf{S} + 1\mathbf{D}$ point doubling algorithm in [1]. The slowing down from $3\mathbf{M} + 4\mathbf{S} + 1\mathbf{D}$ to $4\mathbf{M} + 4\mathbf{S} + 1\mathbf{D}$ will be remedied in Section 4.3 by mixing $\mathcal{E}^e$ with $\mathcal{E}$. A parallel version of the doubling algorithm is given in Section 4.4 for the case $a = -1$.

3.4 More Formulae

Since we have two different addition formulae for computing x_3 and another two for y_3, it is possible to produce hybrid addition formulae from (1) and (3). The hybrid formulae are given by

$$(x_1, y_1) + (x_2, y_2) = \left(\frac{x_1y_1 + x_2y_2}{y_1y_2 + ax_1x_2}, \frac{y_1y_2 - ax_1x_2}{1 - dx_1y_1x_2y_2} \right) = (x_3, y_3), \tag{8}$$

$$(x_1, y_1) + (x_2, y_2) = \left(\frac{x_1y_2 + y_1x_2}{1 + dx_1y_1x_2y_2}, \frac{x_1y_1 - x_2y_2}{x_1y_2 - y_1x_2} \right) = (x_3, y_3). \tag{9}$$

We comment that $\mathcal{E}^e$ analogs of (8) and (9) lead to similar speeds.

4 Applications

We provide further optimizations targeting scalar multiplication operations, nP where n is an integer called the *scalar* and P is the *base point* multiplied by the scalar.

The impact of the new unified addition algorithms in $\mathcal{E}^e$ for preventing side channel attacks is discussed in Section 4.1. Parallel versions of the $8\mathbf{M}+1\mathbf{D}$ unified addition in $\mathcal{E}^e$ are provided in Section 4.2. The speed of scalar multiplication on twisted Edwards curves is increased by mixing $\mathcal{E}^e$ with $\mathcal{E}$ in Section 4.3. A parallel implementation of fast scalar multiplication in $\mathcal{E}^e$ is explained in Section 4.4. When parallelization is desired the algorithms in Section 4.2 and Section 4.4 help to reduce significantly the effective cost of scalar multiplication. Other applications appear in Section 4.5.

4.1 Defeating SPA Attacks

It is well known that a scalar multiplication algorithm can gain SPA protection when unified additions are used as the only group operation, see [10, Section 29.1.2] for instance. From Section 3.4 we know that the unified addition costs $9\mathbf{M}+2\mathbf{D}$ in $\mathcal{E}^e$. For the case $a = -1$ the cost drops to $8\mathbf{M}+1\mathbf{D}$. Both results are faster than all the other unified addition algorithms known to date. Assuming that $\mathbf{S} = 0.8\mathbf{M}$ and $\mathbf{D} \approx 0$, the $8\mathbf{M} + 1\mathbf{D}$ algorithm is approximately 17.5%, 22.5%, 35%, 50%, 55%, 82.5%, 97.5% faster than the best results in [17], [6], [5], [20], [7], [22], [8], respectively. Note, if $\mathbf{S} = \mathbf{M}$ most speedups will be even more significant. Furthermore, both unified addition algorithms are complete for suitably selected parameters, see section 2 for pointers. The completeness is a stronger property than the unification, see [5, p.2].

Another approach to a protected scalar multiplication is using the Montgomery ladder with Montgomery curves or Kummer surfaces. Montgomery's algorithm for Montgomery curves in [23] use $5\mathbf{M} + 4\mathbf{S} + 1\mathbf{D}$ per scalar bit. Gaudry/Lubicz algorithm for Kummer surfaces (genus 1, odd characteristic case) in [16] use $3\mathbf{M} + 6\mathbf{S} + 3\mathbf{D}$ per scalar bit. We will only provide comparisons with Montgomery curves in the rest of the paper. Assuming that an optimized protected scalar multiplication algorithm uses 1.2 unified additions per scalar bit, scalar multiplication using the $8\mathbf{M} + 1\mathbf{D}$ algorithm then requires $(8\mathbf{M} + 1\mathbf{D}) \times 1.2 = 9.6\mathbf{M} + 1.2\mathbf{D}$ per scalar bit. Assuming that $0.67\mathbf{M} \leq \mathbf{S} \leq \mathbf{M}$ and $0 < \mathbf{D} \leq \mathbf{M}$, this will be approximately 6% to 25% slower[2] than Montgomery curves. However, we will show in Section 4.2 that the $8\mathbf{M} + 1\mathbf{D}$ algorithm can be faster on parallel implementations. When designing the parallel algorithms we try exploiting all inherent parallelism. If an $\mathbf{M}$ is performed in parallel with a $\mathbf{D}$ and/or an $\mathbf{S}$ then the cost is counted as an effective $1\mathbf{M}$.

4.2 Defeating SPA Attacks in Parallel Environments

A useful feature of the $8\mathbf{M} + 1\mathbf{D}$ unified addition algorithm is that it is highly parallelizable. In this section, targeting parallel environments, we explain how a protected scalar multiplication using the $8\mathbf{M} + 1\mathbf{D}$ unified addition in $\mathcal{E}^e$ can perform faster than a protected scalar multiplication based on the Montgomery ladder [23]. For details on the ladder algorithm and Montgomery curves, we refer the reader to [23] and [21]. See [18] and [15] for preventing side channel attacks in parallel environments using general elliptic curves.

The Montgomery curve $E_{M,A,B}$ is defined by $By^2 = x^3 + Ax^2 + x$ with $B(A^2 - 4) \neq 0$. Given the projective coordinates of two points $(X_m \colon Z_m)$ and $(X_n \colon Z_n)$ and also $(X_{m-n} \colon Z_{m-n}) = (X_m \colon Z_m) - (X_n \colon Z_n)$; $(X_{m+n}, \colon Z_{m+n}) = (X_m \colon Z_m) + (X_n \colon Z_n)$ is given in [23] by

$$X_{m+n} = Z_{m-n}((X_m - Z_m)(X_n + Z_n) + (X_m + Z_m)(X_n - Z_n))^2,$$
$$Z_{m+n} = X_{m-n}((X_m - Z_m)(X_n + Z_n) - (X_m + Z_m)(X_n - Z_n))^2.$$

Dedicated doubling formulae (which can be faster than the addition) are used to compute X_{2n} and Z_{2n} given in [23] by

$$4X_nZ_n = (X_n + Z_n)^2 - (X_n - Z_n)^2,$$
$$X_{2n} = (X_n + Z_n)^2(X_n - Z_n)^2,$$
$$Z_{2n} = (4X_nZ_n)((X_n - Z_n)^2 + ((A + 2)/4)(4X_nZ_n)).$$

[2] The ratios $\mathbf{S}/\mathbf{M}$ and $\mathbf{D}/\mathbf{M}$ are fixed equally for both cases.

The doubling algorithm uses $2\mathbf{M} + 2\mathbf{S} + 1\mathbf{D}$ and the addition algorithm uses $3\mathbf{M} + 2\mathbf{S}$ assuming that $Z_{m-n} = 1$. The total cost of a doubling and an addition is then $5\mathbf{M}+4\mathbf{S}+1\mathbf{D}$. In a sequential environment it is convenient to consider the addition and doubling operations as a single composite operation. This approach is given in [4]. To follow the same notation rename

$$[(A+2)/4, X_{m-n}, Z_{m-n}, X_m, Z_m, X_n, Z_n, X_{2n}, Z_{2n}, X_{m+n}, Z_{m+n}]$$

as $[a24, X_1, Z_1, X_2, Z_2, X_3, Z_3, X_4, Z_4, X_5, Z_5]$. Assuming that $Z_1 = 1$, a $5\mathbf{M} + 4\mathbf{S} + 1\mathbf{D}$ Montgomery differential-addition-and-doubling algorithm is given in [4, mladd-1987-m] by

$$\begin{gathered}
A \leftarrow X_2 + Z_2, \quad AA \leftarrow A^2, \quad B \leftarrow X_2 - Z_2, \quad BB \leftarrow B^2, \\
E \leftarrow AA - BB, \quad C \leftarrow X_3 + Z_3, \quad D \leftarrow X_3 - Z_3, \quad DA \leftarrow D \cdot A, \\
CB \leftarrow C \cdot B, \quad X_5 \leftarrow (DA + CB)^2, \quad Z_5 \leftarrow X_1 \cdot (DA - CB)^2, \\
X_4 \leftarrow AA \cdot BB, \quad Z_4 \leftarrow E \cdot (BB + a24E).
\end{gathered}$$

2-Processor Montgomery addition and doubling. In [21], it is observed that the doubling and the addition phases of the Montgomery ladder algorithm can be performed independently. From this, it is clear that one of the processors needs $2\mathbf{M} + 2\mathbf{S} + 1\mathbf{D}$ and the other needs $3\mathbf{M} + 2\mathbf{S}$ to perform doubling and addition, respectively. Since $3\mathbf{M} + 2\mathbf{S} \geq 2\mathbf{M} + 2\mathbf{S} + 1\mathbf{D}$ we conclude that one round of computing a doubling and an addition can be done in an effective $3\mathbf{M}+2\mathbf{S}$. Alternatively, we can parallelize the "mladd-1987-m" algorithm in [4]. This approach also yields an effective $3\mathbf{M} + 2\mathbf{S}$. See Appendix-B. The ladder algorithm then uses $3\mathbf{M} + 2\mathbf{S}$ per scalar bit.

2-Processor twisted Edwards ($a = -1$) unified addition in $\mathcal{E}^e$. We now investigate the $8\mathbf{M} + 1\mathbf{D}$ unified addition algorithm. We can split the computational task into 9 steps with a full utilization of 2 processors. The unified addition can then be performed with an effective $4\mathbf{M} + 1\mathbf{D}$ algorithm.

Cost	Step	Processor 1	Processor 2
	1	$R_1 \leftarrow Y_1 - X_1$	$R_2 \leftarrow Y_2 - X_2$
	2	$R_3 \leftarrow Y_1 + X_1$	$R_4 \leftarrow Y_2 + X_2$
1M	3	$R_5 \leftarrow R_1 \cdot R_2$	$R_6 \leftarrow R_3 \cdot R_4$
1M	4	$R_7 \leftarrow T_1 \cdot T_2$	$R_8 \leftarrow Z_1 \cdot Z_2$
1D	5	$R_7 \leftarrow kR_7$	$R_8 \leftarrow 2R_8$
	6	$R_1 \leftarrow R_6 - R_5$	$R_2 \leftarrow R_8 - R_7$
	7	$R_3 \leftarrow R_8 + R_7$	$R_4 \leftarrow R_6 + R_5$
1M	8	$X_3 \leftarrow R_1 \cdot R_2$	$Y_3 \leftarrow R_3 \cdot R_4$
1M	9	$T_3 \leftarrow R_1 \cdot R_4$	$Z_3 \leftarrow R_2 \cdot R_3$

Assuming that an optimized SPA protected scalar multiplication algorithm uses 1.2 unified additions per scalar bit, we have the cost estimate $(4\mathbf{M}+1\mathbf{D}) \times 1.2 = 4.8\mathbf{M} + 1.2\mathbf{D}$ per scalar bit (for each of 2 processors). The fastest system is determined by the ratios $\mathbf{S}/\mathbf{M}$ and $\mathbf{D}/\mathbf{M}$. For instance, if $\mathbf{S} = \mathbf{M}$ and $\mathbf{D} \approx 0$ then twisted Edwards ($a = -1$) curves are approximately 4.2% faster than Montgomery curves. On the other hand, using Montgomery curves still seems to be preferable since the ladder algorithm needs less memory and it is not affected by changes in the ratio $\mathbf{D}/\mathbf{M}$. Note also that $\mathbf{S} < \mathbf{M}$ in some applications.

We omit details for the 3-processor case which can be derived with similar approaches.

4-Processor Montgomery addition and doubling. The Montgomery addition and doubling does not *nicely* fit the 4-processor setting. For instance the "mladd-1987-m" algorithm in [4] seems to be *quite* uncompetitive even if we exploit all inherent parallelism. A quick investigation shows that we can perform a doubling and addition in an effective $2\mathbf{M} + 2\mathbf{S}$. See Appendix-B. The ladder algorithm then uses $2\mathbf{M} + 2\mathbf{S}$ per scalar bit.

4-Processor twisted Edwards ($a = -1$) unified addition in $\mathcal{E}^e$. We can split the computational task into 5 sequential steps among 4 processors. The unified addition can then be performed with an effective $2\mathbf{M} + 1\mathbf{D}$ algorithm.

Cost	Step	Processor 1	Processor 2	Processor 3	Processor 4
	1	$R_1 \leftarrow Y_1 - X_1$	$R_2 \leftarrow Y_2 - X_2$	$R_3 \leftarrow Y_1 + X_1$	$R_4 \leftarrow Y_2 + X_2$
1M	2	$R_5 \leftarrow R_1 \cdot R_2$	$R_6 \leftarrow R_3 \cdot R_4$	$R_7 \leftarrow T_1 \cdot T_2$	$R_8 \leftarrow Z_1 \cdot Z_2$
1D	3	*idle*	*idle*	$R_7 \leftarrow kR_7$	$R_8 \leftarrow 2R_8$
	4	$R_1 \leftarrow R_6 - R_5$	$R_2 \leftarrow R_8 - R_7$	$R_3 \leftarrow R_8 + R_7$	$R_4 \leftarrow R_6 + R_5$
1M	5	$X_3 \leftarrow R_1 \cdot R_2$	$Y_3 \leftarrow R_3 \cdot R_4$	$T_3 \leftarrow R_1 \cdot R_4$	$Z_3 \leftarrow R_2 \cdot R_3$

Following the assumption from the 2-processor case we have the cost estimate $(2\mathbf{M}+1\mathbf{D}) \times 1.2 = 2.4\mathbf{M}+1.2\mathbf{D}$ per scalar bit. If $\mathbf{S} = \mathbf{M}$ and $\mathbf{D} \approx 0$ then twisted Edwards ($a = -1$) curves are approximately 66.7% faster than Montgomery curves. If $\mathbf{S} = 0.8\mathbf{M}$ and $\mathbf{D} = 0.25\mathbf{M}$ then twisted Edwards ($a = -1$) curves are approximately 33.3% faster. If $\mathbf{S} = 0.8\mathbf{M}$ and $\mathbf{D} = \mathbf{M}$ then twisted Edwards ($a = -1$) curves are approximately 5.9% faster.

Assuming $\mathbf{D} \approx 0$, we estimate that a "256-bit, sliding window, 4-NAF" scalar multiplication on twisted Edwards ($a = -1$) curves will require approximately $602\mathbf{M}$ for each of 4 processors, depending on the analysis in [5, Section 5].

Consider the field multiplication operation kR_7 in Step 3. The finite field arithmetic can be implemented building on integer arithmetic. Treating field elements k as a $4n$-bit integer and R_7 as an integer, we fix $k_1, k_2, k_3, k_4 \in [0, 2^n - 1]$ such that $k = k_0 + 2^n k_1 + 2^{2n} k_2 + 2^{3n} k_3$. Now, kR_7 can be obtained as $k_0 R_7 + 2^n (k_1 R_7) + 2^{2n}(k_2 R_7) + 2^{3n}(k_3 R_7)$ by computing $k_i R_7$ in parallel. The rest of the computation for obtaining kR_7 can be practically negligible (depending on the application). Here, the 3 additions to obtain kR_7 and $R_8 \leftarrow 2R_8$ can be put in a new parallel step. Furthermore if $\#K$ is a special prime allowing very fast modular reduction (such as NIST primes) then the cost of casting the integer kR_7 to K (i.e. the modular reduction) can also be practically negligible (depending on the application). This method leads to a better utilization of processors and can be used for decreasing $\mathbf{D}$. Even if k is of the *full size* (i.e. $\mathbf{D} = \mathbf{M}$), this technique fixes each k_i to a quarter of the size of k (i.e. $\mathbf{D}$ is close to $0.25\mathbf{M}$ if schoolbook multiplication and fast reduction are being used). Alternatively, fixing n to the word size of the underlying hardware (or maybe to the size of a compiler-supported data type) can be advantageous in some applications. The same method can be adapted to the 2-processor case.

The parallel implementation of $\mathcal{E}^e \leftarrow \mathcal{E}^e + \mathcal{E}^e$ is easier than the Montgomery case because all processors perform similar tasks at each step. In addition, the implementation does not require a special field squaring circuit to gain better timings.

2×2-Processor Montgomery addition and doubling. If the doubling operation is assigned to a team of two processors and the addition operation is assigned to another team of two processors, the $2\mathbf{M}+2\mathbf{S}$ figure can be improved to $2\mathbf{M}+1\mathbf{S}$. See Appendix-B. Here, we make the assumption that the addition-team and the doubling-team work in an unsynchronized fashion and perform the synchronization at the end (of each round); we are not claiming that the implementation of this is easy. Even with this assumption twisted Edwards ($a=-1$) curves can still be faster. For instance, if $\mathbf{S}=\mathbf{M}$ and $\mathbf{D}\approx 0$ then twisted Edwards ($a=-1$) curves are approximately 25% faster than Montgomery curves.

4.3 Fast Scalar Multiplication

In [11], Cohen, Miyaji, and Ono introduced the modified Jacobian coordinates and studied other systems in the literature, namely affine, projective, Jacobian, and Chudnovsky Jacobian coordinates. To gain better timings they proposed a technique of carefully mixing these coordinates. We follow a similar approach. Note, the notations $\mathcal{E}$ and $\mathcal{E}^e$ follow the notation introduced in [11].

On twisted Edwards curves, the speed of scalar multiplications which involve point doublings can be increased by mixing $\mathcal{E}^e$ with $\mathcal{E}$. The following technique replaces (slower) doublings in $\mathcal{E}^e$ with (faster) doublings in $\mathcal{E}$. In the execution of a scalar multiplication:

(i) If a point doubling is followed by another point doubling, use $\mathcal{E}\leftarrow 2\mathcal{E}$.
(ii) If a point doubling is followed by a point addition, use
  1. $\mathcal{E}^e\leftarrow 2\mathcal{E}$ for the point doubling step; followed by,
  2. $\mathcal{E}\leftarrow \mathcal{E}^e+\mathcal{E}^e$ for the point addition step.

$\mathcal{E}\leftarrow 2\mathcal{E}$ is performed using $3\mathbf{M}+4\mathbf{S}+1\mathbf{D}$ doubling algorithm in [1]. The details of the other operations are given below.

$\mathcal{E}^e\leftarrow 2\mathcal{E}$ using (7):

(i) In Section 3 it was noted that passing from $(X:Y:Z)$ to $(X:Y:T:Z)$ (i.e. passing from $\mathcal{E}$ to $\mathcal{E}^e$) can be performed in $3\mathbf{M}+1\mathbf{S}$. From this, it might seem at the first glance that computing $\mathcal{E}^e\leftarrow 2\mathcal{E}$ will more costly than expected. However, the doubling algorithm for (7) does not use the input T_1 and so it can be used for $\mathcal{E}^e\leftarrow 2\mathcal{E}$ without modification.
(ii) Theorem 1 implies that Z_1 and Z_3 are always nonzero if the base point is of odd order. Alternatively, careful selection of a and d also guarantees that Z_1 and Z_3 are always nonzero regardless of the order of the base point, see [1].

$\mathcal{E}\leftarrow \mathcal{E}^e+\mathcal{E}^e$ based on (either) (5) or (6):

(i) Observe that one field multiplication can be saved by not computing T_3. This can be regarded as a remedy to the extra field multiplication which appears in $\mathcal{E}^e\leftarrow 2\mathcal{E}$ while computing T_3.
(ii) If (6) is used (without computing T_3), scalar multiplication is independent of d. Indeed $\mathcal{E}\leftarrow 2\mathcal{E}$ (see [1]) and $\mathcal{E}^e\leftarrow 2\mathcal{E}$ (see Section 3.3) are also independent of d. Formulae (6) save time if $\mathbf{D}$ is large. In addition, Corollary 1 implies that Z_1, Z_2 and Z_3 are always nonzero if the base point is of odd order.

(iii) If (5) is used (without computing T_3), the curve constant d will be involved in the calculations. Using the concept of readdition discussed in Section 3.4, one can also achieve similar performance in comparison to the case of (6). In addition, Theorem 1 implies that Z_1, Z_2 and Z_3 are always nonzero if the base point is of odd order. Alternatively, careful selection of a and d also guarantees that Z_1, Z_2 and Z_3 are always nonzero regardless of the order of the base point, see [1].

In Table 1, a comparison is made for the speeds that can be achieved under different $\mathbf{S}/\mathbf{M}$ and $\mathbf{D}/\mathbf{M}$ scenarios. These estimates are based on the analysis in [5, Section 5]. To gain the best speed, we assume that ($a = -1$). To make the cost estimation easier (without sacrificing the accuracy), we can consider the cost of $\mathcal{E}^e \leftarrow 2\mathcal{E}$ as $3\mathbf{M}+4\mathbf{S}$ by pushing the extra multiplication to the operation count of $\mathcal{E} \leftarrow \mathcal{E}^e + \mathcal{E}^e$. In this case, the relevant costs for various additions based on the formulae (6) are as follows. Addition: $8\mathbf{M}$; readdition: $8\mathbf{M}$; readdition with $Z_2 = 1$: $7\mathbf{M}$; mixed addition (i.e. addition with $Z_2 = 1$ reasonably denoted by $\mathcal{E} \leftarrow \mathcal{E}^e + \mathcal{A}^e$): $7\mathbf{M}$. As a special case, we also include cost estimates for the Montgomery ladder [23] which require $5\mathbf{M} + 4\mathbf{S} + 1\mathbf{D}$ per scalar bit. The rows are sorted with respect to the column (.8, 0) in descending order. The headers (e.g. (.8, .5)) of columns 2 to 7 fix the ratios $\mathbf{S}/\mathbf{M}$ and $\mathbf{D}/\mathbf{M}$, respectively. (Of course, $\mathbf{D}/\mathbf{M} = 0$ should be regarded as $\mathbf{D}/\mathbf{M} \approx 0$ when it appears.)

Table 1. Cost estimates (**M**) for fast scalar multiplication, 256-bit. (The Montgomery ladder algorithm for Montgomery curves and "sliding window, 4-NAF" method for Edwards, inverted Edwards, and mixed twisted Edwards coordinates)

System	(1,1)	(.8, 1)	(1, .5)	(.8, .5)	(1, 0)	(.8, 0)
Montgomery Ladder, [23]	2560	2355	2432	2227	2304	2099
Edwards, [5]	2351	2139	2326	2115	2301	2090
Inverted Edwards, [6]	2552	2341	2402	2191	2251	2040
Twisted Edwards ($a = -1$), mixed	2152	1951	2152	1951	2152	1951

It is also convenient to consider $\mathcal{E}^e \leftarrow 2\mathcal{E}$ followed by $\mathcal{E} \leftarrow \mathcal{E}^e + \mathcal{E}^e$ as a single composite operation as $\mathcal{E} \leftarrow 2\mathcal{E} + \mathcal{E}^e$ where $\mathcal{E}^e$ is the base point. See [14] for a similar approach in affine Weierstrass coordinates.

4.4 Fast Scalar Multiplication in Parallel Environments

It is natural to ask whether the speed of the protected scalar multiplication discussed in Section 4.2 can be increased by using a fast dedicated doubling algorithm. Unfortunately mixing $\mathcal{E}^e$ with $\mathcal{E}$ does not seem to be helpful in parallel environments for increasing the speed. Nevertheless, $\mathcal{E}^e \leftarrow 2\mathcal{E}^e$ can be performed with an effective $1\mathbf{M} + 1\mathbf{S}$ algorithm, as follows.

Cost	Step	Processor 1	Processor 2	Processor 3	Processor 4
	1	*idle*	*idle*	*idle*	$R_1 \leftarrow X_1 + Y_1$
1S	2	$R_2 \leftarrow X_1^2$	$R_3 \leftarrow Y_1^2$	$R_4 \leftarrow Z_1^2$	$R_5 \leftarrow R_1^2$
	3	$R_6 \leftarrow R_2 + R_3$	$R_7 \leftarrow R_2 - R_3$	$R_4 \leftarrow 2R_4$	*idle*
	4	*idle*	$R_1 \leftarrow R_4 + R_7$	*idle*	$R_2 \leftarrow R_6 - R_5$
1M	5	$X_3 \leftarrow R_1 \cdot R_2$	$Y_3 \leftarrow R_6 \cdot R_7$	$T_3 \leftarrow R_2 \cdot R_6$	$Z_3 \leftarrow R_1 \cdot R_7$

This is essentially the same algorithm as in Section 3.3. It is easy to deduce that the 2-processor point doubling needs an effective 2**M** + 2**S**. Point addition $\mathcal{E}^e \leftarrow \mathcal{E}^e + \mathcal{E}^e$ can be performed with an effective 2**M** algorithm, as follows.

Cost	Step	Processor 1	Processor 2	Processor 3	Processor 4
	1	$R_1 \leftarrow Y_1 - X_1$	$R_2 \leftarrow Y_2 + X_2$	$R_3 \leftarrow Y_1 + X_1$	$R_4 \leftarrow Y_2 - X_2$
1M	2	$R_5 \leftarrow R_1 \cdot R_2$	$R_6 \leftarrow R_3 \cdot R_4$	$R_7 \leftarrow Z_1 \cdot T_2$	$R_8 \leftarrow T_1 \cdot Z_2$
	3	*idle*	*idle*	$R_7 \leftarrow 2R_7$	$R_8 \leftarrow 2R_8$
	4	$R_1 \leftarrow R_8 + R_7$	$R_2 \leftarrow R_6 - R_5$	$R_3 \leftarrow R_6 + R_5$	$R_4 \leftarrow R_8 - R_7$
1M	5	$X_3 \leftarrow R_1 \cdot R_2$	$Y_3 \leftarrow R_3 \cdot R_4$	$T_3 \leftarrow R_1 \cdot R_4$	$Z_3 \leftarrow R_2 \cdot R_3$

This is essentially the same algorithm as in Section 3.2. It is easy to deduce that the 2-processor point doubling needs an effective 4**M**. One may prefer using the parallel version of the addition formulae (1) which comes at the expense of multiplication by d. See the discussions about readdition in Section 3.4 and partitioning k in Section 4.2. Assuming **S** = 0.8**M** and **D** ≈ 0, we estimate that "256-bit, sliding window, 4-NAF" scalar multiplication using $\mathcal{E}^e$ will require approximately 552**M** for each of 4 processors, depending on the analysis in [5, Section 5].

4.5 Other Applications

Point addition intensive operations bring out the full power of the new addition algorithms. Therefore, we will consider the batch signature verification algorithm in this section.

There is a vast literature on the optimization of special exponentiation techniques. A general references is [10]. An example to the case of scalar multiplication is computing $\sum n_i P_i$ with fixed base point(s) or fixed scalar(s). In [5, Section 7], cost estimations for selected applications about $\sum n_i P_i$ are provided for several curve models. The expected increases in speed for twisted Edwards curves can be deduced from [5] by simply substituting the new operation counts. For instance, the batch signature verification technique in [24] attributed to Bos-Coster is summarized in [5, Section 5] for one variant of the ElGamal signature system. The cost estimates for this operation are given in Table 2 in comparison to Edwards coordinates and inverted Edwards coordinates.

Table 2. Cost estimates (**M**) for batched verification of 100 ElGamal signatures, 256-bit

System	(1,1)	(.8, 1)	(1, .5)	(.8, .5)	(1, 0)	(.8, 0)
Edwards, [5]	302	297	289	284	276	271
Inverted Edwards, [6]	276	271	264	259	251	246
Twisted Edwards ($a = -1$), $\mathcal{E}^e$	201	201	201	201	201	201

5 Conclusion

In this work, a new point representation $\mathcal{E}^e$ is introduced for twisted Edwards curves. We derive efficient and highly parallel group operations and discuss alternative ways of preventing exceptional cases. We then provide performance estimates and comparisons for different implementation scenarios.

Defeating SPA Attacks. We provide two fast unified addition algorithms which cost $9\mathbf{M} + 2\mathbf{D}$ and $8\mathbf{M} + 1\mathbf{D}$. The latter case is at least 22% faster than all the other unified addition methods stated in the literature. These formulae are even 17.5% faster than our preliminary result in [17].

Defeating SPA Attacks in Parallel Environments. We provide an effective $2\mathbf{M} + 1\mathbf{D}$ unified point addition algorithm on a 4-processor environment. We further showed that twisted Edwards ($a = -1$) curves can be faster up to 66.7% than Montgomery curves in this parallel environment.

Fast Scalar Multiplication. We first handle single-scalar multiplication. We explain how to perform fast scalar multiplication by mixing $\mathcal{E}^e$ with twisted Edwards coordinates $\mathcal{E}$, improving the current relevant literature bounds by approximately 4%-18%. We then point out that multi-scalar multiplications profit even more from the faster point additions in $\mathcal{E}^e$.

Fast Scalar Multiplication in Parallel Environments. We also point to the parallel versions of fast scalar multiplication offering a speed increase by a factor of 3.54 (using 4 processors) over the optimized sequential case.

In conclusion, we have pushed the recent speed limits of Elliptic Curve Cryptography forward in a wide range of applications. Building on our observations we recommend using $\mathcal{E}^e$ (and mixing $\mathcal{E}^e$ with $\mathcal{E}$ when useful) for speeding up the scalar multiplication in several different settings.

Acknowledgement

The authors thank Tanja Lange and anonymous referees for very useful comments and suggestions.

References

1. Bernstein, D.J., Birkner, P., Joye, M., Lange, T., Peters, C.: Twisted Edwards curves. In: Vaudenay, S. (ed.) AFRICACRYPT 2008. LNCS, vol. 5023, pp. 389–405. Springer, Heidelberg (2008)
2. Bernstein, D.J., Birkner, P., Lange, T., Peters, C.: Optimizing double-base elliptic-curve single-scalar multiplication. In: Srinathan, K., Rangan, C.P., Yung, M. (eds.) INDOCRYPT 2007. LNCS, vol. 4859, pp. 167–182. Springer, Heidelberg (2007)
3. Bernstein, D.J., Birkner, P., Lange, T., Peters, C.: ECM using Edwards curves. Cryptology ePrint Archive, Report 2008/016 (2008), `http://eprint.iacr.org/`
4. Bernstein, D.J., Lange, T.: Explicit-formulas database (2007), `http://www.hyperelliptic.org/EFD`
5. Bernstein, D.J., Lange, T.: Faster addition and doubling on elliptic curves. In: Kurosawa, K. (ed.) ASIACRYPT 2007. LNCS, vol. 4833, pp. 29–50. Springer, Heidelberg (2007)
6. Bernstein, D.J., Lange, T.: Inverted Edwards coordinates. In: Boztaş, S., Lu, H.-F. (eds.) AAECC 2007. LNCS, vol. 4851, pp. 20–27. Springer, Heidelberg (2007)

7. Billet, O., Joye, M.: The Jacobi model of an elliptic curve and side-channel analysis. In: Fossorier, M.P.C., Høholdt, T., Poli, A. (eds.) AAECC 2003. LNCS, vol. 2643, pp. 34–42. Springer, Heidelberg (2003)
8. Brier, E., Joye, M.: Weierstraß elliptic curves and side-channel attacks. In: Naccache, D., Paillier, P. (eds.) PKC 2002. LNCS, vol. 2274, pp. 335–345. Springer, Heidelberg (2002)
9. Brier, E., Joye, M.: Fast point multiplication on elliptic curves through isogenies. In: Fossorier, M.P.C., Høholdt, T., Poli, A. (eds.) AAECC 2003. LNCS, vol. 2643, pp. 43–50. Springer, Heidelberg (2003)
10. Cohen, H., Frey, G. (eds.): Handbook of Elliptic and Hyperelliptic Curve Cryptography. CRC Press, Boca Raton (2005)
11. Cohen, H., Miyaji, A., Ono, T.: Efficient elliptic curve exponentiation using mixed coordinates. In: Ohta, K., Pei, D. (eds.) ASIACRYPT 1998. LNCS, vol. 1514, pp. 51–65. Springer, Heidelberg (1998)
12. Doche, C., Icart, T., Kohel, D.R.: Efficient scalar multiplication by isogeny decompositions. In: Yung, M., Dodis, Y., Kiayias, A., Malkin, T.G. (eds.) PKC 2006. LNCS, vol. 3958, pp. 191–206. Springer, Heidelberg (2006)
13. Edwards, H.M.: A normal form for elliptic curves. Bulletin of the AMS 44(3), 393–422 (2007)
14. Eisenträger, K., Lauter, K., Montgomery, P.L.: Fast elliptic curve arithmetic and improved Weil pairing evaluation. In: Joye, M. (ed.) CT-RSA 2003. LNCS, vol. 2612, pp. 343–354. Springer, Heidelberg (2003)
15. Fischer, W., Giraud, C., Knudsen, E.W., Seifert, J.P.: Parallel scalar multiplication on general elliptic curves over $\mathbb{F}_p$ hedged against non-differential side-channel attacks. Cryptology ePrint Archive, Report 2002/007 (2002), `http://eprint.iacr.org/`
16. Gaudry, P., Lubicz, D.: The arithmetic of characteristic 2 Kummer surfaces. Cryptology ePrint Archive, Report 2008/133 (2008), `http://eprint.iacr.org/`
17. Hisil, H., Wong, K., Carter, G., Dawson, E.: Faster group operations on elliptic curves. Cryptology ePrint Archive, Report 2007/441 (2007), `http://eprint.iacr.org/`
18. Izu, T., Takagi, T.: A fast parallel elliptic curve multiplication resistant against side channel attacks. In: Naccache, D., Paillier, P. (eds.) PKC 2002. LNCS, vol. 2274, pp. 280–296. Springer, Heidelberg (2002)
19. Izu, T., Takagi, T.: Exceptional procedure attack on elliptic curve cryptosystems. In: Desmedt, Y.G. (ed.) PKC 2003. LNCS, vol. 2567, pp. 224–239. Springer, Heidelberg (2003)
20. Joye, M., Quisquater, J.J.: Hessian elliptic curves and side-channel attacks. In: Koç, Ç.K., Naccache, D., Paar, C. (eds.) CHES 2001. LNCS, vol. 2162, pp. 402–410. Springer, Heidelberg (2001)
21. Joye, M., Yen, S.M.: The Montgomery powering ladder. In: Kaliski Jr., B.S., Koç, Ç.K., Paar, C. (eds.) CHES 2002. LNCS, vol. 2523, pp. 291–302. Springer, Heidelberg (2003)
22. Liardet, P.Y., Smart, N.P.: Preventing SPA/DPA in ECC systems using the Jacobi form. In: Koç, Ç.K., Naccache, D., Paar, C. (eds.) CHES 2001. LNCS, vol. 2162, pp. 391–401. Springer, Heidelberg (2001)
23. Montgomery, P.L.: Speeding the Pollard and elliptic curve methods of factorization. Mathematics of Computation 48(177), 243–264 (1987)
24. de Rooij, P.: Efficient exponentiation using precomputation and vector addition chains. In: De Santis, A. (ed.) EUROCRYPT 1994. LNCS, vol. 950, pp. 389–399. Springer, Heidelberg (1995)

A Proof of Lemma 1

Proof. Note that the points at infinity are of even order, see [1]. Assume that $P = (x_1, y_1)$ is of odd order. Thus, P is not one of the points at infinity. Assume that $Q \in S_x \cup S_y - \{P\}$. If Q were one of the points at infinity it would have even order and the claim follows. Note also that $P \neq Q$ and $P \neq -Q$ since $P, -P \notin S_x \cup S_y - \{P\}$. Instead of a further case by case analysis on $S_x \cup S_y - \{P\}$, we will prove the lemma with a general approach. The proof has two parts.

In the first part we will prove that all points in S_x are of even order. Assume that $Q = (x_2, y_2)$ is an element of S_x. By Theorem 2, $ax_1x_2 + y_1y_2 = 0$.

Suppose that $x_1 = 0$. Since P is of odd order $P \neq (0, -1)$ and consequently $P = (0, 1)$. By Theorem 2, $Q = (\pm 1/\sqrt{a}, 0)$. Since $4(\pm 1/\sqrt{a}, 0) = (0, 1)$, Q is of even order as desired.

Assume from now on that $x_1 \neq 0$. We can write $x_2 = -y_1y_2/(ax_1)$ since x_1 is nonzero. Let $M = 2P$ and $N = 2Q$. Since P is of odd order, so is M. Therefore, M is not one of the points at infinity. We can assume that N is not one of the points at infinity; for otherwise Q is of even order as desired. Using the relation $x_2 = -y_1y_2/(ax_1)$ and formula (3) for computing x_3 we get

$$x(N) = \frac{2x_2y_2}{y_2^2 + ax_2^2} = \frac{2(-y_1y_2/(ax_1))y_2}{y_2^2 + a(-y_1y_2/(ax_1))^2} = -\frac{2x_1y_1}{y_1^2 + ax_1^2} = -x(M).$$

The denominators $y_1^2 + ax_1^2$ and $y_2^2 + ax_2^2$ must be nonzero since M and N are not points at infinity. By the curve definition we have

$$y = \pm\sqrt{(1 - ax^2)/(1 - dx^2)}.$$

So $y(M) = \pm y(N)$ since $|x(M)| = |x(N)|$.

$y(M) = -y(N)$ implies that $M - N = (0, -1)$, a point of order 2. Then $2(M - N) = 2(2P - 2Q) = 4(P - Q) = (0, 1)$. So $P - Q$ is a point of order 4.

$y(M) = y(N)$ implies that $M + N = (0, 1)$, the identity. Then $M + N = 2P + 2Q = 2(P + Q) = (0, 1)$. So $P + Q$ is a point of order 2 since $P \neq -Q$.

In conclusion, we have $P \pm Q$ of even order for all situations. Since P is of odd order, $Q \in S_x$ must be of even order.

In the second part of the proof we will prove that all points in $S_y - \{P\}$ are of even order. Assume that $Q = (x_2, y_2)$ is an element of $S_y - \{P\}$. By Theorem 2, $x_1y_2 - y_1x_2 = 0$.

Suppose that $x_1 = 0$. Since P is of odd order $P \neq (0, -1)$ and consequently $P = (0, 1)$. By Theorem 2, $Q = (0, -1)$. Then Q is of even order as desired.

Assume from now on that $x_1 \neq 0$. We can write $y_2 = y_1x_2/x_1$ since x_1 is nonzero. Let $M = 2P$ and $N = 2Q$. Since P is of odd order, so is M. Therefore, M is not one of the points at infinity. We can assume that N is not one of the points at infinity; for otherwise Q is of even order as desired. Using the relation $y_2 = y_1x_2/x_1$ and formula (3) for computing x_3 we get

$$x(N) = \frac{2x_2y_2}{y_2^2 + ax_2^2} = \frac{2x_2(y_1x_2/x_1)}{(y_1x_2/x_1)^2 + ax_2^2} = \frac{2x_1y_1}{y_1^2 + ax_1^2} = x(M).$$

The denominators $y_1^2 + ax_1^2$ and $y_2^2 + ax_2^2$ must be nonzero since M and N are not points at infinity. By the curve definition $y(M) = \pm y(N)$ since $|x(M)| = |x(N)|$.

$y(M) = -y(N)$ implies that $M + N = (0, -1)$, a point of order 2. Then $2(M + N) = 2(2P + 2Q) = 4(P + Q) = (0, 1)$. So $P + Q$ is a point of order 4.

$y(M) = y(N)$ implies that $M - N = (0, 1)$, the identity. Then $M - N = 2P - 2Q = 2(P - Q) = (0, 1)$. So $P - Q$ is a point of order 2 since $P \neq Q$.

In conclusion, we have $P \pm Q$ of even order for all situations. Since P is of odd order, $Q \in S_y - \{P\}$ must be of even order.

In summary, all points in $S_x \cup S_y - \{P\}$ are of even order provided that P is of odd order. □

B Parallel Algorithms

This appendix contains parallel algorithms for Montgomery addition and doubling discussed in Section 4.2.

2-processor Montgomery differential-addition-and-doubling. Effective $3\mathbf{M} + 2\mathbf{S}$, assumption $Z_1 = 1$, adapted from [4, mladd-1987-m].

Cost	Step	Processor 1	Processor 2
	1	$R_1 \leftarrow X_2 + Z_2$	$R_2 \leftarrow X_2 - Z_2$
	2	$R_3 \leftarrow X_3 + Z_3$	$R_4 \leftarrow X_3 - Z_3$
1S	3	$R_5 \leftarrow R_1^2$	$R_6 \leftarrow R_2^2$
	4	$R_7 \leftarrow R_5 - R_6$	*idle*
1M	5	$R_1 \leftarrow R_1 \cdot R_4$	$R_2 \leftarrow R_2 \cdot R_3$
	6	$R_3 \leftarrow R_1 + R_2$	$R_4 \leftarrow R_1 - R_2$
1S	7	$X_5 \leftarrow R_3^2$	$R_2 \leftarrow R_4^2$
1M	8	$R_8 \leftarrow a24R_7$	$X_4 \leftarrow R_5 \cdot R_6$
	9	$R_8 \leftarrow R_6 + R_8$	*idle*
1M	10	$Z_4 \leftarrow R_7 \cdot R_8$	$Z_5 \leftarrow X_1 \cdot R_2$

4-processor Montgomery differential-addition-and-doubling. Effective $2\mathbf{M} + 2\mathbf{S}$, adapted from [4, mladd-1987-m].

Cost	Step	Processor 1	Processor 2	Processor 3	Processor 4
	1	$R_1 \leftarrow X_2 + Z_2$	$R_2 \leftarrow X_2 - Z_2$	$R_3 \leftarrow X_3 + Z_3$	$R_4 \leftarrow X_3 - Z_3$
1S	2	$R_5 \leftarrow R_1^2$	$R_6 \leftarrow R_2^2$	*idle*	*idle*
	3	$R_7 \leftarrow R_5 - R_6$	*idle*	*idle*	*idle*
1M	4	$R_1 \leftarrow R_1 \cdot R_4$	$R_2 \leftarrow R_2 \cdot R_3$	$R_8 \leftarrow a24R_7$	*idle*
	5	$R_3 \leftarrow R_1 + R_2$	$R_4 \leftarrow R_1 - R_2$	$R_8 \leftarrow R_6 + R_8$	*idle*
1S	6	$X_5 \leftarrow R_3^2$	$R_2 \leftarrow R_4^2$	*idle*	*idle*
1M	7	$Z_5 \leftarrow X_1 \cdot R_2$	$X_4 \leftarrow R_5 \cdot R_6$	$Z_4 \leftarrow R_7 \cdot R_8$	*idle*

2×2-processor Montgomery differential-addition and Montgomery doubling. Effective $2\mathbf{M} + 1\mathbf{S}$. Using the notation from [4].

2-processor Montgomery Addition

Cost	Step	Processor 1	Processor 2
	1	$R_0 \leftarrow X_2 - Z_2$	$R_1 \leftarrow X_3 + Z_3$
	2	$R_2 \leftarrow X_2 + Z_2$	$R_3 \leftarrow X_3 - Z_3$
1M	3	$R_0 \leftarrow R_0 \cdot R_1$	$R_2 \leftarrow R_2 \cdot R_3$
	4	$R_1 \leftarrow R_0 + R_2$	$R_3 \leftarrow R_0 - R_2$
1S	5	$R_0 \leftarrow R_1^2$	$R_2 \leftarrow R_3^2$
1M	6	$X_5 \leftarrow Z_1 \cdot R_0$	$Z_5 \leftarrow X_1 \cdot R_2$

2-processor Montgomery Doubling

Cost	Step	Processor 1	Processor 2
	1	$R_4 \leftarrow X_2 + Z_2$	$R_5 \leftarrow X_2 - Z_2$
1S	2	$R_4 \leftarrow R_4^2$	$R_5 \leftarrow R_5^2$
	3	$R_6 \leftarrow R_4 - R_5$	*idle*
1D	4	$R_7 \leftarrow a24R_6$	*idle*
	5	$R_7 \leftarrow R_5 + R_7$	*idle*
1M	6	$X_4 \leftarrow R_4 \cdot R_5$	$Z_4 \leftarrow R_6 \cdot R_7$

The effective cost of addition is $2\mathbf{M}+1\mathbf{S}$ (even if $Z_1 = 1$). The effective cost of doubling is $1\mathbf{M} + 1\mathbf{S} + 1\mathbf{D}$. Since $2\mathbf{M} + 1\mathbf{S} \geq 1\mathbf{M} + 1\mathbf{S} + 1\mathbf{D}$ the overall effective cost is $2\mathbf{M} + 1\mathbf{S}$ depending on the assumption in Section 4.2.

On the Validity of the Φ-Hiding Assumption in Cryptographic Protocols

Christian Schridde and Bernd Freisleben

Department of Mathematics and Computer Science, University of Marburg
Hans-Meerwein-Str. 3, D-35032 Marburg, Germany
{schriddc,freisleb}@informatik.uni-marburg.de

Abstract. Most cryptographic protocols, in particular asymmetric protocols, are based on assumptions about the computational complexity of mathematical problems. The Φ-Hiding assumption is such an assumption. It states that if p_1 and p_2 are small primes exactly one of which divides $\varphi(N)$, where N is a number whose factorization is unknown and φ is Euler's totient function, then there is no polynomial-time algorithm to distinguish which of the primes p_1 and p_2 divides $\varphi(N)$ with a probability significantly greater than 1/2. In this paper, it will be shown that the Φ-Hiding assumption is not valid when applied to a modulus $N = PQ^{2e}$, where $P, Q > 2$ are primes, $e > 0$ is an integer and P hides the prime in question. This indicates that cryptographic protocols using such moduli and relying on the Φ-Hiding assumption must be handled with care.

Keywords: Φ-Hiding assumption, Jacobi symbol, Euler's totient function.

1 Introduction

The Φ-Hiding assumption as defined by Cachin, Micali and Stadler [3] is an assumption about the difficulty of finding small factors of $\varphi(N)$, where N is a number whose factorization is unknown, and $\varphi(\cdot)$ is Euler's totient function, i.e. the number of positive integers less than or equal to N that are coprime to N. The security of several cryptosystems is based on the presumed difficulty of solving this problem [2,5,6,7]. In this paper, it will be shown how information about the unknown factors of $\varphi(N)$ can be obtained when the modulus N is chosen as $N = PQ^{2e}$, where $P, Q > 2$ are primes, $e > 0$ is an integer and P hides the prime in question, such that the Φ-Hiding assumption is not valid in this case. Moduli of the form $N = PQ^{2e}$ are called *Multi-Power RSA* moduli and are used to speed up cryptographic operations [1]. In addition, it will be shown that if two random composite integers instead of two primes are used, the probability of choosing the integer that divides $\varphi(N)$ reaches 99% if the integers have at least 7 prime factors. Furthermore, the paper suggests an approach to get more information about $\varphi(N)$ without knowing the factorization of N.

The paper is organized as follows. In Section 2, two definitions of the Φ-Hiding assumption are given. Our approach to show that the Φ-Hiding assumption is

J. Pieprzyk (Ed.): ASIACRYPT 2008, LNCS 5350, pp. 344–354, 2008.
© International Association for Cryptologic Research 2008

not valid in certain circumstances is presented in Section 3. Section 4 concludes the paper and outlines areas for future research.

2 The Φ-Hiding Assumption

The Φ-Hiding assumption [3] can be defined in two different ways. The first definition illustrates the computational problem the assumption is based on.

Definition 1 (Φ-Hiding assumption (1)). *Given an integer N with unknown factorization, it is computationally hard to decide whether a prime p_i with $2 < p_i << N^{1/4}$ divides $\varphi(N)$ or not.*[1]

The second definition represents a special case of the assumption, since it is assumed that exactly one of two given integers divides $\varphi(N)$.

Definition 2 (Φ-Hiding assumption (2)). *If p_1 and p_2 are two random, small primes and N is constructed such that exactly one of these primes divides $\varphi(N)$, then there is no polynomial-time algorithm to distinguish which of the primes $p_1 > 2$ and $p_2 > 2$ divides $\varphi(N)$ with a probability significantly greater than 0.5, if N is an integer with unknown factorization. If p_i divides $\varphi(N)$, it is said that $\varphi(N)$ hides p_i.*

In cryptographic protocols, Definition 2 of the Φ-Hiding assumption is used, since in this case some previous knowledge is involved (i.e. which of the two primes divides $\varphi(N)$), that can be used to create a necessary backdoor for asymmetric cryptography. To the best of our knowledge, no attack on the Φ-Hiding assumption has been published until now. In the next section, we present our approach to show that the Φ-Hiding assumption is not valid when Multi-Power RSA moduli are used.

3 The Φ-Hiding Assumption Revisited

The Φ-Hiding assumption is only valid when it is applied to a composite number that cannot be completely factored in feasible time, since otherwise it would be trivial to decide whether a prime divides $\varphi(N)$ or not. Our approach to decide whether a prime divides $\varphi(N)$ for a composite number N uses the Jacobi symbol. It can be evaluated efficiently, even for composite numbers with unknown factorization [4]. The Jacobi symbol $\mathrm{J}_P(r)$, for P prime, generalizes the Legrende symbol and states information about quadratic residues: If $a^2 \equiv r \pmod{P}$ for given integers r and P has a solution in a, then $\mathrm{J}_P(r) = 1$, otherwise $\mathrm{J}_P(r) = -1$

[1] Following the remarks of the original paper of Cachin, Micali and Stadler [3], N can be efficiently factored when a prime $> N^{1/4}$ of $\varphi(N)$ is known, thus the Φ-Hiding assumption asks for very small primes. Even if it is known which small primes p_i divide $\varphi(N)$, if $log\ p_i$ is significantly smaller than $(log\ N)^c$, for a constant c between 0 and 1, N cannot be factored significantly faster.

(if $\gcd(P,r) > 1$, then $\mathrm{J}_P(r) = 0$). For composite odd integers, the Jacobi symbol is defined as $\mathrm{J}_N(r) = \prod_{j=1}^{m} \mathrm{J}_{P_j}(r)^{\nu_j}$, if $N = P_1^{\nu_1} \dots P_m^{\nu_m}$. Furthermore, a particular $2k$-th root of unity is used to show that the values of the Jacobi symbol are related to factors of $\varphi(N)$, and that the Jacobi symbol adopts *non-random* values when the evaluated integer r is a divisor of $\varphi(N)$. Thus, the novel idea to use the existence and the non-existence of $2k$-th roots of unity in finite fields/rings allows us to gain knowledge about the divisors of $\varphi(N)$, which in some cases can be used to make the decision whether a given integer divides $\varphi(N)$ or not. These results will be used to show that the Φ-Hiding assumption as defined by Cachin, Micali and Stadler [3] is not valid when applied to a modulus $N = PQ^{2e}$, where $P, Q > 2$ are primes, $e > 0$ is an integer and P hides the prime in question. Lemma 1 is central for our approach:

Lemma 1. *Let ξ_{2k} be any fixed primitive $2k$-th root of unity and $k \in \mathbb{N}^+$, then:*

$$i^{1-k} \prod_{j=1}^{k-1} \left(\xi_{2k}^{j} - \xi_{2k}^{-j} \right) = k \tag{1}$$

Proof (of Lemma 1). The polynomial $f(X) = (X^k - 1)/(X-1) = X^{k-1} + X^{k-2} + \dots + 1$ has ξ_k^j for $j = 1, \dots, k-1$ as its roots, where ξ_k is any fixed primitive kth root of unity. Writing $f(X)$ in factored form $f(X) = \prod_{j=1}^{k-1}(X - \xi_k^j)$, we obtain $f(1) = \prod_{j=1}^{k-1}(1 - \xi_k^j) = k$. Since

$$i^{1-k} \prod_{j=1}^{k-1} (\xi_{2k}^{j} - \xi_{2k}^{-j}) = i^{1-k} \prod_{j=1}^{k-1} \xi_{2k}^{j} \prod_{j=1}^{k-1} (1 - \xi_{k}^{-j}) = i^{1-k} k \prod_{j=1}^{k-1} \xi_{2k}^{j} \tag{2}$$

and since $\prod_{j=1}^{k-1} \xi_{2k}^{j} = \xi_{2k}^{(k-1)k/2} = \xi_4^{k-1} = i^{k-1}$, the product $i^{1-k} \prod_{j=1}^{k-1} \xi_{2k}^{j}$ vanishes and we get

$$i^{1-k} \prod_{j=1}^{k-1} (\xi_{2k}^{j} - \xi_{2k}^{-j}) = k \tag{3}$$

which proves the lemma. □

We now rewrite the $(k-1)$ terms covered by the product symbol in equation (1), such that it contains a large square:

Lemma 2 (Square Lemma). *Let $k \in \mathbb{Z}^+$ and $k > 2$. Then:*
1. If k is odd:

$$\prod_{j=1}^{k-1} \left(\xi_{2k}^{j} - \xi_{2k}^{-j} \right) = \prod_{j=1}^{(k-1)/2} \left(\xi_{2k}^{j} + \xi_{2k}^{k-j} \right)^2 \tag{4}$$

2. If k is even:

$$\prod_{j=1}^{k-1} \left(\xi_{2k}^{j} - \xi_{2k}^{-j} \right) = 2i \prod_{j=1}^{(k-2)/2} \left(\xi_{2k}^{j} + \xi_{2k}^{k-j} \right)^2 \tag{5}$$

Proof (of Lemma 2)
1. k is odd: Since k is odd, the j-th and the $(k-j)$-th factor for $1 \leq j \leq k-1$ can be paired. The result is:

$$\begin{aligned}
&(\xi_{2k}^{j} - \xi_{2k}^{-j}) \cdot (\xi_{2k}^{k-j} - \xi_{2k}^{-(k-j)}) = (\xi_{2k}^{j} - \xi_{2k}^{-j}) \cdot (\xi_{2k}^{k-j} + \xi_{2k}^{j}) \\
&= \xi_{2k}^{j}\xi_{2k}^{k-j} + \xi_{2k}^{j}\xi_{2k}^{j} - \xi_{2k}^{-j}\xi_{2k}^{k-j} - \xi_{2k}^{-j}\xi_{2k}^{j} = -1 + \xi_{2k}^{2j} - \xi_{2k}^{k-2j} - 1 \\
&= \xi_{2k}^{2j} - 2 - \xi_{2k}^{k-2j} = \xi_{2k}^{2j} - 2 + \xi_{2k}^{k}\xi_{2k}^{k-2j} \\
&= \xi_{2k}^{2j} - 2 + \xi_{2k}^{2(k-j)} = (\xi_{2k}^{j} + \xi_{2k}^{k-j})^2
\end{aligned}$$

The pairing contains a square. Since $k-1$ is even, no term is left and a product of $(k-1)/2$ squares is generated, which proves the case for odd values of k.
2. k is even: Since k is even, the jth and the $(k-j)$th factor for $1 \leq j < k/2$ and $k/2 < j \leq k-1$ can be paired, which leads to the same terms as in case 1. The difference is that the factor $\left(\xi_{2k}^{j} - \xi_{2k}^{-j}\right)$ with $j = k/2$ remains. For this factor, $\xi_{2k}^{k/2} - \xi_{2k}^{-k/2} = (-1)^{1/2} - (-1)^{-1/2} = i - i^{-1} = i(1 - 1/i^2) = 2i$, which proves the case for even values of k. □

By Lemma 2, the product in Equation (1) is transformed to a product with a perfect square and the factor i^{1-k} (k odd) and $2i^{2-k}$ (k even), respectively.

3.1 Application to Finite Fields and Rings

In this section, the results are applied to finite fields $\mathbb{F}_P$ with P being a prime number. We distinguish between two cases. In the first case, we assume that a $\xi_{2k} \in \mathbb{F}_P$ does not exist, and in the second case, we assume that a $\xi_{2k} \in \mathbb{F}_P$ exists.

Case 1: A $\xi_{2k} \in \mathbb{F}_P$ does not exist. In this case, it is assumed that $\mathbb{F}_P$ does not contain a $2k$-th root of unity. As a consequence, there is no integer of order $2k$ and thus the factors $\left(\xi_{2k}^{j} + \xi_{2k}^{k-j}\right)$ are not defined properly in $\mathbb{F}_P$. Thus, it cannot be assumed that the product $\prod_{j=1}^{(k-1)/2}\left(\xi_{2k}^{j} + \xi_{2k}^{k-j}\right)^2$ forms a valid square in $\mathbb{F}_P$ and vanishes from the Jacobi symbol. The integer k, which nevertheless exists, has no defined counterpart on the left side of Equation 1. In this case, $\mathrm{J}_P(k)$ cannot be distinguished from a random coin flip between 1 and -1.

Case 2: A $\xi_{2k} \in \mathbb{F}_P$ exists. This leads to the fact that the square $\prod_{j=1}^{(k-1)/2}\left(\xi_{2k}^{j} + \xi_{2k}^{k-j}\right)^2$ obtained from Lemma 2 is valid in $\mathbb{F}_P$, since each ξ_{2k} is defined properly. Therefore, equation (1) can be written as a well defined congruence in $\mathbb{F}_P$. Corollary 1 shows the outcome when the Jacobi symbol is applied to this congruence and the square obtained from Lemma 2 is inserted.

Corollary 1. *Let P be an odd prime number, $k \in \mathbb{F}_P$. Assume that a $\xi_{2k} \in \mathbb{F}_P$ exists, then:*

1. *If k is odd:*

$$\mathrm{J}_P\left((-1)^{(1-k)/2}\prod_{j=1}^{(k-1)/2}\left(\xi_{2k}^{j}+\xi_{2k}^{k-j}\right)^2\right)=J_P((-1)^{(1-k)/2})=J_P(k) \qquad (6)$$

2. *If k is even:*

$$\mathrm{J}_P\left(2(-1)^{1-k/2}\prod_{j=1}^{(k-2)/2}\left(\xi_{2k}^{j}+\xi_{2k}^{k-j}\right)^2\right)=J_P(2(-1)^{1-k/2})=J_P(k) \qquad (7)$$

After the square has vanished from the Jacobi symbol, a simple congruence is left. This congruence indicates a relationship between the value of the Jacobi symbol and the divisors of $\varphi(P)$, because Corollary 1 is only valid if $2k$ divides $\varphi(P)$. Again, this implicitly shows that it is important to distinguish between the two cases of divisibility introduced above, since the square vanishes only if it is defined properly. Otherwise, the Jacobi symbol of an arbitrary integer k would always be equal to $\mathrm{J}_P((-1)^{(1-k)/2})$ or $\mathrm{J}_P(2(-1)^{1-k/2})$, respectively, which obviously is wrong.

EXAMPLE: Let $P = 31$ with $\varphi(31) = 30$. By setting $k = 5$ due to $(2{\cdot}5)|30$, there must be an integer of order 10, e.g. 23 or 15. It does not matter which of them is chosen here, since it disappears after applying the Jacobi symbol. Now, calculate $(-1)^{(1-5)/2} = (-1)^{-2} = 1$. Since k is odd, $\mathrm{J}_{31}((-1)^{(1-5)/2}) = \mathrm{J}_{31}(1) = \mathrm{J}_{31}(5)$ must hold, which is true since both sides are equal to 1.

Next, a Theorem is stated that describes the relationship between $\mathrm{J}_P(k)$ and ξ_{2k}.

Theorem 1. *Let P be an odd prime number, $k \in \mathbb{F}_P$. $\mathrm{J}_P(k)$ and the divisors of $\varphi(P)$ are connected via the following implications:*

1. *If k is odd, then:*

$$\begin{array}{rcl} \textit{If } \xi_{2k} \in \mathbb{F}_P \textit{ exists} & \Rightarrow & \mathrm{J}_P((-1)^{(1-k)/2}) = \mathrm{J}_P(k). \\ \textit{If } \mathrm{J}_P((-1)^{(1-k)/2}) \neq \mathrm{J}_P(k) & \Rightarrow & \xi_{2k} \in \mathbb{F}_P \textit{ does not exist.} \end{array}$$

2. *If k is even, then:*

$$\begin{array}{rcl} \textit{If } \xi_{2k} \in \mathbb{F}_P \textit{ exists} & \Rightarrow & \mathrm{J}\left(2(-1)^{1-k/2}\right) = \mathrm{J}_P(k). \\ \textit{If } \mathrm{J}\left(2(-1)^{1-k/2}\right) \neq \mathrm{J}_P(k) & \Rightarrow & \xi_{2k} \in \mathbb{F}_P \textit{ does not exist.} \end{array}$$

Proof (of Theorem 1)
The proof of the theorem follows directly from Corollary 1. □

Theorem 1 indicates that either a divisor k of $\varphi(P)$ must be known to conclude that the corresponding Jacobi symbols $\mathrm{J}_P(k)$ and $\mathrm{J}_P((-1)^{(1-k)/2})$ (or $\mathrm{J}\left(2(-1)^{1-k/2}\right)$) are equal, or it must be tested whether the two Jacobi symbols $\mathrm{J}_P(k)$ and $\mathrm{J}_P((-1)^{(1-k)/2})$ (or $\mathrm{J}\left(2(-1)^{1-k/2}\right)$) are different in order to get the information that k cannot be a divisor of $\varphi(P)$. In the two other cases, no information can be obtained. The reason is that either the k-th root of -1 is not defined, or from the equality of the Jacobi symbols it cannot be concluded that k divides $\varphi(P)$.

To summarize, if $2k$ divides $\varphi(P)$, the Jacobi symbol of k adopts non-random values. Furthermore, Corollary 1 shows that the resulting congruences $\mathrm{J}_P((-1)^{(1-k)/2}) \equiv \mathrm{J}_P(k)$ and $\mathrm{J}_P(2(-1)^{1-k/2}) \equiv \mathrm{J}_P(k)$ for odd and even values of k are *independent* of the chosen ξ_{2k}. Thus, it is only essential that a ξ_{2k} exists in $\mathbb{F}_P$, but it is not necessary to know it.

3.2 Leakage Corollaries

In this section, we present tables for special composite integers N that contain the values the Jacobi symbol must adopt to leak information about the divisors of $\varphi(N)$. For composite integers N with unknown factorization, we do not know the order of an arbitrary integer a, but we can compute the Jacobi symbol $\mathrm{J}_N(a)$. Thus, we are only able to use the first implication of item 1 and and the second implication of item 2 of Theorem 1. For clarity, the following corollary divides these items further with respect to different residue classes of a prime P and an integer k.

Corollary 2 (Leakage corollary for prime numbers). *Let P be an odd prime number, $k \in \mathbb{F}_P$. In any of the following six cases, there does not exist a $\xi_{2k} \in \mathbb{F}_P$.*
If $P \equiv 1 \pmod 4$:

If k is odd: If $\mathrm{J}_P\left(i^{1-k}\right) = 1 \neq -1 = \mathrm{J}_P(k)$.
If k is even: If $\mathrm{J}_P\left(2i^{2-k}\right) = (-1)^{(p^2-1)/8} \neq \mathrm{J}_P(k)$.

If $P \equiv 3 \pmod 4$:

If $k \equiv 0 \pmod 4$: If $\mathrm{J}_P\left(2(-1)^{1-k/2}\right) = (-1)^{(P^2+7)/8} \neq \mathrm{J}_P(k)$.
If $k \equiv 1 \pmod 4$: If $\mathrm{J}_P\left((-1)^{(1-k)/2}\right) = 1 \neq \mathrm{J}_P(k)$.
If $k \equiv 2 \pmod 4$: If $\mathrm{J}_P\left(2(-1)^{1-k/2}\right) = (-1)^{(P^2-1)/8} \neq \mathrm{J}_P(k)$.
If $k \equiv 3 \pmod 4$: If $\mathrm{J}_P\left((-1)^{(1-k)/2}\right) = -1 \neq \mathrm{J}_P(k)$.

Corollary 2 states which two Jacobi symbols must differ to be sure that the integer k is not a divisor of $\varphi(P)$. Thus, in some cases, the access to the Jacobi symbol is sufficient to decide whether a prime divides $P-1$ or not. Next, the corollary is extended to composite integers N being the product of two distinct prime numbers P and Q. This leads to the tables shown in Figure 1. The tables must be read in the following way: The four tables handle the four different residues of k modulo 4. Furthermore, the first two tables (horizontal direction) show the 64 combinations of the 8 different residues of P and Q modulo 16 ($P, Q > 2$) for even residues of k. The third tables was reduced to one a single row since it contains 64 values of -1. The fourth table shows the 64 combinations of the 8 different residues of P and Q modulo 16 ($P, Q > 2$) for $k \equiv 3 \pmod 4$. The entries for each combination of P and Q illustrate which

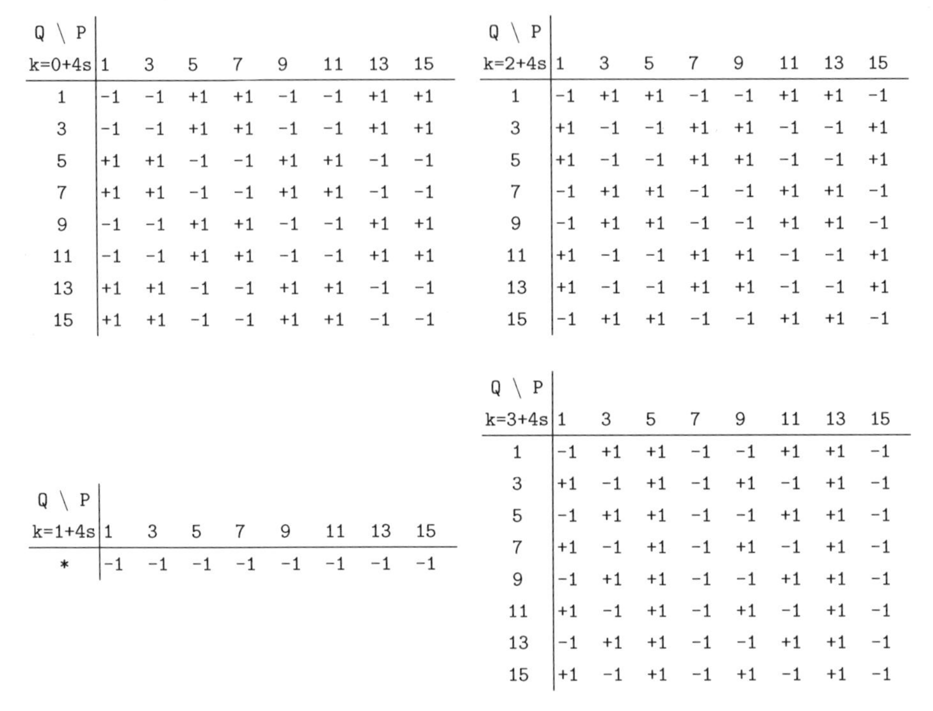

Q \ P k=0+4s	1	3	5	7	9	11	13	15
1	-1	-1	+1	+1	-1	-1	+1	+1
3	-1	-1	+1	+1	-1	-1	+1	+1
5	+1	+1	-1	-1	+1	+1	-1	-1
7	+1	+1	-1	-1	+1	+1	-1	-1
9	-1	-1	+1	+1	-1	-1	+1	+1
11	-1	-1	+1	+1	-1	-1	+1	+1
13	+1	+1	-1	-1	+1	+1	-1	-1
15	+1	+1	-1	-1	+1	+1	-1	-1

Q \ P k=2+4s	1	3	5	7	9	11	13	15
1	-1	+1	+1	-1	-1	+1	+1	-1
3	+1	-1	-1	+1	+1	-1	-1	+1
5	+1	-1	-1	+1	+1	-1	-1	+1
7	-1	+1	+1	-1	-1	+1	+1	-1
9	-1	+1	+1	-1	-1	+1	+1	-1
11	+1	-1	-1	+1	+1	-1	-1	+1
13	+1	-1	-1	+1	+1	-1	-1	+1
15	-1	+1	+1	-1	-1	+1	+1	-1

Q \ P k=1+4s	1	3	5	7	9	11	13	15
*	-1	-1	-1	-1	-1	-1	-1	-1

Q \ P k=3+4s	1	3	5	7	9	11	13	15
1	-1	+1	+1	-1	-1	+1	+1	-1
3	+1	-1	+1	-1	+1	-1	+1	-1
5	-1	+1	+1	-1	-1	+1	+1	-1
7	+1	-1	+1	-1	+1	-1	+1	-1
9	-1	+1	+1	-1	-1	+1	+1	-1
11	+1	-1	+1	-1	+1	-1	+1	-1
13	-1	+1	+1	-1	-1	+1	+1	-1
15	+1	-1	+1	-1	+1	-1	+1	-1

Fig. 1. Entries: $\mathrm{J}_{PQ}(k)$. Tables for $N = PQ$ for different residues of P and Q modulo 16.

value of the Jacobi symbol $\mathrm{J}_N(k)$ reveals that there is no integer of order $2k$ for at least one of the primes P and Q. For example, the first entry of -1 in the upper left table represents the case $k \equiv 0 \pmod 4$ and $P \equiv Q \equiv 1 \pmod{16}$. Applying Corollary 2 to this combination yields $\mathrm{J}_P\left(2i^{2-k}\right) = \mathrm{J}_Q\left(2i^{2-k}\right) = 1$. The corresponding table entry of -1 shows that $\mathrm{J}_N(k)$ must be -1, therefore at least for one of the primes P or Q, there is no integer of order $2k$.

The conclusion is too weak to obtain knowledge regarding the Φ-Hiding assumption, since $\phi(N)$ could still be divisible by $2k$. Some integers, even with unknown factorization, allow to obtain more information about the divisors of $\varphi(N)$. These are integers of the form $N = PQ^{2e}$, since one of the two involved primes is a square, which is ignored by the Jacobi symbol. In this way, the Jacobi symbol leaks information about the other prime involved. If N has the form $N = PQ^{2e}$, then for the Jacobi symbol and a co-prime integer $k > 2$, $\mathrm{J}_N(k) = \mathrm{J}_{PQ^{2e}}(k) = \mathrm{J}_P(k) \cdot \mathrm{J}_Q(k)^{2e} = \mathrm{J}_P(k)$.

Using this fact, the tables displayed in Figure 2 show the values the Jacobi symbol $\mathrm{J}_N(k)$ must adopt such that $2k$ does not divide $\varphi(P)$.

Example: Suppose $N = 1323801442080750176044871$ and N is of the form $N = PQ^{2e}$, $e > 0$. Suppose we want to test whether $k = 41$ divides $P - 1$. Since $k \equiv 1 \pmod 4$, the third table must be used. Thus, $\mathrm{J}_N(41) = -1$. The table shows that whenever the Jacobi symbol of k is negative, k can not divide $P - 1$.

Q \ P k=0+4s	1	3	5	7	9	11	13	15
*	-1	-1	+1	+1	-1	-1	+1	+1

Q \ P k=2+4s	1	3	5	7	9	11	13	15
*	-1	+1	+1	-1	-1	+1	+1	-1

Q \ P k=1+4s	1	3	5	7	9	11	13	15
*	-1	-1	-1	-1	-1	-1	-1	-1

Q \ P k=3+4s	1	3	5	7	9	11	13	15
*	-1	+1	-1	+1	-1	+1	-1	+1

Fig. 2. Entries: $\mathrm{J}_{PQ^{2e}}(k)$. Tables for $N = PQ^{2e}$ for different residues of P and Q modulo 16.

In the next section, the last two tables in Figure 2 are used to invalidate the Φ-Hiding assumption when using moduli of the form $N = PQ^{2e}$ and choosing P to hide the prime number in question.

3.3 Application to the Φ-Hiding Assumption

In both Definitions 1 and 2 of Section 2, it is only required that N is a composite integer with unknown factorization. By applying our results from the previous sections, we show that this requirement is not sufficient. If the Φ-Hiding assumption is applied to a modulus of the form PQ^{2e}, where the integer P is constructed in such a way that P hides a given prime, then the Φ-Hiding assumption is violated with non-negligible probability. Moduli of this form, mostly with $e = 1$, are used by several cryptographic protocols, as described by Boneh and Shacham [1] and used, e.g., by Poupard and Stern [8], to speed up some computations that profit from the form PQ^{2e} with $e > 0$ instead of PQ. Using the results of the previous sections, the following theorem can be stated:

Theorem 2. *Let $N = PQ^{2e}$ and suppose that P hides p. Then, the Φ-Hiding assumption from Definition 2 can be violated. An attacker can choose the hidden prime with an average success probability of $\frac{3}{4}$.*

The following notation is used: N is again of the form $N = PQ^2$ and $\mathrm{T}(N, k)$ is the value of the corresponding table entry of Figure 2.

Proof (of Theorem 2). Suppose that either p_1 or p_2 divides $\varphi(N)$ and an attacker has to decide which of them divides $\varphi(N)$. Without loss of generality, we assume that p_1 is the prime that is hidden by P. For this prime, $\mathrm{J}_N(p_1) \neq \mathrm{T}(N, p_1)$ holds, because it divides $P - 1$ (see Theorem 1). Thus, the attacker will find *at least* one matching Jacobi symbol concerning the primes p_1 and p_2. From the attackers point of view, the probability that a prime p_i, $i \in \{1, 2\}$ divides $\varphi(N)$ is

$$Prob[p_i|\varphi(N)] = \begin{cases} 0, & \mathrm{J}_N(p_i) = \mathrm{T}(N, p_i) \\ 1, & \mathrm{J}_N(\overline{p}_i) = \mathrm{T}(N, \overline{p}_i) \\ \frac{1}{2}, & \mathrm{J}_N(p_i) = \mathrm{J}_N(\overline{p}_i) \end{cases} \quad (8)$$

where $\overline{p}_i$ denotes the other one of the two primes. Note the factorization of N is not needed to construct the tables in figure 2. They are universally valid for moduli of the form $N = PQ^{2e}$ and thus known to the attacker. Whenever the Jacobi symbol $\mathrm{J}_N(p_i)$ is equal to $\mathrm{T}(N, p_i)$, Theorem 1 states that p_i cannot be a divisor of $\varphi(N)$, thus the probability is $Prob[p_i|\varphi(N)] = 0$. Consequently, the Jacobi symbol $\mathrm{J}_N(\overline{p}_i)$ must be not equal to $T(N, \overline{p}_i)$, which indicates that it is the hidden prime. If both Jacobi symbols do not match the table entry, no information is leaked and the attacker cannot argue in any direction. Thus, in this case the probability is $Prob[p_i|\varphi(N)] = \frac{1}{2}$. Since the primes p_i are chosen randomly, it can be assumed that the Jacobi symbol $\mathrm{J}_N(p_2)$ adopts random values of -1 and $+1$. The calculation of the total probability for the attacker to choose the hidden prime correctly is as follows: Whenever a Jacobi symbol evaluates to a value unequal to the table entry, it cannot be the prime that is hidden by P, so the attacker chooses the other one, the hidden one, with a probability of 1. When both Jacobi symbols evaluate to $\neq \mathrm{T}(N, \cdot)$, the attacker chooses the right one with a probability of $\frac{1}{2}$. Thus, in total there is an average probability of $\frac{1}{2} \cdot 1 + \frac{1}{2} \cdot \frac{1}{2} = \frac{3}{4}$ to choose the correct prime, which proves Theorem 2. □

Composite Integers. The situation is even worse when the Φ-Hiding assumption is used with composite integers n_1 and n_2 instead of the primes p_1 and p_2, as done, for example, by Gentry et al. [5]. Assume that there is a modulus of the form $N = PQ^2$ and we want to determine whether the composite integer n_i, which is the product of m distinct primes greater than 2, divides $\varphi(N)$. Suppose the Jacobi symbol is applied and the result does not allow to decide whether n_i divides $\varphi(N)$ or not. In this case, we can proceed with the prime factors of n_i. Since n_i is $\prod_{j=1}^{m} p_j$, the Jacobi symbol can simply be evaluated for all of its prime factors. If there is a prime p_j with a Jacobi symbol that leaks the required information, we know that n_i cannot divide $\varphi(N)$, since from $n_i|\varphi(N)$ it follows that $p_j|\varphi(N)$ must also hold. If the integers in question consist only of 7 prime numbers, there already is a success probability of $\approx 99\%$ to choose the right integer.

Corollary 3. *If $n_1 = \prod_{j=1}^{l_1} p_i$ and $n_2 = \prod_{j=1}^{l_2} q_j$ are two random, composite integers that are odd and square free and n_1 is the hidden integer, then an attacker has a success probability of $(1 - \frac{1}{2^{l_2}})$ to choose the hidden integer.*

Proof. Let $n_1 = \prod_{j=1}^{l_1} p_j$ and $n_2 = \prod_{j=1}^{l_2} q_j$ be two odd, square free integers. If $N = PQ^{2e}$ and exactly one of the two integers n_1 and n_2 divides $\varphi(N)$, the probability to choose the right one of the two possibilities is as follows. The case $l_1 = l_2 = 1$ was already addressed in the paper; it has a success probability of $\frac{3}{4}$. Note that if $n_i|\varphi(N)$, then also each divisor of n_i is a divisor of N. Thus, if we find a divisor of n_i that does not divide $\varphi(N)$, we can conclude that n_i is not the integer hidden by $\varphi(N)$. Since the same argument applies to all divisors that are prime numbers, it is sufficient to check all prime factors of n_i whether they are divisors of $\varphi(N)$ or not.

Table 1. Success Probability

$l_1 = l_2$	1	2	3	4	5	6	7
	0.5	0.75	0.875	0.938	0.969	0.984	0.992

Without loss of generality, we assume that n_1 is the integer hidden by $\varphi(N)$. For each of its l_1 prime factors p_i, $\mathrm{J}_N(p_i) \neq \mathrm{T}(N, p_i)$ must hold. For the other integer n_2, it follows that for each of its l_2 prime factors q_i it holds with a probability of $\frac{1}{2}$ that $\mathrm{J}_N(q_i) \neq \mathrm{T}(N, q_i)$ and with a probability of $\frac{1}{2}$ that $\mathrm{J}_N(q_i) = \mathrm{T}(N, q_i)$. Whenever the first case occurs, no knowledge is gained. But whenever the latter case occurs, the information that n_2 cannot be a divisor of $\varphi(N)$ is gained, so n_1 is the hidden number. The method fails if for all prime factors $\mathrm{J}_N(q_i) \neq \mathrm{T}(N, q_i)$ is obtained, which occurs with a probability of $\prod_{i=1}^{l_2} Prob[\mathrm{J}_N(q_i) \neq \mathrm{T}(N, q_i)] = \frac{1}{2^{l_2}}$. Thus, the success probability of choosing the right integer is $(1 - \frac{1}{2^{l_2}})$. □

Table 1 illustrates the success probability of choosing the right prime for different numbers of prime factors.

3.4 Discussion

In the previous section we have shown that in some circumstances it can be efficiently decided whether a given prime p divides $\varphi(N)$ or not. A necessary condition is that moduli of the form PQ^{2e} with $e > 1$ are used and P hides p. If someone implements a cryptographic protocol based on the Φ-Hiding assumption and uses such moduli, an attacker has an average probability of $\frac{3}{4}$ to choose the right prime, if the primes the attacker can choose from are selected randomly. In cases when it is desired to ask which composite number n_i is hidden by P, the success probability would be even greater than $\frac{3}{4}$, since for each prime factor of n the attacker has the success probability of $\frac{3}{4}$.

There are two possible countermeasures to the presented attack. First, moduli of the form $PQ^{2e}, e > 1$ should not be used in conjunction with the Φ-Hiding assumption. Second, the primes a user can choose from should not be selected randomly, but only those primes that have a positive Jacobi symbol regarding N should be used. Thus, the assumption as stated in the original form should be adapted to avoid its vulnerability to the presented attack.

4 Conclusions

In this paper, it was shown that by utilizing an identity of $2k$-th roots in $\mathbb{Z}_N$ and the Jacobi symbol, it is possible to gain knowledge about the unknown factors of Euler's totient function $\varphi(N)$ even if N is computationally hard to factorize. This knowledge was used to invalidate the Φ-Hiding assumption as defined by Cachin, Micali and Stadler [3] for moduli of the form $N = PQ^{2e}$ with P hiding the prime in question, since the Jacobi symbol adopts non-random values when being applied to a factor of $\varphi(N)$. Our results are important for evaluating the security of cryptographic protocols that use the Φ-Hiding assumption and exemplify the situation when it has to handled with care.

There are several areas for future work. For example, an interesting issue is to examine the case when the integer k does not divide $\varphi(N)$. In this case, the identity is not well defined. Thus, it should be investigated whether there are methods to bypass this problem to obtain further relationships between the Jacobi symbol and the factors of $\varphi(N)$. Since the approach makes use of an identity of $2k$-th roots in $\mathbb{Z}_N$ and this identity is only one of many, future work should be directed to analyze other results of such identities that may offer attack possibilities on the Φ-Hiding assumption.

Acknowledgements. The authors would like to thank Frederik Vercauteren for his excellent comments to improve the presentation of the material contained in this paper.

References

1. Boneh, D., Shacham, H.: Fast Variants of RSA. CryptoBytes 5(1) (Winter/Spring 2002)
2. Cachin, C.: Efficient Private Bidding and Auctions with an Oblivious Third Party. In: ACM Conference on Computer and Communications Security, pp. 120–127 (1999)
3. Cachin, C., Micali, S., Stadler, M.: Computationally Private Information Retrieval with Polylogarithmic Communication. In: Stern, J. (ed.) EUROCRYPT 1999. LNCS, vol. 1592, pp. 402–407. Springer, Heidelberg (1999)
4. Eikenberry, S.M., Sorenson, J.P.: Efficient Algorithms for Computing the Jacobi Symbol. Journal of Symbolic Computation 26(4), 509–523 (1998)
5. Gentry, C., Mackenzie, P., Ramzan, Z.: Password Authenticated Key Exchange Using Hidden Smooth Subgroups. In: Proceedings of the 12th ACM Conference on Computer and Communications Security, pp. 299–309. ACM Press, New York (2005)
6. Gentry, C., Ramzan, Z.: Single-Database Private Information Retrieval with Constant Communication Rate. In: Proceedings of the 32nd International Colloquium on Automata, Languages and Programming, Lisbon, Portugal, pp. 803–815 (2005)
7. Hemenway, B., Ostrovsky, R.: Public Key Encryption which is Simultaneously a Locally-Decodable Error-Correcting Code. In: Electronic Colloquium on Computational Complexity, Report No. 21 (2007)
8. Poupard, G., Stern, J.: Fair Encryption of RSA Keys. In: Preneel, B. (ed.) EUROCRYPT 2000. LNCS, vol. 1807, pp. 172–189. Springer, Heidelberg (2000)

Chosen Ciphertext Security with Optimal Ciphertext Overhead

Masayuki Abe[1], Eike Kiltz[2,*], and Tatsuaki Okamoto[1]

[1] NTT Information Sharing Platform Laboratories, NTT Corporation, Japan
[2] CWI Amsterdam, The Netherlands

Abstract. Every public-key encryption scheme has to incorporate a certain amount of randomness into its ciphertexts to provide semantic security against chosen ciphertext attacks (IND-CCA). The difference between the length of a ciphertext and the embedded message is called the *ciphertext overhead*. While a generic brute-force adversary running in 2^t steps gives a theoretical lower bound of t bits on the ciphertext overhead for IND-CPA security, the best known IND-CCA secure schemes demand roughly $2t$ bits even in the random oracle model. Is the t-bit gap essential for achieving IND-CCA security?

We close the gap by proposing an IND-CCA secure scheme whose ciphertext overhead matches the generic lower bound up to a small constant. Our scheme uses a variation of a four-round Feistel network in the random oracle model and hence belongs to the family of OAEP-based schemes. Maybe of independent interest is a new efficient method to encrypt long messages exceeding the length of the permutation while retaining the minimal overhead.

1 Introduction

1.1 Background

MOTIVATION. Ever since Goldwasser and Micali introduced the concept of "probabilistic encryption" [16] it is well understood that every public-key encryption scheme has to incorporate a certain amount of randomness into their ciphertexts in order to achieve semantic security. Thus a ciphertext c must be longer than the embedded message m and the difference $\ell_{\mathsf{oh}} := |c| - |m|$ is called the *ciphertext overhead*. In order to achieve stronger security properties, the ciphertext overhead tends to be even larger due to the use of extended randomness or extra integrity checking mechanisms. In this paper we are asking for the minimal possible ciphertext overhead to protect against adaptive chosen ciphertext attacks (IND-CCA security).

A GENERIC LOWER BOUND. A ciphertext overhead of ℓ_{oh} bits means that at most ℓ_{oh} bits of randomness can be incorporated into a ciphertext. A brute-force

* Supported by the research program Sentinels (http://www.sentinels.nl). Sentinels is being financed by Technology Foundation STW, the Netherlands Organization for Scientific Research (NWO), and the Dutch Ministry of Economic Affairs.

J. Pieprzyk (Ed.): ASIACRYPT 2008, LNCS 5350, pp. 355–371, 2008.
© International Association for Cryptologic Research 2008

Table 1. Upper bounds on the ciphertext overhead (up to small additive constants) in OAEP variants for $(2^{\varepsilon}, 2^{-t})$-adversaries. The lower bound is $\ell_{\mathsf{oh}} \geq t + \varepsilon$. OW: one-wayness. SPD-OW: set partial domain one-wayness.

Scheme	Ciphertext Overhead	Assumption on TDP	#Feistel rounds
OAEP [4, 15]	$\ell_{\mathsf{oh}} \leq 3t + 2\varepsilon$	SPD-OW	2
OAEP+ [25]	$\ell_{\mathsf{oh}} \leq 3t + 2\varepsilon$	OW	2
PSS-E [10]	$\ell_{\mathsf{oh}} \leq 2t + 2\varepsilon$	SPD-OW	2
PSP2 S-Pad [14]	$\ell_{\mathsf{oh}} \leq 2t + 2\varepsilon$	OW	4
OAEP-3R [23]	$\ell_{\mathsf{oh}} \leq 2t + \varepsilon$	OW	3
OAEP-4X (ours)	$\ell_{\mathsf{oh}} = t + \varepsilon$	OW	4

adversary in the IND-CPA experiment can exhaustively search for the randomness used for the challenge ciphertext. After encrypting one of the challenge messages up to 2^t times, it has an advantage of $\Omega(2^t/2^{\ell_{\mathsf{oh}}})$. Requiring the advantage to be smaller than $2^{-\varepsilon}$ (and ignoring small additive constants), it must hold that

$$\ell_{\mathsf{oh}} \geq t + \varepsilon .$$

Accordingly, $t+\varepsilon$ bits are a lower bound on the ciphertext overhead with respect to adversaries running in 2^t steps and having a success probability of at most $2^{-\varepsilon}$, by counting encryption as one step. (We refer to Section 2 for a more formal treatment.) We say that the ciphertext overhead is *optimal* if it matches the lower bound up to a (small) constant term, i.e., if $\ell_{\mathsf{oh}} \leq t+\varepsilon+O(1)$. Since every IND-CPA adversary is also an IND-CCA adversary, the above lower bound also applies to IND-CCA secure schemes.

For a number of schemes the ciphertext overhead primarily depends on the size of the underlying number-theoretic primitive, which often suffers from more sophisticated attacks. For example, ciphertexts of ElGamal-type schemes contain at least one group element of overhead which must be longer than $2t + \varepsilon$ bits due to the generic square-root bounds on the discrete-logarithm problem. Hence, the ciphertext overhead of such schemes can never match the generic lower bound.

Upper Bounds from Existing Schemes. Among the cryptosystems based on trapdoor permutations, there are ones whose ciphertext overhead is essentially independent of the size of the underlying permutation. We focus on such schemes for the rest of the paper. An example with optimal ciphertext overhead is the basic version of OAEP [4], which omits the zero padding and therefore only offers IND-CPA security. Considering IND-CCA security, however, OAEP loses its optimal ciphertext overhead as exemplified in Section 2.2. On the other hand, concrete security proofs for existing schemes provide upper bounds on the ciphertext overhead with which the desired level of security is attained. Table 1 summarizes the ciphertext overhead of existing schemes. Its content is discussed in the rest of this section.

IND-CCA SECURITY VIA VALIDITY CHECKING. As in OAEP, a common approach [25, 19, 21, 10, 20, 14] to achieve IND-CCA security is to attach a deterministic *validity string* (such as zero-padding or a hash of the message, etc) to the message (or the ciphertext) so that decryption can verify and reject almost all invalid ciphertexts. The ciphertext overhead is thus determined by the size of the randomness and the validity string. OAEP and the schemes in [25, 19] require randomness of $2t+\varepsilon$ bits plus a validity string of $t+\varepsilon$ bits. (See Section 2.2 for details on how to compute these values.) Their ciphertext overhead is thus $\ell_{\mathsf{oh}} = 3t+2\varepsilon$. The schemes in [10, 14] have a better security reduction and achieve $\ell_{\mathsf{oh}} = 2t + 2\varepsilon$, which seems the best one can expect as long as encryption incorporates a validity string into the ciphertexts.

VALIDITY-FREE ENCRYPTION. A considerable step towards minimizing the ciphertext overhead was the *validity-free* approach introduced by Phan and Pointcheval [22, 23]. In their scheme (called 3-round OAEP) decryption never rejects but returns a randomly looking message if a given ciphertext was not properly created with the encryption algorithm. Since no validity string is needed, the ciphertext overhead only depends on the randomness. As we shall discuss later, their security reduction however forces the ciphertext overhead to be $\ell_{\mathsf{oh}} = k_r = 2t + \varepsilon$ bits because of a "quadratic term" $q_h q_d / 2^{k_r}$ that appears in the success probability of their reduction. A more recent scheme in [13] suffers from the same problem. In summary, these schemes successfully eliminate the validity string but instead demand an extended randomness to prove IND-CCA security.

ENCRYPTING LONG MESSAGES. The problem of getting optimal overhead becomes even more difficult when considering longer messages. Notice that all above schemes limit the messages to the size of the permutation minus the overhead. To encrypt long inputs, [4, 17] suggest to stretch the width of the Feistel network to cover the entire message and apply the permutation only to a part of the output. But no general and formal treatment has been given to this methodology and it is unclear if and how it affects the ciphertext overhead. Furthermore, for schemes that use several Feistel rounds, this approach is expensive in computation as every internal hash function has to deal with a long input or output. A number of methods for constructing hybrid encryption are available (e.g., [12, 8, 9, 1, 6]), but they all increase the ciphertext overhead mainly because a one-time session-key is being encrypted.

1.2 Our Contribution

Our main contribution is an IND-CCA-secure public-key encryption scheme with optimal ciphertext overhead based on arbitrary family of trapdoor one-way permutation in the random oracle model. We follow the validity-free approach of 3-round OAEP [22] but instead use a 4-round Feistel network. (See Figure 1 in Section 3 for a diagram.) We stress that the essential difference is not the increased number of rounds; it is rather the way we bind the message to the randomness in the first round of the Feistel network while most of OAEP variants separately input the message and the randomness. (See Section 1.3 for more intuition.)

Our contribution is mostly theoretical; Our scheme demonstrates that lower and upper bounds on the ciphertext overhead with respect to IND-CCA security can match up to a small additive constant in the random oracle model. The design approach that binds the message to the randomness and the security proof may be of technical interest, too. In practice, when implemented with an 1024-bit RSA permutation (80-bit security), our scheme encrypts 943-bit and longer messages while it is 863 bits for a known best scheme, which is at most 9% increase of the message space. Though such a t-bit saving may have limited practical impact in general, the scheme could find applications with edgy requirements in bandwidth.

We also introduce a novel method to securely combine simple passively secure symmetric encryption with the Feistel network to encrypt long messages while retaining the optimal ciphertext overhead. While the construction is interesting in that it suggests a new variant of a KEM that allows partial message recovery, it is interesting also in a theoretical sense as it illustrates the difference in the properties of the round functions in a 4-round Feistel network as it will be discussed later.

1.3 Technical Overview

ACHIEVING OPTIMAL OVERHEAD. We explain the technical details in 3-round OAEP that seem to make it difficult to prove an optimal ciphertext overhead. The extended randomness of size $k_r \geq 2t + \varepsilon$ stems from a quadratic term $q_h\, q_d/2^{k_r}$ in the success probability of the security reduction. Since an adversary running in time 2^t can make at most $q_h \leq 2^t$ hash oracle queries and $q_d \leq 2^t$ decryption queries, we must assume that $q_h\, q_d \approx (2^t)^2$. Requiring $q_h\, q_d/2^{k_r} \leq 2^{-\varepsilon}$ results in $k_r \geq 2t + \varepsilon$.

Where does this quadratic loss in the reduction actually come from? In the security proof, every time the simulated decryption oracle receives a ciphertext that was not legitimately generated by asking the random oracles, it returns a random plaintext. Later, it patches the hash table for the simulated randomness so that the hash output looks consistent. The patching fails if the randomness has already been asked to the random oracle. This happens with probability at most $q_h/2^{k_r}$ since there are at most q_h hash queries. Throughout the attack, there are at most q_d decryption queries and hence the error probability of the patching is bounded by $q_h\, q_d/2^{k_r}$.

Our main technical contribution is to provide a security analysis for our scheme where only linear terms of the form $q_h/2^{k_r}$ or $q_d/2^{k_r}$ appear. We overcome the problem observed in 3-round OAEP by feeding the randomness *together* with a part of the input message (say m_1) into the hash function, i.e., by computing $H_1(r \,\|\, m_1)$. This link between the randomness and the message allows the reduction to partition hash queries by m_1 and therefore reducing the error probability in patching the hash table to $q_{h,m_1}/2^{k_r}$, where q_{h,m_1} is the number of hash queries with respect to m_1. By summing up the probabilities for all m_1 returned from the decryption oracle, the error probability is bounded by $\sum_{m_1} q_{h,m_1}/2^{k_r} \leq q_h/2^{k_r}$. The quadratic term is thus eliminated. The fourth round of the Feistel network is then needed to cover m_1.

ENCRYPTING LONG MESSAGES. In order to encrypt long messages exceeding the size of the permutation (while retaining the optimal overhead), we incorporate the idea of the Tag-KEM/DEM framework [1] that allows to use a simple passively secure length-preserving symmetric cipher. The exceeding part of the message is encrypted with the symmetric cipher whose key is derived from the randomness used in the asymmetric part of encryption. The symmetric part is then tied to the asymmetric part of the ciphertext by feeding it back into one of the hash function used in the Feistel network. Conceptually, our approach is similar to Tag-KEMs with partial ciphertext recovery [6] but in our case the message can be directly recovered. Namely, the main part of our construction can be used as *a Tag-KEM with partial message recovery.*

A concrete technical difficulty is how and where to include the feedback from the symmetric part. Including it in the F-function (random oracle) in every round of the 4-round Feistel network should work but may be redundant. Is it then secure if the feedback is given only to one of the F-functions? Which one? [24] showed that the inner two rounds have different properties than the outer two ones. Does that also apply to our case? Our result shows that it is sufficient to give the feedback to one of the inner two hash functions. We remark that when including the feedback only in the outer hash functions then either our security proof does no longer hold or there is a concrete attack. We refer to Section 3.3 for further details.

1.4 Related Work

IN OTHER MODELS. [22] constructed a simple scheme with optimal ciphertext overhead in the ideal full-domain permutation model. Looking at the construction and the security proof, however, one can see that the model is very strong and has little difference from idealizing the encryption function itself. Recently it is shown that ideal full-domain permutation can be constructed using random oracles [11] but the reduction is very costly and a *tight reduction* needed to retain the optimal overhead is highly unlikely. Note that [22] could only present a non-optimal scheme in the random oracle model, which shows the difficulty of achieving the optimality.

FOR SHORT MESSAGES. Schemes based on general one-way permutations can never offer the optimal overhead for messages shorter than the size of the permutation. For the state of art in this issue, we refer to [2] which presents a scheme that offers non-optimal but $\ell_{\mathsf{oh}} \geq 2t + \varepsilon$ that is currently the shortest overhead for messages of arbitrary (small) length. It is left as another open problem to construct a scheme with optimal overhead for arbitrary message size.

2 Lower Bound of Ciphertext Overhead

We follow the standard definition of public-key encryption $\mathsf{PKE} = (\mathcal{G}, \mathcal{E}, \mathcal{D})$ and indistinguishability against chosen plaintext attacks (IND-CPA) and adaptive chosen ciphertext attacks (IND-CCA). For formal definitions, we refer to the full version [3].

2.1 General Argument

Let $\mathsf{PKE} = (\mathcal{G}, \mathcal{E}, \mathcal{D})$ be a public-key encryption scheme and let $\mathcal{M}$ and $\mathcal{R}$ be the message and randomness space associated to a public-key pk. For $(pk, sk) \leftarrow \mathcal{G}(1^k)$ and $M \in \mathcal{M}$, let $C(M)$ denote the set of ciphertexts that recover message M. The ciphertext overhead ℓ^k_{oh} with respect to k is defined by $\ell^k_{\mathsf{oh}} = |\mathcal{E}_{pk}(M;r)| - |M|$. To obtain a simple form of the lower bound, we restrict ourselves to PKE where ℓ^k_{oh} is a fixed positive constant for any $pk \in \mathcal{G}(1^k)$, $M \in \mathcal{M}$ and $r \in \mathcal{R}$.

Let A be an adversary that runs in 2^t steps and breaks the semantic (IND-CPA) security of PKE with advantage at most $2^{-\varepsilon}$. To study the relation between the adversary's ability and the ciphertext overhead, we treat t, ε independently from k and represent the bounds of the ciphertext overhead as a function $\ell^k_{\mathsf{oh}}(t, \varepsilon)$. In the following argument, we count every encryption as one step. A launches the following attack.

1. Given pk generated by $(pk, sk) \leftarrow \mathcal{G}(1^k)$, pick arbitrary M_0 and M_1 of the same length from $\mathcal{M}$. Send (M_0, M_1) to the challenger and receive $c^* = \mathcal{E}_{pk}(M_b)$ where $b \leftarrow \{0, 1\}$.
2. Repeat the following up to 2^t times.
   - $r \leftarrow \mathcal{R}$, $c = \mathcal{E}_{pk}(M_0; r)$.
   - If $c = c^*$, output $\tilde{b} = 0$ and stop.
3. Output $\tilde{b} = 1$.

For a string c, let $p(c)$ denote the probability that $c = \mathcal{E}_{pk}(M_0; r)$ happens for uniformly chosen r. Similarly, let $p'(pk)$ denote the probability that pk is selected by $\mathcal{G}(1^k)$. The advantage of adversary A in breaking the semantic security with respect to pk is

$$\begin{aligned} \mathbf{Adv}_{\mathsf{A},pk} &= |\Pr[\tilde{b} = 0 \mid b = 0] - \Pr[\tilde{b} = 0 \mid b = 1]| \\ &= \Pr[\tilde{b} = 0 \mid b = 0] - 0 \\ &= \sum_{c \in C(M_0)} p(c)(1 - (1 - p(c))^{2^t}). \end{aligned} \tag{1}$$

Let η be the min-entropy with respect to the ciphertexts in $C(M_0)$ in bits. Since $p(c) \geq \frac{1}{2^\eta}$ for any $c \in C(M_0)$,

$$\mathbf{Adv}_{\mathsf{A},pk} \geq \sum_{c \in C(M_0)} p(c)(1 - (1 - \frac{1}{2^\eta})^{2^t}) \geq \frac{2^t}{2^\eta} - \frac{2^t - 1}{2^{2\eta}}. \tag{2}$$

Since $\eta \leq \ell_{\mathsf{oh}}^k$, we have

$$\begin{aligned}\mathbf{Adv}_{\mathsf{A}}(k) &= \sum_{pk\in\mathcal{G}(k)} p'(pk)\cdot \mathbf{Adv}_{\mathsf{A},pk} \\ &\geq \sum_{pk\in\mathcal{G}(k)} p'(pk)\cdot\left(\frac{2^t}{2^{\ell_{\mathsf{oh}}^k}} - \frac{2^t-1}{2^{2\ell_{\mathsf{oh}}^k}}\right) \\ &\geq \frac{1}{2}\cdot\frac{2^t}{2^{\ell_{\mathsf{oh}}^k}}. \end{aligned} \tag{3}$$

Since we require $\mathbf{Adv}_{\mathsf{A}}(k) \leq 2^{-\varepsilon}$, it holds that $2^{-\varepsilon} \geq \frac{1}{2}\cdot\frac{2^t}{2^{\ell_{\mathsf{oh}}^k}}$ for $t, \varepsilon \geq 1$. Thus we have the lower bound:

$$\ell_{\mathsf{oh}}^k(t,\varepsilon) \geq t + \varepsilon - 1\ . \tag{4}$$

If $c \leftarrow \mathcal{E}_{pk}(M; r)$ is bijective with respect to c and r, the adversary can search r one by one without duplication and the advantage for this case is $\mathbf{Adv}_{\mathsf{A},pk} = \frac{2^t}{2^\eta}$, which results in $\ell_{\mathsf{oh}}^k(t,\varepsilon) \geq t + \varepsilon$.

In the above discussion we used the simplified argument to count one encryption as one single time unit. More generally, one should count each fundamental cryptographic operation (such as hashing, group operation, etc.) as one step. Hence the value 2^t is understood as the total number of times the adversary performs the fundamental cryptographic operations. A precise assessment is possible by incorporating an adequate scaling factor that represent the exact number of steps (depending on the computational model).

2.2 Example: Ciphertext Overhead of OAEP

OAEP includes randomness of size k_r and zero-padding of size k_v. These parameters define the ciphertext overhead as $\ell_{\mathsf{oh}} = k_r + k_v$. Together with the size of permutation, n, they are provided as a security parameter $k = (n, k_r, k_v)$. According to [15, Th. 1], the advantage of an adversary A against the IND-CCA security of OAEP, making up to q decryption and hash queries is upper bounded by

$$\mathbf{Adv}_{\mathsf{A}}^{\mathrm{cca}}(k) \leq \epsilon_{\mathrm{spd}}(n) + \frac{c\,q^2}{2^{k_r}} + \frac{c'q}{2^{k_r}}\ , \tag{5}$$

where $\epsilon_{\mathrm{spd}}(n)$ is the probability of breaking set partial one-wayness of the underlying trapdoor permutation of size n, and $c, c' \geq 1$ are two (small) constants.

Consider an $(2^t, 2^{-\varepsilon})$ adversary that can make at most $q \leq 2^t$ oracle queries. Since parameter n can be chosen essentially independently from k_r and k_v, we can safely assume that $\epsilon_{\mathrm{spd}}(n)$ is small enough. Assuming $\epsilon_{\mathrm{spd}}(n) \leq c''\,2^{-\varepsilon}$ with a constant $0 < c'' \leq \frac{1}{2}$ for concreteness, each of the remaining two terms in (5) must be smaller than $2^{-\varepsilon} - \epsilon_{\mathrm{spd}}(n) \geq (1 - c'')\,2^{-\varepsilon}$. Namely,

$$\frac{c\,2^{2t}}{2^{k_r}} \leq (1-c'')\,2^{-\varepsilon} \quad \text{and} \quad \frac{c'2^t}{2^{k_r}} \leq (1-c'')\,2^{-\varepsilon} \tag{6}$$

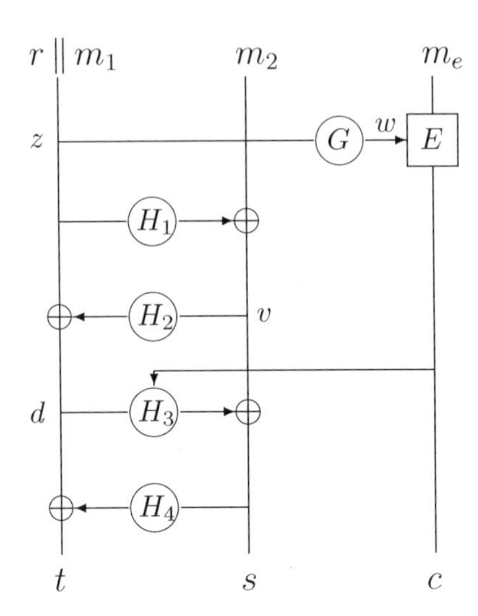

Fig. 1. The diagram of (a part of) encryption. Input message is $m = m_1 \,\|\, m_2 \,\|\, m_e \in \{0,1\}^{k_{m1}} \times \{0,1\}^{k_{m2}} \times \{0,1\}^*$ and the randomness is $r \in \{0,1\}^{k_r}$. The actual ciphertext is (u, c) where $u = f(t \,\|\, s)$.

must hold. Accordingly, in order to attain the desired security level, it is *sufficient* to choose

$$k_r = 2t + \varepsilon \quad \text{and} \quad k_v = t + \varepsilon \tag{7}$$

plus some small positive constants. As a result, the ciphertext overhead of OAEP is upper bounded by

$$k_r + k_v = 3t + 2\varepsilon + O(1). \tag{8}$$

3 Proposed Scheme

3.1 Description

Our construction requires a symmetric-key encryption scheme $\mathsf{SE}_{k_e} = (\mathsf{E}, \mathsf{D})$ and a trapdoor permutation family $\mathcal{P}_n$ as building blocks. The symmetric encryption scheme SE must be length-preserving and passively secure (indistinguishable against passive attacks), and the trapdoor permutation family must be one-way. For formal definitions, we refer to the full version [3].

Let (n, k_e, k_r) be a set of security parameters where n represents the bit-length of the trapdoor permutation, k_e is the key size of the symmetric-key encryption, and k_r is the size of randomness incorporated into the ciphertext. The proposed scheme $\mathsf{PKE} = (\mathcal{G}, \mathcal{E}, \mathcal{D})$ is the following. See also Figure 1 for a diagram of encryption.

Key Generation $\mathcal{G}$: Given a security parameter $k = (n, k_e, k_r)$ for $n \geq 6k_r$, set parameters k_{m_1} and k_{m_2} so that

$$k_{m_1} \geq 2k_r, \quad k_{m_2} \geq 3k_r, \quad n = k_r + k_{m_1} + k_{m_2} \tag{9}$$

are fulfilled. Then select $(f, f^{-1}) \leftarrow \mathcal{P}_n$ (the trapdoor permutation generator) and hash functions G and H_i for $i = 1, 2, 3, 4$ such that

$$\begin{array}{ll} G : \{0,1\}^{k_r+k_{m1}} \rightarrow \{0,1\}^{k_e}, & H_1 : \{0,1\}^{k_r+k_{m1}} \rightarrow \{0,1\}^{k_{m2}}, \\ H_2 : \{0,1\}^{k_{m2}} \rightarrow \{0,1\}^{k_r+k_{m1}}, & H_3 : \{0,1\}^{*} \rightarrow \{0,1\}^{k_{m2}}, \\ H_4 : \{0,1\}^{k_{m2}} \rightarrow \{0,1\}^{k_r+k_{m1}}. & \end{array}$$

The private-key is f^{-1}. The public-key includes f, SE_{k_e}, and the hash functions with associated parameters.

Encryption $\mathcal{E}$: Given a plaintext $m \in \{0,1\}^*$, first chop it into three blocks, m_1, m_2, and m_e such that

$$m = m_1 \,\|\, m_2 \,\|\, m_e \in \{0,1\}^{k_{m1}} \times \{0,1\}^{k_{m2}} \times \{0,1\}^{*}.$$

Then choose random $r \leftarrow \{0,1\}^{k_r}$ and compute

$$\begin{array}{llll} z = r \,\|\, m_1, & w = G(z), & c = \mathsf{E}_w(m_e), & \\ h_1 = H_1(z), & v = h_1 \oplus m_2, & h_2 = H_2(v), & d = h_2 \oplus z, \\ h_3 = H_3(d \,\|\, c), & s = h_3 \oplus v, & h_4 = H_4(s), & t = h_4 \oplus d, \end{array}$$

and $u = f(t \,\|\, s)$. The ciphertext is $(u, c) \in \{0,1\}^n \times \{0,1\}^*$.

Decryption $\mathcal{D}$: Given a ciphertext $(u, c) \in \{0,1\}^n \times \{0,1\}^{k_e}$, compute $y = f^{-1}(u)$ and parse y as $y = t \,\|\, s \in \{0,1\}^{k_r+k_{m1}} \times \{0,1\}^{k_{m2}}$. Then compute the following values:

$$\begin{array}{llll} h_4 = H_4(s), & d = h_4 \oplus t, & h_3 = H_3(d \,\|\, c), & v = h_3 \oplus s, \\ h_2 = H_2(v), & z = h_2 \oplus d, & h_1 = H_1(z), & m_2 = h_1 \oplus v, \\ w = G(z), & m_e = \mathsf{D}_w(c), & & \end{array}$$

and parse $z = r \,\|\, m_1 \in \{0,1\}^{k_r} \times \{0,1\}^{k_{m1}}$. The output is $m_1 \,\|\, m_2 \,\|\, m_e$.

3.2 Security and Optimality

The following theorems hold for PKE described in the previous section. A proof sketch is in Section 4 and the complete proof is in [3].

Theorem 1 (Chosen Ciphertext Security). *Suppose* A *is an adversary that runs in time τ with at most q_h hash queries and q_d decryption queries. Then there exist an adversaries* B *that runs in time at most $\tau + O(q_h^2)$ and an adversary* C *that runs in time at most $\tau + O(1)$ with*

$$\mathbf{Adv}_{\mathsf{A}}^{\mathrm{cca}}(k) \leq \mathbf{Adv}_{\mathsf{C},\mathsf{SE}}^{\text{ind-pa}}(k_e) + 2\mathbf{Adv}_{\mathsf{B},\mathcal{P}}^{\mathrm{owp}}(n) + O(\frac{q_h + q_d}{2^{k_r}}) .$$

Note that the number of hash queries includes the ones made through the decryption queries. In an asymptotic sense, Theorem 1 states that the above scheme is semantically secure against adaptive chosen message attacks in the random oracle model if the trapdoor permutation $\mathcal{P}$ is one-way and SE is passively secure.

As it is the case for most OAEP variants, our security reduction includes a quadratic factor q_h^2 in the running time of the adversary against the one-way permutation. It results in demanding larger n which increases the minimal length of the message the scheme can encrypt attaining the optimal overhead. The approach from [19, 14] helps achieving a linear running time if desired.

Theorem 2 (Optimality in Ciphertext Overhead). *If* $\mathbf{Adv}^{\text{ind-pa}}_{\mathsf{C},\mathsf{SE}}(k_e) + 2\mathbf{Adv}^{\text{owp}}_{\mathsf{B},\mathcal{P}}(n) \leq 2^{-(\varepsilon+1)}$ *holds for all adversaries* C *and* B *running in time* 2^t, *then* $k_r = \ell_{oh} = t + \varepsilon + 4$ *is sufficient for messages of size equal or larger than* $n - k_r$ *bits.*

Note that parameters k_e and n are independent of the overhead and can be set arbitrary to fulfill the condition.

3.3 Notes on Variations

Why Not 3 Rounds? Consider the 3-round version of our scheme obtained by removing H_4 and simply letting $t = d$. We show that the 3-round version is not simulatable, at least with the technique that constructs a plaintext extractor from the queries to the random oracles. Since the following argument holds regardless of the presence of the extended part c, let us ignore it.

Suppose that the adversary creates two ciphertexts u and u' by randomly choosing t, s, t' and computing $s' = H_3(t) \oplus s \oplus H_3(t')$, $u = f(t \,\|\, s)$, and $u' = f(t' \,\|\, s')$. Since $H_3(t) \oplus s = H_3(t') \oplus s'$, decrypting u and u' yield the same v. However, such a relation between u and u' can not be detected by the simulator since $H_2(v)$ is not asked. Accordingly the decryption oracle must return random $m_1 \,\|\, m_2$ and ${m_1}' \,\|\, {m_2}'$ to answer to the queries on u and u', respectively. Then the adversary asks $H_2(v)$ and obtains h_2. For consistency, it must hold that $h_2 = (r \,\|\, m_1) \oplus t = (r' \,\|\, {m_1}') \oplus t'$. However, since m_1 and ${m_1}'$ are randomly chosen before the simulator sees t and t', such a relation can be fulfilled only by chance. The adversary can notice the inconsistency by checking the relation and the simulation should fail.

Including c Into a Hash Other than H_3. We discuss on the variants that includes c into one of the hash functions rather than H_3. In summary, only the inner two hash functions, H_2 and H_3, are the right choice.

- Case of $H_1(z \,\|\, c)$. This is clearly a wrong choice since (u^*, c^*) and (u^*, c) yield the same m_1.
- Case of $H_2(v \,\|\, c)$. It is possible to modify the proof of Theorem 1 to show that this variant is also secure.
- Case of $H_4(s \,\|\, c)$. For this case, we can show that a (powerful) adversary can distinguish the simulation from the reality. The underlying idea is that, given a challenge ciphertext (u^*, c^*), the adversary builds a ciphertext (u, c) that

yields the same plaintext without making queries to H_3. Suppose that the adversary finds (t^*, s^*). It obtains $h_4^* = H_4(s^* \| c^*)$ and $d^* = h_4^* \oplus t^*$. It then selects arbitrary c and asks $h_4 = H_4(s^* \| c)$. Note that c must be different from c^*. It further computes $t = d^* \oplus h_4$ and $u = f(t \| s^*)$. Observe that (u, c) recovers d^* and v^* since $d = t \oplus H_4(s^* \| c) = d^* \oplus h_4 \oplus H_4(s^* \| c) = d^* \oplus h_4 \oplus h_4 = d^*$ and $v = s^* \oplus H_3(d) = s^* \oplus H_3(d^*) = v^*$. Therefore, the selected challenge message is returned if (u, c) is asked to the real decryption oracle. However, since $H_3(d^*)$ has only been defined implicitly and was never directly asked by the adversary, the simulated decryption oracle cannot detect such a case and returns a random message which is noticed by the adversary.

4 Proofs

4.1 Proof of Theorem 1 (Sketch)

We proceed in games. Let X_i denote the event that adversary A outputs $\tilde{b} = b$ in Game i.

Game 0. The original CCA game. By definition, we have

$$\Pr[X_0] = \frac{1}{2} \cdot \mathbf{Adv}_{\mathsf{A}}^{\text{cca}}(k) + \frac{1}{2}. \tag{10}$$

Game 1. Modify the challenge oracle so that it returns random u^* that is independent from the challenge messages as follows.

Challenge Oracle (M_0, M_1).
C.1 Choose $u^* \leftarrow \{0,1\}^n$.
C.2 Choose $b \leftarrow \{0,1\}$ and split M_b into $m_1{}^*$, $m_2{}^*$ and $m_e{}^*$, accordingly. Then choose $w^* \leftarrow \{0,1\}^{k_e}$ and compute $c^* = \mathsf{E}_{w^*}(m_e{}^*)$.
C.3 Return (u^*, c^*).

For u^*, c^* and w^*, let $(t^*, s^*, d^*, v^*, z^*, h_4^*, h_3^*, h_2^*, h_1^*)$ be a consistent internal state. Let AskH_3^+ denote an event such that $(d^* \| c^*)$ is asked to H_3 *after* s^* is asked to H_4. The following bound can be shown.

$$|\Pr[X_0] - \Pr[X_1]| \leq \frac{q_g}{2^{k_r}} + \frac{q_{h_1}}{2^{k_r}} + \frac{q_{h_2}}{2^{k_{m_2}}} + \frac{q_{h_3}}{2^{k_r + k_{m_1}}} + \Pr[\mathsf{AskH}_3^+] \tag{11}$$

It is straightforward to see that distinguishing b breaks the passive security of the symmetric encryption since only the symmetric part is related to b in Game 1. We thus have

$$\Pr[X_1] \leq \frac{1}{2} + \frac{1}{2} \cdot \mathbf{Adv}_{\mathsf{C,SE}}^{\text{ind-pa}}(k_e) , \tag{12}$$

for some suitable adversary C that has similar running time as A.

To bound $\Pr[\mathsf{AskH}_3^+]$, we initiate a new series of sub-games starting from Game 1. In the following games, each random oracle X is simulated with an independent list L_X that is initially empty. When X is first asked on fresh input a, output b is uniformly selected and (a, b) is stored in L_X. If a has been asked before, the corresponding b is read from L_X and returned. By $(a, [b]) \in L_X$, we mean that table L_X includes an entry whose first element is a. If such entry exists, the second element is denoted by b. List L_X is *consistent* for oracle X if every input a is unique in L_X. By $F_{1.i}$ we denote the same event in the following sub-games Game 1.i.

Game 1.0. This game is the same as Game 1. Since this is just a change of notation, we have

$$\Pr[\mathsf{AskH}_3^+] = \Pr[F_{1.0.}] . \tag{13}$$

Game 1.1. The game is modified so that it *immediately* stops at the moment AskH_3^+ happens. To capture event AskH_3^+, hash oracle H_3 is modified so that it checks whether the query $d \,\|\, c$ equals the value $d^* \,\|\, c^*$ by searching L_{H4} for corresponding s^*.

Hash Oracle $H_3(d \,\|\, c)$.
A.1 If $(d \,\|\, c, [h_3]) \in L_{H3}$, return h_3.
A.2 Choose $h_3 \leftarrow \{0,1\}^{k_{m2}}$ and add $(d \,\|\, c, h_3)$ to L_{H3}.
A.3 Repeat the following for every entry (h_4, s) in L_{H4}.
 (a) Compute $t = d \oplus h_4$, $u = f(t \,\|\, s)$.
 (b) If $u = u^*$, abort the game. (event: $F_{1.1.}$).
A.4 Return h_3.

Since this modification does not change the view of the adversary unless AskH_3^+ happens, we have

$$\Pr[F_{1.0.}] = \Pr[F_{1.1.}] . \tag{14}$$

Game 1.2. Modify the decryption oracle so that it returns a *random message* when a decryption query is made on a ciphertext whose associated $d \,\|\, c$ was not yet asked to H_3. Modify H_3 for consistency, too.

Decryption Oracle $\mathcal{D}(u, c)$.
D.1 Compute $t \,\|\, s = f^{-1}(u)$.
D.2 $h_4 \leftarrow H_4(s)$.
D.3 Let $d = t \oplus h_4$. If $(d \,\|\, c, [h_3]) \notin L_{H3}$, go to the next step. Otherwise, return $m_1 \,\|\, m_2 \,\|\, m_e$ computed normally by using t, s, d, and h_3.
D.4 Return $m_1 \,\|\, m_2 \,\|\, m_e$ computed as follows.
 (a) Select m_1, m_2, and w uniformly and compute $m_e = \mathsf{D}_w(c)$.
 (b) Add (u, c, w, m_1, m_2) to L_{watch}.

Hash Oracle $H_3(d \,\|\, c)$.
A.1 If $(d \,\|\, c, [h_3]) \in L_{H3}$, return h_3.
A.2 Choose $h_3 \leftarrow \{0,1\}^{k_{m2}}$ and put $(d \,\|\, c, h_3)$ to L_{H3}.
A.3 Repeat the following for every entry (h_4, s) in L_{H4}.
   (a) Compute $t = d \oplus h_4$, $u = f(t \,\|\, s)$, $v = h_3 \oplus s$.
   (b) If $u = u^*$, abort the game. (event: $F_{1.2.}$).
   (c) If $(u, c, [w], [m_1], [m_2]) \in L_{\mathsf{watch}}$, do as follows.
      – Select $r \leftarrow \{0,1\}^{k_r}$ and compute $z = r \,\|\, m_1$, $h_2 = d \oplus z$, $h_1 = m_2 \oplus v$.
      – Add (z, w), (z, h_1), and (v, h_2) to L_G, L_{H1}, and L_{H2}, respectively.
      – Remove entry (u, c, w, m_1, m_2) from L_{watch}.
A.4 Return h_3.

The following bound can be shown.

$$|\Pr[F_{1.1.}] - \Pr[F_{1.2.}]| \leq \frac{q_d^2}{2^{k_{m_1}}} + \frac{q_{h_1} + q_g}{2^{k_r}} + \frac{q_{h_2}\, q_d}{2^{k_{m_2}}}. \tag{15}$$

Game 1.3. Modify the decryption oracle so that it also returns a *random message* when a decryption query is made on a ciphertext whose associated s was not yet asked to H_4.

Decryption Oracle $\mathcal{D}(u, c)$.
D.1 Compute $t \,\|\, s = f^{-1}(u)$.
D.2 If $(s, [h_4]) \in L_{H4}$ and $(d \,\|\, c, [h_3]) \in L_{H3}$ for $d = t \oplus h_4$, then return $m_1 \,\|\, m_2 \,\|\, m_e$ computed normally by using t, s, d, and h_3.
D.3 Otherwise, return $m_1 \,\|\, m_2 \,\|\, m_e$ computed as follows.
   (a) Select m_1, m_2, and w uniformly and compute $m_e = \mathsf{D}_w(c)$.
   (b) Add (u, c, w, m_1, m_2) to L_{watch}.

The following bound can be shown.

$$|\Pr[F_{1.2.}] - \Pr[F_{1.3.}]| \leq \frac{q_d\, q_{h_3}}{2^{k_r + k_{m_1}}}. \tag{16}$$

Game 1.4. Modify the decryption oracle so that it uses a lookup table instead of computing $t \,\|\, s = f^{-1}(u)$.

Decryption Oracle $\mathcal{D}(u, c)$.
D.1 If $(u, c, [t], [s]) \in L_X$, then continue the normal decryption procedure by using t and s and return the obtained message.
D.2 Otherwise, return random $m_1 \,\|\, m_2 \,\|\, m_e$ computed as follows.
   (a) Select m_1, m_2, and w uniformly and compute $m_e = \mathsf{D}_w(c)$.
   (b) Add (u, c, w, m_1, m_2) to L_{watch} and return $m_1 \,\|\, m_2 \,\|\, m_e$.

Hash Oracle $H_3(d \,\|\, c)$.
A.1 If $(d \,\|\, c, [h_3]) \in L_{H3}$, return h_3.
A.2 Choose $h_3 \leftarrow \{0,1\}^{k_{m_2}}$ and put $(d \,\|\, c, h_3)$ to L_{H3}.
A.3 Repeat the following for every entry (h_4, s) in L_{H4}.
(a) Compute $t = d \oplus h_4$, $u = f(t \,\|\, s)$, $v = h_3 \oplus s$.
(b) If $u = u^*$, abort the game with status 1 (event: $F_{1.4.}$).
(c) If $(u, c, [w], [m_1], [m_2]) \in L_{\mathsf{watch}}$, do as follows
- Select $r \leftarrow \{0,1\}^{k_r}$ and compute $z = r \,\|\, m_1$, $h_2 = d \oplus z$, $h_1 = m_2 \oplus v$.
- Add (z, w), (z, h_1), and (v, h_2) to L_G, L_{H1}, and L_{H2}, respectively.
- Remove entry (u, c, w, m_1, m_2) from L_{watch}.
(d) Put (u, c, t, s) to L_X.
A.4 Return h_3.

Hash Oracle $H_4(s)$.
B.1 If $(s, [h_4]) \in L_{H4}$, return h_4.
B.2 Choose $h_4 \leftarrow \{0,1\}^{k_r + k_{m_1}}$ and put (s, h_4) to L_{H4}.
B.3 Repeat the following for every entry $([d], [c], [h_3])$ in L_{H3}.
(a) Let $t = d \oplus h_4$, $v = s \oplus h_3$, and $u = f(t \,\|\, s)$.
(b) Put (u, c, t, s) to L_X.
B.4 Return h_4.

Since the adversary's view is not influenced by this modification, we have

$$\Pr[F_{1.3.}] = \Pr[F_{1.4.}]. \tag{17}$$

Game 1.4. does not use f^{-1} and any *-marked internal values at all. Challenge u^* is a random element in $\{0,1\}^n$, and $s^* \,\|\, t^*$ such that $f(s^* \,\|\, t^*) = u^*$ can be extracted if $F_{1.4.}$ happens. It is thus straightforward to construct adversary B that computes f^{-1} using adversary A that causes $F_{1.4.}$. We thus have

$$\Pr[F_{1.4.}] \leq \mathbf{Adv}^{\mathrm{owp}}_{\mathsf{B},f}(k)\ . \tag{18}$$

The running time of B is bounded by that of A plus $O(q_h^2)$.

From (11), (14), (16), (17), and (18), we have

$$\mathbf{Adv}^{\mathrm{cca}}_{\mathsf{A}}(k) \leq \mathbf{Adv}^{\mathrm{ind\text{-}pa}}_{\mathsf{C,SE}}(k_e) + 2 \cdot \mathbf{Adv}^{\mathrm{owp}}_{\mathsf{B},\mathcal{P}}(n) + \frac{4(q_{h_1} + q_g)}{2^{k_r}} + \frac{2q_d^2}{2^{k_{m_1}}} + \frac{2q_{h_2}(q_d + 1)}{2^{k_{m_2}}} + \frac{2q_{h_3}(q_d + 1)}{2^{k_r + k_{m_1}}}\ .$$

Finally, using $k_{m_1} \geq 2k_r$, $k_{m_2} \geq 3k_r$ and setting $q_h = q_{h_1} + q_{h_2} + q_{h_3} + q_{h_4} + q_g$, this simplifies to the claimed form in the theorem as follows.

$$\begin{aligned}\mathbf{Adv}^{\mathrm{cca}}_{\mathsf{A}}(k) &\leq \mathbf{Adv}^{\mathrm{ind\text{-}pa}}_{\mathsf{C,SE}}(k_e) + 2 \cdot \mathbf{Adv}^{\mathrm{owp}}_{\mathsf{B},\mathcal{P}}(n) + \frac{4q_h}{2^{k_r}} + \frac{2q_d^2}{2^{2k_r}} + \frac{2q_h(q_d+1)}{2^{3k_r}} \\ &\leq \mathbf{Adv}^{\mathrm{ind\text{-}pa}}_{\mathsf{C,SE}}(k_e) + 2 \cdot \mathbf{Adv}^{\mathrm{owp}}_{\mathsf{B},\mathcal{P}}(n) + O(\frac{q_h + q_d}{2^{k_r}})\end{aligned} \tag{19}$$

4.2 Proof of Theorem 2

Fix ε and t. We require $\mathbf{Adv}_{\mathsf{A}}^{\text{cca}}(k) \leq 1/2^{\varepsilon}$ for adversaries A running time in 2^t. Using the explicit bound (19) from the proof of Theorem 1, it is sufficient to set k_r so that

$$\mathbf{Adv}_{\mathsf{C,SE}}^{\text{ind-pa}}(k_e) + 2 \cdot \mathbf{Adv}_{\mathsf{B},\mathcal{P}}^{\text{owp}}(n) + \frac{4q_h}{2^{k_r}} + \frac{2q_d^2}{2^{2k_r}} + \frac{2q_h(q_d+1)}{2^{3k_r}} = \frac{1}{2^{\varepsilon}} \tag{20}$$

is fulfilled. By assuming that k_e and n are set to satisfy

$$\mathbf{Adv}_{\mathsf{C,SE}}^{\text{ind-pa}}(k_e) + 2 \cdot \mathbf{Adv}_{\mathsf{B},\mathcal{P}}^{\text{owp}}(n) \leq 1/2^{\varepsilon+1},$$

it is sufficient to choose k_r such that

$$\frac{4q_h}{2^{k_r}} + \frac{2q_d^2}{2^{2k_r}} + \frac{2q_h(q_d+1)}{2^{3k_r}} \leq \frac{1}{2^{\varepsilon+1}}. \tag{21}$$

To achieve semantic security, $q_h/2^{k_r} \leq 1$ and $q_d/2^{k_r} \leq 1$ must hold. Since 2^t upper bounds the running time, $q_h \leq 2^t$ and $q_d \leq 2^t$ must hold, too. By using these bounds, the left side of (21) simplifies to

$$\frac{1}{2^{k_r}}(4q_h + 2q_d + q_h + 1) \leq \frac{8 \cdot 2^t}{2^{k_r}}. \tag{22}$$

Thus we have

$$\frac{8 \cdot 2^t}{2^{k_r}} \leq \frac{1}{2^{\varepsilon+1}},$$

which results in $t + \varepsilon + 4 \leq k_r$. Since $\ell_{\mathsf{oh}} = k_r$ holds for all messages of size equal or larger than $n - k_r$ bits, $\ell_{\mathsf{oh}} = k_r = t + \varepsilon + 4$ is sufficient. It matches the lower bound up to the constant term.

5 Conclusion and Open Problems

We propose a variant of OAEP that attains an optimal overhead in the random oracle model and thereby proved that the lower bound of ciphertext overhead is tight even with respect to IND-CCA security. Open problems include:

- Show the bound without random oracles. In the standard model, the schemes in [7, 18] have the shortest known ciphertext overhead consisting of two group elements that results in $\ell_{\mathsf{oh}} \geq 4t + 2\varepsilon$ bits. It remains as a very interesting open question whether or not the optimality can be achieved without random oracles.
- Optimal ciphertext overhead for shorter messages. We refer to [2] whose (DH-based) schemes offer $\ell_{\mathsf{oh}} \geq 2t + \varepsilon$ for short messages.
- Show that 4-round is *necessary* (or not) in our construction.

References

[1] Abe, M., Gennaro, R., Kurosawa, K.: Tag-KEM/DEM: A new framework for hybrid encryption. Journal of Cryptology 21(1), 97–130 (2008)

[2] Abe, M., Kiltz, E., Okamoto, T.: Compact CCA-secure encryption for arbitrary messages (Unpublished Manuscript Available from the authors) (2007)

[3] Abe, M., Kiltz, E., Okamoto, T.: Chosen ciphertext security with optimal overhead. IACR ePrint Archive 2008/374, September 2 (2008)

[4] Bellare, M., Rogaway, P.: Optimal asymmetric encryption. In: De Santis, A. (ed.) EUROCRYPT 1994. LNCS, vol. 950, pp. 92–111. Springer, Heidelberg (1995)

[5] Bellare, M., Rogaway, P.: Code-based game-playing proofs and the security of triple encryption. In: Vaudenay, S. (ed.) EUROCRYPT 2006. LNCS, vol. 4004, pp. 409–426. Springer, Heidelberg (2006); Full version available from IACR ePrint Archive 2004/331

[6] Bjørstad, B., Dent, A., Smart, N.: Efficient KEMs with partial message recovery. In: Galbraith, S.D. (ed.) Cryptography and Coding 2007. LNCS, vol. 4887, pp. 233–256. Springer, Heidelberg (2007)

[7] Boyen, X., Mei, Q., Waters, B.: Direct chosen ciphertext security from identity-based techniques. In: ACM Conference on Computer and Communications Security, pp. 320–329. ACM, New York (2005); Also available at IACR e-print 2005/288

[8] Coron, J., Handschuh, H., Joye, M., Paillier, P., Pointcheval, D., Tymen, C.: GEM: A generic chosen-ciphertext secure encryption method. In: Preneel, B. (ed.) CT-RSA 2002. LNCS, vol. 2271, pp. 263–276. Springer, Heidelberg (2002)

[9] Coron, J., Handschuh, H., Joye, M., Paillier, P., Pointcheval, D., Tymen, C.: Optimal chosen-ciphertext secure encryption of arbitrary-length messages. In: Naccache, D., Paillier, P. (eds.) PKC 2002. LNCS, vol. 2274, pp. 17–33. Springer, Heidelberg (2002)

[10] Coron, J.S., Joye, M., Naccache, D., Paillier, P.: Universal padding schemes for RSA. In: Yung, M. (ed.) CRYPTO 2002. LNCS, vol. 2442, pp. 226–241. Springer, Heidelberg (2002)

[11] Coron, J.S., Patarin, J., Seurin, Y.: The random oracle model and the ideal cipher model are equivalent. In: Wagner, D. (ed.) CRYPTO 2008. LNCS, vol. 5157, pp. 1–20. Springer, Heidelberg (2008)

[12] Cramer, R., Shoup, V.: Design and analysis of practical public-key encryption schemes secure against adaptive chosen ciphertext attack. SIAM Journal on Computing 33(1), 167–226 (2003)

[13] Cui, Y., Kobara, K., Imai, H.: A generic conversion with optimal redundancy. In: Menezes, A. (ed.) CT-RSA 2005. LNCS, vol. 3376, pp. 104–117. Springer, Heidelberg (2005)

[14] Dodis, Y., Freedman, M., Jarecki, S., Walfish, S.: Versatile padding schemes for joint signature and encryption. In: ACM CCS 2004. ACM Press, New York (2004)

[15] Fujisaki, E., Okamoto, T., Pointcheval, D., Stern, J.: RSA-OAEP is secure under the RSA assumption. In: Kilian, J. (ed.) CRYPTO 2001. LNCS, vol. 2139, pp. 260–274. Springer, Heidelberg (2001)

[16] Goldwasser, S., Micali, S.: Probabilistic encryption. Journal of Computer and System Sciences 28, 270–299 (1984)

[17] Jonsson, J.: An OAEP variant with a tight security proof. IACR e-print Archive 2002/034 (2002)

[18] Kiltz, E.: Chosen-ciphertext security from tag-based encryption. In: Halevi, S., Rabin, T. (eds.) TCC 2006. LNCS, vol. 3876, pp. 581–600. Springer, Heidelberg (2006)
[19] Kobara, K., Imai, H.: OAEP++: A very simple way to apply OAEP to deterministic OW-CPA primitives. IACR ePrint archive, 2002/130 (2002)
[20] Komano, Y., Ohta, K.: Efficient universal padding schemes for multiplicative trapdoor one-way permutation. In: Boneh, D. (ed.) CRYPTO 2003. LNCS, vol. 2729, pp. 366–382. Springer, Heidelberg (2003)
[21] Okamoto, T., Pointcheval, D.: REACT: Rapid enhanced-security asymmetric cryptosystem transform. In: Naccache, D. (ed.) CT-RSA 2001. LNCS, vol. 2020, pp. 159–174. Springer, Heidelberg (2001)
[22] Phan, D.H., Pointcheval, D.: Chosen-ciphertext security without redundancy. In: Laih, C.-S. (ed.) ASIACRYPT 2003. LNCS, vol. 2894, pp. 1–18. Springer, Heidelberg (2003)
[23] Phan, D.H., Pointcheval, D.: OAEP 3-round: A generic and secure asymmetric encryption padding. In: Lee, P.J. (ed.) ASIACRYPT 2004. LNCS, vol. 3329, pp. 63–78. Springer, Heidelberg (2004)
[24] Ramzan, Z., Reyzin, L.: On the round security of symmetric-key cryptographic primitives. In: Bellare, M. (ed.) CRYPTO 2000. LNCS, vol. 1880, pp. 376–393. Springer, Heidelberg (2000)
[25] Shoup, V.: OAEP reconsidered. In: Kilian, J. (ed.) CRYPTO 2001. LNCS, vol. 2139, pp. 239–259. Springer, Heidelberg (2001)

Concurrently Secure Identification Schemes Based on the Worst-Case Hardness of Lattice Problems

Akinori Kawachi, Keisuke Tanaka, and Keita Xagawa

Department of Mathematical and Computing Sciences, Tokyo Institute of Technology, Japan
{kawachi,keisuke,xagawa5}@is.titech.ac.jp

Abstract. In this paper, we show that two variants of Stern's identification scheme [IEEE Transaction on Information Theory '96] are provably secure against concurrent attack under the assumptions on the *worst-case* hardness of lattice problems. These assumptions are weaker than those for the previous lattice-based identification schemes of Micciancio and Vadhan [CRYPTO '03] and of Lyubashevsky [PKC '08]. We also construct efficient ad hoc anonymous identification schemes based on the lattice problems by modifying the variants.

Keywords: Lattice-based cryptography, identification schemes, concurrent security, ad hoc anonymous identification schemes.

1 Introduction

Many researchers have so far developed cryptographic schemes based on combinatorial problems related to knapsacks, codes, and lattices, due to the intractability of the underlying problems, the efficiency of primitive operations, and the threat of quantum computers to number-theoretic schemes.

The cryptographic schemes based on combinatorial problems usually assume the *average-case* hardness of the underlying problem because they have to deal with randomly generated cryptographic instances such as keys, plaintexts, and ciphertexts. This implies security risk in such schemes since it is generally hard to show their average-case hardness. In fact, several attacks against such schemes, e.g., [25], were found in practical settings. The cryptographic schemes based only on the average-case hardness are more likely to be at risk of these kinds of attacks.

It is therefore significant to guarantee the security under the worst-case hardness. Ajtai [1] showed that the average-case hardness of some lattice problem is equivalent to its worst-case hardness. His seminal result opened the way to cryptographic schemes based on the worst-case hardness of lattice problems. Several lattice-based schemes were proposed such as public-key encryption schemes, e.g., by Ajtai and Dwork [2], and hash functions [1,11,19].

Among varieties of lattice-based cryptographic schemes, there are very few results on the identification (ID) schemes based on the worst-case hardness of lattice problems. For example, Micciancio and Vadhan proposed ID schemes based on the worst-case hardness of lattice problems, such as the gap versions of the Shortest Vector Problem. These schemes are obtained from their statistical zero-knowledge protocol with efficient

J. Pieprzyk (Ed.): ASIACRYPT 2008, LNCS 5350, pp. 372–389, 2008.
© International Association for Cryptologic Research 2008

provers [20]. Recently, Lyubashevsky also constructed lattice-based ID schemes secure against active attack [14]. Unfortunately, the approximation factors of the underlying problems in their schemes are large for practical use as noted in [14, Sec. 5] since security parameters for ID schemes should be large in order to achieve the required hardness. Therefore, it is necessary to construct the schemes based on weaker assumptions, i.e., the assumptions on lattice problems with smaller approximation factors.

1.1 Our Contributions

In this paper, we propose two variants, which we call $\mathrm{S}^{+}_{\mathrm{GL}}$ and $\mathrm{S}^{+}_{\mathrm{C/IL}}$, of Stern's ID scheme [26]. These variants are secure against *concurrent* attack[1] under the assumptions on the *worst-case* hardness of lattice problems, while Stern's original scheme assumes the *average-case* hardness of certain decoding problem in coding theory and the existence of a collision-resistant hash function, and its security is only against *passive* attack. The underlying problems of $\mathrm{S}^{+}_{\mathrm{GL}}$ and $\mathrm{S}^{+}_{\mathrm{C/IL}}$ are the gap version of the Shortest Vector Problem with approximation factor $\tilde{O}(n)$ ($\mathrm{GapSVP}^{2}_{\tilde{O}(n)}$) and the Shortest Vector Problem for ideal lattices with approximation factor $\tilde{O}(n)$ ($\Lambda(f)\text{-}\mathrm{SVP}^{\infty}_{\tilde{O}(n)}$), respectively, where $\tilde{O}(g(n)) = O(g(n) \operatorname{poly} \log g(n))$ for a function g in n, The assumptions are weaker than those for the previous lattice-based ID schemes [20,14]. We stress that such weaker assumptions will take a step for practical use of lattice-based ID schemes.

Moreover, we show that our variants yield efficient ad hoc anonymous identification schemes (AID schemes). In an AID scheme, which introduced by Dodis, Kiayias, Nicolosi, and Shoup [7], the protocol is done by two parties, a prover and verifier, but we implicitly suppose an ad hoc group. Given public keys of all members of the group to the verifier (and the prover), the goal is to convince the verifier that the prover belongs to the group, without being specified who the prover is of the group, if and only if the prover is an actual member of the group. We formally define a concurrent version of the security notion, the security against impersonation under concurrent chosen-group attack, and prove that our AID schemes satisfy this security notion. Our schemes are based on the worst-case hardness of $\mathrm{GapSVP}^{2}_{\tilde{O}(n)}$ and $\Lambda(f)\text{-}\mathrm{SVP}^{\infty}_{\tilde{O}(n)}$. To authors' best knowledge, this is the first non-trivial construction under the assumption of the worst-case hardness of lattice problems.

1.2 Main Ideas

In this section, we only discuss the ID scheme $\mathrm{S}^{+}_{\mathrm{GL}}$ based on GapSVP. We first construct a string commitment scheme based on the lattice problem which will be used in ID schemes. Then we will describe the idea of the proof on concurrent security of the variant. Finally, we give a sketch of our construction method of an AID scheme.

Before giving the overview, we review the underlying problem GapSVP_{γ} and the fundamental problem, the Small Integer Solution Problem ($\mathrm{SIS}_{q,m,\beta}$), on which our

[1] In *active attack*, an adversary could interact with the prover prior to impersonation. In *concurrent attack*, an adversary could interact with many different prover "clones" concurrently prior to impersonation. Each clone has the same secret key, but has independent random coins and maintains its own state. After interacting with many clones, the adversary tries impersonation.

variants are directly based. The informal definitions and the relationship of two problems are given as follows:

- $\mathrm{SIS}_{q,m,\beta}$: Given a random n-by-m matrix $\mathbf{A}$ whose elements are in $\mathbb{Z}_q$, the problem is finding an m-dimensional integral non-zero vector z such that $Az \equiv \mathbf{0} \pmod q$ and $\|z\|_2 \leq \beta$.
- GapSVP^2_γ: Given an n-dimensional lattice L and a rational number d, the problem is outputting YES if there exists a non-zero vector $\boldsymbol{v} \in L$ such that $\|\boldsymbol{v}\|_2 \leq d$, or NO if for any non-zero vector $\boldsymbol{v} \in L$ $\|\boldsymbol{v}\|_2 > \gamma d$.
- ([19]) For suitable q and m, if there exists a probabilistic polynomial-time algorithm which solves $\mathrm{SIS}_{q,m,\beta}$ on the average then there exists a probabilistic polynomial-time algorithm which solves $\mathrm{GapSVP}^2_{\tilde{O}(\beta n^{1/2})}$ in the worst case.

As in Lyubashevsky's result [14], we use the above relationship for our security reduction. Hence we mainly deals with SIS instead of GapSVP.

We simply obtain the lattice-based hash functions as in [11]: Choose a random matrix $\mathbf{A} \in \mathbb{Z}_q^{n\times m}$. For any $\boldsymbol{x} \in \{0,1\}^m$, a hash value is $f_{\mathbf{A}}(\boldsymbol{x}) := \mathbf{A}\boldsymbol{x} \bmod q$. A collision $(\boldsymbol{x}, \boldsymbol{x}')$ of the hash function $f_{\mathbf{A}}$ implies a solution $z = \boldsymbol{x} - \boldsymbol{x}'$ of $\mathrm{SIS}_{q,m,\sqrt{m}}$. Thus, the security of the hash functions is based on the worst-case hardness of $\mathrm{GapSVP}^2_{\tilde{O}(\sqrt{nm})}$.

String commitment schemes: We construct a string commitment scheme from lattice-based hash functions. General constructions of string commitment schemes from collision-resistant hash functions were shown by Damgård, Pedersen, and Pfizmann [4] and Halevi and Micali [12]. Stern also constructed a string commitment scheme from collision-resistant hash functions in [26, Sec. III-A]: Let h be a hash function. Given a string s and a random string ρ, a commitment is $h(\rho \circ (\rho \oplus s))$, where $\circ$ and $\oplus$ denote the concatenation and XOR operators, respectively. However, its hiding property was not shown. We construct a string commitment scheme by a more direct and simpler way than the general one and Stern's one: Given s and ρ, a commitment is $h(\rho \circ s)$, where h is a lattice-based hash function. The binding property simply follows from the collision-resistance property of h. We derive its hiding property from ϵ-regularity of h for some negligible function ϵ (see, e.g., [16, Sec. 4.1]). As mentioned in the above, we have collision-resistant lattice-based hash functions based on the worst-case hardness of GapSVP, while Stern assumed the existence of collision-resistant hash functions.

Our ID scheme and its concurrent security: In Stern's scheme and our variant, a prover has a binary vector $\boldsymbol{x}$ with fixed Hamming weight as his/her secret key. We also feed to the prover and the verifier a matrix $\mathbf{A}$ as a system parameter and a vector $\boldsymbol{y}$ as the public key corresponding to $\boldsymbol{x}$. The task of the prover is to convince the verifier that he/she knows a correct secret key $\boldsymbol{x}$ satisfying a relation $\mathbf{A}\boldsymbol{x} = \boldsymbol{y}$ and $\boldsymbol{x}$ has a valid weight.

In Stern's protocol [26], the prover computes three commitments and sends them to the verifier. The verifier sends a random challenge to the prover. The prover reveals two of three commitments corresponding to the challenge. He constructed the knowledge extractor which computes a collision of a hash function in a string commitment scheme or a secret key corresponding to the target public key if a passive adversary responds correctly to any challenges after sending commitments.

One of standard strategies to achieve concurrent security is to prove that a public key corresponds to multiple secret keys and that the protocol is witness indistinguishable

(WI) [8] and proof-of-knowledge: The reduction algorithm generates *sk* and *pk* and runs the adversary on *pk* by simulating the prover with *sk*. Using the knowledge extractor of the protocol, the algorithm obtains another *sk'* corresponding to *pk* with probability at least 1/2 since the protocol is WI. The algorithm then solves the underlying problem by using *pk*, *sk*, and *sk'*.

In our reduction, when the algorithm is given $\mathbf{A}$, it generates a secret key $\boldsymbol{x}$ and a public key $\boldsymbol{y} = \mathbf{A}\boldsymbol{x}$, and feeds $\mathbf{A}$ and $\boldsymbol{y}$ to the adversary. Note that the algorithm can simulate the prover with $\mathbf{A}$ and $\boldsymbol{x}$ that the adversary concurrently accesses. Using the knowledge extractor for the adversary in Stern's proof, the algorithm obtains a collision of a string commitment scheme or a secret key $\boldsymbol{x}'$ such that $\boldsymbol{x}' \neq \boldsymbol{x}$ and $\mathbf{A}\boldsymbol{x}' = \boldsymbol{y}$, differently from the general strategy. In the former case, the algorithm outputs the collision $(\boldsymbol{s}, \boldsymbol{s}')$ of a hash function $h_{\mathbf{A}}$ in the string commitment scheme. Thus, the solution for SIS is obtained by $\boldsymbol{z} = \boldsymbol{s} - \boldsymbol{s}'$. In the latter case, the condition $\boldsymbol{x} \neq \boldsymbol{x}'$ will be satisfied with probability at least 1/2 by witness indistinguishability of Stern's protocol. Thus, the algorithm has the solution $\boldsymbol{z} = \boldsymbol{x} - \boldsymbol{x}'$ for SIS. The ℓ_2 norm of both solutions is at most $\sqrt{m} = \tilde{O}(n^{1/2})$. From the relationship between SIS and GapSVP the assumption is the worst-case hardness of $\mathrm{GapSVP}^2_{\tilde{O}(n)}$.

AID schemes: Our construction for AID schemes also has the following structure: Each of l members in the ad hoc group has a vector $\boldsymbol{x}_i$ $(i = 1, \ldots, l)$. Then, the common inputs of the scheme are a system parameter $\mathbf{A}$ and a set of public keys $\boldsymbol{y}_1, \ldots, \boldsymbol{y}_l$ of the members, which satisfy $\boldsymbol{y}_i = \mathbf{A}\boldsymbol{x}_i$ $(i = 1, \ldots, l)$. We can show that, by Stern's protocol, the prover can anonymously convince the verifier that the prover knows $\boldsymbol{x}_i$ corresponding to one of $\boldsymbol{y}_1, \ldots, \boldsymbol{y}_l$, since he/she knows a new vector $\boldsymbol{x}'$ such that $[\mathbf{A}\, \boldsymbol{y}_1 \, \ldots \, \boldsymbol{y}_l]\boldsymbol{x}' = \mathbf{0}$. (This idea is due to Wu, Chen, Wang, and Wang [27], who presented an AID scheme from certain combinatorial problem.) Additionally, we force the prover to prove that the positions of +1 and −1 in $\boldsymbol{x}'$ are proper by modifying Stern's protocol. We succeed to give security proof for the scheme, while Wu et al. gave no formal proof on the security of their scheme.

1.3 Comparison with Other Lattice-Based Schemes

ID schemes: In [20], Micciancio and Vadhan proposed a statistical zero-knowledge and proof-of-knowledge protocol for GapSVP. Combining it with lattice-based hash functions, we obtain an ID scheme which is secure against *passive attack* based on $\mathrm{SIS}_{q,m,\tilde{O}(n)}$, which can be reduced from $\mathrm{GapSVP}^2_{\tilde{O}(n^{1.5})}$.

In the scheme, the prover and the verifier are given a matrix $\mathbf{A}$ as a common input, and the prover has a binary vector $\boldsymbol{x}$ as secret information. The task of the prover is to convince the verifier that he/she knows $\boldsymbol{x}$ satisfying the relations that $\mathbf{A}\boldsymbol{x} = \mathbf{0}$ and $\boldsymbol{x}$ is relatively short. It seems difficult to directly simulate the prover since a simulator has to prepare a dummy short vector $\boldsymbol{x}'$ satisfying $\mathbf{A}\boldsymbol{x}' = \mathbf{0}$, which is the task of SIS itself. Thus, we cannot straightforwardly prove the concurrent security for their ID scheme.

By a simple modification, we can construct a concurrently secure ID scheme ($\mathrm{MV}^{+}_{\mathrm{GL}}$ for short) based on the worst-case hardness of lattice problems by Micciancio and Vadhan's ID scheme as noted in [20, Sec. 5]. In particular, applying techniques of De Santis, Di Crescenzo, Persiano, and Yung [6] and of Feige and Shamir [8], a modification of

Table 1. Comparisons among ID schemes and AID schemes. A secret key sk is $\boldsymbol{x} \in \{0,1\}^m$. The factor n denotes the security parameter. We denote the Hamming weight of $\boldsymbol{x}$ by $w_H(\boldsymbol{x})$. Assume that the protocols are repeated t times in parallel for reducing errors. In the table for AID schemes, l denotes the number of the members in the group. Note that the parameters in ideal-lattice-based versions are almost same as those in general-lattice-based versions.

ID schemes ($\mathbf{A}_0, \mathbf{A}_1, \mathbf{A} \in \mathbb{Z}_q^{n\times m}$)						
	Param.	Public key	Relation	γ in GapSVP^2_γ	Comm. cost	Errors
$\mathrm{MV}^+_{\mathrm{GL}}$ [20]	–	$\mathbf{A}_0, \mathbf{A}_1$	$\mathbf{A}_0\boldsymbol{x} = \mathbf{0}$ or $\mathbf{A}_1\boldsymbol{x} = \mathbf{0}$	$\tilde{O}(n^{1.5})$	$t \cdot \tilde{O}(n)$	1-sided
L_{GL} [14]	(A)	$\mathbf{A}, \boldsymbol{y}$	$\mathbf{A}\boldsymbol{x} = \boldsymbol{y}$	$\tilde{O}(n^2)$	$t \cdot \tilde{O}(n)$	2-sided
$\mathrm{S}^+_{\mathrm{GL}}$	A	$\boldsymbol{y}$	$\mathbf{A}\boldsymbol{x} = \boldsymbol{y}$ and $w_H(\boldsymbol{x}) = m/2$	$\tilde{O}(n)$	$t \cdot \tilde{O}(n)$	1-sided

AID schemes ($\mathbf{A}_{i,0}, \mathbf{A}_{i,1}, \mathbf{A} \in \mathbb{Z}_q^{n\times m}$)						
Base	Param.	Set of pks	Relation	γ in GapSVP^2_γ	Comm. cost	Errors
$\mathrm{MV}^+_{\mathrm{GL}}$ [20]	–	$\{\mathbf{A}_{i,0}, \mathbf{A}_{i,1}\}_{i=1,\ldots,l}$	$\mathbf{A}_{i,0}\boldsymbol{x} = \mathbf{0}$ or $\mathbf{A}_{i,1}\boldsymbol{x} = \mathbf{0}$	$\tilde{O}(n^{1.5})$	$tl \cdot \tilde{O}(n)$	1-sided
L_{GL} [14]	A	$\boldsymbol{y}_1, \ldots, \boldsymbol{y}_l$	$\mathbf{A}\boldsymbol{x} = \boldsymbol{y}_i$	$\tilde{O}(n^2)$	$tl \cdot \tilde{O}(n)$	2-sided
$\mathrm{S}^+_{\mathrm{GL}}$	A	$\boldsymbol{y}_1, \ldots, \boldsymbol{y}_l$	$\mathbf{A}\boldsymbol{x} = \boldsymbol{y}_i$ and $w_H(\boldsymbol{x}) = m/2$	$\tilde{O}(n)$	$t \cdot \tilde{O}(l+n)$	1-sided

the ID scheme can be proven to have concurrent security[2] based on the same problem as that in the original scheme.

Recently, Lyubashevsky proposed new concurrently secure ID schemes based on lattice problems [14]; we call it L_{GL} for short. In his protocol, the prover proves, given $\mathbf{A}$ and $\boldsymbol{y}$, he/she has $\boldsymbol{x} \in \{0,1\}^m$ such that $\mathbf{A}\boldsymbol{x} = \boldsymbol{y}$. Using an active adversary, his knowledge extractor obtains another vector $\boldsymbol{x}'$ such that $\mathbf{A}\boldsymbol{x}' = \boldsymbol{y}$ and the length of $\boldsymbol{x}'$ is at most $O(m^{1.5}) = \tilde{O}(n^{1.5})$. Thus, in the L_{GL} scheme, the underlying problem is $\mathrm{SIS}_{q,m,\tilde{O}(n^{1.5})}$, which can be reduced from $\mathrm{GapSVP}^2_{\tilde{O}(n^2)}$.

As mentioned in the previous section, the assumption of $\mathrm{S}^+_{\mathrm{GL}}$ is the worst-case hardness of $\mathrm{GapSVP}^2_{\tilde{O}(n)}$, which is weaker than those of $\mathrm{MV}^+_{\mathrm{GL}}$ and L_{GL}. This improvement is obtained by the condition that the knowledge extractor outputs another secret key $\boldsymbol{x}'$ whose length is at most $\sqrt{m} = \tilde{O}(\sqrt{n})$. Our schemes has 1-sided error (perfect completeness and soundness error), while L_{GL} has 2-sided error (completeness and soundness errors). As a summary, see Table 1.

AID schemes: By taking OR of l statements [6], we can straightforwardly obtain $\mathrm{MV}^+_{\mathrm{GL}}$-based and L_{GL}-based AID schemes, whose security are based on the worst-case hardness of lattice problems. We feed only $pk_1, \ldots, pk_l$ as the common inputs to the prover and the verifier. In this case, the prover convinces the verifier that he/she has a secret key corresponding to one of public keys, pk_i.

However, each of these simple modifications requires a large overhead cost involving the size of the ad hoc group. Let l be the number of the members of the group and n the security parameter. The protocol is run in t times in parallel to reduce the errors. The

[2] Combining ORing technique by De Santis et al. [6] and discarding technique by Feige and Shamir [8], we derive a construction technique for ID schemes secure against active attack. Moreover, we can construct concurrently secure ID schemes by the same technique as a folklore says.

communication costs of the MV^{+}_{GL}-based and L_{GL}-based schemes are $tl \cdot \tilde{O}(n)$. The size of a set of the public keys is $l \cdot \tilde{O}(n^2)$ and $\tilde{O}(n^2) + l \cdot \tilde{O}(n)$ in the modified versions of MV^{+}_{GL} and L_{GL}, respectively.

On AID schemes, the L_{GL}-based and our schemes require many *vectors* proportional to the size of the group, while the MV^{+}_{GL}-based scheme requires many *matrices* proportional to the size of the group (see Table 1). Additionally, the communication cost of our schemes is $t \cdot \tilde{O}(n + l)$, while those in the MV^{+}_{GL}-based and L_{GL}-based schemes are $tl \cdot \tilde{O}(n)$. This shows the advantage of our scheme on the efficiency.

1.4 Organization

The rest of this paper is organized as follows. In Section 2, we review basic notations and notions, and the cryptographic schemes we consider. In Section 3, we review lattice-based hash functions and give a commitment scheme based on the lattice-based hash functions for our ID and AID schemes. In Section 4, we construct the ID scheme by combining the framework of Stern's scheme with our string commitment scheme. We present the AID scheme in Section 5.

In this paper, due to lack of space, we only describe the schemes based on GapSVP since the construction on $\Lambda(f)$-SVP follows from a similar strategy to that on GapSVP. We discuss the constructions on $\Lambda(f)$-SVP in the full paper.

2 Preliminaries

Basic notions and notations: We denote by n the security parameter of cryptographic schemes throughout this paper, which corresponds to the rank of the underlying lattice problems. We say that a problem is hard in the worst case if there exists no probabilistic polynomial-time algorithm solves the problem in the worst case with non-negligible probability. We sometimes use $\tilde{O}(g(n))$ for any function g in n as $O(g(n) \cdot \mathrm{polylog}(g(n)))$. We assume that all random variables are independent and uniform. For a positive integer n, let $[n]$ denote a set $\{1, 2, \ldots, n\}$.

For any $p \geq 1$, the ℓ_p norm of a vector $\boldsymbol{x} = {}^t(x_1, \ldots, x_n) \in \mathbb{R}^n$, denoted by $\|\boldsymbol{x}\|_p$, is $(\sum_{i \in [n]} x_i^p)^{1/p}$. For ease of notation, we define $\|\boldsymbol{x}\| := \|\boldsymbol{x}\|_2$. The ℓ_∞ norm is defined as $\|\boldsymbol{x}\|_\infty = \lim_{p \to \infty} \|\boldsymbol{x}\|_p = \max_{i \in [n]} |x_i|$. Let $w_H(\boldsymbol{x})$ denote the Hamming weight of $\boldsymbol{x}$, i.e., the number of non-zero elements in $\boldsymbol{x}$. Let $\mathrm{B}(m, w)$ denote the set of binary vectors in $\{0, 1\}^m$ whose Hamming weights are exactly equal to w, i.e., $\mathrm{B}(m, w) := \{\boldsymbol{x} \in \{0, 1\}^m \mid w_H(\boldsymbol{x}) = w\}$. We denote the concatenation of two vectors or strings $\boldsymbol{v}_1$ and $\boldsymbol{v}_2$ by $\boldsymbol{v}_1 \circ \boldsymbol{v}_2$.

We omit the definitions of zero-knowledge arguments and witness-indistinguishable protocols. For formal definitions, see textbooks, e.g., by Goldreich [10].

Hash functions: We briefly review the definition of collision-resistant hash function families. Let $\mathcal{H}_n = \{h_k : M_n \to D_n \mid k \in K_n\}$ be a family of hash functions, where M_n, D_n, and K_n denote a space of messages, digests, and indices, respectively. Let $\mathcal{H} = \{\mathcal{H}_n\}_{n \in \mathbb{N}}$. Roughly speaking, if $\mathcal{H}$ is collision resistant, any polynomial-time adversary cannot, on input a random index k, output a collision of the hash function indexed by k. For a formal definition, see, e.g., the textbook by Katz and Lindell [13, Sec. 4.6.1].

String commitment schemes: We consider a string commitment scheme in the trusted setup model. The trusted setup model is often required to construct practically efficient cryptographic schemes such as non-interactive string commitment schemes. In this model, we assume that a trusted party $\mathcal{T}$ honestly sets up a system parameter for the sender and the receiver.

First $\mathcal{T}$ distributes the index k of a commitment function to the sender and the receiver. Both parties then share a common function Com_k by a given k. The scheme runs in two phase, called committing and revealing phases. In the committing phase, the sender commits his/her decision, say a string s, to a commitment string $c = \mathrm{Com}_k(s;\rho)$ with a random string ρ and sends c to the receiver. In the revealing phase, the sender gives the receiver the decision s and the random string ρ. The receiver verifies the validity of c by computing $\mathrm{Com}_k(s;\rho)$.

We require two security notions of the string commitment schemes, statistically-hiding and computationally-binding properties. Intuitively, we say that the commitment scheme is statistically hiding, if any computationally unbounded adversarial receiver cannot distinguish two commitment strings generated from two distinct strings. Also, it is computationally binding, if any polynomial-time adversarial sender cannot change the committed string after sending the commitment. See, e.g., [12] for the formal definition.

Canonical identification schemes: Let $\mathcal{SI} = (\mathsf{SetUp}, \mathsf{KG}, \mathsf{P}, \mathsf{V})$ be an identification scheme, where SetUp is the setup algorithm which on input 1^n outputs *param*, KG is the key-generation algorithm which on input *param* outputs (pk, sk), P is the prover algorithm taking input sk, V is the verifier algorithm taking inputs *param* and pk. We say $\mathcal{SI}$ is a canonical identification scheme if it is a public-coin 3-move protocol.

We are interested in concurrent attack, which is stronger than active and passive attack. We employ the definition of concurrent security in [3]. In concurrent attack, the adversary will play the role of a cheating verifier prior to impersonation and can interact many different prover clones concurrently. Each clone has the same secret key, but has independent random coins and maintains its own state. We say $\mathcal{SI}$ is secure against impersonation under concurrent attack, if any polynomial-time adversary cannot, given a random public key of a legitimate prover, impersonate the legitimate prover. For the formal definition, see [3].

Ad hoc anonymous identification schemes: An AID scheme allows a user to anonymously prove his/her membership in a group if and only if the user is an actual member of the group, where the group is formed in an ad hoc fashion without help of the group manager. We then assume that every user registers his/her public key to the public key infrastructure.

We define the algorithms in AID schemes. An AID scheme is four tuple $\mathcal{AID} = (\mathsf{SetUp}, \mathsf{Reg}, \mathsf{P}, \mathsf{V})$, where SetUp is the setup algorithm which on input 1^n outputs *param*, Reg is the key generation and registration algorithm which on input *param* outputs (pk, sk), P is the prover algorithm taking inputs *param*, a set of public keys $R = (pk_1, \ldots, pk_l)$, and one of the secret keys sk_i such that $pk_i \in R$, and V is the verifier algorithm taking inputs *param* and R. For more formal definition, see [7].

There are two goals for security of AID schemes: security against impersonation and anonymity. Dodis et al. formally defined security against impersonation under passive attack. They mentioned the definition of security against impersonation under concurrent attack. However, they did not give the formal definition (see [7, Sec. 3.2]). Thus, we define the security notion with respect to concurrent attack. In the setting of chosen-group attack, the adversary could force the prover to prove the membership in an arbitrary group if the prover is indeed a member of the group. Additionally, concurrent attack allows the cheating verifier to interact with the clones of any provers. Also, they allow the cheating prover to interact with the clones of provers, but prohibit it from interacting with the target provers. We say $\mathcal{AID}$ is secure against impersonation under concurrent chosen-group attack, if any polynomial-time adversary cannot impersonate the legitimate prover in the above settings.

The security notion, anonymity against full key exposure, captures the property that an adversary cannot distinguish two transcripts even if the adversary has the secret keys of all the members. We say $\mathcal{AID}$ is anonymous against full key exposure if any polynomial-time adversary cannot distinguish two provers with a common set of public keys even though the adversary generates all keys of the set. The formal definitions of two notions are in the full paper.

3 Main Tools

In this section, we review main tools, lattices, lattice problems, and lattice-based hash functions, and construct string commitment schemes.

Lattices and lattice problems: We first review fundamental notions of lattices, well-known lattice problems, and a related problem.

An n-dimensional lattice in $\mathbb{R}^m$ is the set $L(\boldsymbol{b}_1, \ldots, \boldsymbol{b}_n) = \{\sum_{i \in [n]} \alpha_i \boldsymbol{b}_i \mid \alpha_i \in \mathbb{Z}\}$ of all integral combinations of n linearly independent vectors $\boldsymbol{b}_1, \ldots, \boldsymbol{b}_n \in \mathbb{R}^m$. The sequence of vectors $\boldsymbol{b}_1, \ldots, \boldsymbol{b}_n$ is called a *basis* of the lattice L and denoted by $\mathbf{B}$. For more details on lattices, see the textbook by Micciancio and Goldwasser [18].

We give the definitions of well-known lattice problems, the Shortest Vector Problem (SVP^p) and its approximation version (SVP^p_γ): The problem SVP^p is, given a basis $\mathbf{B}$ of a lattice L, finding the shortest non-zero vector $\boldsymbol{v}$ in L in the ℓ_p norm. The problem SVP^p_γ is, given a basis $\mathbf{B}$ of a lattice L, finding a non-zero vector $\boldsymbol{v}$ in L such that for any non-zero vector $\boldsymbol{x}$ in L, $\|\boldsymbol{v}\|_p \leq \gamma \|\boldsymbol{x}\|_p$.

We next give the definition of the gap version of SVP^p_γ, which is the underlying problem of lattice-based hash functions.

Definition 3.1 (GapSVP^p_γ **[18]).** *For a gap function γ, an instance of* GapSVP^p_γ *is a pair $(\mathbf{B}, d)$ where $\mathbf{B}$ is a basis of a lattice L and d is a rational number. In YES input there exists a vector $\boldsymbol{v} \in L \setminus \{\mathbf{0}\}$ such that $\|\boldsymbol{v}\|_p \leq d$. In NO input, for any vector $\boldsymbol{v} \in L \setminus \{\mathbf{0}\}$, $\|\boldsymbol{v}\|_p > \gamma d$.*

We also define the Small Integer Solution problem SIS (in the ℓ_p norm), which is often considered in the context of average-case/worst-case connections and a source of lattice-based hash functions as we see later.

Definition 3.2 ($\mathrm{SIS}^p_{q,m,\beta}$ **[19]).** *For a fixed integer q and a real β, given a matrix $\mathbf{A} \in \mathbb{Z}_q^{n \times m}$, the problem is finding a non-zero integer vector $z \in \mathbb{Z}^m$ such that $\mathbf{A}z \equiv \mathbf{0} \pmod{q}$ and $\|z\|_p \leq \beta$.*

The relation between SIS and GapSVP is reviewed in the next paragraph.

Lattice-based hash functions: We review the lattice-based hash functions. For a prime $q = q(n) = n^{O(1)}$ and an integer $m = m(n) > n \log q(n)$, we define a family of hash functions,

$$\mathcal{H}(q,m) = \{f_{\mathbf{A}} : \{0,1\}^m \to \mathbb{Z}_q^n \mid \mathbf{A} \in \mathbb{Z}_q^{n\times m}\},$$

where $f_{\mathbf{A}}(\boldsymbol{x}) = \mathbf{A}\boldsymbol{x} \bmod q$.

Originally, Ajtai [1] showed that the worst-case hardness of GapSVP_γ^2 for some polynomial $\gamma(n)$ is reduced to the average-case hardness of $\mathrm{SIS}_{q,m,n}^2$ for suitable $q(n)$ and $m(n)$. It is known that $\mathcal{H}(q,m)$ is indeed collision resistant for suitably chosen q and m by Goldreich, Goldwasser, and Halevi [11]. They observed that finding a collision $(\boldsymbol{x}, \boldsymbol{x}')$ for $f_{\mathbf{A}} \in \mathcal{H}(q,m)$ implies finding a short non-zero vector $z = \boldsymbol{x} - \boldsymbol{x}'$ such that $\|z\| \leq \sqrt{m}$ and $\mathbf{A}z \equiv \mathbf{0} \pmod q$, i.e., solving $\mathrm{SIS}_{q,m,\sqrt{m}}^2$. Recently, Micciancio and Regev showed that $\mathcal{H}(q,m)$ is collision resistant under the assumption that $\mathrm{GapSVP}_{\tilde{O}(n)}^2$ is hard in the worst case [19].

Theorem 3.1 ([19]). *For any polynomially bounded functions $\beta = \beta(n)$, $m = m(n)$, $q = q(n)$, with $q \geq 4\sqrt{m}n^{3/2}\beta$ and $\gamma = 14\pi\sqrt{n}\beta$, there exists a probabilistic polynomial-time reduction from solving GapSVP_γ^2 in the worst case to solving $\mathrm{SIS}_{q,m,\beta}^2$ on the average with non-negligible probability.*

There were another reductions from the gap version of the covering radius problem GapCRP_γ, the shortest independent vector problem SIVP_γ, and the guaranteed distance decoding problem GDD_γ by adjusting the parameters [19]. It is worth that we note the results following the above results: Peikert [22] showed the reductions from the same problems in any ℓ_p norms for $p \geq 2$. The recent paper [9, Sec. 9] by Gentry, Peikert, and Vaikuntanathan showed that the modulus q in SIS can be $\tilde{O}(n)$.

A string commitment scheme: General constructions of statistically-hiding and computationally-binding string commitment schemes are known from a family of collision-resistant hash functions [4,12]. Their constructions used universal hash functions for the statistically-hiding property.

Here, we give a more direct and simpler construction from the lattice-based hash functions without the universal hash functions. The input of the commitment function is an m-bit vector $\boldsymbol{x}$ obtained by concatenating a random string $\rho = (\rho_1, \ldots, \rho_{m/2})$ and a message string $s = (s_1, \ldots, s_{m/2})$, i.e., $\boldsymbol{x} = \rho \circ s$. We then define the commitment function on inputs s and ρ as

$$\mathrm{Com}_{\mathbf{A}}(s;\rho) := \mathbf{A}\boldsymbol{x} \bmod q = \mathbf{A}^t(\rho_1, \ldots, \rho_{m/2}, s_1, \ldots, s_{m/2}) \bmod q.$$

Lemma 3.1. *For $m > 10n\log q$, if $\mathrm{SIS}_{q,m,\sqrt{m}}$ is hard on the average, then $\mathrm{Com}_{\mathbf{A}}$ is a statistically-hiding and computationally-binding string commitment scheme in the trusted set up model. In particular, for any polynomially bounded functions $m = m(n)$, $q = q(n)$, $\gamma = \gamma(n)$, with $q \geq 4mn^{3/2}$, $\gamma = 14\pi\sqrt{nm}$, and $m > 10n\log q$, $\mathrm{Com}_{\mathbf{A}}$ is a statistically-hiding and computationally-binding string commitment scheme in the trusted setup model if GapSVP_γ^2 is hard in the worst case.*

Before the proof, we review a definition of statistical distances: Given two probability density functions ϕ_1 and ϕ_2 on a finite set S, we define the statistical distance between them as $\Delta(\phi_1, \phi_2) := \frac{1}{2}\sum_{x\in S} |\phi_1(x) - \phi_2(x)|$.

Proof. The computationally-binding property immediately follows from the collision-resistant property. We now show the statistically-hiding property.

Let $\mathbf{A} = [\boldsymbol{a}_1 \cdots \boldsymbol{a}_m]$. We then have $\mathrm{Com}_{\mathbf{A}}(s;\rho) = \sum_{i=1}^{m/2} \rho_i \boldsymbol{a}_i + \sum_{i=1}^{m/2} s_i \boldsymbol{a}_{i+m/2}$. The following claim in [24] says that a random subset sum of $\boldsymbol{a}_i$ is statistically close to the uniform distribution for almost all choices of $\boldsymbol{a}_i$.

Claim ([24]). Let G be some finite Abelian group and let l be some integer. For any l elements $g_1, \ldots, g_l \in G$, consider $\Delta(\sum_{i\in[l]} a_i g_i, u)$, where u and a_i is chosen uniformly at random from G and $\{0,1\}$, respectively. Then the expectation of this statistical distance over a uniform choice of $g_1, \ldots, g_l \in G$ is at most $\sqrt{|G|/2^l}$. In particular, the probability that this statistical distance is more than $(|G|/2^l)^{1/4}$ is at most $(|G|/2^l)^{1/4}$.

In our proof, we consider $\mathbb{Z}_q^n$ as a finite Abelian group G. Since $m > 10n \log q$, $(|G|/2^{m/2})^{1/4} \leq q^{-n}$. Thus, for all but an at most q^{-n} fraction of $\mathbf{A} = [\boldsymbol{a}_1, \ldots, \boldsymbol{a}_m] \in \mathbb{Z}_q^{n\times m}$, we have that $\Delta(\boldsymbol{u}, \sum_{i\in[m/2]} \rho_i \boldsymbol{a}_i) \leq q^{-n}$, where $\boldsymbol{u} \in \mathbb{Z}_q^n$ is uniform random variable. Assume that we have such $\mathbf{A}$. So, we have $\Delta(\boldsymbol{u}, \mathrm{Com}_{\mathbf{A}}(0^{m/2};\rho)) \leq q^{-n}$. By the definition of $\mathrm{Com}_{\mathbf{A}}$, for any $s \in \{0,1\}^{m/2}$, we have $\Delta(\boldsymbol{u}, \mathrm{Com}_{\mathbf{A}}(s;\rho)) \leq q^{-n}$. By the triangle inequality, we obtain

$$\Delta(\mathrm{Com}_{\mathbf{A}}(s_1;\rho_1), \mathrm{Com}_{\mathbf{A}}(s_2;\rho_2)) \leq \Delta(\boldsymbol{u}, \mathrm{Com}_{\mathbf{A}}(s_1;\rho_2)) + \Delta(\boldsymbol{u}, \mathrm{Com}_{\mathbf{A}}(s_2;\rho_2)) \leq 2q^{-n},$$

for any message s_1 and s_2. This shows that, for all but negligible fraction of choice of $\mathbf{A}$, the distributions of two commitments are statistically close. □

Using the Merkle-Damgård technique, we obtain a string commitment scheme whose commitment function is $\mathrm{Com}_{\mathbf{A}} : \{0,1\}^* \times \{0,1\}^{m/2} \to \mathbb{Z}_q^n$ rather than $\mathrm{Com}_{\mathbf{A}} : \{0,1\}^{m/2} \times \{0,1\}^{m/2} \to \mathbb{Z}_q^n$ as the following.

Assume that $m = 2r$. Let $\mathbf{A} = [\mathbf{B}\,\mathbf{C}]$, where $\mathbf{B}, \mathbf{C} \in \mathbb{Z}_q^{n\times r}$. For $\mathbf{X} \in \mathbb{Z}_q^{n\times l}$, we define $f_{\mathbf{X}} : \{0,1\}^l \to \mathbb{Z}_q^n$ as the hash function $f_{\mathbf{X}}(\boldsymbol{s}) = \mathbf{X}\boldsymbol{s} \bmod q$. Let l be $\lceil n \log q \rceil$ and let $t : \mathbb{Z}_q^n \to \{0,1\}^l$ be some one-to-one function that we can compute t and t^{-1} efficiently. Let $\mathsf{pad} : \{0,1\}^* \to \{0,1\}^*$ be a padding function for the Merkle-Damgård construction. Applying the Merkle-Damgård construction to $f_{\mathbf{C}}$, we obtain a new hash function $h_{\mathbf{C}} : \{0,1\}^* \to \mathbb{Z}_q^n$. The precise definition of $h_{\mathbf{C}}$ is as follows:

Hash function $h_{\mathbf{C}}$:

1. On input s, obtain a padded message $S \leftarrow \mathsf{pad}(s)$.
2. Chop it into $(S_0, \ldots, S_k)$, where $S_i \in \{0,1\}^{r-l}$.
3. Let $H_0 = \mathbf{0}$ (more generally, some fixed H_0 can be used).
4. For $i = 1$ to $k+1$ do $H_i \leftarrow f_{\mathbf{C}}(t(H_{i-1}) \circ S_{i-1})$.
5. Output H_{k+1}.

Our new commitment scheme is defined as follows: for $s \in \{0,1\}^*$ and $\rho \in \{0,1\}^r$,

$$\mathrm{Com}_{\mathbf{A}}(s;\rho) := h_{\mathbf{C}}(s) + f_{\mathbf{B}}(\rho) \bmod q.$$

Lemma 3.2. *If there exists a polynomial-time machine outputting a collision for* $\mathrm{Com}_{\mathbf{A}}$, *then there exists a polynomial-time machine outputting a collision for* $f_{\mathbf{A}}$.

Proof. Let us assume that we obtain a collision $(s, \rho), (\tilde{s}, \tilde{\rho}) \in \{0,1\}^* \times \{0,1\}^r$ for $\mathrm{Com}_{\mathbf{A}}$. By the assumption, we have

$$h_{\mathbf{C}}(s) + f_{\mathbf{B}}(\rho) \equiv h_{\mathbf{C}}(\tilde{s}) + f_{\mathbf{B}}(\tilde{\rho}) \pmod q.$$

If $\rho = \tilde{\rho}$, we have $s \neq \tilde{s}$ and $h_{\mathbf{C}}(s) = h_{\mathbf{C}}(\tilde{s})$. Using the reduction for the Merkle-Damgård construction (see e.g., [13, Thm. 4.14]), we obtain $u \neq \tilde{u} \in \{0,1\}^r$ such that $f_{\mathbf{C}}(u) = f_{\mathbf{C}}(\tilde{u})$. Thus, we have a collision $u \circ \rho, \tilde{u} \circ \rho \in \{0,1\}^{2r}$ for $f_{\mathbf{A}}$.

Next, we assume that $\rho \neq \tilde{\rho}$. Let S and $\tilde{S}$ be padded messages of s and $\tilde{s}$, respectively. Assume that S and $\tilde{S}$ are chopped into $(S_0, \ldots, S_k)$ and $(\tilde{S}_0, \ldots, \tilde{S}_{k'})$, respectively. Let H_k and $\tilde{H}_{k'}$ be inner hash values for s and $\tilde{s}$ in the algorithm, respectively. By the definition of H_k and $\tilde{H}_{k'}$, we obtain

$$h_{\mathbf{C}}(s) = f_{\mathbf{C}}(t(H_k) \circ S_k),$$
$$h_{\mathbf{C}}(\tilde{s}) = f_{\mathbf{C}}(t(\tilde{H}_{k'}) \circ \tilde{S}_{k'}).$$

Combining the above equations with the assumption, we obtain

$$f_{\mathbf{A}}(t(H_k) \circ S_k \circ \rho) = f_{\mathbf{A}}(t(\tilde{H}_{k'}) \circ \tilde{S}_{k'} \circ \tilde{\rho}).$$

So, we have a collision $t(H_k) \circ S_k \circ \rho$ and $t(\tilde{H}_{k'}) \circ \tilde{S}_{k'} \circ \tilde{\rho} \in \{0,1\}^{2r}$ for $f_{\mathbf{A}}$. □

We use this commitment scheme in the rest of the paper. We often abuse the notation of $\mathrm{Com}_{\mathbf{A}}$. For example, $\mathrm{Com}_{\mathbf{A}}(v_1, v_2; \rho)$ denotes $\mathrm{Com}_{\mathbf{A}}(\mathrm{string}(v_1) \circ \mathrm{string}(v_2); \rho)$, where $\mathrm{string}(v)$ is a binary representation of v.

4 An Identification Scheme

Our variant $\mathrm{S}^{+}_{\mathrm{GL}}$ is obtained by replacing the string commitment scheme in Stern's ID scheme [26] with our lattice-based one. Stern's protocol deals with the decoding problem on binary codewords called the Syndrome Decoding Problem[3]. He also proposed that an analogous scheme in $\mathbb{Z}_q$, where q is extremely small (typically 3, 5, or 7) [26, Sec. VI]. We adjust this parameter to connect his framework to our assumptions of the lattice problems.

We now describe the protocol $\mathrm{S}^{+}_{\mathrm{GL}}$ below. Obviously, it has perfect completeness, and at most 2/3 soundness error. By parallelizing each step of this protocol in $t = \omega(\log n)$ times, the soundness error becomes negligibly small. To simplify the notations, we write Com instead of $\mathrm{Com}_{\mathbf{A}}$ and we do not write random strings in Com explicitly.

SetUp: The setup algorithm, on input 1^n, outputs a random matrix $\mathbf{A} \in \mathbb{Z}_q^{n \times m}$.

KG: The key-generation algorithm, on input $\mathbf{A}$, chooses a random vector $\boldsymbol{x} \in \mathrm{B}(m, m/2)$ and computes $\boldsymbol{y} := \mathbf{A}\boldsymbol{x} \bmod q$. It outputs $(pk, sk) = (\boldsymbol{y}, \boldsymbol{x})$.

P, V: The common inputs are $\mathbf{A}$ and $\boldsymbol{y}$. The prover's auxiliary input is $\boldsymbol{x}$. They interact as follows:

[3] The Syndrome Decoding Problem is defined as follows: Given $\mathbf{A} \in \mathbb{Z}_2^{n \times m}$, $\boldsymbol{y} \in \mathbb{Z}_2^n$, and $w \in \mathbb{N}$, the problem is finding a vector $\boldsymbol{x} \in \mathrm{B}(m, w)$ such that $\mathbf{A}\boldsymbol{x} \equiv \boldsymbol{y} \bmod 2$. We can consider this problem as a restricted version of $\mathrm{SIS}_{q,m,\beta}$.

Step P1: Choose a random permutation π over $[m]$ and a random vector $\boldsymbol{r} \in \mathbb{Z}_q^m$ and send commitments c_1, c_2, and c_3 computed as

- $c_1 = \mathrm{Com}(\pi, \mathbf{A}\boldsymbol{r})$,
- $c_2 = \mathrm{Com}(\pi(\boldsymbol{r}))$,
- $c_3 = \mathrm{Com}(\pi(\boldsymbol{x} + \boldsymbol{r}))$.

Step V1 Send a random challenge $Ch \in \{1, 2, 3\}$ to P.

Step P2

- If $Ch = 1$, reveal c_2 and c_3. So, send $\boldsymbol{s} = \pi(\boldsymbol{x})$ and $\boldsymbol{t} = \pi(\boldsymbol{r})$.
- If $Ch = 2$, reveal c_1 and c_3. Send $\phi = \pi$ and $\boldsymbol{u} = \boldsymbol{x} + \boldsymbol{r}$.
- If $Ch = 3$, reveal c_1 and c_2. Send $\psi = \pi$ and $\boldsymbol{v} = \boldsymbol{r}$.

Step V2

- If $Ch = 1$, check that $c_2 = \mathrm{Com}(\boldsymbol{t})$, $c_3 = \mathrm{Com}(\boldsymbol{s} + \boldsymbol{t})$, and $\boldsymbol{s} \in \mathrm{B}(m, m/2)$.
- If $Ch = 2$, check that $c_1 = \mathrm{Com}(\phi, \mathbf{A}\boldsymbol{u} - \boldsymbol{y})$ and $c_3 = \mathrm{Com}(\phi(\boldsymbol{u}))$.
- If $Ch = 3$, check that $c_1 = \mathrm{Com}(\psi, \mathbf{A}\boldsymbol{v})$ and $c_2 = \mathrm{Com}(\psi(\boldsymbol{v}))$.

Output $Dec = 1$ if all checks are passed, otherwise output $Dec = 0$.

4.1 Statistical Zero-Knowledge Property

The proof of the zero-knowledge property of the original protocol is in [26, Thm. 4]. Stern left completion of the proof as the problem for reader. Thus, we give the whole proof that Stern's protocol is statistically zero knowledge when Com is a statistically-hiding and computationally-binding string commitment scheme.

Theorem 4.1. *The protocol is statistically zero knowledge when* Com *is a statistically-hiding and computationally-binding string commitment scheme.*

Proof. Following the definition, we construct a simulator $\mathcal{S}$ which on input $\mathbf{A}$ and $\boldsymbol{y}$ and given oracle access to a cheating verifier $\mathcal{CV}$, outputs a simulated transcript. A real transcript between P and $\mathcal{CV}$ on input $\mathbf{A}$ and $\boldsymbol{y}$ is denoted by $\langle \mathsf{P}, \mathcal{CV} \rangle (\mathbf{A}, \boldsymbol{y})$.

First, $\mathcal{S}$ chooses a random value $\bar{c}$ from $\{1, 2, 3\}$ which is a prediction what value the cheating verifier $\mathcal{CV}$ will *not* choose. Next, it chooses a random tape of $\mathcal{CV}$, denoted by r'. We remark that, by the assumption on the commitment, the distributions of a challenge from $\mathcal{CV}$ in the real interaction and in the simulation are statistically close.

Case $\bar{c} = 1$: $\mathcal{S}$ computes $\boldsymbol{x}' \in \mathbb{Z}_q^m$ such that $\mathbf{A}\boldsymbol{x}' = \boldsymbol{y}$ by using linear algebra. Next, it chooses a random permutation π' over $[m]$, a random vector $\boldsymbol{r}' \in \mathbb{Z}_q^m$, and random strings ρ_1', ρ_2', and ρ_3'. So, it computes

- $c_1' := \mathrm{Com}(\pi', \mathbf{A}\boldsymbol{r}'; \rho_1')$,
- $c_2' := \mathrm{Com}(\pi'(\boldsymbol{r}'); \rho_2')$,
- $c_3' := \mathrm{Com}(\pi'(\boldsymbol{x}' + \boldsymbol{r}'); \rho_3')$.

It sends them to $\mathcal{CV}$. Since the commitment scheme is statistically hiding, the distribution of a challenge from $\mathcal{CV}$ is statistically close to the real distribution. Receiving a challenge Ch from $\mathcal{CV}$, the simulator $\mathcal{S}$ computes a transcript as follows:

- If $Ch = 1$, $\mathcal{S}$ outputs $\perp$ and halts.
- If $Ch = 2$, it outputs $(r'; (c_1', c_2', c_3'), 2, (\pi', \boldsymbol{x}' + \boldsymbol{r}', \rho_1', \rho_3'))$.
- If $Ch = 3$, it outputs $(r'; (c_1', c_2', c_3'), 3, (\pi', \boldsymbol{r}', \rho_1', \rho_2'))$.

We analyze the case $Ch = 2$. In this case, we obtain that

$$\langle \mathsf{P}, \mathcal{CV}\rangle(\mathbf{A}, \boldsymbol{y}) = (r; (c_1, c_2, c_3), 2, (\pi, \boldsymbol{x} + \boldsymbol{r}, \rho_1, \rho_3),$$
$$\mathcal{S}(\mathbf{A}, \boldsymbol{y}) = (r'; (c_1', c_2', c_3'), 2, (\pi', \boldsymbol{x}' + \boldsymbol{r}', \rho_1', \rho_3')).$$

Assume that $(\pi', \boldsymbol{r}', \rho_1', \rho_3') = (\pi, \boldsymbol{r} + \boldsymbol{x} - \boldsymbol{x}', \rho_1, \rho_3)$. By this equation, we have that $c_1' = c_1$, $c_3' = c_3$, and the responses from the simulator equal to the responses from the prover. Since the commitment is statistically hiding, we have the distributions of c_2 and c_2' are statistically close. Thus, we conclude that the both distributions of the simulated transcript and the real transcript are statistically close.

It is straightforward to show it in the case $Ch = 3$ by using the equation $(\pi', \boldsymbol{r}') = (\pi, \boldsymbol{r})$. Thus, we omit this part from the proof.

Case $\bar{c} = 2$: $\mathcal{S}$ chooses a random permutation π' over $[m]$, two random vectors $\boldsymbol{r}' \in \mathbb{Z}_q^m$, $\boldsymbol{x}' \in \mathrm{B}(m, m/2)$, and random strings ρ_1', ρ_2', and ρ_3'. $\mathcal{S}$ computes commitments

- $c_1' := \mathrm{Com}(\pi', \mathbf{A}\boldsymbol{r}'; \rho_1')$,
- $c_2' := \mathrm{Com}(\pi'(\boldsymbol{r}'); \rho_2')$,
- $c_3' := \mathrm{Com}(\pi'(\boldsymbol{x}' + \boldsymbol{r}'); \rho_3')$.

It sends them to $\mathcal{CV}$. Receiving a challenge Ch, the simulator computes a transcript as follows:

- If $Ch = 1$, then $\mathcal{S}$ outputs $(r'; (c_1', c_2', c_3'), 1, (\pi'(\boldsymbol{x}'), \pi'(\boldsymbol{r}'), \rho_2', \rho_3'))$.
- If $Ch = 2$, then it outputs $\perp$ and halts.
- If $Ch = 3$, then it outputs $(r'; (c_1', c_2', c_3'), 3, (\pi', \boldsymbol{r}', \rho_1', \rho_2'))$.

We analyze the case $Ch = 1$. In this case, we have that

$$\langle \mathsf{P}, \mathcal{CV}\rangle(\mathbf{A}, \boldsymbol{y}) = (r; (c_1, c_2, c_3), 1, (\pi(\boldsymbol{x}), \pi(\boldsymbol{r}), \rho_2, \rho_3),$$
$$\mathcal{S}(\mathbf{A}, \boldsymbol{y}) = (r'; (c_1', c_2', c_3'), 1, (\pi'(\boldsymbol{x}'), \pi'(\boldsymbol{r}'), \rho_2', \rho_3')).$$

Let χ be a permutation over $[m]$ such that $\chi(\boldsymbol{x}') = \boldsymbol{x}$. In this case, we set $(\pi', \boldsymbol{r}', \rho_2', \rho_3') = (\pi \circ \chi^{-1}, \chi(\boldsymbol{r}), \rho_2, \rho_3)$. By this equation, we have that $\pi(\boldsymbol{x}) = \pi'(\boldsymbol{x}')$, $\pi(\boldsymbol{r}) = \pi'(\boldsymbol{r}')$, $c_2' = c_2$, and $c_3' = c_3$, that is, the responses from the simulator equal to the responses from the prover. Since the commitment scheme is statistically hiding, the distributions of the real transcript and the output of the simulator are statistically close.

We omit the proof of the case $Ch = 3$, since it is trivial.

Case $\bar{c} = 3$: $\mathcal{S}$ chooses a random permutation π over $[m]$, two random vectors $\boldsymbol{r} \in \mathbb{Z}_q^m$, $\boldsymbol{x}' \in \mathrm{B}(m, m/2)$, and random strings ρ_1, ρ_2, and ρ_3. $\mathcal{S}$ computes

- $c_1 := \mathrm{Com}(\pi, \mathbf{A}(\boldsymbol{x}' + \boldsymbol{r}) - \boldsymbol{y}; \rho_1)$,
- $c_2 := \mathrm{Com}(\pi(\boldsymbol{r}); \rho_2)$,
- $c_3 := \mathrm{Com}(\pi(\boldsymbol{x}' + \boldsymbol{r}); \rho_3)$.

It sends them to $\mathcal{CV}$.

- If $Ch = 1$, then $\mathcal{S}$ outputs $(r'; (c_1, c_2, c_3), 1, (\pi(\boldsymbol{x}'), \pi(\boldsymbol{r}), \rho_2, \rho_3)$.
- If $Ch = 2$, then it outputs $(r'; (c_1, c_2, c_3), 2, (\pi, \boldsymbol{x}' + \boldsymbol{r}'))$.
- If $Ch = 3$, it outputs $\perp$ and halts.

In the case $Ch = 1$, we consider the equation $(\pi', \boldsymbol{r}', \rho'_2, \rho'_3) = (\pi \circ \chi^{-1}, \chi(\boldsymbol{r}), \rho_2, \rho_3)$, where χ denotes a permutation over $[m]$ such that $\chi(\boldsymbol{x}') = \boldsymbol{x}$. The remaining part of proof is the same as that in the case $\bar{c} = 2$ and $Ch = 1$. In the case $Ch = 2$, we let $(\pi', \boldsymbol{r}', \rho'_1, \rho'_3) = (\pi, \boldsymbol{r} + \boldsymbol{x} - \boldsymbol{x}', \rho_1, \rho_3)$. The remaining part of proof is the same as that in the case $\bar{c} = 1$ and $Ch = 2$.

The probability that the simulator $\mathcal{S}$ outputs $\perp$ is at most $1/3 + \epsilon(n) \leq 1/2$ where ϵ is some negligible function. Additionally, by the above arguments, the distribution of the output of $\mathcal{S}$ conditioned on it is not $\perp$ is statistically close to the distribution of the real transcript. Therefore, we have constructed the simulator and completed the proof. □

Since the protocol is statistically zero knowledge for $t = 1$, it has a witness-indistinguishable property. Witness-indistinguishable property is closed under the parallel composition [8]. Thus, the above protocol is witness indistinguishable for $t = \omega(\log n)$ if a statistically-hiding string commitment scheme is used.

4.2 Security of the Protocol

We show the theorem of the security on our ID protocol, which concerns impersonation under concurrent attack.

Theorem 4.2. *For any $m(n) = \Theta(n \log n)$, there exist $q(n) = O(n^{2.5} \log n)$ and $\gamma(n) = O(n\sqrt{\log n})$ such that $m \geq 10n \log q$ and $q^n / |\mathrm{B}(m, m/2)|$ is negligible in n and the above ID scheme is secure against impersonation under concurrent attack if* GapSVP^2_γ *is hard in the worst case.*

Before the proof of security, we need to mention the following trivial lemma.

Lemma 4.1. *For any fixed* $\mathbf{A}$, *let $Y := \{\boldsymbol{y} \in \mathbb{Z}_q^n \mid |\{\boldsymbol{x} \in \mathrm{B}(m, m/2) \mid \mathbf{A}\boldsymbol{x} = \boldsymbol{y}\}| = 1\}$, i.e., a set of vectors $\boldsymbol{y}$ such that the preimage $\boldsymbol{x}$ of $\boldsymbol{y}$ is uniquely determined for $\mathbf{A}$. If $q^n / |\mathrm{B}(m, m/2)|$ is negligible in n, then the probability that, if we obtain $(\boldsymbol{y}, \boldsymbol{x}) \leftarrow \mathsf{KG}(\mathbf{A})$, then $\boldsymbol{y} \in Y$ is negligible in n.*

We now prove Theorem 4.2. The part of the proof is similar to that in [26].

Proof (Proof of Theorem 4.2). Since there exists average-case/worst-case reduction from GapSVP^2_γ to $\mathrm{SIS}^2_{q,m,\sqrt{m}}$ (Theorem 3.1), we only construct $\mathcal{A}$ solving $\mathrm{SIS}^2_{q,m,\sqrt{m}}$ on the average from an impersonator $\mathcal{I} = (\mathcal{CV}, \mathcal{CP})$ which succeeds impersonation under concurrent attack with non-negligible probability ϵ.

For the clarity, we write the transcript of interaction by (Cmt, Ch, Rsp, Dec). Since the protocol is parallelized, each Cmt, Ch, and Rsp is an ordered list which contains t elements. For example, $Cmt = (Cmt_1, \ldots, Cmt_t)$.

Given $\mathbf{A}$, $\mathcal{A}$ chooses a random secret key $\boldsymbol{x} \in \mathrm{B}(m, m/2)$ and computes $\boldsymbol{y} = \mathbf{A}\boldsymbol{x}$. Using the secret key, it can simulate the prover oracle perfectly. $\mathcal{A}$ runs $\mathcal{CV}$ on input $(\mathbf{A}, \boldsymbol{y})$ and obtains a state for $\mathcal{CP}$. $\mathcal{A}$ feeds the state to $\mathcal{CP}$ and acts as a legitimate verifier. Receiving commitments Cmt, $\mathcal{A}$ chooses three challenges $Ch^{(1)}$, $Ch^{(2)}$, and $Ch^{(3)}$ from $\{1, 2, 3\}^t$ uniformly at random. Rewinding with three challenges, $\mathcal{A}$ obtains three transcripts $(Cmt, Ch^{(i)}, Rsp^{(i)}, Dec^{(i)})$ for $i = 1, 2, 3$ as the results of the interactions.

By the Heavy Row Lemma [21], the probability that all $Dec^{(i)}$ are 1 is at least $(\epsilon/2)^3$. Meanwhile, we have

$$\Pr\left[\exists j \in [t] : \{Ch_j^{(1)}, Ch_j^{(2)}, Ch_j^{(3)}\} = \{1,2,3\}\right] = 1 - (7/9)^t$$

by a simple calculation. Thus the probability that $\mathcal{A}$ has three transcripts $(Cmt, Ch^{(i)}, Rsp^{(i)}, Dec^{(i)})$ for $i = 1,2,3$ such that $Dec^{(i)} = 1$ for all i, and $\{Ch_j^{(1)}, Ch_j^{(2)}, Ch_j^{(3)}\} = \{1,2,3\}$ for some $j \in [t]$ is at least $(\epsilon/2)^3 - (7/9)^t$, which is non-negligible since ϵ is non-negligible and $t = \omega(\log n)$.

We next show how $\mathcal{A}$ obtains a secret key or finds a collision of the hash functions in the string commitment scheme by using three good transcripts. Assume that $\mathcal{A}$ has three transcripts $(Cmt^{(i)}, Ch^{(i)}, Rsp^{(i)}, Dec^{(i)})$ for $i = 1,2,3$ such that $Cmt^{(1)} = Cmt^{(2)} = Cmt^{(3)}$, $Dec^{(i)} = 1$ for all i, and $\{Ch_j^{(1)}, Ch_j^{(2)}, Ch_j^{(3)}\} = \{1,2,3\}$ for some $j \in [t]$. Without loss of generality, we assume that $Ch_j^{(i)} = i$. We parse $Rsp_j^{(i)}$ as in Step V2. We have following equations (We omit j for simplification):

$$\begin{aligned}
c_1 &= \mathrm{Com}_{\mathbf{A}}(\phi, \mathbf{A}\boldsymbol{u} - \boldsymbol{y}; \rho_1^{(2)}) &&= \mathrm{Com}_{\mathbf{A}}(\psi, \mathbf{A}\boldsymbol{v}; \rho_1^{(3)}),\\
c_2 &= \mathrm{Com}_{\mathbf{A}}(\boldsymbol{t}; \rho_2^{(1)}) &&= \mathrm{Com}_{\mathbf{A}}(\psi(\boldsymbol{v}); \rho_2^{(3)}),\\
c_3 &= \mathrm{Com}_{\mathbf{A}}(\boldsymbol{s} + \boldsymbol{t}; \rho_3^{(1)}) &&= \mathrm{Com}_{\mathbf{A}}(\phi(\boldsymbol{u}); \rho_3^{(2)}),\\
\boldsymbol{s} &\in \mathrm{B}(m, m/2).
\end{aligned}$$

If there exists a distinct pair of arguments of $\mathrm{Com}_{\mathbf{A}}$, $\mathcal{A}$ obtains a collision for $\mathbf{A}$ and solves $\mathrm{SIS}_{q,m,\sqrt{m}}$.

Next, we suppose that there exist no distinct pairs of the arguments of $\mathrm{Com}_{\mathbf{A}}$. Let π denote the inverse permutation of ϕ. From the first equation, we have $\pi^{-1} = \phi = \psi$. Thus, we obtain $\boldsymbol{u} = \pi(\boldsymbol{s} + \boldsymbol{t})$ from the third equation. Combining it with the first equation, we have $\mathbf{A}\boldsymbol{v} = \mathbf{A}(\pi(\boldsymbol{s}) + \pi(\boldsymbol{t})) - \boldsymbol{y}$. Since $\boldsymbol{v} = \phi^{-1}(\boldsymbol{t}) = \pi(\boldsymbol{t})$ from the second equation, we obtain $\boldsymbol{y} = \mathbf{A} \cdot \pi(\boldsymbol{s})$. Since $\boldsymbol{s} \in \mathrm{B}(m, m/2)$, so $\pi(\boldsymbol{s})$ also is in $\mathrm{B}(m, m/2)$. Therefore, $\mathcal{A}$ sets $\boldsymbol{x}' := \pi(\boldsymbol{s})$.

We now have to show that $\boldsymbol{x}' \neq \boldsymbol{x}$ with probability at least 1/2. By Lemma 4.1, there must be another secret key $\boldsymbol{x}'$ corresponding to $\boldsymbol{y}$ with overwhelming probability. Recall that the protocol is statistically witness indistinguishable. Hence, $\mathcal{I}$'s view is independent of $\mathcal{A}$'s choice of $\boldsymbol{x}$ with overwhelming probability. Thus we have $\boldsymbol{x}' \neq \boldsymbol{x}$ with probability at least 1/2. In this case $\mathcal{A}$ outputs $\boldsymbol{z} = \boldsymbol{x} - \boldsymbol{x}'$ and solves $\mathrm{SIS}_{q,m,\sqrt{m}}$. □

We note that the above proof is extended into multi-user settings as in the proof of Lyubashevsky [14].

5 An Ad Hoc Anonymous Identification Scheme

We next construct our AID scheme based on GapSVP. First, we sketch a basic idea for our construction: Let $\mathbf{A}$ be a system parameter. Each user has a secret key $\boldsymbol{x}_i \in \mathrm{B}(m, w)$ and a public key $\boldsymbol{y}_i = \mathbf{A}\boldsymbol{x}_i$. In the AID scheme, a group is specified by a set of public keys $(\boldsymbol{y}_1, \ldots, \boldsymbol{y}_l)$ of the members. Let $\boldsymbol{e}_{i,l}$ denote an l-dimensional vector ${}^t(0, \ldots, 0, 1, 0, \ldots, 0)$ whose i-th element is 1. The prover in the group, who has a secret key $\boldsymbol{x}_i$, wants convinces the verifier that he/she knows that $\boldsymbol{x}' := \boldsymbol{x}_i \circ -\boldsymbol{e}_{i,l}$ such that $[\mathbf{A}\, \boldsymbol{y}_1 \ldots \boldsymbol{y}_l]\boldsymbol{x}' = \mathbf{0}$

and $x_i \in \mathrm{B}(m, m/2)$. Changing the parameters and using Stern's protocol, the prover can convinces the verifier that he/she has x' such that $[\mathbf{A}\ y_1\ \dots\ y_l]x' = \mathbf{0}$, the numbers of $+1$ in x' is $m/2$, and the numbers of -1 in x' is 1. Additionally, we force the prover to prove that x' is in the form $x' = x_i \circ -e_{i,l}$. To do so, we divide a permutation π in Step P1 into two permutations.

Let π_h be a permutation over $[m]$ and π_t be a permutation over $[l]$. For a permutation π over $[m+l]$, we denote $\pi = \pi_h \odot \pi_t$ if

$$\pi = \begin{pmatrix} 1 & 2 & \cdots & m \\ \pi_h(1) & \pi_h(2) & \cdots & \pi_h(m) \end{pmatrix} \cdot \begin{pmatrix} m+1 & m+2 & \cdots & m+l \\ m+\pi_t(1) & m+\pi_t(2) & \cdots & m+\pi_t(l) \end{pmatrix}.$$

For any π_h and π_t, we have $(\pi_h \odot \pi_t)^{-1} = \pi_h^{-1} \odot \pi_t^{-1}$. For any $x_h \in \mathbb{Z}^m$ and $x_t \in \mathbb{Z}^l$, if $\pi = \pi_h \odot \pi_t$ then $\pi(x_h \circ x_t) = \pi_h(x_h) \circ \pi_t(x_t)$.

We here construct an AID scheme based on GapSVP. Similarly to the ID scheme in Section 4, the protocol is repeated $t = \omega(\log n)$ times in parallel to achieve exponentially small soundness error. As in the previous section, we hide randomness in $\mathrm{Com}_{\mathbf{A}}$.

SetUp: Same as SetUp of the protocol in Section 4.
Reg: Same as KG of the protocol in Section 4.
P, V: The common inputs are $\mathbf{A}$ and $(y_1, \dots, y_l)$. The prover's auxiliary input is x_i for some $i \in [l]$. Let $\mathbf{A}' := [\mathbf{A}\ y_1\ \dots\ y_l]$ and $x := x_i \circ -e_{i,l}$. We write Com instead of $\mathrm{Com}_{\mathbf{A}}$ for ease of notation. They interact as follows:

Step P1: Choose random permutations π_h over $[m]$ and π_t over $[l]$. Let $\pi = \pi_h \odot \pi_t$. Choose a random vector $r \in \mathbb{Z}_q^{m+l}$. Send commitments c_1, c_2, and c_3 as
- $c_1 = \mathrm{Com}(\pi_h, \pi_t, \mathbf{A}'r)$,
- $c_2 = \mathrm{Com}(\pi(r))$,
- $c_3 = \mathrm{Com}(\pi(x + r))$.

Step V1 Send a random challenge $Ch \in \{1, 2, 3\}$ to P.

Step P2
- If $Ch = 1$, reveal c_2 and c_3. Send $s = \pi(x)$ and $t = \pi(r)$.
- If $Ch = 2$, reveal c_1 and c_2. Send $\phi_h = \pi_h$, $\phi_t = \pi_t$, and $u = x + r$.
- If $Ch = 3$, reveal c_1 and c_3. Send $\psi_h = \pi_h$, $\psi_t = \pi_t$, and $v = r$.

Step V2
- If $Ch = 1$, check that $c_2 = \mathrm{Com}(t)$, $c_3 = \mathrm{Com}(s + t)$, and s is in the form $s_h \circ -e_{j,l}$ for some j and $s_h \in \mathrm{B}(m, m/2)$.
- If $Ch = 2$, check that $c_1 = \mathrm{Com}(\phi_h, \phi_t, \mathbf{A}'u)$ and $c_3 = \mathrm{Com}((\phi_h \odot \phi_t)(u))$.
- If $Ch = 3$, check that $c_1 = \mathrm{Com}(\psi_h, \psi_t, \mathbf{A}')$ and $c_2 = \mathrm{Com}((\psi_h \odot \psi_t)(v))$.

Output $Dec = 1$ if all checks are passed, otherwise output $Dec = 0$.

The security of the above protocol is stated as follows. We omit the proof, since it is similar to the proof of Theorem 4.2.

Theorem 5.1. *Let $m = m(n)$ and $q = q(n)$ be polynomially bounded functions satisfying the conditions that $m \geq 10n \log q$ and $q^n / |\mathrm{B}(m, m/2)|$ is negligible in n. Assume that there exists an impersonator $\mathcal{I}$ that succeeds impersonation under concurrent chosen-group attack with non-negligible probability. Then there exists a probabilistic polynomial-time algorithm $\mathcal{A}$ that solves $\mathrm{SIS}^2_{q,m,\sqrt{m}}$.*

Combining Theorem 5.1 with Theorem 3.1, we obtain the following theorem.

Theorem 5.2. *For any $m(n) = \Theta(n \log n)$, there exist $q(n) = O(n^{2.5} \log n)$ and $\gamma(n) = O(n \sqrt{\log n})$ such that $q^n / |\mathrm{B}(m, m/2)|$ is negligible in n and the above scheme is secure against impersonation under concurrent chosen-group attack if* GapSVP^2_γ *is hard in the worst case.*

The statistical anonymity of the above scheme follows from witness indistinguishability of the protocol.

Acknowledgement

The third author thanks Eiichiro Fujisaki for his inspiring question, which motivated us to combine Stern's protocol with lattice-based hash functions. We thank the anonymous referees for their helpful comments and their suggestions on editorial problems. This work is partly supported by Grant-in-Aid for JSPS Fellows 19-55201, and by the Ministry of Education, Science, Sports and Culture, Grant-in-Aid for Young Scientist (B) No.17700007, 2005 and for Scientific Research (B) No.18300002, 2006.

References

1. Ajtai, M.: Generating hard instances of lattice problems (extended abstract). In: STOC 1996, pp. 99–108 (1996)
2. Ajtai, M., Dwork, C.: A public-key cryptosystem with worst-case/average-case equivalence. In: STOC 1997, pp. 284–293 (1997)
3. Bellare, M., Palacio, A.: GQ and Schnorr identification schemes: Proofs of security against impersonation under active and concurrent attacks. In: Yung, M. (ed.) CRYPTO 2002. LNCS, vol. 2442, pp. 162–177. Springer, Heidelberg (2002)
4. Damgård, I.B., Pedersen, T.P., Pfizmann, B.: On the existence of statistically hiding bit commitment schemes and fail-stop signatures. Journal of Cryptology 10(3), 163–194 (1997)
5. Damgård, I.B., Pedersen, T.P., Pfizmann, B.: Statistical Secrecy and Multibit Commitments. IEEE Transactions on Information Theory 44(3), 1143–1151 (1998)
6. De Santis, A., Di Crescenzo, G., Persiano, G., Yung, M.: On monotone formula closure of SZK. In: FOCS 1994, pp. 454–465 (1994)
7. Dodis, Y., Kiayias, A., Nicolosi, A., Shoup, V.: Anonymous identification in ad hoc groups. In: Cachin, C., Camenisch, J.L. (eds.) EUROCRYPT 2004. LNCS, vol. 3027, pp. 609–626. Springer, Heidelberg (2004)
8. Feige, U., Shamir, A.: Witness indistinguishable and witness hiding protocols. In: STOC 1990, pp. 416–426 (1990)
9. Gentry, C., Peikert, C., Vaikuntanathan, V.: Trapdoors for hard lattices and new cryptographic constructions. In: STOC 2008, pp. 197–206 (2008)
10. Goldreich, O.: Foundations of Cryptography: Volume I – Basic Tools. Cambridge University Press, Cambridge (2001)
11. Goldreich, O., Goldwasser, S., Halevi, S.: Collision-free hashing from lattice problems. ECCC 3(42) (1996)
12. Halevi, S., Micali, S.: Practical and provably-secure commitment scheme from collision-free hashing. In: Koblitz, N. (ed.) CRYPTO 1996. LNCS, vol. 1109, pp. 201–215. Springer, Heidelberg (1996)
13. Katz, J., Lindell, Y.: Introduction to Modern Cryptography. Chapman & Hall/CRC (2007)

14. Lyubashevsky, V.: Lattice-based identification schemes secure under active attacks. In: Cramer, R. (ed.) PKC 2008. LNCS, vol. 4939, pp. 162–179. Springer, Heidelberg (2008)
15. Lyubashevsky, V., Micciancio, D.: Generalized compact knapsacks are collision resistant. In: Bugliesi, M., Preneel, B., Sassone, V., Wegener, I. (eds.) ICALP 2006, Part II. LNCS, vol. 4052, pp. 144–155. Springer, Heidelberg (2006)
16. Lyubashevsky, V., Micciancio, D., Peikert, C., Rosen, A.: SWIFFT: A modest proposal for FFT hashing. In: Nyberg, K. (ed.) FSE 2008. LNCS, vol. 5086, pp. 54–72. Springer, Heidelberg (2008)
17. Micciancio, D.: Generalized compact knapsacks, cyclic lattices, and efficient one-way functions. Computational Complexity 16, 365–411 (2007)
18. Micciancio, D., Goldwasser, S.: Complexity of Lattice Problems: a cryptographic perspective. Kluwer Academic Publishers, Dordrecht (2002)
19. Micciancio, D., Regev, O.: Worst-case to average-case reductions based on Gaussian measures. SIAM Journal on Computing 37(1), 267–302 (2007)
20. Micciancio, D., Vadhan, S.: Statistical zero-knowledge proofs with efficient provers: Lattice problems and more. In: Boneh, D. (ed.) CRYPTO 2003. LNCS, vol. 2729, pp. 282–298. Springer, Heidelberg (2003)
21. Ohta, K., Okamoto, T.: On concrete security treatment of signatures derived from identification. In: Krawczyk, H. (ed.) CRYPTO 1998. LNCS, vol. 1462, pp. 354–369. Springer, Heidelberg (1998)
22. Peikert, C.: Limits on the hardness of lattice problems in l_p norms. Computational Complexity 17(2), 300–351 (2008)
23. Peikert, C., Rosen, A.: Efficient collision-resistant hashing from worst-case assumptions on cyclic lattices. In: Halevi, S., Rabin, T. (eds.) TCC 2006. LNCS, vol. 3876, pp. 145–166. Springer, Heidelberg (2006)
24. Regev, O.: On lattices, learning with errors, random linear codes, and cryptography. In: STOC 2005, pp. 84–93 (2005)
25. Shamir, A.: A polynomial-time algorithm for breaking the basic Merkle-Hellman cryptosystem. IEEE Transactions on Information Theory 30(5), 699–704 (1984)
26. Stern, J.: A new paradigm for public key identification. IEEE Transactions on Information Theory 42(6), 749–765 (1996)
27. Wu, Q., Chen, X., Wang, C., Wang, Y.: Shared-key signature and its application to anonymous authentication in ad hoc group. In: Zhang, K., Zheng, Y. (eds.) ISC 2004. LNCS, vol. 3225, pp. 330–341. Springer, Heidelberg (2004)

Rigorous and Efficient Short Lattice Vectors Enumeration

Xavier Pujol[1] and Damien Stehlé[1,2,⋆]

[1] LIP Arénaire, CNRS/INRIA/ENS Lyon/UCBL/Université de Lyon
[2] University of Macquarie/University of Sydney
xavier.pujol@ens-lyon.fr, damien.stehle@gmail.com

Abstract. The Kannan-Fincke-Pohst enumeration algorithm for the shortest and closest lattice vector problems is the keystone of all strong lattice reduction algorithms and their implementations. In the context of the fast developing lattice-based cryptography, the practical security estimates derive from floating-point implementations of these algorithms. However, these implementations behave very unexpectedly and make these security estimates debatable. Among others, numerical stability issues seem to occur and raise doubts on what is actually computed. We give here the first results on the numerical behavior of the floating-point enumeration algorithm. They provide a theoretical and practical framework for the use of floating-point numbers within strong reduction algorithms, which could lead to more sensible hardness estimates.

Keywords: Lattices, SVP, lattice cryptanalysis, numerical stability.

1 Introduction

A *lattice* L is a discrete subgroup of some $\mathbb{R}^n$. It can always represented by a *basis*, i.e., some $d \leq n$ linearly independent vectors $\boldsymbol{b}_1, \ldots, \boldsymbol{b}_d \in \mathbb{R}^n$ such that $L = \sum \mathbb{Z}\boldsymbol{b}_i$. A given lattice has infinitely many bases as soon as $d \geq 2$. One is most often interested in bases made of rather short/orthogonal vectors, which are generically called reduced. They provide a more tractable description of the lattice. Since a lattice is discrete, it contains a vector of smallest non-zero Euclidean length: this length λ is called the *lattice minimum*. The most famous problem related to lattices is the *Shortest Vector Problem* (SVP), which aims at finding a lattice vector of length λ from an arbitrary basis. SVP is known to be NP-hard under randomized reductions [2]. Another popular lattice problem is the *Closest Vector Problem* (CVP): given a lattice basis and a target vector in $\mathbb{R}^n$, find a lattice vector that is closest to the target. This non-homogeneous version of SVP is NP-hard [7]. Since these problems are costly to solve for large dimensions, one is often satisfied with weaker variants. E.g., in γ-SVP one asks for a non-zero lattice vector no longer than $\gamma \cdot \lambda$.

⋆ This work is part of the Australian Research Council Discovery Project on Lattices and their Theta Series.

J. Pieprzyk (Ed.): ASIACRYPT 2008, LNCS 5350, pp. 390–405, 2008.
© International Association for Cryptologic Research 2008

Lattice reduction algorithms range between two extremes. On one side, the LLL algorithm [20] provides a basis with relatively poor properties, in polynomial time. On the opposite, the Hermite-Korkine-Zolotarev (HKZ) reduction provides an excellent basis but requires a huge computational effort. Schnorr [31] was the first to devise hierarchies of algorithms ranging from LLL to HKZ, depending on a parameter k. Schnorr's algorithms make use, in a LLL fashion, of HKZ reductions in projections of sublattices of dimension $O(k)$. When k increases, the cost increases as well, but the quality of the bases improves. The recent hierarchies [9,10] achieve better trade-offs but follow the same general strategy. In practice, the Schnorr-Euchner BKZ algorithm [32] seems to be the best, at least for small values of k. The HKZ reduction uses the Kannan-Fincke-Pohst (KFP) enumeration of short lattice vectors [19,8]. KFP may be replaced by the probabilistic algorithm of [4], but the latter seems slower in practice [28].

Lattices appeared for the first time in cryptology at the beginning of the 80's, when the renowned LLL algorithm [20] was used to break knapsack cryptosystems [29]. For many years lattices were mostly used as a cryptanalytic tool [18]. The landscape changed dramatically in the mid-90's with the invention of several lattice-based encryption schemes, among which Ajtai-Dwork [3], NTRU [17] and GGH [13]. Their securities provably/heuristically rely on the hardness of relaxed variants of SVP and CVP. For example, in the GGH/NTRU framework, the hardness of recovering the secret key from the public key is related to SVP and the hardness of maliciously deciphering a message is related to CVP. A recent but very active and promising trend consists in building other cryptographic schemes whose securities provably reduce to the assumed worst-case hardness of Poly(d)-SVP for special lattices (called ideal). This includes hashing [23], signatures [22] and public-key identification [21]. Gentry, Peikert and Vaikuntanathan [12] introduced other elaborate schemes, including a signature and an identity-based cryptosystem. We refer to [24] for more details. Besides cryptology, lattice reduction and in particular KFP is used in many areas, including number theory [6] and communications theory [25,15], in which the present results may prove useful as well.

Despite the high-speed development of lattice-based cryptography, its practical security remains to be assessed (see [11] for a first step in that direction). Contrary to factorization and discrete logarithm in finite fields and in elliptic curves, the practical limits for solving SVP and CVP and their relaxed variants are essentially unknown, implying that the practicality of the schemes above is debatable. It could be that the suggested key sizes are below what they should, as what happened to be the case with GGH [26]. They may also be too large and then unnecessarily sacrifice efficiency. No significant computational project has ever been undertaken. The main reason is that the algorithmic facet of lattice reduction remains mysterious. In particular, the theoretically best algorithms [9,10] seem to remain slower than heuristic ones such as [32], whose practical behaviors are themselves suspicious. Let us discuss NTL's BKZ routine [33] which implements [32] and is the only publicly available such implementation: when the so-called block-size k is around 30, the number of internal calls to SVP in dimension k seems to explode suddenly (although the corresponding quantity decreases with k in the theoretical algorithms); when k increases, BKZ seems to require more precision for the underlying floating-point computations, although the considered bases should become more orthogonal, which implies a better conditioning with respect to numerical computations. The latter raises doubts on what is actually computed and thus on the practical security estimates of lattice-based cryptography.

Classically, to obtain correctness guarantees, the lattices under study should be in $\mathbb{Q}^n$ and the KFP enumeration should rely on a rational arithmetic. However, the rationals may have huge bit-sizes (though polynomial in the bit-size of the input basis). The bit-size of the rationals is a polynomial factor of the overall enumeration cost (between linear and quadratic, depending on the integer arithmetic). Keeping a rational arithmetic would decrease the efficiency of KFP significantly. In practice, e.g., in NTL, these rational numbers are always replaced by small precision floating-point numbers. Finding a small lattice vector corresponds to disclosing an integer linear combination of vectors whose coordinates are small, i.e., for which any coordinate is a cancellation of integer multiples of initial coordinates. However, floating-point computations are notoriously inadequate when cancellations occur since it often implies huge losses of precision (and thus a possibly dramatic growth of relative errors). Moreover, the precision is rather low (usually 53 bits), though the number of operations performed may be exponential with the dimension. If the operations reuse the variables sequentially, then one may run out of precision simply because of the accumulation of the errors. Finally, there is no efficient way to check the optimality of a solution but to re-run the whole algorithm in rational arithmetic: by comparing the length of the output vector with the lattice determinant, one can check that it looks reasonable, but it could be that (much) better solutions have been missed.

In the present paper, we give the first analysis of the influence of floating-point arithmetic within the KFP enumeration algorithm. More precisely, we show that if it is called on an LLL-reduced basis of a lattice made of integer vectors and uses floating-point arithmetic with a precision that is $\Omega(d)$ (the constant being explicit), then it finds the desired solution, i.e., a vector reaching the lattice minimum λ. Moreover, if the lattice is known only approximately (which may be the case for the projected sublattices in BKZ-style algorithms), then it finds a close to optimal solution. Finally, we also prove that the floating-point enumeration involves essentially the same number of arithmetic operations as the rational one. The results hold in a broad context: the technique can be adapted to fixed-point arithmetic, a weak condition is required for the input basis (if the input basis is not LLL-reduced, then the cost of the enumeration would grow dramatically), and the input may not be known exactly. Furthermore, the worst-case precision may be provably and adaptively decreased to a usually much smaller sufficient precision that can be computed efficiently from a given input basis. Double precision seems to suffice for KFP for all computationally tractable dimensions.

For the result to be valid, KFP has to be slightly modified (essentially, the initial upper bound has to be enlarged). The proof relies on a subtle analysis of the floating-point variant with respect to the rational enumeration: because of internal tests whose outcomes may differ due to inaccuracies, the execution of the floating-point variant may not mimic at all the ideal one. After working around that difficulty, the proof reduces to standard error analysis. To obtain a low sufficient precision, we heavily use the LLL-reducedness of the input basis.

Our result complements the Nguyen-Stehlé floating-point LLL [27]. By combining these two results, the use of floating-point arithmetic in all practical lattice algorithms may be made rigorous. Providing tight conditions leading to guarantees for the enumeration algorithm is likely to lead to significantly faster algorithms. Since the possible troubles coming from the use of floating-point arithmetic are better understood, one may work around them in the cheapest valid way rather than using unnecessarily large precisions. Like LLL [34], one may hope to design combinations of reduction algorithms whose arithmetic handling is oblivious to the user, that are guaranteed and as fast as possible. A good understanding of the underlying numerical stability issues provides a firm ground to study other questions. Furthermore, the knowledge of a small sufficient precision for the enumeration algorithm is an invaluable ingredient for hardware-based enumeration: in software, one should not use a precision cruder than the processor double precision; in hardware, however, the smaller the precision the faster. Overall, the floating-point analysis of the enumeration algorithm is a step towards intense cryptanalytic computations.

ROAD-MAP. In Section 2, we give the necessary background on lattices and floating-point arithmetic. In Section 3, we precisely describe the algorithm under scope and describe our results. We give elements of the proofs in Section 4, the more technical details being postponed to the appendix of the full version. In Section 5, we discuss the practicality of our results.

NOTATIONS. If $x \in \mathbb{R}$, we denote by $\lfloor x \rceil$ its closest integer (if there are two possibilities, the even one is chosen). A variable $\bar{x}$ is supposed to approximate the corresponding x, and we define $\Delta x = |\bar{x} - x|$.

REMARKS. For simplicity, we will only consider SVP. The results can be extended to CVP. Many variables occur in the text. This is due to the combined technicalities of floating-point arithmetic and LLL. This also comes from the will to provide explicit bounds, which is necessary to actually derive rigorous implementations. Here is a heuristic glossary for a first reading: the LLL-parameters $\delta, \eta, \alpha, \rho$ are essentially $1, 1/2, \sqrt{4/3}, \sqrt{3}$; the variables $C_1, C_2, \ldots$ are $\tilde{O}(1)$; the variables ϵ and ϵ' quantify inaccuracies and are negligible, whereas K is close to 1.

2 Reminders on Lattices and Floating-Point Arithmetic

We give some quick reminders on floating-point arithmetic and lattices. For more details, we respectively refer to [16] and [6].

FLOATING-POINT ARITHMETIC. A precision t *floating-point number* is a triple $(s, e, m) \in \{0, 1\} \times \mathbb{Z} \times (\mathbb{Z} \cap [2^{t-1}, 2^t - 1])$. It represents the real $(-1)^s \cdot m \cdot$

2^{e-t+1}. The *unit in the last place* is $\epsilon = 2^{-t+1}$. If $a \in \mathbb{R}$, we denote by $\diamond(a)$ the floating-point number that is closest to a (the one with an even m if there are two solutions). We have $|a - \diamond(a)| \leq \epsilon/2 \cdot |a|$. If a and b are two floating-point numbers, we define $a \oplus b$, $a \ominus b$ and $a \otimes b$ by $\diamond(a+b)$, $\diamond(a-b)$ and $\diamond(a \cdot b)$. The *double precision* $t = 53$ is a common choice as $\oplus$, $\ominus$ and $\otimes$ are implemented at the processor level in most computers. In practice, and for the KFP enumeration in particular, one should use double precision as much as possible. However, asymptotically with respect to the growing lattice dimension d, we will need $t = \Omega(d)$.

Gram-Schmidt orthogonalization. Let $\boldsymbol{b}_1, \ldots, \boldsymbol{b}_d$ be linearly independent vectors. We define their Gram-Schmidt orthogonalization by $\boldsymbol{b}_i^* = \boldsymbol{b}_i - \sum_{j<i} \mu_{i,j} \boldsymbol{b}_j^*$ with $\mu_{i,j} = \frac{\langle \boldsymbol{b}_i, \boldsymbol{b}_j^* \rangle}{\|\boldsymbol{b}_j^*\|^2}$ for $i > j$. We define $r_i = \|\boldsymbol{b}_i^*\|^2$. The $\mu_{i,j}$'s and r_i's. are the *Gram-Schmidt coefficients*. The $\boldsymbol{b}_i^*$'s are pairwise orthogonal. If the $\boldsymbol{b}_i$'s are integral, then the $\mu_{i,j}$'s are rational and can be computed in polynomial time with the formula above.

LLL-reduction. Let $\eta \in [1/2, 1)$ and $\delta \in (\eta^2, 1)$. Consider a lattice basis $\boldsymbol{b}_1, \ldots, \boldsymbol{b}_d$ and its corresponding $\boldsymbol{b}_i^*$'s and $\mu_{i,j}$'s. The basis is said to be (δ, η)-*LLL-reduced* if for all $i > j$ we have $|\mu_{i,j}| \leq \eta$ and $\delta \|\boldsymbol{b}_{i-1}^*\|^2 \leq \|\boldsymbol{b}_i^* + \mu_{i,i-1} \boldsymbol{b}_{i-1}^*\|^2$. This directly implies that the lengths of the $\boldsymbol{b}_i^*$'s cannot decrease too fast: if $\alpha := (\delta - \eta^2)^{-1/2}$ then $\alpha^2 r_i \geq r_{i-1}$. In this paper, we will further assume that $\delta > \eta^2 + (1+\eta)^{-2}$. This assumption is reasonable, since before starting an enumeration one should always LLL-reduce the lattice with δ close to 1 and η close to 1/2. Our analysis can be adapted to the general case, but this complicates the exposure for a useless situation. Lenstra, Lenstra and Lovász [20] gave an algorithm that computes an LLL-reduced basis from an arbitrary integral basis in time $O(d^5 n \log^3 B)$ where B is the maximum of the lengths of the input vectors. Using (low precision) floating-point arithmetic for the Gram-Schmidt computations, Nguyen and Stehlé [27] decreased that complexity to $O(d^4 n(d + \log B) \log B)$. Their algorithm requires $\eta > 1/2$. They rely on floating-point approximation to the Gram-Schmidt orthogonalization, which is much cheaper to obtain than computing the exact one. As an intermediate result, they show that if the input basis is LLL-reduced and if the computations are based on the exact Gram matrix (the matrix of the pairwise scalar products of the basis vectors), then this approximation is accurate even with low precision (linear with respect to the dimension).

Theorem 1 ([27]). *Let $\boldsymbol{b}_1, \ldots, \boldsymbol{b}_d \in \mathbb{Z}^n$ be a (δ, η)-LLL-reduced basis, with $\eta \in [1/2, 1)$ and $\delta \in (\eta^2, 1)$. Let $u \in (0, 1/16)$ and $\rho = (1 + \eta + u)(\delta - \eta^2)^{-1/2}$. Let t be such that $C_1 \rho^{2d} \epsilon < u$ where $\epsilon = 2^{-t+1}$ and $C_1 = 32d^2$. Starting from the Gram matrix of the $\boldsymbol{b}_i$'s and using precision t floating-point arithmetic, one can compute some $\bar{r}_i$'s and $\bar{\mu}_{i,j}$'s such that:*

$$\forall i > j, |\bar{\mu}_{i,j} - \mu_{i,j}| \leq C_1 \rho^{2j} \epsilon \ \ \textit{and} \ \ \forall i, |\bar{r}_i - r_i| \leq C_1 \rho^{2i} \epsilon \cdot r_i.$$

3 Floating-Point Lattice Enumeration

The usual method to solve SVP and CVP relies on the KFP enumeration [19,8]. We refer to [1] for a comprehensive survey. Here we will consider the variant due to Schnorr and Euchner [32] since it is the fastest and the one used in NTL. After describing the algorithm, we explain how to use floating-point arithmetic and finally give our main results.

3.1 The Enumeration Algorithm

The KFP algorithm for SVP takes as input a lattice basis and returns a shortest non-zero lattice vector. For this, it considers some A and finds all solutions $(x_1, \ldots, x_d) \in \mathbb{Z}^d$ to the equation

$$\left\| \sum_{i=1}^{d} x_i \boldsymbol{b}_i \right\|^2 \leq A. \quad (1)$$

If $A \geq \|\boldsymbol{b}_1\|^2$, then the set of solutions is non-trivial and SVP is solved by keeping the best one. Equation (1) is equivalent to

$$\sum_{i=1}^{d} \left(x_i + \sum_{j=i+1}^{d} \mu_{j,i} x_j \right)^2 r_i \leq A. \quad (2)$$

We let $c_i = -\sum_{j=i+1}^{d} \mu_{j,i} x_j$ and perform the change of variable $y_i := x_i - c_i$. This corresponds to applying to $\boldsymbol{x}$ the triangular matrix whose diagonal coefficients are 1 and whose off-diagonal coefficients are the $\mu_{i,j}$'s. Any sequence $(y_i, \ldots, y_d)$ corresponds to a unique sequence $(x_i, \ldots, x_d)$. Equation (2) becomes $\sum_{i=1}^{d} y_i^2 r_i \leq A$, which implies that:

$$\begin{aligned} y_d^2 r_d &\leq A, \\ y_{d-1}^2 r_{d-1} &\leq A - y_d^2 r_d, \\ &\ldots \\ y_1^2 r_1 &\leq A - \sum_{j=2}^{d} y_j^2 r_j. \end{aligned}$$

KFP finds all y_d's satisfying the first equation, then all (y_{d-1}, y_d)'s satisfying the second equation, etc. until it discloses all $(y_1, \ldots, y_d)$'s satisfying the last equation. Let $i < d$. Suppose that $y_{i+1}, \ldots, y_d$ are already set. Then there is a finite number of possibilities for y_i since y_i belongs to a bounded interval and is the fixed shift (by c_i) of the integer variable x_i. The number of possibilities for y_i is $\leq 1 + 2\sqrt{A/r_i}$. This shows that the bigger the r_i's, the faster the enumeration. We will see that big r_i's also help decreasing the required floating-point precision needed for the computations. Overall, KFP consists in trying to build solution vectors $\sum_{i=1}^{d} x_i \boldsymbol{b}_i$ to Equation (1) by successively looking at the projections orthogonally to the spans of $(\boldsymbol{b}_1, \ldots, \boldsymbol{b}_i)$ for a decreasing i. For a given choice of $(x_{i+1}, \ldots, x_d)$, the variable x_i belongs to an interval centered in c_i. Its length is $\sqrt{\frac{A-\ell_{i+1}}{r_i}}$, where $\ell_{i+1} := \sum_{j>i} y_j^2 r_j$.

Schnorr and Euchner improved KFP as follows. Suppose $(x_{i+1}, \ldots, x_d)$ is set. Instead of looking at the possible x_i's in a straight increasing fashion, they are chosen from the center of the interval to its borders: the first value is $\lfloor c_i \rceil$, then the integer that is second closest to c_i, etc. This has the effect of sorting the ℓ_i's by increasing order, and thus of maximizing the likelihood of quickly finding a solution to Equation (1). Once a solution is found, the value of A may be decreased, which possibly cuts off many branches of the execution tree. In Figure 1, we give a detailed description of the enumeration algorithm using the Schnorr-Euchner zig-zag path. The vector $\boldsymbol{sol}$ stores the non-zero vector $\boldsymbol{x}$ that is currently thought as minimizing $\| \sum_{i \leq d} x_i \boldsymbol{b}_i \|$. It remains $\mathbf{0}$ as long as no length below $\sqrt{A}$ has been found. The Δx_i's and $\Delta^2 x_i$'s are used to implement the zig-zag path.

Input: A bound A. Approximations $\bar{\mu}_{i,j}$'s and $\bar{r}_i$'s to the Gram-Schmidt coefficients of a possibly unknown basis $\boldsymbol{b}_1, \ldots, \boldsymbol{b}_d$.
Output: A coordinate vector $\boldsymbol{x} \in \mathbb{Z}^d \setminus \{\mathbf{0}\}$ such that $\sum_{i=1}^d x_i \boldsymbol{b}_i$ is likely to reach the lattice minimum.

1. $\boldsymbol{x} := (1, 0, \ldots, 0)$; $\boldsymbol{\Delta x} := (1, 0, \ldots, 0)$; $\boldsymbol{\Delta^2 x} := (1, -1, \ldots, -1)$; $\boldsymbol{sol} := \mathbf{0}$.
2. $\boldsymbol{c}, \boldsymbol{\ell}, \boldsymbol{y} := \mathbf{0}$.
3. $i := 1$. Repeat
4. $\quad y_i := |x_i - c_i|$; $\ell_i := \ell_{i+1} + y_i^2 r_i$.
5. $\quad$ If $\ell_i \leq A$ and $i = 1$, then $(\boldsymbol{sol}, A) :=$ `update`$(\boldsymbol{sol}, A, \boldsymbol{x}, \ell_1)$.
6. $\quad$ If $\ell_i \leq A$ and $i > 1$, then $i := i - 1$ and
7. $\qquad c_i := -\sum_{j=i+1}^d x_j \mu_{j,i}$.
8. $\qquad x_i := \lfloor c_i \rceil$; $\Delta x_i := 0$; if $c_i < x_i$ then $\Delta^2 x_i := 1$ else $\Delta^2 x_i := -1$.
9. $\quad$ Else if $\ell_i > A$ and $i = d$ return $\boldsymbol{sol}$ and stop.
10. $\quad$ Else $i := i + 1$ and
11. $\qquad \Delta^2 x_i := -\Delta^2 x_i$; $\Delta x_i := -\Delta x_i + \Delta^2 x_i$; $x_i := x_i + \Delta x_i$.

Fig. 1. The Schnorr-Euchner variant of the KFP enumeration algorithm

The algorithm of Figure 1 calls an `update` routine. In the ideal case, i.e., with correct input Gram-Schmidt coefficients and exact computations, we simply take $\texttt{update}_1(\boldsymbol{sol}, A, \boldsymbol{x}, \ell_1) = (\boldsymbol{x}, \ell_1)$. If we use floating-point arithmetic, however, this strategy may lead us to cut off branches of the tree that could contain the minimal non-zero length: if the computed approximation to ℓ_1 under-estimates it and if the lattice minimum is between both values and has not been reached yet, it will be missed. One can avoid this pitfall when floating-point arithmetic is used but the lattice is perfectly known, i.e., the genuine $\boldsymbol{b}_i$'s or the correct Gram-Schmidt quantities are given. In that situation, it is useful to consider $\texttt{update}_2$ defined as follows: $\texttt{update}_2(\boldsymbol{sol}, A, \boldsymbol{x}, \ell_1) = (\boldsymbol{x}, A)$ when $\boldsymbol{sol} = \mathbf{0}$ or $\| \sum_i x_i \boldsymbol{b}_i \| \leq \| \sum_i sol_i \boldsymbol{b}_i \|$ (exactly), and $\texttt{update}_2(\boldsymbol{sol}, A, \boldsymbol{x}, \ell_1) = (\boldsymbol{sol}, A)$ otherwise.

When using floating-point arithmetic, it is crucial to specify the order in which the operations are performed. At Step 4, we will evaluate the term $y_i^2 r_i$ as: $\bar{r}_i \otimes (\bar{y}_i \otimes \bar{y}_i)$. At Step 7, we will evaluate $\sum_{j=i+1} x_j \mu_{j,i}$ as $(x_{i+1} \otimes \bar{\mu}_{i+1,i}) \oplus [(x_{i+2} \otimes \bar{\mu}_{i+2,i}) \oplus [\ldots \oplus (x_d \otimes \bar{\mu}_{d,i}) \ldots]]$. Finally, notice that the x_i's, Δx_i's, $\Delta^2 x_i$'s and sol_i's remain integers.

An iteration of the loop is uniquely determined by the values of i and $(x_i, \ldots, x_d)$ at the beginning of the iteration. We say that the state is $\sigma = (i, [x_i, \ldots, x_d])$. Let $i \leq d$ and $x_i, \ldots, x_d \in \mathbb{Z}$. The floating-point algorithm and the exact algorithm do not necessarily perform the same iterations, and even if they do they may not be performed in the same order. It is thus impossible to compare the values of the variables for a given loop iteration. However, one may compare the values of the variables for a given state of the loop. In both the exact and floating-point variants, the values of the c_i's, y_i's and ℓ_i's do not depend on the iteration, but only on the state. Furthermore, these values are well-defined even if they are not actually computed: they do not depend on the initial bound A, nor on the existence of an iteration with the right state, nor in the order in which the states are visited. Consider a variable of the algorithm. We use the notation v to represent its value at a given state with exact computations and $\bar{v}$ its value at the same state with floating-point computations.

3.2 Main Results

We consider a lattice basis $\boldsymbol{b}_1, \ldots, \boldsymbol{b}_d$ that is (δ, η)- LLL-reduced with $\eta \in [1/2, 1)$ and $\eta^2 + \frac{1}{(1+\eta)^2} < \delta < 1$. We let $\alpha = \frac{1}{\sqrt{\delta - \eta^2}}$ and $\rho = (1+\eta)\alpha$. The minimum of the lattice spanned by the $\boldsymbol{b}_i$'s is denoted by λ. Below, when using KFP, the basis may not be known. In that case, its Gram-Schmidt coefficients or approximations thereof are known. The former situation may arise if one knows only the Gram matrix of the basis. The latter is typical of BKZ-style algorithms: one tries to reduce a large-dimensional lattice basis $\boldsymbol{b}_1, \ldots, \boldsymbol{b}_d$ by enumerating short vectors of lattices spanned by the projections of the vectors $\boldsymbol{b}_{i+1}, \ldots, \boldsymbol{b}_{i+k}$ orthogonally to $\boldsymbol{b}_1, \ldots, \boldsymbol{b}_i$, for some i and k; usually, one only knows approximations to the Gram-Schmidt coefficients of the projected k-dimensional basis.

Suppose we use floating-point arithmetic in the enumeration procedure, as described above. We denote by ϵ the unit in the last place and we define $K = 1 + \epsilon/2 \approx 1$. We allow the input Gram-Schmidt coefficients to be incorrect. For this purpose, we define:

$$\kappa = \max\left(\max_{i>j} \frac{\Delta \mu_{i,j}}{\epsilon}, \max_i \frac{\Delta r_i}{r_i \cdot \epsilon} \right).$$

If the Gram-Schmidt coefficients are exactly known and then rounded, we have $\kappa \leq 1$. They can also be computed as mentioned in Theorem 1, in which case we have $\kappa \leq C_1 \rho^{2d} (1+u')^{2d}$ for some small $u' > 0$.

To simplify the theorems below, we introduce some notation. We define $R = (1+\kappa\epsilon)\cdot\max_i r_i$: it bounds all the r_i's as well as the $r_i + \Delta r_i$'s. We also define:

$$C_2 = \frac{\kappa+2}{\alpha-1} + \frac{\kappa+4}{\rho-1}, \qquad C_3 = \frac{2\alpha(2+\kappa+2C_2)}{1+\eta-\alpha},$$
$$\epsilon' = 2\frac{R}{r_1}\left[(1+\kappa)\alpha^{2d} + (2C_2+C_3)\rho^d\right]\cdot\epsilon, \qquad C_4 = C_3\frac{K+d\epsilon}{1-\epsilon'}(2+d\epsilon).$$

The following theorem shows that when some exact knowledge of the lattice is provided then the floating-point enumeration solves SVP, if the precision is $\Omega(d)$ and the initial length upper bound is slightly increased in order to take care of the inaccuracies. In particular, in the most usual case where the r_i's decrease, one can choose $A = r_1\cdot(1+(2d+C_3\rho^d)\epsilon)$, which is only slightly larger than r_1. If the r_i's do not decrease, they can still be assumed of the same order of magnitude (up to a factor $2^{O(d)}$), thanks to the LLL-reducedness of the input basis, and the a priori knowledge that larger r_i's will not be used in vectors reaching the minimum.

Theorem 2. *Consider the floating-point KFP algorithm described in Subsection 3.1. Suppose that either the $\mathbf{b}_i$'s are known or that the Gram-Schmidt quantities are correct, and that the* `update`$_2$ *function is used. We assume that $C_2\rho^d\cdot\epsilon \leq 0.01$ and $A \geq (1+2d\epsilon)\cdot\lambda^2 + C_3\rho^d\epsilon\cdot R$. Then the returned coordinates* **sol** *satisfy $\|\sum_{i\leq d} sol_i\mathbf{b}_i\| = \lambda$.*

In the theorem above, we do not cut off branches of the computation once a short vector has been found: we keep the initial bound A. It is possible to decrease A each time a significantly shorter vector is found. Suppose a vector of exact squared norm $A' < A$ has been found. Then we can set $A = \min(A, A'(1+\epsilon''))$, for a well chosen ϵ'' that can be made explicit. This takes care of possible slight over-estimates of internal ℓ_i's which could erroneously lead to the removal of useful loop iterations. For the sake of simplicity, we do not consider this variant here.

Within BKZ-style algorithms, one may only know approximations to the Gram-Schmidt coefficients of the input basis, making Theorem 2 useless in such situations. Furthermore, due to the input uncertainty, one may not be able to decide which is the shortest between two vectors of close-by lengths: one cannot do better than finding a vector which is not much longer than λ. Of course, if there is a sufficient gap between λ and the length of any lattice vector different that does not reach the minimum, then an optimal solution will be found. The theorem below shows that finding a close to optimal vector is actually possible.

Theorem 3. *Consider the floating-point KFP algorithm described in Subsection 3.1, with the* `update`$_1$ *function. Let $\gamma = \|\sum_i sol_i\mathbf{b}_i\|$ be the norm of the found solution. If $A \geq \bar{r}_1$ and $\epsilon' < 0.01$, then:*

$$\lambda^2 \leq \gamma^2 \leq (1+4d\epsilon)\cdot\lambda^2 + C_4\max\left(1,\frac{A}{r_1}\right)\rho^d\epsilon\cdot R.$$

It should be noted that floating-point variants of BKZ cannot solve their internal SVP instantiations exactly: the best they can do is to solve $(1+\epsilon")$-SVP instantiations instead, for some small $\epsilon"$. However, with a small enough $\epsilon"$, this does not change significantly the overall quality of the output bases.

The two results above provide as good as could be expected correctness guarantees to the floating-point enumeration. However, since the algorithm is not the rational one, the complexity analyzes do not hold anymore. The following theorem shows that the overhead of the floating-point enumeration with respect to the rational one is small.

Theorem 4. *Consider the floating-point KFP algorithm described in Subsection 3.1 with either of the update functions and either the knowledge of the basis or the Gram-Schmidt coefficients or only approximations thereof. Let* $\gamma = \|\sum_i sol_i \boldsymbol{b}_i\|$ *be the norm of the found solution. We suppose that* $\epsilon' < 0.01$. *Then the number of loop iterations is lower than the number of loop iterations of the rational algorithm given the genuine basis and an input bound* $A' = (1+d\epsilon)\cdot A + C_4 \max\left(1, \frac{A}{r_1}\right)\rho^d \epsilon \cdot R$.

As a consequence of Theorems 2 and 4, the cost of Kannan's algorithm [19] can be decreased from $\text{Poly}(n, \log B)\cdot d^{\frac{d}{2e}(1+o(1))}$ (see [14]) to $\left(d^{\frac{d}{2e}} + \text{Poly}(n, \log B)\right)\cdot d^{o(d)}$: it suffices to use rationals everywhere but in the enumerations which should be performed with precision $\Theta(d)$.

4 Error Analysis of the Floating-Point Enumeration

We now turn to the proofs of Theorems 2, 3 and 4. We proceed by proving that the computed lengths $\bar{\ell}_i$ of the projected vectors are accurate. Lemma 1 means that $\bar{\ell}_1$ cannot be much larger than ℓ_1, which suffices for Theorem 3. For the other results, we need the converse: Lemma 2 means that the true ℓ_i cannot be much larger than the computed one. The proofs of Lemmata 1 and 2 are explained in Subsection 4.2.

As mentioned in Section 3, an $\bar{\ell}_i$ computed by the floating-point algorithm may not correspond to any ℓ_i computed by the rational one with the same bound A, and vice-versa. To be rigorous, we need the following definitions. For $\boldsymbol{x} \in \mathbb{Z}^d$, we let $n(\boldsymbol{x}) = \|\sum_{i=1}^d x_i \boldsymbol{b}_i\|^2$ and $\bar{n}(\boldsymbol{x})$ its approximation as would be computed by the enumeration were the state $(1, [x_1, \ldots, x_d])$ visited. We use the notations and hypotheses of Subsection 3.1.

Lemma 1. *Suppose that* $C_2\rho^d \cdot \epsilon < 0.01$. *Let* $\boldsymbol{x} \in \mathbb{Z}^d$. *If* $n(\boldsymbol{x}) \leq r_1$, *then:*

$$\bar{n}(\boldsymbol{x}) \leq (1+2d\epsilon)\cdot n(\boldsymbol{x}) + C_3\rho^d\epsilon \cdot R.$$

Lemma 2. *Suppose that* $\epsilon' < 0.01$. *Let* $\boldsymbol{x} \in \mathbb{Z}^d$ *and* $i \leq d$. *We consider the state* $(i, [x_i, \ldots, x_d])$. *Then*

$$\ell_i \leq (1+d\epsilon)\cdot \bar{\ell}_i + C_3 \max\left(1, \frac{\bar{\ell}_i(K+d\epsilon)}{r_1(1-\epsilon')}\right)\rho^d\epsilon \cdot R.$$

4.1 Using Lemmata 1 and 2 to Prove the Theorems

Let us first prove Theorem 2 from Lemma 1. Let $(x_1, \ldots, x_d)$ be the coordinates of a shortest vector. If the state $(1, \boldsymbol{x})$ is considered by the floating-point algorithm with $A \geq (1 + 2d\epsilon) \cdot \lambda^2 + C_3 \rho^d \epsilon \cdot R$, then a shortest vector will be found. Making sure that $(1, \boldsymbol{x})$ is indeed considered is the purpose of the following lemma. It relies on subtle properties of the floating-point model, in particular that the rounding is a non-decreasing function.

Lemma 3. *If one uses the* $\texttt{update}_1$ *function within the enumeration, then all coordinate vectors* $\boldsymbol{x}$ *such that* $\bar{n}(\boldsymbol{x}) \leq A$ *will indeed be considered during the execution.*

Proof. Let $\boldsymbol{x} \in \mathbb{Z}^d$ with $\bar{n}(\boldsymbol{x}) \leq A$. We show by induction on decreasing i that $(i, [x_i, \ldots, x_d])$ is considered and that at this moment the test $\bar{\ell}_i \leq A$ is satisfied. Let $i \leq d$. We consider the sequence $(\sigma_1, \ldots, \sigma_\tau)$ of considered states $(i, [X, x_{i+1}, \ldots, x_d])$ with $X \in \mathbb{Z}$. It is non-empty if $i = d$, and it is also non-empty if $i < d$ by induction hypothesis.

The sequence $\left(\bar{\ell}_i(\sigma_t)\right)_t$ is non-decreasing. The first integer $X = x_i(\sigma_1)$ is exactly $\lfloor \bar{c}_i \rceil$. The computation of $x_i(\sigma_t)$ from $x_i(\sigma_{t-1})$ is exact, and the distance between $x_i(\sigma_t)$ and $\bar{c}_i$ is non-decreasing. Since the rounding function is non-decreasing, the sequence $(\bar{y}_i(\sigma_t))_t$ is also non-decreasing. For the same reason, the sequence $\left(\bar{\ell}_i(\sigma_t)\right)_t$ is non-decreasing.

Consider the value $\bar{\ell}$ of $\bar{\ell}_i$ were it computed with $(x_i, \ldots, x_d)$. We have $\bar{\ell} \leq \bar{n}(\boldsymbol{x}) \leq A$. Since $\bar{\ell}_i(\sigma_\tau) > A$, there must exist t such that $x_i(\sigma_t) = x_i$ and the test $\bar{\ell}_i \leq A$ is satisfied for that state σ_t. □

We now prove Theorem 3. If we use $\texttt{update}_2$, the bound A may decrease during the execution, to finally reach a value A_{end}. The final output would have been the same if we had started with $A = A_{end}$. We consider that it is the case, which implies that A is not modified during the execution. Let $\boldsymbol{x} \in \mathbb{Z}^d$ such that $n(\boldsymbol{x}) = \lambda^2$. Lemma 1 implies that $\bar{n}(\boldsymbol{x}) \leq (1 + 2d\epsilon) \cdot \lambda^2 + C_3 \rho^d \epsilon \cdot R$. We must have $A \leq (1 + 2d\epsilon) \cdot \lambda^2 + C_3 \rho^d \epsilon \cdot R$ since otherwise A would have been decreased after $\boldsymbol{x}$ was found. Applying Lemma 2 with $\boldsymbol{sol}$ and using the above bound on A provides the result.

For Theorem 4, consider a state $(i, [x_i, \ldots, x_d])$ with a successful test $\bar{\ell}_i \leq A$. Lemma 2 gives $\ell_i \leq (1 + d\epsilon) \cdot A + C_3 \max\left(1, \frac{A(K+d\epsilon)}{r_1(1-\epsilon')}\right) \rho^d \epsilon \cdot R \leq A'$. Therefore, the exact algorithm with the bound A' would have considered this state and the corresponding test would have been successful as well. Moreover, there are as many failed loop iterations with $i < d$ as successful loop iterations with $i > 1$. This completes the proof.

4.2 Proving Lemmata 1 and 2

The proofs of Lemmata 1 and 2 rely on standard techniques of floating-point error analysis. We simultaneously bound the errors and the variables, which leads us to use an induction on the decreasing index i. Within the induction step, we rely on three basic facts whose proofs are tedious but straightforward. They are given in the appendix of the full version.

Lemma 4. *Suppose that $C_2\rho^d\epsilon \leq 0.01$. Suppose we are at the end of Step 4 of some loop iteration with state $(i, [x_i, \ldots, x_d])$. If there exists a constant $\nu \geq 1$ such that for any $j > i$ we have $y_j \leq \nu\alpha^{j-1}$, then*

$$\Delta c_i \leq C_2\nu\alpha^d(1+\eta)^{d-i}\epsilon \quad \textit{and} \quad \Delta y_i \leq y_i\epsilon/2 + KC_2\nu\alpha^d(1+\eta)^{d-i}\epsilon.$$

Lemma 5. *At Step 4 of the floating-point algorithm, we have:*

$$|(\bar{y}_i \otimes \bar{y}_i) \otimes \bar{r}_i - r_i y_i^2| \leq RK^2[(\kappa+1)y_i^2\epsilon + (2y_i + \Delta y_i)\Delta y_i]$$

Lemma 6. *Suppose that $C_2\rho^d\epsilon \leq 0.01$. Suppose we are at the end of Step 4 of some loop iteration with state $(i, [x_i, \ldots, x_d])$. If there exists a constant $\nu \geq 1$, such that for any $j \geq i$ we have $y_i \leq \nu\alpha^{i-1}$, then:*

$$\Delta\ell_i \leq d\epsilon \cdot \ell_i + C_3\nu^2\rho^d\epsilon \cdot R \quad \textit{and} \quad \Delta\ell_i \leq d\epsilon \cdot \bar{\ell}_i + C_3\nu^2\rho^d\epsilon \cdot R.$$

We can now prove Lemma 1. Let $\boldsymbol{x} \in \mathbb{Z}^d$ such that $n(\boldsymbol{x}) \leq r_1$. Since the basis is LLL-reduced, the y_i's corresponding to $\boldsymbol{w}$ satisfy $y_i \leq \sqrt{n(\boldsymbol{x})/r_i} \leq \sqrt{r_1/r_i} \leq \alpha^{i-1}$. The first part of Lemma 6 with $\nu = 1$ provides the result.

Finally, we prove Lemma 2. Let $\boldsymbol{x} \in \mathbb{Z}^d$ and $i \leq d$. We show by induction on j decreasing from d to i that the bound on Δc_j of Lemma 4 holds and that we have $y_j \leq \nu\alpha^{j-1}$, with $\nu = \max\left(1, \sqrt{\frac{\bar{\ell}_i(K+d\epsilon)}{r_1(1-\epsilon')}}\right)$. Lemma 2 will then follow from the second part of Lemma 6. Let $j \geq i$. By induction, we have $y_k < \nu\alpha^{k-1}$ for any $k > j$, so that the bounds of Lemma 4 hold. It remains to see that $y_j \leq \nu\alpha^{j-1}$. Lemmata 5 and 6 provide:

$$\begin{aligned} r_j y_j^2 \leq \ell_j \leq \bar{\ell}_j + \Delta\ell_j &\leq K\bar{\ell}_j + \Delta\ell_{j+1} + \left|(\bar{y}_j \otimes \bar{y}_j) \otimes \bar{r}_j - r_j y_j^2\right| \\ &\leq K\bar{\ell}_j + d\epsilon\bar{\ell}_j + C_3\nu^2\rho^d\epsilon R + RK^2\left[(\kappa+1)y_j^2\epsilon + (2y_j + \Delta y_j)\Delta y_j\right]. \end{aligned}$$

We use Lemma 4 to bound Δy_j in the equation above. This leads $P(y_j) \leq 0$, where P is the degree-2 polynomial with coefficients:

$$P_0 = -\bar{\ell}_j(K + d\epsilon) - C_3\nu^2R\rho^d\epsilon - RK^4(C_2\nu\rho^d\epsilon)^2,$$
$$P_1 = -2RK^4C_2\nu\alpha^{d-j}(1+\eta)^d\epsilon \quad \text{and} \quad P_2 = r_j - 2RK^3(\kappa+1)\epsilon.$$

The fact that $\epsilon' < 0.01$ implies that $P_2 > 0$ and thus that y_j is below the positive root of P. It can be checked that $P(\nu\alpha^{j-1}) \geq 0$, which implies that $y_j \leq \nu\alpha^{j-1}$. This completes the proof.

5 Practical Considerations

The algorithm described in Section 3 has been implemented in C++ and is freely distributed within `fplll-3.0` [5]. The code does not use the worst-case bounds above but remains guaranteed, as explained below. We also explain how our results may be used within BKZ-style algorithms.

5.1 Guaranteeing the Computations with Smaller Precision

The worst-case bounds given in Section 3 are very pessimistic for generic instantiations. This is due to the facts that all $|\mu_{i,j}|$'s (resp. r_{i-1}/r_i's) are bounded by their worst-case value η (resp. α^2) and all floating-point errors are considered to be always maximal and in the worst direction. Although they might occur, cases where all these bounds are tight are unlikely. In the worst-case analysis, we also use loose bounds to simplify the technicalities, though they do not modify the terms that are exponential with d. For $(\delta, \eta) = (0.99, 0.51)$, if the Gram-Schmidt coefficients are correct up to their last bit ($\kappa \leq 1$), the provably sufficient precision for a d-dimensional enumeration is $\approx 0.8 \cdot d$ (when d grows to infinity). To take advantage of the machine instructions, one is tempted to use double precision, i.e., $\epsilon = 2^{-52}$. In that case, the enumeration is guaranteed up to dimension ≈ 45 (for an output relative error $\leq 1\%$).

In practice, one should rather turn the worst-case error analysis into an algorithm. One can use the values the actual Gram-Schmidt coefficients rather than general upper bounds. If they are known approximately, one should take into consideration their intrinsic inaccuracies. The adaptive precision computation uses $O(d^2)$ arithmetic operations: Lemmata 4 and 6 are applied $O(d)$ times each and both perform $O(d)$ operations. This computation is thus dominated by the enumeration. The error computations are themselves performed in floating-point arithmetic, but one should be cautious with the rounding modes: since we try to upper bound a quantity, the default rounding to nearest should be replaced by roundings towards infinities and zero. In the code, we used MPFR [30] for that purpose.

The table below illustrates the above technique. Each entry corresponds to 10 samples of the following experiment. A $(d+1) \times d$ matrix B is sampled: for any i, $B[1, i]$ is a random integer with $100 \cdot d$ bits, $B[i+1, i]$ is 1 and the other entries are 0. The columns of the matrix B are then $(0.99, 0.51)$-LLL-reduced. Then the adaptive precision computation is performed. The precision is computed so that the algorithm is guaranteed to solve 1.01-SVP. One observes that double precision suffices for dimensions up to 90, which is higher than what is currently handleable in practice.

Dimension d	20	30	40	50	60	70	80
Worst-case required precision (Theorem 3)	33	41	49	57	66	74	82
Adaptively computed required precision (worst-case over the samples)	20	25	29	33	38	42	47

5.2 Enumerating within BKZ-Style Algorithms

With the floating-point LLL of Nguyen and Stehlé [27] and the present results, one may use floating-point arithmetic within BKZ-style algorithms in a guaranteed way. However, it is not clear yet how to maximize the efficiency while doing this. As a target, double precision should be used as much as possible, since multi-precision arithmetic is significantly slower.

A first solution consists in performing all operations with the same provably sufficient precision, provided by the bounds given in Section 3 after replacing κ by the bounds of Theorem 1 and R by $2\alpha^{2d} \cdot r_1$ (the vectors whose r_i's are $> \alpha^{2d} r_1$ cannot be used to create a vector of minimal non-zero length). Though the precision remains $O(d)$, it will be fairly large and slow multi-precision arithmetic will be necessary. It can be checked that the required precision can be decreased by a constant factor by noticing that in Theorem 1 the errors on $\mu_{i,j}$ and r_j depends on j.

Another possibility is to use a Gram-Schmidt orthogonalization with very high precision and then use the adaptive precision estimate described above. Double precision is likely to be sufficient for all reasonable values of the hierarchy parameter k, making the computed approximations to the Gram-Schmidt coefficients correct up to relative error $\approx 2^{-53}$. Since the enumerations are likely to dominate the overall cost, it is worth using multi-precision arithmetic to compute accurate Gram-Schmidt coefficients in order to be allowed double precision within the enumerations.

If the Gram-Schmidt computations are not negligible with respect to the enumerations, then one could try using double precision in all computations. This may be done by relying with the following strategy:

- Run the floating-point LLL algorithm with double precision for the Gram-Schmidt computations, with infinite loop detection (see [34]).
- If the double precision seemed to suffice (i.e., the execution terminated without an infinite loop detection), compute a posteriori accuracy bounds as described by Villard in [35].
- Run the adaptive precision computation to see if double precision suffices for the enumeration.

6 Concluding Remarks

We proved strong numerical properties of the KFP enumeration algorithm, which gives a stronger insight about the use of floating-point arithmetic within lattice reduction algorithms. To obtain a full hierarchy of reduction algorithms ranging from LLL to HKZ that efficiently relies on floating-point arithmetic, it only remains to see how to combine our new results with those on floating-point LLL from [27]. It would also be interesting to devise new techniques to decrease the required precision in order to be able to use double precision as often as possible.

However, we answered only one of the two main troubles related to BKZ-style algorithms: it is still unknown how to best use small dimensional lattice enumeration within a large dimensional reduction. It would be desirable to have an algorithm which is theoretically at least as good as the best current one [10], that would beat BKZ in practice and whose behavior would be perfectly understood. Once this will be done, there will remain to mount massive computational projects to assess the limits of current computers against lattice-based cryptography. It will then make sense to run the enumeration on hardware. Our analysis extends to fixed-point arithmetic, which is the natural arithmetical choice in hardware.

References

1. Agrell, E., Eriksson, T., Vardy, A., Zeger, K.: Closest point search in lattices. IEEE Transactions on Information Theory 48(8), 2201–2214 (2002)
2. Ajtai, M.: The shortest vector problem in l_2 is NP-hard for randomized reductions (extended abstract). In: Proc. of the 30th Symposium on the Theory of Computing (STOC 1998), pp. 284–293. ACM Press, New York (1998)
3. Ajtai, M., Dwork, C.: A public-key cryptosystem with worst-case/average-case equivalence. In: Proc. of the 29th Symposium on the Theory of Computing (STOC 1997), pp. 284–293. ACM Press, New York (1997)
4. Ajtai, M., Kumar, R., Sivakumar, D.: A sieve algorithm for the shortest lattice vector problem. In: Proc. of the 33rd Symposium on the Theory of Computing (STOC 2001), pp. 601–610. ACM Press, New York (2001)
5. Cadé, D., Pujol, X., Stehlé, D.: fplll-3.0, a floating-point LLL implementation, `http://perso.ens-lyon.fr/damien.stehle`
6. Cohen, H.: A Course in Computational Algebraic Number Theory, 2nd edn. Springer, Heidelberg (1995)
7. van Emde Boas, P.: Another NP-complete partition problem and the complexity of computing short vectors in a lattice. Technical report 81-04, Mathematisch Instituut, Universiteit van Amsterdam (1981)
8. Fincke, U., Pohst, M.: A procedure for determining algebraic integers of given norm. In: van Hulzen, J.A. (ed.) ISSAC 1983 and EUROCAL 1983. LNCS, vol. 162, pp. 194–202. Springer, Heidelberg (1983)
9. Gama, N., Howgrave-Graham, N., Koy, H., Nguyen, P.: Rankin's constant and blockwise lattice reduction. In: Dwork, C. (ed.) CRYPTO 2006. LNCS, vol. 4117, pp. 112–130. Springer, Heidelberg (2006)
10. Gama, N., Nguyen, P.: Finding short lattice vectors within Mordell's inequality. In: Proc. of the 40th Symposium on the Theory of Computing (STOC 2008), pp. 207–216. ACM Press, New York (2008)
11. Gama, N., Nguyen, P.: Predicting lattice reduction. In: Smart, N.P. (ed.) EUROCRYPT 2008. LNCS, vol. 4965, pp. 31–51. Springer, Heidelberg (2008)
12. Gentry, C., Peikert, C., Vaikuntanathan, V.: Trapdoors for hard lattices and new cryptographic constructions. In: Proc. of the 40th Symposium on the Theory of Computing (STOC 2008), pp. 197–206. ACM Press, New York (2008)
13. Goldreich, O., Goldwasser, S., Halevi, S.: Public-key cryptosystems from lattice reduction problems. In: Kaliski Jr., B.S. (ed.) CRYPTO 1997. LNCS, vol. 1294, pp. 112–131. Springer, Heidelberg (1997)
14. Hanrot, G., Stehlé, D.: Improved analysis of Kannan's shortest lattice vector algorithm (extended abstract). In: Menezes, A. (ed.) CRYPTO 2007. LNCS, vol. 4622, pp. 170–186. Springer, Heidelberg (2007)
15. Hassibi, A., Boyd, S.: Integer parameter estimation in linear models with applications to GPS. IEEE Transactions on signal process 46(11), 2938–2952 (1998)
16. Higham, N.: Accuracy and Stability of Numerical Algorithms. SIAM Publications, Philadelphia (2002)
17. Hoffstein, J., Pipher, J., Silverman, J.H.: NTRU: a ring based public key cryptosystem. In: Buhler, J.P. (ed.) ANTS 1998. LNCS, vol. 1423, pp. 267–288. Springer, Heidelberg (1998)

18. Joux, A., Stern, J.: Lattice reduction: a toolbox for the cryptanalyst. Journal of Cryptology 11(3), 161–185 (1998)
19. Kannan, R.: Improved algorithms for integer programming and related lattice problems. In: Proc. of the 15th Symposium on the Theory of Computing (STOC 1983), pp. 99–108. ACM Press, New York (1983)
20. Lenstra, A.K., Lenstra Jr., H.W., Lovász, L.: Factoring polynomials with rational coefficients. Mathematische Annalen 261, 513–534 (1982)
21. Lyubashevsky, V.: Lattice-based identification schemes secure under active attacks. In: Cramer, R. (ed.) PKC 2008. LNCS, vol. 4939, pp. 162–179. Springer, Heidelberg (2008)
22. Lyubashevsky, V., Micciancio, D.: Asymptotically efficient lattice-based digital signatures. In: Canetti, R. (ed.) TCC 2008. LNCS, vol. 4948, pp. 37–54. Springer, Heidelberg (2008)
23. Lyubashevsky, V., Micciancio, D., Peikert, C., Rosen, A.: SWIFFT: a modest proposal for FFT hashing. In: Nyberg, K. (ed.) FSE 2008. LNCS, vol. 5086, pp. 54–72. Springer, Heidelberg (2008)
24. Micciancio, D., Regev, O.: Lattice-based cryptography. In: Buchmann, J., Ding, J. (eds.) PQCrypto 2008. LNCS, vol. 5299. Springer, Heidelberg (2008)
25. Mow, W.H.: Maximum likelihood sequence estimation from the lattice viewpoint. IEEE Transactions on Information Theory 40, 1591–1600 (1994)
26. Nguyen, P.: Cryptanalysis of the Goldreich-Goldwasser-Halevi cryptosystem from Crypto 1997. In: Wiener, M. (ed.) CRYPTO 1999. LNCS, vol. 1666, pp. 288–304. Springer, Heidelberg (1999)
27. Nguyen, P., Stehlé, D.: Floating-point LLL revisited. In: Cramer, R. (ed.) EUROCRYPT 2005. LNCS, vol. 3494, pp. 215–233. Springer, Heidelberg (2005)
28. Nguyen, P., Vidick, T.: Sieve algorithms for the shortest vector problem are practical. Journal of Mathematical Cryptology (to appear, 2008)
29. Odlyzko, A.M.: The rise and fall of knapsack cryptosystems. In: Proc. of Cryptology and Computational Number Theory. In: Proc. of Symposia in Applied Mathematics, vol. 42, pp. 75–88. American Mathematical Society (1989)
30. The SPACES Project. MPFR, a LGPL-library for multiple-precision floating-point computations with exact rounding, `http://www.mpfr.org/`
31. Schnorr, C.P.: A hierarchy of polynomial lattice basis reduction algorithms. Theoretical Computer Science 53, 201–224 (1987)
32. Schnorr, C.P., Euchner, M.: Lattice basis reduction: improved practical algorithms and solving subset sum problems. Mathematics of Programming 66, 181–199 (1994)
33. Shoup, V.: NTL, Number Theory Library, `http://www.shoup.net/`
34. Stehlé, D.: Floating-point LLL: theoretical and practical aspects. In: Proc. of the LLL+25 conference (to appear)
35. Villard, G.: Certification of the QR factor R, and of lattice basis reducedness. In: Proc. of the 2007 International Symposium on Symbolic and Algebraic Computation (ISSAC 2007). ACM Press, New York (2007)

Solving Linear Equations Modulo Divisors: On Factoring Given Any Bits*

Mathias Herrmann and Alexander May

Horst Görtz Institute for IT-Security
Faculty of Mathematics
Ruhr Universität Bochum, Germany
mathias.herrmann@rub.de, alex.may@rub.de

Abstract. We study the problem of finding solutions to linear equations modulo an unknown divisor p of a known composite integer N. An important application of this problem is factorization of N with given bits of p. It is well-known that this problem is polynomial-time solvable if at most half of the bits of p are unknown and if the unknown bits are located in one *consecutive* block. We introduce an heuristic algorithm that extends *factoring with known bits* to an arbitrary number n of blocks. Surprisingly, we are able to show that $\ln(2) \approx 70\%$ of the bits are sufficient for any n in order to find the factorization. The algorithm's running time is however exponential in the parameter n. Thus, our algorithm is polynomial time only for $n = \mathcal{O}(\log\log N)$ blocks.

Keywords: Lattices, small roots, factoring with known bits.

1 Introduction

Finding solutions to polynomial modular equations is a central mathematical problem and lies at the heart of almost any cryptanalytic approach. For instance, most symmetric encryption functions can be interpreted as polynomial transformations from plaintexts to ciphertexts. Solving the corresponding polynomial equations yields the secret key.

Among all polynomial equations the linear equations $f(x_1, \ldots, x_n) = a_1x_1 + a_2x_2 + \cdots + a_nx_n$ play a special role, since they are often easier to solve. Many problems already admit a linear structure. For instance, the subset sum problem for finding a subset of $s_1, \ldots, s_n$ that sums to t asks for a 0,1-solution $(y_1, \ldots, y_n)$ of the linear equation $s_1x_1 + \cdots + s_nx_n - t = 0$. Special instances of this problem can be solved by lattice techniques [CJL+92].

Although many problems are inherently of non-linear type, solution strategies for these problems commonly involve some linearization step. In this work, we address the problem of solving *modular* linear equations $f(x_1, \ldots, x_n) = 0 \bmod N$ for some N with unknown factorization. Note that modular equations usually

* This research was supported by the German Research Foundation (DFG) as part of the project MA 2536/3-1.

J. Pieprzyk (Ed.): ASIACRYPT 2008, LNCS 5350, pp. 406–424, 2008.
© International Association for Cryptologic Research 2008

have many solutions $(y_1, \ldots, y_n) \in \mathbb{Z}_N^n$. An easy counting argument however shows that one can expect a unique solution whenever the product of the unknowns is smaller than the modulus - provided the coefficients a_i are uniformly distributed in $\mathbb{Z}_N$. More precisely, let X_i be upper bounds such that $|y_i| \leq X_i$ for $i = 1 \ldots n$. Then one can roughly expect a unique solution whenever the condition $\prod_i X_i \leq N$ holds.

It is folklore knowledge that under the same condition $\prod_i X_i \leq N$ the unique solution $(y_1, \ldots, y_n)$ can heuristically be recovered by computing a shortest vector in an n-dimensional lattice. In fact, this approach lies at the heart of many cryptanalytic results (see e.g. [GM97, NS01, Ngu04, BM06]). If in turn we have $\prod_i X_i \geq N^{1+\epsilon}$ then the linear equation usually has N^ϵ many solutions, which is exponential in the bit-size of N. So there is no hope to find efficient algorithms that in general improve on this bound, since one cannot even output all roots in polynomial time.

In the late 80's, Hastad [Has88] and Toffin, Girault, Vallée [GTV88] extended the lattice-based approach for linear equations to modular univariate monic polynomials $f(x) = a_0 + a_1 x + \cdots + a_{\delta-1}x^{\delta-1} + x^\delta$. In 1996, Coppersmith [Cop96b] further improved the bounds of [Has88, GTV88] to $|x_0| \leq N^{\frac{1}{\delta}}$ for lattice-based solutions that find small roots of $f(x)$. For modular univariate polynomials $f(x)$ there are again counting arguments that show that this bound cannot be improved in general. Even more astonishing than the improved bound is the fact that Coppersmith's method does neither rely on a heuristic nor on the computation of a shortest vector, but provably provides all roots smaller than this bound and runs in polynomial time using the L^3 algorithm [LLL82].

In the same year, Coppersmith [Cop96a] formulated another rigorous method for bivariate polynomials $f(x, y)$, see also [Cor07]. This method has several nice applications, most notably the problem of *factoring with high bits known* and also an algorithm that shows the deterministic polynomial time equivalence of factoring and computing the RSA secret key [May04, CM07]. In the *factoring with high bits known* problem, one is given an RSA modulus $N = pq$ and an approximation $\tilde{p}$ of p. This enables to compute an approximation $\tilde{q}$ of q, which leads to the bivariate polynomial equation $f(x, y) = (\tilde{p} + x)(\tilde{q} + y) - N$. Finding the unique solution in turn enables to factor. Coppersmith showed that this can be done in polynomial time given 50% of the bits of p and thereby improved upon a result from Rivest and Shamir [RS85], who required 60% of the bits of p. Using an oracle that answers arbitrary questions instead of returning bits of the prime factor, Maurer [Mau95] presented a probabilistic algorithm based on elliptic curves, that factors an integer N in polynomial time making at most $\epsilon \log N$ oracle queries for any $\epsilon > 0$.

In 2001, Howgrave-Graham [HG01] gave a reformulation of the *factoring with high bits known* problem, showing that the remaining bits of p can be recovered if $\gcd(\tilde{p} + x, N)$ is sufficiently large. This can also be stated as finding the root of the linear monic polynomial $f(x) = \tilde{p} + x \bmod p$ where $p \geq N^\beta$ for some $0 < \beta \leq 1$. Later, this was generalized by May [May03] to arbitrary monic modular polynomials of degree δ which results in the bound $|x_0| \leq N^{\frac{\beta^2}{\delta}}$. The result for *factoring with high bits known* follows for the choice $\beta = \frac{1}{2}$, $\delta = 1$.

Notice that in the *factoring with high bits known* problem, the unknown bits have to be in one consecutive block of bits. This variant of the factorization problem is strongly motivated by side-channel attacks that in most cases enable an attacker to recover some of the bits of the secret key. The attacker is then left with the problem of reconstructing the whole secret out of the obtained partial information. Unfortunately, the unknown part is in general not located in one consecutive bit block but widely spread over the whole bit string. This raises the question whether we can sharpen our tools to this general scenario.

Our contribution: We study the problem of finding small roots of linear modular polynomials $f(x_1, \ldots, x_n) = a_1x_1 + a_2x_2 + \cdots + a_nx_n + a_{n+1} \bmod p$ for some unknown $p \geq N^\beta$ that divides the known modulus N. This enables us to model the problem of *factoring with high bits known* to an arbitrary number n of unknown blocks. Namely, if the k-th unknown block starts in the ℓ-th bit position we choose $a_k = 2^\ell$.

We are able to show an explicit bound for the product $\prod_i X_i = N^\gamma$, where γ is a function in β and n. For the special case in which $p = N$, i.e. $\beta = 1$ and the modulus p is in fact known, we obtain the previously mentioned folklore bound $\prod_i X_i \leq N$. Naturally, the larger the number n of blocks, the smaller is the bound for $\prod_i X_i$ and the larger is the running time of our algorithm. In other words, the larger the number of blocks, the more bits of p we do have to know in the *factoring with known bits* problem. What is really surprising about our lattice-based method is that even for an arbitrary number n of blocks, our algorithm still requires only a constant fraction of the bits of p. More precisely, a fraction of $\ln(2) \approx 70\%$ of p is always sufficient to recover p.

Unfortunately, the running time for our algorithm heavily depends on n. Namely, the dimension of the lattice basis that we have to L^3-reduce grows exponentially in n. Thus, our algorithm is polynomial time only if $n = \mathcal{O}(\log \log N)$. For larger values of n, our algorithm gets super-polynomial. To the best of our knowledge state-of-the-art general purpose factorization algorithms like the GNFS cannot take advantage of extra information like given bits of one of the prime factors. Thus, our algorithm still outperforms the GNFS for the *factoring with known bits* problem provided that $n = o(\log^{\frac{1}{3}} N \log\log^{\frac{2}{3}} N)$.

We would like to notice that our analysis for arbitrary n yields a bound $\prod_i X_i \leq N^\gamma$ that holds no matter how the size of the unknowns are distributed among the X_i. In case the X_i are of strongly different sizes, one might even improve on the bound N^γ. For our starting point $n = 2$, we sketch such a general analysis for arbitrary sizes of X_1, X_2. The analysis shows that the bound for the product X_1X_2 is minimal when $X_1 = X_2$ and that it converges to the known Coppersmith result $N^{\frac{1}{4}}$ in the extreme case, where one of the X_i is set to $X_i = 1$.

Notice that if one of the upper bounds is set to $X_i = 1$ then the bivariate linear equation essentially collapses to a univariate equation. In this case, we also obtain the bound $N^{\frac{1}{4}}$ for the *factoring with known bits* problem. Thus, our algorithm does not only include the folklore bound as a special case but also the Coppersmith bound for univariate linear modular equations.

As our lattice-based algorithm eventually outputs multivariate polynomials over the integers, we are using a well-established heuristic [Cop97, BD00] for extracting the roots. We show experimentally that this heuristic works well in practice and always yielded the desired factorization. In addition to previous papers that proposed to use resultant or Gröbner basis computations, we use the multidimensional Newton method from numerical mathematics to efficiently extract the roots.

The paper is organized as follows. Section 2 recalls basic lattice theory. In Section 3 we give the analysis of bivariate linear equations modulo an unknown divisor. As noticed before, we prove a general bound that holds for all distributions of X_1, X_2 as well as sketch an optimized analysis for strongly unbalanced X_1, X_2. Section 4 generalizes the analysis to an arbitrary number n of variables. Here, we also establish the $\ln(2) \approx 70\%$ result for *factoring with known bits*. We experimentally verify the underlying heuristic in Section 5.

2 Preliminaries

Let $b_1, \ldots, b_k$ be linearly independent vectors in $\mathbb{R}^n$. Then the *lattice* spanned by $b_1, \ldots, b_k$ is the set of all integer linear combinations of $b_1, \ldots, b_k$. We call $b_1, \ldots, b_k$ a basis of L. The integer k is called the *dimension* or *rank* of the lattice and we say that the lattice has *full rank* if $k = n$.

Every nontrivial lattice in $\mathbb{R}^n$ has infinitely many bases, therefore we seek for *good* ones. The most important quality measure is the length of the basis vectors which corresponds to the basis vectors' orthogonality. A famous theorem of Minkowski [Min10] relates the length of the shortest vector in a lattice to the determinant:

Theorem 1 (Minkowski). *In an ω-dimensional lattice, there exists a non-zero vector v with*

$$\|v\| \leq \sqrt{\omega} \det(L)^{\frac{1}{\omega}}. \tag{1}$$

In lattices with fixed dimension we can efficiently find a shortest vector, but for arbitrary dimensions, the problem of computing a shortest vector is known to be NP-hard under randomized reductions [Ajt98]. The L^3 algorithm, however, computes in polynomial time an approximation of the shortest vector, which is sufficient for many applications. The basis vectors of an L^3-reduced basis fulfill the following property (for a proof see e.g. [May03]).

Theorem 2 (L^3). *Let L be an integer lattice of dimension ω. The L^3 algorithm outputs a reduced basis spanned by $\{v_1 \ldots, v_\omega\}$ with*

$$\|v_1\| \leq \|v_2\| \leq \ldots \leq \|v_i\| \leq 2^{\frac{\omega(\omega-i)}{4(\omega+1-i)}} \det(L)^{\frac{1}{\omega+1-i}}, \quad i = 1, \ldots, \omega \tag{2}$$

in polynomial time.

The underlying idea of Coppersmith's method for finding small roots of polynomial equations is to reduce the problem of finding roots of $f(x_1,\ldots,x_n) \bmod p$ to finding roots over the integers. Therefore, one constructs a collection of polynomials that share a common root modulo p^m for some well-chosen integer m. Then one finds an integer linear combination which has a sufficiently small norm. The search for such a small norm linear combination is done by defining a lattice basis via the polynomials' coefficient vectors. An application of L^3 yields a small norm coefficient vector that corresponds to a small norm polynomial.

The following lemma due to Howgrave-Graham [HG97] gives a sufficient condition under which modular roots are also roots over $\mathbb{Z}$ and quantifies the term *sufficiently small.*

Lemma 1. *Let $g(x_1,\ldots,x_n) \in \mathbb{Z}[x_1,\ldots,x_n]$ be an integer polynomial with at most ω monomials. Suppose that*

1. $g(y_1,\ldots,y_n) = 0 \bmod p^m$ *for* $|y_1| \leq X_1,\ldots,|y_n| \leq X_n$ *and*
2. $\|g(x_1X_1,\ldots,x_nX_n)\| < \frac{p^m}{\sqrt{\omega}}$

Then $g(y_1,\ldots,y_n) = 0$ holds over the integers.

Our approach relies on heuristic assumptions for computations with multivariate polynomials.

Assumption 1. *Our lattice-based construction yields algebraically independent polynomials. The common roots of these polynomials can be efficiently computed using numerical methods.*

The first part of Assumption 1 assures that the constructed polynomials allow for extracting the common roots, while the second part assures that we are able to compute these common roots efficiently. We would like to point out that our subsequent complexity considerations solely refer to our lattice-based construction, that turns a linear polynomial $f(x_1,\ldots,x_n) \bmod p$ into n polynomials over the integers. We assume that the running time for extracting the desired root out of these n polynomials is negligible compared to the time complexity of the lattice construction. We verify this experimentally in Section 5. Usually, our method yields more than n polynomials, so one can make use of additional polynomials as well.

3 Bivariate Linear Equations

The starting point of our analysis are bivariate linear modular equations $f(x_1,x_2) = a_1x_1 + a_2x_2 + a_3 \bmod p$. The parameter p is unknown, we only know a multiple N of p, and the parameter β that quantifies the size relation $p \geq N^\beta$. Let X_1, X_2 be upper bounds on the desired solution y_1, y_2, respectively. Moreover, we require that our linear polynomial is monic with respect to one of the variables, i.e. either $a_1 = 1$ or $a_2 = 1$. This is usually not a restriction, since we could e.g. multiply $f(x_1,x_2)$ by $a_1^{-1} \bmod N$. If this inverse does not exist, we can factorize N.

In the following theorem, we give an explicit bound on X_1X_2 under which we can find two polynomials $g_1(x_1,x_2)$ and $g_2(x_1,x_2)$ that evaluate to zero at all small points (y_1,y_2) with $|y_1y_2| \leq X_1X_2$. Under the heuristic that g_1 and g_2 are *algebraically independent*, all roots smaller than X_1X_2 can be recovered by standard methods over the integers.

Theorem 3. *Let $\epsilon > 0$ and let N be a sufficiently large composite integer with a divisor $p \geq N^\beta$. Furthermore, let $f(x_1,x_2) \in \mathbb{Z}[x_1,x_2]$ be a linear polynomial in two variables. Under Assumption 1, we can find all solutions (y_1,y_2) of the equation $f(x_1,x_2) = 0 \bmod p$ with $|y_1| \leq N^\gamma$ and $|y_2| \leq N^\delta$ if*

$$\gamma + \delta \leq 3\beta - 2 + 2(1-\beta)^{\frac{3}{2}} - \epsilon \tag{3}$$

The algorithm's time and space complexity is polynomial in $\log N$ and ϵ^{-1}.

Before we provide a proof for Theorem 3, we would like to interpret its implications. Notice that Theorem 3 yields in the special case $\beta = 1$ the bound $X_1X_2 \leq N^{1-\epsilon}$ that corresponds to the folklore bound for linear equations. Since we are unaware of a good reference for the folklore method in the cryptographic literature, we briefly sketch the derivation of this bound in Appendix A. Thus, our result generalizes the folklore method to more general moduli.

On the other hand, we would like to compare our result with the one of Coppersmith for *factoring with high bits known* when p, q are of equal bit-size, i.e. $\beta = \frac{1}{2}$. Coppersmith's result allows a maximal size of $N^{0.25}$ for one unknown block. Our result states a bound of $N^{0.207}$ for the product of two blocks. The best that we could hope for was to obtain a total of $N^{0.25}$ for two blocks as well. However, it seems quite natural that the bound decreases with the number n of blocks. On the other hand, we are able to show that if the unknown blocks are significantly unbalanced in size, then one can improve on the bound $N^{0.207}$. It turns out that the more unbalanced X_1, X_2 are, the better. In the extreme case, we obtain $X_1 = N^{0.25}, X_2 = 1$. Notice that in this case, the variable x_2 vanishes and we indeed obtain the univariate result $N^{0.25}$ of Coppersmith. Hence, our method contains the Coppersmith-bound as a special case as well. We give more details after the following proof of Theorem 3.

Proof. Define $X_1X_2 := N^{3\beta-2+2(1-\beta)^{\frac{3}{2}}-\epsilon}$ and fix $m = \left\lceil \frac{3\beta(1+\sqrt{1-\beta})}{\epsilon} \right\rceil$.

We define a collection of polynomials which share a common root modulo p^t by

$$g_{k,i}(x_1,x_2) := x_2^i f^k(x_1,x_2) N^{\max\{t-k,0\}} \tag{4}$$

for $k = 0,...,m; i = 0,...,m-k$ and some $t = \tau m$, that will be optimized later.

We can define the following polynomial ordering for our collection. Let $g_{k,i}, g_{l,j}$ be two polynomials. If $k < l$ then $g_{k,i} < g_{l,j}$, if $k = l$ then $g_{k,i} < g_{l,j} \Leftrightarrow i < j$. If we sort the polynomials according to that ordering, every subsequent polynomial in the ordering introduces exactly one new monomial. Thus, the corresponding coefficient vectors define a lower triangular lattice basis, like in Figure 1.

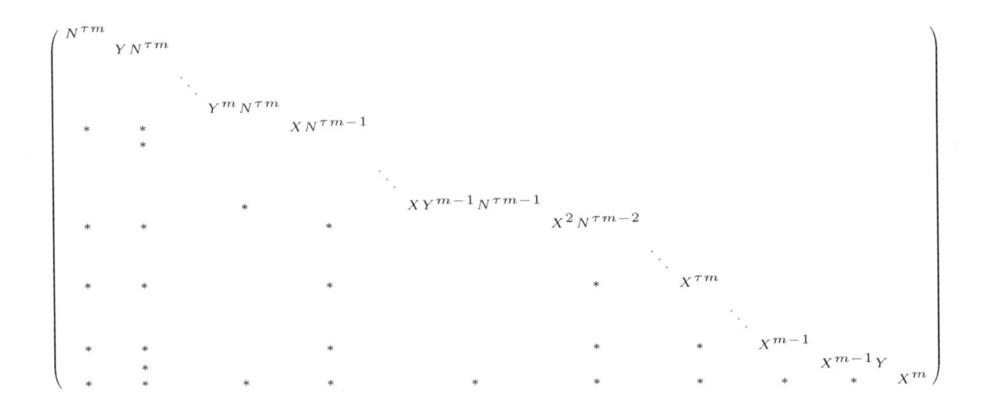

Fig. 1. Basis Matrix in Triangular Form

From the basis matrix we can easily compute the determinant as the product of the entries on the diagonal as $\det(L) = X^{s_x} Y^{s_y} N^{s_N}$, where

$$s_x = s_y = \frac{1}{6}(m^3 + 3m^2 + 2m), \quad s_N = \sum_{i=0}^{\tau m} (m + 1 - i)(\tau m - i) \tag{5}$$

Now we apply L^3 basis reduction to the lattice basis. Our goal is to find two coefficient vectors whose corresponding polynomials contain all small roots over the integer. Theorem 2 gives us an upper bound on the norm of a second-to-shortest vector in the L^3-reduced basis. If this bound is in turn smaller than the bound in Howgrave-Graham's lemma (Lemma 1), we obtain the desired two polynomials. I.e., we have to satisfy the condition

$$2^{\frac{d(d-1)}{4(d-1)}} \det(L)^{\frac{1}{d-1}} < d^{-\frac{1}{2}} N^{\beta \tau m}, \tag{6}$$

where d is the dimension of the lattice L, which in our case is $d = \frac{1}{2}(m^2+3m+2)$. If we plug in the value for the determinant and use the fact that $s_x = \frac{md}{3}$, we obtain the condition

$$X_1 X_2 < 2^{-\frac{3(d-1)}{4m}} d^{-\frac{3(d-1)}{2md}} N^{\frac{3\beta\tau m(d-1)}{md} - \frac{3s_N}{md}}. \tag{7}$$

Setting $\tau = 1 - \sqrt{1-\beta}$, the exponent of N can be lower bounded by

$$3\beta - 2 + 2(1-\beta)^{\frac{3}{2}} - \frac{3\beta\left(1 + \sqrt{1-\beta}\right)}{m}. \tag{8}$$

[Details can be found in Appendix B.]

Comparing this with the value of $X_1 X_2$, which we defined in the beginning, we can express how m depends on the error term ϵ:

$$m \geq \frac{3\beta\left(1 + \sqrt{1-\beta}\right)}{\epsilon}. \tag{9}$$

which holds for our choice of m. Therefore, the required condition is fulfilled.

It remains to show that the algorithm's complexity is polynomial in $\log(N)$ and ϵ^{-1}. The running time is dominated by L^3 reduction, which is polynomial in the dimension of the lattice and in the bitsize of the entries. Recall that our lattice's dimension is $\mathcal{O}(m^2)$ and therefore polynomial in ϵ^{-1}. For the matrix entries we notice that the power f^k in the $g_{k,i}$'s can be reduced modulo N^k, since we are looking for roots modulo N^k. Thus, the coefficients of $f^k N^{max(\tau m-k,0)}$ have bitsize $\mathcal{O}(m\log(N))$. Powers of X_2 appear only with exponents up to m and therefore their bitsize can also be upper bounded by $\mathcal{O}(m\log(N))$. Thus, the coefficients' bitsize is $\mathcal{O}(\epsilon^{-1}\log(N))$.

Remark: We also analyzed the bivariate modular instance as a trivariate equation over the integers, which is modelled by

$$(a_1x_1 + a_2x_2 + a_3)y - N = 0, \tag{10}$$

where y stands for $\frac{N}{p}$. It turns out that we obtain the same bounds as in the modular case.

Theorem 3 holds for any bounds X_1, X_2 within the proven bound for the product X_1X_2. As pointed out before, the analysis can be improved if one of the bounds is significantly smaller than the other one, say $X_1 \ll X_2$. Then one should employ additional extra shifts in the smaller variable, which intuitively means that the smaller variable gets stronger weight since it causes smaller costs.

We do not give the exact formulas for this optimization process. Instead, we show in Figure 2 the resulting graph that demonstrates how the result converges to the known bound $N^{0.25}$ for unbalanced block-sizes.

Notice that the result from Theorem 3 is indeed optimal not only for equal block-sizes $X_1 = X_2$ but for most of the possible splittings of block-sizes. Only in extreme cases a better result can be achieved. In the subsequent chapter, we generalize Theorem 3 to an arbitrary number n of blocks. In the generalization however, we will not consider the improvement that can be achieved for strongly unbalanced block-sizes.

Naturally, the bounds $N^{0.25}$ for $n = 1$ and $N^{0.207}$ for $n = 2$ get worse for arbitrary n. But surprisingly, we will show that for $n \to \infty$ the bound does not converge to N^0 as one might expect, but instead to $N^{0.153}$. To illustrate this result: If N is a 1000-bit modulus and p, q are 500 bit each. Then 153 bit can be recovered given the remaining 347 bits, or 69.4% of p, in any *known* positions. However as we will see in the next section, the complexity heavily depends on the number of unknown blocks.

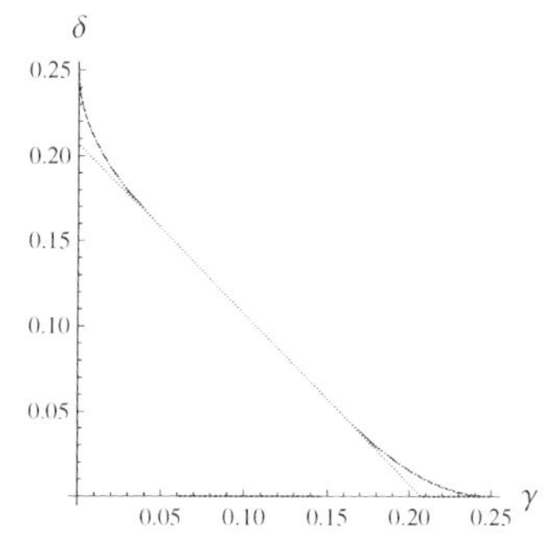

Fig. 2. Optimized Result

4 Extension to More Variables

In this section, we generalize the result of Section 3 from bivariate linear equations with $n = 2$ to an arbitrary number n of variables $x_1, \ldots, x_n$.

Let $X_1, X_2, \ldots, X_n$ be upper bounds for the variables $x_1, x_2, \ldots, x_n$. As in Theorem 3, we will focus on proving a general upper bound for the product $X_1 X_2 \ldots X_n$ that is valid for any $X_1, X_2, \ldots, X_n$. Similar to the reasoning in Section 3 it is possible to achieve better results for strongly unbalanced X_i by giving more weight to variables x_i with small upper bounds. Although we did not analyze it, we strongly expect that in the case $X_1 = N^{0.25}$, $X_2 = \cdots = X_n = 1$ everything boils down to the univariate case analyzed by Coppersmith/Howgrave-Graham – except that we obtain an unnecessarily large lattice dimension.

Naturally, we achieve an inferior bound than $N^{0.25}$. But in contrast, our bound holds no matter how the sizes of the unknowns are distributed among the upper bounds X_i. Let us state our main theorem.

Theorem 4. *Let $\epsilon > 0$ and let N be a sufficiently large composite integer with a divisor $p \geq N^{\beta}$. Furthermore, let $f(x_1, \ldots, x_n) \in \mathbb{Z}[x_1, \ldots, x_n]$ be a monic linear polynomial in n variables. Under Assumption 1, we can find all solutions $(y_1, \ldots, y_n)$ of the equation $f(x_1, \ldots, x_n) = 0 \bmod p$ with $|y_1| \leq N^{\gamma_1}, \ldots, |y_n| \leq N^{\gamma_n}$ if*

$$\sum_{i}^{n} \gamma_i \leq 1 - (1-\beta)^{\frac{n+1}{n}} - (n+1)(1 - \sqrt[n]{1-\beta})(1-\beta) - \epsilon \tag{11}$$

The time and space complexity of the algorithm is polynomial in $\log N$ and $(\frac{e}{\epsilon})^n$, where e is Euler's constant.

We will prove Theorem 4 at the end of this section. Let us first discuss the implications of the result and the consequences for the *factoring with known bits problem*. First of all, the algorithm's running time is exponential in the number n of blocks. Thus in order to obtain a polynomial complexity one has to restrict

$$n = \mathcal{O}\left(\frac{\log \log N}{1 + \log(\frac{1}{\epsilon})}\right).$$

This implies that for any constant error term ϵ, our algorithm is polynomial time whenever $n = \mathcal{O}(\log \log N)$.

The proof of the following theorem shows that the bound for $X_1 \ldots X_n$ in Theorem 4 converges for $n \to \infty$ to $N^{\beta + (1-\beta)\ln(1-\beta)}$. For the *factoring with known bits* problem with $\beta = \frac{1}{2}$ this yields the bound $N^{\frac{1}{2}(1-\ln(2))} \approx N^{0.153}$. This means that we can recover a $(1 - \ln(2)) \approx 0.306$-fraction of the bits of p, or in other words an $\ln(2) \approx 0.694$-fraction of the bits of p has to be known.

Theorem 5. *Let $\epsilon > 0$. Suppose N is a sufficiently large composite integer with a divisor $p \geq N^{\beta}$. Further, suppose we are given an*

$$\left(1 - \frac{1}{\beta}\right) \cdot \ln(1-\beta) + \epsilon \quad \textit{fraction} \tag{12}$$

of the bits of p. Then, under Assumption 1, we can compute the unknown bits of p in time polynomial in $\log N$ and $(\frac{e}{\epsilon})^n$, where e is Euler's constant.

Proof. From Theorem 4 we know, that we can compute a solution to the equation

$$a_1x_1 + a_2x_2 + \ldots + a_nx_n + a_{n+1} = 0 \bmod p$$

as long as the product of the unknowns is smaller than N^γ, where $\gamma = \sum_i^n \gamma_i$ is upper-bounded as in Inequality (11). As noticed already, the bound for γ actually converges for $n \to \infty$ to a value different from zero. Namely,

$$\lim_{n\to\infty} \left(1 - (1-\beta)^{\frac{n+1}{n}} - (n+1)(1 - \sqrt[n]{1-\beta})(1-\beta)\right) = \beta + (1-\beta)\ln(1-\beta) \tag{13}$$

Hence, this is the portion of p we can at least compute, no matter how many unknowns we have.

Conversely, once we have $((\beta-1)\ln(1-\beta)+\epsilon)\log(N)$ bits of p given together with their positions, we are able to compute the missing ones. Since $\log N \leq \frac{\log p}{\beta}$, we need at most an $((1-\frac{1}{\beta})\ln(1-\beta)+\epsilon)$-fraction of the bits of p.

Theorem 5 implies a polynomial-time algorithm for the *factoring with known bits* problem whenever the number of unknown bit-blocks is $n = \mathcal{O}(\log\log N)$. However, the algorithm can be applied for larger n as well. As long as n is sub-polynomial in the bit-size of N, the resulting complexity will be sub-exponential in the bit-size of N.

It remains to prove our main theorem.

Proof of Theorem 4

Define $\prod_{i=1}^n X_i := N^{1-(1-\beta)^{\frac{n+1}{n}} - (n+1)(1-\sqrt[n]{1-\beta})(1-\beta)-\epsilon}$. Let us fix

$$m = \left\lceil \frac{n(\frac{1}{\pi}(1-\beta)^{-0.278465} - \beta\ln(1-\beta))}{\epsilon} \right\rceil \tag{14}$$

We define the following collection of polynomials which share a common root modulo p^t

$$g_{i_2,\ldots,i_n,k} = x_2^{i_2} \ldots x_n^{i_n} f^k N^{\max\{t-k,0\}} \tag{15}$$

where $i_j \in \{0,\ldots,m\}$ such that $\sum_{j=2}^n i_j \leq m-k$. The parameter $t = \tau m$ has to be optimized. Notice that the set of monomials of $g_{i_2,\ldots,i_n,k}$ defines an n-dimensional simplex.

It is not hard to see that there is an ordering of the polynomials in such a way that each new polynomial introduces exactly one new monomial. Therefore the lattice basis constructed from the coefficient vectors of the $g_{i_2,\ldots,i_n,k}$'s has triangular form, if they are sorted according to the order. The determinant $\det(L)$ of the corresponding lattice L is then simply the product of the entries on the diagonal:

$$\det(L) = \prod_{i=1}^n X_i^{s_{x_i}} N^{s_N}, \tag{16}$$

with $s_{x_i} = \binom{m+n}{m-1}$ and $s_N = md\tau - \binom{m+n}{m-1} + \binom{m(1-\tau)+n}{m(1-\tau)-1}$, where $d = \binom{m+n}{m}$ is the dimension of the lattice.

Now we ensure that the vectors from L^3 are sufficiently small, so that we can apply the Lemma of Howgrave-Graham (Lemma 1) to obtain a solution over $\mathbb{Z}$. We have to satisfy the condition

$$2^{\frac{d(d-1)}{4(d-n+1)}} \det(L)^{\frac{1}{d-n+1}} < d^{-\frac{1}{2}} N^{\beta\tau m}$$

Using the value of the determinant in (16) and the fact that $s_{x_i} = \frac{md}{n+1}$ we obtain

$$\prod_{i=1}^{n} X_i \leq 2^{-\frac{(d-1)(n+1)}{4m}} d^{-\frac{(n+1)(d-n+1)}{2md}} N^{(\beta m\tau(d-n+1)-dm\tau+\binom{m+n}{m-1}-\binom{m(1-\tau)+n}{m(1-\tau)-1})\frac{n+1}{md}}.$$

In Appendix C we show how to derive a lower bound on the right-hand side for the optimal value $\tau = 1-(1-\beta)^{\frac{1}{n}}$. Using $X_i = N^{\gamma_i}$ the condition reduces to

$$\sum_{i=1}^{n} \gamma_i \leq 1-(1-\beta)^{\frac{n+1}{n}} - (n+1)(1-\sqrt[n]{1-\beta})(1-\beta) - \frac{n\frac{1}{\pi}(1-\beta)^{-0.278465}}{m} + \beta\ln(1-\beta)\frac{n}{m}.$$

Comparing this to the initial definition of $\prod_{i=1}^{n} X_i$, we obtain for the error term ϵ

$$-\frac{n\frac{1}{\pi}(1-\beta)^{-0.278465}}{m} + \beta\ln(1-\beta)\frac{n}{m} \geq -\epsilon$$

$$\Leftrightarrow m \geq \frac{n(\frac{1}{\pi}(1-\beta)^{-0.278465} - \beta\ln(1-\beta))}{\epsilon} = \mathcal{O}(\frac{n}{\epsilon})$$

which holds for our choice of m.

To conclude the proof, we notice that the dimension of the lattice is $d = \mathcal{O}(\frac{m^n}{n!}) = \mathcal{O}(\frac{n^n e^n}{\epsilon^n n^n}) = \mathcal{O}(\frac{e^n}{\epsilon^n})$. For the bitsize of the entries in the basis matrix we observe that we can reduce the coefficients of f^i in g modulo N^i. Thus the product $f^k N^{\max\{\tau m-k,0\}}$ is upper bounded by $B = m\log(N)$. Further notice that the bitsize of $X_2^{i_2} \ldots X_n^{i_2}$ is also upper bounded by $m\log(N)$ since $\sum_{i=2}^{n} i_j \leq m$ and $X_i \leq N$.

The running time is dominated by the time to run L^3-lattice reduction on a basis matrix of dimension d and bit-size B. Thus, the time and space complexity of our algorithm is polynomial in $\log N$ and $(\frac{e}{\epsilon})^n$. □

5 Experimental Results

We implemented our lattice-based algorithm using the L^2-algorithm from Nguyen, Stehlé [NS05]. We tested the algorithm for instances of the *factoring with known bits* problem with $n = 2, 3$ and 4 blocks of unknown bits. Table 1 shows the experimental results for an 512-bit RSA modulus N with divisor p of size $p \geq N^{\frac{1}{2}}$.

For given parameters m, t we computed the number of bits that one should theoretically be able to recover from p (column *pred* of Table 1). For each bound we made two experiments (column *exp*). The first experiment splits the bound into n equally sized pieces, whereas the second experiment unbalancedly splits

Table 1. Experimental Results

n	m	t	$\dim(L)$	pred (bit)	exp (bit)	time (min)
2	15	4	136	90	45/45	25
2	15	4	136	90	87/5	15
3	7	1	120	56	19/19/19	0.3
3	7	1	120	56	52/5/5	0.3
3	10	2	286	69	23/23/23	450
3	10	2	286	69	57/6/6	580
4	5	1	126	22	7/6/6/6	3
4	5	1	126	22	22/2/2/2	4.5

the bound in one large piece and $n-1$ small ones. In the unbalanced case, we were able to recover a larger number of bits than theoretically predicted. This is consistent with the reasoning in Section 3 and 4.

In all of our experiments, we successfully recovered the desired small root, thereby deriving the factorization of N. We were able to extract the root both by Gröbner basis reduction as well as by numerical methods in a fraction of a second.

For Gröbner basis computations, it turns out to be useful that our algorithm actually outputs more sufficiently small norm polynomials than predicted by the L^3-bounds. This in turn helps to speed up the computation a lot.

As a numerical method, we used multidimensional Newton iteration on the starting point $\frac{1}{2}(X_1,\ldots,X_n)$. Usually this did already work. If not, we were successful with the vector of upper-bounds $(X_1,\ldots,X_n)$ as a starting point. Although this approach worked well and highly efficient in practice, we are unaware of a starting point that provably lets the Newton method converge to the desired root.

Though Assumption 1 worked perfectly for the described experiments, we also considered two pathological cases, where one has to take special care.

First, a problem arises when we have a prediction of k bits that can be recovered, but we use a much smaller sum of bits in our n blocks. In this case, the smallest vector lies in a sublattice of small dimension. As a consequence, we discovered that then usually all of our small norm polynomials shared $f(x)$ as a common divisor. When we removed the gcd, the polynomials were again algebraically independent and we were able to retrieve the root. Notice that removing $f(x)$ does not eliminate the desired root, since $f(x)$ does not contain the root over the integers (but $\mod p$).

A second problem may arise in the case of two closely adjacent unknown blocks, e.g. two blocks that are separated by one known bit only. Since in comparison with the n-block case the case of $n-1$ blocks gives a superior bound, it turns out to be better in some cases to merge two closely adjacent blocks into one variable. That is what implicitly seems to happen in our approach. The computations then yield the desired root only in those variables which are sufficiently separated. The others have to be merged before re-running the algorithm in order to obtain all the unknown bits. Alternatively, we confirmed experimentally that merging the nearby blocks from the beginning immediately yields the desired root.

Both pathological cases are no failure of Assumption 1, since one can still easily extract the desired root. All that one has to do is to either remove a gcd or to merge variables.

6 Conclusion and Open Problems

We proposed a heuristic lattice-based algorithm for finding small solutions of linear equations $a_1x_1 + \cdots + a_nx_n + a_{n+1} = 0 \bmod p$, where p is an unknown divisor of some known N. Our algorithm gives a solution for the *factoring with known bits* problem given $\ln(2) \approx 70\%$ of the bits of p in any locations.

Since the time and space complexity of our algorithm is polynomial in $\log N$ but exponential in the number n of variables, we obtain a polynomial time algorithm for $n = \mathcal{O}(\log \log N)$ and a subexponential time algorithm for $n = o(\log N)$. This naturally raises the question whether there exists some algorithm with the same bound having complexity polynomial in n. This would immediately yield a polynomial time algorithm for *factoring with* 70% *bits given*, independently of the given bit locations *and* the number of consecutive bit blocks. We do not know whether such an algorithm can be achieved for polynomial equations with *unknown divisor*. On the other hand, we feel that the complexity gap between the folklore method for *known divisors* with complexity linear in n and our method is quite large, even though the folklore method relies on much stronger assumptions.

Notice that in the *factoring with known bits* problem, an attacker is given the location of the given bits of p and he has to fill in the missing bits. Let us give a crude analogy for this from coding theory, where one is given the codeword p with erasures in some locations. Notice that our algorithm is able to correct the erasures with the help of the redundancy given by N. Now a challenging question is whether there exist similar algorithms for *error-correction* of codewords p. I.e., one is given p with a certain percentage of the bits flipped. Having an algorithm for this problem would be highly interesting in situations with error-prone side-channels.

We would like to thank the anonymous reviewers and especially Robert Israel for helpful comments and ideas.

References

[Ajt98] Ajtai, M.: The Shortest Vector Problem in L_2 is NP-hard for Randomized Reductions (Extended Abstract). In: STOC, pp. 10–19 (1998)

[BD00] Boneh, D., Durfee, G.: Cryptanalysis of RSA with private key d less than $N^{0.292}$. IEEE Transactions on Information Theory 46(4), 1339 (2000)

[BM06] Bleichenbacher, D., May, A.: New Attacks on RSA with Small Secret CRT-Exponents. In: Public Key Cryptography, pp. 1–13 (2006)

[CJL+92] Coster, M.J., Joux, A., LaMacchia, B.A., Odlyzko, A.M., Schnorr, C.-P., Stern, J.: Improved Low-Density Subset Sum Algorithms. Computational Complexity 2, 111–128 (1992)

[CM07] Coron, J.-S., May, A.: Deterministic Polynomial-Time Equivalence of Computing the RSA Secret Key and Factoring. J. Cryptology 20(1), 39–50 (2007)

[Cop96a] Coppersmith, D.: Finding a Small Root of a Bivariate Integer Equation; Factoring with High Bits Known. In: Maurer, U.M. (ed.) EUROCRYPT 1996. LNCS, vol. 1070, pp. 178–189. Springer, Heidelberg (1996)

[Cop96b] Coppersmith, D.: Finding a Small Root of a Univariate Modular Equation. In: Maurer, U.M. (ed.) EUROCRYPT 1996. LNCS, vol. 1070, pp. 155–165. Springer, Heidelberg (1996)

[Cop97] Coppersmith, D.: Small Solutions to Polynomial Equations, and Low Exponent RSA Vulnerabilities. J. Cryptology 10(4), 233–260 (1997)

[Cor07] Coron, J.-S.: Finding Small Roots of Bivariate Integer Polynomial Equations: A Direct Approach. In: Menezes, A. (ed.) CRYPTO 2007. LNCS, vol. 4622, pp. 379–394. Springer, Heidelberg (2007)

[GM97] Girault, M., Misarsky, J.-F.: Selective Forgery of RSA Signatures Using Redundancy. In: Fumy, W. (ed.) EUROCRYPT 1997. LNCS, vol. 1233, pp. 495–507. Springer, Heidelberg (1997)

[GTV88] Girault, M., Toffin, P., Vallée, B.: Computation of approximate L-th roots modulo n and application to cryptography. In: Goldwasser, S. (ed.) CRYPTO 1988. LNCS, vol. 403, pp. 100–117. Springer, Heidelberg (1990)

[Has88] Hastad, J.: Solving Simultaneous Modular Equations of Low Degree. SIAM Journal on Computing 17(2), 336–341 (1988)

[HG97] Howgrave-Graham, N.: Finding Small Roots of Univariate Modular Equations Revisited. In: Proceedings of the 6th IMA International Conference on Cryptography and Coding, pp. 131–142 (1997)

[HG01] Howgrave-Graham, N.: Approximate Integer Common Divisors. In: Silverman, J.H. (ed.) CaLC 2001. LNCS, vol. 2146, pp. 51–66. Springer, Heidelberg (2001)

[LLL82] Lenstra, A.K., Lenstra, H.W., Lovász, L.: Factoring Polynomials with Rational Coefficients. Mathematische Annalen 261(4), 515–534 (1982)

[Mau95] Maurer, U.M.: On the Oracle Complexity of Factoring Integers. Computational Complexity 5(3/4), 237–247 (1995)

[May03] May, A.: New RSA Vulnerabilities Using Lattice Reduction Methods. PhD thesis, University of Paderborn (2003)

[May04] May, A.: Computing the RSA Secret Key Is Deterministic Polynomial Time Equivalent to Factoring. In: Franklin, M. (ed.) CRYPTO 2004. LNCS, vol. 3152, pp. 213–219. Springer, Heidelberg (2004)

[Min10] Minkowski, H.: Geometrie der Zahlen. Teubner (1910)

[Ngu04] Nguyen, P.Q.: Can We Trust Cryptographic Software? Cryptographic Flaws in GNU Privacy Guard v1.2.3. In: Cachin, C., Camenisch, J.L. (eds.) EUROCRYPT 2004. LNCS, vol. 3027, pp. 555–570. Springer, Heidelberg (2004)

[NS01] Nguyen, P.Q., Stern, J.: The Two Faces of Lattices in Cryptology. In: Silverman, J.H. (ed.) CaLC 2001. LNCS, vol. 2146, pp. 146–180. Springer, Heidelberg (2001)

[NS05] Nguyen, P.Q., Stehlé, D.: Floating-Point LLL Revisited. In: Cramer, R. (ed.) EUROCRYPT 2005. LNCS, vol. 3494, pp. 215–233. Springer, Heidelberg (2005)

[RS85] Rivest, R.L., Shamir, A.: Efficient Factoring Based on Partial Information. In: Pichler, F. (ed.) EUROCRYPT 1985. LNCS, vol. 219, pp. 31–34. Springer, Heidelberg (1986)

A Linear Equations with Known Modulus

We briefly sketch the folklore method for finding small roots of linear modular equations $a_1x_1 + \cdots + a_nx_n = 0 \bmod N$ with *known* modulus N. Further, we assume that $\gcd(a_i, N) = 1$ for some i, wlog $\gcd(a_n, N) = 1$. Let X_i be upper bounds on $|y_i|$. We can handle inhomogeneous modular equations by introducing a term $a_{n+1}x_{n+1}$, where $|y_{n+1}| \le X_{n+1} = 1$.

We would like to point out that the heuristic for the folklore method is quite different compared to the one taken in our approach. First of all, the method requires to solve a shortest vector problem in a certain lattice. This problem is known to be NP-hard for general lattices. Second, one assumes that there is only *one* linear independent vector that fulfills the Minkowski bound (Theorem 1) for the shortest vector.

We will show under this heuristic assumption that the shortest vector yields the unique solution $(y_1, \ldots, y_n)$ whenever

$$\prod_{i=1}^{n} X_i \le N.$$

We multiply our linear equation with $-a_n^{-1}$ and obtain

$$b_1x_1 + b_2x_2 + \ldots + b_{n-1}x_{n-1} = x_n \bmod N \quad \text{,where } b_i = a_n^{-1}a_i \tag{17}$$

For a solution $(y_1, \ldots, y_n)$ of (17) we know $\sum_{i=1}^{n-1} b_iy_i = y_n - yN$ for some $y \in \mathbb{Z}$. Consider the lattice L generated by the row vectors of the following matrix

$$B = \begin{pmatrix} Y_1 & 0 & 0 & \ldots & 0 & Y_nb_1 \\ 0 & Y_2 & 0 & & 0 & Y_nb_2 \\ \vdots & & \ddots & & \vdots & \vdots \\ \vdots & & & & Y_{n-1} & Y_nb_{n-1} \\ 0 & 0 & 0 & \ldots & 0 & Y_nN \end{pmatrix}$$

with $Y_i = \frac{N}{X_i}$. By construction,

$$v = (y_1, \ldots, y_{n-1}, y) \cdot B = (Y_1y_1, \ldots, Y_ny_n)$$

is a vector of L. We show, that this is a short vector which fulfills the Minkowski bound from Theorem 1. If we assume that v is actually the shortest vector, then we can solve an SVP instance.

Since $Y_iy_i = \frac{y_i}{X_i}N \le N$ we have $\|v\| \le \sqrt{n}N$. Further, the determinant of the lattice L is

$$\det(L) = N\prod_{i=1}^{n} Y_i = N\prod_{i=1}^{n}\frac{N}{X_i} = N^{n+1}\prod_{i=1}^{n}\frac{1}{X_i}.$$

The vector v thus fulfills the Minkowski bound, if

$$\sqrt{n}N \le \sqrt{n}\det(L)^{\frac{1}{n}} \quad \Leftrightarrow \quad \prod_{i=1}^{n} X_i \le N.$$

B Lower Bound in Theorem 2

Starting with

$$X_1X_2 < 2^{-\frac{3(d-1)}{4m}} d^{-\frac{3(d-1)}{2md}} N^{\frac{3\beta\tau m(d-1)}{md} - \frac{3s_N}{md}}$$

we wish to derive a lower bound of the right-hand side. First we notice that for sufficiently large N the powers of 2 and d are negligible. Thus, we only examine the exponent of N. We use the values $d = \frac{1}{2}(m^2 + 3m + 2)$ and $s_N = \sum_{i=0}^{\tau m}(m+1-i)(\tau m - i)$ and get

$$\tau\left(3\beta - 3\tau + \tau^2\right) + \frac{-\tau - 6\beta\tau + \tau^3}{1+m} - \frac{2\left(\tau - 3\beta\tau - 3\tau^2 + 2\tau^3\right)}{2+m}.$$

For τ we choose $1 - \sqrt{(1-\beta)}$, resulting in

$$\begin{aligned}&-2 + 2\sqrt{1-\beta} + 3\beta - 2\beta\sqrt{1-\beta}\\ &\quad - \frac{3\sqrt{1-\beta}}{1+m} + \frac{6\sqrt{1-\beta}}{2+m} + \frac{7\beta\sqrt{1-\beta}}{1+m} - \frac{10\beta\sqrt{1-\beta}}{2+m} + \frac{6(-1+2\beta)}{2+m} - \frac{3(-1+3\beta)}{1+m}.\end{aligned}$$

Now we combine the terms that change their sign in the possible β-range, such that we obtain a term which is either positive or negative for all $\beta \in (0,1)$

$$\begin{aligned}-\frac{3\sqrt{1-\beta}}{1+m} - \frac{3(-1+3\beta)}{1+m} + \frac{7\beta\sqrt{1-\beta}}{1+m} &= \frac{3 - 3\sqrt{1-\beta} - 9\beta + 7\beta\sqrt{1-\beta}}{1+m} < 0\\ \frac{6(-1+2\beta)}{2+m} + \frac{6\sqrt{1-\beta}}{2+m} &= \frac{6\left(-1+\sqrt{1-\beta}+2\beta\right)}{2+m} > 0 \text{ for all } \beta \in (0,1).\end{aligned}$$

Finally, we approximate the positive terms by $\frac{*}{2m}$ and the negative ones by $\frac{*}{m}$ and obtain

$$2^{-\frac{3(d-1)}{4m}} d^{-\frac{3(d-1)}{2md}} N^{\frac{3\beta\tau m(d-1)}{md} - \frac{3s_N}{md}} \geq N^{-2+2\sqrt{1-\beta}+3\beta-2\beta\sqrt{1-\beta}-\frac{3\beta(1+\sqrt{1-\beta})}{m}}. \quad (18)$$

C Lower Bound in Theorem 3

We derive a lower bound of

$$2^{-\frac{(d-1)(n+1)}{4m}} d^{-\frac{(n+1)(d-n+1)}{2md}} N^{\left(\beta m\tau(d-n+1) - dm\tau + \binom{m+n}{m-1} - \binom{m(1-\tau)+n}{m(1-\tau)-1}\right)\frac{n+1}{md}}.$$

For sufficiently large N, the powers of 2 and d are negligible and thus we consider in the following only the exponent of N

$$\begin{aligned}&\left(\beta m\tau(d-n+1) - dm\tau + \binom{m+n}{m-1} - \binom{m(1-\tau)+n}{m(1-\tau)-1}\right)\frac{n+1}{md}\\ &= \beta\tau(n+1) - \frac{\beta\tau(n-1)(n+1)}{d} - \tau(n+1) + 1 - \frac{\prod_{k=0}^{n}(m(1-\tau)+k)}{n!md}.\end{aligned}$$

With $d = \binom{m+n}{m} = \frac{(m+n)!}{m!n!} = \frac{\prod_{k=1}^{n}(m+k)}{n!}$ we have

$$\beta\tau(n+1) - \tau(n+1) + 1 - \frac{\beta\tau(n-1)(n+1)!}{\prod_{k=1}^{n}(m+k)} - \frac{\prod_{k=0}^{n}(m(1-\tau)+k)}{\prod_{k=0}^{n}(m+k)}.$$

We now analyze the last two terms separately. For the first one, if we choose $\tau = 1 - \sqrt[n]{(1-\beta)}$ we obtain

$$\frac{\beta\tau(n-1)(n+1)!}{\prod_{k=1}^{n}(m+k)} \leq \frac{\beta(1-\sqrt[n]{1-\beta})(n-1)(n+1)!}{(m+1)\prod_{k=2}^{n}k} \leq \frac{\beta(1-\sqrt[n]{1-\beta})n^2}{m}.$$

Fact 1

$$n(1-\sqrt[n]{1-\beta}) \leq -\ln(1-\beta) \tag{19}$$

Using this approximation, we obtain

$$\frac{\beta\tau(n-1)(n+1)!}{\prod_{k=1}^{n}(m+k)} \leq -\ln(1-\beta)\beta\frac{n}{m}.$$

The analysis of the second term $\frac{\prod_{k=0}^{n}(m(1-\tau)+k)}{\prod_{k=0}^{n}(m+k)}$ is a bit more involved. We use its partial fraction expansion to show an upper bound.

Lemma 2. *For* $\tau = 1-(1-\beta)^{\frac{1}{n}}$ *we have*

$$\frac{\prod_{k=0}^{n}(m(1-\tau)+k)}{\prod_{k=0}^{n}(m+k)} \leq (1-\beta)^{\frac{n+1}{n}} + \frac{1}{\pi}(1-\beta)^{-0.278465}\frac{n}{m}. \tag{20}$$

Proof. First notice that

$$\frac{\prod_{k=0}^{n}(m(1-\tau)+k)}{\prod_{k=0}^{n}(m+k)} = (1-\tau)^{n+1} + \frac{\prod_{k=0}^{n}(m(1-\tau)+k) - (1-\tau)^{n+1}\prod_{k=0}^{n}(m+k)}{\prod_{k=0}^{n}(m+k)}.$$

We analyze the second part of this sum. Its partial fraction expansion is

$$\frac{\prod_{k=0}^{n}(m(1-\tau)+k) - (1-\tau)^{n+1}\prod_{k=0}^{n}(m+k)}{\prod_{k=0}^{n}(m+k)} = \frac{c_0}{m} + \frac{c_1}{m+1} + \ldots + \frac{c_n}{m+n}. \tag{21}$$

Our goal is to determine the values c_i. Start by multiplying with $\prod_{k=0}^{n}(m+k)$:

$$\prod_{k=0}^{n}(m(1-\tau)+k) - (1-\tau)^{n+1}\prod_{k=0}^{n}(m+k) = \sum_{i=0}^{n} c_i \prod_{\substack{k=0\\k\neq i}}^{n}(m+k).$$

Now we successively set m equal to the roots of the denominator and solve for c_i. For the i-th root $m = -i$ we obtain

$$\prod_{k=0}^{n}(-i(1-\tau)+k) = c_i \prod_{\substack{k=0\\k\neq i}}^{n}(k-i)$$

$$c_i = \frac{\prod_{k=0}^{n}(-i(1-\tau)+k)}{\prod_{\substack{k=0\\k\neq i}}^{n}(k-i)}.$$

We can rewrite this in terms of the Gamma function as

$$c_i = (-1)^i \frac{\Gamma(-i(1-\tau)+n+1)}{\Gamma(i+1)\Gamma(n-i+1)\Gamma(-i(1-\tau))}.$$

Using the identity $\Gamma(-z) = -\frac{\pi}{\sin(\pi z)\Gamma(z+1)}$, we obtain

$$c_i = (-1)^{i+1} \frac{\Gamma(-i(1-\tau)+n+1)\Gamma(i(1-\tau)+1)}{\Gamma(i+1)\Gamma(n-i+1)} \frac{\sin(\pi i(1-\tau))}{\pi}.$$

In the following we use $Q := \frac{\Gamma(-i(1-\tau)+n+1)\Gamma(i(1-\tau)+1)}{\Gamma(i+1)\Gamma(n-i+1)}$.

We now give an upper bound on the absolute value of c_i. Start by using the value $\tau = 1 - \sqrt[n]{1-\beta}$ and let $1-\beta = e^{-c}$ for some $c > 0$. Consider

$$\ln \frac{\Gamma(ie^{-\frac{c}{n}}+1)}{\Gamma(i+1)} = \ln(\Gamma(ie^{-\frac{c}{n}}+1)) - \ln(\Gamma(i+1)) = -\int_0^{i-ie^{-\frac{c}{n}}} \Psi(1+i-t)dt \quad \text{and}$$

$$\ln \frac{\Gamma(-ie^{-\frac{c}{n}}+n+1)}{\Gamma(n-i+1)} = \ln(\Gamma(-ie^{-\frac{c}{n}}+n+1)) - \ln(\Gamma(n-i+1)) = \int_0^{i-ie^{-\frac{c}{n}}} \Psi(n-i+1+t)dt.$$

Therefore

$$\ln Q = \int_0^{i-ie^{-\frac{c}{n}}} \Psi(n-i+1+t) - \Psi(1+i-t)dt.$$

The Digamma function Ψ is increasing and thus the integrand is increasing and we get the approximation

$$\ln Q \le (i - ie^{-\frac{c}{n}})(\Psi(n+1-ie^{-\frac{c}{n}}) - \Psi(1+ie^{-\frac{c}{n}})).$$

Let $i = tn$. Then for fixed t the expression on the right-hand side converges for $n \to \infty$ to

$$\lim_{n\to\infty} (i - ie^{-\frac{c}{n}})(\Psi(n+1-ie^{-\frac{c}{n}}) - \Psi(1+ie^{-\frac{c}{n}})) = ct\ln(\frac{1}{t}-1).$$

By numeric computation, the maximum of $t\ln(\frac{1}{t}-1)$ in the range $0 < t < 1$ is 0.278465. Thus,

$$\begin{aligned} \ln Q &\le 0.278465c \\ Q &\le (1-\beta)^{-0.278465}. \end{aligned}$$

Putting things together, we have

$$c_i \le (-1)^{i+1}(1-\beta)^{-0.278465} \frac{\sin(\pi i(1-\tau))}{\pi} \le \frac{1}{\pi}(1-\beta)^{-0.278465}.$$

The initial problem of estimating the partial fraction expansion from equation (21) now states

$$\begin{aligned} \frac{\prod_{k=0}^{n}(m(1-\tau)+k) - (1-\tau)^{n+1}\prod_{k=0}^{n}(m+k)}{\prod_{k=0}^{n}(m+k)} &= \frac{c_0}{m} + \frac{c_1}{m+1} + \ldots + \frac{c_n}{m+n} \\ &\le \frac{\sum c_i}{m} \\ &\le \frac{n\frac{1}{\pi}(1-\beta)^{-0.278465}}{m}. \end{aligned}$$

Now that we have bounds on the individual terms, we can give a bound on the complete expression

$$
\begin{aligned}
&2^{-\frac{(d-1)(n+1)}{4m}} d^{-\frac{(n+1)(d-n+1)}{2md}} N^{\left(\beta m\tau(d-n+1)-dm\tau+\binom{m+n}{m-1}-\binom{m(1-\tau)+n}{m(1-\tau)-1}\right)\frac{n+1}{md}} \\
&\geq N^{\beta\tau(n+1)-\tau(n+1)+1-(1-\tau)^{n+1}-\frac{n^{\frac{1}{\pi}}(1-\beta)^{-0.278465}}{m}+\ln(1-\beta)\beta\frac{n}{m}} .
\end{aligned}
$$

An Infinite Class of Balanced Functions with Optimal Algebraic Immunity, Good Immunity to Fast Algebraic Attacks and Good Nonlinearity*

Claude Carlet[1] and Keqin Feng[2]

[1] Department of Mathematics, University of Paris 8 (MAATICAH), 93526 - Saint-Denis cedex 02, France
claude.carlet@inria.fr

[2] Department of Mathematical Sciences, Tsinghua University, Beijing China 100084
kfeng@math.tsinghua.edu.cn

Abstract. After the improvement by Courtois and Meier of the algebraic attacks on stream ciphers and the introduction of the related notion of algebraic immunity, several constructions of infinite classes of Boolean functions with optimum algebraic immunity have been proposed. All of them gave functions whose algebraic degrees are high enough for resisting the Berlekamp-Massey attack and the recent Rønjom-Helleseth attack, but whose nonlinearities either achieve the worst possible value (given by Lobanov's bound) or are slightly superior to it. Hence, these functions do not allow resistance to fast correlation attacks. Moreover, they do not behave well with respect to fast algebraic attacks. In this paper, we study an infinite class of functions which achieve an optimum algebraic immunity. We prove that they have an optimum algebraic degree and a much better nonlinearity than all the previously obtained infinite classes of functions. We check that, at least for small values of the number of variables, the functions of this class have in fact a very good nonlinearity and also a good behavior against fast algebraic attacks.

Keywords: Algebraic attack, Boolean function, Stream cipher.

1 Introduction

Before this century, the Boolean functions used in the combiner and filter models of stream ciphers (see description *e.g.* in [9]) had mainly to be balanced, to have a high algebraic degree, a high nonlinearity and, in the case of the combiner model, a high correlation immunity (in the case of the filter model, a correlation immunity of order 1 is commonly considered as sufficient; in most cases, it is easily achieved without losing the other properties, by replacing the function by a linearly equivalent one). These properties could be satisfied by functions of about 10 variables. But the algebraic attacks introduced by Courtois and Meier [15] (or more properly speaking improved by them, since the idea of algebraic

* Supported by the NSFC grant 60433050 and the 973 grant of China 2004CB 3180004.

J. Pieprzyk (Ed.): ASIACRYPT 2008, LNCS 5350, pp. 425–440, 2008.
© International Association for Cryptologic Research 2008

attacks comes already from Shannon), which have allowed cryptanalysing several stream ciphers [1,12,13,15,25] have led to more constraints on the functions, and obliged to increase the number of variables up to at least 13 variables and in practice much more (maybe 20). The property needed for resisting the standard algebraic attack of Courtois and Meier [15] is a high algebraic immunity [33]: for a given Boolean function f on n variables, any nonzero Boolean function g such that $f * g = 0$ or $(1 + f) * g = 0$ should have high algebraic degree, where $*$ is the multiplication of functions inherited from multiplication in $\mathbb{F}_2$, the finite field with two elements. The best possible algebraic immunity of n-variable functions is $\lceil \frac{n}{2} \rceil$ [15]. It has been proved in [19] that, for all $a < 1$, when n tends to infinity, $AI(f)$ is almost surely greater than $\frac{n}{2} - \sqrt{\frac{n}{2} \ln \left(\frac{n}{a \ln 2}\right)}$. Hence, random functions behave well with respect to the algebraic immunity (but this does not mean that functions with good algebraic immunity are easy to construct).

Having a high algebraic immunity is not sufficient for resisting the fast algebraic attacks introduced by Courtois in [13]: if one can find g of low degree and $h \neq 0$ of reasonable degree such that $f * g = h$, then a fast algebraic attack (FAA) is feasible. No result is known on the behavior of random functions against FAA.

Even a high resistance to fast algebraic attacks is not sufficient, since algebraic attacks on the augmented function [23] can be efficient when fast algebraic attacks are not. The resistance to these attacks is not properly speaking a property of the function used in a cipher and studying the resistance of the cipher to them obliges to consider all possible update functions (of the linear part of the pseudo-random generator).

It is a difficult challenge to find functions achieving all of the necessary criteria and the research of such functions has taken a significant delay with respect to cryptanalyses. The research of Boolean functions that can resist algebraic attacks, the Berlekamp-Massey attack and the fast correlation attacks has not given fully satisfactory results: we know that functions achieving optimal or suboptimal algebraic immunity and in the same time balancedness, high algebraic degree and high nonlinearity must exist thanks to the results of [19,37]. Such functions have been found with sufficient numbers of variables thanks to Algorithm 1 of [2] (others can be found by using the algorithm of [20]). But the functions given in [2] belong to classes which have not, potentially, a good asymptotic algebraic immunity (see [35]), and there remains to see whether these functions behave well against fast algebraic attacks. No infinite class of functions with good algebraic immunity and good nonlinearity has been exhibited so far.

There are, up to now, two main infinite classes of Boolean functions achieving optimum algebraic immunity. The first one contains functions in even numbers n of variables and is obtained by an iterative construction. The constructed functions have been further studied in [10], where it is shown that their algebraic degrees are close to n but their nonlinearity is $2^{n-1} - \binom{n-1}{\frac{n}{2}}$, which is insufficient. Moreover, they are not balanced (but it is possible to build balanced functions from these ones) and are weak against fast algebraic attacks [2,18]. The second class contains symmetric functions (whose values depend only on the Hamming weight of the input vectors) [3,18] or functions whose values depend on the Hamming weight of the input vectors except for a few inputs [7]. The nonlinearities of these functions are often not exceeding $2^{n-1} - \binom{n-1}{\lfloor \frac{n}{2} \rfloor}$ and when they do, they are not much greater than this number, see [11]. They are still weaker against fast algebraic attacks [2]. The functions constructed in [28,29] seem to have worse nonlinearity than those of [7]. Apart from these infinite classes, some power functions with sub-optimal algebraic immunity, in at most 20 variables, have been exhibited in [2, Table 1]. The behavior of these functions against fast algebraic attacks has not been investigated so far.

In the present paper, we show that an infinite class of balanced functions with optimal algebraic immunity, which has been considered in [22] for showing the tightness of bounds on the algebraic immunity of vectorial functions, has potentially a good nonlinearity. We give a very simple proof of the optimal algebraic immunity of these functions. We show that they have also optimal algebraic degree and we prove a lower bound on their nonlinearities which is much larger than the best nonlinearities of the infinite classes of functions with optimal algebraic immunity found so far. However, this bound is not enough for saying these functions have good nonlinearities. We compute for small values of n the exact values of the nonlinearity, which are very good and much bigger than the lower bound, and we also check for these values of n that the functions behave well against fast algebraic attacks. This is the first time a function (and moreover a whole infinite class of functions) seems able to satisfy all of the main criteria for being used as a filtering function in a stream cipher.

The rest of the paper is organized as follows. In Section 2, we recall the necessary background. In Section 3, we give a simple proof that the functions of the class have optimal algebraic immunity. In Section 4, we calculate the univariate representation of the functions and deduce their algebraic degree. We prove a lower bound on their nonlinearity. We give also the exact values of the nonlinearity for small values of n. In Section 5, we give the results of computer investigations suggesting a good immunity of the functions against fast algebraic attacks.

2 Preliminaries

Let $\mathbb{F}_2^n$ be the n-dimensional vector space over $\mathbb{F}_2$, and B_n the set of n-variable (Boolean) functions from $\mathbb{F}_2^n$ to $\mathbb{F}_2$. The basic representation of a Boolean function $f(x_1, \cdots, x_n)$ is by the output column of its truth table, i.e., a binary string of length 2^n,

$$[f(0,0,\cdots,0), f(1,0,\cdots,0), f(0,1,\cdots,0), f(1,1,\cdots,0), \cdots, f(1,1,\cdots,1)].$$

The *Hamming weight* $\mathrm{wt}(f)$ of a Boolean function $f \in B_n$ is the weight of this string, that is, the size of the support $\mathrm{Supp}(f) = \{x \in \mathbb{F}_2^n \mid f(x) = 1\}$ of the function. The *Hamming distance* $\mathrm{d_H}(f, g)$ between two Boolean functions f and g is the Hamming weight of their difference $f + g$ (by abuse of notation, we use $+$ to denote the addition on $\mathbb{F}_2$, i.e., the XOR). We say that a Boolean function

f is *balanced* if its truth table contains an equal number of 1's and 0's, that is, if its Hamming weight equals 2^{n-1}.

Any Boolean function has a unique representation as a multivariate polynomial over $\mathbb{F}_2$, called the *algebraic normal form* (ANF), of the special form:

$$f(x_1, \cdots, x_n) = \sum_{I \subseteq \{1,2,\cdots,n\}} a_I \prod_{i \in I} x_i.$$

The *algebraic degree*, $\deg(f)$, is the global degree of this polynomial, that is, the number of variables in the highest order term with non zero coefficient. A Boolean function is affine if it has degree at most 1. The set of all affine functions is denoted by A_n.

We shall need another representation of Boolean functions, by univariate polynomials over the field $\mathbb{F}_{2^n}$. We identify the field $\mathbb{F}_{2^n}$ and the vector space $\mathbb{F}_2^n$: this field being an n-dimensional $\mathbb{F}_2$-vector space, we can choose a basis $(\beta_1, \cdots, \beta_n)$ and identify every element $x = \sum_{i=1}^n x_i \beta_i \in \mathbb{F}_{2^n}$ with the n-tuple of its coordinates $(x_1, \cdots, x_n) \in \mathbb{F}_2^n$. Every function $f : \mathbb{F}_{2^n} \to \mathbb{F}_{2^n}$ (and in particular every Boolean function $f : \mathbb{F}_{2^n} \to \mathbb{F}_2$) can then be uniquely represented as a polynomial $\sum_{j=0}^{2^n-1} a_j x^j$ where $a_j \in \mathbb{F}_{2^n}$. Indeed, the mapping which maps every such polynomial to the corresponding function from $\mathbb{F}_{2^n}$ to itself is $\mathbb{F}_{2^n}$-linear, injective (since a non-zero polynomial of degree at most $2^n - 1$ over a field cannot have more than $2^n - 1$ zeroes in this field) and therefore surjective since the $\mathbb{F}_{2^n}$-vector spaces of these polynomials and of the functions from $\mathbb{F}_{2^n}$ to itself have the same dimension 2^n. The function is Boolean if and only if the functions $f(x)$ and $(f(x))^2$ are represented by the same polynomial, that is, if $a_0, a_{2^n-1} \in \mathbb{F}_2$ and, for every $i = 1, \cdots, 2^n - 2$, we have $a_{2j} = (a_j)^2$, where $2j$ is taken mod $2^n - 1$. Then the algebraic degree of the function equals the maximum *2-weight* $w_2(j)$ of j such that $a_j \neq 0$, where the 2-weight of j equals the number of 1's in its binary expansion. We briefly recall why, since the algebraic degree is an important parameter and we will need this when studying the functions. Writing $j = \sum_{s=0}^{n-1} j_s 2^s$, we have the equalities:

$$\begin{aligned} f(x) &= \sum_{j=0}^{2^n-1} a_j \left(\sum_{i=1}^n x_i \beta_i \right)^j \\ &= \sum_{j=0}^{2^n-1} a_j \left(\sum_{i=1}^n x_i \beta_i \right)^{\sum_{s=0}^{n-1} j_s 2^s} \\ &= \sum_{j=0}^{2^n-1} a_j \prod_{s=0}^{n-1} \left(\sum_{i=1}^n x_i \beta_i^{2^s} \right)^{j_s} ; \end{aligned}$$

expanding these products, simplifying and decomposing again over the basis $(\beta_1, \ldots, \beta_n)$ gives the ANF of F; this proves that the algebraic degree is upper bounded by the number $\max\{w_2(j);\ a_j \neq 0\}$, and it cannot be strictly smaller, because the number of those functions from $\mathbb{F}_{2^n}$ to itself of algebraic degrees at most d equals the number of those univariate polynomials $\sum_{j=0}^{2^n-1} a_j x^j$, $a_j \in \mathbb{F}_{2^n}$, such that $\max_{j=0,\ldots,2^n-1/\, a_j \neq 0} w_2(j) \leq d$.

In this representation, the elements of A_n are all the functions $tr(ax)$, $a \in \mathbb{F}_{2^n}$, where tr is the trace function: $tr(x) = x + x^2 + x^{2^2} + \cdots + x^{2^{n-1}}$.

Any Boolean function should have high algebraic degree to allow the cryptosystem resisting the Berlekamp-Massey attack [21].

Boolean functions used in cryptographic systems must have high nonlinearity to withstand fast correlation attacks (see e.g. [6,34]). The *nonlinearity* of an n-variable function f is its distance to the set of all n-variable affine functions, i.e.,

$$nl(f) = \min_{g \in A_n} (\mathrm{d_H}(f, g)).$$

This parameter can be expressed by means of the Walsh transform. Let $x = (x_1, \cdots, x_n)$ and $\lambda = (\lambda_1, \cdots, \lambda_n)$ both belong to $\mathbb{F}_2^n$ and $\lambda \cdot x$ be the usual inner product in $\mathbb{F}_2^n$: $\lambda \cdot x = \lambda_1 x_1 + \cdots + \lambda_n x_n \in \mathbb{F}_2$, or any other inner product in $\mathbb{F}_2^n$. Let $f(x)$ be a Boolean function in n variables. The *Walsh transform* (depending on the choice of the inner product) of $f(x)$ is the integer valued function over $\mathbb{F}_2^n$ defined as

$$W_f(\lambda) = \sum_{x \in \mathbb{F}_2^n} (-1)^{f(x) + \lambda \cdot x}.$$

If we identify the vector space $\mathbb{F}_2^n$ with the field $\mathbb{F}_{2^n}$, then we can take for inner product: $\lambda \cdot x = tr(\lambda x)$.

A Boolean function f is balanced if and only if $W_f(0) = 0$. The nonlinearity of f can also be given by

$$nl(f) = 2^{n-1} - \frac{1}{2} \max_{\lambda \in \mathbb{F}_2^n} |W_f(\lambda)|.$$

For every n-variable function f we have $nl(f) \leq 2^{n-1} - 2^{n/2-1}$.

Algebraic attacks have been introduced recently (see [15]). They recover the secret key, or at least the initialization of the cipher, by solving a system of multivariate algebraic equations. The idea that the key bits can be characterized as the solutions of such a system comes from C. Shannon [39]. In practice, for cryptosystems which are robust against the usual attacks, this system is too complex to be solved (its equations being highly nonlinear). In the case of stream ciphers, we can get a very overdefined system (i.e. a system with a number of linearly independent equations much greater than the number of unknowns). In the combiner or the filter model, with a linear part of size N and with an n-variable Boolean function f as combining or filtering function, there exists a linear permutation $L : \mathbb{F}_2^N \mapsto \mathbb{F}_2^N$ and a linear mapping $L' : \mathbb{F}_2^N \mapsto \mathbb{F}_2^n$ such that, denoting by $u_1, \cdots, u_N$ the initialisation and by $(s_i)_{i \geq 0}$ the pseudo-random sequence output by the generator, we have, for every $i \geq 0$:

$$s_i = f(L' \circ L^i(u_1, \cdots, u_N)).$$

The number of equations can then be much larger than the number of unknowns. This makes less complex the resolution of the system by using Groebner basis, and even allows linearizing the system (i.e. obtaining a system of linear equations by replacing every monomial of degree greater than 1 by a new unknown); the resulting linear system has however too many unkwnowns and cannot be solved. Courtois and Meier have had a simple but very efficient idea. Assume that there exist functions $g \neq 0$ and h of low degrees (say, of degrees at most d) such that $f * g = h$. We have then, for every $i \geq 0$:

$$s_i \, g(L' \circ L^i(u_1, \cdots, u_N)) = h(L' \circ L^i(u_1, \cdots, u_N)).$$

This equation in $u_1, \cdots, u_N$ has degree at most d, since L and L' are linear, and the system of equations obtained after linearization can then be solved by Gaussian elimination. Low degree relations have been shown to exist for several well known constructions of stream ciphers, which were immune to all previously known attacks.

It has been shown [15,33] that the existence of such relations is equivalent to that of non-zero functions g of low degrees such that $f * g = 0$ or $(f+1) * g = 0$. This led to the following definition.

Definition 1. *For $f \in B_n$, we define $AN(f) = \{g \in B_n \,|\, f * g = 0\}$. Any function $g \in AN(f)$ is called an annihilator of f. The algebraic immunity (AI) of f is the minimum degree of all the nonzero annihilators of f and of all those of $f+1$. We denote it by $AI(f)$.*

Note that $AI(f) \leq \deg(f)$, since $f * (1+f) = 0$. Note also that the algebraic immunity, as well as the nonlinearity and the degree, is affine invariant (i.e. is invariant under composition by an affine automorphism). As shown in [15], we have $AI(f) \leq \lceil \frac{n}{2} \rceil$.

The complexity of the standard algebraic attack on the combiner model or the filter model using a nonlinear function f equals roughly $O(D^3)$ in time and $O(D)$ in data, where $D = \sum_{i=0}^{AI(f)} \binom{N}{i}$, where N is the size of the linear part of the pseudo-random generator.

If a function has optimal algebraic immunity $\lceil \frac{n}{2} \rceil$ with n odd, then it is balanced (see e.g. [10]). Whatever is n, a high value of $AI(f)$ automatically implies that the nonlinearity is not very low: M. Lobanov has obtained in [31] the following tight lower bound:

$$nl(f) \geq 2 \sum_{i=0}^{AI(f)-2} \binom{n-1}{i}.$$

However, this bound does not assure that the nonlinearity is high enough:

- For n even and $AI(f) = \frac{n}{2}$, it gives $nl(f) \geq 2^{n-1} - 2\binom{n-1}{n/2-1} = 2^{n-1} - \binom{n}{n/2}$ which is much smaller than the best possible nonlinearity $2^{n-1} - 2^{n/2-1}$ and, more problematically, much smaller than the asymptotic almost sure nonlinearity of Boolean functions, which is, when n tends to ∞, located in the neighbourhood of $2^{n-1} - 2^{n/2-1}\sqrt{2n \ln 2}$ (see [37]); the nonlinearity reached by the known functions with optimal AI is equal to (or is close to) that of the majority function

which maps an input vector $x \in \mathbb{F}_2^n$ to 1 if its weight is not smaller (resp. is strictly greater) than $n/2$ and 0 otherwise (the two versions are affinely equivalent) and of the iterative construction recalled in [10] : $2^{n-1} - \binom{n-1}{n/2} = 2^{n-1} - \frac{1}{2}\binom{n}{n/2}$; it is a little better than what gives Lobanov's bound but it is insufficient. Some functions exhibited in [11,28,29] have better nonlinearities but the increasement is not quite significant.
• For n odd and $AI(f) = \frac{n+1}{2}$, Lobanov's bound gives $nl(f) \geq 2^{n-1} - \binom{n-1}{(n-1)/2} \simeq 2^{n-1} - \frac{1}{2}\binom{n}{(n-1)/2}$ which is a little better than in the n even case, but still far from the average nonlinearity of Boolean functions; the nonlinearity of the majority function matches this bound; here again, some functions exhibited in [11,28,29] have better nonlinearities but the increasement is not sufficient.

A high algebraic immunity is a necessary but not sufficient condition for robustness against all kinds of algebraic attacks. Indeed, if one can find g of low degree and $h \neq 0$ of reasonable degree such that $f * g = h$, then a fast algebraic attack is feasible, see [13,1,24] (note however that fast algebraic attacks need more data than standard ones). This has been exploited in [14] to present an attack on SFINKS [4] and we can say that with this attack, which comes in addition to the standard algebraic attack, Courtois has made very difficult the work of the designer. Since $f * g = h$ implies $f * h = f * f * g = f * g = h$, we see that h is then an annihilator of $f + 1$ and if $h \neq 0$, then its degree is at least equal to the algebraic immunity of f. So summarizing, we shall say that the function behaves well with respect to fast algebraic attacks if there exists k (which can be small with respect to n, but not too small) such that, for every nonzero function g of algebraic degree at most k, the function $h = f * g$ has algebraic degree significantly greater than $\lceil \frac{n}{2} \rceil$. It has been shown in [13] that when $e + d \geq n$, there must exist g of degree at most e and h of degree at most d such that $f * g = h$. Hence, an n-variable function f can be considered as optimal with respect to fast algebraic attacks if there do not exist two functions $g \neq 0$ and h such that $f * g = h$ and $\deg(g) + \deg(h) < n$ with $\deg(g) < n/2$. The question of the existence of such functions was completely open until the present paper.

The pseudo-random generator must also resist algebraic attacks on the augmented function [23], that is, on the vectorial function $F(x)$ whose coordinate functions are $f(x), f(L(x)), \cdots, f(L^{m-1}(x))$, where L is the (linear) update function of the linear part of the generator. Algebraic attacks can be more efficient when applied to the augmented function rather than to the function f itself. The efficiency of the attack depends not only on the function f, but also on the update function (and naturally also on the choice of m), since for two different update functions L and L', the vectorial functions $F(x)$ and $F'(x) = (f(x), f(L'(x)), ..., f({L'}^{m-1}(x))$ are not linearly equivalent (neither equivalent in the more general sense called CCZ-equivalence, that is, affine equivalence of the graphs of the functions). Testing the behavior of a function with respect to this attack is therefore a long term work (all possible update functions have to be investigated).

A new version of algebraic attack has been found recently by S. Rønjom and T. Helleseth [38] and is very efficient. Its time complexity is roughly $O(\mathcal{D})$, where $\mathcal{D} = \sum_{i=0}^{\deg(f)} \binom{N}{i}$, where N is the size of the linear part of the pseudo-random generator. But it needs much more data than standard algebraic attacks: $O(\mathcal{D})$ also! When f has degree close to n and algebraic immunity close to $\frac{n}{2}$, this is the square of what is needed by standard algebraic attacks. However, this attack obliges the designer to choose a function with very high degree.

The functions used in the combiner model must be additionally highly resilient (that is, balanced and correlation immune of a high order; see definition *e.g.* in [9]) to withstand correlation attacks. It seems quite difficult to achieve all of the necessary criteria including this one, and for this reason, the filter generator seems more appropriate.

3 The Infinite Class and Its Algebraic Immunity

We shall show that, for every n, the Boolean function on $\mathbb{F}_{2^n}$ whose support equals $\{0\} \cup \{\alpha^i;\ i = 0, \cdots, 2^{n-1}-2\}$, where α is a primitive element of $\mathbb{F}_{2^n}$, has optimal algebraic immunity. This function (or more precisely its complement) makes thinking of the majority function but we shall see that it is in fact quite different since it has much better nonlinearity and it behaves much better with respect to fast algebraic attacks too.

Theorem 1. *Let n be any integer such that $n \geq 2$ and α a primitive element of the field F_{2^n}.*

Let f be the Boolean function on $\mathbb{F}_{2^n}$ whose support is $\{0, 1, \alpha, \cdots, \alpha^{2^{n-1}-2}\}$. Then f has optimal algebraic immunity $\lceil n/2 \rceil$.

Proof
Let g be any Boolean function of algebraic degree at most $\lceil n/2 \rceil - 1$. Let $g(x) = \sum_{i=0}^{2^n-1} g_i x^i$ be its univariate representation in the field $\mathbb{F}_{2^n}$, where $g_i \in \mathbb{F}_{2^n}$ is null if the 2-weight $w_2(i)$ of i is at least $\lceil n/2 \rceil$ (which implies in particular that $g_{2^n-1} = 0$).

If g is an annihilator of f, then we have $g(\alpha^i) = 0$ for every $i = 0, \cdots, 2^{n-1}-2$, that is, the vector $(g_0, \cdots, g_{2^n-2})$ belongs to the Reed-Solomon code over $\mathbb{F}_{2^n}$ of zeroes $1, \alpha, \cdots, \alpha^{2^{n-1}-2}$ (the Reed-Solomon code of zeroes $\alpha^\ell, \cdots, \alpha^{\ell+r}$ equals by definition the set of vectors $(g_0, \cdots, g_{2^n-2})$ of $\mathbb{F}_{2^n}^{2^n-1}$ such that these elements are zeroes of the polynomial $\sum_{i=0}^{2^n-2} g_i X^i$, see [32]; there exists an equivalent definition where Reed-Solomon codes are given by evaluating polynomials at points but we shall not need it).

According to the BCH bound, if g is non-zero, then the vector $(g_0, \cdots, g_{2^n-2})$ has Hamming weight at least 2^{n-1}. The general proof of this lower bound can be found in [32] as well. For self-completeness, we briefly recall how it can be simply proved in our framework. By definition, we have:

$$\begin{pmatrix} g(1) \\ g(\alpha) \\ g(\alpha^2) \\ \vdots \\ g(\alpha^{2^n-2}) \end{pmatrix} = \begin{pmatrix} 1 & 1 & 1 & \cdots & 1 \\ 1 & \alpha & \alpha^2 & \cdots & \alpha^{2^n-2} \\ 1 & \alpha^2 & \alpha^4 & \cdots & \alpha^{2(2^n-2)} \\ \vdots & \vdots & \vdots & \cdots & \vdots \\ 1 & \alpha^{2^n-2} & \alpha^{2(2^n-2)} & \cdots & \alpha^{(2^n-2)(2^n-2)} \end{pmatrix} \times \begin{pmatrix} g_0 \\ g_1 \\ g_2 \\ \vdots \\ g_{2^n-2} \end{pmatrix}$$

which implies (since for every $0 \leq i, j \leq 2^n - 2$, the sum $\sum_{k=0}^{2^n-2} \alpha^{(i-j)k}$ equals 1 if $i = j$ and 0 otherwise):

$$\begin{pmatrix} g_0 \\ g_1 \\ g_2 \\ \vdots \\ g_{2^n-2} \end{pmatrix} = \begin{pmatrix} 1 & 1 & 1 & \cdots & 1 \\ 1 & \alpha^{-1} & \alpha^{-2} & \cdots & \alpha^{-(2^n-2)} \\ 1 & \alpha^{-2} & \alpha^{-4} & \cdots & \alpha^{-2(2^n-2)} \\ \vdots & \vdots & \vdots & \cdots & \vdots \\ 1 & \alpha^{-(2^n-2)} & \alpha^{-2(2^n-2)} & \cdots & \alpha^{-(2^n-2)(2^n-2)} \end{pmatrix} \times \begin{pmatrix} g(1) \\ g(\alpha) \\ g(\alpha^2) \\ \vdots \\ g(\alpha^{2^n-2}) \end{pmatrix}$$

$$= \begin{pmatrix} 1 & 1 & \cdots & 1 \\ \alpha^{-(2^{n-1}-1)} & \alpha^{-2^{n-1}} & \cdots & \alpha^{-(2^n-2)} \\ \vdots & \vdots & \cdots & \vdots \\ \alpha^{-(2^{n-1}-1)(2^n-2)} & \alpha^{-2^{n-1}(2^n-2)} & \cdots & \alpha^{-(2^n-2)(2^n-2)} \end{pmatrix} \times \begin{pmatrix} g(\alpha^{2^{n-1}-1}) \\ g(\alpha^{2^{n-1}}) \\ \vdots \\ g(\alpha^{2^n-2}) \end{pmatrix}$$

Suppose that at least 2^{n-1} of the g_i's are null. Then, $g(\alpha^{2^{n-1}-1}), \cdots, g(\alpha^{2^n-2})$ satisfy a homogeneous system of linear equations whose matrix is a $2^{n-1} \times 2^{n-1}$ Vandermonde matrix and whose determinant is therefore non-null. This implies that $g(\alpha^{2^{n-1}-1}), \cdots, g(\alpha^{2^n-2})$ and therefore g must then be null, a contradiction. Hence the vector $(g_0, \cdots, g_{2^n-2})$ has weight at least 2^{n-1}.

Moreover, suppose that this vector has Hamming weight 2^{n-1} exactly. Then $g(x) = \sum_{\substack{0 \leq i \leq 2^n-2 \\ w_2(i) \leq (n-1)/2}} x^i$ and n is odd (so that $g(x)$ can have 2^{n-1} terms); but this contradicts the fact that $g(0) = 0$. We deduce that the vector $(g_0, \cdots, g_{2^n-2})$ has Hamming weight strictly greater than 2^{n-1}, leading to a contradiction with the fact that g has algebraic degree at most $\lceil n/2 \rceil - 1$, since the number of integers of 2-weight at most $\lceil n/2 \rceil - 1$ is not strictly greater than 2^{n-1}.

Let g be now a non-zero annihilator of $f + 1$. The vector $(g_0, \cdots, g_{2^n-2})$ belongs then to the Reed-Solomon code over $\mathbb{F}_{2^n}$ of zeroes $\alpha^{2^{n-1}-1}, \cdots, \alpha^{2^n-2}$. According to the BCH bound (which can be proven similarly as above), this vector has then Hamming weight strictly greater than 2^{n-1}. We arrive to the same contradiction. Hence, there does not exist a non-zero annihilator of f or $f + 1$ of algebraic degree at most $\lceil n/2 \rceil - 1$ and f has then (optimal) algebraic immunity $\lceil n/2 \rceil$. $\square$

Remark

1. We have proved in fact that f admits no non-zero annihilator whose univariate representation has at most 2^{n-1} non-zero coefficients.

2. The same proof shows that, for every even n, denoting $D = \sum_{i=0}^{n/2-1} \binom{n}{i} = 2^{n-1} - \binom{n-1}{n/2}$, if the support of f contains $\{0, \alpha^i, \alpha^{i+1}, \cdots \alpha^{i+D-2}\}$ and if the support of $f+1$ contains $\{\alpha^j, \alpha^{i+1}, \cdots \alpha^{j+D-1}\}$ for suitable parameters i, j, then the function f also has optimal AI. Moreover, for every n and every positive integer D, if $\mathrm{Supp}(f) \supseteq \{0, \alpha^i, \alpha^{i+1}, \cdots \alpha^{i+D-2}\}$ and $\mathrm{Supp}(f+1) \supseteq \{\alpha^j, \alpha^{i+1}, \cdots \alpha^{j+D-1}\}$ for suitable parameters i, j, then the function f has AI at least k such that $D \geq \sum_{i=0}^{k-1} \binom{n}{i}$. Hence, we can build functions with sub-optimal algebraic immunity. Sub-optimality is sometimes better than optimality in cryptography, when it allows avoiding a too strong structure of the function. Here, this allows constructing a balanced function of algebraic immunity $\lceil \frac{n}{2} \rceil - 1$ (for instance) and whose support is not made exclusively of consecutive powers of a primitive element.
3. Note that the function of Theorem 1 is not *a priori* linearly equivalent to the Boolean function whose support equals the set of the binary expansions of the integers in the range $[0; 2^{n-1} - 1]$. Indeed, for general $i = \sum_{k=0}^{n-1} 2^{i_k}$, $j = \sum_{k=0}^{n-1} 2^{j_k}$ there is no bilinear relationship between $tr(\alpha^{i+j})$ and $i_0 j_0 + \cdots + i_{n-1} j_{n-1}$. This means that the inner products in both frameworks are not linearly linked.

4 Algebraic Degree and Nonlinearity of the Function

We shall see now that the algebraic degree of the function of Theorem 1 is cryptographically quite satisfactory and that its nonlinearity is provably much better than for the previously known functions with optimal algebraic immunity. However, the lower bound we obtain gives a value which is not high enough for saying that the function has good nonlinearity. Nevertheless, for the values of n for which we could compute the exact value of the nonlinearity, it is quite satisfactory too.

Theorem 2. *The univariate representation of the function f of Theorem 1 equals*

$$1 + \sum_{i=1}^{2^n-2} \frac{\alpha^i}{(1+\alpha^i)^{1/2}} x^i \tag{1}$$

where $u^{1/2} = u^{2^{n-1}}$. Hence, f has algebraic degree $n-1$ (which is optimal for a balanced function).

Proof. Let $f(x) = \sum_{i=0}^{2^n-1} f_i x^i$ be the univariate representation of f. We have $f_0 = f(0) = 1$, $f_{2^n-1} = 0$ (since f has even Hamming weight and therefore algebraic degree at most $n-1$) and for every $i \in \{1, \cdots, 2^n - 2\}$:

$$f_i = \sum_{j=0}^{2^n-2} f(\alpha^j) \alpha^{-ij} = \sum_{j=0}^{2^{n-1}-2} \alpha^{-ij} = \frac{1+\alpha^{-i(2^{n-1}-1)}}{1+\alpha^{-i}} =$$

$$\left(\frac{1+\alpha^{-i(2^n-2)}}{1+\alpha^{-2i}}\right)^{1/2} = \left(\frac{1+\alpha^i}{1+\alpha^{-2i}}\right)^{1/2} = \frac{\alpha^i}{(1+\alpha^i)^{1/2}}.$$

This proves Relation (1). We can see that $f_{2^n-2} \neq 0$ and therefore f has algebraic degree $n-1$. □

Remark. Computing the expression of Theorem 2 has high complexity. Actually, the complexity of computing $f(x)$ is comparable to computing the discrete log since the latter can be obtained by computing n outputs to f (with a dichotomic method).

Theorem 3. *Let f be defined as in Theorem 1, then:*

$$nl(f) \geq 2^{n-1} + \frac{2^{n/2+1}}{\pi} \ln\left(\frac{\pi}{4(2^n-1)}\right) - 1 \approx 2^{n-1} - \frac{2\ln 2}{\pi} n\, 2^{n/2}.$$

Proof.

$$\begin{aligned} nl(f) &= 2^{n-1} - \frac{1}{2}\max_{\lambda\in\mathbb{F}_2^n} |W_f(\lambda)| \qquad (2)\\ &= 2^{n-1} - \frac{1}{2}\max_{0\neq\lambda\in\mathbb{F}_2^n} |W_f(\lambda)| \quad (\text{since } W_f(0)=0)\\ &= 2^{n-1} - \max_{\lambda\in\mathbb{F}_{2^n}^*}\left|\sum_{x\notin supp(f)} (-1)^{tr(\lambda x)}\right|\\ &\quad (\text{since } (-1)^f = 2(f+1)-1 \text{ and } \sum_{x\in\mathbb{F}_2^n}(-1)^{\lambda\cdot x}=0)\\ &= 2^{n-1} - \max_{\lambda\in\mathbb{F}_{2^n}^*} |S_\lambda| \end{aligned}$$

where

$$S_\lambda = \sum_{i=2^{n-1}-1}^{2^n-2} (-1)^{tr(\lambda\alpha^i)} \quad (\lambda\in\mathbb{F}_{2^n}^*) \qquad (3)$$

Let $\zeta = e^{\frac{2\pi\sqrt{-1}}{2^n-1}}$ be a primitive (2^n-1)-th root of 1 in the complex field $\mathbb{C}$, χ be the multiplicative character of $\mathbb{F}_{2^n}$ defined by $\chi(\alpha^j) = \zeta^j$ $(0 \leq j \leq 2^n-2)$ and $\chi(0) = 0$. We define the Gauss sum:

$$G(\chi^\mu) = \sum_{x\in\mathbb{F}_{2^n}^*} \chi^\mu(x)(-1)^{tr(x)} \quad (0 \leq \mu \leq 2^n-2)$$

It is well-known (see [30]) that $G(\chi^0) = -1$ and $|G(\chi^\mu)| = 2^{\frac{n}{2}}$ for $1 \leq \mu \leq 2^n-2$. By Fourier transformation we have

$$(-1)^{tr(\alpha^j)} = \frac{1}{2^n-1}\sum_{\mu=0}^{2^n-2} G(\chi^\mu)\overline{\chi}^\mu(\alpha^j) \quad (0 \leq j \leq 2^n-2)$$

Let $\lambda = \alpha^l$ $(0 \leq l \leq 2^n - 2)$ and $q = 2^n$. Then $\overline{\chi}^\mu(\lambda\alpha^i) = \zeta^{-\mu(l+i)}$ and by (3),

$$S_\lambda = \frac{1}{q-1}\sum_{\mu=0}^{q-2} G(\chi^\mu) \sum_{i=\frac{q}{2}-1}^{q-2} \overline{\chi}^\mu(\lambda\alpha^i)$$
$$= \frac{1}{q-1}\sum_{\mu=0}^{q-2} G(\chi^\mu) \sum_{i=\frac{q}{2}-1}^{q-2} \zeta^{-\mu(l+i)}$$
$$= \frac{1}{q-1}\left(\sum_{\mu=1}^{q-2} G(\chi^\mu)\zeta^{-\mu l}\,\frac{\zeta^{-\mu(\frac{q}{2}-1)}-1}{1-\zeta^{-\mu}} - \frac{q}{2}\right)$$

Therefore, for $\lambda \in \mathbb{F}_q^*$,

$$|S_\lambda| \leq \frac{1}{q-1}\left(\sum_{\mu=1}^{q-2} |G(\chi^\mu)| \cdot \frac{\left|\sin\frac{\pi\mu(\frac{q}{2}-1)}{q-1}\right|}{\sin\frac{\pi\mu}{q-1}} + \frac{q}{2}\right)$$
$$\leq \frac{1}{q-1}\left(\sum_{\mu=1}^{q-2} |G(\chi^\mu)| \cdot \frac{1}{\sin\frac{\pi\mu}{q-1}} + \frac{q}{2}\right)$$
$$= \frac{1}{q-1}\left(2\sqrt{q}\sum_{\mu=1}^{\frac{q}{2}-1}\left(\sin\frac{\pi\mu}{q-1}\right)^{-1} + \frac{q}{2}\right)$$

since $\sin(\pi - u) = \sin(u)$. By convexity of the function $\frac{1}{\sin t}$, we have, for $0 \leq \theta < t$ and $t + \theta \leq \pi$:

$$\frac{1}{\sin(t-\theta)} + \frac{1}{\sin(t+\theta)} \geq \frac{2}{\sin\ t}.$$

Then we deduce

$$\int_{t-\frac{\theta}{2}}^{t+\frac{\theta}{2}} \frac{du}{\sin\ u} \geq \frac{\theta}{\sin\ t}$$

and taking $\theta = \frac{\pi}{q-1}$:

$$\sum_{\mu=1}^{\frac{q}{2}-1}\left(\sin\frac{\pi\mu}{q-1}\right)^{-1} \leq \frac{q-1}{\pi}\sum_{\mu=1}^{\frac{q}{2}-1}\int_{\frac{\pi\mu}{q-1}-\frac{\pi}{2(q-1)}}^{\frac{\pi\mu}{q-1}+\frac{\pi}{2(q-1)}} \frac{du}{\sin\ u}$$
$$= \frac{q-1}{\pi}\int_{\frac{\pi}{2(q-1)}}^{\frac{\pi}{2}} \frac{du}{\sin\ u}.$$

Set $t(x) = \tan(x/2)$. We have $\sin x = \frac{2t(x)}{1+t^2(x)}$ and therefore $\frac{1}{\sin x} = \frac{t'(x)}{t(x)}$. Hence a primitive of $1/\sin x$ equals $\ln(|\tan(x/2)|)$. This implies

$$\begin{aligned}
2^{n-1} - \max S_\lambda &\geq 2^{n-1} - \left(2^{n/2+1}\left[\frac{1}{\pi}\ln(\tan(x/2))\right]_{\frac{\pi}{2(2^n-1)}}^{\frac{\pi}{2}} + 1\right) \\
&= 2^{n-1} + \frac{2^{n/2+1}}{\pi}\ln\left(\tan\left(\frac{\pi}{4(2^n-1)}\right)\right) - 1 \\
&\geq 2^{n-1} + \frac{2^{n/2+1}}{\pi}\ln\left(\frac{\pi}{4(2^n-1)}\right) - 1 \\
&\quad \left(\text{since } \tan x \geq x;\ \forall x \in \left[0; \frac{\pi}{2}\right[\right) \\
&\approx 2^{n-1} - \frac{2\ln 2}{\pi} n\, 2^{n/2}.
\end{aligned}$$

□

Remarks
1. The lower bound given by Theorem 3 shows that the nonlinearity of our function f is provably considerably better (at least asymptotically) than those of the previously found functions. Moreover, we checked for small values of n that the exact value of $nl(f)$ is much better than what gives this lower bound and better than the nonlinearity of random functions and that it seems quite sufficient for resisting fast correlation attacks (for these small values of n, it behaves as $2^{n-1} - 2^{n/2}$). We give in Table 1 below, for n ranging from 6 to 11, the values of the nonlinearity of f compared with Lobanov's lower bound (when applied with optimal algebraic immunity), with the best nonlinearities of those functions with optimal AI known before the present paper, with the lower bound of Theorem 3, and with the upper bound $2^{n-1} - 2^{n/2-1}$.
2. We have seen that the computation of the value of $f(x)$ has high complexity. The power functions seen in [2, Table 1] may be better in practice for being used with a high number of variables, if their behavior against fast algebraic attacks can be proved good. Our construction might be useful with different designs, using less variables. It would be nice to find other infinite classes with the same qualities and which would be more easily computable.

Table 1. The values of the nonlinearity of f compared with Lobanov's lower bound and with the upper bound $2^{n-1} - 2^{n/2-1}$

n	6	7	8	9	10	11
Lobanov's bound	12	44	58	186	260	772
Best nl of fcts with optimal AI known before	22	48	98	196	400	798
The bound of Theorem 3	10	28	70	163	366	798
The values of the nl of fct f of Theorem 1	*24*	*54*	*112*	*232*	*478*	*980*
The upper bound $2^{n-1} - 2^{n/2-1}$	28	58	120	244	496	1001

5 Immunity against Fast Algebraic Attacks

Computer investigations made using [2, Algorithm 2] suggest the following properties of the class of functions of Theorem 1:

- No nonzero function g of degree at most e and no function h of degree at most d exist such that $f * g = h$, when $(e, d) = (1, n - 2)$ for n odd and $(e, d) = (1, n - 3)$ for n even. This has been checked for $n \leq 12$ and we conjecture it for every n.
- For $e > 1$, pairs (g, h) of degrees (e, d) such that $e + d < n - 1$ were never observed. Precisely, the non-existence of such pairs could be checked exhaustively for $n \leq 9$ and $e < n/2$, for $n = 10$ and $e \leq 3$ and for $n = 11$ and $e \leq 2$. This suggests that this class of functions, even if not always optimal against fast algebraic attacks, has a very good behavior.

The instance with $n = 9$ turns out to be optimal. To the best of our knowledge, this is the first time where a function with optimal immunity against FAA's can be observed.

6 Conclusion

The functions of Theorem 1 seem to gather all the properties needed for allowing the stream ciphers using them as filtering functions to resist all the main attacks (the Berlekamp-Massey and Rønjom-Helleseth attacks, fast correlation attacks, standard and fast algebraic attacks). They are the only functions of this kind found so far.

Acknowledgement

The authors thank Sihem Mesnager for her help in computing the nonlinearities of the function and for a nice improvement of a first version of our lower bound on the nonlinearity. We thank also Simon Fischer for his precious help in computing the immunities to fast algebraic attacks of the function and Xiangyong Zeng for a nice observation.

References

1. Armknecht, F.: Improving fast algebraic attacks. In: Roy, B., Meier, W. (eds.) FSE 2004. LNCS, vol. 3017, pp. 65–82. Springer, Heidelberg (2004)
2. Armknecht, F., Carlet, C., Gaborit, P., Künzli, S., Meier, W., Ruatta, O.: Efficient computation of algebraic immunity for algebraic and fast algebraic attacks. In: Vaudenay, S. (ed.) EUROCRYPT 2006. LNCS, vol. 4004, pp. 147–164. Springer, Heidelberg (2006)
3. Braeken, A., Preneel, B.: On the algebraic immunity of symmetric Boolean functions. In: Maitra, S., Veni Madhavan, C.E., Venkatesan, R. (eds.) INDOCRYPT 2005. LNCS, vol. 3797, pp. 35–48. Springer, Heidelberg (2005), `http://homes.esat.kuleuven.be/~abraeken/thesisAn.pdf`

4. Braeken, A., Lano, J., Mentens, N., Preneel, B., Verbauwhede, I.: SFINKS: A Synchronous stream cipher for restricted hardware environments. In: SKEW - Symmetric Key Encryption Workshop (2005)
5. Canteaut, A.: Open problems related to algebraic attacks on stream ciphers. In: Ytrehus, Ø. (ed.) WCC 2005. LNCS, vol. 3969, pp. 120–134. Springer, Heidelberg (2006)
6. Canteaut, A., Trabbia, M.: Improved fast correlation attacks using parity-check equations of weight 4 and 5. In: Preneel, B. (ed.) EUROCRYPT 2000. LNCS, vol. 1807, pp. 573–588. Springer, Heidelberg (2000)
7. Carlet, C.: A method of construction of balanced functions with optimum algebraic immunity. Cryptology ePrint Archive, `http://eprint.iacr.org/2006/149`; Proceedings of the Wuyi Workshop on Coding and Cryptology. Published by World Scientific Publishing Co. Its series of Coding and Cryptology (to appear)
8. Carlet, C.: On the higher order nonlinearities of algebraic immune functions. In: Dwork, C. (ed.) CRYPTO 2006. LNCS, vol. 4117, pp. 584–601. Springer, Heidelberg (2006)
9. Carlet, C.: The monography Boolean Methods and Models. In: Crama, Y., Hammer, P. (eds.) Boolean Functions for Cryptography and Error Correcting Codes. Cambridge University Press, Cambridge (to appear), `http://www-rocq.inria.fr/codes/Claude.Carlet/pubs.html`
10. Carlet, C., Dalai, D.K., Gupta, K.C., Maitra, S.: Algebraic immunity for cryptographically significant Boolean functions: analysis and construction. IEEE Trans. Inform. Theory 52(7), 3105–3121 (2006)
11. Carlet, C., Zeng, X., Li, C.: Further properties of several classes of Boolean functions with optimum algebraic immunity (preprint), IACR e-print archive 2007/370
12. Cho, J.Y., Pieprzyk, J.: Algebraic attacks on SOBER-t32 and SOBER-128. In: Roy, B., Meier, W. (eds.) FSE 2004. LNCS, vol. 3017, pp. 49–64. Springer, Heidelberg (2004)
13. Courtois, N.: Fast algebraic attacks on stream ciphers with linear feedback. In: Boneh, D. (ed.) CRYPTO 2003. LNCS, vol. 2729, pp. 176–194. Springer, Heidelberg (2003)
14. Courtois, N.: Cryptanalysis of SFINKS. In: ICISC 2005. Cryptology ePrint Archive Report 2005/243 (2005), `http://eprint.iacr.org/`
15. Courtois, N., Meier, W.: Algebraic attacks on stream ciphers with linear feedback. In: Biham, E. (ed.) EUROCRYPT 2003. LNCS, vol. 2656, pp. 345–359. Springer, Heidelberg (2003)
16. Courtois, N., Pieprzyk, J.: Cryptanalysis of block ciphers with overdefined systems of equations. In: Zheng, Y. (ed.) ASIACRYPT 2002. LNCS, vol. 2501, pp. 267–287. Springer, Heidelberg (2002)
17. Dalai, D.K., Gupta, K.C., Maitra, S.: Cryptographically significant Boolean functions: construction and analysis in terms of algebraic immunity. In: Gilbert, H., Handschuh, H. (eds.) FSE 2005. LNCS, vol. 3557, pp. 98–111. Springer, Heidelberg (2005)
18. Dalai, D.K., Maitra, S., Sarkar, S.: Basic theory in construction of Boolean functions with maximum possible annihilator immunity. Des. Codes Cryptogr. 40(1), 41–58 (2006)
19. Didier, F.: A new upper bound on the block error probability after decoding over the erasure channel. IEEE Transactions on Information Theory 52, 4496–4503 (2006)
20. Didier, F.: Using Wiedemann's algorithm to compute the immunity against algebraic and fast algebraic attacks. In: Barua, R., Lange, T. (eds.) INDOCRYPT 2006. LNCS, vol. 4329, pp. 236–250. Springer, Heidelberg (2006)

21. Ding, C., Xiao, G., Shan, W. (eds.): The Stability Theory of Stream Ciphers. LNCS, vol. 561. Springer, Heidelberg (1991)
22. Feng, K., Liao, Q., Yang, J.: Maximal values of generalized algebraic immunity. Designs, Codes and Cryptography (to appear)
23. Fischer, S., Meier, W.: Algebraic Immunity of S-boxes and Augmented Functions. In: Biryukov, A. (ed.) FSE 2007. LNCS, vol. 4593, pp. 366–381. Springer, Heidelberg (2007)
24. Hawkes, P., Rose, G.: Rewriting Variables: The Complexity of Fast Algebraic Attacks on Stream Ciphers. In: Franklin, M. (ed.) CRYPTO 2004. LNCS, vol. 3152, pp. 390–406. Springer, Heidelberg (2004)
25. Lee, D.H., Kim, J., Hong, J., Han, J.W., Moon, D.: Algebraic attacks on summation generators, Fast Software Encryption. In: Roy, B., Meier, W. (eds.) FSE 2004. LNCS, vol. 3017, pp. 34–48. Springer, Heidelberg (2004)
26. Li, N., Qi, W.F.: Construction and analysis of Boolean functions of $2t+1$ variables with maximum algebraic immunity. In: Lai, X., Chen, K. (eds.) ASIACRYPT 2006. LNCS, vol. 4284, pp. 84–98. Springer, Heidelberg (2006)
27. Li, N., Qi, W.F.: Symmetric Boolean functions depending on an odd number of variables with maximum algebraic immunity. IEEE Transactions on Information theory 52(5), 2271–2273 (2006)
28. Li, N., Qi, W.-Q.: Construction and analysis of Boolean functions of $2t+1$ variables with maximum algebraic immunity. In: Lai, X., Chen, K. (eds.) ASIACRYPT 2006. LNCS, vol. 4284, pp. 84–98. Springer, Heidelberg (2006)
29. Li, N., Qu, L., Qi, W.-F., Feng, G., Li, C., Xie, D.: On the construction of Boolean functions with optimal algebraic immunity. IEEE Transactions on Information Theory 54(3), 1330–1334 (2008)
30. Lidl, R., Niederreiter, H.: Finite Fields, Encyclopedia of Mathematics and its Applications, vol. 20. Addison-Wesley, Reading (1983)
31. Lobanov, M.: Tight bound between nonlinearity and algebraic immunity. Paper 2005/441 (2005), `http://eprint.iacr.org/`
32. MacWilliams, F.J., Sloane, N.J.: The Theory of Error-Correcting Codes. North-Holland, Amsterdam (1977)
33. Meier, W., Pasalic, E., Carlet, C.: Algebraic attacks and decomposition of Boolean functions. In: Cachin, C., Camenisch, J.L. (eds.) EUROCRYPT 2004. LNCS, vol. 3027, pp. 474–491. Springer, Heidelberg (2004)
34. Meier, W., Staffelbach, O.: Fast correlation attacks on stream ciphers. In: Günther, C.G. (ed.) EUROCRYPT 1988. LNCS, vol. 330, pp. 301–314. Springer, Heidelberg (1988)
35. Nawaz, Y., Gong, G., Gupta, K.: Upper Bounds on Algebraic Immunity of Power Functions. In: Robshaw, M.J.B. (ed.) FSE 2006. LNCS, vol. 4047, pp. 375–389. Springer, Heidelberg (2006)
36. Qu, L., Li, C., Feng, K.: Note on symmetric Boolean functions with maximum algebraic immunity in odd number of variables. IEEE Transactions on Information theory 53(8), 2908–2910 (2007)
37. Rodier, F.: Asymptotic nonlinearity of Boolean functions. Designs, Codes and Cryptography 40(1), 59–70 (2006)
38. Rønjom, S., Helleseth, T.: A new attack on the filter generator. IEEE Trans. Inform. Theory 53(5), 1752–1758 (2007)
39. Shannon, C.E.: Communication theory of secrecy systems. Bell system technical journal 28, 656–715 (1949)

An Improved Impossible Differential Attack on MISTY1

Orr Dunkelman[1,*] and Nathan Keller[2,**]

[1] 'Ecole Normale Supérieure
Département d'Informatique,
CNRS, INRIA
45 rue d'Ulm, 75230 Paris, France
orr.dunkelman@ens.fr

[2] Einstein Institute of Mathematics, Hebrew University
Jerusalem 91904, Israel
nkeller@math.huji.ac.il

Abstract. MISTY1 is a Feistel block cipher that received a great deal of cryptographic attention. Its recursive structure, as well as the added FL layers, have been successful in thwarting various cryptanalytic techniques. The best known attacks on reduced variants of the cipher are on either a 4-round variant with the FL functions, or a 6-round variant without the FL functions (out of the 8 rounds of the cipher).

In this paper we combine the generic impossible differential attack against 5-round Feistel ciphers with the dedicated Slicing attack to mount an attack on 5-round MISTY1 with all the FL functions with time complexity of $2^{46.45}$ simple operations. We then extend the attack to 6-round MISTY1 with the FL functions present, leading to the best known cryptanalytic result on the cipher. We also present an attack on 7-round MISTY1 without the FL layers.

1 Introduction

MISTY1 [10] is a 64-bit block cipher with presence in many cryptographic standards and applications. For example, MISTY1 was selected to be in the CRYPTREC e-government recommended ciphers in 2002 and in the final NESSIE portfolio of block ciphers, as well as an ISO standard (in 2005).

MISTY1 has a recursive Feistel structure, where the round function is in itself (very close to) a 3-round Feistel construction. To add to the security of the cipher, after every two rounds (and before the first round), an FL function is applied to each of the halves independently. The FL functions are key-dependent

* The first author was supported by the France Telecome Chaire. Some of the work presented in this paper was done while the first author was staying at K.U. Leuven, Belgium and supported by the IAP Programme P6/26 BCRYPT of the Belgian State (Belgian Science Policy).

** The second author is supported by the Adams Fellowship Program of the Israel Academy of Sciences and Humanities.

J. Pieprzyk (Ed.): ASIACRYPT 2008, LNCS 5350, pp. 441–454, 2008.
© International Association for Cryptologic Research 2008

linear functions which play the role of whitening layers (even in the middle of the encryption).

MISTY1 has withstood extensive cryptanalytic efforts. The most successful attacks on it are an impossible differential attack on 4 rounds (when the FL layers are present) [8], an integral attack on 5 rounds (when all but the last FL layers are present) [6], and an impossible differential attack on 6 rounds (without FL layers) [9].

In this paper we show that the generic impossible differential attack against 5-round Feistel constructions [2,5] can be combined with the dedicated slicing attack [8] to yield an attack on 5-round MISTY1 with all the FL functions. The data complexity of the attack is 2^{38} chosen plaintexts, and the time complexity is $2^{46.45}$ simple operations. The main idea behind this attack is to actually attack the FL functions themselves as these functions are keyed linear transformations.

After presenting the 5-round attack, we extend it by one more round, and show that by using key schedule considerations and a delicately tailored attack algorithm, it is possible to attack 6 rounds of MISTY1 with all the FL functions present. The 6-round attack requires 2^{51} chosen plaintexts and has a running time of $2^{123.4}$ encryptions.

Finally, we present an impossible differential attack on 7-round MISTY1 when the FL layers are omitted. The attack uses $2^{50.2}$ known plaintexts, and has a running time of $2^{114.1}$ encryptions. We summarize our results along with previously known results on MISTY1 in Table 1.

Table 1. Summary of the Attacks on MISTY1

Attack	Rounds	FL functions	Complexity	
			Data	Time
Impossible Differential [7]	4	Most	2^{23} CP	$2^{90.4}$
Impossible Differential [7]	4	Most	2^{38} CP	2^{62}
Collision Search [7]	4	Most	2^{20} CP	2^{89}
Collision Search [7]	4	Most	2^{28} CP	2^{76}
Slicing Attack [8]	$4^{\dagger}$	All	$2^{22.25}$ CP	2^{45}
Slicing Attack & Impossible Differential [8]	4	All	$2^{27.2}$ CP	$2^{81.6}$
Impossible Differential [8]	4	All	$2^{27.5}$ CP	2^{116}
Integral [6]	5	Most	$2^{10.5}$ CP	$2^{22.11}$
Impossible Differential (Section 3)	$5^{\dagger}$	All	2^{38} CP	$2^{46.45}$
Impossible Differential (Section 4)	6	All	2^{51} CP	$2^{123.4}$
Higher-Order Differential [1]	5	None	$2^{10.5}$ CP	2^{17}
Impossible Differential [7]	6	None	2^{54} CP	2^{61}
Impossible Differential [7]	6	None	2^{39} CP	2^{106}
Impossible Differential [9]	6	None	2^{39} CP	2^{85}
Impossible Differential (Section 5)	7	None	$2^{50.2}$ KP	$2^{114.1}$

KP – Known plaintext, CP – Chosen plaintext.

† – the attack retrieves 41.36 bits of information about the key.

This paper is organized as follows: In Section 2 we give a brief description of the structure of MISTY1. We present our 5-round attack in Section 3, and discuss its extension to 6 rounds in Section 4. In Section 5 we present a 7-round attack which can be applied when there are no FL layers. Section 6 concludes the paper.

2 The MISTY1 Cipher

MISTY1 [10] is a 64-bit block cipher that has a key size of 128 bits. Since its introduction it withstood several cryptanalytic attacks [1,6,7,8,9], mostly due to its very strong round function (which accepts 32-bit input and 112-bit subkey[1]) and the FL layers (keyed linear transformations) which are applied every two rounds. The security of MISTY1 was acknowledged several times, when it was selected to the NESSIE portfolio, the CRYPTREC's list of recommended ciphers, and as an ISO standard.

MISTY1 has a recursive structure. The general structure of the cipher is a 8-round Feistel construction, where the round function, FO, is in itself close to a 3-round Feistel construction. The input to the FO function is divided into two halves. The left one is XORed with a subkey, enters a keyed permutation FI, and the output is XORed with the right half. After the XOR the two halves are swapped, and the same process (including the swap) is repeated two more times. After that, an additional swap and an XOR of the left half with a subkey are performed.

The FI in itself also has a Feistel-like structure. The 16-bit input is divided into two unequal halves — one of 9 bits, and the second of 7 bits. The left half (which contains 9 bits) enters an S-box, $S9$, and the output is XORed with the 7-bit half (after padding the 7-bit value with two zeroes). The two halves are swapped, the 7-bit half enters a different S-box, $S7$, and the output is XORed with 7 bits out of the 9 of the right half. The two halves are then XORed with a subkey, and swapped again. The 9-bit value again enters $S9$, and the output is XORed with the 7-bit half (after padding). The two halves are then swapped for the last time.

Every two rounds, starting before the first one, the two 32-bit halves enter an FL layer. The FL layer is a simple transformation. The input is divided into two halves of 16 bits each, the AND of the left half with a subkey is XORed to the right half, and the OR of the updated right half with another subkey is XORed to the left half. We outline the structure of MISTY1 and its parts in Figure 1.

The key schedule of MISTY1 takes the 128-bit key, and treats it as eight 16-bit words $K_1, K_2, \ldots, K_8$. From this set of subkeys, another eight 16-bit words are generated according to $K'_i = FI_{K_{i+1}}(K_i)$.[2]

[1] In [7] it was observed that the round function has an equivalent description that accepts 105 equivalent subkey bits.

[2] In case the index of the key j is greater than 8, the used key word is $j - 8$. This convention is used throughout the paper.

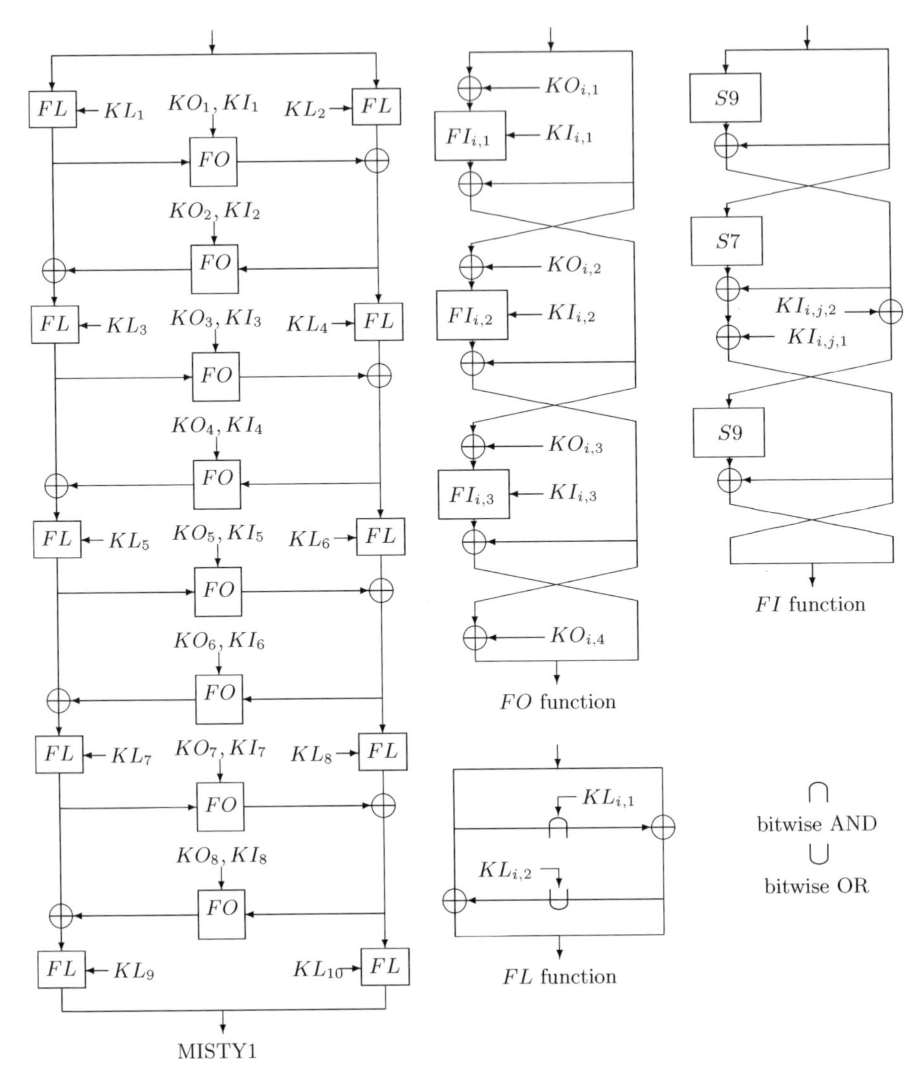

Fig. 1. Outline of MISTY1

In each round, seven words are used as the round subkey, and each of the FL functions accepts two subkey words. We give the exact key schedule of MISTY1 in Table 2.

3 An Impossible Differential Attack on 5-Round MISTY1

Our attack on 5-round MISTY1 with all the FL functions is based on the generic impossible differential attack against 5-round Feistel constructions with a bijective round function [2,5] and on the dedicated slicing attack [8] on reduced-round MISTY1.

Table 2. The Key Schedule Algorithm of MISTY1

$KO_{i,1}$	$KO_{i,2}$	$KO_{i,3}$	$KO_{i,4}$	$KI_{i,1}$	$KI_{i,2}$	$KI_{i,3}$	$KL_{i,1}$	$KL_{i,2}$
K_i	K_{i+2}	K_{i+7}	K_{i+4}	K'_{i+5}	K'_{i+1}	K'_{i+3}	$K_{\frac{i+1}{2}}$ (odd i) $K'_{\frac{i}{2}+2}$ (even i)	$K'_{\frac{i+1}{2}+6}$ (odd i) $K_{\frac{i}{2}+4}$ (even i)

3.1 The New 5-Round Impossible Differential

The generic attack on 5-round Feistel constructions is based on the following impossible differential:

Observation 1 ([2], page 136). *Let $E : \{0,1\}^{2n} \to \{0,1\}^{2n}$ be a 5-round Feistel construction with a bijective round function $f : \{0,1\}^n \to \{0,1\}^n$. Then for all non-zero $\alpha \in \{0,1\}^n$, the differential $(0, \alpha) \to (0, \alpha)$ through E is impossible.*

Our proposition is based on the fact that a similar impossible differential can be constructed even if FL layers are added to the construction, as in MISTY1. Note that since for a given key the FL layers are linear, we can define $FL(\alpha)$ for a difference α as the unique difference β such that $(x \oplus y = \alpha) \Rightarrow (FL(x) \oplus FL(y) = \beta)$.

Proposition 1. *Let E denote a 5-round variant of MISTY1, with all the FL functions present (including an FL layer after round 5). If for the given secret key we have $FL8(FL6(FL4(FL2(\alpha)))) = \beta$, where FLn is FL with the key KL_n, then the differential $(0, \alpha) \to (0, \beta)$ through E is impossible.*

Proof. If the plaintext difference is $(0, \alpha)$, then after the first FL layer, the difference becomes $(0, FL2(\alpha))$. This difference evolves after two rounds (including the second FL layer) to $(x, FL4(FL2(\alpha)))$, where $x \neq 0$ due to the bijectiveness of the round function of MISTY1.

On the other hand, if the output difference is $(0, \beta)$ such that $\beta = FL8(FL6(FL4(FL2(\alpha))))$, then before the last FL layer, the difference is $(0, FL6(FL4(FL2(\alpha))))$, and thus the input difference to round 5 is also $(0, FL6(FL4(FL2(\alpha))))$. Thus, the difference before the third FL layer is $(0, FL4(FL2(\alpha)))$.

However, if the input difference to round 3 is $(x, FL4(FL2(\alpha)))$ and the output difference of round 4 (before the FL layer) is $(0, FL4(FL2(\alpha)))$, then the output difference of the FO function in round 3 is zero. This is impossible since the input difference to this FO function is $x \neq 0$, and the FO function is bijective.

Hence, the differential $(0, \alpha) \to (0, \beta)$ is indeed impossible. □

We note that a similar approach is used in the slicing attack on 4-round MISTY1 [8]. The slicing attack is based on the generic 3-round impossible differential $(0, \alpha) \to (0, \beta)$ for all non-zero α, β which holds for every 3-round Feistel construction with a bijective round function.

3.2 The Structure of the FL Functions

A straightforward way to use the new impossible differential to attack 5-round MISTY1 is to encrypt many pairs with difference $(0, \alpha)$ for non-zero α, consider the pairs whose ciphertext difference is of the form $(0, \beta)$, and discard subkeys of the FL layers for which $FL8(FL6(FL4(FL2(\alpha)))) = \beta$. However, since the subkeys used in $FL2, FL4, FL6$, and $FL8$ are determined by 96 key bits, this approach is very time consuming. Instead, we examine the structure of the FL functions in order to find an efficient way to find the instances for which $FL8(FL6(FL4(FL2(\alpha)))) = \beta$, for a given pair (α, β). We use a series of observations, most of which were first presented in [8].

In the rest of this section, the function $FL8 \circ FL6 \circ FL4 \circ FL2$ is denoted by G.

1. For each $0 \leq i \leq 15$, the i-th bits of both halves of the input to an FL function and the i-th bits of both halves of the subkey used in the FL function, influence only the i-th bits of both halves of the output of the function. As a result, each FL function can be represented as a parallel application of 16 functions $f_i : \{0,1\}^2 \rightarrow \{0,1\}^2$ keyed by two different subkey bits each.
2. Each f_i is linear and invertible.
3. The two observations above hold also for a series of FL functions applied sequentially. In particular, the function $G = FL8 \circ FL6 \circ FL4 \circ FL2$ can be represented as a parallel application of 16 functions $g_i : \{0,1\}^2 \rightarrow \{0,1\}^2$ keyed by eight subkey bits each. The g_i's are all linear and invertible, and hence, can realize only six possible functions.[3] Thus, there are only $6^{16} = 2^{41.36}$ possible G functions.
4. Since each g_i is invertible, the differentials $0 \rightarrow a$ and $a \rightarrow 0$ through g_i are impossible, for each non-zero $a \in \{0,1\}^2$. As a result, most of the differentials of the form $\alpha \rightarrow \beta$ through G are impossible, regardless of the subkeys used in the FL functions. In each of the g_i-s, only 10 out of the 16 possible input/output pairs are possible. Hence, only $(10/16)^{16} = 2^{-10.85}$ of the input/output pairs for G are possible.
5. Assume that $G(\alpha) = \beta$, for fixed α and β. We want to find how many functions of the form G (out of the possible $2^{41.36}$ functions) satisfy this condition. For each g_i, there are 10 possible input/output pairs (the other six pairs are impossible for any subkey). For the $0 \rightarrow 0$ pair, all the six possible g_i functions satisfy this condition. For each of the 9 remaining pairs, two of the six functions satisfy the condition. Since the g_i functions are independent, the expected number of functions satisfying the conditions for all the g_i-s is:

$$\sum_{j=0}^{16} \binom{16}{j} \cdot \left(\frac{9}{10}\right)^j \cdot \left(\frac{1}{10}\right)^{16-j} \cdot 2^j \cdot 6^{16-j} = 2^{20.2}.$$

[3] Since we are interested only in differences, we treat two functions that differ by an additive constant as the same function. The total number of functions for each f_i is actually 24.

The $2^{41.36}$ possible G functions can be enumerated in such a way that the functions satisfying the condition for each (α, β) pair can be found efficiently.

Using these observations on the structure of the FL functions, we are ready to present our attack.

3.3 The New Attack

1. Ask for the encryption of 64 structures of 2^{32} plaintexts each, such that in each structure, the left half of all the plaintexts is equal to some random value A, while the right half obtains all possible values. (As a result, the difference between two plaintexts in the same structure is of the form $(0, \alpha)$).
2. For each structure, find the pairs whose output difference is of the form $(0, \beta)$.
3. For each pair with input difference $(0, \alpha)$ and output difference $(0, \beta)$ check whether $\alpha \to \beta$ is an impossible differential for the function G (as described in Section 3.2). Discard pairs which fail this test.
4. For each remaining pair, find all the G functions satisfying the condition $G(\alpha) = \beta$ and discard them from the list of all possible G functions.
5. After analyzing all the remaining pairs, output the list of remaining G functions.

Step 2 of the algorithm can be easily implemented by a hash table, resulting in about 2^{31} pairs from each structure. Step 3 can be easily performed by evaluating a simple Boolean function on the input and the output (as we are concerned with cases of a zero input causing a non-zero output or vice versa).[4]

As noted in Section 3.2, out of the 2^{31} pairs, about $2^{31} \cdot 2^{-10.85} = 2^{20.15}$ pairs remain from each structure at this point. Each of these pairs discards about $2^{20.2}$ possible values of G on average (as shown in Section 3.2), and thus, each structure is expected to discard about $2^{40.35}$ G functions. The identification of the discarded functions can be performed very efficiently.

Thus, after analyzing about 64 structures, we are left only with the right G function.[5] The time complexity of the attack is about $64 \cdot 2^{20.15} \cdot 2^{20.2} = 2^{46.35}$ simple operations, and the information retrieved by the attacker is equivalent to 41.36 key bits. In many situations, this is considered a break of the system and the attack terminates.

[4] The exact Boolean expression is as follows: Let the input difference of G be (x_1, x_2) and the output difference of G be (y_1, y_2). Also let $\overline{t}$ be the bitwise NOT of t, let & be a bitwise AND, and | be a bitwise OR. If $\overline{x_1} \& \overline{x_2} \& (y_1 | y_2)$ is non-zero then there is a zero input difference transformed to a non-zero output difference. It is also required to check whether the output difference is zero and the input difference is non-zero, which is done by evaluating: $\overline{y_1} \& \overline{y_2} \& (x_1 | x_2)$.

[5] We expect $2^{40.35} \cdot 64 = 2^{46.35}$ functions to be discarded (with overlap). Thus, the probability that a specific function remains after the analysis is

$$\left(1 - 2^{-41.36}\right)^{2^{46.35}} \approx e^{-32} = 2^{-46.2}.$$

3.4 Retrieving the Rest of the Secret Key

If the attacker wants to retrieve the actual value of the key, she can use the G function found in the attack to retrieve the value of the subkeys used in the G function. A naive approach is to try the possible 2^{96} subkeys which affect the functions $FL2, FL4, FL6$, and $FL8$, and check (for each subkey) whether it yields the correct G function. A more efficient algorithm is to guess the values of the subkeys K_3', K_4, K_5, K_6, and K_7, and check whether they induce the correct transformation from the input of G to the right half of the output of G. If this is the case, the attacker can retrieve the suggested value for K_8 efficiently, and if the suggestion is consistent with the correct G function, the attacker obtains a candidate for 96 bits of the key (the knowledge of K_3' and K_4 allows computing K_3). The time complexity of this approach is roughly 2^{80} evaluations of four FL functions, and the attacker gets a list of $2^{96} \cdot 2^{-41.36} = 2^{54.64}$ 96-bit subkeys. Retrieving the rest of the key by exhaustive search leads to a total time complexity of $2^{86.64}$ encryptions.

We note that possibly this part of the attack can be performed much more efficiently using some different attack technique and exploiting the key information obtained so far.[6]

4 Extending the Attack to 6 Rounds

The simplest way to extend a 5-round attack to 6 rounds is to guess the subkey of the last round, peel the last round off, and apply the 5-round attack. In MISTY1, this requires guessing the key of the last FL layer, as well as 112 subkey bits which enter the sixth FO function. Thus, we need to use a more careful analysis and key schedule considerations to present this attack.

In our attack we guess the subkey of the last FL layer (composed of 64 bits), and examine only ciphertext pairs with a special structure in order to reduce the amount of subkey material in the sixth FO we need to handle. Finally, we repeat the five round attack, taking into consideration the already known subkey material.

The special structure of the pairs examined in the attack is based on the following observation, presented in [7]:

Observation 2. *([7]) Assume that the input values to the function FO_i are known. The question whether the output difference of FO_i is of the form (δ, δ), for a 16-bit value δ, depends only on the 50 subkey bits $KO_{i,1}, KO_{i,2}, KI_{i,1,2}$, and $KI_{i,2,2}$.*

4.1 The Attack's Algorithm

1. Take m structures (generated just like in the 5-round attack).
2. For each guess of the subkey used in the last FL layer (subkeys K_2', K_4, K_6', and K_8), partially decrypt all the ciphertexts.

[6] We note that a similar problem is discussed in [8], and several techniques applicable in special cases (e.g., if the attacker can use both chosen plaintext and chosen ciphertext queries) are presented.

3. Find all pairs with plaintext difference $(0, \alpha)$ and ciphertext difference $((\delta, \delta), (x, y))$,[7] such that differential $\alpha \to (x, y)$ through $FL6 \circ FL4 \circ FL2$ is not impossible (see Section 3.2).
4. **Analysis of Round 6:** For each such pair, with difference $((\delta, \delta), (x, y))$, perform the following steps:
   (a) Given $KO_{6,2} = K_8$ compute the actual values just before the key addition with $KI_{6,2}$ for the pair. If the difference in the 7 left bits does not fit the corresponding 7 difference bits of δ — discard the pair.
   (b) Using the input and output differences of the second $S9$ S-box of the function $FI_{6,2}$, find the pairs of actual input values satisfying this difference relation.[8] From the actual input values obtain (on average) one candidate for the 9 bits of $KI_{6,2,2}$.
   (c) For each possible guess of $KI_{6,2,1}$ (i.e., the remaining unknown bits of K'_7) compute $KO_{6,1} = K_6$, and check whether the difference in the 7 left bits before the key addition in the first FI is equal to the difference in the 7 left bits of y.
   (d) Similarly to Step (4b), deduce $KI_{6,1,2}$ using the input/output differences of the second $S9$ in the function $FI_{6,1}$, suggested by the pair.
5. **Application of the 5-Round Attack:** For each guess of the 89 subkey bits (i.e., $K'_2, K_4, K'_6, K'_7, K_8, KI_{6,1,2}$) and for each pair corresponding to this subkey guess, perform the following:
   (a) Guess the 9 least significant bits of K_5 and use the key schedule to compute bits 7, 8 of K'_4 and K'_5. Check whether the relation $FL6(FL4(FL2(\alpha))) = (x, y)$ holds at bits 7, 8 of the left and the right halves of α and β (note that all the subkey bits involved in this relation are already known). If no, discard the pair.
   (b) Guess the remainder of K_5, and compute the full values of K'_4 and K'_5. Check whether the pair can achieve $\alpha \to (x, y)$, and retrieve the suggested value for the 7 remaining bits of K'_3.
   (c) If at this stage, for a given key guess there are remaining pairs, discard the subkey guess (as it suggests an impossible event). Otherwise, retrieve the remaining key bits by exhaustive search.

4.2 Analysis of the Attack

Starting with m structures, for each guess of the subkey used in the last FL layer (64 bits), about $m \cdot 2^{63} \cdot 2^{-16} \cdot 2^{-10.85} = m \cdot 2^{36.15}$ pairs are expected to enter Step (4). Each of these pairs has probability 2^{-7} to satisfy the differential condition of Step (4a), leaving $m \cdot 2^{29.15}$ pairs for each guess of the first 64 subkey bits. Then, in Step (4b) we obtain (for each pair) one candidate on average for 9 additional subkey bits, reducing the number of pairs associated with a given subkey guess (of 73 bits) to $m \cdot 2^{20.15}$ pairs. These two operations (a 7-bit filtering

[7] The reader is advised that we give the values without the swap operation, to be consistent with our figure describing MISTY1.

[8] This can be done easily by examining the difference distribution table of $S9$.

and a 9-bit subkey suggestion) are performed again in Steps (4c,4d) for each guess of 7 additional subkey bits. As a result, $m \cdot 2^{20.15} \cdot 2^{-16} = m \cdot 2^{4.15}$ pairs are expected to enter Step (5), for each of the 89-bit subkey guesses.

In Step (5), we guess a total of 16 additional key bits, and discard all the pairs for which $FL6(FL4(FL2(\alpha))) \neq (x, y)$. Since all the pairs for which the differential $\alpha \to (x, y)$ through $FL6 \circ FL4 \circ FL2$ is impossible were discarded in Step (3) of the attack, the probability of a pair to pass the filtering of Step (5) is $2^{-21.15}$. Hence, the number of pairs remaining after Step (5) for each subkey guess is $m \cdot 2^{4.15} \cdot 2^{-21.15} = m \cdot 2^{-17}$. As a result, the probability that a subkey guess is not discarded is $e^{-m \cdot 2^{-17}}$. Thus, the time complexity of Step (5c) is $2^{128} \cdot e^{-m \cdot 2^{-17}}$ encryptions.

We note that the number of pairs entering Step (5b) is $m \cdot 2^{1.5}$ for each subkey guess. Indeed, in Step (5a) we discard the pairs for which $FL6(FL4(FL2(\alpha))) \neq (x, y)$ in four bits. It may seem that the probability of a pair to pass this filtering is 2^{-4}. However, since the pairs for which the differential $\alpha \to (x, y)$ through $FL6 \circ FL4 \circ FL2$ is impossible were already discarded before, the probability of a pair to pass the filtering[9] is $2^{-2.65}$, and hence the number of remaining pairs is indeed $m \cdot 2^{1.5}$ for each subkey guess.

The two most time consuming steps of the attack are Steps (5b) and (5c). Step (5b) takes $3 \cdot m \cdot 2^{1.5} \cdot 2^{105} = m \cdot 2^{108.1}$ evaluations of FL. We take the moderate assumption that the time complexity of three FL evaluations is not greater than 1/8 of the time required for a 6-round encryption. Hence, the time complexity of Step(5b) is about $m \cdot 2^{103.5}$ MISTY1 encryptions. Step (5c) takes $2^{128} \cdot e^{-m \cdot 2^{-17}}$ encryptions.

The least overall time complexity is achieved when both terms are the same, i.e., when $m \cdot 2^{103.5} = 2^{128} \cdot e^{-m \cdot 2^{-17}}$. Solving this equation numerically, suggests that $m = 2^{18.945}$ is the optimal value. Thus, the data complexity of the attack is $m \cdot 2^{32} \approx 2^{51}$ chosen plaintexts, and the time complexity is $2^{123.4}$ encryptions.

5 Attack on 7-Round MISTY1 with No FL Layers

In this section we show that if the FL layers are removed from the structure of MISTY1, then the generic impossible differential for 5-round Feistel constructions [2,5] can be used to mount an attack on a 7-round variant of the cipher. The attack is based on examining pairs with input difference (α, x) and output difference (α, y), and discarding all the subkeys which lead to the impossible differential $(\alpha, 0) \to (\alpha, 0)$ in rounds 2–6. However, since each of the FO functions uses 112 key bits, trying all the possible subkeys is infeasible. Instead, we use differential properties of the FO function, along with key schedule considerations, in order to discard the possible subkeys efficiently.

[9] As noted earlier, in the filtering in Step 3, the attacker discards (for a given pair of bits) 6 out of 16 possible values. Hence, in this step, the attacker discards 9 out of the remaining 10 values.

5.1 Differential Properties of the *FO* Function

We start with an observation presented in [7].

Observation 3 ([7]). *Given a pair of input values to the function FO_i, the corresponding output difference depends only on the equivalent of 75 subkey bits. These bits are the subkeys $KO_{i,1}, KO_{i,2}, KI_{i,1,2}, KI_{i,2,2}$, and $KI_{i,3,2}$, and the equivalent subkey*

$$AKO_{i,3} = KO_{i,3} \oplus KI_{i,1,1}||00||KI_{i,1,1},$$

where || denotes concatenation.

We refer the reader to [7] for the complete proof of this observation.

Our next proposition is a novel observation concerning MISTY1:

Proposition 2. *Assume that the input values and the output difference of the function FO_i are known, along with one of the following sets of subkey bits:*

1. $KO_{i,1}, KI_{i,1,2}, KI_{i,2,2}, KI_{i,3,2}$, *or*
2. $KO_{i,2}, KI_{i,1,2}, KI_{i,2,2}, KI_{i,3,2}$.

Consider the remaining 32 key bits that influence the output difference (i.e., $KO_{i,2}$ or $KO_{i,1}$, respectively, along with $AKO_{i,3}$). There exists one value of these 32 bits on average which satisfies the input/output condition, and this value can be found efficiently (using only several simple operations).

Proof. Consider the case when the bits of Set (1) are known. The knowledge of bits $KO_{i,1}$ and $KI_{i,1,2}$ allows to encrypt the pair through the first FI layer and (using the output difference of FO_i) obtain the output difference of $FI_{i,2}$. The input difference to $FI_{i,2}$ can be computed from the input of FO_i. Given the input and the output differences to the function $FI_{i,2}$ and the subkey $KI_{i,2,2}$, there exists one pair of inputs on average which satisfies the input/output difference condition. This pair of actual values, along with the input to FO_i, suggests a unique value for the subkey $KO_{i,2}$. Similarly, since the input and output differences to $FI_{i,3}$ and the subkey $KI_{i,3,2}$ are known, they suggest one value of the subkey $AKO_{i,3}$ on average which satisfies these differences.

In the second case, when the bits of Set (2) are known, the knowledge of bits $KO_{i,2}$ and $KI_{i,2,2}$ allows to encrypt the pair through the second FI layer and (using the output difference of FO_i) obtain the output difference of $FI_{i,1}$. The input difference to $FI_{i,1}$ can be computed from the input of FO_i. This input/output difference pair suggests a single value of the subkey $KO_{i,1}$ on average. The single suggestion for $AKO_{i,3}$ can be retrieved as in the first case.

In order to obtain the suggested subkeys efficiently, it is sufficient to pre-compute the full difference distribution table [4] of the FI function (i.e., a table containing also the actual values which satisfy each input/output difference condition), for each possible value of $KI_{i,j,2}$. Each such table requires about 2^{34} bytes of memory. In the on-line phase of the attack, given the input/output differences to an FI function, along with the corresponding subkey $KI_{i,j,2}$, the possible actual values of the input can be found using a single table look-up. Hence, the suggested values for the 32 subkey bits can be found using only several simple operations. □

Now we are ready to present the attack.

5.2 The Attack Algorithm

The attack algorithm is as follows:

1. Ask for the encryption of $2^{50.2}$ known plaintexts.
2. Find all pairs (P_1, P_2) and their corresponding ciphertexts (C_1, C_2), respectively, such that $P_1 \oplus P_2 = (\alpha, x)$ and $C_1 \oplus C_2 = (\alpha, y)$ for some x, y and α. The expected number of pairs remaining after this stage is $(2^{50.2})^2/2 \cdot 2^{-32} = 2^{67.4}$.
3. **Examining round 1:** For each of the remaining pairs, perform the following:
   (a) Guess the subkey K_1 and the 9 least significant bits of the subkeys K'_2, K'_4, K'_6 (which compose the subkeys $KO_{1,1}, KI_{1,1,2}, KI_{1,2,2}$, and $KI_{1,3,2}$). Use Proposition 2 to find the suggested value for the subkeys $KO_{1,2} = K_3$ and $AKO_{1,3}$.
   (b) Guess the remaining bits of K'_6 (which are the bits of $KI_{1,1,1}$), and use the value of $AKO_{1,3}$ to obtain the value $KO_{1,3} = K_8$.
   (c) For each value of the subkeys K_3 and K_8, store the list of all the pairs which suggested this value. The expected number of such pairs is $2^{67.4} \cdot 2^{16+9+9+9} \cdot 2^7/2^{82} = 2^{35.4}$.
4. **Examining round 7:** For each possible value of the 82 bits of the key considered in Step 3 (subkeys K_1, K_3, K'_6, K_8, and the 9 least significant bits of K'_2 and K'_4), and for each of the pairs corresponding to each subkey value, perform the following:
   (a) Use Proposition 2 to find the values $KO_{7,1} = K_7$ and $AKO_{7,3}$ (note that the values $KO_{7,2}, KI_{7,1,2}, KI_{7,2,2}$, and $KI_{7,3,2}$ are known at this stage).
   (b) Use the key schedule to find the value of K_6. Use the knowledge of $AKO_{7,3}$ and $KO_{7,3} = K_6$ to get the value of $KI_{7,1,1}$, along with a 9-bit filtering condition (only pairs for which $AKO_{7,3} \oplus KO_{7,3}$ is of the form $a||00||a$, for some 7-bit value a, remain, and suggest the value $KI_{7,1,1} = a$).
5. Discard the values of the 105 examined key bits ($K_1, K_3, K'_4, K_6, K_7, K_8$, and the 9 least significant bits of K'_2) suggested by at least one pair. The expected number of pairs suggesting each subkey value is $2^{35.4} \cdot 2^{-9}/2^{23} = 10.56$. As the number of pairs suggesting a subkey value has a Poisson distribution, a subkey remains (i.e., is not suggested by any pairs) with probability $e^{-10.56} = 2^{-15.23}$. Hence, the expected number of remaining 105-bit subkeys is $2^{105} \cdot 2^{-15.23} = 2^{89.77}$.

6. For the remaining possibilities of the 105-bit subkey, exhaustively search all possible keys, until the right key is found.

The data complexity of the attack is $2^{50.2}$ known plaintexts. Its time complexity is mostly dominated by Step (4) and Step (6). Step (4) is repeated $2^{35.4} \cdot 2^{82} = 2^{117.4}$ times. Each such key deduction is expected to take one FI application, two memory accesses, and a few XOR operations. For sake of simplicity we assume that this is equal to 1/16 of 7-round MISTY1 encryption, and thus, Step (4) takes a total of $2^{113.4}$ encryptions. Step (6) takes $2^{128} \cdot 2^{-15.23} = 2^{112.8}$ trial encryptions. Therefore, the total time complexity of the attack is $2^{114.1}$ encryptions.

6 Summary and Conclusions

In this paper we presented several new impossible differential attacks on MISTY1. While previous attacks were applicable only up to 4 rounds of the cipher (including the FL layers), we presented a 5-round attack with time complexity of $2^{46.45}$ simple operations, and extended it to an attack on a 6-round variant faster than exhaustive key search. We also presented a 7-round attack on a variant of the cipher without FL functions. The best previously known attacks against this variant were on 6 rounds.

It seems interesting to compare between the attacks on reduced-round variants of MISTY1 including the FL functions, and the attacks on the variant without the FL functions. If the FL functions do not exist, much simpler impossible differential attacks can be mounted, and as a result, the attacks extend to one more round, compared to the case where the FL-s are present. On the other hand, when the FL functions are present, their linear structure can be exploited in order to reduce significantly the time complexity of impossible differential attacks.

Thus, we conclude that while the FL functions do contribute to the security of the full MISTY1 with respect to impossible differential attacks, they may reduce the practical security of reduced variants with a relatively small number of rounds.[10]

References

1. Babbage, S., Frisch, L.: On MISTY1 higher order differential cryptanalysis. In: Won, D. (ed.) ICISC 2000. LNCS, vol. 2015, pp. 22–36. Springer, Heidelberg (2001)
2. Biham, E., Biryukov, A., Shamir, A.: Miss in the Middle Attacks on IDEA and Khufu. In: Knudsen, L.R. (ed.) FSE 1999. LNCS, vol. 1636, pp. 124–138. Springer, Heidelberg (1999)

[10] It seems that the extremely low time complexity of the impossible differential attack on 5-round MISTY1 with the FL layers cannot be achieved if the FL layers are absent (even for a 5-round variant), due to the big amount of subkey bits affecting each FO function. As a result, the practical security of 5-round MISTY1 w.r.t. impossible differential attacks is *reduced* if the FL layers are present. A similar observation regarding a 4-round variant of MISTY1 was made in [8].

3. Biham, E., Biryukov, A., Shamir, A.: Cryptanalysis of Skipjack Reduced to 31 Rounds. In: Stern, J. (ed.) EUROCRYPT 1999. LNCS, vol. 1592, pp. 12–23. Springer, Heidelberg (1999)
4. Biham, E., Shamir, A.: Differential Cryptanalysis of the Data Encryption Standard. Springer, Heidelberg (1993)
5. Knudsen, L.R.: The Security of Feistel Ciphers with Six Rounds or Less. Journal of Cryptology 15(3), 207–222 (2002)
6. Knudsen, L.R., Wagner, D.: Integral Cryptanalysis. In: Daemen, J., Rijmen, V. (eds.) FSE 2002. LNCS, vol. 2365, pp. 112–127. Springer, Heidelberg (2002)
7. Kühn, U.: Cryptanalysis of Reduced-Round MISTY. In: Pfitzmann, B. (ed.) EUROCRYPT 2001. LNCS, vol. 2045, pp. 325–339. Springer, Heidelberg (2001)
8. Kühn, U.: Improved cryptanalysis of MISTY1. In: Daemen, J., Rijmen, V. (eds.) FSE 2002. LNCS, vol. 2365, pp. 61–75. Springer, Heidelberg (2002)
9. Lu, J., Kim, J., Keller, N., Dunkelman, O.: Improving the Efficiency of Impossible Differential Cryptanalysis of Reduced Camellia and MISTY1. In: Malkin, T.G. (ed.) CT-RSA 2008, vol. 4964, pp. 370–386. Springer, Heidelberg (2008)
10. Matsui, M.: Block encryption algorithm MISTY. In: Biham, E. (ed.) FSE 1997. LNCS, vol. 1267, pp. 64–74. Springer, Heidelberg (1997)
11. Tanaka, H., Hisamatsu, K., Kaneko, T.: Strength of MISTY1 without FL function for higher order differential attack. In: Fossorier, M.P.C., Imai, H., Lin, S., Poli, A. (eds.) AAECC 1999. LNCS, vol. 1719, pp. 221–230. Springer, Heidelberg (1999)

Generalized Identity Based and Broadcast Encryption Schemes*

Dan Boneh and Michael Hamburg

Stanford University
{dabo,mhamburg}@cs.stanford.edu

Abstract. We provide a general framework for constructing identity-based and broadcast encryption systems. In particular, we construct a general encryption system called *spatial encryption* from which many systems with a variety of properties follow. The ciphertext size in all these systems is independent of the number of users involved and is just three group elements. Private key size grows with the complexity of the system. One application of these results gives the first *broadcast HIBE* system with short ciphertexts. Broadcast HIBE solves a natural problem having to do with identity-based encrypted email.

1 Introduction

In this paper we develop a general framework for constructing identity-based encryption (IBE) [17,4] and broadcast encryption [9] with constant-size ciphertexts. This framework enables one to easily combine different encryption properties via a product rule and to obtain encryption systems supporting multiple properties. For example, a multi-authority, forward-secure, broadcast encryption system (with constant-size ciphertexts) is easily derived by taking the "product" of three systems. One new concept constructed using our framework is broadcast hierarchical IBE. We discuss this concept at the end of the section and explain its importance to secure email.

We start with an informal description of the framework; a precise definition is given in the next section. Rather than an IBE or a broadcast system we consider a higher level abstraction.

- Let $\mathcal{P}$ be a finite set of *policies*. Roughly speaking, a message m can be encrypted to any policy π in $\mathcal{P}$.
- Let $\mathcal{R}$ be a finite set of *roles*. Each decryptor has a role ρ in $\mathcal{R}$ and can obtain a private key K_ρ corresponding to its role ρ.
- We allow for an arbitrary predicate called open on the set $\mathcal{R}\times\mathcal{P}$ that specifies which roles in $\mathcal{R}$ can open what policies in $\mathcal{P}$.

A key K_ρ can decrypt ciphertexts encrypted for policy π if and only if role ρ opens policy π, i.e. $\mathsf{open}(\rho, \pi)$ is true.

To continue with the abstraction, we provide a notion of delegation which is useful in hierarchical IBE (HIBE) [13,11]. To support delegation we assume

---

* Supported by NSF and the Packard Foundation.

J. Pieprzyk (Ed.): ASIACRYPT 2008, LNCS 5350, pp. 455–470, 2008.
© International Association for Cryptologic Research 2008

there is a partial order $\succeq$ defined on the set of roles $\mathcal{R}$. The idea is that given the key K_{ρ_1} there is a delegation algorithm that can be used to generate the key K_{ρ_2}, whenever $\rho_1 \succeq \rho_2$. Naturally, we require that the open relation respect delegation, meaning that if role ρ_2 opens policy π and $\rho_1 \succeq \rho_2$ then ρ_1 also opens π.

Given the sets $\mathcal{P}, \mathcal{R}$ and relations open and $\succeq$, one obtains a very general notion of identity-based encryption. It generalizes HIBE, broadcast encryption, attribute-based encryption [12], predicate encryption [6,14] and other variants. We refer to such schemes as generalized IBE, or GIBE. In the next section we define GIBE schemes more precisely along with their associated security games.

Spatial encryption. In Section 3 we study an important instance of GIBE called *spatial encryption* in which policies are points in $\mathbb{Z}_p^n$ and roles are affine subspaces of $\mathbb{Z}_p^n$. The delegation relation $\succeq$ on roles is defined by subspace inclusion: role $\rho_1 \succeq \rho_2$ if ρ_1's affine space contains ρ_2's space.

As we will see, spatial encryption enables us to build a host of identity-based and broadcast encryption schemes. In particular, it supports a product rule that lets us combine encryption properties such as forward security, multiple authorities, and others.

In Section 4 we construct an efficient spatial encryption system with constant-size ciphertext. Our starting point is an HIBE construction of Boneh, Boyen, and Goh [2]. We are able to extend their system to obtain a spatial encryption system. However, the proof of security is more difficult and requires the BDDHE assumption introduced in [5] (the proof in [2] used the slightly weaker BDHI assumption). We describe various extensions of the system at the end of Section 4.

Our initial motivation: email encryption. Suppose user A wishes to send an encrypted email to users $B_1, \ldots, B_n$. User A knows the identities of all recipients, but does not know which private key generators (PKGs) issued their private keys. Moreover, user A only trusts PKGs $P_1, \ldots, P_\ell$. She wishes to encrypt the email so that user B_j can decrypt it if and only if B_j has a private key issued by one of the ℓ trusted PKGs. Using basic IBE this will require ciphertext of size $O(n \cdot \ell)$. Our goal is to construct a system whose ciphertext size is constant, that is, independent of n and ℓ.

This natural email encryption problem can be modeled as a GIBE and constructed using the product of two instances of our spatial encryption scheme. Here each PKG has a role which can delegate to a key for any user; a (possibly distributed) dealer holds the master key $K_\top$. We obtain a system that precisely solves the problem described above, with ciphertext size independent of n and ℓ. However, in our current construction the private key size is linear in $n + \ell$.

Similarly, we also construct a broadcast HIBE. Roughly speaking, in a broadcast HIBE there is a tree-like hierarchy of identities and private keys as in HIBE. An encryptor picks a set S of nodes in the hierarchy and encrypts a message m to this set S. We let c be the resulting ciphertext. As in a broadcast system, any user in S can decrypt c, but (proper) coalitions outside of S cannot. We say that the system has constant-size ciphertext if the size of c is independent of the size of S. Broadcast HIBE applies naturally to hierarchical email systems where messages can have many recipients.

Broadcast HIBE can be easily modeled as a GIBE and constructed from our spatial encryption system. This expands on the features of previous constant-size broadcast systems such as Boneh et al. [5] and Sakai and Furukawa [16], albeit at the cost of increased private-key size.

2 Generalized Identity-Based Encryption (GIBE)

A *Generalized Identity-Based Encryption Scheme*, or GIBE, allows a participant to encrypt a message under a certain *policy*, in some set $\mathcal{P}$ of allowable policies. We will enforce no structure on the allowed policies. To decrypt, users may hold secret keys corresponding to *roles*. Roles are organized in a partially-ordered set $\mathcal{R}$, that is, a set endowed with a reflexive, transitive, antisymmetric relation $\succeq$.

A GIBE may be parameterized in some way. For example, a system may have a limited number of identities, hierarchy levels, time periods or the like. We call such choices the *setup parameters* SP. As SP varies, $\mathcal{P}$ and $\mathcal{R}$ will generally also vary. Similarly, $\mathcal{P}$ and $\mathcal{R}$ may depend on the security parameter λ or on randomness chosen at setup. We encode these choices into a policy parameter χ generated at setup, and use policies $\mathcal{P}_\chi$ and roles $\mathcal{R}_\chi$. For brevity, we will omit χ when it is unambiguous.

For a policy π and a role ρ, we write $\mathsf{open}(\rho, \pi)$ if a user with a secret key for ρ is allowed to decrypt a message encrypted under π. We require this relation to be monotone, meaning that if $\rho \succeq \rho'$ and $\mathsf{open}(\rho', \pi)$ then $\mathsf{open}(\rho, \pi)$. For simplicity, we require that $\mathcal{R}$ contains a top element $\top$, such that $\top \succeq \rho$ for all $\rho \in \mathcal{R}$, and $\mathsf{open}(\top, \pi)$ for all $\pi \in \mathcal{P}$. Informally, greater roles open more messages, and the greatest role, $\top$, can open them all. Obviously, only a highly-trusted authority should hold the secret key $K_\top$.

A GIBE consists of four randomized algorithms:

- *Setup*(λ, SP) takes as input a security parameter λ and setup parameters SP. It returns public parameters PP (which include the policy parameter χ) and a master secret key $K_\top$.
- *Delegate*$(\mathrm{PP}, \rho, K_\rho, \rho')$ takes the secret key K_ρ for role ρ and returns a secret key $K_{\rho'}$ for ρ', where $\rho \succeq \rho'$.
- *Encrypt*(PP, π, m) encrypts a message m under a policy π.
- *Decrypt*$(\mathrm{PP}, \rho, K_\rho, \pi, c)$ decrypts a ciphertext c using a secret key K_ρ. Decryption may fail. However, we require that decryption succeeds when $\mathsf{open}(\rho, \pi)$, so that:

$$\mathit{Decrypt}(\ \mathrm{PP},\ \rho,\ K_\rho,\ \pi,\ \mathit{Encrypt}(\mathrm{PP},\ \pi,\ m)\ ) = m$$

for all PP generated by *Setup*, for all policies π and roles ρ, and for all keys K_ρ for ρ delegated directly or indirectly from $K_\top$.

We require that the algorithms *Setup*, *Delegate*, *Encrypt*, *Decrypt* and the predicates open and $\succeq$ all run in expected polynomial time in λ. We also require that delegation is independent of the path taken; that is, if $\rho_1 \succeq \rho_2 \succeq \rho_3$, then

$$\mathit{Delegate}(\mathrm{PP}, \rho_1, K_{\rho_1}, \rho_3)$$

should produce the same distribution as

$$\mathit{Delegate}(\mathrm{PP}, \rho_2, \mathit{Delegate}(\mathrm{PP}, \rho_1, K_{\rho_1}, \rho_2), \rho_3)$$

2.1 Security

We define the security of a GIBE $\mathcal{I}$ in terms of a family of security games between a challenger and an adversary. The system parameters SP are fixed, and the adversary is allowed to depend on them. We define the full, CCA_2, anonymous game first (anonymity here refers to the property that the ciphertext leaks no information about the policy used to create it [1]).

Setup: The challenger runs $\mathit{Setup}(\lambda, \mathrm{SP})$ and sends PP to the adversary.

First query phase: The adversary makes several delegation queries ρ_i to the challenger, which runs $\mathit{Delegate}(\mathrm{PP}, \top, K_\top, \rho_i)$ and returns the resulting K_{ρ_i}. The adversary may also make decryption queries (ρ_i, π_i, c_i) to the challenger, where $\mathsf{open}(\rho_i, \pi_i)$. The challenger runs $K_{\rho_i} \leftarrow \mathit{Delegate}(\mathrm{PP}, \top, K_\top, \rho_i)$, then runs $\mathit{Decrypt}(\mathrm{PP}, \rho_i, K_{\rho_i}, \pi_i, c_i)$ and returns the resulting m_i (or fails).

Challenge: The adversary chooses messages m_0 and m_1 and policies π_0^* and π_1^*, and sends them to the challenger. We require that the adversary has not been given decryption keys for these policies, that is, $\neg\mathsf{open}(\rho_i, \pi_j^*)$ for all delegation queries ρ_i in the first query phase, and for $j \in \{0, 1\}$.

The challenger chooses a random $b \xleftarrow{\mathrm{R}} \{0, 1\}$, runs $\mathit{Encrypt}(\mathrm{PP}, \pi_b^*, m_b)$, and returns the resulting *challenge ciphertext* c^* to the adversary.

Second query phase: The second query phase is exactly like the first, except that the adversary may not issue decryption queries for c^*, and the adversary may not make delegation queries for roles that open π_j^* for $j \in \{0, 1\}$.

Guess: The adversary outputs a bit $b' \in \{0, 1\}$. The adversary wins if $b' = b$, and otherwise it loses.

There are several important variants on the above game:

- In a CCA_1 game, the adversary may not issue decryption queries during the second query phase.
- In a CPA game, the adversary may not issue decryption queries at all.
- In a non-anonymous game, we require that $\pi_0^* = \pi_1^*$.
- In a selective game, the setup phase is modified. The challenger sends the policy parameter χ to the adversary. The adversary chooses in advance its π_0^* and π_1^* and sends them to the challenger. Then the challenger sends the rest of the public parameters PP.

We define adversary $\mathcal{A}$'s advantage in game variant $\mathcal{V}$ (when $\mathcal{A}$ is attacking the GIBE system $\mathcal{I}$ with parameter SP) to be

$$\mathcal{V}\mathrm{Adv}_{\mathcal{A}\leftrightarrow(\mathcal{I},\mathrm{SP})}(\lambda) := |\ \Pr[\mathcal{A} \text{ wins } \mathcal{V}] - \Pr[\mathcal{A} \text{ loses } \mathcal{V}]\ |$$

We say that a GIBE $\mathcal{I}$ is $\mathcal{V}$-secure if for all setup parameters SP and all probabilistic polynomial-time adversaries $\mathcal{A}$, the function $\mathcal{V}\mathrm{Adv}_{\mathcal{A}\leftrightarrow(\mathcal{I},\mathrm{SP})}(\lambda)$ is a negligible function of λ.

In this paper we will primarily focus on the simplest security model, namely selective-security, non-anonymous, against a CPA adversary. We denote the adversary's advantage in this model by $(\mathrm{NonAnon}, \mathrm{Sel}, \mathrm{CPA})\mathrm{Adv}_{\mathcal{A}\leftrightarrow(\mathcal{I},\mathrm{SP})}(\lambda)$.

2.2 Example GIBE Instances

Many instances of GIBE already appear in the literature:

- In traditional IBE [17,4] the policies are simply identities and the roles are identities or $\top$. A message encrypted to an identity I can be decrypted only with a key for I or for $\top$. There is no delegation except from $\top$.
- In broadcast IBE [9] the policies are sets of identities and the roles are identities or $\top$. A message to a set S of identities can be decrypted only with a key for $I \in S$, or for $\top$. There is no delegation except from $\top$.
- In attribute-based encryption (ABE) [12], the policies are subsets of a set S of attributes, and the roles are upwardly closed subsets of $\top := 2^S$. A message to a set S of attributes can be decrypted with a key for any set containing S. [12] does not define a delegation model for attribute-based encryption, but the circuit-based implementation permits delegation by widening a k-of-n threshold gate into a $k+1$-of-$n+1$ threshold gate.
- In hierarchical IBE [13,11] the policies are identities and the roles are points in the hierarchy, with $\top$ at the root of the hierarchy. Here the key for a point x can either delegate to or decrypt from any point y below x.
- In forward-secure [7] systems, the roles and policies include a time t. Roles can be delegated by increasing the time t, and cannot decrypt messages with an earlier t.

The games used to define the security of these instances are special cases of the GIBE games. In the next section we will show that most of these instances can be constructed from a GIBE we call spatial encryption. These generic constructions for IBE and HIBE are competitive with the best known hand-tailored constructions. For broadcast IBE and forward-secure IBE our generic construction has short ciphertexts, but the private key is longer than the best known tailor-made constructions [7,16,5].

2.3 Embedding Lemmas

It is clear that some GIBEs can be used to construct other GIBEs. For example, it is obvious that any broadcast IBE can also function as a traditional IBE.

In particular, suppose that we have a GIBE $\mathcal{I}$ with policies $\mathcal{P}_\chi$ and roles $\mathcal{R}_\chi$, and we wish to define a GIBE with policies $\mathcal{P}'_\chi$ and roles $\mathcal{R}'_\chi$. Suppose that we are given an efficient injective map $f_\mathcal{P} : \mathcal{P}'_\chi \to \mathcal{P}_\chi$ and an efficient embedding $f_\mathcal{R} : \mathcal{R}'_\chi \to \mathcal{R}_\chi$ which satisfy $\mathsf{open}(f_\mathcal{R}(\rho), f_\mathcal{P}(\pi)) \iff \mathsf{open}(\rho, \pi)$ and $f_\mathcal{R}(\top) = \top$. Then we can define a GIBE $\mathcal{I}'$ with policies $\mathcal{P}'_\chi$ and roles $\mathcal{R}'_\chi$ simply by applying all $f_\mathcal{P}$ to all policies and $f_\mathcal{R}$ to all roles.

Lemma 1 (Embedding Lemma). *Let $\mathcal{I}$ and $\mathcal{I}'$ be GIBEs as defined above. For any GIBE adversary $\mathcal{A}$ against $\mathcal{I}'$, there is a GIBE adversary $\mathcal{B}$ against $\mathcal{I}$, running in about the same time as $\mathcal{A}$, such that*

$$\mathcal{V}\mathrm{Adv}_{\mathcal{A}\leftrightarrow(\mathcal{I},SP)}(\lambda) = \mathcal{V}\mathrm{Adv}_{\mathcal{B}\leftrightarrow(\mathcal{I},SP)}(\lambda)$$

Similarly, we can sometimes use collision-resistant hashing to construct new GIBEs. Suppose we have a GIBE $\mathcal{I}$ in which policies and roles are lists of elements of some set $\mathcal{X}$, and in which open and $\succeq$ are decided in a monotone fashion by comparing certain elements for equality. Suppose also that we have an efficient collision-resistant hash $H : \mathcal{X}' \to \mathcal{X}$ on some other set $\mathcal{X}'$. Then we can define a GIBE $\mathcal{I}'$ which is identical to $\mathcal{I}$ except that its policies and roles are lists over $\mathcal{X}'$ instead of $\mathcal{X}$, and all operations apply H pointwise to the policies and roles.

Lemma 2 (Hashed Embedding Lemma). *Let $\mathcal{I}$ and $\mathcal{I}'$ be GIBEs as defined above. For any GIBE adversary $\mathcal{A}$ against $\mathcal{I}'$, there is a GIBE adversary $\mathcal{B}_1$ against $\mathcal{I}$ and a collision-resistance adversary $\mathcal{B}_2$ against H, each running in about the same time as $\mathcal{A}$, such that*

$$\mathcal{V}\mathrm{Adv}_{\mathcal{A}\leftrightarrow(\mathcal{I},SP)}(\lambda) \le \mathcal{V}\mathrm{Adv}_{\mathcal{B}_1\leftrightarrow(\mathcal{I},SP)}(\lambda) + \mathrm{CRAdv}_{\mathcal{B}_2\leftrightarrow H}(\lambda)$$

The proofs of these lemmas are immediate and are omitted.

3 Spatial Encryption: An Important Instance of GIBE

The building block for systems in our paper will be *spatial encryption*, a new GIBE. In spatial encryption, the policies $\mathcal{P}$ are the points of an n-dimensional affine space $\mathbb{Z}_q^n$. The roles $\mathcal{R}$ are all subspaces W of $\mathbb{Z}_q^n$ ordered by inclusion, and $\mathsf{open}(W, \pi) \iff W \ni \pi$.

3.1 Systems Derived from Spatial Encryption

To demonstrate the power of spatial encryption, we show that many other GIBEs are embedded in it.

Hierarchical IBE. Hierarchical IBE is trivially embeddable in spatial encryption. Here the path components are elements of $\mathbb{Z}_q$, and the paths are limited to length at most n. This extends easily to hierarchical IBE where the path components are strings by using the Hashed Embedding Lemma. This is not the only embedding of hierarchical IBE in spatial encryption, however.

Inclusive IBE. In *inclusive IBE*, the policies are subsets of size at most n of a set of identities. The roles are also subsets of size at most n, where $\rho \succeq \rho' \iff \rho \subseteq \rho'$; that is, one can delegate by adding elements to a set. We say that $\mathsf{open}(\rho, \pi)$ iff $\rho \subseteq \pi$; that is, a message to a set can be decrypted with a key for any subset.

We can embed inclusive IBE in a spatial system of dimension $n + 1$. Here the identities are elements of $\mathbb{Z}_q$, but this extends to inclusive IBE with strings as identities using the Hashed Embedding Lemma. We encode a policy $\pi \subset \mathbb{Z}_q$ as the coefficients of the polynomial $\hat{\pi}(t) := \prod_{c \in \pi} x - c$; this polynomial has degree at most n and therefore has at most $n + 1$ coefficients. We encode a role $\rho \subset \mathbb{Z}_q$ as the vector subspace of coefficients of polynomials which are divisible by $\prod_{c \in \rho} x - c$.

Inclusive IBE seems almost as powerful as spatial encryption; nearly all the applications in this paper use inclusive IBE rather than using spatial encryption directly.

Inclusive IBE can be built using attribute-based encryption, but this construction is less efficient than spatial encryption. In particular, the ciphertext has size $O(n)$. Our construction gives constant size ciphertext.

Co-inclusive IBE. Co-inclusive IBE is the dual of inclusive IBE. Policies and roles (other than $\top$) are sets of at most n identities, where $r \succeq r' \iff r \supseteq r'$; that is, one can delegate by removing elements from a set. We say that $\mathsf{open}(\rho, \pi)$ iff $\rho \supseteq \pi$; that is, a message to a set can be decrypted with a key for any set which contains it.

We can embed co-inclusive IBE in a spatial system of dimension $2n$. For a role ρ, we assign the span of $\{\boldsymbol{v}_i : i \in \rho\}$, where $\boldsymbol{v}_i = (1, i, i^2, \ldots, i^{2n-1})$ is the Vandermonde vector for i. To encrypt to a policy π, we encrypt to $\boldsymbol{v}_\pi := \sum_{i \in \pi} \boldsymbol{v}_i$. It is clear that $\boldsymbol{v}_\pi$ is not contained in the subspace for any role $\rho' \not\supseteq \pi$, for then we would have expressed $\boldsymbol{v}_\pi$ as a sum of at most $2n$ linearly independent vectors in two different ways.

Co-inclusive IBE can be built using attribute-based encryption, but this construction is less efficient than spatial encryption. Once again, the ciphertext has size $O(n)$. Our construction gives constant size ciphertext.

Broadcast Hierarchical IBE. Broadcast HIBE (and therefore also vanilla broadcast IBE [16]) is embeddable in inclusive IBE. The role for a path `a/b/c/...` in the hierarchy is the set $\{$`a`, `a/b`, `a/b/c`, $\ldots\}$. The policy for a set of nodes in the hierarchy is the union of their roles. The scheme can broadcast to a set of points S in the hierarchy if the number of distinct path prefixes in S is less than the dimension n.

For a useful broadcast system, short ciphertexts are required. Our spatial encryption has constant-size ciphertexts, so our broadcast HIBE does as well.

Product Schemes. For GIBEs $\mathcal{I}_1, \mathcal{I}_2$ with roles $\mathcal{R}_1, \mathcal{R}_2$ and policies $\mathcal{P}_1, \mathcal{P}_2$, respectively, we define a product scheme $\mathcal{I}_1 \otimes \mathcal{I}_2$. This scheme's roles are $\mathcal{R}_1 \times \mathcal{R}_2$ and its policies are $\mathcal{P}_1 \times \mathcal{P}_2$. Here $\mathsf{open}((\rho_1, \rho_2), (\pi_1, \pi_2))$ if and only if $\mathsf{open}(\rho_1, \pi_1)$

and $\mathsf{open}(\rho_2, \pi_2)$, and similarly $(\rho_1, \rho_2) \succeq (\rho_1', \rho_2')$ if and only if $\rho_1 \succeq \rho_1'$ and $\rho_2 \succeq \rho_2'$. Note that this is different from what can be accomplished with double encryption, for here the recipient needs to be able to decrypt both components using a single key $K_{(\rho_1,\rho_2)}$. For instance, in the forward-secure encryption system that follows, a recipient decrypts with a key for a role ρ issued before time t, not a key for ρ and another key issued before time t.

Using the vector space $\mathbb{Z}_q^{n_1+n_2} \cong \mathbb{Z}_q^{n_1} \times \mathbb{Z}_q^{n_2}$, we can embed two instances of spatial encryption with dimensions n_1 and n_2 in one of dimension n_1+n_2. Therefore, if two schemes $\mathcal{I}_1$ and $\mathcal{I}_2$ are embeddable in spatial systems of dimensions n_1 and n_2, their product $\mathcal{I}_1 \otimes \mathcal{I}_2$ is embeddable in a spatial system of dimension $n_1 + n_2$. Similarly, we can construct product schemes in inclusive IBE. Here the policies are of the form $\pi_1 \uplus \pi_2$ and the roles are of the form $\rho_1 \uplus \rho_2$, where $\uplus$ denotes a disjoint union.

Multiple Authorities. A common limitation in IBE systems is the need to trust a single central authority. The central authority has the ability to decrypt any message sent using the system, but equally importantly, the central authority must correctly decide to whom it will issue keys for a given role. The human element of this authentication problem makes it less amenable to technical solutions.

Product schemes are a step toward a solution to this problem. Let $\mathcal{I}_a$ be a broadcast system whose identities are the names of authorities, and let $\mathcal{I}_s$ be any GIBE. Then the product system $\mathcal{I}_a \otimes \mathcal{I}_s$ is a multi-authority version of $\mathcal{I}_s$. A (possibly distributed) central dealer gives each authority a the decryption key for the role $(a, \top)$. Then if a user wishes to encrypt a message to some policy $\pi \in \mathcal{P}_s$, and trusts a set A of authorities, she encrypts the message to (A, π). This can be decrypted only by a user who holds the key for (a, ρ) where $a \in A$ and $\mathsf{open}(\rho, \pi)$, that is, one whom a has certified for a role which opens π.

Forward Security. There are already constructions of forward-secure IBE from HIBE, so we already know that forward-secure encryption is embeddable in spatial encryption [7]. We show a trivial forward-secure system from spatial encryption that will be useful in constructing product schemes. Set the policy for a time t to be the vector of t ones followed by $n - t$ zeros, and the role for a range of times $[t_1, t_2]$ to be the affine subspace of t_1 ones, followed by any $t_2 - t_1$ components, followed by $n - t_2$ zeros.

A similar construction works for forward-secure IBE based on inclusive IBE. These constructions require many more dimensions than [7], but they require the user to store only one secret key for a given range of times. This makes them more efficient for use in product schemes.

CCA$_2$ Security. Following [3], we can use a MAC and a commitment scheme to create a CCA$_2$-secure encryption $\mathcal{I}'$ scheme from a scheme $\mathcal{I}$ which is merely CPA-secure. To encrypt a message m to a policy π, we choose a random MAC key k, a commitment com to k and the decommitment dec. We encrypt $c :=$ *Encrypt*$(\mathrm{PP}, (\pi, \mathrm{com}), (m, \mathrm{dec}))$ using the product $\mathcal{I} \otimes \mathrm{IBE}$, and set the ciphertext as $(\mathrm{com}, c, \mathrm{MAC}(k, c))$. The resulting scheme is anonymous if $\mathcal{I}$ is, and fully secure if $\mathcal{I}$ is. The proof is exactly as in [3].

Email Encryption. We have now solved the motivating example of practical email encryption: by composing the above constructions, we can easily build a forward-secure, multiple-authority, CCA_2-secure broadcast hierarchical encryption system. This system can encrypt a message to n_r (path prefixes of) recipients, trusting in n_a authorities, with t time periods in a single key. The ciphertexts have constant size, and the private keys have size $O(n_a + n_r + t)$.

Short Identity-Based Ring Signatures. We can convert a GIBE $\mathcal{I}$ to an identity-based signature scheme using the product scheme $\mathcal{I} \otimes \mathrm{IBE}$. The signing key for a role ρ is $K_{(\rho,\top)}$, and a signature of a message m under a role ρ is $K_{(\rho,H(m))}$, where H is a collision-resistant hash. This construction has the curious property that a signature by ρ on a message m can be delegated to produce a signature by ρ' on m for any $\rho' \preceq \rho$. If this property is undesirable, delegation can be prevented by using $H((\rho, m))$ instead of $H(m)$ above.

If the construction of $\mathcal{I} \otimes \mathrm{IBE}$ is fully secure, then this signature scheme will be unforgeable; if it is selectively secure, then the signature scheme will be selectively unforgeable in the random oracle model for H.

If we choose $\mathcal{I}$ to be inclusive IBE, then this construction gives an identity-based ring signature system [15,8,18], in which a user A can sign messages anonymously on behalf of any set of users containing A. A straightforward implementation using spatial encryption would result in long signatures, but the length results from the ability to delegate signatures further. By removing this ability, we can build constant-length identity-based ring signatures. We give the details in the full version of the paper.

4 Constructing a Spatial Encryption System

We now turn to the construction of a selectively-secure n-dimensional spatial encryption system with constant-size ciphertext. Our construction is inspired by the construction of a constant size HIBE given in [2]. Our proof of security, however, requires a slightly stronger complexity assumption, namely the BDDHE assumption previously used in [5].

4.1 Notation

Vectors in this paper are always column vectors. When writing them inline, we transpose them to save space. We will be working with vectors of group elements, so we will adopt a convenient notation. For a vector $\boldsymbol{v} = (v_1, v_2, \ldots, v_n)^\top \in \mathbb{Z}_p^n$ of field elements, we use $g^{\boldsymbol{v}}$ to denote the vector of group elements

$$g^{\boldsymbol{v}} := (g^{v_1}, g^{v_2}, \ldots, g^{v_n})^\top \in \mathbb{G}^n$$

In many cases, we will manipulate these without knowing the actual vector $\boldsymbol{v}$. For example, given $g^{\boldsymbol{v}}$ and $\boldsymbol{w}$, we can easily compute $g^{\langle \boldsymbol{v}, \boldsymbol{w} \rangle}$, where $\langle \boldsymbol{v}, \boldsymbol{w} \rangle := \boldsymbol{v}^\top \boldsymbol{w}$ is the usual dot product on $\mathbb{Z}_p^n$.

We will write $\mathrm{Aff}(M, \boldsymbol{a}) \subseteq \mathbb{Z}_p^n$ for the d-dimensional affine space $\{M\boldsymbol{x} + \boldsymbol{a} : \boldsymbol{x} \in \mathbb{Z}_p^d\}$.

4.2 The System

The system parameters for our spatial encryption system will be a prime p (where $\log p$ is approximately the security parameter λ) and two groups $\mathbb{G}$ and $\mathbb{G}_T$ of order p, with a bilinear pairing $e : \mathbb{G} \times \mathbb{G} \to \mathbb{G}_T$. Additionally, the public parameters will include group elements $g, g^{a_0}, t \in \mathbb{G}_T$ and a vector $g^{\boldsymbol{a}} \in \mathbb{G}^n$.

A secret key for an affine space $V := \mathrm{Aff}(M, \boldsymbol{x})$ will have the form

$$\left(g^r,\ g^{b + r a_0 + r \langle \boldsymbol{x}, \boldsymbol{a} \rangle},\ g^{r M^\top \boldsymbol{a}} \right)$$

where b is the master secret and r is random in $\mathbb{Z}_p$.

The four GIBE algorithms work as follows:

- *Setup*(λ, n) generates the system parameters $p, \mathbb{G}, \mathbb{G}_T$. It then chooses parameters
$$g \xleftarrow{\mathrm{R}} \mathbb{G}^*, \qquad a_0 \xleftarrow{\mathrm{R}} \mathbb{Z}_p, \qquad \boldsymbol{a} \xleftarrow{\mathrm{R}} \mathbb{Z}_p^n$$
and secret parameter $b \xleftarrow{\mathrm{R}} \mathbb{G}$, then computes $t := e(g, g)^b$. It outputs public parameters
$$\mathrm{PP} := (\ p,\ \mathbb{G},\ \mathbb{G}_T;\quad g,\ g^{a_0},\ g^{\boldsymbol{a}},\ t\ )$$
and master secret key
$$K_\top := (\ g,\ g^b,\ g^{\boldsymbol{a}}\ )\ \in \mathbb{G}^{n+2}$$
- *Delegate*$(\mathrm{PP}, V_1, K_{V_1}, V_2)$ takes two subspaces $V_1 := S(M_1, \boldsymbol{x}_1)$ and $V_2 := S(M_2, \boldsymbol{x}_2)$. Since V_2 is a subspace of V_1, we must have $M_2 = M_1 T$ and $\boldsymbol{x}_2' = \boldsymbol{x}_1 + M_1 \boldsymbol{y}$ for some (efficiently computable) matrix T and vector $\boldsymbol{y}$. We can then compute a key
$$\begin{aligned} \hat{K}_{V_2} &:= \left(g^r,\ g^{b + r a_0 + r \langle \boldsymbol{x}_1, \boldsymbol{a} \rangle} \cdot g^{r \boldsymbol{y}^\top M_1^\top \boldsymbol{a}},\ g^{r T^\top M_1^\top \boldsymbol{a}} \right) \\ &= \left(g^r,\ g^{b + r a_0 + r \langle \boldsymbol{x}_2, \boldsymbol{a} \rangle},\ g^{r M_2^\top \boldsymbol{a}} \right) \end{aligned}$$
for V_2. However, we also need to re-randomize it. To do this, we pick a random $s \xleftarrow{\mathrm{R}} \mathbb{Z}_p$ and compute
$$\begin{aligned} K_{V_2} &:= \left(g^r \cdot g^s,\ g^{b + r(a_0 + \langle \boldsymbol{x}_2, \boldsymbol{a} \rangle)} \cdot g^{s(a_0 + \langle \boldsymbol{x}_2, \boldsymbol{a} \rangle)},\ g^{r M_2^\top \boldsymbol{a}} \cdot g^{s M_2^\top \boldsymbol{a}} \right) \\ &= \left(g^{r+s},\ g^{b + (r+s)(a_0 + \langle \boldsymbol{x}_2, \boldsymbol{a} \rangle)},\ g^{(r+s) M_2^\top \boldsymbol{a}} \right) \end{aligned}$$
Notice that V_1 and V_2 may be the same subspace. In that case, this formula translates the secret key between different forms for V_1 and re-randomizes it. As a result, we are free to choose whatever representation of V we wish.

- *Encrypt*$(\mathrm{PP}, \boldsymbol{x}, m)$, where m is encoded as an element of the target group $\mathbb{G}_T$, picks a random $s \xleftarrow{R} \mathbb{Z}_p$ and computes a ciphertext

$$\left(g^s,\ g^{s(a_0+\langle \boldsymbol{x}, \boldsymbol{a} \rangle)},\ m \cdot t^s \right)$$

- *Decrypt*$(\mathrm{PP}, V, K_V, \boldsymbol{x}, c)$ where $c = (c_1, c_2, c_3)$ is the above ciphertext, first delegates K_V to obtain the key $K_{\{\boldsymbol{x}\}} = (k_1, k_2) := \left(g^r, g^{b+r(a_0+\langle \boldsymbol{x}, \boldsymbol{a} \rangle)} \right)$. It then recovers

$$\frac{c_3 \cdot e(c_2, k_1)}{e(c_1, k_2)} = \frac{m \cdot t^s \cdot e(g,g)^{rs(a_0+\langle \boldsymbol{x}, \boldsymbol{a} \rangle)}}{e(g,g)^{sb+rs(a_0+\langle \boldsymbol{x}, \boldsymbol{a} \rangle)}} = m$$

4.3 Bilinear Decision Diffie-Hellman Exponent

To prove security we use a generalization of bilinear Diffie-Hellman first proposed in [5]. Let $\mathbb{G}$ be a group of prime order p, and let g be a generator of g. Let $e : \mathbb{G} \times \mathbb{G} \to \mathbb{G}_T$ be a bilinear map, and let n be a positive integer. We define the notation $g^{\alpha^{[a,b]}}$ for integers $a \le b$ as

$$g^{\alpha^{[a,b]}} := \left(g^{\alpha^a}, g^{\alpha^{a+1}}, \ldots, g^{\alpha^b} \right)^{\top}$$

We then define distributions

$$\begin{aligned} \mathcal{P}_{\mathrm{BDDHE}} := \text{choose: } & g \xleftarrow{R} \mathbb{G}^*, \alpha \xleftarrow{R} \mathbb{Z}_p, h \xleftarrow{R} \mathbb{G}^*, z \leftarrow e(g,h)^{\alpha^n} \\ \text{output: } & \left(g^{\alpha^{[0,n-1]}},\ g^{\alpha^{[n+1,2n]}},\ h,\ z \right) \end{aligned}$$

$$\begin{aligned} \mathcal{R}_{\mathrm{BDDHE}} := \text{choose: } & g \xleftarrow{R} \mathbb{G}^*, \alpha \xleftarrow{R} \mathbb{Z}_p, h \xleftarrow{R} \mathbb{G}^*, z \xleftarrow{R} \mathbb{G}_T \\ \text{output: } & \left(g^{\alpha^{[0,n-1]}},\ g^{\alpha^{[n+1,2n]}},\ h,\ z \right) \end{aligned}$$

We define the BDDHE-advantage of a randomized algorithm $\mathcal{A} : \mathbb{G}^{2n+1} \times \mathbb{G}_T \to \{0,1\}$ as

$$\begin{aligned} \mathrm{BDDHE\,Adv}_{\mathcal{A},n}(\lambda) := \Big| & \Pr\left[\mathcal{A}(x) = 1 : x \xleftarrow{R} \mathcal{P}_{\mathrm{BDDHE}} \right] \\ & - \Pr\left[\mathcal{A}(x) = 1 : x \xleftarrow{R} \mathcal{R}_{\mathrm{BDDHE}} \right] \Big| \end{aligned}$$

4.4 Proof of Selective Security

Call the spatial encryption system above $\mathcal{S}$. To make the proof more readable we abstract away re-randomization terms in the main proof of security. To do so, we divide the proof into two steps:

- First, we show in Observation 1 that if the system $\mathcal{S}$ is insecure then so is a system with rigged randomization parameters (i.e. a system where $a_0, \boldsymbol{a}, b, r$ and s are chosen non-uniformly). This step is straightforward.

– Second, we show in Theorem 1 that a specific rigging of the randomization parameters in $\mathcal{S}$ is secure. The combination of these two steps implies that $\mathcal{S}$ is secure.

We believe that hiding re-randomization terms in the main simulation makes the proof easier to understand.

Observation 1 (Rigged parameters). *Let $\mathcal{S}'$ be identical to $\mathcal{S}$ except that $a_0, \boldsymbol{a}, b$, the r in delegation queries and the s in the challenge ciphertext are chosen by some algorithm rather than uniformly at random. Then for any $\mathcal{V}$-adversary $\mathcal{A}$ against $\mathcal{S}$, there is is a $\mathcal{V}$-adversary $\mathcal{B}$ against $\mathcal{S}'$, running in about the same time as $\mathcal{A}$, such that*

$$\mathcal{V}\mathrm{Adv}_{\mathcal{A}\leftrightarrow(\mathcal{S},n)}(\lambda) = \mathcal{V}\mathrm{Adv}_{\mathcal{B}\leftrightarrow(\mathcal{S}',\backslash)}(\lambda)$$

Proof. The adversary $\mathcal{B}$ runs $\mathcal{A}$, but re-randomizes $\mathcal{A}$'s queries and the simulator's responses. More concretely, at setup time $\mathcal{B}$ chooses uniformly random $a_0' \xleftarrow{R} \mathbb{Z}_p, \boldsymbol{a}' \xleftarrow{R} \mathbb{Z}_p^n, b' \xleftarrow{R} \mathbb{Z}_p$. It sends $\mathcal{A}$ the public parameters

$$\left(p,\ \mathbb{G},\ \mathbb{G}_T;\quad g,\ g^{a_0+a_0'},\ g^{\boldsymbol{a}+\boldsymbol{a}'},\ t \cdot e(g,g)^{b'} \right)$$

$\mathcal{B}$ then adjusts $\mathcal{A}$'s queries to match these public parameters. For example, when $\mathcal{A}$ makes a delegation query, $\mathcal{B}$ passes the query through directly to the challenger. Given the response

$$\left(g^r,\ g^{b+ra_0+r\langle \boldsymbol{x},\boldsymbol{a}\rangle},\ g^{rM^\top \boldsymbol{a}} \right)$$

$\mathcal{B}$ computes a new key

$$\left(g^r,\ g^{b+ra_0+r\langle \boldsymbol{x},\boldsymbol{a}\rangle} \cdot g^{b'} \cdot (g^r)^{a_0'+\langle \boldsymbol{x},\boldsymbol{a}'\rangle},\ g^{rM^\top \boldsymbol{a}} \cdot (g^r)^{M^\top \boldsymbol{a}'} \right)$$

$\mathcal{B}$ re-randomizes it using *Delegate*, and returns it to $\mathcal{A}$.

Because $\mathcal{A}$'s view of the parameters is uniformly random, it is attacking the system $\mathcal{S}$. At the end, $\mathcal{B}$ will win its $\mathcal{S}'$-game if and only if $\mathcal{A}$ wins its $\mathcal{S}$-game, so

$$\mathcal{V}\mathrm{Adv}_{\mathcal{A}\leftrightarrow(\mathcal{S},n)}(\lambda) = \mathcal{V}\mathrm{Adv}_{\mathcal{B}\leftrightarrow(\mathcal{S}',\backslash)}(\lambda)$$

as claimed.

We now proceed to the selective-security game. Here we prove that spatial encryption is selectively CPA secure so long as the BDDHE-problem is hard on $\mathbb{G}$.

Theorem 1. *Let $\mathcal{A}$ be any non-anonymous, selective CPA adversary against $\mathcal{S}$. Then there is a BDDHE-adversary $\mathcal{B}$, running in about the same time as $\mathcal{A}$, such that:*

$$\text{BDDHE Adv}_{\mathcal{B},n+1}(\lambda) = \frac{1}{2} \cdot (\text{NonAnon}, \text{Sel}, \text{CPA})\text{Adv}_{\mathcal{A}\leftrightarrow(\mathcal{S},n)}(\lambda)$$

Proof. We first use the above observation to construct an $\mathcal{S}'$-adversary $\mathcal{A}'$ with the same advantage as $\mathcal{A}$. Our proof then follows by direct reduction. The simulator $\mathcal{B}$ takes $p, \mathbb{G}, \mathbb{G}_T$ and $(g^{\alpha^{[0,n]}}, g^{\alpha^{[n+2,2n+2]}}, h, z)$ from the BDDHE problem above. For the setup phase, $\mathcal{B}$ passes to $\mathcal{A}'$ the policy parameters $\chi = (p, \mathbb{G}, \mathbb{G}_T, n)$. Upon receiving the intended target policy $\boldsymbol{v}$, the simulator sets

$$\boldsymbol{a} = \alpha^{[1,n]}, \qquad a_0 = -\langle \boldsymbol{v}, \boldsymbol{a} \rangle, \qquad b = \alpha^{n+1}$$

Note that while $\mathcal{B}$ cannot efficiently compute $\boldsymbol{a}, a_0$ or b, it can compute $g^{\boldsymbol{a}}, g^{a_0}$ and $e(g,g)^b$ which are all it needs to present the public parameters to $\mathcal{A}'$.

To answer delegation queries for a subspace $V = \mathrm{Aff}(M, \boldsymbol{x})$, the simulator finds a vector $\boldsymbol{u} = (u_1, u_2, \ldots, u_n)^\top$ such that $M^\top \boldsymbol{u} = 0$, but $\langle \boldsymbol{x} - \boldsymbol{v}, \boldsymbol{u} \rangle \neq 0$. Such a $\boldsymbol{u}$ must exist since $\boldsymbol{v} \notin V$, and it can easily be found by the Gram-Schmidt process. The simulator then formally sets

$$r = \frac{u_1\alpha^n + u_2\alpha^{n-1} + \ldots + u_n\alpha}{\langle \boldsymbol{x} - \boldsymbol{v}, \boldsymbol{u} \rangle}$$

Note that while $\mathcal{B}$ cannot efficiently compute r, it can compute g^r. Now, for any vector y, the coefficient of the missing term α^{n+1} in $r\langle \boldsymbol{y}, \boldsymbol{a} \rangle$ is exactly $\langle \boldsymbol{y}, \boldsymbol{u} \rangle / \langle \boldsymbol{x} - \boldsymbol{v}, \boldsymbol{u} \rangle$. Therefore, $rM^\top \boldsymbol{a}$ is a vector of polynomials in α of degree at most $2n$, and the coefficient of α^{n+1} is zero by the choice of $\boldsymbol{u}$. Therefore $\mathcal{B}$ can compute $g^{rM^\top \boldsymbol{a}}$ efficiently from $g^{\alpha^{[0,n]}}$ and $g^{\alpha^{[n+2,2n]}}$. Similarly, $\mathcal{B}$ can compute

$$\begin{aligned} g^{b + r(a_0 + \langle \boldsymbol{x}, \boldsymbol{a} \rangle)} &= g^{\alpha^n + r\langle \boldsymbol{v} - \boldsymbol{x}, \boldsymbol{a} \rangle} \\ &= g^{\alpha^n + P(\alpha) + \langle \boldsymbol{v} - \boldsymbol{x}, \boldsymbol{u} \rangle \alpha^n / \langle \boldsymbol{x} - \boldsymbol{v}, \boldsymbol{u} \rangle} \\ &= g^{P(\alpha)} \end{aligned}$$

where $P(\alpha)$ has degree $2n$ and a zero coefficient on the α^{n+1} term. $\mathcal{B}$ uses this technique to answer delegation queries during both query phases.

To construct a challenge ciphertext for the message m_i, the simulator formally sets $s = \log_g h$, returning $c = (h, z \cdot m)$.

$\mathcal{B}$ returns 1 if $\mathcal{A}'$ guesses correctly, and 0 otherwise. Now, if $z = e(g, h)^{\alpha^{n+1}}$, this is a valid challenge ciphertext, so $\mathcal{A}'$ wins with probability

$$\frac{1}{2} + \frac{1}{2} \cdot (\mathrm{NonAnon, Sel, CPA})\mathrm{Adv}_{\mathcal{A}' \leftrightarrow (\mathcal{S}', \backslash)}(\lambda)$$

On the other hand, if z is random, then so is c and $\mathcal{A}'$ wins with probability $\frac{1}{2}$. As a result,

$$\begin{aligned} \mathrm{BDDHE\ Adv}_{\mathcal{B}, n+1}(\lambda) &= \frac{1}{2} \cdot (\mathrm{NonAnon, Sel, CPA})\mathrm{Adv}_{\mathcal{A}' \leftrightarrow (\mathcal{S}', \backslash)}(\lambda) \\ &= \frac{1}{2} \cdot (\mathrm{NonAnon, Sel, CPA})\mathrm{Adv}_{\mathcal{A} \leftrightarrow (\mathcal{S}, n)}(\lambda) \end{aligned}$$

as claimed.

4.5 Extensions to Spatial Encryption

Short Public Parameters. The public parameters g, g^{a_0} and $g^{\boldsymbol{a}}$ in spatial encryption consist of uniformly random elements of $\mathbb{G}$ (with the caveat that $g \neq 1$). Therefore, given a random-oracle hash $H : [1, n+2] \to \mathbb{G}$, these parameters can be omitted.

Policy Delegation. It may be desirable to re-encrypt a message from a policy π to a more restrictive policy π'. A simple model of this is to make $\mathcal{P} \uplus \mathcal{R}$ into a partially-ordered set. We say that $\pi \succeq \pi'$ if π' can be delegated to π, and $\rho \succeq \pi$ if $\mathsf{open}(\rho, \pi)$. The bottom $\bot \in \mathcal{P}$ of the partially ordered set represents plaintext or plaintext-equivalent, i.e. a policy which anyone can decrypt. Then encryption becomes a special case of policy delegation, just as key generation is a special case of delegation.

We can implement policy delegation in spatial encryption by allowing encryptions to any affine subspace $W = \mathrm{Aff}(M, \boldsymbol{x}) \subset \mathbb{Z}_p^n$. This can be decrypted by a key K_V if and only if $V \cap W \neq \emptyset$. The encryptions look much like the private keys in Section 4.2:

$$\left(g^s,\ g^{s(a_0 + \langle \boldsymbol{x}, \boldsymbol{a} \rangle)},\ g^{sM^\top \boldsymbol{a}},\ m \cdot t^s \right)$$

This allows us to construct dual systems for many of the systems in Section 3, in which policies and roles are transposed. It also enables us to turn co-inclusive encryption into a k-of-n threshold system.

However, ciphertexts for the policy-delegated systems are no longer constant-size: their size is instead proportional to the dimension of the policy as a subspace of $\mathbb{Z}_p^n$. Furthermore, while the proof given in Section 4.4 still holds, the limitations of selective security seem much stronger: the adversary must choose a subspace to attack ahead of time.

5 Future Work

The biggest drawback of cryptosystems derived from spatial encryption is that our proof only shows selective security. We leave as a significant open problem the construction of a fully-secure spatial encryption system under a compact, refutable assumption (preferably one simpler than our BDDHE assumption). Since most of the systems derived in this paper can be constructed through inclusive IBE, a fully-secure inclusive system would be almost as strong a result. We note that Gentry's recent fully-secure "key-randomizable broadcast IBE" [10] is nearly identical to our inclusive IBE, except that Gentry's adversary is only allowed to issue delegation requests for singleton identities. This result suggests that a fully-secure inclusive IBE system is within reach.

Another important challenge is to reduce the the size of the secret keys. Our current construction requires users to store $O(n \log \lambda)$ bits of sensitive information in memory and on disk, which may be challenging in some scenarios.

6 Conclusions

We presented GIBE, a general framework for viewing identity-based and broadcast encryption systems. We also constructed a spatial encryption system, which is an important instance of GIBE. Spatial encryption supports a product rule which enables us to easily construct systems with various encryption properties. One result of spatial encryption is broadcast HIBE with short ciphertexts.

A natural open problem is to constuct a spatial encryption system where both ciphertexts and private keys are short. Perhaps the techniques in [5] or [16] can be used towards this goal.

Acknowledgement

Special thanks to Adam Barth for helpful discussions on multi-authority email encryption.

References

1. Abdalla, M., Bellare, M., Catalano, D., Kiltz, E., Kohno, T., Lange, T., Malone-Lee, J., Neven, G., Paillier, P., Shi, H.: Searchable encryption revisited: Consistency properties, relation to anonymous IBE, and extensions. In: Shoup, V. (ed.) CRYPTO 2005. LNCS, vol. 3621, pp. 205–222. Springer, Heidelberg (2005)
2. Boneh, D., Boyen, X., Goh, E.: Hierarchical identity based encryption with constant size ciphertext. In: Cramer, R. (ed.) EUROCRYPT 2005. LNCS, vol. 3494, pp. 440–456. Springer, Heidelberg (2005)
3. Boneh, D., Canetti, R., Halevi, S., Katz, J.: Chosen-ciphertext security from identity-based encryption. SIAM J. of Computing (SICOMP) 36(5), 915–942 (2006)
4. Boneh, D., Franklin, M.: Identity-based encryption from the Weil pairing. SIAM Journal of Computing 32(3), 586–615 (2003); Extended abstract in Crypto 2001
5. Boneh, D., Gentry, C., Waters, B.: Collusion resistant broadcast encryption with short ciphertexts and private keys. In: Shoup, V. (ed.) CRYPTO 2005. LNCS, vol. 3621, pp. 258–275. Springer, Heidelberg (2005)
6. Boneh, D., Waters, B.: Conjunctive, subset, and range queries on encrypted data. In: Vadhan, S.P. (ed.) TCC 2007. LNCS, vol. 4392, pp. 535–554. Springer, Heidelberg (2007)
7. Canetti, R., Halevi, S., Katz, J.: A forward-secure public-key encryption scheme. Journal of Cryptology 20(3), 265–294 (2007); Early version in Eurocrypt 2003
8. Dodis, Y., Kiayias, A., Nicolosi, A., Shoup, V.: Anonymous identification in adhoc groups. In: Cachin, C., Camenisch, J.L. (eds.) EUROCRYPT 2004. LNCS, vol. 3027, pp. 609–626. Springer, Heidelberg (2004)
9. Fiat, A., Naor, M.: Broadcast encryption. In: Stinson, D.R. (ed.) CRYPTO 1993. LNCS, vol. 773, pp. 480–491. Springer, Heidelberg (1994)
10. Gentry, C.: Hierarchial identity based encryption with polynomially many levels. Personal communications (2008)
11. Gentry, C., Silverberg, A.: Hierarchical ID-based cryptography. In: Zheng, Y. (ed.) ASIACRYPT 2002. LNCS, vol. 2501, pp. 548–566. Springer, Heidelberg (2002)

12. Goyal, V., Pandey, O., Sahai, A., Waters, B.: Attribute-based encryption for fine-grained access control of encrypted data. In: Proceedings of ACM CCS 2006 (2006)
13. Horwitz, J., Lynn, B.: Towards hierarchical identity-based encryption. In: Knudsen, L.R. (ed.) EUROCRYPT 2002. LNCS, vol. 2332, pp. 466–481. Springer, Heidelberg (2002)
14. Katz, J., Sahai, A., Waters, B.: Predicate encryption supporting disjunctions, polynomial equations, and inner products. In: Smart, N.P. (ed.) EUROCRYPT 2008. LNCS, vol. 4965, pp. 146–162. Springer, Heidelberg (2008)
15. Rivest, R., Shamir, A., Tauman, Y.: How to leak a secret. In: Boyd, C. (ed.) ASIACRYPT 2001. LNCS, vol. 2248, pp. 552–565. Springer, Heidelberg (2001)
16. Sakai, R., Furukawa, J.: Identity-based broadcast encryption (2007), `http://eprint.iacr.org/2007/217`
17. Shamir, A.: Identity-based cryptosystems and signature schemes. In: Blakely, G.R., Chaum, D. (eds.) CRYPTO 1984. LNCS, vol. 196, pp. 47–53. Springer, Heidelberg (1985)
18. Zhang, F., Kim, K.: ID-based blind signature and ring signature from pairings. In: Zheng, Y. (ed.) ASIACRYPT 2002. LNCS, vol. 2501, pp. 533–547. Springer, Heidelberg (2002)

Speeding Up the Pollard Rho Method on Prime Fields

Jung Hee Cheon, Jin Hong, and Minkyu Kim

ISaC and Department of Mathematical Sciences
Seoul National University, Seoul 151-747, Korea
{jhcheon,jinhong,minkyu97}@snu.ac.kr

Abstract. We propose a method to speed up the r-adding walk on multiplicative subgroups of the prime field. The r-adding walk is an iterating function used with the Pollard rho algorithm and is known to require less iterations than Pollard's original iterating function in reaching a collision. Our main idea is to follow through the r-adding walk with only partial information about the nodes reached.

The trail traveled by the proposed method is a normal r-adding walk, but with significantly reduced execution time for each iteration. While a single iteration of most r-adding walks on $\mathbf{F}_p$ require a multiplication of two integers of $\log p$ size, the proposed method requires an operation of complexity only linear in $\log p$, using a pre-computed table of size $O((\log p)^{r+1} \cdot \log\log p)$. In practice, our rudimentary implementation of the proposed method increased the speed of Pollard rho with r-adding walks by a factor of more than 10 for 1024-bit random primes p.

Keywords: Pollard rho, r-adding walk, discrete logarithm problem, prime field.

1 Introduction

Let G be a finite cyclic group of order q generated by g. Given $h \in G$, the discrete logarithm problem (DLP) over G is to find the smallest non-negative integer x such that $g^x = h$. The answer x is called the discrete logarithm of h to the base g, and is denoted by $\log_g h$. Along with the integer factorization problem, the DLP is one of two most important mathematical primitives in public key cryptography and its hardness is the basis of various cryptosystems such as Diffie-Hellman key agreement protocol [3], ElGamal cryptosystem [6], and signature schemes [5, 6].

Many of these systems, including the Digital Signature Standard [5], are implemented on a multiplicative subgroup G of prime order q of a prime field $\mathbf{F}_p$. In such a setting, the index calculus method [1] determines the size of p to be used, but the size of q is set by the Pollard rho method [14].

In this work, we use the r-adding walk style of iterating function for the Pollard rho method, which is known to require less iterations before collision than Pollard's original iterating function. In an r-adding walk, a set $\mathcal{M}$ of r

J. Pieprzyk (Ed.): ASIACRYPT 2008, LNCS 5350, pp. 471–488, 2008.
© International Association for Cryptologic Research 2008

random elements from G is first fixed. Given the i-th element $g_i \in G$ of the walk, the $(i+1)$-th element g_{i+1} is defined to be the product of g_i and an element $M_s \in \mathcal{M}$, whose choice is given by the *index* $s = s(g_i)$, a function of g_i. Our idea is to define the index function s in such a way that $s(g_{i+1})$ can be computed from g_i and $M_{s(g_i)}$, without fully computing the product $g_{i+1} = g_i M_{s(g_i)}$. In the next iteration, $g_{i+2} = g_{i+1} \cdot M_{s(g_{i+1})}$ is considered as a product of g_i and $M_{s(g_i)} M_{s(g_{i+1})}$, with the second term taken from a pre-computed table of products among $\mathcal{M}$ elements. Thus, $s(g_{i+2})$ is computed without fully computing g_{i+2}. More generally, we prepare a table $\mathcal{M}_\ell := (\mathcal{M} \cup \{1\})^\ell$ of ℓ products from $\mathcal{M}$ and a full product computation is done when we reach ℓ iterations. Our method can be used with the distinguished points [15] collision detection method, and hence allows efficient parallelization, as with the original Pollard rho, i.e., n times speedup with n processors [12].

The proposed method produces a normal r-adding walk trail, and hence should reach a collision and solve DLP in the same number of steps as with any other r-adding walk, but the execution time of each iteration is significantly reduced. For 1024-bit random primes p the proposed algorithm replaces a multiplication of two 1024-bit words by 64 multiplications between a 16-bit word and a 32-bit word, and our rudimentary implementation of the proposed method was faster than the usual r-adding walks by a factor of more than 10. An incremental use of this algorithm will reduce each iteration of the original r-adding walk on $G \subset \mathbf{F}_p^\times$ from one multiplication of integers of $\log p$ size to an operation of complexity linear in $\log p$, using a pre-computed table of size $O((\log p)^{r+1} \cdot \log\log p)$.

Previous Works. The fastest algorithm for the DLP on a finite field $\mathbf{F}_p$ is the index calculus method whose complexity is sub-exponential in the size of the base field [1]. Since the performance of the method depends on the size of the base field, this method has the same performance on any subgroup of $\mathbf{F}_p^\times$. If the subgroup has a composite order, we can use Pohlig and Hellman [13] algorithm to reduce the DLP in the subgroup to the DLP in its prime order subgroups.

For prime-order cyclic groups G, including multiplicative subgroups of sufficiently large finite fields, the first non-trivial algorithm solving the DLP was the Baby-Step Giant-Step method suggested by Shanks [20]. It requires $O(\sqrt{q})$ operations and memory to work on an abelian group of order q. Pollard [14] proposed a probabilistic algorithm, called the Pollard rho method, with the same complexity, but requiring only small size of memory. There have been several variants proposing different collision detection methods [2, 11, 19] and iterating functions [17, 23]. An efficient parallelization of Pollard rho was developed by van Oorschot and Wiener [12] using distinguished points. For (hyper-)elliptic curves with fast endomorphisms, more efficient variants of Pollard rho methods are known [4, 7, 25].

Organization. In Section 2, we introduce the Pollard rho method, r-adding walks, and the distinguished point collision detection method. In Section 3, we propose *Tag Tracing*, a method to speed up Pollard rho. In Section 4, we apply this to prime fields and analyze its complexity. Also, we present some implementation result for 1024-bit primes. In Section 5, we estimate the asymptotic complexity of our algorithm when it is used incrementally. Section 6 concludes this paper. Tag tracing on binary fields is briefly treated in Appendix B.

2 Pollard Rho Algorithm

To set the basis of our discussion and fix notation, we will quickly review variants of the Pollard rho method in this section. Readers should consult the original papers for any detail. Throughout this paper $G = \langle g \rangle$ will be a finite cyclic group of prime order q, on which we wish to solve a discrete logarithm problem.

2.1 Function Iteration and Collision

Given any function $f : G \to G$, we can create a sequence $(g_i)_{i \geq 0}$ by iteratively defining

$$g_{i+1} = f(g_i) \quad (i \geq 0),$$

starting from a random starting point $g_0 \in G$. Because G is a finite set, this sequence is eventually periodic. The smallest integers $\mu \geq 0$ and $\lambda \geq 1$ satisfying $g_{\lambda+\mu} = g_\mu$ are said to be the pre-period and period of the sequence $(g_i)_{i \geq 0}$, respectively.

When the function f is chosen uniformly at random from the set of all functions sending G to G, the value $\lambda + \mu$ is expected to be $\sqrt{\pi q/2} \sim 1.253\sqrt{q}$. Each variant of Pollard rho method provides an *iterating function* f and a method to detect a *collision*, i.e., the happening of $g_i = g_j$ with $i \neq j$.

Suppose we are trying to solve for $\log_g h$. Given any element $y \in G$, there are many ways to write it in the exponent form $y = g^a h^b$. Let us say that a function $f : G \to G$ is *exponent traceable*, or *allows exponent tracing*, with respect to g and h, if it is possible to express the function in the form

$$f(g^a h^b) = g^{f_g(a,b)} h^{f_h(a,b)},$$

with some (simple) functions f_g and f_h of the exponents. For example, if f was the squaring function on G, we could set $f_g(a, b) = 2a$ and $f_h(a, b) = 2b$.

The iterating function of a Pollard rho algorithm variant is always chosen in such a way that it is exponent traceable. Thus, starting from $g_0 = g^{a_0} h^{b_0}$, with randomly chosen, but known, (a_0, b_0), we can always keep track of the exponents (a_i, b_i) satisfying $g_i = g^{a_i} h^{b_i}$. Then, when a collision $g_i = g_j$ is detected, setting $x = \log_g h$, we know $g^{a_i}(g^x)^{b_i} = g^{a_j}(g^x)^{b_j}$, so we can use

$$a_i + x \cdot b_i \equiv a_j + x \cdot b_j \pmod{q}$$

to solve for x.

2.2 Iterating Functions

An iterating function is taken to be of good design if the number of iterations it takes to reach a collision is close to $\sqrt{\pi q/2}$, the value expected of a random function.

Pollard. Pollard [14] originally targeted the DLP on $(\mathbf{Z}/p\mathbf{Z})^*$, but his iterating function, which we shall denote as $f_P : G \to G$, can be modified for use on any cyclic group. Let $G = T_0 \cup T_1 \cup T_2$ be a partition of G into nearly equal sized subsets. The iterating function is defined as follows.

$$f_P(y) = \begin{cases} gy, & \text{if } y \in T_0, \\ y^2, & \text{if } y \in T_1, \\ hy, & \text{if } y \in T_2. \end{cases}$$

It is clear that this allows exponent tracing. For example, when $g^a h^b \in T_0$, we have $(f_P)_g(a,b) = a+1$ and $(f_P)_h(a,b) = b$. Tests have shown that it takes more than $\sqrt{\pi q/2}$ iterations for f_P to reach a collision [23, 24], so that f_P is not an optimal choice for an iterating function.

***r*-adding walks.** Let $3 \le r \le 100$ be a small positive integer and let $G = T_0 \cup \cdots \cup T_{r-1}$ be a partition of G into r-many subsets of roughly the same size. The index function $s : G \to \{0, 1, \ldots, r-1\}$ is defined by setting $s(y) = s$ for $y \in T_s$. For each $s = 0, \ldots, r-1$, randomly choose integers $m_s, n_s \in \mathbf{Z}/q\mathbf{Z}$ and set the multipliers to $M_s = g^{m_s} h^{n_s}$. The iterating function is given by

$$f_T(y) = yM_{s(y)}.$$

That is, one of the r-many fixed elements $M_s \in G$ is multiplied, depending on which subset T_s the input belongs to. This is clearly exponent traceable, with the exponent functions being addition by m_s and n_s. The name r-adding refers to the additions.

This method was introduced in [17] and the work [16] shows that any $r \ge 8$ will suffice for cyclic groups. Testing [24] on cyclic elliptic curve groups show that 20-adding walks perform very close to a random function.

2.3 Collision Detection

The main issues with collision detection is to detect a collision with minimal number of additional iterating function applications after collision occurs, and with a small amount of memory. There have been several proposals on collision detection methods by Floyd [9], Brent [2], Sedgewick-Szymanski-Yao [19], Quisquater-Delescaille [15], and Nivasch [11].

Among them, the method using distinguished points by Quisquater and Delescaille [15] is regarded as the most efficient one. This was originally an idea for use with time-memory trade-off techniques. Distinguished points are those elements of G that satisfy a certain condition, which is easy to check. For example, with a fixed encoding for G, we may set them to be those elements with a certain number of starting bits equal to zero.

After each application of the iterating function, the current g_i is stored in a table, if it is a distinguished point. The algorithm terminates when a collision is found among the distinguished points. The distinguished points should be defined so that this table is of manageable size.

Let θ be the fraction of elements in G which satisfy the distinguishing property. The algorithm is expected to terminate with a collision after $\sqrt{\pi q/2} + 1/\theta$ applications of the iterating function.

This method has the advantage that it can lead to n-times speedup with n-processor parallelization [12].

3 Tag Tracing

Let us recall the r-adding walk iterating function f_T. Given an input $g_i \in G$, it first determines the index $s = s(g_i)$, and produces $g_{i+1} = g_i M_s \in G$ as the output. Occasionally, the output $g_i M_s$ is placed in a table of small size.

Notice that the storing operation is not very frequent. So, one may question whether computing the product $g_i M_s$ is really necessary at every iteration. Of course, iterated applications of f_T require current g_i to be available, but this is avoidable if we have a pre-computed table of suitably many products of M_s. Then it suffices for one to compute just the index at each iteration. We shall explore this line of reasoning in this section.

3.1 Preparation

As in the r-adding walk, we fix an index set $\mathcal{S} = \{0, 1, \ldots, r-1\}$ for some small r and let $\mathcal{M} = \{M_s = g^{m_s} h^{n_s}\}_{s \in \mathcal{S}}$ be a multiplier set for the r-adding walk. Fix a small positive number ℓ and consider the product set $\mathcal{M}_\ell = (\mathcal{M} \cup \{1\})^\ell$, i.e., the set of products of at most ℓ-many M_s. Notice that we know how to write each element of $\mathcal{M}_\ell$ in the form $g^m h^n$. We shall treat the set $\mathcal{M}_\ell$ as a table of elements of G, listed together with their respective exponent forms.

For our tag tracing approach to the DLP, we want to pre-compute $\mathcal{M}_\ell$ before going into the actual r-adding walk, and the following two lemmas show the range of r and ℓ one may choose, depending on the resources available.

Lemma 1. *The size of $\mathcal{M}_\ell$ is at most $\binom{\ell+r}{r}$.*

Proof. The size of $\mathcal{M}_\ell$ is bounded above by the number of combinations with repetitions, where one chooses ℓ times from the set $\mathcal{M} \cup \{1\}$ of size $r+1$. The bound is reached only if all product elements produced are distinct. □

Lemma 2. *The set $\mathcal{M}_\ell$ can be constructed in $\binom{\ell+r}{r} - 1$ multiplications in G.*

Proof. Consider the complete r-ary tree structure of depth ℓ. Label each edge with an index from $\mathcal{S}$ in such a way that from each node, the r edges extending to its children nodes are labeled with different indices. We label the root node with $1 \in G$ and label each node below with the element of $\mathcal{M}_\ell$ which is the product of multipliers labeled by edges on its way down.

The nodes of the complete r-ary tree will contain multiple copies of $\mathcal{M}_\ell$. It is clear that if we collect just the nodes with paths to the root that are labeled in non-decreasing order, then we will obtain one copy of $\mathcal{M}_\ell$. As the number of edges leading to these nodes is one less than the number of these nodes, and since each edge corresponds to one multiplication used in creation of node labels, we arrive at our claim. □

Some example values would be $\binom{84}{20} \sim 2^{63.2}$ for 20-adding walks with $\ell = 64$, and $\binom{72}{8} \sim 2^{33.4}$ for 8-adding walks with $\ell = 64$. As $\mathcal{M}_\ell$ can only be computed after h, whose discrete logarithm we are looking for, is given, we do not want these pre-computation complexities to go over our main attack complexity.

We now fix a *tag set* $\mathcal{T}$ together with three functions.

$$\begin{aligned}
&\tau : G \to \mathcal{T}.\\
&\bar{\tau} : G \times \mathcal{M}_\ell \to \mathcal{T} \cup \{\text{fail}\}.\\
&\sigma : \mathcal{T} \to \mathcal{S} = \{0, 1, \ldots, r-1\}.
\end{aligned}$$

The first function τ is named the *tag function*. We define the index function $s : G \to \mathcal{S}$ to be $s = \sigma \circ \tau$ and also consider the function $\bar{s} = \sigma \circ \bar{\tau} : G \times \mathcal{M}_\ell \to \mathcal{S} \cup \{\text{fail}\}$. The three functions above are to be chosen so that they satisfy the following condition.

1. The index function $s = \sigma \circ \tau$ is surjective and roughly pre-image uniform, i.e., grouping G according to its image points under s partitions G into subsets of roughly the same size.
2. When $\bar{s}(g, M) \in \mathcal{S}$, we have $\bar{s}(g, M) = s(g \cdot M)$. In particular, any successful output of $\bar{s}$ depends only on the product of its inputs.

So we are looking for a function τ that resembles a normal index function, but with a larger image set, and also another way $\bar{\tau}$ to evaluate τ on product of group elements.

The situation we have in mind concerning τ and $\bar{\tau}$ is as follows. Given a random $M \in \mathcal{M}_\ell$ and $g \in G$, the expected time for calculation of $\bar{\tau}(g, M)$ is smaller than the time needed for computation of the product $M \cdot g$. The general thought behind this is that it should take less effort to obtain some partial information about a product than the full product itself. For example, consider the case $G \subset \mathbf{F}_p^\times$ and define $\tau(g)$ to be the most significant k bits of $g \in G$. Intuitively, computing k bits out of the $\log p$ bits of product gM may take as little as $\frac{k}{\log p}$ of the time for full product computation. If some of the product bits were easier to calculate than others, the time could be even shorter.

We shall denote the expected time for $\bar{s}(g, M)$ evaluation by $|\bar{s}|$.

3.2 Iterating Function

The iterating function of our tag tracing algorithm will follow the usual r-adding walks. We have already fixed an index set $\mathcal{S} = \{0, 1, \ldots, r-1\}$ with an appropriate index function $s = \sigma \circ \tau$ and a multiplier set $\mathcal{M}$ during the preparation phase.

We start with a random $g_0 \in G$ and the first index $s_0 = s(g_0)$ is computed. We set $g_1 = g_0 M_{s_0}$, exactly as in the normal r-adding walk process, but the product $g_0 M_{s_0}$ is *not* computed. Instead, $\bar{s}(g_0, M_{s_0})$ is computed in time $|\bar{s}|$. If $\bar{s}(g_0, M_{s_0}) \in \mathcal{S}$, we have computed

$$s_1 = s(g_1) = s(g_0 M_{s_0}) = \bar{s}(g_0, M_{s_0}).$$

We have not fully computed g_1, but can set $g_2 = g_1 M_{s_1} = g_0 M_{s_0} M_{s_1}$, which is, once again, not computed.

Now, since $M_{s_0} M_{s_1} \in \mathcal{M}_\ell$ is an element which has been pre-computed, we can evaluate $\bar{s}(g_0, M_{s_0} M_{s_1})$ in time $|\bar{s}|$. This leads us to index value s_2 and we can continue as before.

If we come across the situation $\bar{s}(g_0, M_{s_0} \cdots M_{s_k}) \notin \mathcal{S}$, or arrive at ℓ iterations of the above process, we do a full product computation. That is, we compute $g_{k+1} = g_0 M_{s_0} \cdots M_{s_k}$ and let this replace the role g_0 has taken up to that iteration. Notice that this full product requires just one multiplication, since $M_{s_0} \cdots M_{s_k} \in \mathcal{M}_\ell$ has been pre-computed.

Notice that since the set $\mathcal{M}_\ell$ is a table of elements of G, listed together with its respective exponent forms, the above process is fully exponent traceable.

3.3 Collision Detection

To complete the description of the tag tracing method, we need to check if is possible to detect collisions. The distinguished point method is well suited for our tag tracing.

Usually, the distinguished points is defined to be points with a certain number of starting bits equal to zero, under a fixed encoding. With tag tracing, we use this usual definition, but for more efficiency, impose an additional condition to be satisfied. This extra condition is set to depend on the tag value $\tau(g)$ in such a way that it can only be satisfied when $\bar{\tau}(g', M') \notin \mathcal{T}$ for every g' and M' such that $g = g'M'$. Then, whenever there is a chance of some g_i being a distinguished point, we would already have the full form for g_i, and there are no additional full product computations involved in relation to collision detection. With the extra condition on the tag, the original condition can be relaxed to maintain the number of distinguished points.

3.4 Complexity Analysis

Let us make a rough time complexity comparison of our tag tracing with the original r-adding walks. The storage complexity of tag tracing is given by Lemma 1.

We can assume that the various parameters for tag tracing has been chosen so that the time taken for preparation, given by Lemma 2, is insignificant compared to the main function iterations. We shall also not include efforts needed in following through the exponents needed in final computation of the discrete logarithm.

Consider the time taken to do a full product computation in G. This will be almost equal to $|f_T|$, the time taken for one iteration of the r-adding walk. Even though this time will depend on the encoding for G, we shall assume that computation of full product in our tag tracing also requires time $|f_T|$.

Recalling the notation $|\bar{s}|$ introduced earlier, we can restate one of our requirements on $\bar{\tau}$ as $\frac{|\bar{s}|}{|f_T|} < 1$. It is now easy to see that a single iteration of tag tracing is expected to take time

$$|\bar{s}| + \big(\frac{1}{\ell} + P_{\text{fail}}\big)|f_T|, \tag{1}$$

at the most, where P_{fail} is the probability of reaching $\bar{s}$ value not in $\mathcal{S}$.

The expected running time of tag tracing is the above value multiplied by the number of iterations required for a normal r-adding walk style algorithm. Hence the ratio of running time between tag tracing and a normal r-adding walk would be

$$\frac{|\bar{s}|}{|f_T|} + \frac{1}{\ell} + P_{\text{fail}}.$$

If this is less than 1, we have a reduction in discrete logarithm solving time. As discussed earlier, it should be possible to find τ and $\bar{\tau}$ such that $|\bar{s}|$ is much smaller than $|f_T|$, making the above a meaningful reduction in time.

4 Application to Prime Fields and Its Implementation

Throughout this section, p will be a prime and $G = \langle g \rangle \subset \mathbf{F}_p^\times$ will be a cyclic group of order q. We will show how to apply the proposed tag tracing algorithm to G and present some implementation results. Tag tracing on subgroups of the binary field, which is quite similar, is dealt in Appendix B.

4.1 Parameter Setup

We fix the index set size r and the multiplier product pre-computation length ℓ in such a way that the time and storage complexities given by Lemma 1 and Lemma 2 are manageable. The tag set $\mathcal{T} = \{0, 1, 2, \ldots, T-1\}$ is taken to be of size $T = r \cdot b$, a multiple of r. We take a positive integer ε and set $d = \lceil \log_\varepsilon p \rceil$. Then we choose integer $\omega' \geq d(\varepsilon - 1) + 1$. We use the notation $\omega = T\omega'$ and assume that $\omega < p^{\frac{1}{3}}$.

Optimal choice for these parameters will depend on many factor including the size of prime p, resources available, and the speed of large integer multiplications. The parameter set below with $\ell = 128$ may be appropriate for use on a modern PC when primes p is of 1024-bit size. Readers may keep these in mind to facilitate understanding of further material.

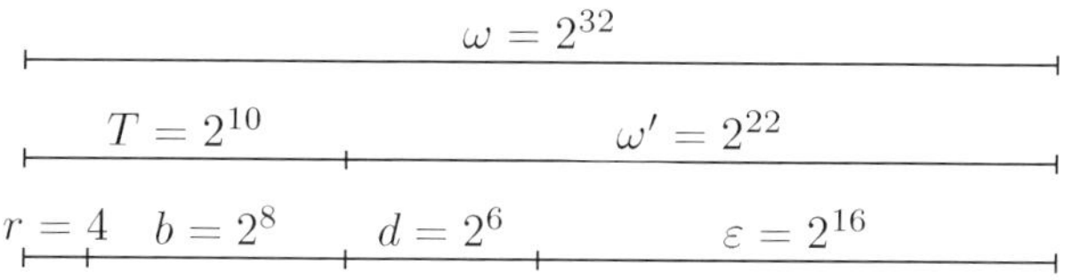

$\omega = 2^{32}$

$T = 2^{10}$ $\omega' = 2^{22}$

$r = 4$ $b = 2^8$ $d = 2^6$ $\varepsilon = 2^{16}$

4.2 Tag Function

Our assumption $\omega < p^{\frac{1}{3}}$ implies that we may always choose integer $B > p^{\frac{2}{3}}$ such that $0 \le \omega B - p < B^{\frac{1}{2}}$. For example, setting $B = \lceil p/\omega \rceil$ should always work. We fix any such B and define the tag function $\tau : G \to \mathcal{T}$ as

$$\tau(g) = \left\lfloor \frac{g \bmod p}{\omega' B} \right\rfloor, \tag{2}$$

where we are using "$x \bmod y$" to denote the unique integer between 0 and $y-1$ that is congruent to x modulo y. Notice that $0 \le \omega B - p$ implies $\frac{p-1}{\omega' B} < T$, so that the above quotient indeed lies in $\mathcal{T} = \{0, 1, \ldots, T-1\}$. We also define $\sigma : \mathcal{T} \to \mathcal{S} = \{0, 1, \ldots, r-1\}$ as $\sigma(x) = \lfloor x/b \rfloor$ and this fixes the index function $s = \sigma \circ \tau : G \to \mathcal{S}$.

The following lemma shows that we can expect τ to be roughly pre-image uniform.

Lemma 3. *If variable* $\mathbf{x}$ *is uniformly distributed over* $\mathbf{F}_p$*, then the probability distribution of* $\tau(\mathbf{x})$ *over* $\mathcal{T}$ *is almost uniform in the sense that*

$$\left|\mathrm{Prob}[\tau(\mathbf{x}) = k] - \mathrm{Prob}[\tau(\mathbf{x}) = k']\right| < \frac{1}{p^{\frac{1}{2}}}$$

for any $k, k' \in \mathcal{T}$.

Proof. We view τ as having been defined on all of $\mathbf{F}_p$. Note that $\bar{p} := T\omega' B - p = \omega B - p < B^{\frac{1}{2}} < \omega' B$. This implies that for each fixed $k = 0, \ldots, T-2$, there are exactly $\omega' B$ elements $0 \le \mathbf{x} < p$ with $\tau(\mathbf{x}) = k$ and that there are $\omega' B - \bar{p}$ elements satisfying $\tau(\mathbf{x}) = T-1$. Thus the maximal difference between pre-image sizes is $\bar{p}$. Notice that the condition $\omega B - p < B^{\frac{1}{2}}$ implies $B < p$. The maximal probability difference can now be seen to be less than $\bar{p}/p < B^{\frac{1}{2}}/p < p^{-\frac{1}{2}}$. □

Since the condition $T = r \cdot b$ makes σ exactly pre-image uniform, the above lemma holds even when $\tau(\mathbf{x})$ is replaced by $s(\mathbf{x})$, and we can state the following.

Proposition 1. *Assuming that the elements of* G *are uniformly distributed over* $\mathbf{F}_p$*, we can expect the index function* s *to be roughly pre-image uniform.*

4.3 Auxiliary Functions

We should now present the auxiliary function $\bar{\tau} : G \times \mathcal{M}_\ell \to \mathcal{T} \cup \{\text{fail}\}$ which is essentially equal to τ.

Given $\mathbf{x}, \mathbf{y} \in \mathbf{F}_p$, we can always write

$$\mathbf{x} = \sum_{i=0}^{d-1} x_i \varepsilon^i \quad (0 \le x_i < \varepsilon)$$

and, for each $0 \le i \le d-1$, we can write

$$\varepsilon^i \mathbf{y} \bmod p = \hat{y}_i B + \check{y}_i \quad (0 \le \hat{y}_i \le \frac{p-1}{B} < \omega,\ 0 \le \check{y}_i < B). \tag{3}$$

Using this notation, we define

$$\bar{\bar{\tau}}(\mathbf{x}, \mathbf{y}) = \left\lfloor \frac{\sum_{i=0}^{d-1} x_i \hat{y}_i \bmod \omega}{\omega'} \right\rfloor. \tag{4}$$

Let us check how close $\bar{\bar{\tau}}(\mathbf{x}, \mathbf{y})$ is to $\tau(\mathbf{xy})$.

Lemma 4. *Given* $\mathbf{x}, \mathbf{y} \in \mathbf{F}_p$, *we have* $\tau(\mathbf{xy}) = \bar{\bar{\tau}}(\mathbf{x}, \mathbf{y})$ *or* $\bar{\bar{\tau}}(\mathbf{x}, \mathbf{y}) + 1$, *unless* $\bar{\bar{\tau}}(\mathbf{x}, \mathbf{y}) = T - 1$.

Proof. Before going into the proof, for easy reference, let us recall some of the conditions that were placed on the parameters: $d(\varepsilon - 1) < \omega'$; $\omega = T\omega'$; $\omega < p^{\frac{1}{3}} < B^{\frac{1}{2}}$; $\omega B - p < B^{\frac{1}{2}}$;

We start by writing

$$\sum_{i=0}^{d-1} x_i \hat{y}_i = a_2 \omega + a_1 \omega' + a_0,$$

where the coefficients a_0, a_1, and a_2 are to be obtained through usual integer divisions. In particular, we have $a_2 \le \frac{d(\varepsilon-1)(\omega-1)}{\omega} < d(\varepsilon - 1) < \omega' < \omega < B^{\frac{1}{2}}$. It should also be noted that $a_1 = \bar{\bar{\tau}}(\mathbf{x}, \mathbf{y})$.

In the above notation, we may write

$$\begin{aligned} \mathbf{xy} = \sum_{i=0}^{d-1} x_i \varepsilon^i \mathbf{y} &\equiv \Big(\sum_{i=0}^{d-1} x_i \hat{y}_i\Big) B + \sum_{i=0}^{d-1} x_i \check{y}_i \pmod p \\ &\equiv a_1 \omega' B + a_0 B + a_2(\omega B - p) + \sum_{i=0}^{d-1} x_i \check{y}_i \pmod p. \end{aligned}$$

The various conditions allow us to bound the lower terms by

$$\begin{aligned} a_0 B &+ a_2(\omega B - p) + \sum_{i=0}^{d-1} x_i \check{y}_i \\ &< (\omega' - 1)B + B^{\frac{1}{2}} \cdot B^{\frac{1}{2}} + d(\varepsilon - 1)(B - 1) \\ &< \omega' B + (\omega' - 1)B = 2\omega' B - B. \end{aligned}$$

Now, if $a_1 = \bar{\bar{\tau}}(\mathbf{x}, \mathbf{y})$ is strictly less than $T-1$, then

$$\begin{aligned} a_1\omega' B + a_0 B + a_2(\omega B - p) + \sum_{i=0}^{d-1} x_i \breve{y}_i \\ < (T-2)\omega' B + 2\omega' B - B = \omega B - B < p. \end{aligned}$$

So, when $\bar{\bar{\tau}}(\mathbf{x}, \mathbf{y}) \neq T-1$, we know

$$\mathbf{xy} \bmod p = a_1\omega' B + \Big\{ a_0 B + a_2(\omega B - p) + \sum_{i=0}^{d-1} x_i \breve{y}_i \Big\}.$$

Finally, since the sum of terms inside the braces is non-negative and strictly less than $2\omega' B$, the quotient of $\mathbf{xy} \bmod p$ divided by $\omega' B$ must be either a_1 or $a_1 + 1$. □

Based on this lemma, we define $\bar{\tau} : G \times \mathcal{M}_\ell \to \mathcal{T} \cup \{\text{fail}\}$ as follows.

$$\bar{\tau}(g, M) = \begin{cases} \text{fail} & \text{if } \bar{\bar{\tau}}(g, M) \bmod b \text{ is either } b-1 \text{ or } b-2, \\ \bar{\bar{\tau}}(g, M) & \text{if otherwise.} \end{cases}$$

Recalling the definitions $\sigma(x) = \lfloor x/b \rfloor$, $s = \sigma \circ \tau$, and $\bar{s} = \sigma \circ \bar{\tau}$, it is now easy to show the following proposition.

Proposition 2. *When $\bar{\tau}(g, M) \in \mathcal{T}$ and hence $\bar{s}(g, M) \in \mathcal{S}$, we have $s(g \cdot M) = \bar{s}(g, M)$ and $\tau(gM) \bmod b \neq b-1$.*

Proof. We note that $T - 1 \bmod b = b - 1$, so that Lemma 4 together with $\bar{\bar{\tau}}(g, M) \bmod b \neq b-1$ implies $\tau(gM) = \bar{\bar{\tau}}(g, M)$ or $\bar{\bar{\tau}}(g, M) + 1$.

Now, this together with the condition that $\bar{\bar{\tau}}(g, M) \bmod b$ is neither $b-1$ nor $b-2$ implies $\tau(gM) \bmod b \neq b-1$.

In addition, we have $\bar{\tau}(g, M) = \bar{\bar{\tau}}(g, M$ and $\bar{\tau}(g, M) \bmod b \neq b-1$ implies

$$\lfloor \bar{\tau}(g, M)/b \rfloor = \lfloor (\bar{\tau}(g, M) + 1)/b \rfloor,$$

which must be $s(g \cdot M)$. □

4.4 Tag Tracing

We are now ready to start tag tracing. Using the proof of Lemma 2 as a hint, we compute a table containing entries (M, m, n) for $M = g^m h^n \in \mathcal{M}_\ell$. We also append the associated vector

$$\mathbf{v}_{\varepsilon,B}(M) = \left(\left\lfloor \frac{\varepsilon^0 M \bmod p}{B} \right\rfloor, \ldots, \left\lfloor \frac{\varepsilon^{d-1} M \bmod p}{B} \right\rfloor \right)$$

to each entry of the table. Notice that these are the $\hat{y}_i$ appearing in equation (3).

We can now follow the discussion of Section 3.2 to compute each iteration of tag tracing. The elements of G are written in ε-ary representation so that we may quickly compute s using equation (4) and Proposition 2. Whenever we reach ℓ iterations or an $\bar{s}$ calculation failure, the complete product g_i is computed using one group multiplication. A point $g \in G$ is defined to be a distinguished points only if $\tau(g) \bmod b = b - 1$ and if it satisfies some additional on g. According to Proposition 2, $g \cdot M$ can be a distinguished point only when $\bar{\tau}(g, M) \notin T$.

4.5 Implementation

We have tested tag tracing with an implementation on a modern PC and compared it with a normal 20-adding walk. Both the tag tracing and 20-adding walks were set to use distinguished points for collision detection. We used the finite field arithmetics provided by the NTL [21] library to implement the 20-adding walk, so as not to be biased. Throughout the test, prime p was taken to be of 1024-bit size, and whenever random primes p, q and $\langle g \rangle \subset \mathbf{F}_p^\times$ of order q was needed, they were generated in the style specified for DSA [5].

After comparing rho lengths of r-adding walks for various r, we opted to use $r = 4$ for tag tracing, as we did not have much memory available. Compared to the 20-adding walk, our tag tracing with $r = 4$ will have approximately 1.3 times longer rho length. This is explained in Appendix A. Other parameters were set to $b = 2^8$, $\varepsilon = 2^{16}$, $T = 2^{10}$, $\omega' = 2^{22}$, and $\omega = 2^{32}$.

For speed comparison, we chose q to be of 160-bit size and ran both the 20-adding walk and tag tracing for 2^{28} iterations. For tag tracing this was done for various choices of ℓ, with a set of randomly chosen primes p and q, group generator g, DLP target h, adding walk multipliers M_s, and initial starting point. Timings are listed in Table 1. The size $\binom{\ell+r}{r} \cdot \big((1 + \|\omega\|/\|\varepsilon\|)\|p\| + 2\|q\|\big)$ of table $\mathcal{M}_\ell$ and its preparation time is also listed, where the $\|\cdot\|$ notation has been used for bit length. The corresponding time, averaged over 10 randomly generated starting points was 1071.4 seconds for the 20-adding walk.

The table shows that the speed of tag tracing iteration can be over 15 times faster than that of a 20-adding walk. Since the rho length of a 4-adding walk is 1.3 times longer than that of a 20-adding walk, this translates to tag tracing being more than 11.5 times faster than a 20-adding walk in solving DLP.

Table 1 is also interesting in that it reflects the complexity estimate given by equation (1). Larger ℓ imply smaller number of full product computation and

Table 1. Tag tracing timing for 2^{28} iterations ($\|q\| = 160$)

ℓ	10	20	30	40	50	60	70	80	90	100
$\lvert\bar{s}\rvert$ (sec)	156.6	91.8	75.0	70.5	70.3	70.0	70.8	71.6	72.6	73.9
$\lvert f_T\rvert/\lvert\bar{s}\rvert$	6.8	11.7	14.3	**15.2**	**15.2**	**15.3**	15.1	15.0	14.8	14.5
$\mathcal{M}_\ell$ size (MB)	0.4	4.3	18.8	54.9	127.9	256.9	465.3	780.2	1233.1	1859.3
$\mathcal{M}_\ell$ comp time (sec)	0.21	2.27	9.90	29.1	68.0	137	245	414	650	983

Table 2. Full running time comparison of tag tracing and 20-adding walks

	$\|q\| = 35$	$\|q\| = 40$	$\|q\| = 45$	$\|q\| = 50$	$\|q\| = 55$
Pollard rho	1.103 sec	6.272 sec	38.738 sec	203.138 sec	1185.578 sec
20-adding	0.940 sec	5.174 sec	29.653 sec	159.977 sec	959.027 sec
tag tracing	0.093 sec	0.441 sec	2.634 sec	13.481 sec	80.785 sec
Pollard rho/tag tracing	11.89	14.24	14.70	15.07	14.68
20-adding/tag tracing	10.14	11.75	11.26	11.87	11.87

this results in steep increased speed for $\ell = 10 \sim 60$. The gradual decrease in speed after that seems to be from two factors. As we are using $b = 2^8$, we have $P_{\text{fail}} = 1/2^7$, and the increasing of ℓ looses effect as we approach $\ell = 2^7$. We have experienced through various tweaks that table lookups to $\mathcal{M}_\ell$ present a considerable fraction of the time taken by a tag tracing iteration. This coupled with our poor use of memory is another reason for decrease in speed at high ℓ. In any case, unlike our primitive testing, large scale implementation of tag tracing will need to use advanced hash table techniques that allow constant time table lookups.

We verified with small q that tag tracing has no problem in solving DLPs. Except for the q size, parameters identical to the above were used with $\ell = 40$. The timings, averaged over 200 randomly generated starting points and multiplier sets, are given in Table 2. The figures do not include the approximately 29 seconds spent on creation of $\mathcal{M}_\ell$. This may seem illogical here, but as table creation time does not change much with q, the speed ratio calculated in this way will reflect what can be expected of the ratio at large q. The data in Table 2 roughly coincides with our prediction of 11.5 factor speedup.

5 Asymptotic Complexity

In this section, we consider the asymptotic complexity of the proposed algorithm for large p. We will use $\text{Mul}(k)$ to denote the cost of multiplication modulo an integer of k-bit size.

Looking at equation (4) and the definition of σ, we can check that the cost of evaluating the auxiliary index function $\bar{s}$ is d multiplications modulo ω, $d-1$ additions modulo ω, and two divisions of integers less than ω. Thus, ignoring the small fixed number of divisions and the relatively cheaper additions, we can say that $\bar{s}$ evaluation costs approximately $d\,\text{Mul}(\|\omega\|)$. Recalling (1), we can write the average cost of a single tag tracing iteration as

$$d\,\text{Mul}(\|\omega\|) + \left(\frac{1}{\ell} + \frac{2}{b}\right)\text{Mul}(\|p\|).$$

If ω is set to grow with p, this complexity would not be linear in $k = \|p\|$. To obtain linear complexity, we perform tag computation in an incremental way, starting with fixed small parameters and recomputing with incrementally larger parameters only when the previous attempt fails. Let us explain the procedure in more detail.

We fix $r \geq 4$ and $b \geq 2$ to be small constants and define $t = \lfloor \log_b k \rfloor$. We take $\ell = \Theta(k)$, fix ε to a positive integer satisfying $\varepsilon^t \leq p^{\frac{1}{3}}/(rk^2)$, and let $d = \lceil \log_\varepsilon p \rceil$, as before. Based on these, we prepare a parameter set for each index $i = 1, \ldots, t$, as follows: $b_i = b^i$, $T_i = rb_i$, $\varepsilon_i = \varepsilon^i$, $\omega'_i = d\varepsilon_i$, $\omega_i = T_i \omega'_i$, $B_t = \lceil p/\omega_t \rceil$, $B_i = (\frac{\omega_t}{\omega_i})B_t$.

Note that ω'_i does not involve $d_i = \lceil \log_{\varepsilon_i} p \rceil$, allowing each ω_i to divide ω_{i+1} and making each B_i an integer. It is possible to check that each set of parameters satisfy all conditions set forth on Section 4.1 and Section 4.2. For example, $\omega_i \leq \omega_t = rb^t d\varepsilon^t < p^{\frac{1}{3}}$ and $\omega'_i = d\varepsilon_i \geq d_i \varepsilon_i \geq d_i(\varepsilon_i - 1) + 1$. We can also use $p^{\frac{2}{3}} < p/\omega_t \leq B_t$ to show $0 \leq \omega_i B_i - p = \omega_t B_t - p < \omega_t < p^{\frac{1}{3}} < B_t^{1/2} \leq B_i^{\frac{1}{2}}$.

For each i, we can define the tag function τ_i and the index function s_i as in Section 4, i.e.,

$$\tau_i(g) = \left\lfloor \frac{g \bmod p}{\omega'_i B_i} \right\rfloor, \quad s_i(g) = \left\lfloor \frac{\tau_i(g)}{b_i} \right\rfloor = \left\lfloor \frac{g \bmod p}{\omega'_i B_i b_i} \right\rfloor. \tag{5}$$

Since $b_i \omega'_i B_i = b_t \omega'_t B_t$, we have $s_i(g) = s_t(g)$, for any i. We already know that this common index function is roughly pre-image uniform.

Let $g \in G$ and $M \in \mathcal{M}_\ell$. For each i, we can define $\bar{\tau}_i(g, M)$ as in Section 4, which is computed in time $d_i \operatorname{Mul}(\|\omega_i\|)$, and is successful in giving $s_i(g \cdot M)$ with probability $1 - 2/b^i$. We now use an incremental approach in computing the common index $s_1(g \cdot M)$. First, $\bar{\tau}_1(g \cdot M)$ is computed. If it returns a failure, we compute $\bar{\tau}_2(g \cdot M)$, and so on. We stop whenever an output $s_i(g \cdot M)$ for $i \leq t$ is successfully obtained and move onto the next iteration of tag tracing. The full product of g and M is computed if all t attempts fail.

Then the time complexity of this incremental approach is

$$d_1 \operatorname{Mul}(\|\omega_1\|) + \frac{2d_2}{b} \operatorname{Mul}(\|\omega_2\|) + \cdots + \frac{2d_t}{b^{t-1}} \operatorname{Mul}(\|\omega_t\|) + \left(\frac{1}{\ell} + \frac{2}{b^t}\right) \operatorname{Mul}(\|p\|)$$

$$\leq 2\left(\frac{b}{b-1}\right)^2 \operatorname{Mul}(\|\omega_1\|) \lceil \log_\varepsilon p \rceil + \left(\frac{1}{\ell} + \frac{2}{b^t}\right) \operatorname{Mul}(\|p\|) = O(\|p\|) = O(k),$$

where we have used the facts $\omega_i < \omega^i$, $d_i \leq \lfloor d/i \rfloor$, $\ell = \Theta(k)$, $b^t = \Theta(k)$, and that $\operatorname{Mul}(k)$ is at most quadratic in k.

The incremental approach requires t tables and since an entry in the i-th table is of $d_i \|\omega_i\| < \log_\varepsilon \omega \log p$ bits, noting that $\prod_{i=1}^{r} \frac{\ell+i}{i} \leq \ell^r = O(k^r)$, we can write the storage requirement as

$$t \binom{\ell + r}{r} (\log_\varepsilon \omega \cdot \log p) = O(k^{r+1} \cdot \log k).$$

It only remains to consider collision detections. A point $g \in G$ is defined to be a distinguished point only if $\tau_t(g) \bmod b^t = b^t - 1$ with possibly some additional conditions on g. Because $\lfloor \tau_i(g)/b^i \rfloor = \lfloor \tau_t(g)/b^t \rfloor$ and $\tau_i(g) = \lfloor \tau_t(g)/b^{t-i} \rfloor$, we have $\tau_t(g) \bmod b^t = \left(\tau_i(g) \bmod b^i\right) b^{t-i} + \tau_t(g) \bmod b^{t-i}$. From this, we see that $\tau_t(g) \bmod b^t = b^t - 1$ implies $\tau_i(g) \bmod b^i = b^i - 1$ for any i. Thus distinguished point candidates can be noticed from any $\bar{\tau}_i(g, M)$.

6 Conclusion

In this paper, we proposed a method to speed up the Pollard rho algorithm on cyclic subgroups of the prime field $\mathbf{F}_p$. The proposed algorithm replaces the multiplication needed in r-adding walks with an operation of linear complexity. As a further work, we would like to generalize our algorithms to elliptic or hyperellipic curves.

Acknowledgments. This work was supported by the Korea Science and Engineering Foundation (KOSEF) grant (No. R01-2008-000-11287-0).

References

1. Adleman, L.: A Subexponential Algorithm for the Discrete Logarithm Problem with Applications to Cryptography. In: Proc. of the IEEE 20th Annual Symposium on Foundations of Computer Science (FOCS), pp. 55–60 (1979)
2. Brent, R.: An improved Monte Carlo Factorization Algorithm. BIT 20, 176–184 (1980)
3. Diffie, W., Hellman, M.: New Directions in Cryptology. IEEE Trans. Inform. Theory 22, 644–654 (1976)
4. Duursma, I., Gaudry, P., Morain, F.: Speeding up the Discrete Log Computation on Curves with Automorphisms. In: Lam, K.-Y., Okamoto, E., Xing, C. (eds.) ASIACRYPT 1999. LNCS, vol. 1716, pp. 103–121. Springer, Heidelberg (1999)
5. Digital Signature Standard, NIST. U.S. Department of Commerce. Federal Information Processing Standards Publication (FIPS PUB) 186 (May 1994)
6. ElGamal, T.: A Public Key Cryptosystem and a Signature Scheme based on Discrete Logarithms. IEEE Trans. Infrom. Theory 31, 469–472 (1985)
7. Gallant, R., Lambert, R., Vanstone, S.: Improving the Parallelized Pollard Lambda Search on Binary Anomalous Curves. Math. Comp. 69, 1699–1705 (2000)
8. Karatsuba, A., Ofman, Y.: Multiplication of Multidigit Numbers on Automata. Soviet Physics-Doklady 7, 595–596 (1963)
9. Knuth, D.: The Art of Computer Programming. Seminumerical Algorithms, vol. II. Addison-Wesley, Reading (1969)
10. Knuth, D.: The Art of Computer Programming. Sorting and Searching, vol. III. Addison-Wesley, Reading (1973)
11. Nivasch, G.: Cycle Detection using a Stack. Information Processing Letters 90, 135–140 (2004)
12. van Oorschot, P., Wiener, M.: Parallel Collision Search with Cryptanalytic Applications. J. Cryptology 12, 1–28 (1999)
13. Pohlig, S., Hellman, M.: An Improved Algorithm for Computing Discrete Logarithms over GF(p) and its Cryptographic Significance. IEEE Trans. Inform. Theory 24, 106–110 (1978)

14. Pollard, J.: A Monte Carlo Method for Index Computation (*modp*). Math. Comp. 32(143), 918–924 (1978)
15. Quisquater, J., Delescaille, J.: How easy is Collision Search? Application to DES. In: Quisquater, J.-J., Vandewalle, J. (eds.) EUROCRYPT 1989. LNCS, vol. 434, pp. 429–434. Springer, Heidelberg (1990)
16. Sattler, J., Schnorr, C.: Generating Random Walks in Groups. Ann. -Univ. -Sci. -Budapest. -Sect. -Comput. 6, 65–79 (1985)
17. Schnorr, C., Lenstra Jr., H.: A Monte Carlo Factoring Algorithm with Linear Storage. Math. Comp. 43(167), 289–311 (1984)
18. Schönhage, A., Strassen, V.: Schnelle Multiplikation Grobner Zahlen. Computing 7, 281–292 (1971)
19. Sedgewick, R., Szymanski, T., Yao, A.: The Complexity of Finding Cycles in Periodic Functions. SIAM Journal on Computing 11(2), 376–390 (1982)
20. Shanks, D.: Class number, a Theory of Factorization and Genera. In: Proc. Symp. Pure Math., vol. 20, pp. 415–440 (1971)
21. Shoup, V.: NTL: A Library for doing Number Theory, Ver 5.4.1, `http://shoup.net/ntl/`
22. Shoup, V.: A Computational Introduction to Number Theory and Algebra. Cambridge University Press, Cambridge (2005)
23. Teske, E.: Speeding up Pollard's rho Method for Computing Discrete Logarithms. In: Buhler, J.P. (ed.) ANTS 1998. LNCS, vol. 1423, pp. 541–554. Springer, Heidelberg (1998)
24. Teske, E.: On Random Walks for Pollard's rho Method. Math. Comp. 70, 809–825 (2001)
25. Wiener, M., Zuccherato, R.: Fast Attacks on Elliptic Curve Cryptosystems. In: Tavares, S., Meijer, H. (eds.) SAC 1998. LNCS, vol. 1556, pp. 190–200. Springer, Heidelberg (1999)

A Performance of 4-Adding Walks

For $r \geq 3$, let us write f_r to denotes the r-adding walk iterating function. We also write f_P for the Pollard's iterating function. Where as the rho length of a function graph on a set of size q is expected to be $\sqrt{\pi q/2}$ for a random function, the actual rho lengths of various f_r and f_P are a small constant multiple of $\sqrt{\pi q/2}$. We shall write C_r and C_P for these constants. In this section, we show experiment results on these values. During the test, size of p was always set to 1024 bits, but varying q sizes were used.

In order to use the iterating functions f_r and f_P we need to define an index function. For each $r \geq 3$, the index function $s_r : \mathbf{F}_p \to \{0, \ldots, r-1\}$ was set to $s_r(g) = \lfloor r \cdot (A \cdot g \mod 1) \rfloor$, where A is a rational approximation of the golden ratio $(\sqrt{5}-1)/2$. When A is of sufficient precision, this is known to bring about uniform looking distribution [10], even on non-uniform inputs. For our experiment, a precision of 1044 binary places for A is sufficient.

Estimates for the constants C_r and C_P were found as follows. Primes p, q and cyclic group generator g of order q in $\mathbf{F}_p^{\times}$ were randomly generated in the DSA style [5], and the multiplier set was randomly selected. Then the iterating function was iterated from a random starting point until the walk intersected

Table 3. Experimental rho length constant for various iterating functions

$\|q\|$	10	15	20	25	30	35	40
C_P	1.244	1.267	1.307	1.289	1.304	1.325	1.312
C_3	1.628	1.830	2.051	2.201	2.408	2.568	2.742
C_4	1.336	1.346	1.328	1.374	1.360	1.368	1.370
C_8	1.092	1.105	1.072	1.087	1.061	1.098	1.058
C_{20}	0.995	1.008	1.036	1.004	1.014	1.047	1.034
C_4/C_{20}	1.342	1.335	1.282	1.369	1.342	1.308	1.325

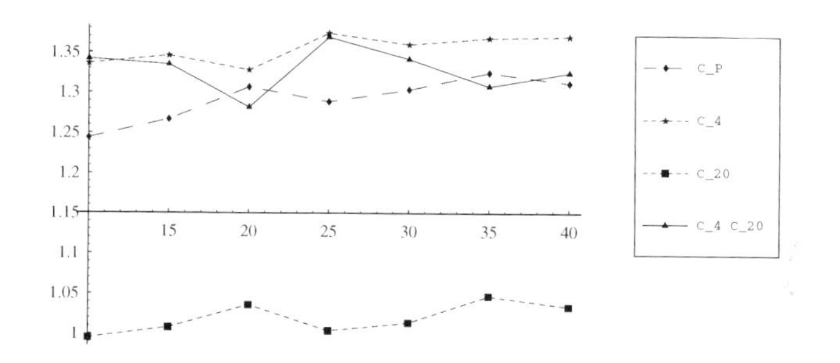

Fig. 1. Expected rho length constants for f_P, f_4, and f_{20}

itself in a rho. The length of the rho was recorded and the process redone with a newly generated g, h and multiplier set. This was repeated 1000 times for each iterating function. The average rho lengths divided by $\sqrt{\pi q/2}$ are the constant C_r and C_P, and this is summarized in Table 3. We have also provided graphs for some of these in Figure 1.

It is clear that our data is not very accurate, but it is good enough for one to conclude that C_4/C_{20} will not be too different from 1.3, even for large q.

B Tag Tracing on Binary Fields

Let us explain how tag tracing can be applied to cyclic subgroups of binary fields. We shall be very brief, as much of this case is quite similar to the prime field case.

Fix the binary field to $\mathbf{F}_{2^m} = \mathbf{F}_2[t]/p(t)$, where $p(t)$ is an irreducible polynomial of degree m, so that elements of the cyclic group $G \subset \mathbf{F}_{2^m}^\times$ may be written in the polynomial basis. Adopting the notation used with integers, we shall write $\lfloor p_1(t)/p_2(t) \rfloor$ and $p_1(t) \bmod p_2(t)$ to denote the quotient and remainder, respectively, resulting from the polynomial division of $p_1(t)$ by $p_2(t)$.

We fix positive integers u and v, such that $v < u \leq \frac{m+1}{2}$, and define the polynomial $B(t) = \lfloor p(t)/t^u \rfloor$. The tag function $\tau : G \to \mathcal{T} = \{f \in \mathbf{F}_2[t] \mid \deg f < u - v\}$ is defined as

$$\tau\big(g(t)\big) = \left\lfloor \frac{g(t) \bmod p(t)}{t^v \cdot B(t)} \right\rfloor. \tag{6}$$

Note that this map is surjective and will be roughly pre-image uniform for usual choices of G.

Given an $\mathbf{x}(t) \in \mathbf{F}_2[t]$, we can write $\mathbf{x}(t) = \sum_i x_i(t) \cdot t^{(v+1)i}$, with $\deg x_i(t) \le v$. Also, given $\mathbf{y}(t) \in \mathbf{F}_2[t]$, we can write, $t^{(v+1)i} \cdot \mathbf{y}(t) \bmod p(t) = \hat{y}_i(t) \cdot B(t) + \check{y}_i(t)$ with $\deg \check{y}_i(t) < m - u$, for each meaningful i. Using this notation, we define the auxiliary tag function as

$$\bar{\tau}\big(\mathbf{x}(t), \mathbf{y}(t)\big) = \left\lfloor \frac{\sum_i x_i(t) \cdot \hat{y}_i(t) \bmod t^u}{t^v} \right\rfloor. \tag{7}$$

Then, through careful counting of degrees and argument similar to the proof of Lemma 4, one can show that

$$\tau\big(\mathbf{x}(t) \cdot \mathbf{y}(t)\big) = \bar{\tau}\big(\mathbf{x}(t), \mathbf{y}(t)\big).$$

We emphasize that this is true for any choice of $\mathbf{x}(t), \mathbf{y}(t) \in \mathbf{F}_2[t]$.

Finally, we view the polynomial set $\mathcal{T}$ as the set of non-negative integers less than $|\mathcal{T}| = 2^{u-v}$ and define $\sigma : \mathcal{T} \to \mathcal{S} = \{0, \dots, r-1\}$ to be division by $\lceil |\mathcal{T}|/r \rceil$. Then, the index function $s = \sigma \circ \tau : G \to \mathcal{S}$ is pre-image uniform for r that is a power of 2. For other r, the probability of reaching each of the indices may differ by at most $1/|\mathcal{T}|$.

In the binary field case, unlike the prime field case, the auxiliary tag function always gives the correct tag value, so one has a better chance of running through the full ℓ-many tag tracing steps with the pre-computed table $\mathcal{M}_\ell$, without fully computing any product. However, the asymptotic complexity of the binary field case remains equal to that of the prime field case.

Sufficient Conditions for Intractability over Black-Box Groups: Generic Lower Bounds for Generalized DL and DH Problems*

Andy Rupp[1], Gregor Leander[1], Endre Bangerter[2], Alexander W. Dent[3], and Ahmad-Reza Sadeghi[1]

[1] Horst Görtz Institute for IT-Security (Germany)
{arupp,sadeghi}@crypto.rub.de, gregor.leander@rub.de
[2] Bern University of Applied Sciences (Switzerland)
endre.bangerter@bfh.ch
[3] Royal Holloway, University of London (United Kingdom)
a.dent@rhul.ac.uk

Abstract. The generic group model is a valuable methodology for analyzing the computational hardness of number-theoretic problems used in cryptography. Although generic hardness proofs exhibit many similarities, still the computational intractability of every newly introduced problem needs to be proven from scratch, a task that can easily become complicated and cumbersome when done rigorously. In this paper we make the first steps towards overcoming this problem by identifying criteria which guarantee the hardness of a problem in an extended generic model where algorithms are allowed to perform any operation representable by a polynomial function.

Keywords: Generic Group Model, Straight-Line Programs, Hardness Conditions, Lower Bounds.

1 Introduction

The *generic group model* was introduced by Nechaev [1] and Shoup [2]. In this model one considers algorithms that given a group G as black box, may only perform a restricted set of operations on the elements of G such as applying the group law, inversion of group elements and equality testing. Since in this model the group is treated as black box, the algorithms cannot exploit any special properties of a concrete group representation.

Many fundamental cryptographic problems were proven to be computationally intractable in the generic model, most notably the discrete logarithm

* The work described in this paper has been supported in part by the European Commission through the IST Programme under Contract IST-2002-507932 ECRYPT. The information in this document reflects only the authors' views, is provided as is and no guarantee or warranty is given that the information is fit for any particular purpose. The user thereof uses the information at its sole risk and liability. The names of the last three authors are in alphabetical order.

J. Pieprzyk (Ed.): ASIACRYPT 2008, LNCS 5350, pp. 489–505, 2008.
© International Association for Cryptologic Research 2008

problem (DLP), the computational and decisional Diffie-Hellman problem (DHP and DDHP) [2], and the root extraction problem (in groups of hidden order) [3]. These intractability results are considered to be evidence supporting cryptographic assumptions of number-theoretic nature which underly the security of a vast number of systems of applied cryptography. Moreover, loosely speaking, it has become considered good practice, when making new intractability assumptions, to prove the underlying problem to be hard in the generic model. Many novel assumptions rely on more complex algebraic settings than the standard assumptions. They involve multiple groups and operations on group elements additional to the basic operations. Examples include the numerous assumptions based on bilinear pairings (e.g., see [4,5]). Since the properties ensuring generic hardness had not been well-studied and formalized before this work, for each novel problem an entire hardness proof had to be done from scratch.

A generic group algorithm can only perform a subset of the operations that can be performed by an algorithm that may exploit specific properties of the representation of group elements. This implies that proving a problem to be intractable in the generic group model is a *necessary, but not sufficient* condition for the problem to be intractable in any concrete group. A generically intractable problem that is easy in any concrete group has been considered in [6].

Our contributions. In a nutshell, we identify the core aspects making cryptographic problems hard in the generic model. We provide a set of conditions, which given the description of a cryptographic problem allow one to check whether the problem at hand is intractable with respect to generic algorithms performing certain operations. In this way we aim at (i) providing means to structure and analyze the rapidly growing set of cryptographic assumptions as motivated in [7] and (ii) making the first steps towards automatically checkable hardness conditions in the generic model.

Related Work. In [8] the author analyzes a generalization of the Diffie-Hellman problem, the P-Diffie-Hellman problem: given group elements (g, g^{x_1}, g^{x_2}) the challenge is to compute $g^{P(x_1,x_2)}$, where P is a (non-linear) polynomial and g is a generator of some group G. Among other results, it is shown there that the computational and decisional variant of this problem class is hard in the generic model. Another general problem class has been introduced in [9] to cover DH related problems over bilinear groups. The authors show that decisional problems belonging to this class are hard in the generic model.

Recent work by Bresson et al. [10] independently analyzes generalized decisional problems over a single prime order group in the plain model. They showed that under several restrictions a so-called (P, Q)-DDH problem is efficiently reducible to the standard DDH problem. However, one important requirement for applying their results is that the P and Q polynomials describing the problem need to be *power-free*, i.e., variables are only allowed to occur with exponents being equal to zero or one.

2 Some Preliminaries

Let $\mathsf{poly}(x)$ denote the class of univariate polynomials in x with non-negative integer coefficients. We call a function f *negligible* if $\forall poly \in \mathsf{poly}(x)\, \exists \kappa_0\, \forall \kappa \geq \kappa_0 : f(\kappa) \leq \frac{1}{poly(\kappa)}$.

Throughout the paper we are concerned with multivariate *Laurent polynomials* over the ring $\mathbb{Z}_n$. Informally speaking, Laurent polynomials are polynomials whose variables may also have negative exponents. More precisely, a Laurent polynomial P over $\mathbb{Z}_n$ in indeterminates $X_1, \ldots, X_\ell$ is a finite sum $P = \sum a_{\alpha_1,\ldots,\alpha_\ell} X_1^{\alpha_1} \cdots X_\ell^{\alpha_\ell}$ where $a_{\alpha_1,\ldots,\alpha_l} \in \mathbb{Z}_n$ and $\alpha_i \in \mathbb{Z}$. The set of Laurent polynomials over $\mathbb{Z}_n$ forms a ring with the usual addition and multiplication. By $\deg(P) = \max\{\sum_i |\alpha_i| \mid a_{\alpha_1,\ldots,\alpha_l} \not\equiv 0 \bmod n\}$ we denote the *(absolute) total degree* of a Laurent polynomial $P \neq 0$. Furthermore, we denote by $\mathfrak{L}_n^{(\ell,c)}$ (where $0 \leq c \leq l$) the subring of Laurent polynomials over $\mathbb{Z}_n$ where only the variables $X_{c+1}, \ldots, X_\ell$ can appear with negative exponents. Note that for any $P \in \mathfrak{L}_n^{(\ell,c)}$ and $\mathbf{x} = (x_1, \ldots, x_\ell) \in \mathbb{Z}_n^c \times (\mathbb{Z}_n^*)^{\ell-c}$ the evaluation $P(\mathbf{x})$ is well-defined.

If $\mathcal{A}$ is a probabilistic algorithm, then $y \xleftarrow{R} \mathcal{A}(x)$ denotes the assignment to y of the output of $\mathcal{A}$'s run on x with fresh random coins. Furthermore, by $[\mathcal{A}(x)]$ we denote the set of all possible outputs of a probabilistic algorithm $\mathcal{A}$ on input of a fixed value x. If S is a set, then $x \xleftarrow{R} S$ denotes the random generation of an element $x \in S$ using the uniform distribution.

3 Problem Classes

In this section we formally define the classes of computational problems under consideration. For our formalization we adapt and extend the framework in [11].

Definition 1 (DL-/DH-type problem). *A* DL-/DH-type problem $\mathcal{P}$ *is characterized by*

- *A tuple of parameters*

$$Param_{\mathcal{P}} = (k, \ell, c, z)$$

  consisting of some constants $k, \ell \in \mathbb{N}, c \in \mathbb{N}_0$ where $c \leq \ell$ and $z \in \mathsf{poly}(x)$.
- *A structure instance generator* $\mathsf{SIGen}_{\mathcal{P}}(\kappa)$ *that on input of a security parameter κ outputs a tuple of the form*

$$((\mathbf{G}, \mathbf{g}, n), (\mathbf{I}, Q)),$$

  where
  - $(\mathbf{G}, \mathbf{g}, n)$ *denotes the* algebraic structure instance *consisting of descriptions of cyclic groups $\mathbf{G} = (G_1, \ldots, G_k)$ of order n and corresponding generators $\mathbf{g} = (g_1, \ldots, g_k)$,*
  - $(\mathbf{I}, Q)$ *denotes the* relation structure instance *consisting of the* input polynomials $\mathbf{I} = (\mathbf{I_1}, \ldots, \mathbf{I_k})$, *with* $\mathbf{I_i} \subset \mathfrak{L}_n^{(\ell,c)}$, $|\mathbf{I_i}| \leq z(\kappa)$, *and the* challenge polynomial $Q \in \mathfrak{L}_n^{(\ell,c)}$.

Then a problem instance of $\mathcal{P}$ consists of a structure instance $((\mathbf{G}, \mathbf{g}, n), (\mathbf{I}, Q)) \stackrel{R}{\leftarrow} \mathsf{SIGen}_{\mathcal{P}}(\kappa)$ *and group elements* $(g_i^{P(\mathbf{x})} | P \in \mathbf{I_i}, 1 \leq i \leq k)$*, where* $\mathbf{x} \stackrel{R}{\leftarrow} \mathbb{Z}_n^c \times (\mathbb{Z}_n^*)^{\ell - c}$ *are secret values. Given such a problem instance, the challenge is to compute*

$$\begin{cases} Q(\mathbf{x}), & \text{for a DL-type problem} \\ g_1^{Q(\mathbf{x})}, & \text{for a DH-type problem} \end{cases}.$$

Numerous cryptographically relevant problems fall into the class of DL-type or DH-type problems. Examples are problems such as the DLP [2], DHP [2], a variant of the representation problem [11], generalized DHP [11], square and inverse exponent problem [11], bilinear DHP [4], w-bilinear DH inversion problem [5], w-bilinear DH exponent problem [9], co-bilinear DHP [4], and many more. In Appendix A we extend our definitions and conditions to also include problems like the w-strong DH and w-strong BDH problem where the challenge is specified by a *rational function.* As an illustration of the definition, we consider the w-BDHI problem in more detail.

Example 1 (w-BDHIP). For the w-BDHI problem we have parameters $Param_{w\text{-BDHI}} = (3, 1, 0, w + 1)$ and a structure instance generator $\mathsf{SIGen}_{w\text{-BDHI}}$ that on input κ returns

$$((\mathbf{G} = (G_1, G_2, G_3), \mathbf{g} = (g_1, g_2, g_3), p),$$
$$(\mathbf{I} = (\mathbf{I_1} = \{1\}, \mathbf{I_2} = \{1, X_1^1, \ldots, X_1^{w(\kappa)}\}, \mathbf{I_3} = \{1\}), Q = X_1^{-1}))$$

such that p is a prime, there exists a non-degenerate, efficiently computable bilinear mapping $e : G_2 \times G_3 \rightarrow G_1$ with $e(g_2, g_3) = g_1$, and an isomorphism $\psi : G_2 \rightarrow G_3$ with $\psi(g_2) = g_3$. A problem instance additionally comprises group elements $(g_i^{P(\mathbf{x})} | P \in \mathbf{I_i}, 1 \leq i \leq 3) = (g_1, g_2, g_2^{x_1}, \ldots, g_2^{x_1^{w(\kappa)}}, g_3)$, where $\mathbf{x} = x_1 \stackrel{R}{\leftarrow} \mathbb{Z}_n^*$, and the task is to compute $g_1^{Q(\mathbf{x})} = g_1^{x_1^{-1}}$.

In the remainder of this paper, we are often only interested in individual parts of the output of $\mathsf{SIGen}_{\mathcal{P}}$. To this end, we introduce the following simplifying notation: By $\$ \stackrel{R}{\leftarrow} \mathsf{SIGen}_{\mathcal{P}}^{\$}(\kappa)$, where \$ is a wildcard character, we denote the projection of $\mathsf{SIGen}_{\mathcal{P}}$'s output to the part \$. For instance, $(n, \mathbf{I}, Q) \stackrel{R}{\leftarrow} \mathsf{SIGen}_{\mathcal{P}}^{(n, \mathbf{I}, Q)}(\kappa)$ denotes the projection of the output to the triple consisting of the group order, the input polynomials, and the challenge polynomial. Furthermore, by $[\mathsf{SIGen}_{\mathcal{P}}^{\$}(\kappa)]$ we denote the set of all possible outputs \$ for a given fixed security parameter κ.

4 Extending Shoup's Generic Group Model

4.1 Generic Operations

For our framework we restrict to consider operations of the form $\circ : G_{s_1} \times \ldots \times G_{s_u} \rightarrow G_d$, where $u \geq 1$, $s_1, \ldots, s_u, d \in \{1, \ldots, k\}$ are some fixed constants (that do not depend on κ). Furthermore, we demand that the action of $\circ$ on the group elements can be represented by a fixed regular polynomial. That means, there exists a fixed $F \in \mathbb{Z}[Y_1, \ldots, Y_u]$ (also not depending on κ) such that for any generators $g_{s_1}, \ldots, g_{s_u}, g_d$ given as part of a problem instance we have that $\circ(a_1, \ldots, a_u) = g_d^{F(y_1,\ldots,y_u)}$ where $a_1 = g_{s_1}^{y_1}, \ldots, a_u = g_{s_u}^{y_u}$. For instance, the bilinear mapping $e : G_2 \times G_3 \to G_1$ which is part of the algebraic setting of the w-BDHIP is such an operation: for any g_2, g_3 and $g_1 = e(g_2, g_3)$ it holds that $e(a_1, a_2) = e(g_2^{y_1}, g_3^{y_2}) = g_1^{F(y_1,y_2)}$ where $F = Y_1Y_2$. In fact, to the best of our knowledge, virtually any deterministic operation considered in the context of the generic group model in the literature so far belongs to this class of operations.

We represent an operation of the above form by a tuple $(\circ, s_1, \ldots, s_u, d, F)$, where the first component is a symbol serving as a unique identifier of the operation. The set of allowed operations can thus be specified by a set of such tuples. The full version of this paper [12] explains how to extend the operation set to include decision oracles.

Example 2 (Operations Set for w-BDHIP). The operations set $\Omega = \{(\circ_1, 1, 1, 1, Y_1 + Y_2), (\circ_2, 2, 2, 2, Y_1 + Y_2), (\circ_3, 3, 3, 3, Y_1 + Y_2), (inv_1, 1, 1, -Y_1), (inv_2, 2, 2, -Y_1), (inv_3, 3, 3, -Y_1), (\psi, 2, 3, Y_1), (e, 2, 3, 1, Y_1 \cdot Y_2)\}$ specifies operations for the group law $(\circ_i)$ and inversion (inv_i) over each group as well as the isomorphism $\psi : G_2 \to G_3$ and the bilinear map $e : G_2 \times G_3 \to G_1$.

4.2 Generic Group Algorithms and Intractability

In this section, we formally model the notion of generic group algorithms for DL-/DH-type problems. We adapt Shoup's generic group model [2] for this purpose.

Let $S_n \subset \{0,1\}^{\lceil \log_2(n) \rceil}$ denote a set of bit strings of cardinality n and Σ_n the set of all bijective functions from $\mathbb{Z}_n$ to S_n. Furthermore, let $\sigma = (\sigma_1, \ldots, \sigma_k) \in \Sigma_n^k$ be a k-tuple of randomly chosen encoding functions for the groups $G_1, \ldots, G_k \cong \mathbb{Z}_n$.

A *generic algorithm* $\mathcal{A}$ is a probabilistic algorithm that is given access to a *generic (multi-) group oracle* $\mathcal{O}_\Omega$ allowing $\mathcal{A}$ to perform operations from Ω on encoded group elements. Since any cyclic group of order n is isomorphic to $(\mathbb{Z}_n, +)$, we will always use $\mathbb{Z}_n$ with generator 1 for the internal representation of a group G_i.

As internal state $\mathcal{O}_\Omega$ maintains two types of lists, namely *element lists* $L_1, \ldots, L_k$, where a $L_i \subset \mathfrak{L}_n^{(\ell,c)}$, and *encoding lists* $E_1, \ldots, E_k$, where $E_i \subset S_n$. For an index j let $L_{i,j}$ and $E_{i,j}$ denote the j-th entry of L_i and E_i, respectively. Each list L_i is initially populated with the corresponding input polynomials given as part of a problem instance of a DL-/DH-type problem $\mathcal{P}$, i.e., $L_i = (P | P \in \mathbf{I}_i)$. A list E_i contains the encodings of the group elements corresponding to the entries of L_i, i.e., $E_{i,j} = \sigma_i(L_{i,j}(\mathbf{x}))$. E_i is initialized with $E_i = (\sigma_i(P(\mathbf{x})) | P \in \mathbf{I}_i)$. $\mathcal{A}$ is given (read) access to all encodings lists. In order to be able to perform operations on the randomly encoded elements, the algorithm may query $\mathcal{O}_\Omega$. Let $(\circ, s_1, \ldots, s_u, d, F)$ be an operation from Ω. Upon receiving a query $(\circ, j_1, \ldots, j_u)$, the oracle computes $P := F(L_{s_1,j_1}, \ldots, L_{s_u,j_u})$, appends P to L_d and $\sigma_d(P(\mathbf{x}))$ to the encoding list E_d. After having issued a number of queries, $\mathcal{A}$ eventually provides its final output. In the case that $\mathcal{P}$ is a DL-type problem, we say that $\mathcal{A}$ has solved the problem instance of $\mathcal{P}$ if its output a satisfies $Q(\mathbf{x}) - a \equiv 0 \bmod n$. In the case that $\mathcal{P}$ is a DH-type problem, $\mathcal{A}$ has solved the problem instance if its output $\sigma_1(a)$ satisfies $Q(\mathbf{x}) - a \equiv 0 \bmod n$.

Let a DL-/DH-type problem over cyclic groups $G_1, \ldots, G_k$ of order n be given. We can write the group order as $n = p^e \cdot s$ with $\gcd(p, s) = 1$ where p be the largest prime factor of n. Then for each i it holds that $G_i \cong G_i^{(p^e)} \times G_i^{(s)}$ where $G_i^{(p^e)}$ and $G_i^{(s)}$ are cyclic groups of order p^e and s, respectively. It is easy to see that solving an instance of a DL-/DH-type over groups G_i of order n is equivalent for a generic algorithm to solving it *separately* over the subgroups $G_i^{(p^e)}$ *and* the subgroups $G_i^{(s)}$. Thus, computing a solution over the groups G_i is a least as hard for generic algorithms as computing a solution over the groups $G_i^{(p^e)}$. In the following we always assume that $\mathsf{SIGen}_{\mathcal{P}}$ on input κ generates groups of prime power order $n = p^e$ with $p > 2^\kappa$ and $e > 0$.

Definition 2 (q-GGA). *A q-GGA is a generic group algorithm that for any $\kappa \in \mathbb{N}$, it receives as part of its input, makes at most $q(\kappa)$ queries to the generic group oracle.*

Definition 3 (GGA-intractability of DL-type Problems). *A DL-type problem $\mathcal{P}$ is (Ω, q, ν)-GGA-intractable if for all q-GGA $\mathcal{A}$ and $\kappa \in \mathbb{N}$ we have*

$$\Pr\left[Q(\mathbf{x}) \equiv a \bmod n \;\middle|\; \begin{array}{l} (n, \mathbf{I}, Q) \stackrel{R}{\leftarrow} \mathsf{SIGen}_{\mathcal{P}}^{(n,\mathbf{I},Q)}(\kappa); \sigma \stackrel{R}{\leftarrow} \Sigma_n^k; \mathbf{x} \stackrel{R}{\leftarrow} \mathbb{Z}_n^c \times (\mathbb{Z}_n^*)^{\ell-c}; \\ a \stackrel{R}{\leftarrow} \mathcal{A}^{\mathcal{O}_\Omega}(\kappa, n, \mathbf{I}, Q, (\sigma_i(P(\mathbf{x}))|P \in \mathbf{I_i})_{1 \le i \le k}) \end{array} \right] \le \nu(\kappa)$$

Definition 4 (GGA-intractability of DH-type Problems). *A DH-type problem $\mathcal{P}$ is (Ω, q, ν)-GGA-intractable if for all q-GGA $\mathcal{A}$ and $\kappa \in \mathbb{N}$ we have*

$$\Pr\left[Q(\mathbf{x}) \equiv a \bmod n \;\middle|\; \begin{array}{l} (n, \mathbf{I}, Q) \stackrel{R}{\leftarrow} \mathsf{SIGen}_{\mathcal{P}}^{(n,\mathbf{I},Q)}(\kappa); \sigma \stackrel{R}{\leftarrow} \Sigma_n^k; \mathbf{x} \stackrel{R}{\leftarrow} \mathbb{Z}_n^c \times (\mathbb{Z}_n^*)^{\ell-c}; \\ \sigma_1(a) \stackrel{R}{\leftarrow} \mathcal{A}^{\mathcal{O}_\Omega}(\kappa, n, \mathbf{I}, Q, (\sigma_i(P(\mathbf{x}))|P \in \mathbf{I_i})_{1 \le i \le k}) \end{array} \right] \le \nu(\kappa)$$

5 Abstract Hardness Conditions: Linking GGA and SLP Intractability

Informally speaking, the grade of intractability of a DL-/DH-type problem with respect to generic algorithms can be "measured" by means of two "quantities":

1. The probability of gaining information about the secret choices $\mathbf{x}$ in the course of a computation by means of non-trivial equalities between group elements. This quantity is called *leak-resistance*.
2. The probability to solve problem instances using a trivial strategy, i.e., by taking actions independently of (in)equalities of computed group elements and thus independent of the specific problem instance. This quantity is called *SLP-intractability*.

For formalizing both quantities, we make use of so-called straight-line program (SLP) generators. Note that SLPs are a very common concept in the field of computational algebra and has also proved its usefulness in the area of cryptography. However, the SLP model and the GGA model have not been explicitly related in the literature so far.

Definition 5 (**(Ω, q)-SLP-generator**). *A (Ω, q)-SLP-generator $\mathcal{S}$ is a probabilistic algorithm that on input $(\kappa, n, \mathbf{I}, Q)$, outputs lists $(L_1, \ldots, L_k)$ where $L_i \subset \mathfrak{L}_n^{(\ell,c)}$. Each list L_i is initially populated with $L_i = (P|P \in \mathbf{I}_i)$. The algorithm can append a polynomial to a list by applying an operation from Ω to polynomials already contained in the lists, i.e., for an operation $(\circ, s_1, \ldots, s_u, d, F) \in \Omega$ and existing polynomials $P_1 \in L_{s_1}, \ldots, P_u \in L_{s_u}$ the algorithm can append $F(P_1, \ldots, P_u)$ to L_d. In this way, the algorithm may add up to $q(\kappa)$ polynomials in total to the lists. The algorithm additionally outputs an element $a \in \mathbb{Z}_n$ in the case of a DL-type problem and a polynomial $P \in L_1$ in the case of DH-type problem, respectively.*

Let us first formalize the leak-resistance of a problem. When do group elements actually leak information due to equality relations? To see this, reconsider the definition of the generic oracle in Section 4.2 and observe that two encodings $E_{i,j}$ and $E_{i,j'}$ are equal if and only if the evaluation $(L_{i,j} - L_{i,j'})(\mathbf{x})$ yields zero. However, it is clear that such an equality relation yields no information about particular choices $\mathbf{x}$ if it holds for *all* elements from $\mathbb{Z}_n^c \times (\mathbb{Z}_n^*)^{\ell-c}$. Thus, denoting the ideal of $\mathfrak{L}_n^{(\ell,c)}$ containing all Laurent polynomials that are effectively zero over $\mathbb{Z}_n^c \times (\mathbb{Z}_n^*)^{\ell-c}$ by

$$\mathcal{I}_n = \{P \in \mathfrak{L}_n^{(\ell,c)} \mid \forall \mathbf{x} \in \mathbb{Z}_n^c \times (\mathbb{Z}_n^*)^{\ell-c} : \; P(\mathbf{x}) \equiv 0 \bmod n\} \tag{1}$$

an equality yields no information at all if $(L_{i,j} - L_{i,j'}) \in \mathcal{I}_n$. Otherwise, a *non-trivial collision* occurred and $\mathcal{A}$ learns that $\mathbf{x}$ is a modular root of $L_{i,j} - L_{i,j'}$.

By Definition 6 we capture the chance that information about the secret choices $\mathbf{x}$ is leaked in the course of a computation due to non-trivial equalities between group elements. For this purpose we can make use of (Ω, q)-SLP-generators since they generate all possible sequences of polynomials that may occur in an execution of a q-GGA.

Definition 6 (**Leak-resistance**). *A DL-/DH-type problem $\mathcal{P}$ is (Ω, q, ν)-leak-resistant if for all (Ω, q)-SLP-generators $\mathcal{S}$ and $\kappa \in \mathbb{N}$ we have*

$$\Pr\left[\begin{array}{c} \exists i \text{ and } P, P' \in L_i \text{ such that} \\ (P - P')(\mathbf{x}) \equiv 0 \bmod n \;\wedge\; P - P' \notin \mathcal{I}_n \end{array} \left| \begin{array}{l} (n, \mathbf{I}, Q) \stackrel{R}{\leftarrow} \mathsf{SIGen}_{\mathcal{P}}^{(n,\mathbf{I},Q)}(\kappa); \\ (L_1, \ldots, L_k) \stackrel{R}{\leftarrow} \mathcal{S}(\kappa, n, \mathbf{I}, Q); \\ \mathbf{x} \stackrel{R}{\leftarrow} \mathbb{Z}_n^c \times (\mathbb{Z}_n^*)^{\ell-c} \end{array}\right.\right] \leq \nu(\kappa)$$

Now assume that no information about $\mathbf{x}$ can be gained. In this case, we can restrict to consider algorithms applying trivial solution strategies to solve instances of a problem. That means, we can restrict our considerations to the subclass of generic algorithms that, when fixing all inputs except for the choice of $\mathbf{x}$, always apply the *same fixed sequence of operations* from Ω and provide the same output in order to solve an arbitrary problem instance. Thus, the algorithm actually acts as a straight-line program in this case.

Definition 7 (SLP-intractability of DL-Type Problems). *A DL-type problem $\mathcal{P}$ is (Ω, q, ν)-SLP-intractable if for all (Ω, q)-SLP-generators $\mathcal{S}$ and $\kappa \in \mathbb{N}$ we have*

$$\Pr\left[Q(\mathbf{x}) \equiv a \bmod n \;\middle|\; \begin{array}{l} (n, \mathbf{I}, Q) \stackrel{R}{\leftarrow} \mathsf{SIGen}_{\mathcal{P}}^{(n,\mathbf{I},Q)}(\kappa); \\ (a, L_1, \ldots, L_k) \stackrel{R}{\leftarrow} \mathcal{S}(\kappa, n, \mathbf{I}, Q); \\ \mathbf{x} \stackrel{R}{\leftarrow} \mathbb{Z}_n^c \times (\mathbb{Z}_n^*)^{\ell-c} \end{array} \right] \leq \nu(\kappa)$$

Definition 8 (SLP-intractability of DH-type Problems). *A DH-type problem $\mathcal{P}$ is (Ω, q, ν)-SLP-intractable if for all (Ω, q)-SLP-generators $\mathcal{S}$ and $\kappa \in \mathbb{N}$ we have*

$$\Pr\left[(P - Q)(\mathbf{x}) \equiv 0 \bmod n \;\middle|\; \begin{array}{l} (n, \mathbf{I}, Q) \stackrel{R}{\leftarrow} \mathsf{SIGen}_{\mathcal{P}}^{(n,\mathbf{I},Q)}(\kappa); \\ (P, L_1, \ldots, L_k) \stackrel{R}{\leftarrow} \mathcal{S}(\kappa, n, \mathbf{I}, Q); \\ \mathbf{x} \stackrel{R}{\leftarrow} \mathbb{Z}_n^c \times (\mathbb{Z}_n^*)^{\ell-c} \end{array} \right] \leq \nu(\kappa)$$

Theorem 1 (GGA-intractability of DL-/DH-type Problems). *If a DL-type problem is (Ω, q, ν_1)-leak-resistant and (Ω, q, ν_2)-SLP-intractable then it is $(\Omega, q, \nu_1 + \nu_2)$-GGA-intractable. If a DH-type problem is (Ω, q, ν_1)-leak-resistant and (Ω, q, ν_2)-SLP-intractable then it is $(\Omega, q, \frac{1}{2^\kappa - (q(\kappa) + z(\kappa))} + \nu_1 + \nu_2)$-GGA-intractable.*

The proof of this theorem is given in the full version of the paper [12].

6 Practical Conditions

In this section, we present easily checkable conditions ensuring that a DL-/DH-type problem is (Ω, q, ν_1)-leak-resistant and (Ω, q, ν_2)-SLP-intractable with q being polynomial and ν_1 and ν_2 being negligible functions in the security parameter. Reviewing the corresponding definitions, we see that the probabilities ν_1 and ν_2 are closely related to the probability of randomly picking roots of certain multivariate Laurent polynomials. Lemma 1 shows in turn that the probability of finding such a root is small for non-zero polynomials in $\mathfrak{L}_n^{(\ell,c)}$ having low total degrees.

Lemma 1. *Let p be a prime, $e \in \mathbb{N}$, $n = p^e$, and let $P \in \mathfrak{L}_n^{(\ell,c)}$ be a non-zero Laurent polynomial of total degree d. Then for $\mathbf{x} \stackrel{R}{\leftarrow} \mathbb{Z}_n^c \times (\mathbb{Z}_n^*)^{\ell-c}$ we have*

$$\Pr[P(\mathbf{x}) \equiv 0 \bmod n] \leq \frac{(\ell - c + 1)d}{p - 1}.$$

6.1 Operations Sets as Graphs: Bounding Polynomial Degrees

We aim at formalizing the class of operations sets that only allow for a small rise in the degrees of polynomials that can be generated by any (Ω, q)-SLP-generator $\mathcal{S}$. Remember, these are the polynomials that can be generated from the input polynomials by applying operations from Ω at most $q(\kappa)$ times. To this end, we introduce a special type of graph, called *operations set graph* (Definition 9), modeling an operations set and reflecting the corresponding rise of degrees.

Definition 9 (Operations Set Graph). *An operations set graph $\mathcal{G} = (V, E)$ is a directed multi-edge multi-vertex graph. There are two types of vertices, namely group and product vertices. The vertex set V contains at least one group vertex. Each group vertex in V is labeled with a unique integer. All product vertices are labeled by Π. Any edge in E may connect two group vertices or a group and a product vertex.*

Let Ω be an operations set involving k groups. Then the operations set graph $\mathcal{G}_\Omega = (V, E)$ corresponding to Ω is constructed as follows: V is initialized with k group vertices representing the k different groups, where these vertices are labeled with the numbers that are used in the specification of Ω, say the numbers 1 to k. For each operation $(\circ, s_1, \ldots, s_u, d, F) \in \Omega$ we add additional product vertices to V and edges to E. Let $F = \sum_i M_i$ be represented as the sum of non-zero monomials. Then for each M_i we do the following:

1. We add a product vertex and an edge from this vertex to the group vertex with label d.
2. For each variable Y_j $(1 \leq j \leq u)$ occurring with non-zero exponent ℓ in M_i we add ℓ edges from the group vertex labeled with the integer s_j to the product vertex just added before.

In order to embed the notion of increasing polynomial degrees by applying operations into the graph model we introduce the following graph terminology: We associate each group vertex in a graph with a number, called *weight*. The weight may change by doing walks through the graph. Taking a *walk* through the graph means to take an arbitrary path that contains exactly two group vertices (that are not necessarily different) where one of these vertices is the start point and the other is the end point of the path. A walk modifies the weight of the end vertex in the following way:

- If the path contains only the two group vertices, the new weight is set to be the maximum of the weights of the start and end vertex.
- If the path contains a product vertex, the new weight is set to be the maximum of the old weight and $\sum_{j=1}^{u} w_j$, where u is the indegree and w_j is the weight of the j-th predecessor of this product vertex.

We define a *free walk* to be a walk through a path that only consists of the two group vertices and no other vertex. A *non-free walk* is a walk through a path containing a product vertex. It is important to observe that

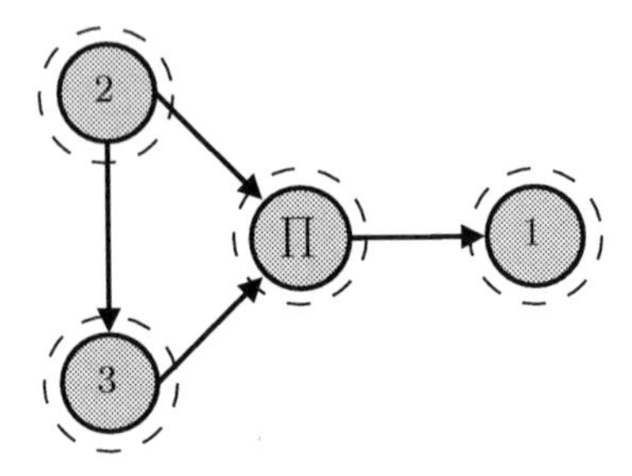

Fig. 1. $\mathcal{G}_\Omega$ for $\Omega = \{(\circ_1, 1, 1, 1, Y_1 + Y_2), (\circ_2, 2, 2, 2, Y_1 + Y_2), (\circ_3, 3, 3, 3, Y_1 + Y_2), (inv_1, 1, 1, -Y_1), (inv_2, 2, 2, -Y_1), (inv_3, 3, 3, -Y_1), (\psi, 2, 3, Y_1), (e, 2, 3, 1, Y_1 \cdot Y_2)\}$ of w-BDHIP. Strongly connected components are marked by dashed borders.

- a non-free walk can actually increase the maximum vertex weight of a graph in contrast to a free-walk.
- after each non-free walk the weight of any vertex can be changed at most finitely many times by doing free walks.

Hence, the following definition of the *q-weight* makes sense: Let q be a fixed positive number. We consider finite sequences of walks through a graph, where each sequence consists of exactly q non-free walks and an arbitrary finite number of free walks. We define the q-weight of a (group) vertex to be the maximum weight of this vertex over all such sequences. Similarly, we define the q-weight of an operations set graph to be the maximum of the q-weights of all its vertices.

Obviously, the q-weights of the vertices $1, \ldots, k$ of an operations set graph $\mathcal{G}_\Omega$ can be used to upper bound the degrees of the output polynomials $L_1, \ldots, L_k$ of any (Ω, q)-SLP-generator $\mathcal{S}$ when setting the initial weight of each group vertex i to the maximal degree of the polynomials in $\mathbf{I_i}$. Similarly, we can bound the maximum positive or negative exponent of a single variable X_j by setting the initial weight of the group vertex i to be the maximum degree of X_j in any polynomial in $\mathbf{I_i}$.

With regard to the definition of the q-weight, we can immediately simplify the structure of operations set graphs: Clearly, we do not change the q-weight of a graph if we remove self-loops and product vertices with indegree 1, where in the latter case the two edges entering and leaving the vertex are replaced by a single edge going from the predecessor vertex to the successor vertex. We call such a graph a *reduced* operations set graph. As an illustrating example, consider the reduced operations set graph depicted in Figure 1, which belongs to the operations set for the w-BDHI problem (cf. Example 2).

The following condition characterizes graphs that do not allow for a super-polynomial grow of vertex weights. Intuitively, it prohibits any kind of repeated doubling. For the q-weight of operations set graphs satisfying Condition 1, it is possible to derive non-trivial upper bounds as given in Theorem 2. The proof is given in the full version of the paper [12].

Condition 1. *Let $\mathcal{G}_\Omega$ be a reduced operations set graph. Then for every strongly connected component*[1] $\mathcal{C}$ *of* $\mathcal{G}_\Omega$ *it holds that every product vertex contained in* $\mathcal{C}$ *has at most one incoming edge from a vertex that is also contained in* $\mathcal{C}$.

Theorem 2. *Let $\mathcal{G}_\Omega$ be a reduced operations set graph satisfying Condition 1. Let n_1 denote the number of product vertices contained in $\mathcal{G}_\Omega$, $u_{\max}$ the maximal indegree of these product vertices, $d_{\max}$ the maximal initial weight of any group vertex, and n_2 the number of SCCs containing at least one product and one group vertex. Then the q-weight of $\mathcal{G}_\Omega$ is upper bounded by*

$$D(n_1, n_2, u_{\max}, d_{\max}, q) = \begin{cases} d_{\max}(u_{\max})^{n_1}, & n_2 = 0 \\ d_{\max}\mathrm{e}^{n_1}, & n_2 > 0 \text{ and } q < \frac{\mathrm{e}}{u_{\max}} n_1 \\ d_{\max}\left(\frac{u_{\max} q}{n_1}\right)^{n_1}, & n_2 > 0 \text{ and } q \geq \frac{\mathrm{e}}{u_{\max}} n_1 \end{cases},$$

where e *denotes Euler's number.*

Example 3. Condition 1 is satisfied for $\mathcal{G}_\Omega$ depicted in Figure 1 since the strongly connected component containing the product vertex contains no other vertices. We have $n_1 = 1$, $n_2 = 0$, and $u_{\max} = 2$. Since the problem instance implies $d_{\max} = w$ we have that the q-weight of the graph is bounded by $2w$.

Note that the factor by which the (maximal) initial weight of the vertices can be increased only depends on the particular operations set graph. Hence, once we have shown that an operations set only allows to increase degrees by a low (i.e., polynomial) factor, this certainly holds for all problems involving this operations set and does not need to be reproven (as it is currently done in the literature).

It is possible to devise a graph algorithm (Algorithm 1) that finds individual bounds on the q-weights of the group vertices which are often tighter than the generic bound from Theorem 2. The principle of the algorithm is simple. We consider the directed acyclic graph that is composed of the SCCs of the operations set graph. We move from the sources to the sinks of the DAG and recursively bound the q-weights of the vertices within each SCC. In the end when all SCCs are labeled with such a bound, the q-weight of a group vertex is simply set to be the q-weight bound of the (unique) SCC in which it is contained.

6.2 Practical Conditions: Leak-Resistance

To provide leak-resistance, we ensure that any difference of two distinct polynomials computable by a (Ω, q)-SLP-generator is of low degree. We do so by demanding that the input polynomials $\mathbf{I}$ of a problem $\mathcal{P}$ have low degrees (Condition 2) and restrict to operations sets Ω only allowing for small increase of degrees (Condition 1). If these conditions are satisfied, we can derive a concrete leak-resistance bound ν for any runtime bound q (Theorem 3).

[1] A strongly connected component of a directed graph $\mathcal{G}_\Omega = (V, E)$ is a maximal set of vertices $U \subset V$ s.t. every two vertices in U are reachable from each other. The strongly connected components of a graph can be computed in time $O(|V| + |E|)$.

Algorithm 1. Computation of the q-weigths of group vertices

Input: q, reduced operations set graph $\mathcal{G}$ satisfying Condition 1, initial weights for the k group vertices in G

Output: q-weights $w_1, \ldots, w_k$ of vertices $1, \ldots, k$

1: Perform a topological sort on the DAG of $\mathcal{G}$, i.e., arrange the SCCs of $\mathcal{G}$ in layers 0 to ℓ such that SCCs in layer j can only receive edges from SCCs contained in layers $i < j$.
2: **for** each layer $j = 0 : \ell$ **do**
3:   **for** each SCC $\mathcal{C}$ in layer j **do**
4:     **if** $\mathcal{C}$ consists only of group vertices **then**
5:       set weight of $\mathcal{C}$ to maximum of weights of vertices contained in $\mathcal{C}$ and weights of SCCs in layers $i < j$ having edges to $\mathcal{C}$
6:     **end if**
7:     **if** $\mathcal{C}$ consists only of a single product vertex **then**
8:       set weight of $\mathcal{C}$ to sum of weights of SCCs in layers $i < j$ having edges to $\mathcal{C}$
9:     **end if**
10:     **if** $\mathcal{C}$ consists of at least one product vertex and one group vertex **then**
11:       let w be the maximum of the weights of group vertices contained in $\mathcal{C}$ and the weights of SCCs in layers $i < j$ having edges to these group vertices
12:       for each product vertex Π in $\mathcal{C}$, compute sum of weights of SCCs in layers $i < j$ having edges to Π, and let v be the maximum of these sums
13:       set weight of $\mathcal{C}$ to $w + qv$
14:     **end if**
15:   **end for**
16: **end for**
17: **for** $i = 1 : k$ **do**
18:   set w_i to weight of SCC containing the group vertex i
19: **end for**

Condition 2. *There exists* $r_1 \in \mathsf{poly}(x)$ *such that for all* $\kappa \in \mathbb{N}$, $\mathbf{I} \in [\mathsf{SIGen}^{\mathbf{I}}_{\mathcal{P}}(\kappa)]$ *we have* $\max_{1 \le i \le k, P \in \mathbf{I_i}} (\deg(P)) \le r_1(\kappa)$

Theorem 3. *Let Ω be an operations set such that Condition 1 is satisfied. Furthermore, let $\mathcal{P}$ be a DL-type or DH-type problem satisfying Condition 2. Then for any $q \in \mathsf{poly}(x)$, the problem $\mathcal{P}$ is (Ω, q, ν)-leak-resistant, where*

$$\nu(\kappa) = 2^{-\kappa} k(q(\kappa) + z(\kappa))^2(\ell - c + 1) D(n_1, n_2, u_{\max}, r_1(\kappa), q(\kappa)) .$$

Example 4 (Leak-resistance for w-BDHIP). The degrees of the input polynomials of the w-BDHI problem are polynomially upper bounded through w by definition. Example 3 showed that Ω satisfies Condition 1 yielding $D(1, 0, 2, w(\kappa), q(\kappa)) = 2w(\kappa)$. Furthermore, for w-BDHIP we have parameters $k = 3$, $\ell = 1$, and $c = 0$. Thus, by Theorem 3 the problem $\mathcal{P}$ is (Ω, q, ν)-leak-resistant, where $\nu(\kappa) = 2^{-\kappa} 12(q(\kappa) + w(\kappa) + 1)^2 w(\kappa)$.

6.3 Practical Conditions: SLP-Intractability of DL-Type Problems

In view of Lemma 1, in order to ensure SLP-intractability for a DL-type problem it suffices to require the challenge polynomial being non-constant (Condition 3) and of low degree (Condition 4).

Condition 3. *There exists* $\kappa_0 \in \mathbb{N}$ *such that for all* $\kappa \geq \kappa_0$, $(n, Q) \in [\mathsf{SIGen}_{\mathcal{P}}^{(n,Q)}(\kappa)]$ *the polynomial* Q *is not a constant in* $\mathfrak{L}_n^{(\ell,c)}$.

Condition 4. *There exists* $r_2 \in \mathsf{poly}(x)$ *such that for all* $\kappa \in \mathbb{N}$, $Q \in [\mathsf{SIGen}_{\mathcal{P}}^{Q}(\kappa)]$ *we have* $\deg(Q) \leq r_2(\kappa)$.

Assuming the above conditions are satisfied for a DL-type problem, Theorem 4 implies that the problem is (Ω, q, ν)-SLP-intractable, where q is an arbitrary polynomial and ν is a negligible function in the security parameter.

Theorem 4. *Let* $\mathcal{P}$ *be a DL-type problem satisfying Condition 4 and Condition 3. Then for any* $q \in \mathsf{poly}(x)$ *and any operations set* Ω, $\mathcal{P}$ *is* (Ω, q, ν)*-SLP-intractable, where*

$$\nu(\kappa) = \begin{cases} 1, & \kappa < \kappa_0 \\ \frac{(\ell - c + 1) r_2(\kappa)}{2^\kappa}, & \kappa \geq \kappa_0 \end{cases}.$$

6.4 Practical Conditions: SLP-Intractability of DH-Type Problems

To ensure SLP-intractability of DH-type problems we formulate similar conditions as in the case of DL-type problems. More precisely, we ensure that the difference polynomials considered in the definition of SLP-intractability (Definition 8) are never zero and of low degree.

The non-triviality condition (Condition 5) states that an efficient SLP-generator can hardly ever compute the challenge polynomial, and thus solve the problem with probability 1.

Condition 5. *For every* $q \in \mathsf{poly}(x)$ *there exists* $\kappa_0 \in \mathbb{N}$ *such that for all* $\kappa \geq \kappa_0$, (Ω, q)*-SLP-generators* $\mathcal{S}$, $(n, \mathbf{I}, Q) \in [\mathsf{SIGen}_{\mathcal{P}}^{(n,\mathbf{I},Q)}(\kappa)]$, *and* $(P, L_1, \ldots, L_k) \in [\mathcal{S}(\kappa, n, \mathbf{I}, Q)]$ *we have* $P \neq Q$ *in* $\mathfrak{L}_n^{(\ell,c)}$.

We note that Condition 5 appears to be more complex compared to the practical conditions seen so far and it is not clear to us how to verify it in its full generality. However, it is usually easy to check in the case of a problem of practical relevance. Usually, one of the following properties is satisfied implying the validity of Condition 5:

- The total degree of $P \in L_1$ is bounded by a value which is smaller than the total degree of Q.
- The positive/negative degree of $P \in L_1$ is bounded by a value which is smaller than the positive/negative degree of Q.

– The positive/negative degree of some variable X_j of $P \in L_1$ is bounded by a value which is smaller than the positive/negative degree of that variable in Q.

Remember, that we can make use of the results from Section 6.1 for proving that a problem satisfies one of these properties.

Moreover, we have to prevent that an (Ω, q)-SLP-generator outputs a polynomial $P \neq Q$ which frequently "collides" with Q and thus constitutes a good interpolation for Q. If P is low degree (Conditions 1 and 2), then it is sufficient to demand that Q is of low degree as well (Condition 4).

Hence, we need the practical conditions for leak-resistance in addition to the ones stated in this section for showing that a DH-type problem is (Ω, q, ν)-SLP-intractable, where ν is a negligible function in the security parameter.

Theorem 5. *Let Ω be an operations set such that Condition 1 is satisfied. Furthermore, let $\mathcal{P}$ be DH-type problem satisfying Condition 2, Condition 4 and Condition 5. Then for any $q \in \mathsf{poly}(x)$, the problem $\mathcal{P}$ is (Ω, q, ν)-SLP-intractable, where*

$$\nu(\kappa) = \begin{cases} 1, & \kappa < \kappa_0 \\ \frac{(\ell - c + 1)(r_2(\kappa) + D(n_1, n_2, u_{\max}, r_1(\kappa), q(\kappa)))}{2^\kappa}, & \kappa \geq \kappa_0 \end{cases}$$

is a negligible function.

Example 5 (SLP-intractability of w-BDHIP). Remember that for this problem the challenge polynomial is fixed to $Q = X_1^{-1}$. Moreover, observe that all variables occurring in the input polynomials only have positive exponents. Thus, any polynomial $P \in L_1$ has only positive exponents in any variable. Hence, Condition 5 is trivially satisfied (independently of the considered operations set Ω).[2] Condition 4 is satisfied since we always have $\deg(Q) = 1 =: r_2(\kappa)$. As we have already seen in the previous sections, Conditions 1 and 2 hold yielding the upper bound $D(1, 0, 2, w(\kappa), q(\kappa)) = 2w(\kappa)$ on the degrees of the polynomials $P \in L_1$. Thus, by Theorem 5 the problem is (Ω, q, ν)-SLP-intractable, where $\nu(\kappa) = 2^{-\kappa}(2 + 4w(\kappa))$.

References

1. Nechaev, V.I.: Complexity of a determinate algorithm for the discrete logarithm. Mathematical Notes 55(2), 165–172 (1994)
2. Shoup, V.: Lower bounds for discrete logarithms and related problems. In: Fumy, W. (ed.) EUROCRYPT 1997. LNCS, vol. 1233, pp. 256–266. Springer, Heidelberg (1997)

[2] Note that $X_1^{-1} \neq X_1^{\phi(n)-1}$ in $\mathfrak{L}_n^{(1,0)}$ but these polynomials evaluate to the same value for all $x_1 \in \mathbb{Z}_n^*$. However, Conditions 1 and 2 ensure that for any efficient SLP-generator there exists κ_0 s.t. for all $\kappa \geq \kappa_0$ the polynomial $X_1^{\phi(n)-1}$ cannot be computed.

3. Damgård, I., Koprowski, M.: Generic lower bounds for root extraction and signature schemes in general groups. In: Knudsen, L.R. (ed.) EUROCRYPT 2002. LNCS, vol. 2332, pp. 256–271. Springer, Heidelberg (2002)
4. Boneh, D., Franklin, M.: Identity-based encryption from the weil pairing. In: Kilian, J. (ed.) CRYPTO 2001. LNCS, vol. 2139, pp. 213–229. Springer, Heidelberg (2001)
5. Boneh, D., Boyen, X.: Efficient selective-ID secure identity based encryption without random oracles. In: Cachin, C., Camenisch, J.L. (eds.) EUROCRYPT 2004. LNCS, vol. 3027, pp. 223–238. Springer, Heidelberg (2004)
6. Dent, A.W.: Adapting the weaknesses of the random oracle model to the generic group model. In: Zheng, Y. (ed.) ASIACRYPT 2002. LNCS, vol. 2501, pp. 100–109. Springer, Heidelberg (2002)
7. Boneh, D.: Number-theoretic assumptions. TCC's Special Session on Assumptions for Cryptography (2007) (Invited Talk)
8. Kiltz, E.: A tool box of cryptographic functions related to the Diffie-Hellman function. In: Pandu Rangan, C., Ding, C. (eds.) INDOCRYPT 2001. LNCS, vol. 2247, pp. 339–350. Springer, Heidelberg (2001)
9. Boneh, D., Boyen, X., Goh, E.: Hierarchical identity based encryption with constant size ciphertext (full paper). Cryptology ePrint Archive, Report 2005/015 (2005), http://eprint.iacr.org/2005/015
10. Bresson, E., Lakhnech, Y., Mazaré, L., Warinschi, B.: A generalization of DDH with applications to protocol analysis and computational soundness. In: Menezes, A. (ed.) CRYPTO 2007. LNCS, vol. 4622, pp. 482–499. Springer, Heidelberg (2007)
11. Sadeghi, A., Steiner, M.: Assumptions related to discrete logarithms: Why subtleties make a real difference. In: Pfitzmann, B. (ed.) EUROCRYPT 2001. LNCS, vol. 2045, pp. 244–260. Springer, Heidelberg (2001)
12. Rupp, A., Leander, G., Bangerter, E., Dent, A.W., Sadeghi, A.R.: Sufficient conditions for intractability over black-box groups: Generic lower bounds for generalized DL and DH problems (full version). Cryptology ePrint Archive, Report 2007/360 (2007), http://eprint.iacr.org/2007/360

A Rational Functions Specifying Problem Challenges

Our framework so far only covers problems where the solution of a problem instance can be represented as a Laurent polynomial. This restriction excludes important problems like the w-strong Diffie-Hellman problem or the w-strong Bilinear Diffie-Hellman problem. Informally speaking, the w-SDH problem can be described as follows: Given group elements $g, g^{x^1}, g^{x^2}, \ldots, g^{x^w}$, where $x \stackrel{R}{\leftarrow} \mathbb{Z}_p^*$, the task is to find an integer $v \in \mathbb{Z}_p^*$ and a group element a such that $a = g^{\frac{1}{x+v}}$. Observe that here the solution is defined by a *rational function* of the secret choices *and* the value v that can be chosen freely. If $\frac{1}{x+v}$ is not defined over $\mathbb{Z}_p$ for particular x and v, then the problem instance is deemed to be not solved.

To let the class of DL-/DH-type problems (Definition 1) cover this problem type we do the following: We first need introduce two additional parameters ℓ' and c' defining the range $\mathbb{Z}_n^{c'} \times (\mathbb{Z}_n^*)^{\ell'-c'}$ from which the algorithm is allowed to choose the value $\mathbf{v}$. Furthermore, we consider structure instance generators $\mathsf{SIGen}_{\mathcal{P}}$ which output two Laurent polynomials Q_1 and Q_2 over $\mathbb{Z}_n$ in the variables $X_1, \ldots, X_\ell, V_1, \ldots V_{\ell'}$, where only the variables $X_{c+1}, \ldots, X_\ell$ and

$V_{c'+1}, \ldots, V_{\ell'}$ may appear with negative exponents. These polynomials represent a rational function

$$R : (\mathbb{Z}_n^c \times (\mathbb{Z}_n^*)^{\ell-c}) \times (\mathbb{Z}_n^{c'} \times (\mathbb{Z}_n^*)^{\ell'-c'}) \to \mathbb{Z}_n\,,$$
$$(\mathbf{x}, \mathbf{v}) \mapsto \frac{Q_1(\mathbf{x}, \mathbf{v})}{Q_2(\mathbf{x}, \mathbf{v})}\,.$$

A problem instance of such an extended DL-/DH-type problem is defined as before. Given a problem instance, the challenge is to output some $\mathbf{v} \in \mathbb{Z}_n^{c'} \times (\mathbb{Z}_n^*)^{\ell'-c'}$ with $Q_2(\mathbf{x}, \mathbf{v}) \in \mathbb{Z}_n^*$ and the element

$$\begin{cases} \frac{Q_1(\mathbf{x},\mathbf{v})}{(Q_2(\mathbf{x},\mathbf{v}))}, & \text{for a DL-type problem} \\ g_1^{\frac{Q_1(\mathbf{x},\mathbf{v})}{(Q_2(\mathbf{x},\mathbf{v}))}}, & \text{for a DH-type problem} \end{cases} . \qquad (2)$$

Adapting most of the framework to the new definition is quite straightforward. In fact, the definition of leak-resistance, the corresponding conditions and theorems stay the same since the definition is completely independent of the challenge polynomial. In the following, we only sketch important differences to the previous version of the conditions.

For this purpose, we need to introduce some new notation: By

$$\mathfrak{F}(\mathfrak{L}_n^{(\ell,c)}) := \left\{ \frac{Q_1}{Q_2} \mid Q_1, Q_2 \in \mathfrak{L}_n^{(\ell,c)},\ Q_2 \text{ is not a zero-divisor} \right\}$$

we denote the ring of fractions of $\mathfrak{L}_n^{(\ell,c)}$. An element $\frac{a}{b} \in \mathfrak{F}(\mathfrak{L}_n^{(\ell,c)})$ with $a, b \in \mathbb{Z}_n$ is called a *constant fraction*. The ring $\mathfrak{L}_n^{(\ell,c)}$ can be seen as a subring of this ring by identifying $Q \in \mathfrak{L}_n^{(\ell,c)}$ with $\frac{Q}{1} \in \mathfrak{F}(\mathfrak{L}_n^{(\ell,c)})$. Note that if we evaluate the fraction $\frac{Q_1}{Q_2}$ with some $\mathbf{v} \in \mathbb{Z}_n^{c'} \times (\mathbb{Z}_n^*)^{\ell'-c'}$ we obtain a fraction $\frac{Q_1(\mathbf{X},\mathbf{v})}{Q_2(\mathbf{X},\mathbf{v})}$ that is not necessarily a well-defined element of $\mathfrak{F}(\mathfrak{L}_n^{(\ell,c)})$. This is because $Q_2(\mathbf{X}, \mathbf{v})$ might be a zero-divisor in $\mathfrak{L}_n^{(\ell,c)}$. However, we can exclude this case, because by choosing such a fraction (i.e., by selecting this particular $\mathbf{v}$) an algorithm can never solve a problem instance.

We stipulate the following definitions for the SLP-intractability of a (extended) DL-type and a DH-type problem, respectively. Note that the SLP-generators now additionally output $\mathbf{v}$ in order to select a specific fraction.

Definition 10 (SLP-intractability of DL-Type Problems). *A DL-type problem $\mathcal{P}$ is (Ω, q, ν)-SLP-intractable if for all (Ω, q)-SLP-generators $\mathcal{S}$ and $\kappa \in \mathbb{N}$ we have*

$$\Pr\left[\begin{array}{c} Q_2(\mathbf{x}, \mathbf{v}) \in \mathbb{Z}_n^* \text{ and} \\ R(\mathbf{x}, \mathbf{v}) \equiv a \bmod n \end{array} \middle| \begin{array}{l} (n, \mathbf{I}, Q_1, Q_2) \stackrel{R}{\leftarrow} \mathsf{SIGen}_{\mathcal{P}}^{(n,\mathbf{I},Q_1,Q_2)}(\kappa); \\ (\mathbf{v}, a, L_1, \ldots, L_k) \stackrel{R}{\leftarrow} \mathcal{S}(\kappa, n, \mathbf{I}, Q_1, Q_2); \\ R \leftarrow \frac{Q_1}{Q_2}; \mathbf{x} \stackrel{R}{\leftarrow} \mathbb{Z}_n^c \times (\mathbb{Z}_n^*)^{\ell-c} \end{array} \right] \leq \nu(\kappa)$$

Definition 11 (SLP-intractability of DH-type Problems). *A DH-type problem $\mathcal{P}$ is (Ω, q, ν)-SLP-intractable if for all (Ω, q)-SLP-generators $\mathcal{S}$ and $\kappa \in \mathbb{N}$ we have*

$$\Pr\left[\begin{array}{c} Q_2(\mathbf{x},\mathbf{v}) \in \mathbb{Z}_n^* \text{ and} \\ (P - R(\mathbf{X},\mathbf{v}))(\mathbf{x}) \equiv 0 \bmod n \end{array} \middle| \begin{array}{l} (n, \mathbf{I}, Q_1, Q_2) \stackrel{R}{\leftarrow} \mathsf{SIGen}_{\mathcal{P}}^{(n,\mathbf{I},Q_1,Q_2)}(\kappa); \\ (\mathbf{v}, P, L_1, \ldots, L_k) \stackrel{R}{\leftarrow} \mathcal{S}(\kappa, n, \mathbf{I}, Q_1, Q_2); \\ R \leftarrow \frac{Q_1}{Q_2}; \mathbf{x} \stackrel{R}{\leftarrow} \mathbb{Z}_n^c \times (\mathbb{Z}_n^*)^{\ell-c} \end{array}\right] \leq \nu(\kappa)$$

The GGA-intractability of a DL-/DH-type problem is still related in the same way to the leak-resistance property and the SLP-intractability of the problem. That means, Theorem 1 holds unchanged for our extension.

To ensure SLP-intractability, we have Condition 6 and 7 for DL-type problems and Condition 6 and 8 for DH-type problems. These conditions imply (Ω, q, ν)-SLP-intractability for the same negligible functions ν as stated in Theorems 4 and 5.

Condition 6. *There exists $r_2 \in \mathsf{poly}(x)$ such that for all $\kappa \in \mathbb{N}$, $(n, Q_1, Q_2) \in [\mathsf{SIGen}_{\mathcal{P}}^{(n,Q_1,Q_2)}(\kappa)]$, and $\mathbf{v} \in \mathbb{Z}_n^{c'} \times (\mathbb{Z}_n^*)^{\ell'-c'}$ we have* $\max\{\deg(Q_1(\mathbf{X}, \mathbf{v})), \deg(Q_2(\mathbf{X}, \mathbf{v}))\} \leq r_2(\kappa)$.

Condition 7. *There exists $\kappa_0 \in \mathbb{N}$ such that for all $\kappa \geq \kappa_0$, $(n, Q_1, Q_2) \in [\mathsf{SIGen}_{\mathcal{P}}^{(n,Q_1,Q_2)}(\kappa)]$, and $\mathbf{v} \in \mathbb{Z}_n^{c'} \times (\mathbb{Z}_n^*)^{\ell'-c'}$ we have that $\frac{Q_1(\mathbf{X},\mathbf{v})}{Q_2(\mathbf{X},\mathbf{v})}$ is not a constant fraction in $\mathfrak{F}(\mathfrak{L}_n^{(\ell,c)})$.*

Condition 8. *For every $q \in \mathsf{poly}(x)$ there exists $\kappa_0 \in \mathbb{N}$ such that for all $\kappa \geq \kappa_0$, (Ω, q)-SLP-generators $\mathcal{S}$, $(n, \mathbf{I}, Q_1, Q_2) \in [\mathsf{SIGen}_{\mathcal{P}}^{(n,\mathbf{I},Q_1,Q_2)}(\kappa)]$, and $(\mathbf{v}, P, L_1, \ldots, L_k) \in [\mathcal{S}(\kappa, n, \mathbf{I}, Q_1, Q_2)]$ we have that $\frac{Q_1(\mathbf{X},\mathbf{v})}{Q_2(\mathbf{X},\mathbf{v})} \neq P$ in $\mathfrak{F}(\mathfrak{L}_n^{(\ell,c)})$.*

Example 6 (SLP-intractability of w-SDHP). For the w-SDH problem we have parameters $Param_{w\text{-SDH}} = (k = 1, \ell = 1, c = 0, z = w + 1, \ell' = 1, c' = 0)$ and a structure instance generator $\mathsf{SIGen}_{w\text{-SDH}}$ that on input κ returns

$$((\mathbf{G} = G_1, \mathbf{g} = g_1, n = p), (\mathbf{I} = \mathbf{I_1} = \{1, X_1^1, \ldots, X_1^{w(\kappa)}\}, Q_1 = 1, Q_2 = X_1 + V_1)).$$

Note that for any $v_1 \in \mathbb{Z}_p^*$, the fraction $\frac{Q_1(\mathbf{X},\mathbf{v})}{Q_2(\mathbf{X},\mathbf{v})} = \frac{1}{X_1+v_1}$ is an element of $\mathfrak{F}(\mathfrak{L}_n^{(\ell,c)})$ but not an element of the subring $\mathfrak{L}_n^{(\ell,c)}$. Hence, Condition 8 is trivially satisfied, since P is always a Laurent polynomial (independently of the considered operations set Ω). Condition 6 is satisfied since we always have $\max\{\deg(Q_1(\mathbf{X}, \mathbf{v}), \deg(Q_2(\mathbf{X}, \mathbf{v}))\} = 1 =: r_2(\kappa)$. As we can easily see, Conditions 1 and 2 hold assuming an operations set containing operations for performing the group law and inversion of elements in G_1, this yields an upper bound $D(0, 0, 0, w(\kappa), q(\kappa)) = w(\kappa)$ on the degrees of the polynomials $P \in L_1$. Thus, the problem is (Ω, q, ν)-SLP-intractable, where $\nu(\kappa) = 2^{-\kappa}(w(\kappa) + 1)$.

OAEP Is Secure under Key-Dependent Messages

Michael Backes[1,2], Markus Dürmuth[1], and Dominique Unruh[1]

[1] Saarland University, Saarbrücken, Germany
{backes,duermuth,unruh}@cs.uni-sb.de
[2] Max-Planck-Institute for Software Systems, Saarbrücken, Germany
backes@mpi-sws.mpg.de

Abstract. Key-dependent message security, short KDM security, was introduced by Black, Rogaway and Shrimpton to address the case where key cycles occur among encryptions, e.g., a key is encrypted with itself. We extend this definition to include the cases of adaptive corruptions and arbitrary active attacks, called adKDM security incorporating several novel design choices and substantially differing from prior definitions for public-key security. We also show that the OAEP encryption scheme (using a partial-domain one-way function) satisfies the strong notion of adKDM security in the random oracle model. The OAEP construction thus constitutes a suitable candidate for implementating symbolic abstractions of encryption schemes in a computationally sound manner under active adversaries.

Keywords: Key-dependent message security, chosen ciphertext attacks, RSA-OAEP.

1 Introduction

Encryption schemes constitute the oldest and arguably the most important cryptographic primitive. Their security was rigorously studied very early, starting with Shannon's work for the information-theoretic case [31]. Computational definitions for public-key encryption were developed over time, in particular in [23,32,30,19]. For symmetric encryption, the first real definitions were, to the best of our knowledge, given in [19,28,8], using the same basic ideas as in public-key encryption. While these definitions seemed to take care of standard usage of encryption schemes, it was soon recognized that larger protocols might pose additional requirements on the encryption schemes, e.g., in multi-party computations with dynamic corruptions as in [7]. It was also recognized that in some cases, symmetric encryption initially seemed to be the appropriate method to use, but upon study other primitives such pseudorandom permutations [10,8] or authenticated encryption [12,9] proved to be better.

A specific additional requirement some larger protocols pose on encryption schemes is the ability to securely encrypt key-dependent messages. One speaks of key-dependent messages if a key K is used to encrypt a message m where m contains or depends on the key K (or the corresponding secret key in the case of public-key encryption). The first concrete use of this case seems to have been

J. Pieprzyk (Ed.): ASIACRYPT 2008, LNCS 5350, pp. 506–523, 2008.
© International Association for Cryptologic Research 2008

in [15], where multiple private keys were used to encrypt one another in order to implement an all-or-nothing property in a credential system to discourage people from transferring individual credentials. Such key cycles also occur in implementations of disk encryption in, e.g., Windows Vista, that can store an encryption of its own secret keys to the disk in some situations. Key cycles also occur in some naively designed key exchange protocols of session keys given master keys shared among the two parties or with a key distribution center, where at the end of the protocol the newly exchanged key is "confirmed" by using it to encrypt or authenticate something that might include the master keys.

Another area that has brought additional requirements on cryptographic primitives, and in particular that of encryption with key cycles, is the use of formal methods or "symbolic cryptography". Here the question is whether simple abstractions of cryptographic primitives exist that can be used by automated proof tools (model checkers or theorem provers) to prove or disprove a wide range of security protocols that use cryptography in a blackbox manner. The original abstractions used by this automation community are term algebras constructed from certain base types and cryptographic operators such as E and D for encryption and decryption. They are often called Dolev-Yao models after the first such abstraction [20]. As soon as one has a multi-user variant of such a model, the keys are terms, and from the term algebra side it is natural that keys can also be encrypted, i.e., most models simply assume that key cycles are allowed. Once cryptographic justification of such models was started in [2], it was recognized that key cycles had to be excluded from the original models to get cryptographic results. The same holds for later results [1,26,6,27,29,4,18,17].

Motivated primarily by symbolic cryptography, a definition of key-dependent message security (*KDM security*) was introduced in [13]. It generalizes the definition from [15] by allowing arbitrary functions of the keys (and not just individual keys) as plaintexts, and by considering symmetric encryption schemes. [13] also presents a definition and a construction (without proof) for the asymmetric case against *passive* attackers. In [5] it was shown that, in the case of symmetric encryption, an extension of the KDM definition that additionally allows for a limited revelation of secret keys of honest users, called DKDM security, is suitable for extending results about the justification of Dolev-Yao models to include protocols with key cycles. Full security in the presence of key-dependent messages has so far only been achieved in the random oracle model. In [24] and [25], the problem of implementing KDM secure symmetric encryption schemes without random oracles is investigated. There, solutions are given for relaxed variants of KDM security, e.g., security against a bounded number of queries or security with respect to a *single* key dependency function. No scheme is known, however, that fulfills any form of full-fledged KDM security (passive or active) without the use of random oracles. In [14], a scheme is presented that is secure if the key dependency functions are guaranteed to be affine. Extensions of KDM security for public-key encryption to active adversaries have not been proposed yet, and establishing meaningful definitions for this case indeed raises non-trivial problems.

Our Contributions. We first propose a new definition of security under key-dependent messages, called *adKDM security*, that captures security against active attackers and adaptive corruptions in the case of public-key encryption. This definition incorporates several novel design choices and substantially differs from prior definitions for public-key security; in particular, it allows the adversary to iteratively construct nested encryptions without necessarily revealing inner encryptions, and it is required to keep track of the knowledge that the adversary maintains in an ideal setting.

We then investigate the OAEP encryption scheme and prove that it satisfies adKDM security in the random oracle model, assuming the partial-domain one-wayness of the underlying trapdoor-permutation. This in particular shows the OAEP construction to constitute a suitable candidate for soundly implementing symbolic abstractions of cryptography (so-called computational soundness). We leave it as an open problem for future work to prove that our definition of adKDM security is sufficient for a computational soundness result.

The need to incorporate key dependencies and the adaptive nature of adKDM security require substantial changes to the CCA2-security proof of OAEP. In particular, adKDM security does not allow for determining in advance which encryptions will be used as challenge encryptions. At the point of construction of these bitstrings, the adversary might not even know the challenge encryptions. Consequently, performing the reduction to the underlying assumption requires us to lazily construct them in order to decide as late as possible which encryption constitutes a challenge encryption.

2 Preliminaries

In this section, we present some definitions and conventions that will be used later on in the paper.

Notation. Let $\oplus$ denote the XOR operation, and let $\|$ denote concatenation. For a probabilistic algorithm B, let $y \leftarrow B(x)$ denote assigning the output of $B(x)$ to y. Let $\Pr[\pi : X]$ denote the probability that π holds after executing the instructions in X (which are of the form $y \leftarrow B(x)$). A function in n is negligible if it is in $n^{-\omega(1)}$. A function is non-negligible if it is not negligible. We formulate all our results for uniform adversaries, but they hold for nonuniform adversaries as well.

Definition 1 (Circuit). *A* circuit *is a Boolean circuit with $n_1 + \cdots + n_t$ input bits ($t \geq 0$) and m output bits. The circuit may have arbitrary fan-in and fan-out, AND-, OR- and NOT-gates, and—in the case of an encryption scheme in the random-oracle model—gates for querying the random oracle(s). We assume that a circuit is always encoded by explicitly specifying all its gates and the numbers $n_1, \dots, n_t, m$. The evaluation $f(x_1, \dots, x_t)$ of a circuit f on bitstrings $x_1, \dots, x_t$*

is defined as follows: Let x'_i be the result of truncating or padding x_i with 0^* to the length n_i. Then $f(x_1, \dots, x_t)$ is the result of evaluating f with input $x'_1 \| \dots \| x'_t$.[1]

Convention: Encryption is length-regular. For any encryption scheme, we impose the following assumption on the output of the encryption function Enc and the decryption function Dec: The length of the output of Enc depends only on the public key and the length of the message. The length of the output of Dec depends only on the *public* key and the length of the ciphertext. This can easily be achieved by suitable padding and encoding.

The OAEP scheme. The optimal asymmetric encryption padding (OAEP) scheme [11] constitutes a widely employed encryption scheme in the random oracle model based on a trapdoor 1-1 function.

Definition 2 (OAEP). *Let k denote the security parameter and let k_0 and k_1 be functions such that $k_0, k_1, k - k_0 - k_1$ are superlogarithmic. Assume a 1-1 trapdoor function f with domain $\{0,1\}^k = \{0,1\}^{k-k_0} \times \{0,1\}^{k_0}$. Let $G : \{0,1\}^{k_0} \to \{0,1\}^{k-k_0}$ and $H : \{0,1\}^{k-k_0} \to \{0,1\}^{k_0}$ denote random oracles. The public and secret key for the OAEP encryption scheme* (Enc, Dec) *consists of a public key and a trapdoor for f. An encryption $c = \mathrm{Enc}(pk, m)$ with $|m| = k - k_0 - k_1$ is computed as $r \leftarrow \{0,1\}^{k_0}$, $s := (m \| 0^{k_1}) \oplus G(r)$, $t := r \oplus H(s)$, $c := f_{pk}(s\|t)$.*

A decryption $\mathrm{Dec}(sk, c)$ is computed as $s\|t := f_{sk}^{-1}(c)$, $r := t \oplus H(s)$, $m\|z := s \oplus G(r)$ with $|s| = k - k_0$, $|t| = k_0$, $|m| = k - k_0 - k_1$ and $|z| = k_1$. If $z = 0^{k_1}$, the plaintext m is returned, otherwise the decryption fails with output $\perp$.

It has been shown in [21] that the OAEP scheme is IND-CCA2 secure in the random oracle model under the assumption that f fulfills the following Definition 3 of partial-domain one-wayness. They further showed that the RSA-trapdoor permutation, which is most commonly used for the OAEP scheme, is partial-domain one-way.

Definition 3 (Partial-Domain One-Wayness). *A 1-1 function $f\colon S \times T \to \mathrm{range}\, f$ with key generation* KeyGen_f *is* partial-domain one-way *if for any polynomial-time adversary A we have that*

$$\Pr\left[s = s' : pk \leftarrow \mathrm{KeyGen}_f, (s,t) \stackrel{\$}{\leftarrow} S \times T, s' \leftarrow A(pk, f_{pk}(s \parallel t))\right]$$

is negligible in k, where $A, \mathrm{KeyGen}_f, f, S, T$ depend on the the security parameter k. We sometimes call this probability the advantage of A.

3 The Definition of adKDM

We now present our definition of adKDM security. Since this definition incorporates several novel design choices and substantially differs from prior security

[1] Not granting a circuit access to the length of its arguments is not a restriction in our case, since this length will always be known in advance.

definitions for public key security, we do not immediately present the definition. Instead, we start with a direct adaption of an existing definition and show using an example why this adaption is not sufficient. We proceed with several plausible approaches for extending this adaption and explain why they fail. We finally present our definition of adKDM security and explain why it solves the problems observed with the tentative definitions discussed before.

Extending DKDM security. In [5] the security notion DKDM was proposed for the case of symmetric key-dependent encryptions. It is the strongest notion of KDM security considered so far; restating it one-to-one in the public-key setting would yield the following definition:[2]

Definition 4 (DKDM, public key setting – sketch). *The DKDM oracle maintains a sequence of key pairs pk_i, sk_i and a random challenge bit b. It answers to the following queries:*

- *$pk(j)$: Return pk_j.*
- *$reveal(j)$ where j has not been used in an $enc(j,\cdot)$ query: Return sk_j.*
- *$enc(j,f)$ where f is a circuit and j has not been used in a $reveal(j)$ query: Compute $m_0 := f(sk_1, sk_2, \dots)$, $m_1 := 0^{|m_0|}$ and encrypt $c := \mathrm{Enc}(pk_j, m_b)$. Return c.*
- *$dec(j,c)$ where c has not been returned by an $enc(j,\cdot)$ query with the same key index j: Return $\mathrm{Dec}(sk_j, c)$.*

A public key encryption scheme $(\mathrm{Enc}, \mathrm{Dec})$ *is DKDM secure if no polynomial time adversary interacting with the DKDM-oracle guesses b with probability non-negligibly greater than $\frac{1}{2}$.*

This definition is an almost immediate generalization of the IND-CCA definition to the multi-session setting (i.e., with several key pairs instead of only one). DKDM extends IND-CCA in two ways: First, the messages that are contained an $enc(\cdot,\cdot)$ encryption query may depend on all secret keys in the system. Second, one can reveal secret keys as long as the corresponding public keys have not been used for encrypting (otherwise one could decrypt a challenge ciphertext so that the definition cannot be met).

Although the notion of DKDM has been shown to be useful for soundness results for a specific class of protocols, it has obvious restrictions on the class of protocols considered. In particular, it is not allowed to reveal a key that has been used for encryption. The following simple protocol illustrates that this indeed constitutes a restriction: Alice holds two secret keys sk_1, sk_2 and a secret message m and sends the following messages to Bob:

$$c_1 := \mathrm{Enc}(pk_1, \mathrm{Enc}(pk_2, m\|sk_1\|sk_2)), \quad c_2 := \mathrm{Enc}(pk_2, \mathrm{Enc}(pk_1, m\|sk_1\|sk_2))$$

Then Bob chooses a value $i = 1, 2$ and Alice sends sk_i to Bob. We would intuitively expect the message m to stay secret since Bob learns at most one of the

[2] We have omitted one condition of their definition, namely that it should not be possible to generate a valid ciphertext without the knowledge of the secret key. This condition is not applicable to the public-key setting.

keys sk_1, sk_2. However, a direct reduction against DKDM security fails. Namely, we have basically four possibilities to construct the messages c_1, c_2 by querying the DKDM oracle (note that *enc* denotes the query to the adKDM oracle while Enc is the encryption algorithm):

(i) $c_1 := enc(1, g_1)$, $c_2 := enc(2, g_2)$ where g_1 and g_2 are circuits computing $\mathrm{Enc}(pk_2, m\|sk_1\|sk_2)$ and $\mathrm{Enc}(pk_1, m\|sk_1\|sk_2)$, respectively (given input (sk_1, sk_2)).
(ii) $c_1 := \mathrm{Enc}(pk_1, enc(2, g))$, $c_2 := \mathrm{Enc}(pk_2, enc(1, g))$ where g computes $m\|sk_1\|sk_2$.
(iii) $c_1 := \mathrm{Enc}(pk_1, enc(2, g))$, $c_2 := enc(2, g_2)$ where g and g_2 are as before.
(iv) $c_1 := enc(1, g_1)$, $c_2 := \mathrm{Enc}(pk_2, enc(1, g))$, where g and g_1 are as before.

Then, depending on the value of i chosen by Bob, we have to issue $reveal(i)$. In cases (i) and (ii), no reveal query is allowed since queries of the forms $enc(1, \cdot)$ and $enc(2, \cdot)$ have been performed which excludes reveal queries $reveal(1)$ and $reveal(2)$ by Definition 4. Similarly, in case (iii) we are not allowed to query $reveal(2)$, and in case (iv) we are not allowed to query $reveal(1)$. Thus in order to perform the first step, we have to know in advance what the value of i will be and to construct c_1, c_2 as in case (iii) or (iv), respectively. Of course, in the present example it is possible to save the reduction proof by guessing i; however, it is easy to thwart this possibility by performing many such games in parallel.[3] A natural approach to extend the definition of DKDM to this case would be to allow to even reveal keys sk_j that are used in encryption queries $enc(j, \cdot)$. However, a query $enc(j, \cdot)$ returns an encryption c of the message m_b. So given the secret key sk_j, we could easily determine m_b from c and therefore the challenge bit b. Therefore, we will have to distinguish between two types of encryption queries: A normal encryption query $enc(j, f)$ will return the encryption of $m_0 := f(sk_1, \dots)$ irrespective of the value of b. A challenge encryption query $challenge(j, f)$ returns m_b where m_0 is as for $enc(j, f)$ and $m_1 := 0^{|m_0|}$. This leads to the following tentative definition:

Definition 5 (KDM security – tentative). *The oracle $\mathcal{T}$ chooses a random bit b and accepts the following queries.*

- *$pk(j)$ and $reveal(j)$: Return pk_j and sk_j, respectively. $dec(j, c)$: Return* $\mathrm{Dec}(sk_j, c)$.
- *$enc(j, f(i_1, \dots, i_t))$ where f is a circuit: Compute $m_0 := f(sk_{i_1}, \dots, sk_{i_t})$ and return* $\mathrm{Enc}(pk_j, m_0)$.
- *$challenge(j, f(i_1, \dots, i_t))$: Compute m_0 as before, $m_1 := 0^{|m_0|}$ and return* $\mathrm{Enc}(pk_j, m_b)$.

---

[3] E.g., Alice sends $m_1^{(1)}, m_2^{(1)}, \dots, m_1^{(n)}, m_2^{(n)}$ with $m_1^{(\mu)} := \mathrm{Enc}(pk_1^{(\mu)}, \mathrm{Enc}(pk_2^{(\mu)}, m\|keys))$, $m_2^{(\mu)} := \mathrm{Enc}(pk_2^{(\mu)}, \mathrm{Enc}(pk_1^{(\mu)}, m\|keys))$ and $keys := sk_1^{(1)}\|sk_2^{(1)}\|\dots\|sk_1^{(n)}\|sk_2^{(n)}$. Then Bob chooses $i_1, \dots, i_n \in \{1, 2\}$ and Alice sends $sk_{i_1}^{(1)}, \dots, sk_{i_n}^{(n)}$. The fact that all keys are contained in each encryption also disables hybrid arguments. To the best of our knowledge, the security of this protocol cannot be reduced to DKDM security.

The oracle aborts in the following cases: reveal(j) *is queried but challenge*$(j,\cdot)$ *has been queried before. challenge*$(j,\cdot)$ *is queried but reveal*(j) *has been queried. dec*(j,c) *is queried but* c *was produced by challenge*$(j,\cdot)$*. A scheme is KDM secure if no polynomial-time adversary guesses* b *with probability noticeably larger than* $\frac{1}{2}$.

This definition might look appealing, but it cannot be met: For example, one could encrypt a challenge plaintext under pk_1 via the query *challenge*$(1,m)$, then encrypt the key sk_1 under pk_2 via $c := enc(2, sk_1)$, and finally reveal sk_2 via *reveal*(2).[4] This sequence of queries is not forbidden by Definition 5. Now we can compute sk_1 from c using sk_2 and then decrypt the challenge encryption using sk_1. This allows us to determine the bit b. Hence no encryption scheme can fulfill Definition 5. We hence have to relax the definition by excluding queries that would trivially allow to decrypt a challenge ciphertext. For this, we have to reject queries to the oracle that would allow the adversary to decrypt the challenge even in an ideal setting. For this, we keep track of the keys that the adversary can deduce from the queries made so far. We call this set *know* (the knowledge of the adversary) because it represents what the adversary knows *ideally.* The set *know* is inductively defined as follows: (a) If *reveal*(j) has been queried, then $j \in know$. (b) If $j \in know$, and a $enc(j, f(i_1, \dots, i_t))$ has been queried, then $i_1, \dots, i_n \in know$. (c) If $enc(j, f(i_1, \dots, i_t))$ has been queried and returned the ciphertext c, and $dec(j, c)$ has subsequently been queried, then $i_1, \dots, i_t \in know$. Roughly, we say that the adversary knows all keys that either were revealed or are contained in ciphertexts it could decrypt using keys it knows. We can now relax Definition 5 by disallowing queries that would allow the adversary to know a secret key for a challenge encryption.

Definition 6 (KDM security – tentative). *KDM security is defined as in Definition 5 except that the oracle* $\mathcal{T}$ *additionally aborts if a query would lead to the following situation: For some* $j \in know$*, a query challenge*$(j,\cdot)$ *has been performed (or is being performed).*

Introducing hidden encryptions. Definition 6, however, is still too weak to allow to adaptively choose which keys to reveal. In particular, the example protocol given above can still not be proven secure: When producing c_1, c_2 in a reduction proof, we have to decide which of the ciphertexts will be created by challenge encryptions (*challenge*$(\cdot,\cdot)$ queries) and which will be created by normal encryptions ($enc(\cdot,\cdot)$). Since we might have to invoke *reveal*(1) later, we may not use *challenge*$(1,\cdot)$ queries, and since we might have to invoke *reveal*(2), we may not use *challenge*$(2,\cdot)$ queries. But if no *challenge*$(\cdot,\cdot)$ query is issued, the oracle $\mathcal{T}$ never uses the bit b and thus the adversary cannot guess b.[5]

Handling adaptive revelations of keys hence requires to further extend our approach. A closer inspection reveals why we failed to prove the security of the example protocol: We had two possible ways to construct the ciphertext

[4] We use the shorthand m and sk_1 for the circuits outputting m and sk_1, respectively.

[5] Again, this problem might be remedied by guessing in advance whether sk_1 or sk_2 will be needed, but see footnote 3 for an example where guessing does not work.

c_1. Either (a) we could ask the oracle to produce $c_1' := \mathrm{Enc}(pk_2, m\|sk_1\|sk_2)$ and encrypt it ourselves using pk_1 to produce c_1. Or (b) we could request the ciphertext c_1 directly by sending to the oracle a circuit f that computes c_1' from sk_1, sk_2. In case (a), we are not allowed to reveal sk_2 since this would allow to decrypt c_1' and thus reveal m. In case (b), if we were to reveal sk_1 this would allow to decrypt c_1. As the plaintext c_1' for c_1 has been produced using a circuit f from sk_1, sk_2 and m, the oracle has no way of knowing that c_1' is actually an encryption of these values (this would require an analysis of the circuit to determine what it does) and thus has to consider the values sk_1, sk_2 and m to be leaked when c_1 is decrypted. Thus in case (b), we have to disallow the revelation of sk_1. This analysis shows that we need a way to send the following instructions to the oracle: "First produce the ciphertext c_1' as an encryption of $m\|sk_1\|sk_2$ (where $m\|sk_1\|sk_2$ is described by a suitable circuit). Do *not* return the value c_1' (as otherwise we would be in case (a)). Then produce the ciphertext c_1 by encrypting c_1'. Return c_1."

Given these instructions, the oracle has enough information to deduce that when revealing sk_1, the message m is still protected by the encryption c_1' using pk_2 (the details of this deduction process are discussed below). And if only sk_2 is revealed instead, c_1 cannot be a decryption and m is protected. Analogous reasoning applies to the construction of c_2.

Hence we have to define an oracle $\mathcal{T}$ that allows us to construct ciphertexts without revealing them. Instead, for each ciphertext we can adaptively decide whether to reveal it or whether we only use it inside other ciphertexts (that again may or may not be revealed). More concretely, whenever a query is issued to $\mathcal{T}$, instead of directly returning the result of that query, it is stored in some register $bits_h$ inside the oracle where h is a handle identifying the register. Only upon a special reveal query, the value $bits_h$ is returned to the adversary. A challenge encryption (i.e., one whose content depends on the challenge bit b) is then produced as follows: First produce a plaintext m (possibly using a circuit and depending on other hidden strings) and assign it to register $bits_{h_1}$. Then, depending on b, assign $bits_{h_1}$ or $0^{|bits_{h_1}|}$, respectively, to register $bits_{h_2}$ (using a special challenge query $h_2 \leftarrow C(h_1)$). Encrypt $bits_{h_2}$ using some key and assign the result to $bits_{h_3}$. Finally (optionally) reveal $bits_{h_3}$.[6]

These considerations lead to the following definition of the adKDM oracle (however, for the definition of adKDM security we will additionally define which sequences of queries are allowed):

Definition 7 (adKDM Oracle). *The* adKDM oracle $\mathcal{T}$ *maintains two partial functions cmd and bits (to increase readability we write $bits_h$ for $bits(h)$ and*

[6] This is, of course, not the only possible way to model challenge encryptions. One could, e.g., use a special command for producing a challenge encryption. However, we believe that the approach of being able to make challenge values out of arbitrary messages allows for more direct reductions in proofs. E.g., in our example protocol we could directly model the fact that m is the value that should remain hidden by using oracle call $h' \leftarrow C(h)$ when $bits_h$ contains m and then using $bits_{h'}$ instead of $bits_h$ in subsequent encryptions.

cmd_h for $cmd(h)$), a set Φ, a sequence of secret/public key pairs sk_i, pk_i ($i \in \mathbb{N}$) (which are generated when first accessed), and a bit b (the challenge bit*). The function cmd will store the structure of previous queries, the function bits will store the corresponding bitstrings, and Φ will keep track of query results that are revealed to the adversary. We will refer to the elements in the domain of cmd and bits as handles in the following. Upon the first activation, b is chosen uniformly from $\{0,1\}$, bits and cmd are initially undefined, and Φ is empty. The oracle responds to the following commands:*

- Encryption: $h' \leftarrow E(j,h)$ *where $cmd_{h'}$ has not been assigned, cmd_h has been assigned, and j is a key index: Set $bits_{h'} := \mathrm{Enc}(pk_j, bits_h)$ and $cmd_{h'} := E(j,h)$.*
- Decryption: $h' \leftarrow D(j,h)$ *where $cmd_{h'}$ has not been assigned, cmd_h has been assigned, and j is a key index: Set $bits_{h'} := \mathrm{Dec}(pk_j, bits_h)$, and $cmd_{h'} := D(j,h)$.*
- Circuit evaluation: $h' \leftarrow F(f, h_1, \dots, h_t)$ *where $cmd_{h'}$ has not been assigned, cmd_{h_i} has been assigned for all i, and f is a circuit with t arguments: Set $bits_{h'} := f(bits_{h_1}, \dots, bits_{h_t})$ and set $cmd_{h'} := F(f, h_1, \dots, h_t)$.*
- *Key request: $h' \leftarrow K(j)$ where $cmd_{h'}$ has not been assigned and j is a key index: Set $cmd_{h'} := K(j)$ and $bits_{h'} := sk_j$.*
- *Challenge: $h' \leftarrow C(h)$ where $cmd_{h'}$ has not been assigned and cmd_h has been assigned: Set $cmd_{h'} := C(h)$. If $b = 1$, set $bits_{h'} := bits_h$, otherwise set $bits_{h'} := 0^{|bits_h|}$.*
- *Reveal: $reveal(h)$ where cmd_h has been assigned: Add h to Φ and return $bits_h$.*
- Public key request: $pk(j)$ *where j is a key index: Return pk_j.*

The above commands in particular allow to assign a constant c to a handle h' by issuing $h' \leftarrow F(f)$ where f is a nullary circuit that returns c. We abbreviate this as $h' \leftarrow F(c)$. Note that the length of every bitstring is always known to the adversary, because Enc, Dec, and all f are length-regular.

The knowledge of the adversary. If $\mathcal{T}$ can be accessed in arbitrary ways, it is easy to determine b, e.g., querying $h_1 \leftarrow F(1)$, $h_2 \leftarrow C(h_1)$, $reveal(h_2)$ will return b. Thus we have to restrict the adversary to queries that will not trivially allow to deduce b. The necessary criteria are given below. In analogy to Definition 6 we do this by deriving a set *know* that characterizes what the adversary would ideally be able to know after the queries it performed. In contrast to Definition 6 the set *know* does not only contain keys, but the handles of all values produced by the oracle that the adversary would be able to know in an ideal setting. Intuitively, the knowledge *know* is defined by the following rules: All handles that the adversary requested (the set Φ) are considered known. If the decryption of a message is known, then that message is considered known.[7] If a circuit evaluation is known, all its arguments are considered known. If a challenge is

[7] It may seem surprising that by learning the result of a decryption we may learn something about the ciphertext. However, in fact we can get a single bit about the ciphertext, namely whether it is valid or not. Combining this with the application of circuits, we can in principle retrieve the full ciphertext.

known, the underlying message is considered known. If a key is known and an encryption of some message under that key is known, the message is considered known. And finally, if a decryption of some handle h_1 is known, and some handle h_2 evaluates to the same bitstring as h_1, and that handle h_2 resulted from an encryption of some message m, then that message m is considered known.

The last rule merits some additional explanation: The adversary may, e.g., construct and reveal an encryption c (assigned to some handle h_2) of some m. Then it constructs a circuit f that evaluates to c (by hard-coding c into f) and assigns $h_1 \leftarrow F(f)$. Now h_1 and h_2 refer to the same bitstring. By revealing the decryption of h_1, the adversary will then learn m. So after this sequence of queries, we have to ensure that m is considered known to the adversary. This is ensured by the last of the above rules. The following definition formally states the definition of the knowledge of the adversary.

Definition 8 (Knowledge). *For partial functions $cmd, bits$ and a set Φ, we define the* knowledge $know = know_{cmd,bits,\Phi}$ of the adversary *to be inductively defined as follows:*

- $\Phi \subseteq know$.
- *If $h' \in know$ and $cmd_{h'} = D(j, h)$ then $h \in know$.*
- *If $h' \in know$ and $cmd_{h'} = F(f, h_1, \dots, h_t)$ then $h_1, \dots, h_t \in know$.*
- *If $h' \in know$ and $cmd_{h'} = C(h)$ then $h \in know$.*
- *If $h' \in know$ and $cmd_{h'} = D(j, h_1)$, $bits_{h_1} = bits_{h_2}$ and $cmd_{h_2} = E(j, h_3)$ then $h_3 \in know$.*
- *If $h'_1, h'_2 \in know$ and $cmd_{h'_1} = K(j)$ and $cmd_{h'_2} = E(j, h)$ then $h \in know$.*

Note that *know* can be efficiently computed given Φ, *cmd*, and *bits* by adding handles to *know* according to the rules in Definition 8 until *know* does not grow any more. We are now ready to state the final definition of adKDM security. Intuitively, an encryption scheme is adKDM secure if the probability that the adversary guesses b correctly without performing a query that would even ideally allow it to retrieve a bitstring constructed using a $C(\cdot)$ query.

Definition 9 (Adaptive KDM Security (adKDM)). *An encryption scheme* (Enc, Dec) *is* adKDM secure *if for any polynomial-time adversary A there is a negligible function μ such that the following holds:*

$$\Pr[\mathsf{Guess} \wedge \neg\mathsf{Invalid}] \leq \tfrac{1}{2} + \mu(k)$$

where the events refer to an execution of A with input 1^k and oracle access to $\mathcal{T}_{(\mathrm{Enc},\mathrm{Dec})}$ and the events are defined as follows:

By Guess *we denote the event that the adversary outputs b where b is the challenge bit.*

By Invalid *we denote the event that $h \in know_{cmd,bits,\Phi}$ with cmd_h being of the form $C(\cdot)$.*

We will show that this definition can be met (at least in the random oracle model) in the next section. Clearly adKDM security implies DKDM security, since if we can only reveal keys that are not used for decrypting, the plaintexts of the challenge encryptions will never be in *know*.

Adaptive KDM security in the random oracle model. As the OAEP construction is formulated in the random oracle model, we need to know how Definition 9 needs to be adapted when used in the random oracle model. In this case, the adversary A is given access to the random oracle, and the circuits f passed to the adKDM oracle are allowed to contain invocations of the random oracle. Furthermore, the key generation, encryption, and decryption algorithms may contain invocations of the random oracle.

On simulation-based notions. We often motivated our design choices above by comparison with an ideal setting in which the adversary knows exactly the bitstrings associated with handles in *know*. This leads to the question whether it is possible to instead directly define security under key-dependent message attacks using a simulation-based definition, i.e., to define an ideal functionality that handles encryption and decryption queries in an ideal fashion. This approach has been successfully used to formulate IND-CCA security in the UC framework [16]. Their approach, however, strongly depends on the fact that the functionality only needs to output public keys and (fake) encryptions (secret keys are only implicitly present due to the ability to use the functionality to decrypt messages).[8] It is currently unclear how this approach could be extended to a functionality that can output secret keys. (It is of course possible to define a functionality that outputs secret keys as long as no encryption queries have been performed for that key, but this lead to a definition that is too weak to handle, e.g., our example protocol and that would roughly correspond to Definition 4.) This difficulty persists if we do not use the strong UC model [16] but instead the weaker stand-alone model as in [22, Chapter 7]. Consequently, although a simulation-based definition of KDM security might be very useful, it is currently unknown how to come up with such a definition.

4 OAEP Is adKDM-Secure

We now prove the adKDM security of the OAEP scheme for a partial-domain one-way function. In particular, since the RSA permutation is partial-domain one-way under the RSA assumption [21], the adKDM security of RSA-OAEP follows.

Theorem 10 (OAEP is adKDM secure). *If f is a partial-domain one-way trapdoor 1-1 function, then the OAEP scheme* $(\mathrm{Enc}, \mathrm{Dec})$ *based on f is adKDM secure in the random oracle model.*

To show this theorem, we first define an alternative characterization of partial-domain one-wayness.

[8] Technically, the reason is that a simulator has to be constructed that chooses the outputs of the functionality. As long as only public keys and ciphertexts are output, fake ciphertexts can be used since they cannot be decrypted. If the simulator had to generate secret keys, the fake ciphertexts could be decrypted and recognized.

Definition 11 (PD-Oracle). *The PD-oracle $\mathcal{P}_f$ for a trapdoor 1-1 function $f : S \times T \to \text{range}\, f$ (that may depend on a security parameter) maintains sequences of public/secret key pairs sk_i, pk_i (generated on first use). It understands the following queries:*

- *$pk(j)$ and $sk(j)$: Return pk_j or sk_j, respectively.*
- *$challenge(h, j)$: If h has already been used, ignore this query. Let $j_h := j$. Choose (s_h, t_h) uniformly from $S \times T$. Set $c_h := f_{pk_{j_h}}(s_h, t_h)$. Return c_h.*
- *$decrypt(h)$: Return (s_h, t_h).*
- *$xdecrypt(c, j)$ where $(c, j) \neq (c_h, j_h)$ for all h. Check whether $f^{-1}_{sk_j}(c) = (s_h, t_h)$ for some h. If so, return (s_h, t_h). Otherwise return $\perp$.*
- *$check(s)$: Return the first h with $s_h = s$. If no such h exists, return $\perp$.*

By PDBreak *we denote the event that a query $check(s)$ is performed such that*

- *The query returns $h \neq \perp$.*
- *No query $sk(j_h)$ and no query $decrypt(h)$ has been performed before the current query.*

Lemma 12. *If f is partial-domain oneway, then for any polynomial-time adversary A querying $\mathcal{P}_f$ we have that* $\Pr[\mathsf{PDBreak}]$ *is negligible in the security parameter.*

The proof is given in Appendix A. We additionally define a variant of the notion of knowledge as defined in Definition 8. We call this variant *lazy knowledge*.

Definition 13 (Lazy knowledge). *For partial functions $cmd, bits$ and a set Φ, we define the* lazy knowledge *$lknow = lknow_{cmd,bits,\Phi}$ of the adversary to be inductively defined as follows:*

- *$\Phi \subseteq lknow$.*
- *If $h' \in lknow$ and $cmd_{h'} = D(j, h)$ then $h \in lknow$.*
- *If $h' \in lknow$ and $cmd_{h'} = F(f, h_1, \dots, h_t)$ then $h_1, \dots, h_t \in lknow$.*
- *If $h' \in lknow$ and $cmd_{h'} = C(h)$ then $h \in lknow$.*
- *If $h', h_1, h_2 \in lknow$, $cmd_{h'} = D(j, h_1)$, $bits_{h_1} = bits_{h_2}$ and $cmd_{h_2} = E(j, h_3)$ then $h_3 \in lknow$.*
- *If $h'_1, h'_2 \in lknow$ and $cmd_{h'_1} = K(j)$ and $cmd_{h'_2} = E(j, h)$ then $h \in lknow$.*

The only change with respect to Definition 8 is that in the fifth rule we require that $h_1, h_2 \in lknow$. In Definition 13 all rules depend only on values $bits_h$ for which $h \in lknow$; thus one can efficiently compute $lknow$ without accessing $bits_h$ for values $h \notin lknow$ by adding handles to $lknow$ according to these rules until $lknow$ does not grow any further. We call this algorithm the *lazy knowledge algorithm*. Note that $lknow \subseteq know$.

Proof sketch (of Theorem 10). To prove Theorem 10 we give a sequence of games that transforms an attack against the adKDM security of the OAEP scheme into an attack against the PD-oracle. This proof sketch only contains the proof structure and highlights selected steps. The full proof is given in the full version [3].

GAME$_1$. The adversary A runs with access to the unmodified adKDM oracle $\mathcal{T}$. We assume that $\mathcal{T}$ invokes an encryption oracle $\mathcal{E}$ for encrypting and a decryption oracle $\mathcal{D}$ for decrypting. In particular, the encryption oracle $\mathcal{E}$ performs the following actions in the i-th query:

$$r \xleftarrow{\$} \{0,1\}^{k_0},\ g := G(r),\ s := (m\|0^{k_1}) \oplus g,\ h := H(s),\ t := r \oplus h,\ c := f_{pk}(s,t).$$

The decryption oracle $\mathcal{D}$ acts as follows, assuming key index j and ciphertext c:

- $(s,t) := f_{pk_j}^{-1}(c)$, $r := t \oplus H(s)$, $(m,z) := s \oplus G(r)$ with $|m| = k - k_1 - k_0$ and $|z| = k_1$.
- If $z = 0^{k_1}$, return m, otherwise return $\bot$.

GAME_2. We change the encryption oracle to first choose the ciphertext c and then compute the values s,t,r,h,t,g from it, i.e., upon the i-th query the encryption oracle does the following:

$$(s,t) \xleftarrow{\$} \{0,1\}^{k-k_0} \times \{0,1\}^{k_0},\ c \xleftarrow{\$} f_{pk}(s,t),\ r \xleftarrow{\$} \{0,1\}^{k_0},\ h := r \oplus t,\ g := (m\|0^{k_1}) \oplus s$$

In particular, the values h and g are not retrieved from the oracles G and H any more. In order to keep the distribution of the values c,s,t,r,h,t,g consistent with the answers of the oracles G and H, the oracles G and H are additionally modified to return the values g and h chosen by the encryption oracle. We show that the probability of a successful attack is modified only by a negligible amount with respect to GAME_1.

GAME_3. We now change the definition of what constitutes a successful attack. In GAME_1–GAME_2, we considered it a successful attack if the adversary guessed the bit b chosen by the adKDM oracle $\mathcal{T}$ without performing queries such that the knowledge in the sense of Definition 8 would contain a handle corresponding to a query of the form $C(\cdot)$; see Definition 9.

Now, in GAME_3, we consider it to be a successful attack if the adversary guessed b without performing queries such that the *lazy* knowledge in the sense of Definition 13 does not contain a handle corresponding to a query $C(\cdot)$. Since the lazy knowledge is a subset of the knowledge, this represents a weakening of the restrictions put on the adversary. Thus the probability of an attack in GAME_3 is upper-bound by the probability of an attack in GAME_2.

GAME_4. This step is arguably the most important step in the proof. In GAME_3, bitstrings $bits_h$ associated to handles h are often computed but never used. For example, the adversary might perform a query $h \leftarrow E(\dots)$ and never use the handle h again. More importantly, however, even if the adversary performs a query $h' \leftarrow E(j,h)$ for that handle h, the value $bits_h$ does not need to be computed due to the following observation: The encryption oracle as introduced in GAME_2 chooses the ciphertext c at random. The value g (which is the only value depending on the plaintext m) is only needed for suitably reprogramming the oracles G (namely such that $G(r) = g$). Thus we can delay the computation of g until G is queried at position r. Thus in case of a query $h' \leftarrow E(j,h)$, the value $m = bits_h$ is not needed for computing $bits_{h'}$. We use this fact to rewrite the whole game GAME_3 such that it only computes a value $bits_h$ when it is actually needed for computing some output sent to the adversary or for computing the lazy knowledge.

The bit b is only used in this game if a value $bits_h$ is computed that corresponds to a query $h \leftarrow C(\cdot)$. If this is not the case, the communication between the adversary and $\mathcal{T}$ is independent of b. Hence, for proving that the probability of attack in the sense of GAME$_3$ is only negligibly larger than $\frac{1}{2}$ (which then shows Theorem 10), it is sufficient to show that only with negligible probability, a value $bits_h$ is computed such that h is not in the lazy knowledge. Namely, as long as no such value $bits_h$ is computed, the adversary cannot have a higher probability in guessing b than $\frac{1}{2}$ unless $h \in lknow$.

GAME$_5$. Now we replace the decryption oracle by a plaintext extractor. More concretely, the decryption oracle performs the following steps when given a ciphertext c:

(a) First, it checks whether $c = f_{pk}(s,t)$ for some pair (s,t) generated by the encryption oracle.[9] Then values (s,t) are known such that $f_{pk}(s,t) = c$, and the oracle can decrypt c without accessing the secret key sk.

(b) Otherwise, it checks whether for some s that has been computed by the encryption oracle, there exists a value t such that $f_{pk}(s,t) = c$. (Doing this efficiently requires the secret key; otherwise we had to iterate over all possible values t.) If so, reject the ciphertext.

(c) Otherwise, for all values s, r that have been generated so far, compute $t := r \oplus H(s)$ and $(m,z) = s \oplus G(r)$. Then check whether $f_{pk}(s,t) = c$ and $z = 0^{k_1}$. If so, return m. Otherwise reject the ciphertext.

We can show that this plaintext extractor is a good simulation of the original decryption oracle (in particular, the adversary is able to produce an s triggering rejection in (b) only if the decryption would fail anyway). Thus the probability that a value $bits_h$ is computed such that h is not in the lazy knowledge does not increase by a non-negligible amount.

GAME$_6$. In this final step, we modify GAME$_5$ not to generate the public/secret key pairs on its own, but to use the PD-oracle $\mathcal{P}$ defined in Definition 11. In particular, we make the following changes:

- When the secret key sk_j is needed (for computing $bits_h$ for a $h \leftarrow K(j)$ query), query $sk(j)$ from $\mathcal{P}$.
- When producing a ciphertext $bits_{h'}$ (that are produced just to be random images of f_{pk}), use $challenge(h', j)$ where j is the corresponding key index.
- In the decryption oracle, for checking the condition (a) in GAME$_5$, we distinguish two cases. If c was produced by the encryption oracle the decryption oracle sends a $decrypt(h)$ to $\mathcal{P}$ where h is the query where c was produced. Otherwise it sends an $xdecrypt(c, j)$ query to $\mathcal{P}$ where j is the index of the key used in the decryption query. In both cases, if the check in (a) would have succeeded, $\mathcal{P}$ will send back a preimage (s,t) of c.
- The check (b) is performed by sending $check(s)$ to $\mathcal{P}$.

A case analysis reveals that if a value $bits_h$ is computed such that h is not in the lazy knowledge, then the event PDBreak (as in Definition 11) occurs. By

[9] This does not imply that c has been generated by the encryption oracle since the encryption oracle might have used a different public key pk at that time.

Lemma 12 this can only happen with negligible probability. Thus no value $bits_h$ is computed such that h is not in the lazy knowledge, and therefore the advantage of the adversary is negligible (as discussed in GAME$_4$). □

References

1. Abadi, M., Jürjens, J.: Formal eavesdropping and its computational interpretation. In: Proc. 4th International Symposium on Theoretical Aspects of Computer Software (TACS), pp. 82–94 (2001)
2. Abadi, M., Rogaway, P.: Reconciling two views of cryptography: The computational soundness of formal encryption. In: Watanabe, O., Hagiya, M., Ito, T., van Leeuwen, J., Mosses, P.D. (eds.) TCS 2000. LNCS, vol. 1872, pp. 3–22. Springer, Heidelberg (2000)
3. Backes, M., Dürmuth, M., Unruh, D.: OAEP is secure under key-dependent messages (2008), `http://www.infsec.cs.uni-sb.de/~unruh/publications/backes08oaep.html`
4. Backes, M., Pfitzmann, B.: Symmetric encryption in a simulatable Dolev-Yao style cryptographic library. In: Proc. 17th IEEE Computer Security Foundations Workshop (CSFW), pp. 204–218 (2004)
5. Backes, M., Pfitzmann, B., Scedrov, A.: Key-dependent message security under active attacks – BRSIM/UC-soundness of symbolic encryption with key cycles. In: Proc. of 20th IEEE Computer Security Foundation Symposium (CSF) (June 2007); Preprint on IACR ePrint 2005/421
6. Backes, M., Pfitzmann, B., Waidner, M.: A composable cryptographic library with nested operations (extended abstract). In: Proc. 10th ACM Conference on Computer and Communications Security, pp. 220–230 (January 2003); Full version in IACR Cryptology ePrint Archive 2003/015
7. Beaver, D., Haber, S.: Cryptographic protocols provably secure against dynamic adversaries. In: Rueppel, R.A. (ed.) EUROCRYPT 1992. LNCS, vol. 658, pp. 307–323. Springer, Heidelberg (1993)
8. Bellare, M., Desai, A., Jokipii, E., Rogaway, P.: A concrete security treatment of symmetric encryption. In: Proc. 38th IEEE Symposium on Foundations of Computer Science (FOCS), pp. 394–403 (1997)
9. Bellare, M., Namprempre, C.: Authenticated encryption: Relations among notions and analysis of the generic composition paradigm. In: Okamoto, T. (ed.) ASIACRYPT 2000. LNCS, vol. 1976, pp. 531–545. Springer, Heidelberg (2000)
10. Bellare, M., Rogaway, P.: Entity authentication and key distribution. In: Stinson, D.R. (ed.) CRYPTO 1993. LNCS, vol. 773, pp. 232–249. Springer, Heidelberg (1994)
11. Bellare, M., Rogaway, P.: Optimal asymmetric encryption. In: De Santis, A. (ed.) EUROCRYPT 1994. LNCS, vol. 950, pp. 92–111. Springer, Heidelberg (1995)
12. Bellare, M., Rogaway, P.: Encode-then-encipher encryption: How to exploit nonces or redundancy in plaintexts for efficient constructions. In: Okamoto, T. (ed.) ASIACRYPT 2000. LNCS, vol. 1976, pp. 317–330. Springer, Heidelberg (2000)
13. Black, J., Rogaway, P., Shrimpton, T.: Encryption-scheme security in the presence of key-dependent messages. In: Proc. 9th Annual Workshop on Selected Areas in Cryptography (SAC), pp. 62–75 (2002)
14. Boneh, D., Halevi, S., Hamburg, M., Ostrovsky, R.: Circular-secure encryption from decision diffie-hellman. In: Wagner, D. (ed.) CRYPTO 2008. LNCS, vol. 5157, pp. 108–125. Springer, Heidelberg (2008)

15. Camenisch, J., Lysyanskaya, A.: An efficient system for non-transferable anonymous credentials with optional anonymity revocation. In: Pfitzmann, B. (ed.) EUROCRYPT 2001. LNCS, vol. 2045, pp. 93–118. Springer, Heidelberg (2001)
16. Canetti, R.: Universally composable security: A new paradigm for cryptographic protocols. In: Proc. 42nd IEEE Symposium on Foundations of Computer Science (FOCS), pp. 136–145 (2001); Extended version in Cryptology ePrint Archive, Report 2000/67
17. Canetti, R., Herzog, J.: Universally composable symbolic analysis of mutual authentication and key exchange protocols. In: Halevi, S., Rabin, T. (eds.) TCC 2006. LNCS, vol. 3876, pp. 380–403. Springer, Heidelberg (2006)
18. Cortier, V., Warinschi, B.: Computationally sound, automated proofs for security protocols. In: Proc. 14th European Symposium on Programming (ESOP), pp. 157–171 (2005)
19. Dolev, D., Dwork, C., Naor, M.: Nonmalleable cryptography. SIAM Journal on Computing 30(2), 391–437 (2000)
20. Dolev, D., Yao, A.C.: On the security of public key protocols. IEEE Transactions on Information Theory 29(2), 198–208 (1983)
21. Fujisaki, E., Okamoto, T., Pointcheval, D., Stern, J.: RSA-OAEP is secure under the RSA assumption. Journal of Cryptology 17(2), 81–104 (2004)
22. Goldreich, O.: Foundations of Cryptography. Basic Applications, vol. 2. Cambridge University Press, Cambridge (May 2004)
23. Goldwasser, S., Micali, S.: Probabilistic encryption. J. Comput. Syst. Sci. 28, 270–299 (1984)
24. Halevi, S., Krawczyk, H.: Security under key-dependent inputs. In: Proc. of the 14th ACM Conference on Computer and Communications Security (to appear, 2007); Preprint on IACR ePrint 2007/315
25. Hofheinz, D., Unruh, D.: Towards key-dependent message security in the standard model (August 2007); Preprint on IACR ePrint 2007/333
26. Laud, P.: Semantics and program analysis of computationally secure information flow. In: Proc. 10th European Symposium on Programming (ESOP), pp. 77–91 (2001)
27. Laud, P.: Symmetric encryption in automatic analyses for confidentiality against active adversaries. In: Proc. 25th IEEE Symposium on Security & Privacy, pp. 71–85 (2004)
28. Luby, M.: Pseudorandomness and Cryptographic Applications. Princeton Computer Society Notes, Princeton (1996)
29. Micciancio, D., Warinschi, B.: Soundness of formal encryption in the presence of active adversaries. In: Naor, M. (ed.) TCC 2004. LNCS, vol. 2951, pp. 133–151. Springer, Heidelberg (2004)
30. Rackoff, C., Simon, D.R.: Non-interactive zero-knowledge proof of knowledge and chosen ciphertext attack. In: Feigenbaum, J. (ed.) CRYPTO 1991. LNCS, vol. 576, pp. 433–444. Springer, Heidelberg (1992)
31. Shannon, C.E.: Communication theory of secrecy systems. Bell System Technical Journal 28(4), 656–715 (1949)
32. Yao, A.C.: Theory and applications of trapdoor functions. In: Proc. 23rd IEEE Symposium on Foundations of Computer Science (FOCS), pp. 80–91 (1982)

A Proof of Lemma 12

Proof. Given an adversary A against the PD-Oracle $\mathcal{P}$ we construct an adversary B against partial-domain one-wayness of the underlying function f as follows.

The machine B that implements the PD-oracle with slight changes: Let q be an upper bound on the number of queries performed by A. Then B gets as input a key pair pk^*, sk^*, values $(s^*, t^*) \in S \times T$ and a value c^*. Let j^* be the i_1-th key index that is used in A's queries, and let h^* the i_2-th handle that is used in a query of the form $challenge(h, j^*)$. Then B answers to A's queries as follows (for simplicity, if we write f_{pk}^{-1} we mean an application of the secret key sk):

- $pk(j)$: If $j = j^*$, return pk^*, otherwise return pk_j.
- $sk(j)$: If $j = j^*$, return sk^*, otherwise return sk_j.
- $challenge(h, j)$: If h has already been used, ignore this query.
  - If $h = h^*$ (and thus also $j = j^*$) then set $c_h := c^*$ and return c_h.
  - If $h \neq h^*$ then choose (s_h, t_h) uniformly from $S \times T$. Set $c_h := f_{pk_{j_h}}(s_h, t_h)$. Return c_h.
- $decrypt(h)$: If $h = h^*$, return (s^*, t^*). Otherwise return (s_h, t_h).
- $xdecrypt(c, j)$ where $(c, j) \neq (c_h, j_h)$ for all h. This is equivalent to the following:
  - If $j \neq j^*$ then check whether $f_{pk_j}^{-1}(c) = (s_h, t_h)$ for some $h \neq h^*$ or $f_{pk_{j^*}}(f_{pk_j}^{-1}(c)) = c_{h^*}$. If so, return $f_{pk_j}^{-1}(c)$. Otherwise, return $\perp$.
  - If $j = j^*$ then test if $f_{pk_j}(s_h, t_h) = c$ for any $h \neq h^*$. If such an h exists, output (s_h, t_h). Otherwise, return $\perp$.
- $check(s)$: If $s = s_h$ for some h, return the first h with $s_h = s$. If $sk(j^*)$ or $decrypt(h^*)$ has been queried, check whether $s = s^*$. If so, return h^*.

We claim that this machine B behaves identically to the PD-oracle $\mathcal{P}$ until the event PDBreak occurs and that A's view is independent of i_1, i_2 until the event PDBreak occurs (assuming that the inputs sk^*, pk^* are an honestly generated key pair, (s^*, t^*) is uniformly distributed on $S \times T$ and $c^* = f_{pk^*}(s^*, t^*)$). For the queries pk, sk, $challenge$, and $decrypt$ this is straightforward. In the case of $xdecrypt$ we distinguish two cases: For $j \neq j^*$, the check performed is equivalent to checking whether $f_{pk_j}^{-1}(c) = (s_h, t_h)$ for some $h \neq h^*$ or $f_{pk_j}^{-1}(c) = (s^*, t^*)$ and then returning h or h^*, respectively. Thus in this case the answer to the query $xdecrypt$ is the same as that the PD-oracle $\mathcal{P}$ would give. For $j = j^*$, in comparison to $\mathcal{P}$, the check whether $f_{pk_j}(s^*, t^*) = c$ is missing. However, if this check held true, we would have that $(c, j) = (c^*, j^*)$ which is excluded. To see that the query $check(s)$ gives the same answers in B and $\mathcal{P}$ until PDBreak occurs, note that the only case where $check(s)$ would give another answer in $\mathcal{P}$ is when $s = s^*$ but neither $sk(j^*)$ nor $decrypt(h^*)$ have been queried. However, in this case h^* would be returned in $\mathcal{P}$, thus PDBreak occurs.[10] So altogether, we have that B behaves identically to $\mathcal{P}$ and A's view is independent of i_1, i_2 until the event PDBreak occurs. By $\mathsf{PDBreak}_{i'_1, i'_2}$ denote the event that $check(s)$ is queried with $s = s_h$ where h is the i'_2-th handle used by A, and no query $sk(j_h)$ or $decrypt(h)$ has been performed where j_h is the i'_1-th key index used by A. Obviously, if

[10] In slight abuse of notation, we denote by PDBreak not the event that $h \neq \perp$ is returned without a query of $sk(j_h)$ or $decrypt(h)$, but that some $check(s)$ is queried such that $s = s_h$ and no query $sk(j_h)$ or $decrypt(h)$ has been performed. Since for $\mathcal{P}$ these are equivalent, it is enough to show the lemma w.r.t. this slightly changed definition.

$\mathsf{PDBreak}$ occurs, then $\mathsf{PDBreak}_{i'_1,i'_2}$ occurs for some $i'_1, i'_2 \in \{1,\dots,q\}$. Since the view of A is independent of i_1, i_2, we have that $\Pr[\mathsf{PDBreak}_{i_1,i_2}] \geq \frac{1}{q^2}\Pr[\mathsf{PDBreak}]$. So it is enough to show that $\Pr[\mathsf{PDBreak}_{i_1,i_2}] =: \varepsilon$ is negligible. Observe that in the description of B, in case of the event $\mathsf{PDBreak}_{i_1,i_2}$ the inputs sk^*, s^*, h^* are never accessed. So if we run B with the inputs sk^*, s^*, h^* set to $\bot$, $\mathsf{PDBreak}_{i_1,i_2}$ still occurs with probability at least ε. Further, $\mathsf{PDBreak}_{i_1,i_2}$ implies that $check(s)$ is called an s satisfying $f^{-1}(c^*) = \bot$. So if let B output one of the values s used in $check(s)$ queries (randomly chosen), we break the partial-domain one-wayness of f with probability at least ε/q. Thus by contradiction, ε must be negligible. Thus $\Pr[\mathsf{PDBreak}]$ is negligible in an execution of B and thus also in one of $\mathcal{P}$. □

Cryptanalysis of Sosemanuk and SNOW 2.0 Using Linear Masks

Jung-Keun Lee, Dong Hoon Lee, and Sangwoo Park

ETRI Network & Communication Security Division,
909 Jeonmin-dong, Yuseong-gu, Daejeon, Korea

Abstract. In this paper, we present a correlation attack on Sosemanuk with complexity less than 2^{150}. Sosemanuk is a software oriented stream cipher proposed by Berbain et al. to the eSTREAM call for stream cipher and has been selected in the final portfolio. Sosemanuk consists of a linear feedback shift register(LFSR) of ten 32-bit words and a finite state machine(FSM) of two 32-bit words. By combining linear approximation relations regarding the FSM update function, the FSM output function and the keystream output function, it is possible to derive linear approximation relations with correlation $-2^{-21.41}$ involving only the keystream words and the LFSR initial state. Using such linear approximation relations, we mount a correlation attack with complexity $2^{147.88}$ and success probability 99% to recover the initial internal state of 384 bits. We also mount a correlation attack on SNOW 2.0 with complexity $2^{204.38}$.

Keywords: stream cipher, Sosemanuk, SNOW 2.0, correlation attack, linear mask.

1 Introduction

Sosemanuk[3] is a software oriented stream cipher proposed by Berbain et al. to the eSTREAM call for stream cipher and has been selected in the final portfolio. The merits of Sosemanuk has been recognized as its considerable security margin and moderate performance[2].

Sosemanuk is based on the stream cipher SNOW 2.0[11] and the block cipher Serpent[1]. Though SNOW 2.0 is a highly reputed stream cipher, it is vulnerable to linear distinguishing attacks using linear masks[14,15]. To strengthen against linear distinguishing attacks, Sosemanuk applies the multiplication modulo 2^{32} with a bit rotation in the FSM update function and a Serpent S-box in bit slice mode in the keystream output function. As of now, there are no known attacks against Sosemanuk with complexity less than 2^{226}[5].

Linear masking has been used in the linear distinguishing attacks on word-based stream ciphers such as SNOW 1.0[9], SNOW 2.0, NLS[7], and Dragon[8]. Coppersmith et al.[9] presented a linear distinguishing attack on SNOW 1.0. They identified linear approximation relations of large correlation involving only the LFSR states and the keystream words. Then using simple bitwise recurrence relations between the LFSR state words, they were able to mount a linear

J. Pieprzyk (Ed.): ASIACRYPT 2008, LNCS 5350, pp. 524–538, 2008.
© International Association for Cryptologic Research 2008

distinguishing attack on SNOW 1.0. Watanabe et al.[15] presented a linear distinguishing attack on SNOW 2.0 and then Nyberg and Wallén[14] refined the attack.

On the other hand, Berbain et al.[4] presented a correlation attack on Grain using linear approximation relations between the initial LFSR state and the keystream bits to recover the initial LFSR state. As to solving systems of linear approximation equations, similar technique was used in [6] and iterative decoding technique was used in [12].

In this paper, combining the linear masking method with the techniques in [4] using fast Walsh transform to recover the initial LFSR state of Grain, we mount a correlation attack on Sosemanuk. The time, data and memory complexity are all less than 2^{150}.

This paper is organized as follows. In Sect. 2, we present a description of Sosemanuk. In Sect. 3, we show how to get approximation relations between the initial LFSR state and the keystream words. In Sect. 4, we describe the attack using the approximation relations. In Sect. 5, we present simulation results. In Sect. 6, we present a correlation attack on SNOW 2.0. We conclude in Sect. 7.

2 Preliminaries

2.1 Notations and Definitions

We define the correlation of a function with respect to masks as follows. Let $f : (\mathrm{GF}(2)^n)^k \rightarrow \mathrm{GF}(2)^n$ be a function and let $\Gamma_0, \Gamma_1, \ldots, \Gamma_k$ be n-bit masks. Then the correlation of f with respect to the tuple $(\Gamma_0; \Gamma_1, \ldots \Gamma_k)$ of masks is defined as

$$\mathrm{c}_f(\Gamma_0; \Gamma_1, \ldots, \Gamma_k) := 2\,\mathrm{Prob}(\Gamma_0 \cdot f(x_1, \ldots, x_k) = \Gamma_1 \cdot x_1 \oplus \ldots \oplus \Gamma_k \cdot x_k) - 1,$$

where $\cdot$ represents the inner product which will be omitted henceforth. We also define the correlation of an approximation relation as

$$2\,\mathrm{Prob}(\text{the approximation holds}) - 1 \ .$$

The following notations will be used in the following sections.

- $\mathrm{wt}(x)$: the Hamming weight of a binary vector or a 32-bit word x
- $\boxplus$: addition modulo 2^{32}
- $\times$: multiplication modulo 2^{32}
- $[i_1, \ldots, i_m]$: the 32-bit linear mask $2^{i_1} + \ldots + 2^{i_m}$ ($i_1, \ldots, i_m$ are distinct integers in between 0 and 31.)
- $\mathrm{c}_+(\Gamma_0; \Gamma_1, \ldots, \Gamma_m)$: the correlation of $f(x_1, \ldots, x_m) = x_1 \boxplus \ldots \boxplus x_m$ with respect to the tuple $(\Gamma_0; \Gamma_1, \ldots \Gamma_k)$ of 32-bit masks
- $\mathrm{c2}_+(\Gamma) := \mathrm{c}_+(\Gamma; \Gamma, \Gamma)$ for 32-bit linear mask Γ
- $\mathrm{c3}_+(\Gamma) := \mathrm{c}_+(\Gamma; \Gamma, \Gamma, \Gamma)$ for 32-bit linear mask Γ
- $\mathrm{c}_T(\Gamma_0; \Gamma_1)$: the correlation of $\mathrm{Trans}(x)$ with respect to the tuple $(\Gamma_0; \Gamma_1)$ of 32-bit masks
- $\mathrm{c2}_T(\Gamma) = \mathrm{c}_T(\Gamma; \Gamma)$ for 32-bit linear mask Γ
- $x_{(j)}$: j-th least significant bit of a nibble, a byte or a 32-bit word x

2.2 Description of Sosemanuk

The structure of Sosemanuk[3] is depicted in Fig. 1. Sosemanuk consists of three main components: a 10-word linear feedback shift register, a 2-word finite state machine, and a nonlinear output function. Sosemanuk is initialized with the key of length in between 128 and 256 and the 128-bit initialization value. The output of the cipher is a sequence of 32-bit keystream words $(z_t)_{t\geq 1}$. The LFSR state at time t is denoted by $LR^t = (s_{t+1}, s_{t+2}, \ldots, s_{t+10})$.($t = 0$ designates the time after initialization.) The LFSR is updated using the recurrence relation

$$s_{t+10} = s_{t+9} \oplus \alpha^{-1} s_{t+3} \oplus \alpha s_t \text{ for all } t \geq 1,$$

where α is a zero of the primitive polynomial

$$P(X) = X^4 + \beta^{23} X^3 + \beta^{245} X^2 + \beta^{48} X + \beta^{239}$$

on GF$(2^8)(X)$ and GF$(2^8) =$ GF$(2)[\gamma]$, where γ is a zero of the primitive polynomial

$$Q(X) = X^8 + X^7 + X^5 + X^3 + 1$$

on GF$(2)(X)$. The FSM state at time t is denoted by $(R1_t, R2_t)$. The FSM is updated as follows.

$$\begin{aligned} R1_t &= R2_{t-1} \boxplus (s_{t+1} \oplus \text{lsb}(R1_{t-1}) s_{t+8}), \\ R2_t &= \text{Trans}(R1_{t-1}) = (M \times R1_{t-1})^{<<<7}, \end{aligned}$$

where $M =$ 0x54655307. The FSM has output

$$f_t = (s_{t+9} \boxplus R1_t) \oplus R2_t \ .$$

The keystream words are obtained as follows.

$$\begin{aligned} &(z_{t+3}, z_{t+2}, z_{t+1}, z_t) = Serpent1(f_{t+3}, f_{t+2}, f_{t+1}, f_t) \oplus (s_{t+3}, s_{t+2}, s_{t+1}, s_t) \\ &(t \equiv 1 (\text{mod } 4)) \end{aligned}$$

where $Serpent1$ denotes the Serpent S-box S_2 applied in bit slice mode. Four words are output per 4 LFSR clockings.

3 Linear Approximations

In this section, we get linear approximation relations involving only the LFSR states and the keystream words with non-negligible correlation by approximating the FSM update functions, the FSM output functions, and the keystream output function using linear masks with non-negligible correlation.

Let $a_t = \text{lsb}(R1_t)$. We consider the following approximations using 32-bit linear masks Γ by replacing all operations (modular additions and the Trans function) by XORs in the FSM update function and the FSM output function:

$$\begin{aligned} \Gamma R1_{t+1} &= \Gamma R2_t \oplus \Gamma(s_{t+2} \oplus a_t s_{t+9}), \\ \Gamma R2_{t+1} &= \Gamma R1_t, \\ \Gamma f_t &= \Gamma s_{t+9} \oplus \Gamma R1_t \oplus \Gamma R2_t, \\ \Gamma f_{t+1} &= \Gamma s_{t+10} \oplus \Gamma R1_{t+1} \oplus \Gamma R2_{t+1} \ . \end{aligned}$$

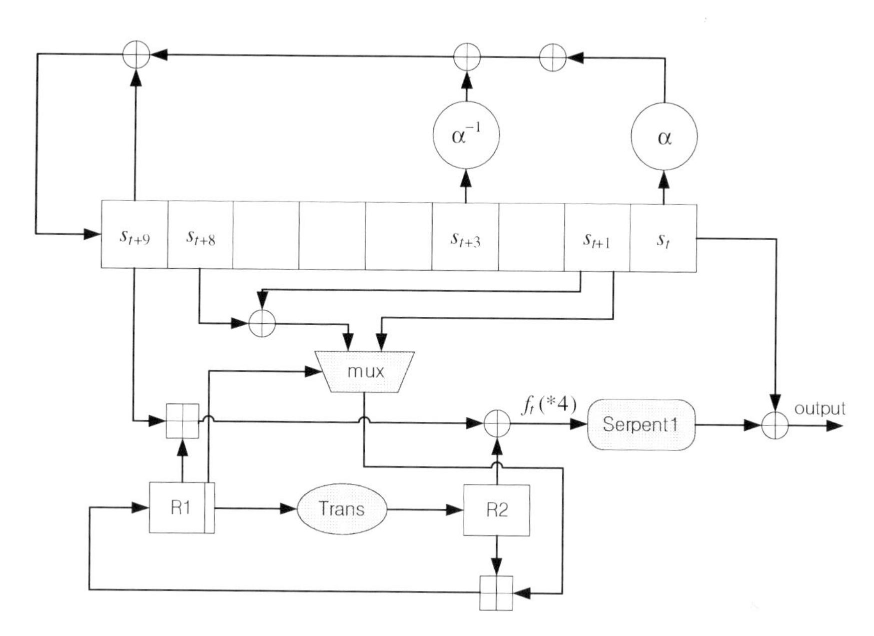

Fig. 1. The Structure of Sosemanuk

XORing the above relations and applying the Piling-Up Lemma, we have the approximation

$$\Gamma(f_t \oplus f_{t+1}) = \Gamma s_{t+2} \oplus a_t \Gamma s_{t+9} \oplus \Gamma s_{t+9} \oplus \Gamma s_{t+10} \tag{1}$$

with correlation $\mathrm{c}2_+(\Gamma)^3 \mathrm{c}2_T(\Gamma)$ *assuming that the four linear approximations are independent.*

However the way of computing the correlation as above is not accurate since the approximation relations have high dependencies. For example, approximations of two modular additions with correlations c_1, c_2 do not necessarily yield an approximation with correlation $c_1 c_2$. So we need to consider approximation relations which do not have obvious dependencies. We have the following equations regarding the internal states and keystream words:

$$\begin{aligned} f_t \oplus R2_t &= s_{t+9} \boxplus \mathrm{Trans}^{-1}(R2_{t+1}), \\ f_{t+1} \oplus R2_{t+1} &= s_{t+10} \boxplus (R2_t \boxplus (s_{t+2} \oplus a_t s_{t+9})) \ . \end{aligned}$$

We consider the following associated approximation relations

$$\begin{aligned} \Gamma f_t \oplus \Gamma R2_t &= \Gamma s_{t+9} \oplus \Gamma R2_{t+1}, \\ \Lambda f_{t+1} \oplus \Lambda R2_{t+1} &= \Lambda s_{t+10} \oplus \Lambda R2_t \oplus \Lambda s_{t+2} \oplus a_t \Lambda s_{t+9} \ . \end{aligned}$$

where Γ and Λ are linear masks as depicted in Fig. 2. The correlations of the above approximations are

$$\sum_{\Phi} \mathrm{c}_+(\Gamma; \Gamma, \Phi) \mathrm{c}_T(\Gamma; \Phi)$$

and $\mathrm{c}3_+(\Lambda)$, respectively.

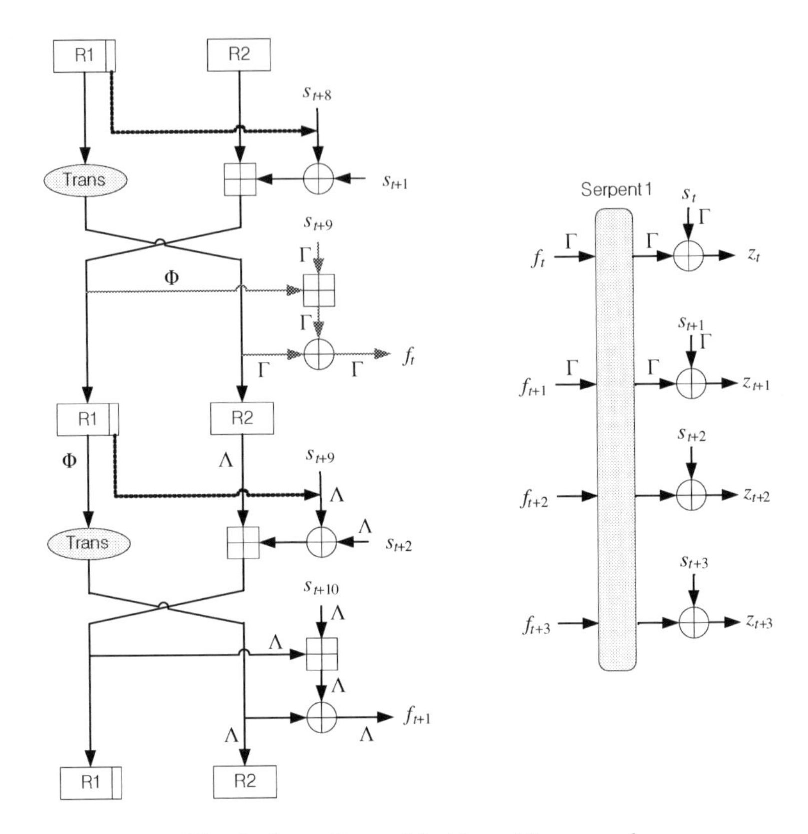

Fig. 2. Some Linear Masking of Sosemanuk

Note that the first correlation is a composite correlation of the function $x_2 = \mathrm{Trans}^{-1}(x_3)$ and the function $y = x_1 \boxplus x_2$ with respect to $(\Gamma; \Gamma, \Gamma)$ which can be computed as a sum of partial correlations [13, Theorem 3][14]. So if we let $\Gamma = \Lambda$, we have the same approximation relation (1) with correlation

$$\mathrm{c3}_{+}(\Gamma) \sum_{\Phi} \mathrm{c}_{+}(\Gamma; \Gamma, \Phi) \mathrm{c}_{T}(\Gamma; \Phi) \ .$$

In order to remove terms involving f_t and f_{t+1} in (1), we will utilize a linear approximation relation regarding the keystream output function that comes from the third S-box S_2 of the block cipher Serpent in bit slice mode.

```
unsigned char S2[16] = {8,6,7,9,3,12,10,15,13,1,14,4,0,11,5,2}
```

S_2 has maximal linear correlation $\frac{1}{2}$. Regarding the function $y = S_2(x)$, we have 8 linear approximation relations with maximal correlation $\frac{1}{2}$ which is of the form

$$x_{(i)} + x_{(i+1)} + (\text{terms involving only } y) = 0 \ .$$

Each of such approximation relations gives linear approximation relations regarding the keystream output function. We will use the relation

$$x_{(0)} + x_{(1)} + y_{(0)} + y_{(3)} = 0$$

which induces the following relation for any $j = 0, \ldots, 31$,

$$(f_t)_{(j)} \oplus (f_{t+1})_{(j)} \oplus (z_t)_{(j)} \oplus (s_t)_{(j)} \oplus (z_{t+3})_{(j)} \oplus (s_{t+3})_{(j)} = 0,$$

with correlation $\frac{1}{2}$ when $t \equiv 1 (\mathrm{mod}\ 4)$. Thus if Γ is a linear mask, then

$$\Gamma(f_t \oplus f_{t+1}) \oplus \Gamma z_t \oplus \Gamma s_t \oplus \Gamma z_{t+3} \oplus \Gamma s_{t+3} = 0, \tag{2}$$

holds with correlation $(\frac{1}{2})^{\mathrm{wt}(\Gamma)}$ when $t \equiv 1 (\mathrm{mod}\ 4)$. Noting that

$$a_t \Gamma s_{t+9} \oplus \Gamma s_{t+9} = 0$$

holds with correlation $\frac{1}{2}$, we have linear approximation (3) involving only LFSR states and keystream words by XORing relations (1) and (2)

$$\Gamma s_t \oplus \Gamma s_{t+2} \oplus \Gamma s_{t+3} \oplus \Gamma s_{t+10} = \Gamma z_t \oplus \Gamma z_{t+3} \tag{3}$$

with correlation

$$C(\Gamma) := (\frac{1}{2})^{\mathrm{wt}(\Gamma)+1} \mathrm{c3}_+(\Gamma) \sum_{\Phi} \mathrm{c}_+(\Gamma; \Gamma, \Phi) \mathrm{c}_T(\Gamma; \Phi)$$

when $t \equiv 1 (\mathrm{mod}\ 4)$, assuming that the approximations are independent. Note that we don't see obvious dependencies between the approximations given above. We check the validity of our estimation by simulations described in Sect. 5.

3.1 Search for Linear Masks

We try to find Γ such that $|C(\Gamma)|$ is as large as possible. Taking into consideration the factor $(\frac{1}{2})^{\mathrm{wt}(\Gamma)}$, we confined the search to masks of weight less than or equal to 5. Furthermore, we have the following observation from many examples though we don't have a proof:

- If $\mathrm{c2}_T(\Gamma) = 0$, then $C(\Gamma) = 0$.

Based on this observation, we compute $C(\Gamma)$ for a given mask Γ in the following way:
If $\mathrm{c2}_T(\Gamma) \neq 0$, then

1. we compute $\mathrm{c3}_+(\Gamma)$ using [14, Theorem 1] regarding correlation of modular addition.
2. We compute $\sum_{\Phi} \mathrm{c}_+(\Gamma; \Gamma, \Phi) \mathrm{c}_T(\Gamma; \Phi)$ using [14, Theorem 1] and fast Walsh transform. Once Γ is fixed, we can compute $\mathrm{c}_+(\Gamma; \Gamma, \Phi)$ for any Φ using the description with finite automaton in [14]. It turns out that for each Γ, $\mathrm{c}_+(\Gamma; \Gamma, \Phi) = 0$ except for most Φ's. Using fast Walsh transform, for each fixed Γ, we can compute $\mathrm{c}_T(\Gamma; \Phi)$ for all Φ with time complexity 2^{37} and memory complexity 2^{32}.

Table 1. Correlations with respect to some linear masks of weight 4

Γ	$\log_2(\lvert c3_+(\Gamma)\rvert)$	$\log_2(\lvert\Sigma_\Phi\rvert)$	$-(\mathrm{wt}(\Gamma)+1)$	$\lvert C(\Gamma)\rvert$
$[25, 14, 13, 0]$	-3.17	-14.33	-5	$2^{-22.50}$
$[25, 24, 14, 0]$	-3.17	-13.24	-5	$2^{-21.41}$
$[25, 22, 18, 0]$	-4.55	-15.13	-5	$2^{-24.68}$

Then we obtain the following results:

- There does not exist a mask Γ of weight 1,2, or 3 such that $|C(\Gamma)| > 2^{-29}$.
- The only masks Γ of weight 2 such that $C(\Gamma) \neq 0$ are $[i, i+25]$ $(i = 0, \ldots, 6)$.
- There exist masks Γ of weight 4 such that $|C(\Gamma)| > 2^{-25}$. Some of them are listed in Table 1.

We also considered some masks Γ of the form $[i, i+25, j, k, l]$, but we could not find one such that $|C(\Gamma)| > 2^{-25}$. Thus the best linear mask we found out is $[25, 24, 14, 0]$, for which the correlation is $-2^{-21.41}$.

4 Correlation Attack on Sosemanuk

In this section, we describe a correlation attack against Sosemanuk recovering the initial internal state. Using the approximation relations (3) involving only LFSR state words and keystream words with non-negligible correlation obtained in the preceding section, we apply the techniques in [4] using fast Walsh transform to mount the attack.

Getting Approximation Relations between Initial LFSR State and Keystream Words. Let Γ be the linear mask $[25, 24, 14, 0]$, $\kappa = C(\Gamma) = -2^{-21.41}$, and $\epsilon = |\kappa/2| = 2^{-22.41}$ throughout this section. Starting with the approximation (3) with correlation κ, we can obtain arbitrarily many linear approximations with correlation κ involving the initial LFSR state $s_1, \cdots, s_{10}$ and the keystream words using the relation

$$\begin{aligned}&(\Gamma_0, \Gamma_1, \cdots, \Gamma_9) \cdot (s_{t+j}, s_{t+j+1}, \cdots, s_{t+j+9}) \\ &= (\mathcal{G}^j(\Gamma_0, \Gamma_1, \cdots, \Gamma_9)) \cdot (s_t, s_{t+1}, \cdots, s_{t+9})\end{aligned}$$

for each $j > 0$, where $\mathcal{G}$ is the "dual" of the LFSR update transformation and is given by

$$\begin{aligned}&\mathcal{G}(\Gamma_0, \Gamma_1, \Gamma_2, \Gamma_3, \Gamma_4, \Gamma_5, \Gamma_6, \Gamma_7, \Gamma_8, \Gamma_9) \\ &= (\alpha^*\Gamma_9, \Gamma_0, \Gamma_1, \Gamma_2 \oplus (\alpha^{-1})^*\Gamma_9, \Gamma_3, \Gamma_4, \Gamma_5, \Gamma_6, \Gamma_7, \Gamma_8 \oplus \Gamma_9),\end{aligned}$$

where $\alpha^*\Gamma$ and $(\alpha^{-1})^*\Gamma$ are 32-bit linear masks such that $(\alpha^*\Gamma)(x) = \Gamma(\alpha x)$ and $((\alpha^{-1})^*\Gamma)(x) = \Gamma(\alpha^{-1}x)$ for each 32-bit x.

To be more explicit, the approximation relations (3) can be rewritten as

$$\begin{aligned}&(\Gamma \oplus \alpha^*\Gamma, 0, \Gamma, \Gamma \oplus (\alpha^{-1})^*\Gamma, 0, 0, 0, 0, 0, \Gamma) \cdot (s_1, \cdots, s_{10}) = \Gamma z_1 \oplus \Gamma z_4 \\ &(\Gamma \oplus \alpha^*\Gamma, 0, \Gamma, \Gamma \oplus (\alpha^{-1})^*\Gamma, 0, 0, 0, 0, 0, \Gamma) \cdot (s_5, \cdots, s_{14}) = \Gamma z_5 \oplus \Gamma z_8 \\ &(\Gamma \oplus \alpha^*\Gamma, 0, \Gamma, \Gamma \oplus (\alpha^{-1})^*\Gamma, 0, 0, 0, 0, 0, \Gamma) \cdot (s_9, \cdots, s_{18}) = \Gamma z_9 \oplus \Gamma z_{12} \\ &\cdots,\end{aligned} \tag{4}$$

which are again equivalent to

$$\begin{array}{l}(\Gamma \oplus \alpha^*\Gamma, 0, \Gamma, \Gamma \oplus (\alpha^{-1})^*\Gamma, 0,0,0,0,0, \Gamma) \cdot (s_1, \cdots, s_{10}) = \Gamma z_1 \oplus \Gamma z_4 \\ \mathcal{F}(\Gamma \oplus \alpha^*\Gamma, 0, \Gamma, \Gamma \oplus (\alpha^{-1})^*\Gamma, 0,0,0,0,0, \Gamma) \cdot (s_1, \cdots, s_{10}) = \Gamma z_5 \oplus \Gamma z_8 \\ \mathcal{F}^2(\Gamma \oplus \alpha^*\Gamma, 0, \Gamma, \Gamma \oplus (\alpha^{-1})^*\Gamma, 0,0,0,0,0, \Gamma) \cdot (s_1, \cdots, s_{10}) = \Gamma z_9 \oplus \Gamma z_{12} \\ \cdots ,\end{array} \quad (5)$$

where $\mathcal{F} = \mathcal{G}^4$. Thus the complexity of getting R relations between the initial LFSR state and the keystream words is comparable to the complexity of getting $128R$ bits of keystream.

Recovering Part of the Initial LFSR State. We apply the "Second LFSR Derivation Technique" in [4]. Let $n = 320$ be the size of the LFSR state in bits and $m < n$. Let $\epsilon' = 2\epsilon^2 = 2^{-43.82}$ and $N = (\frac{2\lambda}{3\epsilon'})^2$, where λ satisfies

$$\frac{1}{\sqrt{2\pi}} \int_{\lambda}^{\infty} e^{-\frac{t^2}{2}} \, dt = 2^{-m}.$$

Let $R = \sqrt{N 2^{n-m+1}}$. Let $u_1, \cdots, u_n$ be the bits of the LFSR initial state $s_1, \cdots, s_{10}$. Suppose we have R linear approximation relations of correlation κ involving u_i's. Let $i_1, \cdots, i_m$ be any integers such that $1 \le i_1 < \ldots < i_m \le n$. XORing pairs of those R equations, we get about $R(R-1)2^{m-n-1} \approx N$ approximation relations with correlation $2\epsilon'$ involving only $u_{i_1}, \ldots, u_{i_m}$ among u_i's. Let these relations be

$$a^j_{i_1} u_{i_1} + \cdots + a^j_{i_m} u_{i_m} = b^j. \; (j = 1, \ldots, N) \quad (6)$$

Let us define the function $\sigma : \mathrm{GF}(2)^m \to \mathbb{Z}$ by

$$\begin{aligned}\sigma(a_1, \cdots, a_m) = & \, |\{j \in \{1, \ldots, N\} : (a^j_{i_1}, \ldots, a^j_{i_m}) = (a_1, \ldots, a_m), \; b^j = 0\}| \\ & - |\{j \in \{1, \ldots, N\} : (a^j_{i_1}, \ldots, a^j_{i_m}) = (a_1, \ldots, a_m), \; b^j = 1\}|\end{aligned}$$

Let W be the fast Walsh transform defined by

$$W(f)(y_1, \ldots, y_m) = \sum_{x_1, \ldots, x_m \in \mathrm{GF}(2)} f(x_1, \ldots, x_m)(-1)^{y_1 x_1 + \ldots + y_m x_m}$$

for $f : \mathrm{GF}(2)^m \to \mathbb{Z}$. Note that, for each $(u_{i_1}, \cdots, u_{i_m})$, $W(\sigma)(u_{i_1}, \cdots, u_{i_m})$ is

$$\begin{array}{l} \text{the number of relations in (6) satisfied by} (u_{i_1}, \cdots, u_{i_m}) \\ - \text{the number of relations in (6) not satisfied by} (u_{i_1}, \cdots, u_{i_m}). \end{array} \quad (7)$$

For the right value of $(u_{i_1}, \cdots, u_{i_m})$, above number follows the normal distribution $\mathrm{N}(2N\epsilon', N(1 - 4\epsilon'^2))$. So, using $N(1 - 4\epsilon'^2) \approx N$, for the right value of $(u_{i_1}, \cdots, u_{i_m})$,

$$\mathrm{Prob}\left(W(\sigma)(u_{i_1}, \cdots, u_{i_m}) < \frac{3}{2} N\epsilon' \right) = \frac{1}{\sqrt{2\pi}} \int_{\frac{\lambda}{3}}^{\infty} e^{-\frac{t^2}{2}} \, dt \; .$$

Table 2. Complexity of the Attack

	with Precomputation			without Precomputation		
	time(unit)	memory(bit)	data(bit)	time(unit)	memory(bit)	data(bit)
Precomputation	$2^{147.47}$	$2^{148.34}$				
Online computation	$2^{144.66}$	$2^{144.55}$	$2^{145.50}$	$2^{147.88}$	$2^{147.10}$	$2^{145.50}$

But for random $(u_{i_1}, \cdots, u_{i_m})$, (7) follows the distribution $\mathrm{N}(0, N)$. So for random $(u_{i_1}, \cdots, u_{i_m})$,

$$\mathrm{Prob}\left(W(\sigma)(u_{i_1}, \cdots, u_{i_m}) > \frac{3}{2}N\epsilon'\right) = \frac{1}{\sqrt{2\pi}}\int_{\lambda}^{\infty} e^{-\frac{t^2}{2}}\, dt = 2^{-m} .$$

Thus, when we use the threshold value $\frac{3}{2}N\epsilon'$ for determining whether a partial LFSR state candidate $(u_{i_1}, \cdots, u_{i_m})$ is the right one, we have non-detection probability less than $\frac{1}{\sqrt{2\pi}}\int_{\frac{\lambda}{3}}^{\infty} e^{-\frac{t^2}{2}}\, dt$ and false alarm rate 2^{-m}.

Complexity of the Attack. The attack can be performed in two ways. One way is to precompute the coefficients $(a^j_{i_1}, \cdots, a^j_{i_m})$ and then perform all other computations in online phase. The other is to perform all the computations online. Complexity of both ways are described below and summarized in Table 2.

Attack with Precomputation. To recover partial bits $u_{i_1}, \cdots, u_{i_m}$ of the initial initial state, in the precomputation phase, we get the coefficients of the left hand sides of the R approximation relations (5) between the LFSR initial states and the keystream words. Store the $(320 + \lceil \log_2(R) \rceil)$-bit values (U_i, i) $(i = 0, \cdots, R-1)$ in a list, where

$$U_i := \mathcal{F}^i(\Gamma \oplus \alpha^* \Gamma, 0, \Gamma, \Gamma \oplus (\alpha^{-1})^* \Gamma, 0, 0, 0, 0, 0, \Gamma)$$

for each i. Then sort the list according to the components in $\{1, \cdots, m\} - \{i_1, \cdots, i_m\}$. For each pair (i, k) such that the components of U_i and U_k in $\{1, \cdots, m\} - \{i_1, \cdots, i_m\}$ coincides, compute $X_{i,k} := (U_i \oplus U_k$ restricted to i_1-th, $\cdots$, i_m-th components), and store $(X_{i,k}, i, k)$ in a list. The list has about N entries of size $m + 2\lceil \log_2(R) \rceil$. In the online phase, set the function $\sigma : \mathrm{GF}(2)^m \to \mathbb{Z}$ as zero. Let $w_i = \Gamma z_{4i+1} + \Gamma z_{4i+4}$ for each $i = 0, \cdots, R-1$. For each $(X_{i,k}, i, k)$ in the list, compute the value $w_i + w_k$ and update σ. (The update rule is that $\sigma(X_{i,k})$ increases by 1 if $w_i + w_k = 0$ and decreases by 1 otherwise.) Perform the fast Walsh transform to σ and check if there is some $(u_{i_1}, \cdots, u_{i_m})$ such that $W(\sigma)(u_{i_1}, \cdots, u_{i_m}) > \frac{3}{2}N\epsilon'$. The complexity of the above attack to recover m bits of the initial LFSR state is as follows. The complexity of the above attack to recover m bits of the initial LFSR state is as follows. We assume the complexity of the basic operations as in Table 3. The precomputation phase has time complexity of about $128R + R\log_2(R)(320 + \lceil \log_2(R) \rceil) + (N + R)(320 + \lceil \log_2(R) \rceil)$ and memory requirement of $R(320 + \lceil \log_2(R) \rceil) + N(m + 2\lceil \log_2(R) \rceil)$ bits if we apply a sorting algorithm of small memory requirement. The online phase

Table 3. Complexity of basic operations

operations	time complexity
XOR of two k-bit words	k
Comparison of two k-bit words	k
Sorting a list with r k-bit entries	$kr\log_2(r)$
Walsh transform for 2^m k-bit integers	$km2^m$

takes $2^m\lceil\log_2(N)\rceil$-bits of memory and time complexity of $8N + m2^m\lceil\log_2(N)\rceil$. The data complexity of the online phase is 2^7R bits. Let $m = 138$. Then $\lambda \approx 13.6$(by e.g. Lemma 1 in the Appendix), $N = 2^{94.00}$ and $R = 2^{138.50}$. For recovery of the whole n bits of the LFSR initial state, we recover $(u_1, \cdots, u_m)$ and $(u_m, \cdots, u_{2m-1})$ using above-mentioned methods. Then restore the remaining 45 bits of the initial LFSR state and 64 initial FSM bits simultaneously using exhaustive search. The precomputation phase takes time complexity of $128R + 2(R\log_2(R)(320 + \lceil\log_2(R)\rceil) + (N + R)(320 + \lceil\log_2(R)\rceil)) = 2^{155.47}$. (The number in the table is $2^{147.47}$ regarding 1 time unit as the time needed to generate 256 bits of keystream which is not greater than the time cost of one trial in the exhaustive search .)The required memory is $2R(320 + \lceil\log_2(R)\rceil) + N(m + 2\lceil\log_2(R)\rceil) = 2^{148.34}$ bits. The online phase has time complexity of $2(8N + m2^m\lceil\log_2(N)\rceil) = 2^{152.66}$, memory requirement of $2^m\lceil\log_2(N)\rceil = 2^{144.55}$ bits, and data complexity of $2^7R = 2^{145.50}$ bits. The non-detection probability is less than $\frac{2}{\sqrt{2\pi}}\int_{\frac{\lambda}{3}}^{\infty} e^{-\frac{t^2}{2}}\,dt \leq 0.01$. We mention that the increased complexity due to sorting was not considered in [4].

Attack without Precomputation. To recover partial bits $u_{i_1}, \cdots, u_{i_m}$ of the initial LFSR state, we first get all the coefficients of the R approximation relations using the keystreams. Store the $(320 + 1)$-bit values (U_i, w_i) $(i = 0, \cdots, R - 1)$. Then sort the list according to the components in $\{1, \cdots, m\} - \{i_1, \cdots, i_m\}$. Set the function σ as zero. For each pair (i, k) such that the components of U_i and U_k in $\{1, \cdots, m\} - \{i_1, \cdots, i_m\}$ coincides, compute $X_{i,k}$ and update the function σ using $(X_{i,k}, w_i + w_k)$. Perform the fast Walsh transform to σ and check if there is some $(u_{i_1}, \cdots, u_{i_m})$ such that $W(\sigma)(u_{i_1}, \cdots, u_{i_m}) > \frac{3}{2}N\epsilon'$. The time complexity is about $128R + R\log_2(R)(n + 1) + N(n + 1) + m2^m\lceil\log_2(N)\rceil$ and memory requirement is about $\lceil\log_2(N)\rceil 2^m + (320 + 1)R$ bits. The data complexity is 2^7R bits. Let $m = 138$. For recovery of the whole n bits of the LFSR initial state, we recover $(u_1, \cdots, u_m)$ and $(u_m, \cdots, u_{2m-1})$ using above-mentioned methods. Then restore the remaining 45 bits of the initial LFSR state and 64 initial FSM bits simultaneously using exhaustive search. The time complexity is $2(128R + R\log_2(R)(n+1) + N(n+1) + m2^m\lceil\log_2(N)\rceil) + 129\cdot 2^{129} = 2^{155.88}$. The memory requirement is $\lceil\log_2(N)\rceil 2^m + (320 + 1)R = 2^{147.10}$ bits, and the data complexity is $2^7R = 2^{145.50}$ bits.

Improving the Attack. We can reduce the data complexity without increasing the time complexity. For the Serpent S-box S_2, we have 8 linear approximations with correlation $\frac{1}{2}$ which is of the form

$$x_{(i)} + x_{(i+1)} + (\text{terms involving only } y) = 0 \ .$$

Using these approximations, we can get 8 linear approximation relations involving the LFSR initial state and keystream words with correlation κ. Thus we can reduce the data complexity at least by the factor of 2^3. We can also reduce the memory requirement of the attack using the "Improved Hybrid Method" [4] without increasing time complexity or data complexity much.

5 Simulations and Results

5.1 Simulations for a Reduced Cipher

We validate our claims by simulating a reduced version of Sosemanuk keystream generator defined as follows. It consists of an LFSR of five bytes and an FSM of two bytes. The LFSR state at time t is $(s_t, s_{t+1}, \ldots, s_{t+5})$. The LFSR is updated using the relation

$$s_{t+5} = s_{t+4} \oplus \beta^{-1} s_{t+3} \oplus \beta s_t,$$

where β is a zero of $x^8 + x^7 + x^5 + x^3 + 1$ in

$$GF(2^8) = GF(2)(\beta) = GF(2)[x]/ < x^8 + x^7 + x^5 + x^3 + 1 >$$

The FSM state at time t is denoted by $(R1_t, R2_t)$. The FSM is updated as follows.

$$R1_t = R2_{t-1} + (s_{t+1} \oplus \text{lsb}(R1_{t-1}) s_{t+3}) (\text{mod } 2^8)$$
$$R2_t = \text{Trans}(R1_{t-1}) = ((M \times R1_{t-1})(\text{mod } 2^8))^{<<<3}$$

where, $M =$ `0x59`. The FSM has output

$$f_t = (s_{t+4} + R1_t)(\text{mod } 2^8) \oplus R2_t.$$

The keystream bytes are obtained as follows.

$$(z_{t+3}, z_{t+2}, z_{t+1}, z_t) = Serpent1(f_{t+3}, f_{t+2}, f_{t+1}, f_t) \oplus (s_{t+3}, s_{t+2}, s_{t+1}, s_t)$$
$$(t \equiv 1 (\text{mod } 4))$$

Then we get a linear approximation relation

$$\Gamma s_t \oplus \Gamma s_{t+2} \oplus \Gamma s_{t+3} \oplus \Gamma s_{t+5} = \Gamma z_t \oplus \Gamma z_{t+3} \ \ (t \equiv 1 (\text{mod } 4))$$

with correlation

$$(\frac{1}{2})^{\text{wt}(\Gamma)+1} \text{c3}_+(\Gamma) \sum_{\Phi} \text{c}_+(\Gamma; \Gamma, \Phi) \text{c}_T(\Gamma; \Phi)$$

when $t \equiv 1(\text{mod } 4)$, for each 8-bit mask Γ. In the simulation, we generate 2^{30} bytes of keystream and observe the actual correlation of the linear approximation regarding the LFSR states and the keystream bytes for various initial internal states. The observed actual correlation is about $-2^{-6.12}$ when $\Gamma = [5, 0]$ and

Table 4. Correlations with respect to linear masks of weight 2

Γ	$\log_2(\lvert c3_+(\Gamma)\rvert)$	$\log_2(\lvert\Sigma_\Phi\rvert)$	$-(\mathrm{wt}(\Gamma)+1)$	correlation
[5, 0]	−1.59	−1.91	−3	$-2^{-6.50}$
[6, 1]	−10	−3	−3	-2^{-16}
[7, 2]	−3.57	−3.36	−3	$-2^{-9.93}$

about $-2^{-10.31}$ when $\Gamma = [7,2]$ regardless of the initial internal state. Using the observed correlation for $\Gamma = [5,0]$, we are able to recover the initial internal state using the method explained in Sect. 4. The parameters are $n = 40$, $m = 24$, $\lambda = 2.83$, $N = 2^{30.31}$ and $R = 2^{23.66}$. We get R approximation relations regarding the n-bit initial LFSR state and the keystream words. Then we get about N approximations regarding the latter m bits of the initial LFSR state. Applying the fast Walsh transform to an array with 2^m entries, we can recover the m bits correctly most of the time. We performed the experiments to recover the latter 24 bits of the initial LFSR state for 100 initial initial internal states as follows.

- LFSR initial states: $(i, i+1, i+2, i+3, i+4)$ $(i = 0, \cdots, 99)$
- FSM initial state: (0,0) (fixed)

With the threshold $\frac{3}{2}N2^{-13.24} = 206382$, we were able to get the right 24-bit value in each case except when $i = 26$. In each case 0–4 false alarms occurred with average 1.18. A few minutes was spent on a Pentium IV 3.4GHz CPU with 1GB RAM for each case. This experimental results corroborate our assertions.

5.2 Simulations with Long Keystreams for Full Sosemanuk

To check if the correlation of relations (3) is correct in another way, we generate long keystreams for Sosemanuk for some initial internal states. We consider the following 2 LFSR initial states and 8 FSM initial states.

- LFSR initial states
  - A: (`0x9000, 0x8000,` $\cdots$ `, 0x1000, 0x0000`)
  - B: (`0x9111, 0x8000,` $\cdots$ `, 0x1000, 0x0111`) (the same as A except for the first and the last word)
- FSM initial states: (`0x0000, 0x0000`), $\cdots$, (`0x7000, 0x7000`)

For each of the 16 initial states, we generate Sosemanuk keystreams of 2^{53} bits and count how many of the 2^{46} induced relations (3) are satisfied for the mask $\Gamma = [25, 24, 14, 0]$ and compute the observed correlation. The results are as in Table 5. In the table, "z-value" represents

$$\frac{(\text{the number of the satisfied among the } 2^{46} \text{ relations}) - (2^{45} + 2^{45}C(\Gamma))}{2^{22}},$$

which is the normalized deviation in the assumed normal distribution. In total, the observed correlation using the 2^{50} relations is $-2^{-21.45}$, which is very close to $C(\Gamma)$. This result also corroborates our assertions.

Table 5. Simulation Result for Long Keystreams

LFSR	FSM	z-value	correlation*	LFSR	FSM	z-value	correlation*
A	0	-0.29	$-2^{-21.28}$	B	0	1.93	$-2^{-22.89}$
	1	-2.13	$-2^{-20.64}$		1	-0.65	$-2^{-21.13}$
	2	0.69	$-2^{-21.79}$		2	-0.15	$-2^{-21.34}$
	3	0.35	$-2^{-21.59}$		3	1.09	$-2^{-22.06}$
	4	0.54	$-2^{-21.70}$		4	-0.95	$-2^{-21.02}$
	5	-0.35	$-2^{-21.25}$		5	-0.16	$-2^{-21.34}$
	6	-0.48	$-2^{-21.20}$		6	0.99	$-2^{-21.99}$
	7	-0.62	$-2^{-21.14}$		7	1.73	$-2^{-22.64}$

*:Observed correlation.

6 Correlation Attack on SNOW 2.0

SNOW 2.0[11] consists of an LFSR consisting of 16 words and an FSM of 2 words. In [14], it was shown that there exists a linear approximation relation of the LFSR bits and keystream bits with bias $2^{-15.496}$ or correlation $\pm 2^{-14.496}$[14, Table 2]. One of such approximation relations is

$$\Lambda s_t + \Lambda s_{t+1} + \Lambda s_{t+5} + \Lambda s_{t+15} + \Lambda s_{t+16} = \Lambda z_t + \Lambda z_{t+1},$$

where $\Lambda = [0, 15, 16]$. Applying the "Second LFSR derivation technique" again with parameter $n = 512$ and $\epsilon = 2^{-15.496}$, we can mount a correlation attack on SNOW 2.0 without precomputation as follows.

Let $m = 192$. Using the same notation as in Sect. 4, $\lambda \approx 16.1$, $N = 2^{66.54}$ and $R = 2^{193.77}$. The time complexity of the attack for recovering m bits is $32R + R\log_2(R)(n+1) + N(n+1) + m2^m\lceil\log_2(N)\rceil$. (The factor 32 comes from the fact that 32 bits of keystreams are needed per one approximation relation.) Memory requirement is about $\lceil\log_2(N)\rceil 2^m + (512+1)R$ bits. The data complexity is $2^5 R$ bits. For recovery of the whole initial LFSR state, recover partial 192 bits of LFSR three times and then recover the initial FSM state by exhaustive search. The total time complexity is $3(32R + R\log_2(R)(n+1) + N(n+1) + m2^m\lceil\log_2(N)\rceil) = 2^{212.38}$. The memory complexity is about $\lceil\log_2(N)\rceil 2^m + (512+1)R = 2^{202.83}$ bits. The data complexity is $2^5 R = 2^{198.77}$ bits. Since the initialization of SNOW 2.0 is a reversible process, we can recover the key from the initial state.

7 Conclusion

We described an attack recovering the initial internal state with time complexity $2^{147.88}$, memory complexity $2^{147.10}$ bits, and data complexity $2^{145.50}$ bits. Though the attack does not threaten the claimed 128-bit security of Sosemanuk, it indicates that using keys longer than 150 bits for Sosemanuk does not guarantee the security level of the key size. The main reason Sosemanuk is vulnerable to the attack described in this paper is that the LFSR state is too small in the presence of a relatively large correlation between the LFSR state and the keystream words. Similar attack of complexity $2^{204.38}$ is valid against SNOW 2.0.

References

1. Anderson, R., Biham, E., Knudsen, L.: Serpent: A Proposal for the Advanced Encryption Standard, http://www.cl.cam.ac.uk/~rja14/serpent.html
2. Babbage, S., et al.: The eSTREAM Portfolio (April 15, 2008), http://www.ecrypt.eu.org/stream/portfolio.pdf
3. Berbain, C., Billet, O., Canteaut, A., Courtois, N., Gilbert, H., Goubin, L., Gouget, A., Granboulan, L., Lauradoux, C., Minier, M., Pornin, T., Sibert, H.: SOSEMANUK, a fast software-oriented stream cipher. eSTREAM Report 2005/027 (2005)
4. Berbain, C., Gilbert, H., Maximov, A.: Cryptanalysis of Grain. In: Robshaw, M.J.B. (ed.) FSE 2006. LNCS, vol. 4047, pp. 15–29. Springer, Heidelberg (2006)
5. Bernstein, D.: Which eSTREAM ciphers have been broken? eSTREAM Report 2008/010 (2008)
6. Chepyzhov, V., Johansson, T., Smeets, B.: A Simple Algorithm for Fast Correlation Attacks on Stream Ciphers. In: Schneier, B. (ed.) FSE 2000. LNCS, vol. 1978, pp. 181–195. Springer, Heidelberg (2001)
7. Cho, J., Pieprzyk, J.: Crossword Puzzle Attack on NLS. In: Biham, E., Youssef, A.M. (eds.) SAC 2006. LNCS, vol. 4356, pp. 249–265. Springer, Heidelberg (2007)
8. Cho, J.: An Improved Estimate of the Correlation of Distinguisher for Dragon. In: Workshop Record of The State of the Art of Stream Ciphers (SASC 2008), pp. 11–20 (2008)
9. Coppersmith, D., Halevi, S., Jutla, C.: Cryptanalysis of stream ciphers with linear masking. In: Yung, M. (ed.) CRYPTO 2002. LNCS, vol. 2442, pp. 515–532. Springer, Heidelberg (2002)
10. Ekdahl, P., Johansson, T.: SNOW - a new stream cipher, http://www.it.ith.se/cryptology/snow/
11. Ekdahl, P., Johansson, T.: A new version of the stream cipher SNOW. In: Nyberg, K., Heys, H.M. (eds.) SAC 2002. LNCS, vol. 2595, pp. 47–61. Springer, Heidelberg (2003)
12. Golic, J., Bagini, V., Morgari, G.: Linear Cryptanalysis of Bluetooth Stream Cipher. In: Knudsen, L.R. (ed.) EUROCRYPT 2002. LNCS, vol. 2332, pp. 238–255. Springer, Heidelberg (2002)
13. Nyberg, K.: Correlation theorems in cryptanalysis. Discrete Applied Mathematics 111, 177–188 (2001)
14. Nyberg, K., Wallén, J.: Improved Linear Distinguishers for SNOW 2.0. In: Robshaw, M.J.B. (ed.) FSE 2006. LNCS, vol. 4047, pp. 144–162. Springer, Heidelberg (2006)
15. Watanabe, D., Biryukov, A., De Canniere, C.: A Distinguishing Attack of SNOW 2.0 with Linear Masking Method. In: Matsui, M., Zuccherato, R.J. (eds.) SAC 2003. LNCS, vol. 3006, pp. 222–233. Springer, Heidelberg (2004)

A An Approximation of the Cumulative Normal Distribution Function

Lemma 1. *For any $0 < a < 1$, we have*

$$\frac{a}{\lambda}e^{-\frac{\lambda^2}{2}} \le \int_{\lambda}^{\infty} e^{-\frac{t^2}{2}}\,dt \le \frac{1}{\lambda}e^{-\frac{\lambda^2}{2}}$$

for any $\lambda \ge 1$ such that $a \le \frac{\lambda^2}{\lambda^2+1}$.

Proof. Let

$$F(x) = \int_{x}^{\infty} e^{-\frac{t^2}{2}}\,dt - \frac{1}{x}e^{-\frac{x^2}{2}} \quad (x > 0)\ .$$

Then $F'(x) = \frac{1}{x}e^{-\frac{x^2}{2}} > 0$ and $\lim_{x\to\infty} F(x) = 0$. Hence $F(x) < 0$ for all $x > 0$. Let

$$G(x) = \int_{x}^{\infty} e^{-\frac{t^2}{2}}\,dt - \frac{a}{x}e^{-\frac{x^2}{2}} \quad (x > 0)\ .$$

Then $G'(x) = (a-1)e^{-\frac{x^2}{2}} + \frac{a}{x^2}e^{-\frac{x^2}{2}}$ so that $G'(x) < 0$ if $a < \frac{x^2}{x^2+1}$. Since $\lim_{x\to\infty} G(x) = 0$, $G(x) > 0$ when $a < \frac{x^2}{x^2+1}$. □

A New Attack on the LEX Stream Cipher

Orr Dunkelman[1,⋆] and Nathan Keller[2,⋆⋆]

[1] École Normale Supérieure
Département d'Informatique,
CNRS, INRIA
45 rue d'Ulm, 75230 Paris, France
orr.dunkelman@ens.fr

[2] Einstein Institute of Mathematics, Hebrew University
Jerusalem 91904, Israel
nkeller@math.huji.ac.il

Abstract. In [6], Biryukov presented a new methodology of stream cipher design, called *leak extraction*. The stream cipher LEX, based on this methodology and on the AES block cipher, was selected to phase 3 of the eSTREAM competition. The suggested methodology seemed promising, and LEX, due to its elegance, simplicity and performance was expected to be selected to the eSTREAM portfolio.

In this paper we present a key recovery attack on LEX. The attack requires about $2^{36.3}$ bytes of key-stream produced by the same key (possibly under many different IVs), and retrieves the secret key in time of 2^{112} simple operations. Following a preliminary version of our attack, LEX was discarded from the final portfolio of eSTREAM.

Keywords: LEX, AES, stream cipher design.

1 Introduction

The design of stream ciphers, and more generally, pseudo-random number generators (PRNGs), has been a subject of intensive study over the last decades. One of the well-known methods to construct a PRNG is to base it on a keyed pseudo-random permutation. A provably secure construction of this class is given by Goldreich and Levin [19]. An instantiation of this approach (even though an earlier one) is the Blum and Micali [11] construction (based on the hardness of RSA). A more efficiency-oriented construction is the BMGL stream cipher [21] (based on the Rijndael block cipher). However, these constructions are relatively slow, and hence are not used in practical applications.

⋆ The first author was supported by the France Telecome Chaire. Some of the work presented in this paper was done while the first author was staying at K.U. Leuven, Belgium and supported by the IAP Programme P6/26 BCRYPT of the Belgian State (Belgian Science Policy).

⋆⋆ The second author is supported by the Adams Fellowship Program of the Israel Academy of Sciences and Humanities.

J. Pieprzyk (Ed.): ASIACRYPT 2008, LNCS 5350, pp. 539–556, 2008.
© International Association for Cryptologic Research 2008

In [6], Biryukov presented a new methodology for constructing PRNGs of this class, called *leak extraction*. In this methodology, the output key stream of the stream cipher is based on parts of the internal state of a block cipher at certain rounds (possibly after passing an additional filter function). Of course, in such a case, the "leaked" parts of the internal state have to be chosen carefully such that the security of the resulted stream cipher will be comparable to the security of the original block cipher.

As an example of the leak extraction methodology, Biryukov presented in [6] the stream cipher LEX, in which the underlying block cipher is AES. The key stream of LEX is generated by applying AES in the OFB (Output Feedback Block) mode of operation and extracting 32 bits of the intermediate state after the application of each full AES round.

LEX was submitted to the eSTREAM competition (see [7]). Due to its high speed (2.5 times faster than AES), fast key initialization phase (a single AES encryption), and expected security (based on the security of AES), LEX was considered a very promising candidate and selected to the third (and final) phase of evaluation.

During the eSTREAM competition, LEX attracted a great deal of attention from cryptanalysts due to its simple structure, but nevertheless, only two attacks on the cipher were reported: A slide attack [23] requiring 2^{61} different IVs (each producing 20,000 keystream bytes), and a generic attack [17] requiring $2^{65.7}$ resynchronizations. Both attacks are applicable only against the original version of LEX presented in [6], but not against the tweaked version submitted to the second phase of eSTREAM [8]. In the tweaked version, the number of IVs used with a single key is bounded by 2^{32}, and hence both attacks require too much data and are not applicable to the tweaked version.

In this paper we present an attack on LEX. The attack requires about $2^{36.3}$ bytes of key stream produced by the same key, possibly under different IVs. The time complexity of the attack is 2^{112} simple operations. Following a preliminary version of our attack, LEX was discarded from the final portfolio of eSTREAM.

Our attack is composed of three steps:

1. **Identification of a special state:** We focus our attention on pairs of AES encryptions whose internal states satisfy a certain difference pattern. While the probability of occurrence of the special pattern is 2^{-64}, the pattern can be observed by a 32-bit condition on the output stream. Thus, the attacker repeats the following two steps for about 2^{32} cases which satisfy this 32-bit condition.
2. **Extracting information on the special state:** By using the special difference pattern of the pair of intermediate values, and guessing the difference in eight more bytes, the attacker can retrieve the actual values of 16 internal state bytes in both encryptions.
3. **Guess-and-Determine attack on the remaining unknown bytes:** Using the additional known byte values, the attacker can mount a guess-and-determine attack that retrieves the key using about 2^{112} simple operations in total.

The second and the third steps of the attack use several observations on the structure of the AES round function and key schedule algorithm.[1] One of them is the following, probably novel, observation:

Proposition 1. *Denote the 128-bit subkey used in the r-th round of AES-128 by k_r, and denote the bytes of this subkey by an 4-by-4 array $\{k_r(i,j)\}_{i,j=0}^{3}$. Then for every $0 \leq i \leq 3$ and r,*

$$k_r(i,1) = k_{r+2}(i,1) \oplus SB(k_{r+1}(i+1,3)) \oplus RCON_{r+2}(i),$$

where SB denotes the SubBytes operation, $RCON_{r+2}$ denotes the round constant used in the generation of the subkey k_{r+2}, and $i+1$ is replaced by 0 for $i = 3$.

It is possible that the observations on the structure of AES presented in this paper can be used not only in attacks on LEX, but also in attacks on AES itself.

This paper is organized as follows: In Section 2 we briefly describe the structures of AES and LEX, and present the observations on AES used in our attack. In Section 3 we show that a specific difference pattern in the internal state can be partially detected by observing the output stream, and can be used to retrieve the actual value of 16 bytes of the internal state (in both encryptions). In Section 4 we leverage the knowledge of these 16 bytes into a complete key recovery attack that requires about 2^{112} simple operations. We give several additional observations that may be useful for further cryptanalysis of LEX in Section 5. We conclude the paper in Section 6.

2 Preliminaries

In this section we describe the structures of AES and LEX, and present the observations on AES used in our attack.

2.1 Description of AES

The advanced encryption standard [14] is an SP-network that supports key sizes of 128, 192, and 256 bits. As this paper deals with LEX which is based on AES-128, we shall concentrate the description on this variant and refer the reader to [22] for a complete detailed description of AES.

A 128-bit plaintext is treated as a byte matrix of size 4x4, where each byte represents a value in $GF(2^8)$. An AES round applies four operations to the state matrix:

- SubBytes (SB) — applying the same 8-bit to 8-bit invertible S-box 16 times in parallel on each byte of the state,

[1] We note that in [6] it was remarked that the relatively simple key schedule of AES may affect the security of LEX, and it was suggested to replace the AES subkeys by 1280 random bits. Our attack, which relies heavily on properties of the AES key schedule, would fail if such replacement was performed. However, some of our observations can be used in this case as well.

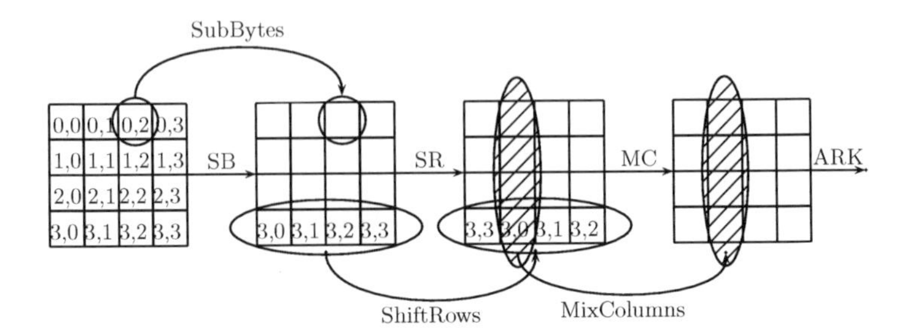

Fig. 1. An AES round

- ShiftRows (SR) — cyclic shift of each row (the i'th row is shifted by i bytes to the left),
- MixColumns (MC) — multiplication of each column by a constant 4x4 matrix over the field $GF(2^8)$, and
- AddRoundKey (ARK) — XORing the state with a 128-bit subkey.

We outline an AES round in Figure 1. Throughout the paper we allow ourselves the abuse of notation $SB(x)$ to denote the application of the S-box to x (whether it is one S-box when x is 8-bit value, or four times when x is 32-bit value). In the first round, an additional AddRoundKey operation (using a whitening key) is applied, and in the last round the MixColumns operation is omitted. We note that in LEX these changes to the first and last round are not applied.

AES-128, i.e., AES with 128-bit keys, has 10 rounds. For this variant, 11 subkeys of 128 bits each are derived from the key. The subkey array is denoted by $W[0,\ldots,43]$, where each word of $W[\cdot]$ consists of 32 bits. The first four words of $W[\cdot]$ are loaded with the user supplied key. The remaining words of $W[\cdot]$ are updated according to the following rule:

- For $i = 4,\ldots,43$, do
  - If $i \equiv 0 \bmod 4$ then $W[i] = W[i-4] \oplus SB(W[i-1] \lll 8) \oplus RCON[i/4]$,
  - Otherwise $W[i] = W[i-1] \oplus W[i-4]$,

where $RCON[\cdot]$ is an array of predetermined constants, and $\lll$ denotes rotation of the word by 8 bits to the left.

2.2 Description of LEX

For the ease of description, we describe only the tweaked version of LEX submitted to the second phase of eSTREAM [8]. The original version of LEX can be found in [6]. We note that our attacks can be easily adopted to the original version as well.

In the initialization step, the publicly known IV is encrypted by AES[2] under the secret key K to get $S = AES_K(IV)$. Then, S is repeatedly encrypted in the

[2] Actually, LEX uses a tweaked version of AES where the AddRoundKey before the first round is omitted, and the MixColumns operation of the last round is present. We allow ourselves the slight abuse of notations, for sake of clarity.

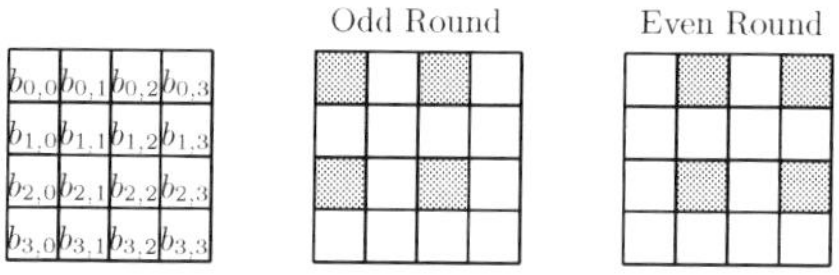

$b_{0,0}$	$b_{0,1}$	$b_{0,2}$	$b_{0,3}$
$b_{1,0}$	$b_{1,1}$	$b_{1,2}$	$b_{1,3}$
$b_{2,0}$	$b_{2,1}$	$b_{2,2}$	$b_{2,3}$
$b_{3,0}$	$b_{3,1}$	$b_{3,2}$	$b_{3,3}$

Odd Round

Even Round

The gray bytes are the output bytes.

Fig. 2. Odd and Even Rounds of LEX

OFB mode of operation under K, where during the execution of each encryption, 32 bits of the internal state are leaked each round. These state bits compose the key stream of LEX. The state bytes used in the key stream are shown in Figure 2. After 500 encryptions, another IV is chosen, and the process is repeated. After 2^{32} different IVs, the secret key is replaced.[3]

2.3 Notations Used in the Paper

As in [6], the bytes of each internal state during AES encryption, as well as the bytes of the subkeys, are denoted by a 4-by-4 array $\{b_{i,j}\}_{i,j=0}^{3}$, where $b_{i,j}$ is the j-th byte in the i-th row. For example, the output bytes in the even rounds are $b_{0,1}, b_{0,3}, b_{2,1}, b_{2,3}$.

2.4 Observations on AES Used in Our Attack

Throughout the paper we use several observations concerning AES.

Observation 1. *For every non-zero input difference to the SubBytes operation, there are 126 possible output differences with probability* 2^{-7} *each (i.e., only a single input pair with the given difference leads to the specified output difference), and a single output difference with probability* 2^{-6}.

As a result, for a randomly chosen pair of input/output differences of the SubBytes operation, with probability 126/256 there is exactly one unordered pair of values satisfying these differences. With probability 1/256 there are two such pairs, and with probability 129/256, there are no such pairs.

We note that while each ordered pair of input/output differences suggests one pair of actual values on average, it actually never suggests *exactly* one pair. In about half of the cases, two (or more) *ordered* pairs are suggested, and in the rest of the cases, no pairs are suggested. In the cases where two (or more) pairs are suggested, the analysis has to be repeated for each of the pairs. On the other hand, if no pairs are suggested, then the input/output differences pair is discarded as a wrong pair and the analysis is not performed at all. Hence,

[3] We note that in the original version of LEX, the number of different IVs used with a single key was not bounded. Following the slide attack presented in [23], the number of IVs used with each key was restricted. This restriction also prevents the attack suggested later in [17] which requires $2^{65.7}$ re-synchronizations.

when factoring both events, it is reasonable to assume that each input/output differences pair suggests one pair of actual values.

Our attack uses this observation in situations where the attacker knows the input and output differences to some SubBytes operation. In such cases, using the observation she can deduce the actual values of the input and the output (for both encryptions). This can be done efficiently by preparing the difference distribution table [4] of the SubBytes operation, along with the actual values of the input pairs satisfying each input/output difference relation (rather than only the number of such pairs). In the actual attack, given the input and output differences of the SubBytes operation, the attacker can retrieve the corresponding actual values using a simple table lookup.

Observation 2. *Since the MixColumns operation is linear and invertible, if the values (or the differences) in any four out of its eight input/output bytes are known, then the values (or the differences, respectively) in the other four bytes are uniquely determined, and can be computed efficiently.*

The following two observations are concerned with the key schedule of AES. While the first of them is known (see [18]), it appears that the second was not published before.

Observation 3. *For each $0 \leq i \leq 3$, the subkeys of AES satisfy the relations:*

$$k_{r+2}(i,0) \oplus k_{r+2}(i,2) = k_r(i,2).$$

$$k_{r+2}(i,1) \oplus k_{r+2}(i,3) = k_r(i,3).$$

Proof. Recall that by the key schedule, for all $0 \leq i \leq 3$ and for all $0 \leq j \leq 2$, we have $k_{r+2}(i,j) \oplus k_{r+2}(i,j+1) = k_{r+1}(i+1,j+1)$. Hence,

$$k_{r+2}(i,0) \oplus k_{r+2}(i,2) = (k_{r+2}(i,0) \oplus k_{r+2}(i,1)) \oplus (k_{r+2}(i,1) \oplus k_{r+2}(i,2)) = k_{r+1}(i,1) \oplus k_{r+1}(i,2) = k_r(i,2),$$

and the second claim follows similarly.

Observation 4. *For each $0 \leq i \leq 3$, the subkeys of AES satisfy the relation:*

$$k_{r+2}(i,1) \oplus SB(k_{r+1}((i+1) \bmod 4, 3)) \oplus RCON_{r+2}(i) = k_r(i,1),$$

Proof. In addition to the relation used in the proof of the previous observation, we use the relation

$$k_{r+2}(i,0) = k_{r+1}(i,0) \oplus SB(k_{r+1}((i+1) \bmod 4, 3)) \oplus RCON_{r+2}(i).$$

Thus,

$$k_{r+2}(i,1) \oplus SB(k_{r+1}((i+1) \bmod 4, 3)) \oplus RCON_{r+2}(i) = (k_{r+2}(i,1) \oplus k_{r+2}(i,0)) \oplus (k_{r+2}(i,0) \oplus SB(k_{r+1}((i+1) \bmod 4, 3)) \oplus RCON_{r+2}(i)) = k_{r+1}(i,1) \oplus k_{r+1}(i,0) = k_r(i,1).$$

These two observations allow the attacker to use the knowledge of bytes of k_{r+2} (and the last column of k_{r+1}) to get the knowledge of bytes in k_r, while "skipping" (some of) the values of k_{r+1}.

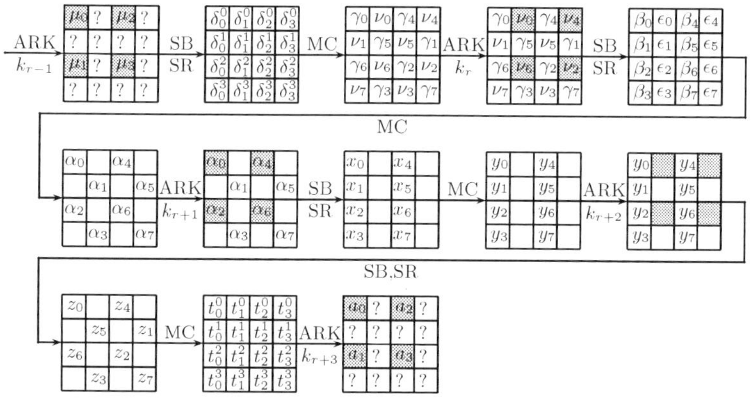

In gray we mark bytes whose value is known from the output key stream.

Fig. 3. The Special Difference Pattern (for Odd Rounds)

3 Observable Difference Pattern in LEX

Our attack is applicable when the special difference pattern starts either in odd rounds or in even rounds. For sake of simplicity of the description, we present the results assuming the difference pattern occurs in the odd rounds, and give in Appendix A the modified attack applicable when the difference pattern is observed in even rounds.

3.1 Detecting the Difference Pattern

Consider two AES encryptions under the same secret key, K. The special difference pattern corresponds to the following event: The difference between the intermediate values at the end of the $(r+1)$-th round is non-zero only in bytes $b_{0,0}, b_{0,2}, b_{1,1}, b_{1,3}, b_{2,0}, b_{2,2}, b_{3,1}$, and $b_{3,3}$. The probability of this event is 2^{-64}. The pattern, along with the evolution of the differences in rounds $r, r+1, r+2$, and $r+3$, is presented in Figure 3.

The difference pattern can be partially observed by a 32-bit condition on the output key stream: If the pattern holds, then all the four output bytes in round $r+2$ (bytes $b_{0,1}, b_{0,3}, b_{2,1}, b_{2,3}$) have zero difference.

Therefore, it is expected that amongst 2^{64} pairs of AES encryptions under the same key, one of the pairs satisfies the difference pattern, and about 2^{32} pairs satisfy the filtering condition. Thus, the following steps of the attack have to be repeated 2^{32} times on average (once for each candidate pair).

We note that if the special difference pattern is satisfied, then by the linearity of the MixColumns operation, there are only 255^2 possible values for the difference in each of the columns before the MixColumns operation of round $r+1$ (denoted by β-s and ϵ-s in Figure 3), and in each of the columns after the MixColumns operation of round $r+2$ (denoted by t-s in Figure 3). This property is used in the second step of the attack to retrieve the actual values of several state bytes.

3.2 Using the Difference Pattern to Retrieve Actual Values of 16 Intermediate State Bytes

In this section we show how the attacker can use the special difference pattern, along with a guess of the difference in eight additional bytes, in order to recover the actual values of 16 intermediate state bytes in both encryptions. We show in detail how the attacker can retrieve the actual value of byte $b_{0,0}$ of the state in the end of round r. The derivation of additional 15 bytes, which is performed in a similar way, is described briefly.

The derivation of the actual value of byte $b_{0,0}$ of the state at the end of round r is composed of several steps (described also in Figure 4):

1. The attacker guesses the differences ν_1, ν_7 and applies the following steps for each such guess.
2. The attacker finds the difference in Column 0 before the MixColumns operation of round $r+1$, i.e., $(\beta_0, \beta_1, \beta_2, \beta_3)$. This is possible since the attacker knows the difference in Column 0 at the end of round $r+1$ (which is $(\alpha_0, 0, \alpha_2, 0)$ where α_0 and α_2 are known from the key stream), and since the AddRoundKey and the MixColumns operations are linear. By performing the inverse ShiftRows operation, the attacker can compute the output difference in byte $b_{0,0}$ after the SubBytes operation of round $r+1$.
3. Given the differences ν_1 and ν_7, there are 255^2 possible differences after the MixColumns of round r in the leftmost column. Using the output bytes $b_{0,0}, b_{2,2}$ of round $r-1$, the attacker knows the difference in two bytes of the same column before the MixColumns operation. Hence, using Observation 2 (the linearity of the MixColumns operation), the attacker retrieves the difference in the whole column, both before and after the MixColumns operation, including the difference γ_0.
4. At this point, the attacker knows the input difference (γ_0) and the output difference (β_0) to the SubBytes operation in byte $b_{0,0}$ of round $r+1$. Hence, using Observation 1 (the property of the SubBytes operation), the attacker finds the actual values of this byte using a single table look-up. In particular, the attacker retrieves the actual value of byte $b_{0,0}$ at the end of round r.

The additional 15 bytes are retrieved in the following way:

1. The value of byte $b_{2,2}$ at the end of round r is obtained in the same way using bytes $b_{0,2}, b_{2,0}$ of the output of round $r-1$ (instead of bytes $b_{0,0}, b_{2,2}$) and examining the third column (instead of the first one).
2. The value of bytes $b_{0,2}$ and $b_{2,0}$ at the end of round r is found by examining α_4, α_6 (instead of α_0, α_2), guessing the differences ν_3, ν_5 (instead of ν_1, ν_7), and repeating the process used in the derivation of bytes $b_{0,0}, b_{2,2}$.
3. In a similar way, by guessing the differences x_1, x_3, x_5, x_7 and using the output bytes of round $r+3$, the attacker can retrieve the actual values of bytes $b_{0,0}, b_{0,2}, b_{2,0}$ and $b_{2,2}$ in the output of round $r+2$.

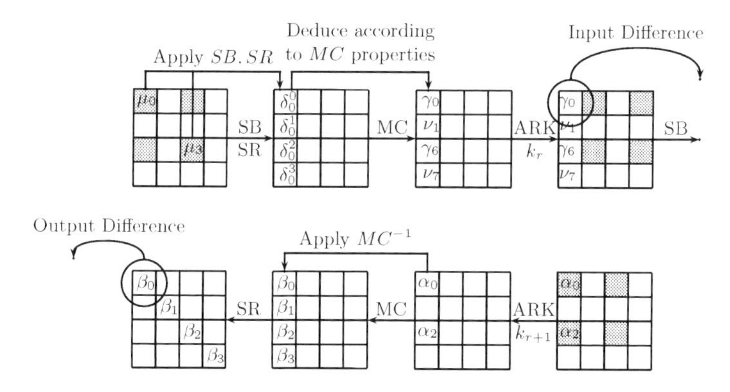

Fig. 4. Deducing the actual value of $b_{0,0}$ in the end of round r

4. Using the output of round r and Observation 2, the attacker can obtain the differences $\alpha_1, \alpha_3, \alpha_5, \alpha_7$. Then, she can use the guessed differences x_1, x_3, x_5, x_7 and Observation 1 to obtain the actual values of bytes $b_{1,1}, b_{1,3}, b_{3,1}$ and $b_{3,3}$ at the end of round $r+1$.
5. Finally, using again the output of round r and Observation 2, the attacker can obtain the differences $\epsilon_1, \epsilon_3, \epsilon_5, \epsilon_7$. Then, using the guessed differences $\nu_1, \nu_3, \nu_5, \nu_7$ and Observation 1, the attacker can obtain the actual values of bytes $b_{1,0}, b_{1,2}, b_{3,0}$, and $b_{3,2}$ at the end of round r.

The bytes whose actual values are known to the attacker at this stage are presented in Figure 5 marked in gray.

4 Retrieving the Key in the Special Cases

The last step of the attack is a guess-and-determine procedure. Given the actual values of the 16 additional state bytes obtained in the second step of the attack, the entire key can be recovered using Observations 2 and 3 (properties of the MixColumns operation and of the key schedule algorithm of AES-128).

The deduction is composed of two phases. In the first phase, presented in Figure 5, no additional information is guessed. We outline in Appendix B the exact steps of the deduction. At the beginning of the second phase, presented in Figure 6, the attacker guesses the value of two additional subkey bytes. We outline in Appendix C the exact steps the attacker performs after guessing these two bytes. In both figures we use gray bytes to mark bytes which are known at the beginning of that deduction phase. Then, if a byte contains a number i it means that this byte is computed in the i-th step of the deduction sequence.

Summarizing the attack, the attacker guesses 10 bytes of information (8 bytes of differences guessed in the second step of the attack, and 2 subkey bytes guessed in the third step of the attack), and retrieves the full secret key. Since all the operations used in the attack are elementary, the attack requires 2^{80} simple operations for each time the attack procedure is applied. Thus, as the attack

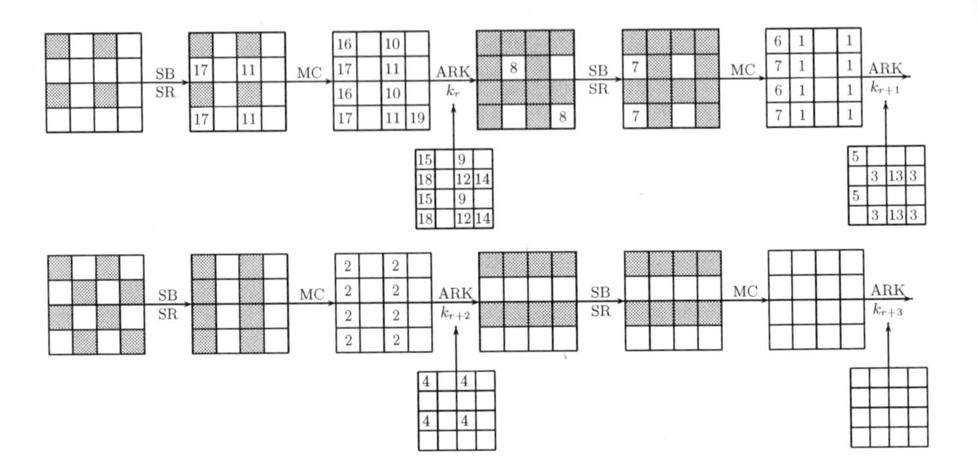

Gray boxes are bytes which are known.
Bytes marked with i, are bytes which are computed in step i.
Transition 9 is based on Observation 3.

Fig. 5. The First Phase of the Guess-and-Determine Attack on LEX (for Odd Rounds)

procedure is repeated 2^{32} times, the total running time of the attack is 2^{112} operations. Since the most time-consuming step of the attack is a guess-and-determine procedure, it is very easy to parallelize the attack, and obtain a speed up equivalent to the number of used CPUs.

4.1 Data Complexity of the Attack

The attack is based on examining special difference patterns. Since the probability of occurrence of a special pattern is 2^{-64}, it is expected that $2^{32.5}$ encryptions under the same key (possibly with different IVs) yield a single pair of encryptions satisfying the special pattern.

However, we note that the attack can be applied for several values of the starting round of the difference pattern. The attack presented above is applicable if r is equal to $1, 3, 5$, or 7, and a slightly modified version of the attack (presented in Appendix A) is applicable if r is equal to $0, 2, 4$, or 6.[4] Hence, $2^{64}/8 = 2^{61}$ pairs of encryptions are sufficient to supply a pair satisfying one of the eight possible difference patterns. These 2^{61} pairs can be obtained from 2^{31} AES encryptions, or equivalently, $2^{36.3}$ bytes of output key stream generated by the same key, possibly under different IVs.

[4] We note that while the attack considers five rounds of the encryption (rounds $r-1$ to $r+3$), it is not necessary that all the five rounds are contained in a single AES encryption. For example, if $r=7$ then round $r+3$ considered in our attack is actually round 0 of the next encryption. The only part of the attack which requires the rounds to be consecutive rounds of the same encryption is the key schedule considerations. However, in these considerations only three rounds (rounds r to $r+2$) are examined.

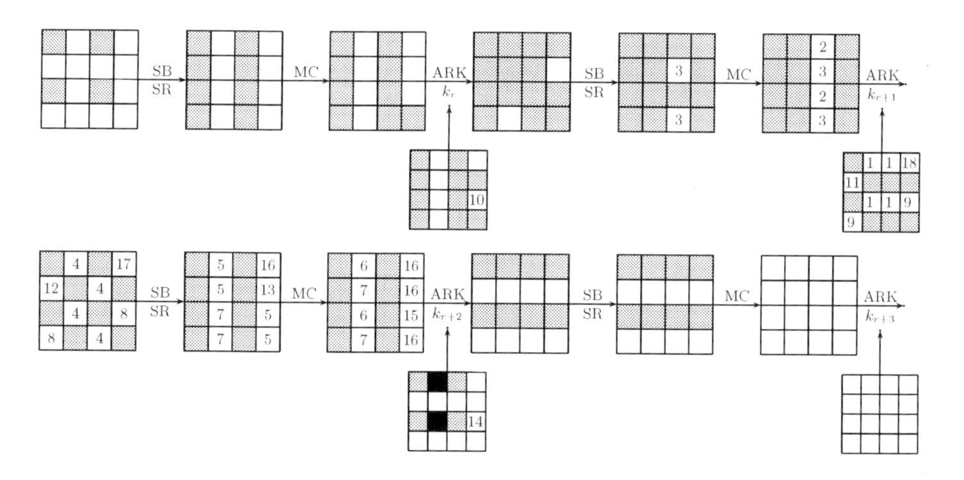

Gray boxes are bytes which are known.
Black boxes are the two bytes guessed in this phase.
Bytes marked with i, are bytes which are computed in step i of the second phase.

Fig. 6. The Second Phase of the Guess-and-Determine Attack on LEX (for Odd Rounds)

5 Further Observations on LEX

In this section we present several observations on the structure of LEX that may be helpful in further cryptanalysis of the cipher.

5.1 Sampling Resistance of LEX

One of the main advantages of LEX, according to the designers (see [6], Section 1), is the small size of its internal state allowing for a very fast key initialization (a single AES encryption). It is stated that the size of the internal state (256 bits) is the minimal size assuring resistance to time-memory-data tradeoff attacks.

Time-memory-data tradeoff (TMDTO) attacks [2,9,10,20] are considered a serious security threat to stream ciphers, and resistance to this class of attacks is a mandatory in the design of stream ciphers (see, for example, [16]). A cipher with an n-bit key is considered (certificationally) secure against TMDTO attacks if any TMDTO attack on the cipher has either data, memory, or time complexity of at least 2^n.

In order to ensure security against conventional TMDTO attacks trying to invert the function (State $\rightarrow$ Key Stream), it is sufficient that the size of the internal state is at least twice the size of the key [10]. LEX satisfies this criterion (the key size is 128 bits and the size of the internal state is 256 bits). As a result, as claimed by the designers (see [6], Sections 3.2 and 5), the cipher is secure with respect to TMDTO attacks.

However, as observed in [1], having the size of the internal state exactly twice larger than the key length is not sufficient if the cipher has a low *sampling resistance*. Roughly speaking, a cipher has a sampling resistance of 2^{-t}, if it is possible to list all internal states which lead to some t-bit output string efficiently. In other words, if it is possible to find a (possibly special) string of t bits, whose "predecessor" states are easily computed, then the cipher has sampling resistance of at most 2^{-t}.

It is easy to see that LEX has maximal sampling resistance of 2^{-32}, as out of the 256 bits of internal state, 32 bits are output directly every round. As a result, using the attack algorithm presented in [10], it is possible to mount a TMDTO attack on LEX with data complexity 2^{88}, and time and memory complexities of 2^{112}. Hence, LEX provides only 112-bit security with respect to TMDTO attacks.

5.2 Loss of Entropy in the Initialization of LEX

The first step in the initialization of LEX is the encryption of IV by AES under the secret key K. When considering $AES_K(IV)$ as a function of K, one can easily see that under reasonable randomness assumptions on AES, this function is a random function of the key K. As a result, the first internal state S used in LEX, does not contain 128 bits of entropy, even when the IV has full entropy. Actually, the expected number of possible S's for a given IV is about 63% of all possible values, i.e., about $2^{127.3}$ possible S's.

Even though our attack does not use this observation, it might still be used in attacks which rely on entropy. Especially, the variant of [15] of time-memory-data tradeoff attacks (trying to invert the function $(key, IV) \to keystream$) might use this observation by trying to invert the function $(key, S) \to keystream$.

5.3 Analysis of the Submitted Reference Implementation of the Original (Untweaked) Version of LEX

After communicating a preliminary version of our attack, we received a request to discuss the implementation of the original (untweaked) version of LEX submitted to eSTREAM. According to a claim made in [3] and verified later by us, the submitted code of the untweaked LEX outputs different bytes than intended and specified (specifically, in the even rounds, $b_{1,1}, b_{1,3}, b_{3,1}$ and $b_{3,3}$ are given as the key stream). Of course, this seems like an unintended typo made in the submission pack (as the fact that it was corrected in the tweaked submission of LEX).

It appears that this variant is much weaker than the intended cipher: First, given the difference in the key stream corresponding to an even round of AES and the consecutive odd round, the difference in two full columns (i.e., four additional internal bytes) can be found easily, without any assumption on the difference between the states. Second, it is possible to devise a simple meet-in-the-middle attack which uses only 256 bits of output stream and retrieves the secret key using 2^{112} simple operations.

The difference between the security of the intended version and that of the actual implementation emphasizes the importance of verifying the implementations of cryptographic primitives very carefully. This importance was first observed in [12] with respect to public key encryption, and adopted in [5] to the symmetric key scenario. While differential fault analysis assumes that the attacker can access both a faulty implementation and a regular implementation, our observations are valid when the attacker has to attack only the faulty implementation.

6 Summary and Conclusions

In this paper we presented a new attack on the LEX stream cipher. We showed that there are special difference patterns that can be easily observed in the output key stream, and that these patterns can be used to mount a key recovery attack.

The attack uses a total of $2^{36.3}$ bytes of key stream produced by a single key (possibly under different IVs) and takes 2^{112} simple operations to implement.

Our results show that for constructions based on the Goldreich-Levin approach (i.e., PRNGs based on pseudo-random permutations), the pseudo-randomness of the underlying permutation is crucial to the security of the resulting stream cipher. In particular, a small number of rounds of a (possibly strong) block cipher cannot be considered random in this sense, at least when a non-negligible part of the internal state is extracted.

References

1. Babbage, S.H., Dodd, M.: Specification of the Stream Cipher Mickey 2.0, submitted to eSTREAM (2006), http://www.ecrypt.eu.org/stream/p3ciphers/mickey/mickey_p3.pdf
2. Babbage, S.H: Improved "exhaustive search" attacks on stream ciphers. In: IEE European Convention on Security and Detection, IEE Conference publication, vol. 408, pp. 161–165. IEE (1995)
3. Bernstein, D.J.: Personal communication (2008)
4. Biham, E., Shamir, A.: Differential Cryptanalysis of the Data Encryption Standard. Springer, Heidelberg (1993)
5. Biham, E., Shamir, A.: Differential Fault Analysis of Secret Key Cryptosystems. In: Kaliski Jr., B.S. (ed.) CRYPTO 1997. LNCS, vol. 1294, pp. 513–525. Springer, Heidelberg (1997)
6. Biryukov, A.: The Design of a Stream Cipher LEX. In: Biham, E., Youssef, A.M. (eds.) SAC 2006. LNCS, vol. 4356, pp. 67–75. Springer, Heidelberg (2007)
7. Biryukov, A.: A New 128-bit Key Stream Cipher LEX, ECRYPT stream cipher project report 2005/013, http://www.ecrypt.eu.org/stream
8. Biryukov, A.: The Tweak for LEX-128, LEX-192, LEX-256, ECRYPT stream cipher project report 2006/037, http://www.ecrypt.eu.org/stream

9. Biryukov, A., Mukhopadhyay, S., Sarkar, P.: Improved Time-Memory Tradeoffs with Multiple Data. In: Preneel, B., Tavares, S. (eds.) SAC 2005. LNCS, vol. 3897, pp. 245–260. Springer, Heidelberg (2006)
10. Biryukov, A., Shamir, A.: Cryptanalytic Time/Memory/Data Tradeoffs for Stream Ciphers. In: Okamoto, T. (ed.) ASIACRYPT 2000. LNCS, vol. 1976, pp. 1–13. Springer, Heidelberg (2000)
11. Blum, M., Micali, S.: How to Generate Cryptographically Strong Sequences of Pseudo-Random Bits. SIAM Journal of Computation 13(4), 850–864 (1984)
12. Boneh, D., DeMillo, R.A., Lipton, R.J.: On the Importance of Checking Cryptographic Protocols for Faults (Extended Abstract). In: Fumy, W. (ed.) EUROCRYPT 1997. LNCS, vol. 1233, pp. 37–51. Springer, Heidelberg (1997)
13. Daemen, J., Rijmen, V.: AES Proposal: Rijndael, NIST AES proposal (1998)
14. Daemen, J., Rijmen, V.: The design of Rijndael: AES — the Advanced Encryption Standard. Springer, Heidelberg (2002)
15. Dunkelman, O., Keller, N.: Treatment of the Initial Value in Time-Memory-Data Tradeoff Attacks on Stream Ciphers. Information Processing Letters 107(5), 133–137 (2008)
16. ECRYPT, Call for Stream Cipher Primitives, version 1.3 (April 12, 2005), `http://www.ecrypt.eu.org/stream/call/`
17. Englund, H., Hell, M., Johansson, T.: A Note on Distinguishing Attacks. In: Preproceedings of State of the Art of Stream Ciphers workshop (SASC 2007), Bochum, Germany, pp. 73–78 (2007)
18. Ferguson, N., Kelsey, J., Lucks, S., Schneier, B., Stay, M., Wagner, D., Whiting, D.: Improved Cryptanalysis of Rijndael. In: Schneier, B. (ed.) FSE 2000. LNCS, vol. 1978, pp. 213–230. Springer, Heidelberg (2001)
19. Goldreich, O., Levin, L.A.: A Hard-Core Predicate for all One-Way Functions. In: Proceedings of 21st STOC, pp. 25–32. ACM, New York (1989)
20. Golic, J.D.: Cryptanalysis of Alleged A5 Stream Cipher. In: Fumy, W. (ed.) EUROCRYPT 1997. LNCS, vol. 1233, pp. 239–255. Springer, Heidelberg (1997)
21. Håstad, J., Näslund, M.: BMGL: Synchronous Key-stream Generator with Provable Security. NESSIE project (submitted, 2000), `http://www.nessie.eu.org`
22. National Institute of Standards and Technology, Advanced Encryption Standard, Federal Information Processing Standards Publications No. 197 (2001)
23. Wu, H., Preneel, B.: Attacking the IV Setup of Stream Cipher LEX, ECRYPT stream cipher project report 2005/059, `http://www.ecrypt.eu.org/stream`

A Special Difference Pattern Starting with an Even Round

In this section we present the modified version of the attack that can be applied if the special difference pattern occurs in the even rounds. The first two steps of the attack (observing the difference pattern and deducing the actual values of 16 additional bytes of the state) are similar to the first two steps of the attack presented in Section 3. The known byte values after these steps are presented in Figure 7, marked in gray. The third step of the attack is slightly different due to the asymmetry of the key schedule, and Observation 4 is used in this step along with Observations 2 and 3. The two phases of this step are presented in

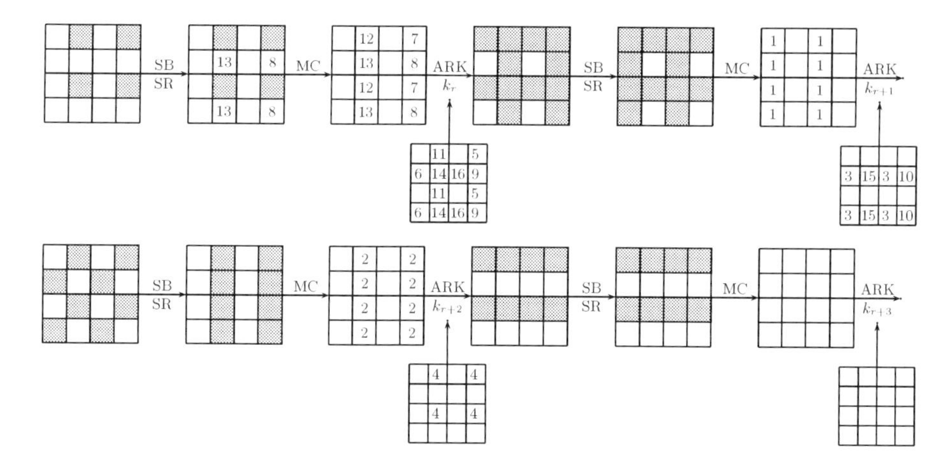

Gray boxes are bytes which are known.
Bytes marked with i, are bytes which are computed in step i.
Step 5 is based on Observation 3, and step 11 is based on Observation 4.

Fig. 7. The First Phase of the Guess-and-Determine Attack on LEX (in Even Rounds)

Figures 7 and 8. The overall time complexity of the attack is 2^{112} operations, like in the case of a difference pattern in the odd rounds.

B Detailed Description of the Steps in the First Deduction Phase

In this section we present the exact deduction steps done during the first phase depicted in Figure 5. The numbers of the steps correspond to the numbers in the figure.

1. The application of MixColumns in round $r+1$ on two columns (second and fourth) gives these bytes.
2. The application of MixColumns in round $r+2$ on two columns (first and third) gives these bytes.
3. The knowledge of the value of four bytes before the XOR with the subkey k_{r+1} and after the XOR, gives the value of the subkey in these bytes.
4. The knowledge of the value of four bytes before the XOR with the subkey k_{r+2} and after the XOR, gives the value of the subkey in these bytes.
5. By the key schedule of AES, the knowledge of byte (0,0) of the subkey k_{r+2} and byte (1,3) of the subkey k_{r+1} gives the value of byte (0,0) of the subkey k_{r+1}. Similarly, the knowledge of byte (2,0) of the subkey k_{r+2} and byte (3,3) of the subkey k_{r+1} gives the value of byte (2,0) of the subkey k_{r+1}.
6. These two bytes are the XOR of the two subkey bytes found in the previous step and known bytes.

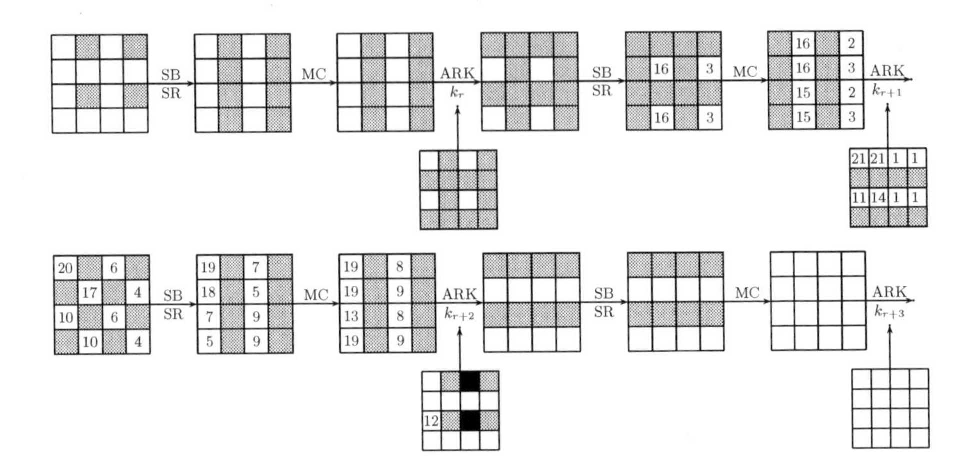

Gray boxes are bytes which are known.
Black boxes are the two bytes guessed in this phase.
Bytes marked with i, are bytes which are computed in step i of the second phase.

Fig. 8. The Second Phase of the Guess-and-Determine Attack on LEX (in Even Rounds)

7. Applying Observation 2 to the first column in the MixColumns operation of round $r+1$ gives these four bytes.
8. The two bytes after the SubBytes and ShiftRows operation are just computed backwards.
9. These bytes are computed using the four bytes found in Step 4, and the application of Observation 3.
10. These bytes are computed by XORing the subkey bytes found in the previous step with known values.
11. Applying Observation 2 to the third column in the MixColumns operation of round r gives these four bytes.
12. The input and output of the AddRoundKey operation of round k_r in these two bytes is known, and allows retrieving these two subkey bytes.
13. By the key schedule of AES, the knowledge of bytes (1,1) and (3,1) of the subkey k_{r+1} and bytes (1,2) and (3,2) of the subkey k_r gives the values of bytes (1,2) and (3,2) of the subkey k_{r+1}, respectively.
14. By the key schedule of AES, the knowledge of bytes (1,2) and (3,2) of the subkey k_r and bytes (1,3) and (3,3) of the subkey k_{r+1} gives the values of bytes (1,3) and (3,3) of the subkey k_r, respectively.
15. By the key schedule of AES, the knowledge of byte (0,0) of the subkey k_{r+1} and byte (1,3) of the subkey k_r gives the value of byte (0,0) of the subkey k_r. Similarly, the knowledge of byte (2,0) of the subkey k_{r+1} and byte (3,3) of the subkey k_r gives the value of byte (2,0) of the subkey k_r.
16. These bytes are computed by XORing the subkey bytes found in the previous step with known values.

17. Applying Observation 2 to the first column in the MixColumns operation of round r gives these four bytes.
18. These bytes are the XOR of the bytes found in the previous step with known bytes.
19. This byte is the XOR of one of the bytes found in Step 14 and a byte found in Step 8.

C Detailed Description of the Steps in the Second Deduction Phase

In this section we present the exact deduction steps performed during the second phase depicted in Figure 6. The numbers of the steps correspond to the numbers in the figure.

1. Using the key schedule algorithm, it is possible to deduce four bytes of k_{r+1} (each is the XOR of two known bytes in k_{r+2}).
2. Decrypting two known bytes using two of the subkey words found in Step 1 gives these two bytes.
3. Applying Observation 2 to the third column in the MixColumns operation of round $r+1$.
4. These four bytes are computed by the XOR of known state bytes and subkey bytes (in k_{r+1}).
5. These four bytes are the application of the SubBytes and ShiftRows operations on the bytes found in the previous step.
6. These bytes are the XOR of known bytes and the subkey bytes that were guessed.
7. Applying Observation 2 to the second column in the MixColumns operation of round $r+2$ gives these four bytes.
8. These two bytes are found by applying the inverse ShiftRows and SubBytes operations to two of the bytes found in the previous step.
9. These two subkey bytes are computed as the XOR of the corresponding bytes before and after the AddRoundKey operation of round $r+1$.
10. By the key schedule of AES, the knowledge of byte (3,0) of the subkey k_{r+1} and byte (3,0) of the subkey k_r gives the value of byte (2,3) of the subkey k_r.
11. By the key schedule of AES, the knowledge of bytes (1,0) and (2,3) of the subkey k_{r+1} gives the value of byte (1,0) of the subkey k_{r+1}.
12. This byte is the XOR of a known state byte with the subkey byte found in the previous step.
13. This byte is computed by applying the SubBytes and ShiftRows operations to the byte found in the previous step.
14. By the key schedule of AES, the knowledge of byte (2,2) of the subkey k_{r+2} and byte (2,3) of the subkey k_{r+1} gives the value of byte (2,3) of the subkey k_{r+2}.
15. This byte is the XOR of a known state byte with the subkey byte found in the previous step.

16. This byte is the partial decryption of a known byte by the byte found in the previous step.
17. Applying Observation 2 to the fourth column in the MixColumns operation of round $r + 2$ gives these four bytes.
18. This byte is found by applying the inverse ShiftRows and SubBytes operations to one of the bytes found in the previous step.
19. This byte is the partial decryption of a known byte by the byte found in the previous step.

Breaking the F-FCSR-H Stream Cipher in Real Time

Martin Hell and Thomas Johansson

Dept. of Electrical and Information Technology, Lund University,
P.O. Box 118, 221 00 Lund, Sweden

Abstract. The F-FCSR stream cipher family has been presented a few years ago. Apart from some flaws in the initial propositions, corrected in a later stage, there are no known weaknesses of the core of these algorithms. The hardware oriented version, called FCSR-H, is one of the ciphers selected for the eSTREAM portfolio.

In this paper we present a new and severe cryptanalytic attack on the F-FCSR stream cipher family. We give the details of the attack when applied on F-FCSR-H. The attack requires a few Mbytes of received sequence and the complexity is low enough to allow the attack to be performed on a single PC within seconds.

1 Introduction

The cryptographic scene include a variety of efficient and trusted block ciphers. However the same does not seem to hold for stream ciphers. The stream ciphers that have received attention through use in various standards tend to have more or less serious security weaknesses. Examples are A5 algorithms used in GSM, the RC4 algorithm used in for example WLAN applications through the WEP protocol and the E0 stream cipher used in Bluetooth.

Based on a belief that a dedicated stream cipher still has a capability of significantly outperforming a block cipher, the eSTREAM project was launched in 2004. The goal of this project was to solicit and evaluate submitted proposals of stream ciphers for future standardization. The main evaluation criteria set up were long-term security, efficiency in terms of performance, flexibility and market requirements.

The eSTREAM project considered two different profiles, one targeting software implemented stream ciphers; and one for hardware implemented stream ciphers (in particular constrained devices). The hardware category received a total of 25 submitted proposals. After three phases of evaluation, the final eSTREAM portfolio recommended four of them. One of them is a design called F-FCSR-H v2.

F-FCSR-H v2 is one of several algorithms in the F-FCSR family of stream ciphers designed by the French researchers F. Arnault, T.P. Berger, and C. Lauradoux. The family of ciphers is based on feedback with carry shift registers (FCSR) together with a filtering function. The idea of using FCSRs to generate sequences for cryptographic applications was initially proposed by Klapper

J. Pieprzyk (Ed.): ASIACRYPT 2008, LNCS 5350, pp. 557–569, 2008.
© International Association for Cryptologic Research 2008

and Goresky in [6]. The F-FCSR family was introduced in [1], proposing four concrete constructions. These proposals were cryptanalyzed in [5]. The initial version submitted to eSTREAM, targeting hardware, was called F-FCSR-H. It was shown in [4] that this construction also had security problems. This lead to a change in the initialization procedure and the resulting algorithm was named F-FCSR-H v2. This paper will focus on the specification of F-FCSR-H v2 given in [2].

The eSTREAM class of hardware stream ciphers (and F-FCSR-H v2 in particular) prescribes a key of length 80 bits. Apart from the initial flaws (on the IV-setup procedure, and a TMD tradeoff attack), there are yet no known weaknesses of the core of these algorithms and the best attack on F-FCSR-H v2 is an exhaustive key search.

In this paper we present a new and severe cryptanalytic attack on the F-FCSR stream cipher family. We give the details of the attack when applied on F-FCSR-H v2. The attack is based on observing that the contribution of nonlinearity comes from the carry bits and that sometimes this contribution is too low and the system can be linearized. The whole attack require a few Mbytes of received sequence and the complexity is low enough to allow the attack to be performed on a single PC within seconds. The attack has been fully implemented using the designers' reference implementation.

In Section 2 we give an overview of the FCSR automaton and the F-FCSR construction. In Section 3 we then discuss the underlying weaknesses giving the attack. In Section 4 we give a description of the attack and in Section 5 we give a more detailed analysis of parts of the attack and we also give the estimated and simulated complexities. In Section 6 we give a rough outline of how the key could be reconstructed from a known state.

2 Recalling the FCSR Automaton and the F-FCSR Construction

Recall that a Feedback with Carry Shift Register (FCSR) is a device that computes the binary expansion of a 2-adic number p/q, where p and q are some integers, with q odd. For simplicity one can assume that $q < 0 < p < |q|$. Following the notation from [2], the size n of the FCSR is the value such that $n+1$ is the bitlength of $|q|$. In the stream cipher construction, p depends on the secret key (and the IV), and q is a public parameter. The choice of q induces some properties of the FCSR. The most important one is that it completely determines the length of the period T of the keystream. The conditions for an optimal choice as used in the F-FCSR family of stream ciphers are: q is a (negative) prime of bitsize $n+1$; the order of 2 modulo q is $|q|-1$; and $T = (|q|-1)/2$ is also prime. Furthermore, set $d = (1+|q|)/2$. Then the Hamming weight $W(d)$ of the binary expansion of d is checked to be not too small, say $W(d) > n/2$.

The FCSR automaton as described in [2] is one way to efficiently implement the generation of the 2-adic expansion sequence. It contains two registers: the main register $\mathbf{M}$ and the carries register $\mathbf{C}$. The main register $\mathbf{M}$ contains n

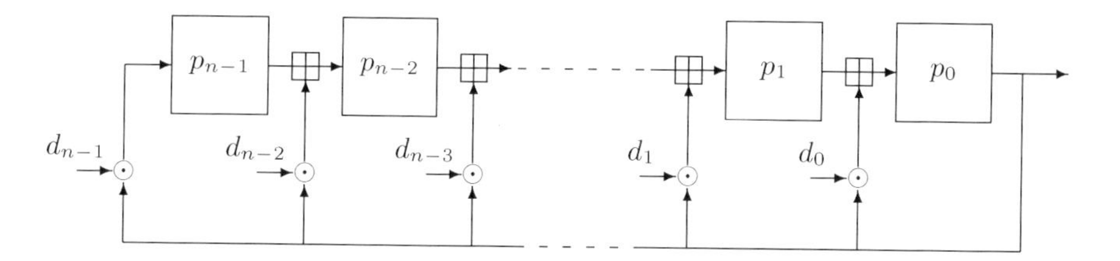

Fig. 1. Automaton to compute the 2-adic expansion of p/q

cells. Let $\mathbf{M} = (m_{n-1}, m_{n-2}, \ldots, m_1, m_0)$ and associate $\mathbf{M}$ to the integer $M = \sum_{i=0}^{n-1} m_i \cdot 2^i$.

Recall the positive integer $d = (1 + |q|)/2$ and its binary representation $d = \sum_{i=0}^{n-1} d_i \cdot 2^i$. The carries register contains l active cells where $l+1$ is the number of nonzero d_i binary digits in d. The active cells are the ones in the interval $0 \leq i \leq n-2$ and $d_{n-1} = 1$ always hold. For this purpose we write the carries register $\mathbf{C}$ as $\mathbf{C} = (c_{n-2}, c_{n-3}, \ldots, c_1, c_0)$ and associate $\mathbf{C}$ to the integer $C = \sum_{i=0}^{n-2} c_i \cdot 2^i$. Note that only l of the bits in $\mathbf{C}$ are active and the remaining ones are set to zero. Let the integer p be written as $p = \sum_{i=0}^{n-1} p_i \cdot 2^i$, where $p_i \in \{0, 1\}$. Then the 2-adic expansion of the number p/q is computed by the automaton given in Figure 1.

The automaton is referred to as the Galois representation and it is very similar to the Galois representation of a usual LFSR. Other representations in connection with F-FCSR were considered in [7]. For all defined variables we also introduce a time index t, and let $\mathbf{M}(t)$ denote the content of $\mathbf{M}$ at time t. Similarly, $\mathbf{C}(t)$ denotes the content of $\mathbf{C}$ at time t.

The addition with carry, denoted $\boxplus$ in Figure 1, has a one bit memory (the carry). It takes three inputs in total, two external inputs and the carry bit. It outputs the XOR of the inputs and it sets the new carry value to one if the integer sum of the three inputs is two or three.

In Figure 2 we give an illustrating example (following [2]). Here $q = -347$ giving $d = 174$ and its binary expansion (10101110). The F-FCSR family of stream ciphers uses this particular automaton as the central part of their construction. So for future considerations in this paper we only need to recall the FCSR

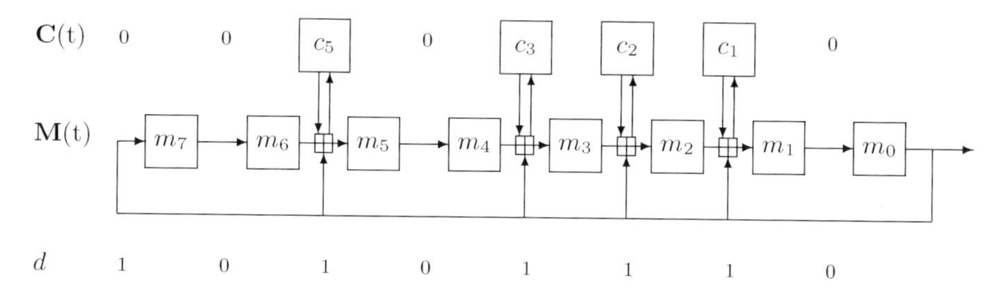

Fig. 2. Example of an FCSR

automaton as implemented in Figure 1 and Figure 2. Important facts are that the FCSR automaton has n bits of memory in the main register and l bits in the carry register, in total $n+l$ bits. If $(\mathbf{M}, \mathbf{C})$ is our *state*, then many states are equivalent in the sense that starting in equivalent states will produce the same output. As the period is $|q|-1 \approx 2^n$ the number of states equivalent to a given state is in the order of 2^l.

2.1 Describing the F-FCSR-H Construction

The F-FCSR family of stream ciphers combines the FCSR automaton with a filtering function. The filtering function extracts keystream bits from the state of the main register in the FCSR automaton. The filter is a simple linear function of bits from the state. In order to increase the throughput, the constructions extract not only one but many bits each clock cycle. The number of extracted bits is eight for F-FCSR-H. Thus there are 8 different filters, now called subfilters, used to extract an 8 bits keystream byte after each transition of the automaton.

A one bit filter F is a bitstring $(f_0, \ldots, f_{n-1})$ of length n. The output bit of the filter is defined to be,

$$F(\mathbf{M}) = \bigoplus_{i=0}^{n-1} f_i m_i,$$

i.e., the scalar product. As F is a known string the output is a linear function (in F_2).

For the 8 bit filter, it consists of 8 such binary functions $F_0, F_1, \ldots, F_7$. However, filter F_j uses only cells m_i in the main register that satisfies $i = j \pmod 8$.

The parameters for F-FCSR-H are now given. The proposal uses key length 80 and IV of bitsize v with $32 \leq v \leq 80$. The core of the F-FCSR-H algorithm has remained identical to the one originally proposed in [1]. Only the key and IV initialization procedure was updated in [2].

The FCSR length (size of the main register) is $n = 160$. The carries register contains $l = 82$ cells. The feedback is determined by the prime

$$q = 1993524591318275015328041611344215036460140087963.$$

This gives

$$d = (1 + |q|)/2 = (\text{AE985DFF 26619FC5 8623DC8A AF46D590 3DD4254E})$$

(hexadecimal notation). So addition boxes and carries cells are present at the positions matching the binary ones in the binary expansion of d. To extract one keystream byte, FCSR-H uses the static filter

$$F = d = (AE985DFF26619FC58623DC8AAF46D5903DD4254E).$$

Using the designers notation, this means that the 8 subfilters (subfilter j is obtained by selecting the bit j in each byte of F) are given by

$$F_0 = (00110111010010101010), F_4 = (01110010001000111100),$$
$$F_1 = (10011010110111000001), F_5 = (10011100010010001010),$$
$$F_2 = (10111011101011101111), F_6 = (00110101001001100101),$$
$$F_3 = (11110010001110001001), F_7 = (11010011101110110100).$$

So the F-FCSR-H generator outputs one byte every time instance and it is simply given as

$$\mathbf{z} = (m_8 + m_{24} + m_{40} + m_{56} + \ldots + m_{136}, m_1 + m_{49} + \ldots, \ldots, m_{23} + \ldots).$$

The key and IV initialization consists of loading key and IV into the main register, clocking 20 times and extracting 20 bytes of output. These 160 bits are used as initial state in the main register of the FCSR automaton and it is clocked 162 times without producing output. More details are given in Section 6.

The second relevant construction in the F-FCSR family, called F-FCSR-16, is constructed in a similar manner. However it has a larger state and extracts 16 bits every clock cycle.

3 Weaknesses of the FCSR Automaton and the F-FCSR Family of Stream Ciphers

As the filtering function is F_2 linear, essentially all the security of the FCSR constructions rely on the FCSR automaton ability to create nonlinearity. It might at first glance look like this is achieved. The nonlinearity lies in the carry bit calculation, and carry bits are quickly spread over the entire main register. They enter new carry bit calculations, thus increasing the degree of nonlinear expressions rapidly. This is probably the first way one tries to analyze the construction, looking at the algebraic expressions created when the automaton is clocked a few times. It looks difficult to find some useful algebraic expression or some correlation between different variables that can be tracked all the way to the keystream symbols.

Instead, we look at the nonlinearity from a different perspective. The main observation we use is the fact that the carry bits in the carries register behave very far from random. The key point is that they all have one common input variable, the feedback bit. Let us look at what happens for a carry bit when the feedback bit is set to zero. We can see that when the feedback bit is zero then a carry bit that is zero must remain zero whereas if the carry bit is one then by probability 1/2 it will turn to zero (assuming random input on the active input). If we now assume that the feedback bit is zero a few consecutive time instances, then it is very likely that the carry bit is pushed to zero.

Actually, the same arguments can be repeated when the feedback bit is one. Then the carry is more likely to be one and by repeatedly having ones on the feedback bit we push the carry value to one. However, for the moment we ignore this case.

Since the feedback bit is a common input to all carries, this has a dramatic effect on the carries vector $\mathbf{C}$. We know that $\mathbf{C}$ has $l = 82$ active cells (carry bits) and we can expect that on average **C** will have a weight of 41. However, the weight is strongly correlated to the values of the feedback bit. Every time the feedback bit is zero all cells in **C** that are zero must remain zero, whereas those with value one has a 50% chance of becoming zero. So a zero feedback bit at time t gives a carries vector at time $t+1$ of roughly half the weight compared to time t. This behavior is easily checked by just running the generator and observing the contents of **C**.

Having found this crucial observation, the attack looks almost trivial. We assume that we have a number of consecutive feedback bits all zero. This would push the carries register to the all zero content. Then 19 more zero feedback bits to keep **C** zero all the time. During this time the generator outputs 20 bytes, or 160 bits. We can thus reconstruct the main register from knowing these values and the fact that **C** is zero. The only problem is that this does not work.

4 Describing the Attack

The underlying ideas of the attack were given in the previous section. However, the assumption that a large number of consecutive zero feedback bits would push the weight of **C** to zero is wrong. By simply running the generator we could see that this never happened. Once you look at the details, there is a simple explanation for this. Look at the FCSR automaton as illustrated in Figure 2, especially the last (least significant) active cell c_1 among the carries. Assume that the feedback bits are zero from time t to $t+t_0$ and the feedback bit at time $t-1$ was one. Now since the feedback bit at time $t-1$ was one and the feedback bits are zero from time t to $t+t_0$ the last carry addition must return zero to the next main register cell. Thus it must set the carry to one. Now, when the carry is one the only way we can have zero output and thus zero feedback is if the main register input to the last carry addition is one. Thus the last carry cell will never be pushed to zero, as we initially hoped. The fact that the carry vector and the feedback will not be zero for several consecutive clock cycles was actually observed in [3]. It was shown that this situation can not occur if the FCSR automata has reached a state of the main cycle, which is the case for all proposed F-FCSR stream ciphers.

However, this is not a problem. We slightly modify our approach and then it will work. As we described above, the all zero feedback sequence can appear if the main register input to the last carry addition is the all one sequence and we start with setting the carry bit to one. Then the all zero feedback will push the weight of **C** to one (the last active carry cell is always one). So it is natural to define the following event.

$$\text{Event } E_{\text{zero}} : \mathbf{C}(t) = \mathbf{C}(t+1) = \ldots = \mathbf{C}(t+19) = (0,0,\ldots,0,1,0).$$

When this happens we know that we have had 20 consecutive zeros in the feedback and that the carry has remained constant for 20 time instances. Using our previous arguments we would think that we need about $\log_2 82 \approx 7$ zeros in the feedback to push the weight of **C** to 1 and then an additional 19 zeros in the feedback to keep $\mathbf{C}$ constant for 20 time instances. Assuming a uniform distribution on the feedback bits this would lead to a probability of very roughly 2^{-26} for the event E_{zero} to happen. As we will see in the next section is it possible to use more information about the state in order to increase the efficiency of the attack. For now, let us just assume that we know how the main register $\mathbf{M}$ at time $t+1, t+2, \ldots, t+19$ depends on $\mathbf{M}(t)$ and that this dependency is linear.

Assuming that event E_{zero} occurs, the remaining part is to recover the main register from the given keystream bytes $\mathbf{z}(t), \mathbf{z}(t+1), \ldots, \mathbf{z}(t+19)$. This will lead to a linear system of equations with 160 equations in 160 unknowns. This could basically be solved through Gaussian elimination, costing something like 160^3 operations. However, we observe that the equations have the special byte structure explained before. There are 20 equations that only include the main register variables $m_0, m_8, m_{16}, \ldots, m_{152}$, there are 20 equations that only include $m_1, m_9, m_{17}, \ldots, m_{153}$, etc. Note that we are only shifting in zeros in $\mathbf{M}$ due to the assumption.

So it is much more efficient to treat each 20 by 20 system of equations independently. Let us describe the received systems of linear equations in more detail. We denote the least significant bit of $\mathbf{z}(t)$ by $\mathbf{z}(t)_0$, the next bit by $\mathbf{z}(t)_1$ etc, i.e., the output byte $\mathbf{z}(t)$ at time t is given by

$$\mathbf{z}(t) = (\underbrace{\mathbf{z}(t)_7}_{MSB}, \mathbf{z}(t)_6, \mathbf{z}(t)_5, \mathbf{z}(t)_4, \mathbf{z}(t)_3, \mathbf{z}(t)_2, \mathbf{z}(t)_1, \underbrace{\mathbf{z}(t)_0}_{LSB}). \tag{1}$$

Then the linear equations involving the main register bits m_i when $i \equiv 0 \mod 8$ at time t can be written as

$$\begin{aligned}
\mathbf{z}(t)_0 &= m_8 \oplus m_{24} \oplus \ldots \oplus m_{136}, \\
\mathbf{z}(t+1)_7 &= m_{24} \oplus m_{40} \oplus \ldots \oplus m_{152}, \\
&\vdots \\
\mathbf{z}(t+19)_5 &= m_{32} \oplus m_{48} \oplus \ldots \oplus m_{152}.
\end{aligned}$$

Similar equations containing only the main register bits m_i such that $i \equiv 1 \mod 8$ can also be listed. The same then goes for equations using only m_i bits when $i \equiv 2 \mod 8$, etc. Altogether, we can for simplicity write

$$\begin{aligned}
\mathbf{W_0} &= (\mathbf{z}(t)_0, \mathbf{z}(t+1)_7, \ldots, \mathbf{z}(t+19)_5), \\
\mathbf{W_1} &= (\mathbf{z}(t)_1, \mathbf{z}(t+1)_0, \ldots, \mathbf{z}(t+19)_6), \\
&\vdots \\
\mathbf{W_7} &= (\mathbf{z}(t)_7, \mathbf{z}(t+1)_6, \ldots, \mathbf{z}(t+19)_4).
\end{aligned}$$

The vector of main register values $m_0, m_8, m_{16}, \ldots, m_{152}$ is denoted $\hat{\mathbf{M}}_\mathbf{0}$. Then we get

$$\mathbf{W_0} = \hat{\mathbf{M}}_\mathbf{0} P_0, \tag{2}$$

where P_0 is a known 20 by 20 matrix (determined from the filter F). Similarly, $\hat{\mathbf{M}}_\mathbf{i}$, $1 \leq i \leq 7$ will denote the main register variables $(m_i, m_{i+8},$

$m_{i+16}, \ldots, m_{i+152}$). With this notation we can write the eight 20 by 20 linear systems of equations as

$$\mathbf{W_0} = \hat{\mathbf{M}}_0 P_0, \mathbf{W_1} = \hat{\mathbf{M}}_1 P_1, \ldots, \mathbf{W_7} = \hat{\mathbf{M}}_7 P_7. \quad (3)$$

Of course, some equations need to have 1 added to them since we have to compensate for the fact that the carry vector is given as $\mathbf{C} = (0, 0, \ldots, 0, 1, 0)$.

The idea is now to precompute, for each linear system, the solution $\hat{\mathbf{M}}_\mathbf{i}$ for each possible value of the vector of keystream bits $\mathbf{W_i}$. This would require 8 tables of size 2^{20} entries, each entry being a 20 bit vector. Though, the real time phase will be more efficient if 20 bytes are stored in each entry, having values only in the bit positions corresponding to the bits in $\hat{\mathbf{M}}_\mathbf{i}$. Then a full candidate state can be found by just ORing together the 8 saved contributions.

Finding the main register content would then require only to compute the vectors $\mathbf{W_i}$, $0 \leq i \leq 7$ from the keystream and then 8 table lookups to get the candidate main register state. The part of a candidate main register state given by $\mathbf{W_i}$ is denoted $\text{TABLE}_i[\mathbf{W_i}]$.

We can note that the P_i matrices are not all of full rank. This means that for our table of solutions, some $\mathbf{W_i}$ values will have no solutions whereas other values will have multiple (a power of two) solutions. This fact will then be combined over all 8 systems of equations, leading to a total number of $S = \prod_{i=0}^{7} s_i$ solutions, where s_i is the number of solutions to the ith system. Thus $\text{TABLE}_i[\mathbf{W_i}]$ returns a set of zero or more solutions.

In our case this property will increase the efficiency of the attack because if we get a value $\mathbf{W_0}$ for which $\text{TABLE}_0[\mathbf{W_0}]$ returns no solutions we can immediately stop and conclude that our assumption of event E_{zero} was wrong.

We now summarize our attack as follows.

0. **for** $t = 1$ to T_{max} do
1. Select the 20 consecutive output bytes $\mathbf{z}(t), \mathbf{z}(t+1), \ldots, \mathbf{z}(t+19)$.
   for $i = 0$ to 7
     Compute $\mathbf{W_i}$
     if $\text{TABLE}_i[\mathbf{W_i}]$ has no solutions
       go to 0.
     else
       store all possible values for $\hat{\mathbf{M}}_\mathbf{i}$.
   end for
3. "Check candidate states": Test all possible values of $(\hat{\mathbf{M}}_0, \hat{\mathbf{M}}_1, \ldots, \hat{\mathbf{M}}_7)$, by checking if a candidate value generates $\mathbf{z}(t+20), \mathbf{z}(t+21), \ldots$.
4. go to 0.

5 Improving the Attack Complexity

In the previous section we assumed that the carry vector was fixed to $\mathbf{C}(t) = \mathbf{C}(t+1) = \ldots = \mathbf{C}(t+19) = (0, 0, \ldots, 0, 1, 0)$ for all considered time instances.

However we note that this is not necessary. As long as we can express the output bits in $\mathbf{z}(t), \mathbf{z}(t+1), \ldots, \mathbf{z}(t+19)$ as linear equations in the main register variables at time t, the attack will work.

Denote the state at time t as $(\mathbf{M}, \mathbf{C})(t)$ and let x represent bits in the state that the output can be expressed as linear combinations of. Let ? represent bits that we do not need to know the value of. Assume that the state $(\mathbf{M}, \mathbf{C})(t)$ is given by

$$(\mathbf{M}, \mathbf{C})(t) = (xx \ldots xx0\underbrace{11\ldots11}_{16}00, 000\ldots0010).$$

Then, the state will be updated as

$$\begin{aligned}
(\mathbf{M}, \mathbf{C})(t+1) &= (xx \ldots xx0\underbrace{11\ldots11}_{15}00, 000\ldots0010),\\
(\mathbf{M}, \mathbf{C})(t+2) &= (xx \ldots xx0\underbrace{11\ldots11}_{14}00, 000\ldots0010),\\
&\vdots\\
(\mathbf{M}, \mathbf{C})(t+15) &= (xxxxxxxx \ldots xx0100, 000\ldots0010),\\
(\mathbf{M}, \mathbf{C})(t+16) &= (xxxxxxxx \ldots xxx000, 000\ldots0010),\\
(\mathbf{M}, \mathbf{C})(t+17) &= (xxxxxxxx \ldots xxxx10, 000\ldots0000),\\
(\mathbf{M}, \mathbf{C})(t+18) &= (xxxxxxxx \ldots xxxxx1, 000\ldots0000),\\
(\mathbf{M}, \mathbf{C})(t+19) &= (xxxxxxxx \ldots xxxxxx, ??????????).
\end{aligned}$$

The only difference from the case presented in the previous section is that we should not compensate for the carry bit when computing the state $(\mathbf{M}, \mathbf{C})(t+18)$ and we need to compensate for the 1 in the feedback when computing the state $(\mathbf{M}, \mathbf{C})(t+19)$. Note that the feedback used when calculating $(\mathbf{M}, \mathbf{C})(t+19)$ will cause the carry vector to be unpredictable. However, only $\mathbf{M}(t+19)$ is used to extract $\mathbf{z}(t+19)$ and knowledge of the carry vector here is not necessary. Using these observations, we can conclude that we only require the carry vector to take the value $(0, 0, \ldots, 0, 1, 0)$ at least 17 consecutive time instances. Thus, we update the definition of E_{zero} to

$$\text{Event } E_{\text{zero}} : \mathbf{C}(t) = \mathbf{C}(t+1) = \ldots = \mathbf{C}(t+16) = (0, 0, \ldots, 0, 1, 0).$$

The probability of E_{zero} has been simulated using in total 2 TB data and 2000 different keys and is estimated to be

$$P(E_{\text{zero}}) = 2^{-25.3}. \tag{4}$$

Thus, we would expect that we need on average $2^{25.3}$ bytes of keystream to recover the state.

The attack using the observations from this section has been fully implemented. The low complexity of the attack allows it to be simulated targeting the full version of F-FCSR-H v2. Using 5000 random keys, the state was recovered using on average $2^{24.7}$ bytes of keystream. The success rate was 100%. The slightly lower amount of keystream which was observed compared to the expected amount can easily be accounted for. For each state there are many equivalent states and sometimes one of these equivalent states is recovered. As an example, if $\mathbf{C}(t) = \mathbf{C}(t+1) = \mathbf{C}(t+15) = (0,0,\ldots,0,1,0)$ but $\mathbf{C}(t-1) \neq (0,0,\ldots,0,1,0)$, then $(\mathbf{M},\mathbf{C})_{t-1}$ can be recovered if it is equivalent to another state $(\mathbf{M}',\mathbf{C}')$ with $\mathbf{C}' = (0,0,\ldots,0,1,0)$. Since the two states will merge after a few clocks, the attack will also recover the real state.

A slight improvement of the attack is achieved by noting that we can also look at the situation when the carry vector is one in all active positions except the last. The required keystream length will be halved, but the attack time will remain unchanged. The same simulation was performed with this improvement and as expected the state was recovered using on average $2^{23.7}$ bytes of keystream.

6 Recovering the Key

We have described a state recovery attack that completely breaks F-FCSR-H. We now outline how we can also derive the key from a known state at any time t. In order to shortly describe this, we recall the initialization from the design document (reference code). Inputs to the initialization are a key K of length 80 bits and an IV of length $v \leq 80$ bits. For simplicity we fix the IV length to 80 bits.

Key+IV setup

1. The main register $\mathbf{M}$ is initialized with key and IV by
$$\mathbf{M} = K + 2^{80}IV = (IV||K),$$
and the carries register $\mathbf{C} = 0$.
2. A loop is iterated 20 times. Each iteration of this loop consists in clocking the FCSR and then extracting a pseudorandom byte $S_i (0 \leq i \leq 19)$ using the filter.
3. The main register $\mathbf{M}$ is reinitialized with these bytes:
$$\mathbf{M} = (S_{19}, S_{18}, \ldots, S_0),$$
and $\mathbf{C} = 0$.
4. The FCSR is clocked 162 times (output is discarded).

Keystream generation

Keystream is produced by first clocking the FCSR, then extracting one pseudorandom byte using filter F as described before.

Let us assume that time $t = 0$ appears directly after **3.** in the initialization above, i.e.,
$$\mathbf{M}(0) = (S_{19}, S_{18}, \ldots, S_0).$$

Recall from Section 2 that every state $(\mathbf{M},\mathbf{C})$ is associated with an integer p, $1 \leq p \leq |q|$, as the state generate the 2-adic expansion of p/q, where $p = M + 2C$. Let us write the value of p at time t as $p(t)$.

Now assume that we have recovered the state $\mathbf{M}$ and the carries register $\mathbf{C}$ at some time t. So $p(t)$ is known. Thus $p(0)$ can be derived since $p(0) = p(t) \cdot 2^t \bmod q$. This gives us knowledge of $\mathbf{M}(0) = (S_{19}, S_{18}, \ldots, S_0)$, since the carries register at time 0 was 0.

Recall that $(S_{19}, S_{18}, \ldots, S_0)$ was the output from F-FCSR-H when the main register was initialized with IV and key bits with $\mathbf{C} = 0$. If we for simplicity assume that $IV = 0$, then the remaining problem is to reconstruct the key bits. We give a rough outline on how such a reconstruction could be done. A more careful analysis might reveal more efficient ways to solve the problem.

The main register starts as $\mathbf{M} = (0^{80} || k_{79}k_{78} \ldots k_1 k_0)$ and $\mathbf{C} = 0$. The FCSR is clocked once before any output.

We start by guessing the first 8 key bits $k_7, k_6, \ldots, k_0$ that control the feedback the first 8 output bytes. With known feedback we can describe how every state bit can be expressed in algebraic form. Note that as long as we have zero feedback the carries register remain zero and we just get linear equations from the output bytes. The nonlinearity starts to grow when feedback is one. So assuming that the first feedback bit is one, we can examine the equations from the output bytes.

Similarly as before, let $\hat{\mathbf{K}}_\mathbf{0} = (k_0, k_8, \ldots, k_{72})$, $\hat{\mathbf{K}}_\mathbf{1} = (k_1, k_9, \ldots, k_{73})$, etc. Let $\mathcal{L}_i(\hat{\mathbf{K}}_\mathbf{i})$ denote some linear function of variables in $\hat{\mathbf{K}}_\mathbf{i}$ and let $\mathcal{C}_i(\hat{\mathbf{K}}_{\mathbf{i_1}}, \hat{\mathbf{K}}_{\mathbf{i_2}}, \ldots, \hat{\mathbf{K}}_{\mathbf{i_n}})$ denote some nonlinear function of variables in $\hat{\mathbf{K}}_{\mathbf{i_1}}, \hat{\mathbf{K}}_{\mathbf{i_2}}, \ldots, \hat{\mathbf{K}}_{\mathbf{i_n}}$. Then the received equations for the first output byte have the form

$$
\begin{aligned}
(S_0)_7 &= \mathcal{L}_0(\hat{\mathbf{K}}_\mathbf{0}), \\
(S_0)_1 &= \mathcal{L}_1(\hat{\mathbf{K}}_\mathbf{1}), \\
\vdots \quad & \quad \vdots \\
(S_0)_6 &= \mathcal{L}_7(\hat{\mathbf{K}}_\mathbf{7}).
\end{aligned}
$$

The next output byte is written

$$
\begin{aligned}
(S_1)_6 &= \mathcal{L}_8(\hat{\mathbf{K}}_\mathbf{0}) + \mathcal{C}_8(\hat{\mathbf{K}}_\mathbf{7}), \\
(S_1)_7 &= \mathcal{L}_9(\hat{\mathbf{K}}_\mathbf{1}) + \mathcal{C}_9(\hat{\mathbf{K}}_\mathbf{0}), \\
\vdots \quad & \quad \vdots \\
(S_1)_5 &= \mathcal{L}_{15}(\hat{\mathbf{K}}_\mathbf{7}) + \mathcal{C}_{15}(\hat{\mathbf{K}}_\mathbf{6}),
\end{aligned}
$$

and then

$$
\begin{aligned}
(S_2)_5 &= \mathcal{L}_{16}(\hat{\mathbf{K}}_\mathbf{0}) + \mathcal{C}_{16}(\hat{\mathbf{K}}_\mathbf{6}, \hat{\mathbf{K}}_\mathbf{7}), \\
(S_2)_6 &= \mathcal{L}_{17}(\hat{\mathbf{K}}_\mathbf{1}) + \mathcal{C}_{17}(\hat{\mathbf{K}}_\mathbf{7}, \hat{\mathbf{K}}_\mathbf{0}), \\
\vdots \quad & \quad \vdots \\
(S_2)_4 &= \mathcal{L}_{23}(\hat{\mathbf{K}}_\mathbf{7}) + \mathcal{C}_{23}(\hat{\mathbf{K}}_\mathbf{5}, \hat{\mathbf{K}}_\mathbf{6}),
\end{aligned}
$$

and so on. The last one we use is

$$\begin{aligned}
(S_7)_0 &= \mathcal{L}_{56}(\hat{\mathbf{K}}_{\mathbf{0}}) + \mathcal{C}_{56}(\hat{\mathbf{K}}_{\mathbf{1}}, \ldots, \hat{\mathbf{K}}_{\mathbf{6}}, \hat{\mathbf{K}}_{\mathbf{7}}), \\
(S_7)_6 &= \mathcal{L}_{57}(\hat{\mathbf{K}}_{\mathbf{1}}) + \mathcal{C}_{57}(\hat{\mathbf{K}}_{\mathbf{2}}, \ldots, \hat{\mathbf{K}}_{\mathbf{7}}, \hat{\mathbf{K}}_{\mathbf{0}}), \\
&\vdots \quad \vdots \\
(S_7)_4 &= \mathcal{L}_{63}(\hat{\mathbf{K}}_{\mathbf{7}}) + \mathcal{C}_{63}(\hat{\mathbf{K}}_{\mathbf{0}}, \ldots, \hat{\mathbf{K}}_{\mathbf{5}}, \hat{\mathbf{K}}_{\mathbf{6}}).
\end{aligned}$$

When $\hat{\mathbf{K}}_{\mathbf{i}}$ appears in the linear expression but not in the nonlinear expression in an equation, we can use the equation to eliminate one variable. Starting with $\hat{\mathbf{K}}_{\mathbf{7}}$ we have 8 such equations. Since we guessed the first key byte $\hat{\mathbf{K}}_{\mathbf{7}}$ contains 9 unknown variables. By leaving or guessing one bit in $\hat{\mathbf{K}}_{\mathbf{7}}$ we can derive the remaining ones as functions $\mathcal{C}(\hat{\mathbf{K}}_{\mathbf{0}}, \ldots, \hat{\mathbf{K}}_{\mathbf{5}}, \hat{\mathbf{K}}_{\mathbf{6}})$. These functions are inserted instead of $\hat{\mathbf{K}}_{\mathbf{7}}$ variables in the remaining equations. Then examining the equations and looking for those with $\hat{\mathbf{K}}_{\mathbf{6}}$ only in the linear part gives 7 more equations that can be used to eliminate $\hat{\mathbf{K}}_{\mathbf{6}}$ variables. Then the same for $\hat{\mathbf{K}}_{\mathbf{5}}$ gives 6 more equations etc. Altogether we can remove 36 variables in this way and we have to do a work effort of trying 2^{44} choices of certain key bits. The algebraic expressions we need to test can be precomputed. Observe that if the first feedback bit is zero (probability $1/2$) the complexity drops to 2^{36}, two zero feedback bits give complexity 2^{28}, etc.

The key recovery part has not been fully implemented but the given arguments show that also key recovery can be done with low complexity.

7 Conclusions

We have given a very strong attack on the F-FCSR-H stream cipher, a cipher that has been selected for the eSTREAM portfolio. The state recovery attack has been fully implemented to attack F-FCSR-H using the designers reference code. It succeeds in a few seconds using on average $2^{23.7}$ bytes (≈ 13 Mbyte) of keystream.

The weakness that was exploited is that the FCSR automata sometimes temporarily (almost) behaves as a regular LFSR. Together with the fact that the output filter is linear, the complete cipher became temporarily linear, which allowed us to recover the internal state.

References

1. Arnault, F., Berger, T.: F-FCSR: Design of a new class of stream ciphers. In: Gilbert, H., Handschuh, H. (eds.) FSE 2005. LNCS, vol. 3557, pp. 83–97. Springer, Heidelberg (2005)
2. Arnault, F., Berger, T., Lauradoux, C.: Update on F-FCSR stream cipher. eSTREAM, ECRYPT Stream Cipher Project, Report 2006/025 (2006), `http://www.ecrypt.eu.org/stream`

3. Arnault, F., Berger, T., Minier, M.: Some results on FCSR automata with applications to the security of FCSR-based pseudorandom generators. IEEE-IT 54(2), 836–840 (2008)
4. Jaulmes, E., Muller, F.: Cryptanalysis of ECRYPT candidates F-FCSR-8 and F-FCSR-H. eSTREAM, ECRYPT Stream Cipher Project, Report 2005/046 (2005), http://www.ecrypt.eu.org/stream
5. Jaulmes, E., Muller, F.: Cryptanalysis of the F-FCSR stream cipher family. In: Preneel, B., Tavares, S. (eds.) SAC 2005. LNCS, vol. 3897, pp. 36–50. Springer, Heidelberg (2006)
6. Klapper, A., Goresky, M.: 2-adic shift registers. In: Anderson, R. (ed.) FSE 1993. LNCS, vol. 809, pp. 174–178. Springer, Heidelberg (1994)
7. Fischer, S., Meier, W., Stegemann, D.: Equivalent representations of the F-FCSR Keystream Generator. In: SASC 2008, Workshop Record, pp. 87–96 (2008)

Author Index